# Medicinal Plants and their Bioactive Compounds in Human Health: Volume 2

Mohammad Azam Ansari
Shoaib Shoaib • Md. Abul Barkat
Editors

# Medicinal Plants and their Bioactive Compounds in Human Health: Volume 2

Springer

*Editors*
Mohammad Azam Ansari
Department of Epidemic Disease Research
Institute for Research and Medical Consultations (IRMC)
Imam Abdulrahman Bin Faisal University
Dammam, Saudi Arabia

Shoaib Shoaib
Department of Genetics
School of Medicine
University of Alabama at Birmingham
Birmingham, AL, USA

Md. Abul Barkat
Department of Pharmaceutics
College of Pharmacy
University of Hafr Al-Batin
Hafr Al-Batin, Saudi Arabia

ISBN 978-981-95-3619-1 ISBN 978-981-95-3620-7 (eBook)
https://doi.org/10.1007/978-981-95-3620-7



This Springer imprint is published by the registered company Springer Nature Singapore Pte Ltd.
The registered company address is: 152 Beach Road, #21-01/04 Gateway East, Singapore 189721, Singapore

*The editors are proud to dedicate this book to their parents, who inspired them to pursue higher studies. We attribute all our achievements to them.*

# Preface

After successful publication of *Medicinal Plants and Their Bioactive Compounds in Human Health: Volume 1* (https://link.springer.com/book/10.1007/978-981-97-6895-0), Volume 2 includes 27 additional medicinal plants for biomedical application in human health. This book highlights the crucial role of plant bioactive compounds in human health, emphasizing their medicinal and nutritional significance. Humans have used the leaves, fruits, roots, bark, flowers, and stems of different plants for a number of reasons since ancient times. Plant-based medicines are also used to treat a wide range of diseases. Remarkably, they are employed as a substitute strategy to combat diabetes, microbiological infections, cardiovascular diseases, neurological diseases, skin and blood disorders, malaria, wound healing, and complications associated to the gastrointestinal and kidney functions. Additionally, plants are thought to be rich in several types of phytoconstituents that are vital for body growth and defense against different diseases. Typically, plants are an excellent source of water, vitamins, minerals, carbohydrates, protein, and other nutrients. Furthermore, the antioxidant, antiproliferative, antibacterial, antiviral, antidiabetic, antimalarial, antifungal, anti-inflammatory, antidepressant, anticoagulant, and anticancer potential of plants and their bioactive components are widely accepted.

Currently, most of the drugs used for various diseases are either synthetic or semi-synthetic chemicals, which exert severe side effects during the course of the treatment. In particular, majority of anticancer and antibacterial drugs are gaining the resistance, which is a significant worry for the clinicians and researchers worldwide. On the other hand, plants have broad scope of the structural diversity of the compounds and plants can serve as a big reservoir of novel and safe drug candidates for treating various diseases as well as their bioactive compounds may enhance the drug efficacy. Additionally, plants and their bioactive compounds can be easily digested and absorbed through the gut wall without exerting significant adverse side effects. Plants and their bioactive compounds can be metabolized easily in human body. Cost-effectiveness, safety profile, fewer side effects, and ease to approach make the plants ideal source for the development and identification of new drug regimens for different diseases. It displays the significant impact of medicinally and nutritionally relevant plants and their products on health and wellness by highlighting their various uses in industries, medicine, cosmetics, and therapies. We sincerely appreciate the contributor's commitment to helping us compile all of our ideas. We are deeply appreciative of the Springer Nature Team's gracious support over the entire book publishing process.

Dammam, Saudi Arabia Mohammad Azam Ansari
Birmingham, AL, USA Shoaib Shoaib
Hafr Al batin, Saudi Arabia Md. Abul Barkat

# Contents

1 **Medicinal and Nutritional Importance of *Azadirachta indica* in Human Health** . . . . . . . . . . . . . . . . . . . . . . . . . . . . . . . . . . . . . . . . . . 1
Sahar Afzal, Shoaib Shoaib, Roshan Jahan, Mo Shadab, Mohammad N. Alomary, and Mohammad Azam Ansari

2 **Medicinal and Nutritional Importance of *Nicotiana* spp. in Human Health** . . . . . . . . . . . . . . . . . . . . . . . . . . . . . . . . . . . . . . . . . . 17
Shoaib Shoaib, Roshan Jahan, Mohammad N. Alomary, Md. Abul Barkat, Saad Alghamdi, and Mohammad Azam Ansari

3 **Nutritional and Medicinal Importance of *Lepidium didymum* in Human Health** . . . . . . . . . . . . . . . . . . . . . . . . . . . . . . . . . . . . . . . . . . 31
Mo Shadab, Farha Aijaz, Nazish Akhtar, Abdul Raheem, Saima Malik, Shoaib Shoaib, and M. B. Siddiqui

4 **Medicinal and Nutritional Importance of *Salvia rosmarinus* in Human Health** . . . . . . . . . . . . . . . . . . . . . . . . . . . . . . . . . . . . . . . . . . 49
Asma Abid, Kamilia Bireche, Nourelhouda Mekhadmi, and Mohammad Azam Ansari

5 **Medicinal and Nutritional Importance of *Rhodomyrtus tomentosa* (Aiton) Hassk** . . . . . . . . . . . . . . . . . . . . . . . . . . . . . . . . . . . . . . . . . . . 71
Dwi Marlina Syukri, Tessa Sjahriani, Tosin Ayodeji Oladosu, Sudarshan Singh, and Mohit Agrawal

6 **Medicinal and Nutritional Importance of *Withania somnifera* in Human Health** . . . . . . . . . . . . . . . . . . . . . . . . . . . . . . . . . . . . . . . . . . 91
Ishani Yadav, Neelu Sinha, and Udita Tiwari

7 **Medicinal and Nutritional Importance of *Bacopa monnieri* in Human Health** . . . . . . . . . . . . . . . . . . . . . . . . . . . . . . . . . . . . . . . . . . 105
Mohd Hanzala Khan, Sabnam Farhat, Ruhi Firdaus, Mohd Ashraf Ashfaq, Md Salik Noorani, and Asghar Ali

8 **Nutritional and Medicinal Importance of *Zingiber officinale* in Human Health** . . . . . . . . . . . . . . . . . . . . . . . . . . . . . . . . . . . . . . . . . . 137
Sabry A. H. Zidan and Hamdy K. Assaf

9 **Medicinal and Nutritional Importance of Garlic (*Allium sativum* L.) Human Health** . . . . . . 161
Gulzar Ahmed Rather, Mohd Yasin Bhat, Mudasir Irfan Dar, Mohmmad Ashaq Sofi, Raman Yadav, Humeraha Nazneen, and Asma Abid

10 **Medicinal and Nutritional Importance of *Prunus dulcis* in Human Health** . . . . . . 183
Aman Safdar, Furqan Choudhary, Mohtashim Rehan, Abdur Rahmaan Akhtar, and Humaira Farooqi

11 **Medicinal and Nutritional Importance of *Cymbopogon citratus* in Human Health** . . . . . . 201
Abdelmonem Siddiq, Manisha Das, Muhammed Shabil, Shubham Shrestha, Swarali Yatin Chodnekar, and Nityanand Jain

12 **Medicinal and Nutritional Importance of *Ziziphus mauritiana* in Human Health** . . . . . . 233
Md. Khokon Miah Akanda, Sanzia Mehjabin, Nazia Hoque, A. H. M. Nazmul Hasan, and G. M. Masud Parvez

13 **Medicinal and Nutritional Importance of *Hypericum perforatum* in Human Health** . . . . . . 251
Pranjal Gujarathi, Meghraj Suryawanshi, and Piyush Gujarathi

14 **Medicinal and Nutritional Importance of *Cannabis sativa* in Human Health** . . . . . . 287
Roy Dinata, Vanlalhruaii Famhawite, Puranjoy Mipun, Sritapa Basu Mallick, and Rinku Baishya

15 **Medicinal and Nutritional Importance of *Prunus persica* in Human Health** . . . . . . 319
Shafia Arshad, Farah Zafar, and Faiza Irshad

16 **Medicinal and Nutritional Importance of *Catharanthus roseus* L. in Human Health** . . . . . . 343
Priyanka Tolambiya, Bhagawatilal Jagetiya, and Sujata Mathur

17 **Medicinal and Nutritional Importance of *Sphaeranthus indicus* in Human Health** . . . . . . 363
Murtaza Vandeliwala, Ruchi Pathak, and Prinsa R. Nagar

18 **Medicinal and Nutritional Importance of *Gloriosa superba* L. in Human Health** . . . . . . 385
Kiran Soni, Yash Pal Sharma, Kapoor, Ankush Moran, and Pooja Sharma

19 **Medicinal and Nutritional Importance of *Gmelina asiatica* in Human Health** . . . . . . 401
Abha R. Vyas, Deep B. Sathwara, and Radhika N. Kachhadiya

**20 Medicinal and Nutritional Importance of *Nigella sativa* on Human Health** . . . . . . 415
Ushakar Mishra, Benazir Fatima, and Abha Singh

**21 Medicinal and Nutritional Importance of *Enhydra fluctuans* in Human Health** . . . . . . 431
Avijit Bej, Nandita Mandal, Anwesha Das, Sanjukta Das, and Arijit Nandi

**22 Medicinal and Nutritional Importance of *Trachyspermum ammi* in Human Health** . . . . . . 453
Kamilia Bireche and Asma Abid

**23 Medicinal and Nutritional Importance of *Nardostachys jatamansi* in Human Health** . . . . . . 479
Sharmistha Ganguly, Dwaipayan Sinha, Ratul Bhattacharya, Dhavalkumar Patel, Abdullah Salik, Srijan Mukhopadhyay, Prajwal Kiran Saldanha, Somnath Kar, Shalini Kumari, and Mohammad Azam Ansari

**24 Medicinal and Nutritional Importance of *Andrographis paniculata* in Human Health** . . . . . . 531
Benazir Fatima, Ushakar Mishra, and Abha Singh

**25 Medicinal and Nutritional Importance of *Mentha piperita* in Human Health** . . . . . . 549
Ayman M. Al-Qaaneh, Mohammad A. Aburumman, Mounir M. Bekhit, Heba A. El-Nashar, Moulay H. Youssoufi, Suriya Rehman, and Shafiq Ur Rahman

**26 Medicinal and Nutritional Importance of *Passiflora incarnata* in Human Health** . . . . . . 573
Sanjana Mishra, Siddharth Singh, Kshitij Patel, Prekshi Garg, and Prachi Srivastava

**27 Medicinal and Nutritional Importance of *Cinnamomum zeylanicum* in Human Health** . . . . . . 593
Akansha Agrwal and Sheetal Mital

# Editors and Contributors

## About the Editors

**Mohammad Azam Ansari** is currently working as Associate Professor at the Institute for Research & Medical Consultation, Imam Abdulrahman Bin Faisal University, Dammam, Saudi Arabia. His research interest is in green synthesis of different kinds of metal and metal oxide nanomaterials by using microbes and various parts of the plants, and characterization of their biomedical and pharmaceuticals applications. His research work mainly focuses on nanoformulation, biomimetic nanomaterials, medicinal plants, and natural products for biomedical, pharmaceutical, and industrial applications, such as the development of antimicrobial, antibiofilm, anti-quorum sensing, and anticancer agents and nanodrugs. These innovative approaches aim to enhance therapeutic efficacy while minimizing side effects, paving the way for more effective treatments. By leveraging the unique properties of nanomaterials, his work contributes significantly to advancing healthcare solutions and improving patient outcomes. He has been consistently recognized among the top 2% of scientists in the world in 2021, 2022, 2023, 2024, and 2025.

**Shoaib Shoaib** is a Postdoctoral Fellow at the University of Alabama at Birmingham, USA, where he conducts cutting-edge research in cancer biology, immunology, and molecular therapeutics. His research spans cancer prevention, drug resistance, phytochemical-based therapeutics, and nanomaterial applications in biomedicine, with specialized expertise in animal models, biochemical assays, advanced microscopy, and molecular biology techniques. His work explores novel cancer therapies, the role of phytochemicals in drug development, and nature-driven solutions for disease treatment.

**Md. Abul Barkat** is an Assistant Professor in the Department of Pharmaceutics at the College of Pharmacy, University of Hafr Al Batin, Saudi Arabia. He has more than a decade of combined teaching and research experience, with a focus on drug delivery science, particularly in the areas of nanomedicine and targeted therapies. His research interests include the development and optimization of herbal/synthetic drug-loaded nanostructured delivery systems, controlled-release drug delivery systems, bio-enhanced drug delivery systems, nanomaterials, nanocomposites for a variety of conditions like burns, cancer, neurological disease, etc.

## Contributors

**Asma Abid** Laboratory of Valorization and Promotion of Saharan Resources (VPRS), Faculty of Mathematics and Matter Sciences, University of Ouargla, Ouargla, Algeria

Higher School of Saharan Agriculture, El Oued, Algeria

**Mohammad A. Aburumman** Department of Nursing, Faculty of Nursing, Al-Balqa Applied University (BAU), Al-Salt, Jordan

**Sahar Afzal** Department of Biotechnology, Jamia Hamdard University, New Delhi, India

**Mohit Agrawal** Department of Pharmacology, School of Medical and Allied Sciences, K.R. Mangalam University, Gurugram, Haryana, India

**Akansha Agrwal** Department of Applied Science and Humanities, KIET Group of Institutions, Ghaziabad, Delhi-NCR, Uttar Pradesh, India

**Farha Aijaz** Allelopathy and Plant Taxonomy Laboratory, Department of Botany, Aligarh Muslim University, Aligarh, India

**Md. Khokon Miah Akanda** Department of Pharmacy, University of Asia Pacific, Dhaka, Bangladesh

**Abdur Rahmaan Akhtar** Department of Biosciences, Faculty of Life Science, Jamia Millia Islamia, New Delhi, India

**Nazish Akhtar** Allelopathy and Plant Taxonomy Laboratory, Department of Botany, Aligarh Muslim University, Aligarh, India

**Saad Alghamdi** Department of Clinical Laboratory Sciences, Faculty of Applied Medical Sciences, Umm Al-Qura University, Makkah, Saudi Arabia

**Asghar Ali** School of Chemical and Life Sciences, Jamia Hamdard, New Delhi, India

**Mohammad N. Alomary** Applied Genomic Technologies Institute, King Abdulaziz City for Science and Technology (KACST), Riyadh, Saudi Arabia

**Ayman M. Al-Qaaneh** Faculty of Allied Medical Sciences, Al-Balqa Applied University (BAU), Al-Salt, Jordan

Department of Pharmaceutical Technology, Faculty of Pharmacy, Jordan University of Science and Technology (JUST), Irbid, Jordan

**Mohammad Azam Ansari** Department of Epidemic Diseases Research, Institute for Research and Medical Consultations (IRMC), Imam Abdulrahman Bin Faisal University, Dammam, Saudi Arabia

**Shafia Arshad** Office of Research Innovation and Commercialization, The Government Sadiq College Women University, Bahawalpur, Pakistan

**Mohd Ashraf Ashfaq** School of Chemical and Life Sciences, Jamia Hamdard, New Delhi, India

**Hamdy K. Assaf** Department of Pharmacognosy, Faculty of Pharmacy, Al-Azhar University, Assiut, Egypt

**Rinku Baishya** Centre for Preclinical Studies (CPS), CSIR-North East Institute of Science and Technology (NEIST), Pulibor, Jorhat, Assam, India

**Md. Abul Barkat** Department of Pharmaceutics, College of Pharmacy, University of Hafr Al-Batin, Hafr Al-Batin, Saudi Arabia

**Avijit Bej** Mahadeva Lal Schroff College of Pharmacy, Aurangabad, Bihar, India

**Mounir M. Bekhit** Department of Pharmaceutics, College of Pharmacy, King Saud University, Riyadh, Saudi Arabia

**Mohd Yasin Bhat** Government Degree College Baramulla (Autonomous), Higher Education Department, Government of Jammu and Kashmir, Jammu and Kashmir, India

**Ratul Bhattacharya** Department of Botany, Raja Ram Mohan Roy Mahavidyalaya, Hooghly, West Bengal, India

**Kamilia Bireche** Laboratory of Valorization and Promotion of Saharan Resources (VPRS), Faculty of Mathematics and Matter Sciences, University of Ouargla, Ouargla, Algeria

**Swarali Yatin Chodnekar** Faculty of Medicine, Teaching University Geomedi LLC, Tbilisi, Georgia

**Furqan Choudhary** Department of Biotechnology, Jamia Hamdard, New Delhi, India

**Mudasir Irfan Dar** Government Degree College Sogam, Higher Education Department, Government of Jammu and Kashmir, Jammu and Kashmir, India

**Anwesha Das** Department of Pharmacy, Sanaka Educational Trust's Group of Institutions (SETGOI), Malandighi, Durgapur, West Bengal, India

**Manisha Das** Faculty of Medicine, Dhaka Medical College, Dhaka University, Dhaka, Bangladesh

**Sanjukta Das** Department of Pharmacy, Birbhum Pharmacy School, Bandersole, Birbhum, West Bengal, India

**Roy Dinata** Animal Science Research Center, Division of Animal Sciences, University of Missouri, Columbia, MO, USA

**Heba A. El-Nashar** Department of Pharmacognosy, Faculty of Pharmacy, Ain Shams University, Cairo, Egypt

**Vanlalhruaii Famhawite** Centre for Preclinical Studies (CPS), CSIR-North East Institute of Science and Technology (NEIST), Pulibor, Jorhat, Assam, India

**Sabnam Farhat** School of Chemical and Life Sciences, Jamia Hamdard, New Delhi, India

**Humaira Farooqi** Department of Biotechnology, Jamia Hamdard, New Delhi, India

**Benazir Fatima** Microbial Biodiversity Lab, Department of Botany, Faculty of Science, Patna University, Patna, Bihar, India

**Ruhi Firdaus** School of Unani Medicine and Surgery, Jamia Hamdard, New Delhi, India

**Sharmistha Ganguly** University Department of Botany, Ranchi University, Ranchi, Jharkhand, India

**Prekshi Garg** BioinfoCore Solutions (OPC) Pvt Ltd, Lucknow, Uttar Pradesh, India

**Piyush Gujarathi** Department of Community Medicine, Vidhyadeep Homeopathic Medical College and Research Centre, Vidhyadeep University, Anita, Surat, Gujarat, India

**Pranjal Gujarathi** Department of Pharmacology, Vidhyadeep Institute of Pharmacy, Vidhyadeep University, Anita, Surat, Gujarat, India

**A. H. M. Nazmul Hasan** Department of Pharmacy, University of Asia Pacific, Dhaka, Bangladesh

**Nazia Hoque** Department of Pharmacy, East West University, Dhaka, Bangladesh

**Faiza Irshad** University College of Conventional Medicine, Faculty of Medicine & Allied Health Sciences, The Islamia University of Bahawalpur, Bahawalpur, Pakistan

**Bhagawatilal Jagetiya** Phytotechnology Research Laboratory, Department of Botany, M.L.V. Government College, Bhilwara, Rajasthan, India

**Roshan Jahan** Department of Botany, Faculty of Lifesciences, Aligarh Muslim University, Aligarh, Uttar Pradesh, India

**Nityanand Jain** Independent Consultant, Chandigarh, India

**Radhika N. Kachhadiya** Department of Pharmaceutical Chemistry, LJ Institute of Pharmacy, LJ University, Ahmedabad, Gujarat, India

**Kapoor** Department of Silviculture and Agroforestry, Dr. Yashwant Singh Parmar University of Horticulture and Forestry, Nauni, Solan, Himachal Pradesh, India

**Somnath Kar** Department of Botany, Holy Cross College, Tripura University, Agartala, Tripura, India

**Mohd Hanzala Khan** School of Chemical and Life Sciences, Jamia Hamdard, New Delhi, India

**Shalini Kumari** University Department of Botany, Ranchi University, Ranchi, Jharkhand, India

**Saima Malik** Allelopathy and Plant Taxonomy Laboratory, Department of Botany, Aligarh Muslim University, Aligarh, India

**Sritapa Basu Mallick** Indian Institute of Science Education and Research Kolkata (IISER Kolkata), Mohanpur, West Bengal, India

**Nandita Mandal** Department of Pharmacy, Sanaka Educational Trust's Group of Institutions (SETGOI), Malandighi, Durgapur, West Bengal, India

**Sujata Mathur** Department of Botany, B.B.D. Government PG College, Jaipur, Rajasthan, India

**Sanzia Mehjabin** Department of Pharmacy, Varendra University, Rajshahi, Bangladesh

**Nourelhouda Mekhadmi** Laboratory of the Development and Technology of Saharan Resources (VTRS), Echahid Hamma Lakhdar El Oued University, El-Oued, Algeria

Faculty of Life and Natural Sciences, Department of Biology, University of El Oued, El-Oued, Algeria

**Puranjoy Mipun** Department of Botany, Bhattadev University, Bajali, Assam, India

**Sanjana Mishra** Amity Institute of Biotechnology, Amity University Uttar Pradesh, Lucknow Campus, Lucknow, Uttar Pradesh, India

**Ushakar Mishra** Microbial Biodiversity Lab, Department of Botany, Faculty of Science, Patna University, Patna, Bihar, India

**Sheetal Mital** Department of Applied Science and Humanities, KIET Group of Institutions, Ghaziabad, Delhi-NCR, Uttar Pradesh, India

**Ankush Moran** Department of Silviculture and Agroforestry, Dr. Yashwant Singh Parmar University of Horticulture and Forestry, Nauni, Solan, Himachal Pradesh, India

**Srijan Mukhopadhyay** Taxonomy and Biosystematic Laboratory, PG Department of Botany, Ramakrishna Mission Vivekananda Centenary College, Rahara, Kolkata, West Bengal, India

**Prinsa R. Nagar** Department of Pharmaceutical Chemistry, L. J. Institute of Pharmacy, L J University, Ahmedabad, India

**Arijit Nandi** Institute of Molecular Bioscience, The University of Queensland, St Lucia, QLD, Australia

Department of Pharmacy, Sanaka Educational Trust's Group of Institutions (SETGOI), Malandighi, Durgapur, West Bengal, India

**Humeraha Nazneen** Nizam College Osmania University, Hyderabad, Telangana, India

**Md Salik Noorani** School of Chemical and Life Sciences, Jamia Hamdard, New Delhi, India

**Tosin Ayodeji Oladosu** Department of Chemistry and Biochemistry, University of Minnesota Duluth, Duluth, MN, USA

**G. M. Masud Parvez** Department of Pharmaceutical and Biomedical Sciences, College of Pharmacy, University of Georgia, Athens, GA, USA

**Dhavalkumar Patel** Department of Microbiology & Biotechnology, University School of Sciences, Gujarat University, Ahmedabad, Gujarat, India

Department of Research & Development, Latambarcem Brewers Private Limited, Dicholi, Goa, India

**Kshitij Patel** Amity Institute of Biotechnology, Amity University Uttar Pradesh, Lucknow Campus, Lucknow, Uttar Pradesh, India

**Ruchi Pathak** Department of Pharmaceutical Chemistry, L. J. Institute of Pharmacy, L J University, Ahmedabad, India

**Abdul Raheem** Department of Botany, University of Lucknow, Lucknow, Uttar Pradesh, India

**Shafiq Ur Rahman** Department of Pharmacy, Shaheed Benazir Bhutto University, Sheringal, Khyber Pakhtunkhwa, Pakistan

**Gulzar Ahmed Rather** AMR and Nanotherapeutics Laboratory, Department of Pharmacology, Saveetha Dental College and Hospitals, Saveetha Institutes of Medical and Technical Sciences (SIMATS), Chennai, Tamil Nadu, India

**Mohtashim Rehan** Paramedical College, Faculty of Medicine, Aligarh Muslim University, Aligarh, Uttar Pradesh, India

**Suriya Rehman** Department of Epidemic Diseases Research, Institute for Research and Medical Consultation, Imam Abdulrahman Bin Faisal University, Dammam, Saudi Arabia

**Aman Safdar** Department of Biotechnology, Jamia Hamdard, New Delhi, India

**Prajwal Kiran Saldanha** Department of Botany, Bangalore University, Jnana Bharathi Campus, Bangalore, Karnataka, India

**Abdullah Salik** Department of Pharmaceutical Sciences, Birla Institute of Technology, Ranchi, Jharkhand, India

**Deep B. Sathwara** Department of Pharmaceutical Chemistry, LJ Institute of Pharmacy, LJ University, Ahmedabad, Gujarat, India

**Muhammed Shabil** Department of Pharmacy Practice, Faculty of Pharmacy, M.S. Ramaiah University of Applied Sciences, Bangalore, Karnataka, India

Global Center for Evidence Synthesis (GCES), Chandigarh, India

**Mo Shadab** Department of Botany, Faculty of Lifesciences, Aligarh Muslim University, Aligarh, Uttar Pradesh, India

Allelopathy and Plant Taxonomy Laboratory, Department of Botany, Aligarh Muslim University, Aligarh, India

**Pooja Sharma** Department of Tree Improvement and Genetic Resources, Dr. Yashwant Singh Parmar University of Horticulture and Forestry, Nauni, Solan, Himachal Pradesh, India

**Yash Pal Sharma** Department of Forest Products, Dr. Yashwant Singh Parmar, University of Horticulture and Forestry, Nauni, Solan, Himachal Pradesh, India

**Shoaib Shoaib** Department of Genetics, School of Medicine, University of Alabama at Birmingham, Birmingham, AL, USA

**Shubham Shrestha** Patan Academy of Health Sciences, Patan Hospital, Lalitpur, Nepal

**Abdelmonem Siddiq** Faculty of Pharmacy, Mansoura University, Mansoura, Egypt

**M. B. Siddiqui** Allelopathy and Plant Taxonomy Laboratory, Department of Botany, Aligarh Muslim University, Aligarh, India

**Abha Singh** Microbial Biodiversity Lab, Department of Botany, Faculty of Science, Patna University, Patna, Bihar, India

**Siddharth Singh** Amity Institute of Biotechnology, Amity University Uttar Pradesh, Lucknow Campus, Lucknow, Uttar Pradesh, India

**Sudarshan Singh** Faculty of Pharmacy, Chiang Mai University, Chiang Mai, Thailand

**Dwaipayan Sinha** Department of Botany, Government General Degree College, Mohanpur, Paschim Medinipur, West Bengal, India

**Neelu Sinha** Department of Biochemistry, School of Life Sciences, Swami Vivekanand Campus, Dr. Bhimrao Ambedkar University, Agra, UP, India

**Tessa Sjahriani** Faculty of Medicine, Malahayati University, Bandar Lampung, Lampung, Indonesia

**Mohmmad Ashaq Sofi** Department of Biomedical Engineering, Sathyabama Institute of Science and Technology, Chennai, Tamil Nadu, India

**Kiran Soni** Department of Forest Products, Dr. Yashwant Singh Parmar University of Horticulture and Forestry, Nauni, Solan, Himachal Pradesh, India

**Prachi Srivastava** Amity Institute of Biotechnology, Amity University Uttar Pradesh, Lucknow Campus, Lucknow, Uttar Pradesh, India

**Meghraj Suryawanshi** Department of Pharmaceutics, Sandip Institute of Pharmaceutical Sciences (SIPS), Affiliated To Savitribai Phule Pune University (SPPU, Pune), Nashik, Maharashtra, India

**Dwi Marlina Syukri** Faculty of Medicine, Malahayati University, Bandar Lampung, Lampung, Indonesia

**Udita Tiwari** Department of Biochemistry, School of Life Sciences, Swami Vivekanand Campus, Dr. Bhimrao Ambedkar University, Agra, UP, India

**Priyanka Tolambiya** Department of Botany, Government PG College, Kekri, Rajasthan, India

**Murtaza Vandeliwala** Department of Pharmaceutical Chemistry, L. J. Institute of Pharmacy, L J University, Ahmedabad, India

**Abha R. Vyas** Department of Pharmaceutical Chemistry, LJ Institute of Pharmacy, LJ University, Ahmedabad, Gujarat, India

**Ishani Yadav** Department of Biochemistry, School of Life Sciences, Swami Vivekanand Campus, Dr. Bhimrao Ambedkar University, Agra, UP, India

**Raman Yadav** Department of Pharmacology School of Pharmacy Noida International University Gautam Budh Nagar, Uttar pradesh, India

**Moulay H. Youssoufi** Laboratory of Applied Chemistry & Environment, Faculty of Sciences, Mohammed First University, Oujda, Morocco

**Farah Zafar** University College of Conventional Medicine, Faculty of Medicine & Allied Health Sciences, The Islamia University of Bahawalpur, Bahawalpur, Pakistan

**Sabry A. H. Zidan** Department of Pharmacognosy, Faculty of Pharmacy, Al-Azhar University, Assiut, Egypt

# Medicinal and Nutritional Importance of *Azadirachta indica* in Human Health

1

Sahar Afzal, Shoaib Shoaib, Roshan Jahan, Mo Shadab, Mohammad N. Alomary, and Mohammad Azam Ansari

**Abstract**

*Azadirachta indica* (Neem) is a well-known traditional medicine in Ayurveda and Unani system and has served as a cornerstone of natural healthcare for decades. Interestingly, every part of neem plant has been utilized for its diverse medicinal properties in the management of diseases such as skin ailments, malaria, diabetes, acne, and bacterial infections. Nutritional aspects of neem plant offer a rich spectrum of proteins, vitamins, minerals, and different bioactive compounds including alkaloids, terpenoids, and flavonoids. Accumulating evidences demonstrate that these compounds have been studied for their potent antioxidant, anti-inflammatory, antimicrobial, and immunomodulatory properties. In

Sahar Afzal and Shoaib Shoaib contributed equally with all other contributors.

S. Afzal
Department of Biotechnology, Jamia Hamdard University, New Delhi, India

S. Shoaib
Department of Genetics, School of Medicine, University of Alabama at Birmingham, Birmingham, AL, USA

R. Jahan · M. Shadab
Department of Botany, Faculty of Lifesciences, Aligarh Muslim University, Aligarh, Uttar Pradesh, India

M. N. Alomary
Applied Genomic Technologies Institute, King Abdulaziz City for Science and Technology (KACST), Riyadh, Saudi Arabia

M. A. Ansari (✉)
Department of Epidemic Diseases Research, Institute for Research and Medical Consultations (IRMC), Imam Abdulrahman Bin Faisal University, Dammam, Saudi Arabia
e-mail: maansari@iau.edu.sa



M. A. Ansari et al. (eds.), *Medicinal Plants and their Bioactive Compounds in Human Health: Volume 2*, https://doi.org/10.1007/978-981-95-3620-7_1

recent years, neem has been evaluated for its crucial role in oral health, displaying comparable efficacy in reducing plaque, gingivitis, and cariogenic bacteria. Studies showed that neem exhibits significant antidiabetic effects, with phytocompounds such as nimbidin, azadirachtin, and stigmasterol modulating glucose metabolism, enhancing insulin secretion, and mitigating oxidative stress. Furthermore, beyond the human health, neem has been evaluated for its biopesticidal compounds and livestock health, aligning with eco-friendly farming practices. Intriguingly, advances in biotechnology have led to the enhanced production of secondary metabolites, paving the way for developing neem-based bioformulations. Taken together, these multifaceted attributes underscore *Azadirachta indica* as a nutritional and medicinal source of immense potential, warranting continued exploration and integration into modern healthcare and agriculture.

**Keywords**

*Azadirachta indica* · Phytocompounds · Health · Nutraceutics · Nutrition · Medicines

## 1.1 Introduction

*Azadirachta indica* (Neem) is a member of Meliaceae family that has been a staple of traditional medicine, especially in Ayurveda and Unani medical systems. Neem is often known as the "Divine Tree" that has long been valued for its wide range of health benefits in curing a variety of illnesses. It has been used for more than 4000 years to cure skin conditions, leprosy, and malaria (Trivedi et al. 2019). Every part of the neem tree including the leaves, bark, seeds, and flowers have been used medicinally in traditional Ayurvedic purposes. For example, neem oil, which is produced from the seeds, has been applied topically for skin ailments and as an antiseptic, while neem leaves have been used to treat fever, respiratory conditions, and dental issues (Zeeshan et al. 2024). In formulations for the treatment of chronic illnesses like diabetes, cancer, and digestive issues, neem has been a crucial component and is considered as a versatile plant due to its antibacterial and anti-inflammatory qualities.

## 1.2 Nutritional and Medicinal Importance of Azadirachta indica

Several medicinally and nutritionally important plants including *Curcuma long*, *Linum usitatissimum*, and *Azadirachta indica* have been investigated for their beneficial properties in India in last few decades (Jahan et al. 2024; Shoaib et al. 2024). Neem's high protein, vitamin, mineral, and phytochemical content emphasizes its nutritional value by supporting a number of bodily functions and has potential medical uses. In addition to vitamins B3 and C, with minor amounts of the other B vitamins as well as trace amounts of copper, magnesium, iron, and zinc, neem stem bark is high in calcium and potassium (Adeoye et al. 2024). The high crude protein content and significant crude lipid levels of neem seeds are especially noteworthy because they are necessary for both bodily processes and energy production. Neem leaves and seeds are also a great source of carbohydrates and nutritional fiber, which promotes digestive health and gives you energy (Obikaonu 2012). Further studies conducted identified all the essential amino acids in this protein while leucine, aspartic acid, lysine, glutamic acid, serine, and alanine appeared with highest percentage (Ilesanmi and Gungula 2016).

Accumulated evidences showed that the plant is rich in bioactive compounds such as terpenoids, alkaloids, flavonoids, and limonoids, which contribute to its diverse health-related properties including antioxidant, anti-inflammatory, antimicrobial, and immunomodulatory activities. Interestingly, diverse properties of phytochemicals make *Azadirachta indica* highly important for human diseases, animal husbandry, and agriculture purposes. *Azadirachta indica* contains a pool of beneficial secondary metabolites, highlighting its medicinal importance in research and pharmaceutical industry. Its compounds can be categorized as isoprenoid (nimbin, azadirone, gedunin, azadirachtin, protomeliacins, and limomoids) and non-isoprenoid (carbohydrates, proteins, sulfur compounds, dihydrochalcones polyphenolics). Many compounds with diverse physical and chemical properties have been isolated from almost every part of the plant, for example, stigmasterol, terpinen-4-ol, sugiol, 4-cymene, nimbiol, α-terpinene, and vitamin E recovered from bark and leaves extracts. The major bioactive compounds found in Neem is shown in Fig. 1.1.

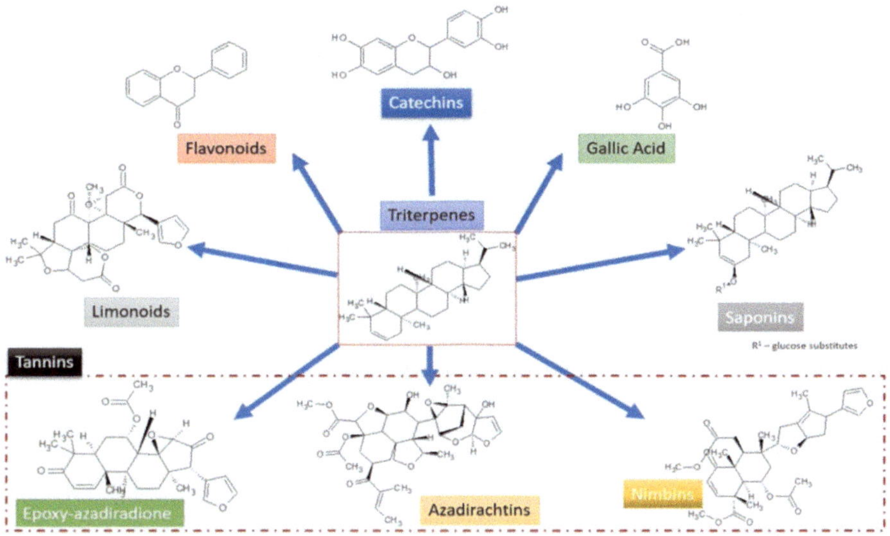


**Fig. 1.1** The major bioactive compounds found in Neem. (Source: Islas et al. 2020)

## 1.3 Pharmacological Activities of Neem

In recent decades hundreds of plants and their bioactive compounds have been explored for their medicinal importance including prevention and treatment of neurodegenerative and skin diseases as well as different cancers (Gahtani et al. 2024; Weir et al. 2023; Shoaib et al. 2023). The bioactive compounds present in neem help to treat diseases by activating antioxidant enzymes, breaking down the cell walls of bacteria, and acting as chemotherapy preventatives via controlling cellular pathways. Various pharmacological activities of bioactive compounds present in neem has been shown in Fig. 1.2.

## 1.4 Antimicrobial Potential of Neem: Application in Oral Hygiene

Neem has gained attention for its potential role in promoting oral health. Studies suggest that neem effectively reduces bacterial plaque, gingivitis, and oral bacteria, presenting it as a promising natural alternative in oral care products (Abhishek et al. 2015). In one study comparing neem and charcoal toothbrushes, both reduced plaques, though charcoal toothbrushes showed slightly higher effectiveness over a 15-day period. Nonetheless, neem toothbrushes still provided notable plaque reduction, highlighting their usefulness in oral hygiene maintenance (Somnathe et al. 2022). Additionally, neem mouthwash has been found to match the effectiveness of

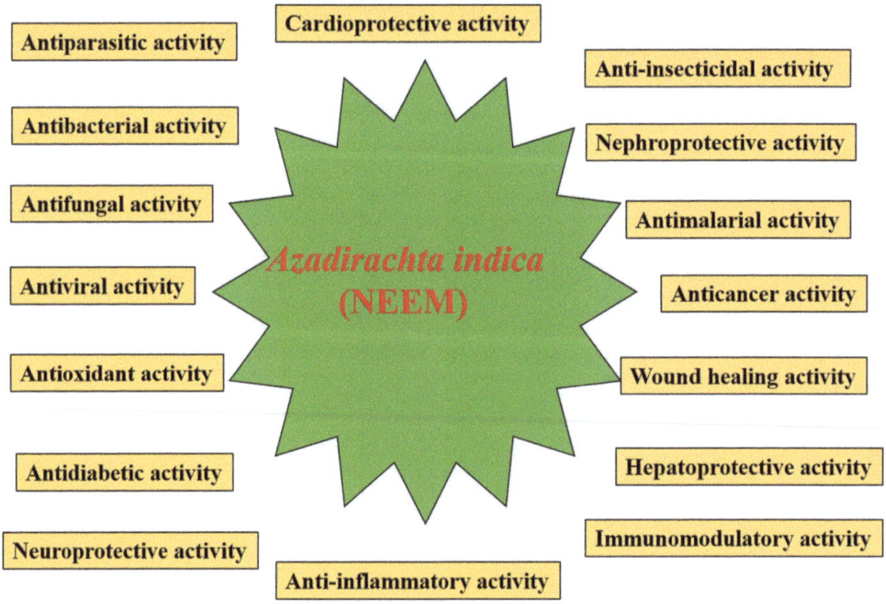


**Fig. 1.2** Pharmacological activities of neem and its bioactive compounds

chlorhexidine, a widely used antiseptic, in lowering plaque and gingivitis. A crossover study revealed that neem mouthwash significantly reduced gingival index scores, indicating its potential as an affordable and accessible option to replace or complement chlorhexidine (Jalaluddin et al. 2017). In a comparative study of 40 chocolate consuming children, it was found that salivary pH was significantly reversed to baseline after chewing leaves of neem and tulsi (*Ocimum sanctum*) (Murthy et al. 2020). A 6-week clinical study assessed the antimicrobial effects of *Azadirachta indica* leaf extract in males aged 20–30 years, comparing a neem extract dental gel to chlorhexidine gluconate mouthwash. The neem gel significantly reduced plaque formation and bacterial counts of *Streptococcus mutans* and *Lactobacilli* species compared to the control group (Pai et al. 2004). Another study found that *Azadirachta indica* mouth rinse was as effective as chlorhexidine in reducing periodontal indices. Both treatments significantly lowered gingival, bleeding, and plaque indices over 21 days compared to placebo. *A. indica* mouth rinse showed comparable efficacy with fewer side effects and may serve as a complementary therapy for plaque-induced gingivitis (Chatterjee et al. 2011). A recent study evaluated the antimicrobial properties of a mouthwash formulated with *Azadirachta indica* and *Ficus benghalensis* against *Staphylococcus aureus*, *Streptococcus mutans*, *Enterococcus faecalis*, and *Candida albicans*. The results showed that the mouthwash exhibited low cytotoxicity and demonstrated strong antimicrobial activity. Therefore, this mouthwash formulation may serve as an effective therapeutic agent for managing oral bacterial infections and as a prophylactic measure (Begum et al. 2025). Another study concluded that a 40% water-soluble azadirachtin formulation is as effective as a 0.2% chlorhexidine mouth rinse in reducing *Streptococcus mutans* levels in dental plaque. This finding suggests that a water-soluble azadirachtin formulation could offer substantial benefits in preventing dental caries, providing a natural and effective alternative for maintaining oral health (Kankariya et al. 2016). The antibacterial, antifungal, antiparasitic, and antiviral properties of different parts of neem against various microbes are illustrated in Fig. 1.3.

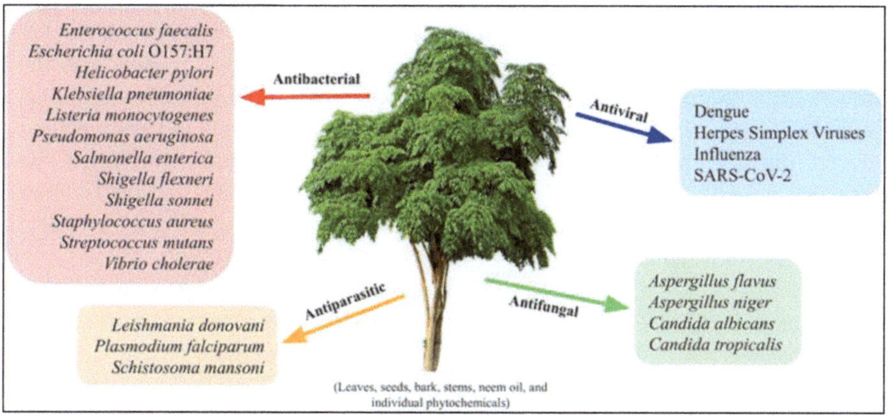


**Fig. 1.3** Representative antimicrobial properties of the different parts (seeds, bark, leaves, oils, and stems) of the *A. indica* tree. (Source: Wylie and Merrell 2022)

## 1.5 Antidiabetic Effects of Neem

Globally, diabetes is one of the most serious health issues and the unavailability of efficient documentation, quality control, and treatment options is the main hurdle in switching to alternative medicines. This comes under the category of diseases that are less curable and more manageable. Therefore, this led to interests in plants for researching phytocompound-based medication for diabetes as well as hypoglycemic compounds. Nimbidin from neem plants at 200 mg/kg effectively delayed post-glucose blood sugar elevation, demonstrating anti-hyperglycemic properties of the plant while another study showed that aqueous neem leaves extract for an equivalent amount of dry leaves in capsules was noted to decrease insulin need by 30–50% in diabetic patients (Khan and Aslam 2008). In a study, the extract of neem helped the pancreas to secrete enough insulin for glucose uptake, and as a result 70% ethanolic alcoholic extract of the root displayed antidiabetic effects owing to glibenclamide, which effectively lowered blood sugar levels in glucose tolerance tests (Patil et al. 2013). Moreover, studies indicated that *Azadirachta indica* enables diabetes mellitus benefits or blood sugar regulation and disease prevention. In comparison to a placebo, neem at dosages of 125, 250, and 500 mg substantially enhanced endothelial function, reduced oxidative stress, and reduced systemic inflammation. It also significantly decreased postprandial blood sugar, fasting blood sugar, and glycosylated hemoglobin (Pingali et al. 2020). In a study, molecular docking simulations identified 3-deacetyl-3-cinnamoyl-azadirachtin as the most promising compound for binding with the diabetes mellitus type 2 enzyme target phosphoenolpyruvate carboxykinase (PEPCK), warranting further in vitro and in vivo validation (Jalil et al. 2013). In a recent in vitro study, butanol fraction of the neem extract showed antioxidant and antidiabetic activities, including superior 2,2-diphenyl-1-picrylhydrazyl (DPPH) scavenging ($IC_{50}$ 0.0154 μg/mL), oxidative stress reduction, and glucose uptake enhancement ($GU_{50}$ 6.22 μg/mL) while the aqueous fraction resulted in the increased ferric reducing antioxidant power (FRAP) activity, and the ethyl acetate and butanol fractions inhibited α-amylase and α-glucosidase, respectively. Gas chromatography-mass spectrometry (GC-MS) analysis identified compounds like sitosterol and stigmasterol, which showed positive molecular docking interactions with AMP-activated protein kinase (α-AMPK), α-amylase, and α-glucosidase (Sanni et al. 2019). The combined extract of *Azadirachta indica* and *Vernonia amygdalina* significantly reduced blood glucose levels in streptozotocin-induced diabetic rats (SDRs) and restored hepatic glucose metabolism and α-amylase activity. While SDR showed decreased serum insulin, T3, and T4 levels, the extract also lowered serum and blood glucose levels in non-diabetic rats (NDRs). After 28 days of treatment, the extract enhanced superoxide dismutase (SOD) activity and alleviated oxidative stress in SDR by reducing glutathione peroxidase (GPx) and catalase (CAT) levels. Histological analysis of the liver and pancreas revealed diabetes-related pathology, which was reversed by the treatment (Atangwho et al. 2012). The effects of water extracts of seeds, stems, flowers, and bark of *Azadirachta indica* on hemoglobin content (Hb) and blood glucose levels in *Rattus norvegicus* were investigated. Varying doses of the extracts were administered every 48 h for 14 days and the results showed a significant reduction in both hemoglobin content and blood glucose levels in all treated groups compared to the control group (Bakr 2012).

## 1.6 Neem in Agriculture and Farming

Neem is a versatile plant with notable agricultural benefits, especially for sustainable farming. Its biopesticidal and soil-enhancing properties, driven by active compounds like azadirachtin, salannin, and nimbin, make it an eco-friendly alternative to chemical pesticides. By improving pest control and soil health, neem supports increased crop yields while reducing environmental impact. Neem products, especially azadirachtin, effectively control insect pests by repelling them, inhibiting feeding, and disrupting their development (Adhikari et al. 2020). These neem-based formulations are biodegradable, safe for non-target organisms, and less likely to cause resistance compared to synthetic pesticides. A study explored the effects of plant growth regulators (PGRs) thidiazuron (TDZ) and 2,4-dichlorophenoxyacetic acid (2,4-D) on callus formation and azadirachtin accumulation in *Azadirachta indica*. Callus induction from both explant types yielded the highest fresh weight, with six-week cultures producing green, brown, and cream-colored callus. Azadirachtin was detected in brown and cream callus, while green callus showed the highest azadirachtin content and was non-toxic to the zebrafish model. These findings highlight the potential of manipulating PGRs in vitro to enhance callus and secondary metabolite production, offering a sustainable approach for producing natural, eco-friendly pesticides (Ashokhan et al. 2020). A study evaluated neem leaf extracts prepared using green technology for their pesticide potential against *Bactrocera dorsalis*. Among the extracts, an eight-hour vinegar extract exhibited the highest insect mortality, though not significantly different from vinegar-fermentation, rice wash, vinegar-maceration, distilled water-fermentation, and rice wash-fermentation extracts while the phytochemical composition of the extracts was influenced by the solvent–extraction interaction (Paragas et al. 2018). The whole neem tree is proven to be extremely beneficial in the field of agriculture. Communities living in both urban and rural areas improve their economic situation by using the short timespan to produce goods from neem seeds and leaves. Neem tree leaves are extensively employed throughout different sectors of agriculture production.

A study evaluated the effectiveness of crude neem leaf powder in goats naturally infected with strongyle parasites. Twenty-four adult goats were treated with neem leaf powder (0.5 g and 1.0 g/kg body weight), resulting in a significant reduction in parasite load. The anthelmintic activity is attributed to neem's alkaloids (Dongre et al. 2015). Further research is recommended to identify the active compounds, determine optimal dosage, establish administration methods, and assess parasite susceptibility, to support neem's potential as a herbal anthelmintic. Phytogenic supplements, such as neem leaf powder, have the potential to enhance rumen fermentation and growth performance in lambs. Compared to monensin, neem leaf powder supplementation increased body weight gain without negatively affecting rumen fermentation, nutrient intake, or digestibility. This suggests that neem leaf powder could serve as an effective natural alternative to synthetic growth promoters in sheep diets (Taethaisong et al. 2023). Another study showed that *Azadirachta indica* could be used as a feed additive that will be safe and inexpensive, as *Azadirachta indica* acts as an anti-methanogenic agent for the lambs and does not compromise the quality of meat (Al-Marzooqi et al. 2022). Another study reported that neem leaf powder served as a potential phytogenic feed for lambs as it increased growth performance without any adverse effects and also promoted body weight in growing lambs (Orzuna-Orzuna et al. 2024). Apart from this, a study on broilers showed increased body weight owing to the positive effects of neem leaf extracts as supplement in comparison to antibiotic growth promoters (Paul et al. 2020). These kinds of studies show huge potential of *Azadirachta indica* to be used as an important supplement for animals as neem could be used to improve their life quality. Supplementing goat diets with sole neem leaf meal instead of concentrate mix and pigeon pea leaf meal yielded similar growth performance. Additionally, neem and pigeon pea leaf supplementation significantly improved dressing percentage based on slaughter and empty body weight (El-Zaiat et al. 2022). Economically, replacing concentrate mix and pigeon pea leaf meal with neem leaf meal reduced feed costs and increased net returns, highlighting its cost-effectiveness.

## 1.7 Antiparasitic and Insecticidal Effects of Neem

Indian Ayurvedic practitioners have used neem decoction in water since time immemorial as oral treatment for malarial fevers. Documents from the Indian medicine literature reveal antimalarial properties in neem from 2000 BC as well as 1500 BC. It was demonstrated that neem successfully inhibited both larval development and organism emergence of malaria, filariasis, and dengue fever vectors (Gunasekaran et al. 2009). A study evaluated the anthelmintic activity of citronella and neem oils on *Fasciola gigantica*. Citronella oil (1%) exhibited a significant in vitro anthelmintic effect, while neem oil showed no significant activity. The study assessed gross motility and used score indices, alongside morphological and histopathological evaluations, to determine treatment effects compared to controls (Jeyathilakan et al. 2010). Coccidiosis and helminthiasis are two parasitic diseases that adversely affect health. A study investigated the antiparasitic potential of neem leaf extracts in murine models of coccidiosis and helminthiasis. The extracts induced paralysis, mortality, and structural deformities in worms, with scanning electron microscopy revealing cuticular shrinkage. In vitro, neem extract significantly inhibited oocyst sporulation and exhibited notable antioxidant activity, highlighting its potential as a natural antiparasitic agent (Abu Hawsah et al. 2023). Neem has most of its active compounds as fatty acids and terpenoids that include the well-known insect repellent effect of azadirachtin alongside multiple other beneficial actions. *Azadirachta indica* has shown strong repellent effects against *Phlebotomus papatasi* (Zatelli et al. 2022). A neem-based insecticide formulation was evaluated for controlling the western cherry fruit fly (*Rhagoletis indifferens*). Field tests showed no significant behavioral changes in adult flies exposed to the insecticide. However, ingestion of the extract immediately after eclosion led to reduced adult survival and egg development, with continuous exposure completely suppressing egg maturation. Applied as a foliar spray at the right time in cherry orchards, the neem-based insecticide could effectively reduce viable egg production and adult survival, offering a promising pest management tool. Neem oil demonstrated high toxicity against pests, with an $LC_{50}$ after eight days, five times more toxic than corn oil and inhibited 50% of adult emergence at 6 ppm. Concentrations above 8 ppm significantly reduced growth indices, pupation, and prolonged larval periods, while corn oil showed no disruption in development (Okumu et al. 2007). Another study assessing neem extracts against *Chrysomya bezziana* larvae found methanol extract to be the most toxic, followed by chloroform, petroleum ether, and ethyl acetate, across both dipping and thin film application methods, though $LC_{50}$ values varied by extract and method (Singh and Kaur 2016). A field study evaluated the repellent efficacy of neem oil (5 mL/L) against *Aedes albopictus* adults. Conducted across 6 locations over 6 months (May–October), data were collected weekly from 900 ovitrap samplings. Repellency was highest during June and July, with monitoring of daily temperature and rainfall confirming seasonal variations in mosquito activity (Macchioni et al. 2019).

## 1.8 Traditional Remedies and Ayurvedic Formulations

Traditional practices give us an account of neem leaves, and their pastes are employed to treat allergic skin responses and viral infections of smallpox and chickenpox (Singh et al. 2021). A significant portion of urban population from Nepal and India along with Bangladesh clean their teeth with neem twigs. People use the leaf juice as both a tonic to enhance appetite and an intestine worm plant-based removal remedy (Darestani and Basheer 2023). A few studies show that neem extract substances efficiently restrict viral replication indicating its possible use in the management of viral diseases such as herpes and dengue (Tiwari et al. 2010). The anti-inflammatory components of neem act to reduce the discomfort produced by skin rashes caused by viral infections (Singh et al. 2021). Medical practitioners have been using neem flower extracts to manage the skin disorders and improve gut microbiota as well as combat viruses and normalize body temperature. This includes utilizing neem flowers to cleanse the liver and boost its functioning (Seriana et al. 2021). The medical traditional usage of this substance includes removal of parasites from the intestines. People use neem flower extracts as both pastes and decoctions as they possess antibacterial and anti-inflammatory properties that help in treating various skin conditions. The extracts made from neem flowers help promote hair growth and fight dandruff while the application of face masks utilizes neem to treat acne along with reducing skin blemishes and cleansing the face (Dua and Sharma 2023; Niharika et al. 2010).

## 1.9 Neem in Pharmaceutics and Biotechnological Applications

As neem possesses antibacterial and anti-inflammatory characteristics, its oil serves as a therapeutic treatment for acne along with eczema and fungal infections of the skin (Arora et al. 2022). It has also been reported that neem bark extract demonstrated exceptional effectiveness against various microbial species through its extensive antimicrobial properties that help quality enrichment of oral health products. A number of cosmetic manufacturers adopt neem in their products as it decreases inflammation while avoiding acne and improving skin conditions (Baby et al. 2022). The effectiveness of neem extract has been documented for livestock treatment of multiple parasitic infections that affect ticks, fleas, and internal worms (Schmahl et al. 2010). The process of developing the neem nutraceuticals relies heavily on in silico methods for structural identification, interaction between receptor and ligand, and drug development processes, reducing both biological and financial expenses. When it comes to delivering nutraceuticals to human body, hydrogel is considered to be one of the best biopolymorphic systems in a number of nutraceuticals carrier. Another known health supplements are neem leaf bitters; as the name suggests, they are made from neem leaves. These bitters are effectively known to manage gut health, involving secretion of digestive enzymes and encouraging healthy bowel movements (Dare 2022). Antioxidants contained in a neem extract enable the ability to counteract free radicals in the body (Shareef and Akhtar 2018). For example, neem leaf extract is full of antioxidants and is considered best for the detoxification, serving as a nutraceutical for detoxifying liver and boosting the immunity (Iman et al. 2022).

Indica derivatives are advantageous agents, as they are environment friendly, biodegradable, and less toxic toward non-target organisms like pollinators and vertebrates. Plant cell/hairy root culture facilitates mass scale production in bioreactors of such compounds of interest in neem. An endophytic fungus obtained from neem plant produces both azadirachtin A and B. The fungus was identified through molecular analysis that involved examining its region of rDNA consisting of ITS1 along with ITS2 and the 5.8S rRNA gene sequence in between them. Endophytic fungal compounds were derived by performing liquid chromatography-high resolution mass spectrometry (LC-HRMS) analysis then comparing their results to authentic reference standards (Kusari et al. 2012). The research on endophytic microflora in neem plants has previously explored the diversity and distribution of endophytic microbes throughout the plants. Extracts isolated from neem plant leaves represent the most accessible biomolecules available for use as attractive building elements when constricting environment-friendly resistive random-access memory (RRAM) devices because they hold important characteristics such as biodegradability, optical transparency, and light weight (Sharma et al. 2023).

## 1.10 Neem as a Functional Food

Scientists have identified neem seeds as the most important plant component because they contain numerous active substances and oil. The presence of certain physiochemical and functionality properties helps us know whether a plant protein can be consumed as an industrial novel food product or not. Such properties are usually influenced by structural characteristics of the proteins. The structure-function property of proteins can moreover be changed by the interaction of proteins with other biomolecules such as lipid-soluble complexes and lipids. The whole neem tree is considered to be extremely beneficial in the field of agriculture. Communities living in both urban and rural areas improve their economic situation by using the short timespan to produce goods from neem seeds and leaves. Neem tree leaves are extensively employed throughout different sectors of agriculture production. Several studies indicate that neem leaf flour contains substantial amounts of crude protein along with dietary fibers except limited quantities of crude fat as well as ash (Andersa et al. 2024). Testing revealed azadirachtin together with alkaloids, sodium nimbinate, nimbin, salannin, nimbidin, and quercetin are the active elements found within neem leaf flour. The carotenoid content in neem flour serves as a primary source for food coloring as it is a pigment, while the presence of glycoproteins is crucial for healthy functioning of human physiology (Sageer et al. 2024). One more observed fact is that incorporating neem leaf flour extends both the life span and superior quality of tomatoes (Hosea et al. 2017). Despite all the mentioned benefits of neem leaf flour, it may cause low toxicity as well as sub-chronic toxicity in humans. The problem can, however, be solved by utilizing lower usage amounts. The active ingredients in flour play crucial roles in supplying nutrients (Boeke et al. 2004). Overall, because of its pharmacological properties, neem is considered to be a potential candidate for biopreservative and therapeutic use. The phytochemical analysis of neem leaves reveals high concentration of alkaloids, triterpenes, flavonoids, and saponins.

## 1.11 Conclusion

*Azadirachta indica* is renowned in traditional medicine in Ayurveda and Unani for its broad therapeutic potential. This chapter highlights neem's pharmacological power, driven by its rich bioactive profile including terpenoids, flavonoids, alkaloids, and limonoids, which underpins its antimicrobial, anti-inflammatory, antioxidant, antidiabetic, and antiparasitic properties. Neem is comparable to chlorhexidine in reducing plaque and gingivitis, improving glycemic control, and controlling insulin sensitivity. Neem's potential extends to pharmaceuticals, nutraceuticals, and eco-friendly materials, contributing to functional foods and natural preservation systems.

## References

Abhishek KN, Supreetha S, Sam G, Khan SN, Chaithanya KH, Abdul N (2015) Effect of neem containing toothpaste on plaque and gingivitis—a randomized double blind clinical trial. J Contemp Dent Pract 16(11):880–883

Abu Hawsah M, Al-Otaibi T, Alojayri G, Al-Shaebi EM, Dkhil MA, Elkhadragy MF et al (2023) In vitro studies for the antiparasitic activities of Azadirachta indica extract. Food Sci Technol 43:e117122

Adeoye AE, Adeoye AM, Mufliat M (2024) Comprehensive analysis of nutrient composition: evaluating vitamins, essential minerals, and trace metals in neem (Azadirachta indica) stem bark extract. Earthline J Chem Sci 11(2):283–296

Adhikari K, Bhandari S, Niraula D, Shrestha J (2020) Use of neem (Azadirachta indica A. Juss) as a biopesticide in agriculture: a review. J Agric Appl Biol 1(2):100–117

Al-Marzooqi W, Sallam SM, Alqaisi O, El-Zaiat HM (2022) Potential of graded doses of neem (Azadirachta indica) seed oil on ruminal fermentation characteristics, degradability, and methane formation in vitro. Ann Anim Sci 22(3):993–999

Andersa KN, Tamiru M, Teka TA, Ali IM, Chane KT, Regasa TK, Ahmed EH (2024) Proximate composition, some phytochemical constituents, potential uses, and safety of neem leaf flour: a review. Food Sci Nutr 12(10):6929–6937

Arora P, Shiveena B, Garg M, Kumari S, Goyal A (2022) Curative potency of medicinal plants in management of eczema: a conservative approach. Phytomed Plus 2(2):100256

Ashokhan S, Othman R, Abd Rahim MH, Karsani SA, Yaacob JS (2020) Effect of plant growth regulators on coloured callus formation and accumulation of azadirachtin, an essential biopesticide in Azadirachta indica. Plants 9(3):352

Atangwho IJ, Ebong PE, Eyong EU, Asmawi MZ, Ahmad M (2012) Synergistic antidiabetic activity of Vernonia amygdalina and Azadirachta indica: biochemical effects and possible mechanism. J Ethnopharmacol 141(3):878–887

Baby AR, Freire TB, Marques GDA, Rijo P, Lima FV, Carvalho JCMD et al (2022) Azadirachta indica (Neem) as a potential natural active for dermocosmetic and topical products: a narrative review. Cosmet 9(3):58

Bakr A (2012) Changes of hemoglobin content and glucose levels in the blood of Rattus norvegicus by water extracts of Azadirachta indica. Chin J Nat Med 10(2):135–137

Begum NF, Reethega L, Abilasha R, Ramani P, Rajeshkumar S, Krishnan RP (2025) Assessment of antimicrobial and cytotoxic activity of Azadirachta indica and Ficus benghalensis herbal formulation for orthodontic patients mouthwash preparation—and invitro study. In: Advances in sports science and technology. CRC Press, Boca Raton, pp 930–936

Boeke SJ, Boersma MG, Alink GM, van Loon JJ, van Huis A, Dicke M, Rietjens IM (2004) Safety evaluation of neem (Azadirachta indica) derived pesticides. J Ethnopharmacol 94(1):25–41

Chatterjee A, Saluja M, Singh N, Kandwal A (2011) To evaluate the antigingivitis and antipalque effect of an Azadirachta indica (neem) mouthrinse on plaque induced gingivitis: a double-blind, randomized, controlled trial. J Indian Soc Periodontol 15(4):398–401

Dare AKO (2022) Effects of neem (Azadirachta indica) and bitter leaf (Vernonia amygdalina) ethanolic extracts as additives on the performance of West African Dwarf Sheep. Master's thesis, Kwara State University, Nigeria

Darestani AS, Basheer MKA (2023) A review of medicinal herbs in the treatment of common human disease. Aust Herb Insight 6(1):1–13

Dongre S, Das G, Nath S, Dixit AK, Agrawal V (2015) Anthelmintic efficacy of Azadirachta indica (neem) against strongyles in goats. Indian J Vet Sci Biotechnol 10(3):18–21

Dua S, Sharma M (2023) Preparation of herbal face pack cum scrub using papaya and neem extracts. B.Tech (Biotechnology) Thesis. Department of Bt & Bi, Jaypee University of Information Technology Waknaghat, Solan-173234, Himachal Pradesh. http://www.ir.juit.ac.in:8080/jspui/handle/123456789/10165

El-Zaiat HM, Elshafie EI, Al-Marzooqi W, Dughaishi KA (2022) Effects of neem (Azadirachta indica) leaf powder supplementation on rumen fermentation, feed intake, apparent digestibility and performance in Omani sheep. Animals 12(22):3146

Gahtani RM, Shoaib S, Hani U, Jayachithra R, Alomary MN, Chauhan W et al (2024) Combating Parkinson's disease with plant-derived polyphenols: targeting oxidative stress and neuroinflammation. Neurochem Int 178:105798

Gunasekaran K, Vijayakumar T, Kalyanasundaram M (2009) Larvicidal & emergence inhibitory activities of NeemAzal T/S 1.2 per cent EC against vectors of malaria, filariasis & dengue. Indian J Med Res 130(2):138–145

Hosea ZY, Liamngee K, Owoicho AL, David T (2017) Effect of neem leaf powder on post harvest shelf life and quality of tomato fruits in storage. Int J Dev Sustain 6(10):1334–1349

Ilesanmi J, Gungula DT (2016) Amino acid composition of cowpea grains preserved with mixtures of neem (Azadirachta indica) and moringa (Moringa oleifera) seed oils. Am J Food Nutr 4(6):150–156

Iman M, Taheri M, Bahari Z (2022) The anti-cancer properties of neem (Azadirachta indica) through its antioxidant activity in the liver: its pharmaceutics and toxic dosage forms. A literature review. J Complement Integr Med 19(2):203–211

Islas JF, Acosta E, Zuca G, Delgado-Gallegos JL, Moreno-Treviño MG, Escalante B, Moreno-Cuevas JE (2020) An overview of Neem (Azadirachta indica) and its potential impact on health. J Funct Foods 74:104171

Jahan R, Malik S, Shoaib S, Alomary MN, Ansari MA, Islam N, Khan S (2024) Medicinal and nutritional importance of Linum usitatissimum L. in human health. In: Medicinal plants and their bioactive compounds in human health, vol 1. Springer Nature Singapore, Singapore, pp 25–45

Jalaluddin M, Rajasekaran UB, Paul S, Dhanya RS, Sudeep CB, Adarsh VJ (2017) Comparative evaluation of neem mouthwash on plaque and gingivitis: a double-blind crossover study. J Contemp Dent Pract 18(7):567–571

Jalil A, Ashfaq UA, Shahzadi S, Javed MR, Rasul I, Rehman SU et al (2013) Screening and design of anti-diabetic compounds sourced from the leaves of neem (Azadirachta indica). Bioinformation 9(20):1031

Jeyathilakan N, Murali K, Anandaraj A, Latha BR, Basith SA (2010) Anthelmintic activity of essential oils of Cymbopogan nardus and Azadirachta indica on Fasciola gigantica. J Vet Anim Sci 6(6):204–209

Kankariya AR, Patel AR, Kunte SS (2016) The effect of different concentrations of water soluble azadirachtin (neem metabolite) on Streptococcus mutans compared with chlorhexidine. J Indian Soc Pedod Prev Dent 34(2):105–110

Khan SA, Aslam J (2008) Study on the effect of neem (Azadirachta indica) leaves smoke in controlling airborne bacteria in residential premises. Curr Res Bacteriol 1(2). https://doi.org/10.3923/crb.2008.64.66

Kusari S, Verma VC, Lamshoeft M, Spiteller M (2012) An endophytic fungus from Azadirachta indica A. Juss. that produces azadirachtin. World J Microbiol Biotechnol 28(3):1287–1294

Macchioni F, Sfingi M, Chiavacci D, Cecchi F (2019) Azadirachta indica (Sapindales: Meliaceae) neem oil as a repellent against Aedes albopictus (Diptera: Culicidae) mosquitoes. J Insect Sci 19(6):12

Murthy GS, Murthy SS, Kavitha M (2020) Effect of chewing Azadirachta indica (neem) and Ocimum sanctum (tulsi) leaves on salivary acidogenicity: a comparative study. J Oral Maxillofac Pathol 24(3):479–483

Niharika A, Aquicio JM, Anand A (2010) Antifungal properties of neem (Azadirachta indica) leaves extract to treat hair dandruff. E-ISRJ 2:244–252

Obikaonu HO (2012) Evaluation of the nutritional value of neem (Azadirachta indica) leaf meal on the performance of finisher broilers. Int J Agric Rural Dev 15(3):1235–1239

Okumu FO, Knols BG, Fillinger U (2007) Larvicidal effects of a neem (Azadirachta indica) oil formulation on the malaria vector Anopheles gambiae. Malar J 6:1–8

Orzuna-Orzuna JF, Lara-Bueno A, Gloria-Trujillo A, Mendoza-Martínez GD, Miranda-Romero LA, Hernández-García PA (2024) Growth performance, dietary energetics, blood metabolites, carcass traits, meat quality, and gene expression of lambs supplemented with a polyherbal phytogenic additive. Vet Sci 11(11):520

Pai MR, Acharya LD, Udupa N (2004) Evaluation of antiplaque activity of Azadirachta indica leaf extract gel—a 6-week clinical study. J Ethnopharmacol 90(1):99–103

Paragas DS, Cruz K, Fiegalan ER (2018) Assessment of green solvents and extraction methods for biopesticide preparation from neem Azadirachta indica leaves against Oriental fruit fly Bactrocera dorsalis (Hendel). Preprints. https://doi.org/10.20944/preprints201805.0179.v1

Patil P, Patil S, Mane A, Verma S (2013) Antidiabetic activity of alcoholic extract of neem (Azadirachta indica) root bark. Natl J Physiol Pharm Pharmacol 3(2):142

Paul TK, Hasan MM, Haque MA, Talukder S, Sarker YA, Sikder MH et al (2020) Dietary supplementation of Neem (Azadirachta indica) leaf extracts improved growth performance and reduced production cost in broilers. Vet World 13(6):1050

Pingali U, Ali MA, Gundagani S, Nutalapati C (2020) Evaluation of the effect of an aqueous extract of Azadirachta indica (Neem) leaves and twigs on glycemic control, endothelial dysfunction and systemic inflammation in subjects with type 2 diabetes mellitus–a randomized, double-blind, placebo-controlled clinical study. Diabetes Metab Syndr Obes 13:4401–4412

Sageer SM, Sharma PK, Bawazir AS (2024) Plant ingredient screening as potential excipients for pharmaceutical dosage forms. Biochem Cell Arch 24(Suppl. 1):4149–4156

Sanni O, Erukainure OL, Chukwuma CI, Koorbanally NA, Ibeji CU, Islam MS (2019) Azadirachta indica inhibits key enzyme linked to type 2 diabetes in vitro, abates oxidative hepatic injury and enhances muscle glucose uptake ex vivo. Biomed Pharmacother 109:734–743

Schmahl G, Al-Rasheid KA, Abdel-Ghaffar F, Klimpel S, Mehlhorn H (2010) The efficacy of neem seed extracts (Tre-san®, MiteStop®) on a broad spectrum of pests and parasites. Parasitol Res 107:261–269

Seriana I, Akmal M, Darusman D, Wahyuni S, Khairan K, Sugito S (2021) Neem leaf (Azadirachta indica A. Juss) ethanolic extract on the liver and kidney function of rats. Sci World J 2021(1):7970424

Shareef M, Akhtar MS (2018) Neem (Azadirachta indica) and its potential for safeguarding health, prevention and treatment of diseases. Matrix Sci Medica 2(1):04–08

Sharma N, Singh K, Tripathi CC, Bera MK (2023) Resistive switching in neem (Azadirachta indica) thin film for a cost-effective and washable biomemristor. J Mater Sci Mater Electron 34(1):50

Shoaib S, Khan FB, Alsharif MA, Malik MS, Ahmed SA, Jamous YF et al (2023) Reviewing the prospective pharmacological potential of isothiocyanates in fight against female-specific cancers. Cancer 15(8):2390

Shoaib S, Jahan R, Rehman AS, Shadab M, Chauhan W, Alomary MN et al (2024) Medicinal and nutritional importance of curcuma longa in human health. In: Medicinal plants and their bioactive compounds in human health, vol 1. Springer Nature Singapore, Singapore, pp 1–24

Singh A, Kaur J (2016) The bioefficacy of crude extracts of Azadirachta indica (Meliaceae) on the survival and development of myiasis-causing larvae of Chrysomya bezziana (Diptera: Calliphoridae). Trop Anim Health Prod 48(1):117–124

Singh V, Roy M, Garg N, Kumar A, Arora S, Malik DS (2021) An insight into the dermatological applications of neem: a review on traditional and modern aspect. Recent Adv Antiinfect Drug Discov 16(2):94–121

Somnathe S, Madhu PP, Chhabra KG, Chapade A, Gothane D (2022) Comparison of plaque removal efficacy of toothbrush with bristle containing neem and charcoal for maintaining oral hygiene: a randomised clinical trial. J Clin Diagn Res 16(10). https://doi.org/10.7860/JCDR/2022/55416.17036

Taethaisong N, Paengkoum S, Kaewwongsa W, Onjai-Uea N, Thongpea S, Paengkoum P (2023) The effect of neem leaf supplementation on growth performance, rumen fermentation, and ruminal microbial population in goats. Animals 13(5):890

Tiwari V, Darmani NA, Yue BY, Shukla D (2010) In vitro antiviral activity of neem (Azardirachta indica L.) bark extract against herpes simplex virus type-1 infection. Phytother Res 24(8):1132–1140

Trivedi A, Fatima N, Husain I, Misra A (2019) An update on the therapeutic potential of neem and its active constituents: a panacea for all diseases. Era's J Med Res 6(1):110–117

Weir SA, Kc K, Shoaib S, Yusuf N (2023) The immunotherapeutic role of type I and III interferons in melanoma and non-melanoma skin cancers. Life 13(6):1310

Wylie MR, Merrell DS (2022) The antimicrobial potential of the neem tree Azadirachta indica. Front Pharmacol 13:891535

Zatelli A, Fondati A, Maroli M, Canine Leishmaniosis Working Group (2022) The knowns and unknowns of the efficacy of neem oil (Azadirachta indica) used as a preventative measure against Leishmania sand fly vectors (Phlebotomus genus). Prev Vet Med 202:105618

Zeeshan A, Amjad W, Masood M, Akram W, Yameen I, Mansoor M et al (2024) Neem's bioactive marvels: a therapeutic review. J Health Rehabil Res 4(1):185–195

# Medicinal and Nutritional Importance of *Nicotiana* spp. in Human Health

2

Shoaib Shoaib, Roshan Jahan, Mohammad N. Alomary, Md. Abul Barkat, Saad Alghamdi, and Mohammad Azam Ansari

**Abstract**

*Nicotiana* genus represents a remarkable reservoir of bioactive compounds with excellent pharmacological potential. Accumulating evidences indicate that nicotine, flavonoids, terpenoids, and glycoalkaloids are among many specific metabolites of the *Nicotiana* species that have been evaluated for their anti-inflammatory, antibacterial, analgesic, antioxidant, and neuropharmacological properties. Flavonoid-mediated antioxidant protection against oxidative stress and cembranoid-based anti-inflammatory properties emphasize the importance of *Nicotiana*. The plant can exhibit outstanding antibacterial and antiparasitic effects through

S. Shoaib
Department of Genetics, School of Medicine, University of Alabama at Birmingham, Birmingham, AL, USA

R. Jahan
Department of Botany, Faculty of Lifesciences, Aligarh Muslim University, Aligarh, Uttar Pradesh, India

M. N. Alomary
Applied Genomic Technologies Institute, King Abdulaziz City for Science and Technology (KACST), Riyadh, Saudi Arabia

M. A. Barkat
Department of Pharmaceutics, College of Pharmacy, University of Hafr Al-Batin, Hafr Al-Batin, Saudi Arabia

S. Alghamdi
Department of Clinical Laboratory Sciences, Faculty of Applied Medical Sciences, Umm Al-Qura University, Makkah, Saudi Arabia

M. A. Ansari (✉)
Department of Epidemic Diseases Research, Institute for Research and Medical Consultations (IRMC), Imam Abdulrahman Bin Faisal University, Dammam, Saudi Arabia
e-mail: maansari@iau.edu.sa

 
M. A. Ansari et al. (eds.), *Medicinal Plants and their Bioactive Compounds in Human Health: Volume 2*, https://doi.org/10.1007/978-981-95-3620-7_2

immunomodulation, enzyme inhibition, and membrane disruption. Although its promising applications concern about nicotine's toxicity, bioactive compounds' standardization and clinical validation still exist. This chapter summarizes medicinal chemistry of *Nicotiana* while highlighting the need of further studies. *Nicotiana* species provide unmet promise for developing novel therapies to address global health issues, such as antibiotic resistance and neurological illnesses, by integrating ethnobotany with contemporary pharmacology.

**Keywords**

*Nicotiana* · Phytocompounds · Health · Disease · Medicine

## 2.1 Introduction

Plants and their products have long been utilized by humans for different purposes including a few as wood, food, medicines, diseases prevention, etc. (Jahan et al. 2024). In recent decades plants and their bioactive compounds have been investigated for the prevention of treatment of various diseases such as neurodegeneration, cancer, oxidative stress, inflammation, pathogenic infections, skin problems, and liver and gastric problems (Gahtani et al. 2024; Weir et al. 2023; Shoaib et al. 2022). *Nicotiana* genus, a member of the Solanaceae, possesses almost 82 species that are native to America; however, a significant portion belongs to Australia's dry regions. *Nicotiana* species exhibit a wide range of morphological characteristics, including variations in leaf, flower structure, and growth patterns. Some species have evolved specialized root systems and/or leaf structures that enable them to endure dry conditions, and many species have evolved to live in arid environments (Poczai et al. 2023). The richness of *Nicotiana* species is reflected in their distribution across multiple continents. The majority of species are found in the Americas, while many endemic species are found in Australia. Additionally, some species of tobacco have naturalized in regions outside of their original range, and the genus has spread throughout the world as weeds or ornamental plants (Berbeć 2024). Different metabolites such as nicotine, nornicotine, and anabasine are produced by *Nicotiana* species that are essential to the ecological interactions of plants, serving as defense systems against different pathogens including viruses and herbivores (Jassbi et al. 2017). These metabolites have biological activity with medicinal potential underlining the molecules' ecological and evolutionary significance. Archaeological evidence from the Wishbone site in North America indicates that tobacco was used as early as 12,300 years ago, which is much earlier than previously recorded. Humans have been utilizing *Nicotiana* species for thousands of years. According to this evidence, the plant was widely employed for a number of functions, including cultural and ceremonial rituals, and was known to the ancient North American population (Duke et al. 2021). In many indigenous civilizations, *Nicotiana* species were highly valued for their cultural and spiritual significance in addition to their hallucinogenic properties. The therapeutic properties of *Nicotiana* species have been recognized by numerous traditional civilizations. In the Peruvian Amazon, *Nicotiana rustica* is utilized to treat mental health conditions; ritualistic use of the herb greatly enhances patient welfare (Berlowitz et al. 2024).

## 2.2 Botanical Description and Phytochemistry of *Nicotiana* Plant

Due to its numerous uses in industry, agriculture, and medicine, *Nicotiana*, and particularly *Nicotiana tabacum* L., is a plant of botanical and phytochemical importance. It is renowned for having a diverse phytochemical profile that includes a range of bioactive substances with possible uses in industry and medicine. The herbaceous plant *Nicotiana tabacum* can grow up to two meters in height. It features tubular blooms that are frequently pink or white in color, as well as with long, wide leaves. *Nicotiana tabacum* roots have long been used for their tonic and sedative properties. Under a microscope, tobacco roots were found to include calcium oxalate crystals, pitted arteries, and cork cells (Sunil et al. 2012). Although *Nicotiana tabacum* contains additional alkaloids that contribute to its pharmacological activities, nicotine is the most well-known alkaloid. Flavonoids and phenols, which are abundant in tobacco, help to explain its antioxidant properties. Virginia tobacco leaf ethanolic extract exhibits strong antioxidant properties and a high phenolic content. Different parts of the plant contain terpenoids and steroids, which have antibacterial and anti-inflammatory properties (Prommaban et al. 2022). The tobacco also contains phytosterols, tannins, and saponins, all of which are believed to contribute to a number of health benefits, such as anti-inflammatory and antibacterial qualities. Tobacco is being investigated for anti-aging and skin health-conscious uses because of its antioxidant and antibacterial qualities (Zhang et al. 2024).

## 2.3 Toxicity and Health Hazards of *Nicotiana* Plant

*Nicotiana tabacum* and *Nicotiana glauca* contain alkaloids, such as nicotine and anabasine, which make them highly toxic and injurious to health. Serious health issues like neurological symptoms and respiratory failure might result from these substances. Nicotine is well known for its addictive qualities and negative health effects, such as cardiovascular and pulmonary conditions. Anabasine, an agonist on nicotinic acetylcholine receptors found in *Nicotiana glauca*, produces symptoms such as tachycardia, paralysis, and respiratory arrest when taken (Kear et al. 2020). Nicotine causes cancer and other major health problems, including harm to the reproductive system. The addictive qualities of nicotine cause long-term health issues associated with tobacco use, including cardiovascular and respiratory conditions, hormonal imbalances, and reproductive health issues, particularly with reference to fertility (Bhardwaj et al. 2025). Studies displayed a significant risk of oral cancers and pancreatic cancer from smokeless tobacco use, especially in areas like India where tobacco is widely used.

## 2.4 Nutritional Properties of *Nicotiana* spp.

The nutritional aspects of the tobacco plant, especially *Nicotiana tabacum*, include its protein content, phytochemical content, and mineral nutrient requirements. Comprehending these components is essential for maximizing tobacco production and investigating its possible health advantages. In addition to its psychotropic qualities, the *Nicotiana* plant has a number of nutritionally significant substances, such as fatty acids, essential amino acids, macronutrients, micronutrients, and secondary metabolites. Even though tobacco is harmful and therefore not utilized as a food, research on its nutritional profile is crucial for the development of plant-based bioproducts, phytochemicals, and biofortification. *Nicotiana tabacum* needs a balanced supply of micronutrients (iron, manganese, and zinc) and macronutrients (nitrogen, phosphorus, and potassium) for optimal growth, yield, and quality. The accumulation of micronutrients occurs primarily in the leaves, where iron is being the most absorbed, followed by manganese, boron, zinc, and copper (Cremonesi et al. 2019). Although tobacco leaves also contains low level of macronutrients, but they are significant from agronomic and phytobiological perspectives (Henry et al. 2018). These nutrients add to the biochemical diversity of the leaves in addition to having an impact on the plant's agronomic performance. The study of tobacco's nutritional makeup is nevertheless crucial for plant science, agronomy, and phytochemical engineering even though it is not directly ingested by people. Precision agriculture, especially in areas where tobacco is grown, depends on an understanding of the macronutrient requirements of tobacco. Increased leaf yield and decreased environmental contamination are two benefits of balanced fertilization (Lisuma et al. 2020).

## 2.5 Medicinal and Pharmacological Properties

Globally, plants and their bioactive compounds have gained tremendous attention due to their health benefiting properties. Their compounds have been evaluated for antiproliferative, antioxidant, anti-inflammatory, antidiabetic, antiparasitic, antibacterial including anti-SARS-CoV-2 (Severe Acute Respiratory Syndrome Coronavirus-2) and *Mycobacterium* as well as their hepatoprotective and cardioprotective effects (Ansari et al. 2023; Shoaib et al. 2023; Azmi et al. 2022). *Nicotiana tabacum* in particular, has many well-known pharmacological and medicinal qualities (Fig. 2.1). Many cultures have long used the plant for its therapeutic qualities, but more recent research has found a number of bioactive compounds that boost the plant's potential for application in pharmaceutical development (Al-Snafi 2022). *Nicotiana tabacum* contains a wide range of chemical components, including flavonoids, terpenoids, alkaloids, and phenolic compounds.

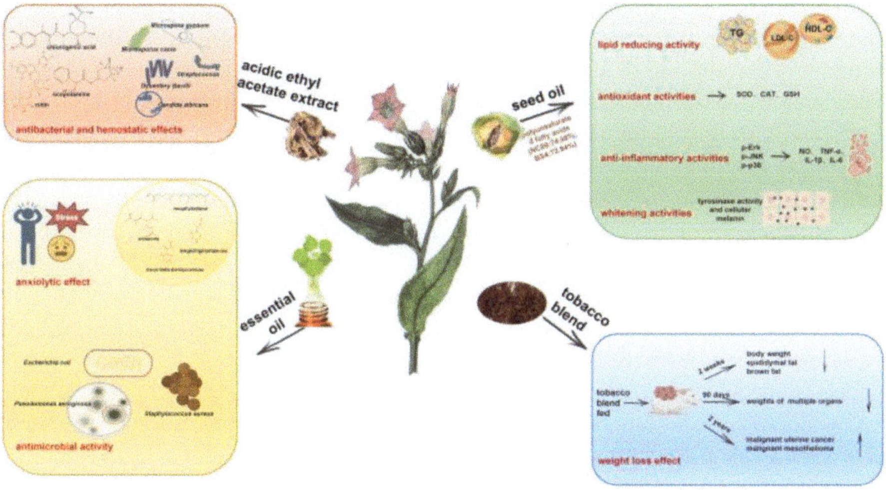


**Fig. 2.1** Biological properties of different parts of tobacco plant extracts. (Source: Zhang et al. 2024)

## 2.6 Anti-inflammatory and Analgesic Effects

Significant anti-inflammatory effects are exhibited by *Nicotiana* spp. through a variety of compounds, and several phytochemical components have been linked to anti-inflammatory effects due to their interactions with a range of biological targets. According to molecular simulations, *Nicotiana tabacum*-derived rutin and chlorogenic acid have significant binding affinities with important inflammatory mediators like cyclooxygenase-1 (COX-1), COX-2, and tumor necrosis factor-alpha (TNF-α), suggesting that they may have anti-inflammatory properties (Patricia et al. 2024). Particularly, rutin exhibits a high binding energy, suggesting that it may be able to reduce inflammation through molecular interactions. Nicotine, an alkaloid with both pro- and anti-inflammatory properties, is found in *Nicotiana*. Through vagus nerve activation and immunological modulation, it helps lessen inflammation in diseases like ulcerative colitis and arthritis (Van Der Zanden et al. 2009). However, the effects of nicotine vary according to the environment and dosage because it can also exacerbate oral inflammations (Zhang et al. 2022). Extracts from *Nicotiana rustica* have demonstrated anti-inflammatory properties in animal experiments that are similar to those of common medications like diclofenac, suggesting potential for use in medicine (Bensalah et al. 2009). *Nicotiana plumbaginifolia* polymethoxyflavone also has analgesic and anti-inflammatory effects by triggering opioids and ATP-sensitive K+ channels (Shajib et al. 2018). Additionally, the high quantities of polyphenols in *Nicotiana tabacum* leafy galls caused by *Rhodococcus fascians* enhance their anti-inflammatory properties and justify their use in traditional medicine. However, depending on the situation and dosage, the plant's bioactive compounds can have both positive and negative effects.

To completely comprehend these dynamics and utilize the medicinal potential of compounds derived from *Nicotiana*, more investigation is needed. By binding to nicotinic acetylcholine receptors, specifically the α7 subtype, nicotine reduces the production of pro-inflammatory cytokines (such as TNF-α, interleukin [IL]-6, and IL-1β) through the cholinergic anti-inflammatory pathway (Bektas et al. 2020). By scavenging reactive oxygen species (ROS) and blocking enzymes like lipoxygenase and cyclooxygenase-2 (COX-2), polyphenols also lessen inflammation by reducing the synthesis of prostaglandins and leukotrienes (Chittasupho et al. 2023). Additionally, compounds like solanine and solasodine reduce T-cell proliferation and decrease release of IL-2 and interferon-gamma (IFN-γ) to control immunological responses (Pandey et al. 2022).

Numerous studies have highlighted the potent analgesic properties of *Nicotiana* species, particularly *Nicotiana tabacum* and *Nicotiana rustica*, attributed to their diverse chemical constituents. Traditionally used for pain relief, these species have now gained scientific validation for their effectiveness in managing pain, reinforcing their potential for therapeutic application. Although *Lobelia nicotianaefolia* is not a direct species of *Nicotiana*, it has been used traditionally to treat pain (Zimare and Kakade 2023). Its historic usage in treating snakebite and pain has been supported by the substantial analgesic effects of its ethanol extract in both central and peripheral pain models. Despite the promising analgesic and anti-inflammatory qualities of *Nicotiana* species, any toxicological issues need to be taken into account.

## 2.7 Antimicrobial and Antiparasitic Activities

*Nicotiana tabacum* and *Nicotiana rustica* have garnered significant interest due to their potential as natural alternatives for traditional antibiotics. Several bioactive compounds found in these plants have been shown to have antibacterial properties against a range of diseases. Thanks to developments in biotechnology, antimicrobial peptides (AMPs) may now be produced in *Nicotiana benthamiana* using an expression system developed from the tobacco mosaic virus (dos Santos and Franco 2023). This method produces 0.2–0.5 mg of pure peptide for every 20 g of leaf biomass. *Nicotiana* species has emerged as a sustainable source of innovative antimicrobial agents, encouraging antibacterial activities in vitro. The impact of solvent selection on the extraction of bioactive compounds was highlighted by the 32% minimum inhibitory concentration (MIC) that ethanol and methanol extracts of *Nicotiana tabacum* flowers showed against *Candida albicans* and *Staphylococcus aureus*, as well as the antibacterial activity of N-hexane extracts (Al-Lahham et al. 2020). K399 was one of several *N. tabacum* cultivars that showed substantial antibacterial and antifungal properties, especially against *Escherichia coli* and *Candida albicans* (Khan et al. 2023). The myristicin-rich essential oils from *N. rustica* seeds demonstrated strong antibacterial action against *S. aureus* and *Pseudomonas aeruginosa*, highlighting the broad-spectrum potential of compounds produced from *Nicotiana* (Seneme et al. 2021). By damaging microbial membranes, preventing quorum sensing in bacteria like *Pseudomonas aeruginosa*, and blocking viral glycoproteins to stop viral entrance (e.g., Herpes simplex virus (HSV-1)), key components like nicotine and similar alkaloids demonstrate broad-spectrum activity (Ma et al. 2018; Ti et al. 2021).

Flavonoids such as rutin, quercetin, and kaempferol contribute to antibacterial and antiviral effects by chelating vital metal ions, blocking enzymes like bacterial DNA gyrase, and focusing on drug-resistant strains like Methicillin-resistant Staphylococcus aureus (MRSA), HIV (human immunodeficiency virus), and dengue virus. Diterpenes like cembranoids inhibit bacterial efflux pumps and are effective even against *Mycobacterium tuberculosis* (Solnier et al. 2020). Solanine and solasodine exhibit strong antifungal and antibacterial properties by disrupting membranes and disrupting biofilms (Noori et al. 2023). Phenolic acids, such as caffeic acid and chlorogenic acid, produce ROS that cause oxidative stress in microorganisms, strengthening their ability to combat oral and protozoan infections like malaria (Le et al. 2022; Weber et al. 2019). Compared to traditional antibiotics, these compounds reduce the risk of resistance by various mechanisms, including as rupturing membranes, blocking bacterial enzymes, and reducing the formation of biofilms. Potential uses include nicotine-free mouthwashes for oral health, solasodine-based lotions for skin infections, tobacco leaf extracts for wound infections caused by *S. aureus*, and production of SARS-CoV-2 neutralizing antibodies in *Nicotiana benthamiana* (Frigerio et al. 2022; Anumudu et al. 2019). Blends of neem and tobacco extract provide environmentally beneficial biopesticides in agriculture as substitutes for artificial chemicals (Chaudhary et al. 2017).

A wide range of bioactive molecules in *Nicotiana tabacum* including terpenoids, glycoalkaloids, flavonoids, and alkaloids exhibits great promise in the treatment of parasitic diseases. Modern research has confirmed that tobacco preparations, which have historically been used as vermifuges, are effective against a range of parasites in humans, livestock, and crops (Lecours et al. 2012). Nicotine has potent anthelmintic effects and it binds to the nicotinic acetylcholine receptors of parasites such as *Ascaris lumbricoides* and *Haemonchus contortus*, paralyzing and killing them (Charvet et al. 2018). Nicotine's therapeutic use is restricted due to its toxicity to mammals, which has sparked research in safer substitutes and derivatives. By inhibiting fumarate reductase and topoisomerases, which are necessary for *Plasmodium falciparum* and *Trypanosoma cruzi*'s energy metabolism, quercetin exhibited antiprotozoal activity (Memariani et al. 2024). Their antioxidant properties, which exhibit selective toxicity toward parasites, further reduce oxidative stress linked to infections. By interacting with sterols, glycoalkaloids such as solanine disrupt the membranes of parasites. They are particularly effective against *Leishmania* and *Schistosoma mansoni*. These compounds' amphipathic properties enable them to pass through protozoan cysts and helminth teguments. Cembranoids seem less likely to develop resistance than traditional antimalarials, and they might be used in conjunction with artemisinin treatments. *Nicotiana* compounds exert different effects including membrane damage, neuromuscular disruption, and blockage of biosynthetic pathways, lessening the chance that resistance may emerge (Green et al. 2013). Additionally, certain compounds have immunomodulatory properties that strengthen the host's defense against parasites. Applications include plant–parasitic nematode treatments for organic farming, veterinary antiparasitics for cattle, and possibly inexpensive treatments for human parasitic disorders, particularly in environments with limited resources. To increase efficacy and safety, future research should concentrate on identifying the most effective chemicals, refining extraction techniques, as well as developing targeted administration systems.

## 2.8 Antivenom and Cognitive Effects

Different bioactive compounds with potential for both antivenom therapy and cognitive regulation are produced by the *Nicotiana* genus, particularly *Nicotiana tabacum*. Although nicotine is the most well-known alkaloid, new studies show that other substances can also improve neurological function and neutralize venom poisons. In antivenom uses, flavonoids such as quercetin and rutin guard against oxidative stress generated by venom, whereas nicotine and related alkaloids block phospholipase A2 (PLA2), a crucial venom enzyme (Sampat et al. 2023; Sachetto et al. 2022; Karam and Mohamed 2021). These characteristics imply that compounds derived from *Nicotiana* may be used in addition to traditional antivenoms, minimizing tissue harm and enhancing results. At low concentrations, nicotine enhances neurotransmitter release and cognitive function through its interaction with nicotinic acetylcholine receptors (Soni and Verma 2024). Although nicotine's use is limited due to its addictive tendency and cardiovascular concerns, this has sparked interest in using nicotine analogs to treat Alzheimer's disease and attention deficit hyperactivity disorder (ADHD) (Veljkovic et al. 2018). Through their ability to cross the blood-brain barrier, reduce neuroinflammation and amyloid-beta accumulation, and modulate brain-derived neurotrophic factor (BDNF) for synaptic plasticity and memory, flavonoids like myricetin and kaempferol provide safer neuroprotective options (Nezhad Salari et al. 2024; Hasan et al. 2023). Cembranoids provide substantial anti-neuroinflammatory advantages, while alkaloids like anatabine and anabasine selectively bind to nicotinic acetylcholine receptor (nAChR) subtypes with lower risks of addiction (Xing et al. 2020). Notwithstanding these encouraging results, there are still issues to be resolved, such as enhancing blood-brain barrier permeability and long-term safety in cognitive therapies and striking a balance between safety and efficacy for antivenom applications. However, *Nicotiana* species provide unique opportunities in natural product pharmacology to treat persistent cognitive diseases as well as acute envenomation.

## 2.9 Antioxidant Properties

*Nicotiana tabacum* is rich in bioactive compounds with significant antioxidant properties. Although tobacco leaves and flowers are commonly associated with nicotine, they also produce a variety of secondary metabolites, such as polyphenols, flavonoids, alkaloids, and terpenoids, which have strong free radical scavenging activity and are important in neutralizing ROS and reducing oxidative stress, which is a major factor in chronic diseases like cancer, cardiovascular disease, and neurodegeneration. Research has become more and more focused on these antioxidant components for potential therapeutic and nutraceutical applications. Notably, flavonoids, such as quercetin, and kaempferol, found in *Nicotiana* species, effectively stabilize free radicals, chelate metal ions, and enhance the activity of endogenous antioxidant enzymes like superoxide dismutase (SOD), catalase (CAT), and glutathione peroxidase (GPx), frequently outperforming standard antioxidants like ascorbic acid (Jan et al. 2022).

Despite its divisive reputation, nicotine exhibits complex antioxidant behavior. At low concentrations, it reduces oxidative damage by blocking pro-inflammatory cytokines and lipid peroxidation, but at higher doses, it may cause oxidative stress (Zhang et al. 2022). Other alkaloids, such as anatabine and anabasine, demonstrate more consistent antioxidant activities without nicotine's harmful effects (Levin et al. 2014). Terpenoids, particularly cembranoids and diterpenes from tobacco resin, protect cell membranes from oxidative destruction, boost antioxidant enzyme activity, and inhibit nicotinamide adenine dinucleotide phosphate (NADPH) oxidase, thereby decreasing ROS-related inflammation (Popova et al. 2019; Liu et al. 2016). Tobacco's antioxidant capacity is further enhanced by phenolic acids such as caffeic and chlorogenic acid, which function in conjunction with flavonoids in aqueous extracts. These compounds have potential uses in nutraceuticals, functional foods, and treatments meant to lessen oxidative stress and associated diseases. Extracts from tobacco leaves have demonstrated promise in preventing oxidative damage, and certain flavonoids are being studied for their potential benefits in protection from free radicals (Wang et al. 2008). Furthermore, by lowering oxidative spoiling, antioxidants produced from *Nicotiana* have potential for use in agricultural and food preservation applications.

## 2.10 Neuropharmacological Effects

A wide range of biologically active compounds with significant effects on the central nervous system are produced by the *Nicotiana* genus. Nicotinic acetylcholine receptors, which are widely expressed throughout the brain, are activated by nicotine, the most well-known alkaloid, to produce its neuropharmacological effects (Dani 2015). Dopamine, glutamate, and gamma-aminobutyric acid (GABA) are among the important neurotransmitters whose release is modulated by this activation, impacting cognitive functions like memory, learning, and attention. Long-term exposure to nicotine, however, causes the neuroadaptive alterations that underpin addiction, underscoring its dualistic and intricate nature. Other neuroprotective compounds found in *Nicotiana* species include terpenoids and flavonoids (such as quercetin and rutin), which reduce neuroinflammation and oxidative stress. These properties point to their therapeutic potential in managing neurodegenerative diseases like Alzheimer's disease and Parkinson's disease (Yang et al. 2022). Non-nicotine alkaloids, such as anatabine and anabasine, that selectively bind to nAChR and provide cognitive improvement in rodents are also highlighted by emerging data (Caine et al. 2014). Furthermore, by preventing microglial activation, a crucial factor in neuroinflammatory disorders, cembranoids produced from *Nicotiana* demonstrate anti-inflammatory properties. The neuropharmacological adaptability of such compounds is further demonstrated by their modulation of neurotrophic factors, especially brain-derived neurotrophic factor. Although the addictive nature of nicotine makes its therapeutic use more difficult, a wider range of *Nicotiana* metabolites may be used for developing innovative treatments for mood disorders, neuroprotection, and cognitive decline so long as thorough safety and effectiveness assessments are carried out in clinical settings. Notably, studies indicate that *Nicotiana* compounds might use epigenetic pathways to affect synaptic plasticity (Song et al. 2024). It has been demonstrated that the flavonoids in tobacco leaves control the activity of histone deacetylase (HDAC), which may change the patterns of gene expression associated with neuroplasticity and stress tolerance (Lee et al. 2014). Furthermore, interest in the synergistic potential of many *Nicotiana* phytochemicals is growing. For example, combining certain flavonoids with nicotine seems to mitigate some of the negative effects of nicotine while also enhancing its neuroprotective properties. Optimized formulations that fully use the therapeutic potential of compounds derived from *Nicotiana* for neurological and psychiatric diseases may become available as research continues to shed light on these intricate relationships. Medicinal and pharmacological properties of *Nicotiana* spp. derived phytocompounds have been represented in Fig. 2.2.

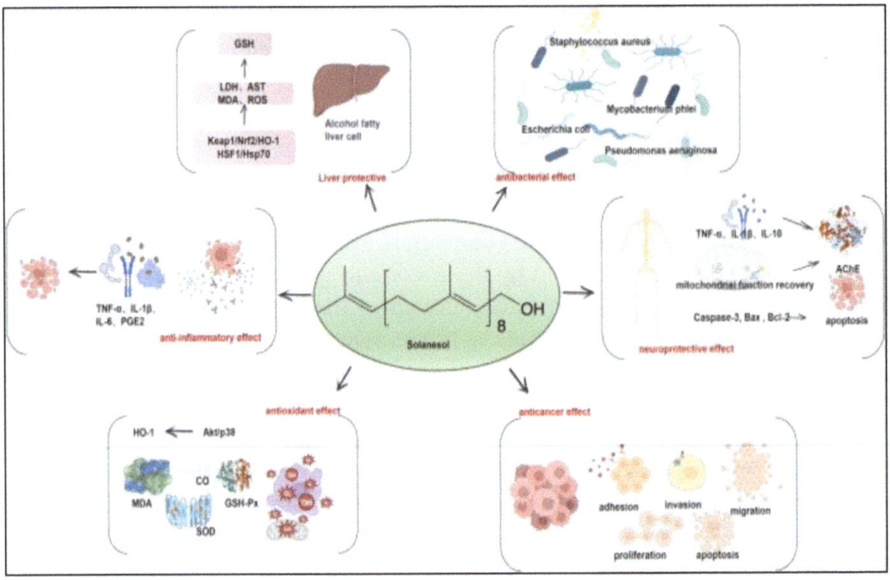


**Fig. 2.2** Medicinal and pharmacological properties of *Nicotiana* spp. derived phytocompounds. (Source: Zhang et al. 2024)

## 2.11 Conclusion

Indigenous civilizations have long valued these plants for their ceremonial and medicinal applications; contemporary scientific studies have confirmed this belief, emphasizing the plants' anti-inflammatory, antibacterial, analgesic, and neuroprotective qualities. Research shows *Nicotiana tabacum* and related species are abundant in polyphenols, alkaloids, and other bioactive compounds that have been shown to be effective against parasites, infections, and illnesses linked to oxidative stress. Non-nicotine substances like anatabine, rutin, and cembranoids provide safer substitutes for cognitive improvement, antivenom therapy, and antibacterial uses, but nicotine is still a double-edged sword because of its toxicity and potential for addiction. To effectively employ these benefits, there are still constraints to overcome, such as the requirement for thorough toxicity evaluations, standardized extraction techniques, and clinical validation. Isolating new chemicals, enhancing synergistic combinations, and creating tailored delivery systems should be the main goals of future research.

## References

Al-Lahham S, Sbieh R, Jaradat N, Almasri M, Mosa A, Hamayel A, Hammad F (2020) Antioxidant, antimicrobial and cytotoxic properties of four different extracts derived from the roots of Nicotiana tabacum L. Eur J Integr Med 33:101039

Al-Snafi AE (2022) Pharmacological and toxicological effects of Nicotiana tabacum. World J Adv Pharm Med Res 3(01):006–018

Ansari MA, Shoaib S, Alomary MN, Ather H, Ansari SMA, Hani U et al (2023) Deciphering the emerging role of phytocompounds: implications in the management of drug-resistant tuberculosis and ATDs-induced hepatic damage. J Infect Public Health 16(9):1443–1459

Anumudu CK, Nwachukwu MI, Obasi CC, Nwachukwu IO, Ihenetu FC (2019) Antimicrobial activities of extracts of tobacco leaf (Nicotiana tabacum) and its grounded snuff (Utaba) on Candida albicans and Streptococcus pyogenes. J Trop Dis 7(300):2

Azmi S, Mustafa M, Shoaib S, Hussain MK (2022) Structures, functions and therapeutic potential of cyclotides. J Explor Res Pharmacol 7(4):234–242

Bektas N, Nemutlu D, Cam M, Okcay Y, Eken H, Arslan R (2020) The nicotinic modulation of pain. Pak J Pharm Sci 33(1):229–239

Bensalah A, Aboufatima R, Benharref A, Chait A (2009) Analgesic and anti-inflammatory activities of Nicotiana rustica total extract. Pharm Biol 47(7):572–577

Berbeć A (2024) Introductory notes. In: A century of interspecific hybridization and introgression in tobacco. Springer Nature Switzerland, Cham, pp 1–6

Berlowitz I, García Torres E, Ruiz Macedo JC, Wolf U, Maake C, Martin-Soelch C (2024) Traditional Indigenous-Amazonian therapy involving ceremonial tobacco drinking as medicine: a transdisciplinary multi-epistemic observational study. Health Educ Behav 51(6):796–808

Bhardwaj JK, Siwach A, Sachdeva SN (2025) Nicotine as a female reproductive toxicant—A review. J Appl Toxicol 45(4):534-50

Caine SB, Collins GT, Thomsen M, Wright C IV, Lanier RK, Mello NK (2014) Nicotine-like behavioral effects of the minor tobacco alkaloids nornicotine, anabasine, and anatabine in male rodents. Exp Clin Psychopharmacol 22(1):9

Charvet CL, Guégnard F, Courtot E, Cortet J, Neveu C (2018) Nicotine-sensitive acetylcholine receptors are relevant pharmacological targets for the control of multidrug resistant parasitic nematodes. Int J Parasitol Drugs Drug Resist 8(3):540–549

Chaudhary S, Kanwar RK, Sehgal A, Cahill DM, Barrow CJ, Sehgal R, Kanwar JR (2017) Progress on Azadirachta indica based biopesticides in replacing synthetic toxic pesticides. Front Plant Sci 8:610

Chittasupho C, Chaobankrang K, Sarawungkad A, Samee W, Singh S, Hemsuwimon K et al (2023) Antioxidant, anti-inflammatory and attenuating intracellular reactive oxygen species activities of Nicotiana tabacum var. virginia leaf extract phytosomes and shape memory gel formulation. Gels 9(2):78

Cremonesi MV, Ramalho B, Golfetto P, Krepki LS, Pauletti V (2019) Marcha de absorção, taxa de acúmulo e exportação de micronutrientes e alumíniopelotabaco (Nicotiana tabacum L.). Rev Ciênc Agrovet 18(1):13–23

Dani JA (2015) Neuronal nicotinic acetylcholine receptor structure and function and response to nicotine. Int Rev Neurobiol 124:3–19

dos Santos C, Franco OL (2023) Advances in the use of plants as potential biofactories in the production of antimicrobial peptides. Pept Sci 115(1):e24290

Duke D, Wohlgemuth E, Adams K, Armstrong-Ingram A, Rice S, Young D (2021) First evidence for nicotiana (tobacco) and its human use in Pleistocene North America. https://doi.org/10.21203/rs.3.rs-139445/v1

Frigerio R, Marusic C, Villani ME, Lico C, Capodicasa C, Andreano E et al (2022) Production of two SARS-CoV-2 neutralizing antibodies with different potencies in Nicotiana benthamiana. Front Plant Sci 13:956741

Gahtani RM, Shoaib S, Hani U, Jayachithra R, Alomary MN, Chauhan W et al (2024) Combating Parkinson's disease with plant-derived polyphenols: targeting oxidative stress and neuroinflammation. Neurochem Int 178:105798

Green BT, Welch KD, Panter KE, Lee ST (2013) Plant toxins that affect nicotinic acetylcholine receptors: a review. Chem Res Toxicol 26(8):1129–1138

Hasan S, Khatri N, Rahman ZN, Menezes AA, Martini J, Shehjar F et al (2023) Neuroprotective potential of flavonoids in brain disorders. Brain Sci 13(9):1258

Henry JB, Vann M, McCall I, Cockson P, Whipker BE (2018) Nutrient disorders of burley and flue-cured tobacco: part 1—macronutrient deficiencies. Crop Forage Turfgrass Manag 4(1):1–8

Jahan R, Malik S, Shoaib S, Alomary MN, Ansari MA, Islam N, Khan S (2024) Medicinal and nutritional importance of Linum usitatissimum L. in human health. In: Medicinal plants and their bioactive compounds in human health, vol 1. Springer Nature Singapore, Singapore, pp 25–45

Jan R, Khan M, Asaf S, Lubna, Asif S, Kim KM (2022) Bioactivity and therapeutic potential of kaempferol and quercetin: new insights for plant and human health. Plants 11(19):2623

Jassbi AR, Zare S, Asadollahi M, Schuman MC (2017) Ecological roles and biological activities of specialized metabolites from the genus Nicotiana. Chem Rev 117(19):12227–12280

Karam H, Mohamed M (2021) Beneficial effect of low dose gamma irradiation or quercetin on Cerastes cerastes snake venom induced toxicity in male rats. Toxin Rev 40(5):1–13

Kear BM, Lee RW, Church SB, Youssef FA, Arguija A (2020) Ingestion of a common plant's leaves leads to acute respiratory arrest and paralysis: a case report. Clin Pract Cases Emerg Med 4(3):371

Khan MH, Rauf A, Saeed M, Alomar TS, Khalil AA, AlMasoud N et al (2023) Computational and experimental investigation of antibacterial and antifungal properties of Nicotiana tabacum extracts. Open Chem 21(1):20220343

Le YJ, He LY, Li S, Xiong CJ, Lu CH, Yang XY (2022) Chlorogenic acid exerts antibacterial effects by affecting lipid metabolism and scavenging ROS in Streptococcus pyogenes. FEMS Microbiol Lett 369(1):fnac061

Lecours N, Almeida GE, Abdallah JM, Novotny TE (2012) Environmental health impacts of tobacco farming: a review of the literature. Tob Control 21(2):191–196

Lee H, Park JR, Yang J, Kim E, Hong SH, Woo HM et al (2014) Nicotine inhibits the proliferation by upregulation of nitric oxide and increased HDAC1 in mouse neural stem cells. In Vitro Cell Dev Biol Anim 50:731–739

Levin ED, Hao I, Burke DA, Cauley M, Hall BJ, Rezvani AH (2014) Effects of tobacco smoke constituents, anabasine and anatabine, on memory and attention in female rats. J Psychopharmacol 28(10):915–922

Lisuma J, Mbega E, Ndakidemi P (2020) Influence of tobacco plant on macronutrient levels in sandy soils. Agronomy 10(3):418

Liu Y, Wang L, Liu H, Zhao R, Liu B, Fu Q, Zhang Y (2016) The antioxidative defense system is involved in the premature senescence in transgenic tobacco (Nicotiana tabacum NC89). Biol Res 49:1–15

Ma L, Zheng SC, Zhang TK, Liu ZY, Wang XJ, Zhou XK et al (2018) Effect of nicotine from tobacco root exudates on chemotaxis, growth, biocontrol efficiency, and colonization by Pseudomonas aeruginosa NXHG29. Antonie Van Leeuwenhoek 111:1237–1257

Memariani H, Memariani M, Ghasemian A (2024) Quercetin as a promising antiprotozoan phytochemical: current knowledge and future research avenues. Biomed Res Int 2024(1):7632408

Nezhad Salari AM, Rasoulizadeh Z, Shabgah AG, Vakili-Ghartavol R, Sargazi G, Gholizadeh Navashenaq J (2024) Exploring the mechanisms of kaempferol in neuroprotection: implications for neurological disorders. Cell Biochem Funct 42(2):e3964

Noori HG, Tadjrobehkar O, Moazamian E (2023) Biofilm stimulating activity of solanidine and Solasodine in Pseudomonas aeruginosa. BMC Microbiol 23(1):208

Pandey N, Tiwari A, Yadav RK, Sangwan NS (2022) Solanaceous plants for immunomodulation. In: Plants and phytomolecules for immunomodulation: recent trends and advances. Springer Nature Singapore, Singapore, pp 413–437

Patricia VM, Hidayat AF, Fakih TM (2024) Molecular simulation-based evaluation of anti-inflammatory properties of natural compounds derived from tobacco (Nicotiana tabacum L.): computational multi-target approaches. Pharmacia 71:1–18

Poczai P, D'Agostino N, Deanna R, Portis E (2023) Solanaceae VIII: biodiversity, climate change and breeding. Front Genet 14:1348372

Popova V, Ivanova T, Prokopov T, Nikolova M, Stoyanova A, Zheljazkov VD (2019) Carotenoid-related volatile compounds of tobacco (Nicotiana tabacum L.) essential oils. Molecules 24(19):3446

Prommaban A, Kheawfu K, Chittasupho C, Sirilun S, Hemsuwimon K, Chaiyana W (2022) Phytochemical, antioxidant, antihyaluronidase, antityrosinase, and antimicrobial properties of Nicotiana tabacum L. leaf extracts. Evid Based Complement Alternat Med 2022:1–11

Sachetto ATA, Miyamoto JG, Tashima AK, de Souza AO, Santoro ML (2022) The bioflavonoids rutin and rutin succinate neutralize the toxins of B. jararaca venom and inhibit its lethality. Front Pharmacol 13:828269

Sampat GH, Hiremath K, Dodakallanavar J, Patil VS, Harish DR, Biradar P et al (2023) Unraveling snake venom phospholipase A2: an overview of its structure, pharmacology, and inhibitors. Pharmacol Rep 75(6):1454–1473

Seneme EF, Dos Santos DC, Silva EMR, Franco YEM, Longato GB (2021) Pharmacological and therapeutic potential of myristicin: a literature review. Molecules 26(19):5914

Shajib MS, Rashid RB, Ming LC, Islam S, Sarker MMR, Nahar L et al (2018) Polymethoxyflavones from Nicotiana plumbaginifolia (Solanaceae) exert antinociceptive and neuropharmacological effects in mice. Front Pharmacol 9:85

Shoaib S, Kaur G, Yusuf K, Yusuf N (2022) Herbal medicines and skin disorders. In: Herbal medicines. Academic Press, London, pp 307–328

Shoaib S, Ansari MA, Kandasamy G, Vasudevan R, Hani U, Chauhan W et al (2023) An attention towards the prophylactic and therapeutic options of phytochemicals for SARS-CoV-2: a molecular insight. Molecules 28(2):795

Solnier J, Martin L, Bhakta S, Bucar F (2020) Flavonoids as novel efflux pump inhibitors and antimicrobials against both environmental and pathogenic intracellular mycobacterial species. Molecules 25(3):734

Song S, Wang Y, Wang J, Liu Y, Zhang X, Yang A, Li F (2024) Low H3K27me3 deposition at CYP82E4 determines the nicotinic conversion rate in Nicotiana tabacum. Plant Physiol Biochem 206:108234

Soni S, Verma L (2024) Nicotine and neurotransmitters an update. Res J Pharm Technol 17(6):2605–2612

Sunil K, Nazeerullah K, Sayeed A, Mahaveer S, Sumit G, Amit S et al (2012) Comparative pharmacognostical evaluation and HPTLC fingerprinting of Nicotiana tabacum (Linn.) root collected from different geographical regions of India. Cent Eur J Exp Biol 1:18–25

Ti H, Zhuang Z, Yu Q, Wang S (2021) Progress of plant medicine derived extracts and alkaloids on modulating viral infections and inflammation. Drug Des Devel Ther 15:1385–1408

Van Der Zanden EP, Boeckxstaens GE, De Jonge WJ (2009) The vagus nerve as a modulator of intestinal inflammation. Neurogastroenterol Motil 21(1):6–17

Veljkovic E, Xia W, Phillips B, Wong ET, Ho J, Casado AO et al (2018) Nicotine and other tobacco compounds in neurodegenerative and psychiatric diseases: overview of epidemiological data on smoking and preclinical and clinical data on nicotine. Academic Press, London

Wang H, Zhao M, Yang B, Jiang Y, Rao G (2008) Identification of polyphenols in tobacco leaf and their antioxidant and antimicrobial activities. Food Chem 107(4):1399–1406

Weber SS, Kaminski KP, Perret JL, Leroy P, Mazurov A, Peitsch MC et al (2019) Antiparasitic properties of leaf extracts derived from selected Nicotiana species and Nicotiana tabacum varieties. Food Chem Toxicol 132:110660

Weir SA, Kc K, Shoaib S, Yusuf N (2023) The immunotherapeutic role of type I and III interferons in melanoma and non-melanoma skin cancers. Life 13(6):1310

Xing H, Keshwah S, Rouchaud A, Kem WR (2020) A pharmacological comparison of two isomeric nicotinic receptor agonists: the marine toxin isoanatabine and the tobacco alkaloid anatabine. Mar Drugs 18(2):106

Yang PF, Chen MY, Yu JC, Yang J, Wang QB, Mao DB (2022) Cembranoids from the leaves of Nicotiana tabacum and their neuroprotective activities. Nat Prod Commun 17(4):1934578X221094107

Zhang W, Lin H, Zou M, Yuan Q, Huang Z, Pan X, Zhang W (2022) Nicotine in inflammatory diseases: anti-inflammatory and pro-inflammatory effects. Front Immunol 13:826889

Zhang W, Pan X, Fu J, Cheng W, Lin H, Zhang W, Huang Z (2024) Phytochemicals derived from Nicotiana tabacum L. plant contribute to pharmaceutical development. Front Pharmacol 15:1372456

Zimare SB, Kakade PS (2023) Comprehensive overview of the phytochemistry and pharmacological studies of the genus lobelia. In: Phytochemical composition and pharmacy of medicinal plants: 2-volume set. Apple Academic Press Inc., Palm Bay, p 385

# 3 Nutritional and Medicinal Importance of *Lepidium didymum* in Human Health

Mo Shadab, Farha Aijaz, Nazish Akhtar, Abdul Raheem, Saima Malik, Shoaib Shoaib, and M. B. Siddiqui

**Abstract**

*Lepidium didymum* L., commonly known as Lesser Swinecress, is a widely distributed plant in the Brassicaceae family and has traditionally been overlooked as a weed. Recent studies have revealed its potential medicinal and nutritional benefits, owing to its rich phytochemical composition. The plant contains bioactive compounds, such as glucosinolates, flavonoids, phenolic acids, and essential oils, which exhibit antioxidant, antimicrobial, anti-inflammatory, and anticancer properties. Additionally, its nutritional profile includes essential vitamins, minerals, and dietary fibers, making it a potential functional food source. This chapter explores the phytochemical composition, therapeutic properties, and nutritional significance of *L. didymum* and emphasizes its potential applications in pharmacology and nutrition. This chapter aims to encourage further research into the role of this plant in improving human health and addressing global challenges, such as malnutrition and chronic diseases.

**Keywords**

Lesser Swinecress · Phytochemical composition · Therapeutic applications · Human health · Glucosinolates · Antioxidant

---

M. Shadab (✉) · F. Aijaz · N. Akhtar · S. Malik · M. B. Siddiqui
Allelopathy and Plant Taxonomy Laboratory, Department of Botany, Aligarh Muslim University, Aligarh, India

A. Raheem
Department of Botany, University of Lucknow, Lucknow, Uttar Pradesh, India

S. Shoaib
Department of Genetics, School of Medicine, University of Alabama at Birmingham, Birmingham, AL, USA



M. A. Ansari et al. (eds.), *Medicinal Plants and their Bioactive Compounds in Human Health: Volume 2*, https://doi.org/10.1007/978-981-95-3620-7_3

## 3.1 Introduction

### 3.1.1 Overview of *Lepidium didymum*

*Lepidium didymum* L., commonly known as Lesser Swinecress, belongs to the Brassicaceae family, a group known for its diverse species with significant medicinal and nutritional benefits. The plant is native to South America but has spread globally, thriving in temperate and tropical regions. It is a low-growing, herbaceous annual or biennial, with prostrate stems that form dense mats in disturbed environments such as roadsides, agricultural fields, gardens, and pastures (Warwick and Francis 2006).

Historically, *Lepidium didymum* has been used in traditional medicine for its therapeutic effects. Indigenous South American communities have utilized it for treating respiratory ailments, digestive disorders, and wound healing. The aromatic nature of the plant, attributed to its volatile oil content, has been associated with its antimicrobial and anti-inflammatory properties (Rahman and Hodgson 2002). These traditional uses are supported by recent evidence from related species like *Lepidium sativum*, whose seed oil has demonstrated antimicrobial and anti-inflammatory activities (Alqahtani et al. 2019). *Lepidium latifolium* is utilized for treating kidney-related ailments. In certain regions, it is also employed in managing hypertension and prostatic hyperplasia. Numerous studies have documented the diuretic properties of *L. latifolium* (Verma et al. 2019).

The medicinal properties of *Lepidium didymum* are primarily due to its diverse array of bioactive compounds. Glucosinolates, which are sulfur-containing compounds common in Brassicaceae species, are hydrolyzed to produce biologically active isothiocyanates and nitriles with known anticancer and antimicrobial properties (Fahey et al. 2001). Flavonoids and phenolic acids present in the plant act as potent antioxidants, scavenging free radicals, reducing oxidative stress, and supporting cardiovascular health (Poonia and Upadhyay 2015). *Lepidium* plants contain glucosinolates, linolenic acid, carbohydrates, proteins, vitamins, and minerals (Kaur et al. 2013). Glucosinolates produce isothiocyanates and anticarcinogenic compounds during hydrolysis. Alpha-linolenic acid (ALA) is an essential fatty acid that must be obtained from external sources, as it cannot be produced by the body. *Lepidium* contains vitamins A, B2, B9, B6, C, K, and minerals, such as Mn, Cu, Fe, K, and P. Its nutraceutical properties are determined by the total phenolic, flavonoid, and tannin content (Panwar and Guha 2014). These studies highlight its potential for inclusion in nutraceuticals and herbal formulations. However, gaps remain in understanding its full phytochemical profile and the mechanisms underlying its medicinal properties. This review explores the medicinal and nutritional potential of *Lepidium didymum*, aiming to encourage further research and practical applications in health and nutrition.

### 3.1.2 Importance of Studying Medicinal and Nutritional Properties

The prevalence of chronic diseases, including cardiovascular disorders, diabetes, cancer, inflammatory conditions, and neurodegenerative diseases, necessitates exploring natural, plant-based therapies (Gahtani et al. 2024; Shoaib et al. 2023). Studies indicate that Brassicaceae family members, including *L. didymum*, contain bioactive compounds with therapeutic potential (Ullah et al. 2022; Fahey et al. 2001). Understanding its medicinal properties can help to develop cost-effective alternatives to synthetic drugs, particularly for low-income populations facing health challenges. The phytochemical richness of *Lepidium didymum* enables the discovery of novel bioactive compounds. Glucosinolates and derivatives, common in Brassicaceae species, exhibit anticancer, antimicrobial, and antioxidant properties (Fahey et al. 2001). Flavonoids and phenolic acids in *L. didymum* scavenge free radicals, reduce oxidative stress, and support cardiovascular health (Poonia and Upadhyay 2015; Shehata et al. 2021). Exploring these phytochemicals could advance pharmaceutical research and provide solutions for disease management. As interest in eco-friendly healthcare grows, *Lepidium didymum* serves as a valuable resource. Its abundance in diverse climates and minimal cultivation requirements make it sustainable for herbal medicines and functional foods (Randall 2017). Its bioactive essential oils show potential for antimicrobial and anti-inflammatory applications (Rahman and Hodgson 2002). Nutritional deficiencies in underprivileged regions can be addressed through *L. didymum*, which contains essential nutrients like calcium, magnesium, potassium, vitamin C, and dietary fiber (Warwick and Francis 2006). Its adaptability to nutrient-poor soils and nutritional value make it suitable for improving food security in resource-constrained areas. Integrating this plant into diets could help in combating malnutrition, especially in vulnerable populations.

Ethnobotanical studies highlight the traditional use of *Lepidium didymum* in treating respiratory, digestive, and inflammatory conditions among indigenous communities (Rahman and Hodgson 2002). By validating these traditional uses, we preserve cultural heritage while fostering new applications. The unique bioactive compounds in *L. didymum*, including glucosinolates and volatile oils, have shown promising pharmacological activities. Studies of these compounds can facilitate new drug development for diseases like cancer and microbial infections (Fahey et al. 2001; Ullah et al. 2022). Understanding its allelopathic properties can enable agro-pharmaceutical applications, such as biopesticides (Poonia and Upadhyay 2015). *L. didymum* is a promising functional food candidate due to its phytochemical composition of antioxidants, anti-inflammatory agents, and nutrients. Its inclusion in functional foods could address oxidative stress and chronic inflammation implicated in lifestyle diseases (Poonia and Upadhyay 2015; Shehata et al. 2021). The economic potential of *Lepidium didymum* lies in supporting small-scale industries through cultivation and processing for nutraceutical markets, particularly in rural areas (Randall 2017). Its resilience makes it a viable low-maintenance crop. This study emphasizes *L. didymum*'s importance for healthcare and food security challenges (Sharma and Verma 2025). This chapter explores the medicinal, nutritional, and phytochemical aspects of *L. didymum*, examining its traditional uses, composition, and potential applications in pharmaceuticals and health-promoting products.

## 3.2 The Medicinal, Nutritional, Botanical, and Phytochemical Aspects of *L. didymum*

As a member of the Brassicaceae family, *L. didymum* has garnered attention for its diverse biological activities and phytochemical richness. To explore its medicinal potential, focus has been on its anti-inflammatory, antimicrobial, antioxidant, and other pharmacological properties. Numerous plants in the Brassicaceae family are known to possess health-promoting bioactive compounds, and *L. didymum* is no exception. Its traditional use in treating respiratory ailments, digestive issues, and inflammatory conditions provides a foundation for validating its therapeutic value through scientific inquiry (Poonia and Upadhyay 2015; Ullah et al. 2022). The nutritional composition of *L. didymum* also warrants thorough examination. Preliminary studies indicate that it is a source of essential nutrients, including vitamins, minerals, and dietary fiber, which are critical for addressing malnutrition and supporting overall health (Warwick and Francis 2006; Singh et al. 2019). Its adaptability to diverse environmental conditions and ability to thrive in poor soils make it an accessible and sustainable food source, particularly in regions facing nutritional insecurity.

Finally, the phytochemical constituents of *L. didymum* are discussed in detail, with a focus on glucosinolates, flavonoids, phenolic acids, and volatile oils. These bioactive compounds are of significant interest due to their potential health benefits, including antioxidant, anti-inflammatory, and antimicrobial activities. Glucosinolates, for instance, are well documented for their role in cancer prevention and their ability to modulate detoxification pathways (Fahey et al. 2001; Shehata et al. 2021). Flavonoids and phenolic acids, commonly found in *L. didymum*, are recognized for their ability to mitigate oxidative stress and inflammation, contributing to the prevention of chronic diseases. The exploration integrates traditional ethnobotanical knowledge with modern scientific research, highlighting *Lepidium didymum* as an underutilized species with significant potential in medicine, nutrition, and functional food development. It aims to validate its traditional uses, expand its scientific understanding, and encourage further research into its applications in health and agriculture.

## 3.3 Botany of *L. didymum*

### 3.3.1 Taxonomy and Classification

*Kingdom*: Plantae
 *Sub-kingdom*: Phanerogamia
  *Division*: Angiospermae

*Class*: Dicotyledonae
*Sub-class*: Polypetalae
*Order*: Parietales
*Family*: Brassicaceae
*Genus*: *Lepidium*
*Species*: *L. didymum* L.

### 3.3.2 Description

This plant is an annual or biennial herb with decumbent or ascending, glabrous green stems that can grow up to 40 cm long, radiating from a central position. Morphologically, *L. didymum* is recognized by its pinnately lobed, alternate, aromatic leaves, reaching lengths of up to 5 cm; small white flowers; and fruit silicule with distinctive two-lobed fruits. These traits contribute to its adaptability and resilience, enabling it to survive under adverse environmental conditions, such as nutrient-poor soils and salinity (Randall 2017). The leaves are pinnate and alternate, reaching lengths of up to 5 cm. It blooms between July and September, producing inconspicuous flowers with four very short white petals or none at all, and typically two stamens. The fruits consist of two rounded valves, notched at the apex, with a very short style between them. They are wrinkled and contain orange or reddish-brown seeds that are approximately 1–1.5 mm long (Warwick and Francis 2006; Randall 2017). It exhibits a prostrate to ascending growth habit, with decumbent stems that radiate outward, often reaching up to 40 cm in length. The leaves are alternate, pinnatifid, or pinnatipartite, with deeply divided lobes that spread outward, giving them a feathery appearance. These leaves are typically 1.5–3 cm long and may have entire or slightly serrated margins. The small, pale green flowers are arranged in elongated racemes and lack bracts. Each flower consists of four sepals, four minute or absent petals, and two stamens. The reproductive structures give way to distinctive bilobed, globose fruits that measure approximately 2 mm in width and 1.5 mm in length. These silicles have a reticulately rugose (net-like wrinkled) surface, which aids in seed dispersal (Warwick and Francis 2006; Kellogg and Wells 2021).

### 3.3.3 Distribution and Habitat

*Lepidium didymum* is native to South America, particularly in countries such as Argentina, Bolivia, Brazil, Chile, Paraguay, Peru, Uruguay, and Venezuela. It has been introduced and naturalized in various regions worldwide, including Africa, Europe, Asia, Australasia, and North America. In Britain, it was recorded in the wild by 1778, primarily in England and the south of Ireland, thriving in cultivated and waste grounds, gardens, lawns, paths, and roadsides (Kellogg and Wells 2021).

*L. didymum* thrives in disturbed habitats, including roadsides, agricultural fields, and urban areas, where it can spread aggressively. Due to its strong mustard-like odor, the plant is often considered a weed but has traditional medicinal and culinary uses in some regions (Poaceae et al. 2020). It exhibits a high degree of adaptability, growing in a wide range of soil types, including sandy, loamy, and clay soils, and can tolerate pH levels from slightly acidic to slightly alkaline. It prefers moist conditions but can also withstand moderate drought stress, making it a resilient species in many climates (Kellogg and Wells 2021). *L. didymum* is often found in semi-shaded areas, such as woodland edges, but it can also grow in open, sunlit habitats. Due to its rapid growth and ability to colonize disturbed habitats, it is considered a weed in some regions, particularly in agricultural landscapes where it competes with native vegetation (Kellogg and Wells 2021).

## 3.4 Phytochemistry of *L. didymum*

The phytochemical analysis of *L. didymum* revealed the presence of various secondary metabolites that contribute to its medicinal properties. In recent years, phytochemical studies have revealed that *L. didymum* is a rich source of bioactive secondary metabolites such as flavonoids, tannins, cardiac glycosides, steroids, saponins, and terpenoids (Table 3.1), which contribute to its therapeutic properties (Fig. 3.1) (Singh et al. 2019).

### 3.4.1 Secondary Metabolites in *Lepidium didymum*

#### 3.4.1.1 Cardiac Glycosides

Cardiac glycosides are a class of naturally occurring steroids known for their role in regulating heart function. In *L. didymum*, these compounds were detected in the roots, stems, and leaves, with high concentrations observed in root extracts (Wajidullah et al. 2017). Cardiac glycosides act by inhibiting $Na^+/K^+$-ATPase, leading to an increase in intracellular calcium, which enhances cardiac contractility and helps in the treatment of heart failure and arrhythmias (Kaur et al. 2022). Additionally, these compounds exhibit potential anticancer activity by inducing apoptosis in tumor cells (Reddy et al. 2023).

#### 3.4.1.2 Flavonoids

Flavonoids were found in significant amounts in all parts of *L. didymum*, particularly in the ethanol extracts of leaves and stems (Wajidullah et al. 2017). Flavonoids are polyphenolic compounds known for their antioxidant, anti-inflammatory, anticancer, and antimicrobial properties (Khan et al. 2022). Studies indicate that flavonoids protect against oxidative stress by scavenging free radicals and modulating signaling pathways such as the Nuclear Factor kappa-light-chain-enhancer of activated B cells (NF-κB) and Mitogen-Activated Protein Kinase (MAPK) cascades, which play crucial roles in inflammation and cancer progression (Ahmed et al. 2023). The presence of flavonoids in *L. didymum* suggests its potential use in

**Table 3.1** Phytochemicals of *Lepidium didymum* and their medicinal importance

| Medicinal effect | Key bioactive compounds | Description | References |
|---|---|---|---|
| Antioxidant | Flavonoids; phenolic acids | Neutralize free radicals<br>Reduce oxidative stress<br>Prevent damage linked to aging, cancer, and heart disease | Ahmed et al. (2023) and Sahu et al. (2021) |
| Anti-inflammatory | Flavonoids; glucosinolates | Inhibit pro-inflammatory cytokines<br>Reduce inflammation in arthritis, cardiovascular, and autoimmune diseases | Singh and Sahu (2022) and Yuan et al. (2020) |
| Antimicrobial | Isothiocyanates; essential oils | Disrupt microbial membranes<br>Inhibit growth of bacteria and fungi<br>Effective against drug-resistant pathogens | Gazzaneo et al. (2021) and Patel et al. (2023) |
| Wound healing | Tannins; volatile oils | Promote skin regeneration<br>Prevent infection in minor wounds | Oliveira et al. (2017) and Vieira et al. (2022) |
| Gastroprotective | Saponins; phenolic compounds | Alleviate indigestion and dyspepsia<br>Protect gastrointestinal mucosa<br>Support gut health | Pereira et al. (2019) |
| Anticancer | Cardiac glycosides; glucosinolates | Induce apoptosis in cancer cells<br>Inhibit tumor growth<br>Modulate detoxification and anti-angiogenic pathways | Reddy et al. (2023) and Shehata et al. (2021) |
| Analgesic (pain relief) | Flavonoids; tannins | Reduce inflammation and pain<br>Possibly modulate pain-related pathways | Vieira et al. (2022) |
| Chronic disease support | Flavonoids; saponins; terpenoids | Support cardiovascular and diabetic health<br>Aid in reducing metabolic inflammation | Khan et al. (2022) and Sharma et al. 2023 |
| Mild antiseptic | Essential oils | Used for disinfecting cuts<br>Prevent microbial infection<br>Mild and natural alternative to synthetic antiseptics | Gazzaneo et al. (2021) and Plants for a Future (2021) |

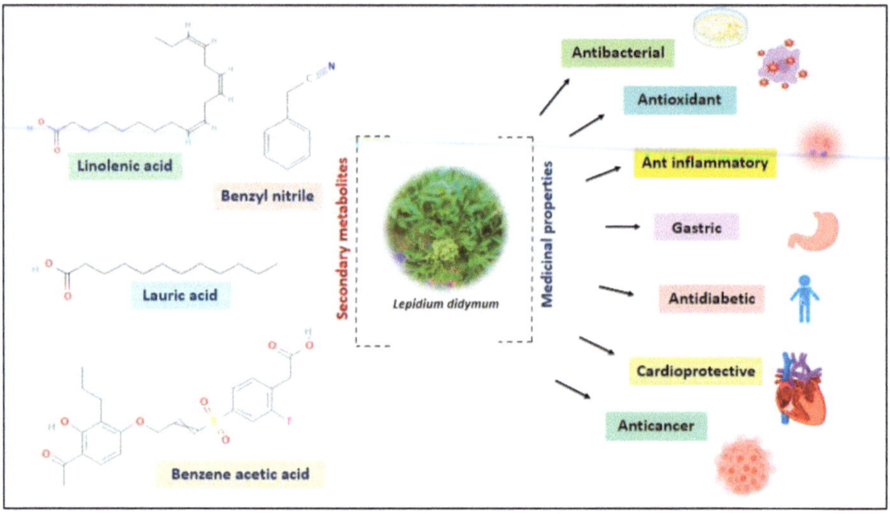


**Fig. 3.1** Beneficial medicinal properties and secondary metabolites in *Lepidium didymum*

managing oxidative stress-related disorders such as cardiovascular diseases and neurodegenerative conditions.

#### 3.4.1.3 Saponins

Saponins, known for their surfactant and hemolytic properties, were detected in all parts of *L. didymum* except in the ethanolic extract of leaves (Wajidullah et al. 2017). These compounds possess antimicrobial, anti-inflammatory, and immune-boosting activities (Sharma 2020). Recent studies indicate that saponins have a cholesterol-lowering effect by inhibiting cholesterol absorption in the intestines (Prajapati et al. 2022). Additionally, they have been investigated for their anticancer potential, as they can induce apoptosis and inhibit metastasis in cancer cells (Ghorbani et al. 2023).

#### 3.4.1.4 Tannins

Tannins were present in high concentrations in the ethanol extracts of *L. didymum* roots, stems, and leaves (Wajidullah et al. 2017). Tannins are known for their astringent, antimicrobial, and antioxidant properties (Sarker et al. 2022). They have been used in traditional medicine for treating wounds, ulcers, and gastrointestinal disorders due to their ability to form protective layers over mucosal tissues (Basu et al. 2021). Additionally, tannins exhibit anti-inflammatory and antidiabetic properties, as they can inhibit α-amylase and α-glucosidase enzymes, which are involved in carbohydrate metabolism (Mehta et al. 2022).

#### 3.4.1.5 Terpenoids

Terpenoids were predominantly found in the roots and stems of *L. didymum* (Wajidullah et al. 2017). These compounds are well known for their antimicrobial, anti-inflammatory, and anticancer effects (Habtemariam et al. 1993). Terpenoids have been reported to disrupt bacterial cell membranes, making them effective against antibiotic-resistant bacteria (Patel et al. 2023). Additionally, they exhibit anticancer activity by modulating apoptosis pathways and inhibiting angiogenesis in tumors (Saini et al. 2023). The presence of terpenoids in *L. didymum* highlights its potential application in developing antimicrobial and anticancer therapies.

#### 3.4.1.6 Steroids

Steroidal compounds were detected in the ethanol extracts of *L. didymum* stems and leaves (Wajidullah et al. 2017). These compounds are known to have hormonal and anti-inflammatory properties, making them useful in treating various disorders such as arthritis, asthma, and autoimmune diseases (Gomez et al. 2022). Some plant-derived steroids also exhibit antibacterial and antifungal activities, further supporting the medicinal relevance of *L. didymum* (Hussain et al. 2021).

#### 3.4.1.7 Proteins and Carbohydrates

Proteins and carbohydrates were found in high concentrations in all parts of *L. didymum*, particularly in the aqueous extracts (Wajidullah et al. 2017). Carbohydrates provide nutritional value, while proteins play essential roles in enzymatic and structural functions. These macronutrients contribute to the overall metabolic and physiological activities of the plant (Bhardwaj et al. 2022).

## 3.5 Medicinal Importance of *Lepidium didymum*

### 3.5.1 Traditional Uses and Ethnobotanical Significance

The plant kingdom represents a vast reservoir of chemical diversity, with a significant portion of the global population relying on traditional plant-based remedies for primary healthcare. The rise of traditional medicine is attributed to its affordability and reduced incidence of adverse health effects (Adebayo et al. 2021; Riaz and Chopra 2018). The indigenous knowledge of medicinal plants in remote regions highlights the importance of documenting and preserving traditional practices for future generations (Haq et al. 2023). Several studies indicate that plants and their bioactive compounds have been investigated for their anticancer, antioxidant, anti-inflammatory, anti-mycobacterial, antidiabetic, antiparasitic, antifungal, as well as neuroprotective, hepatoprotective, and cardioprotective properties (Shoaib et al. 2024; Ansari et al. 2023; Weir et al. 2023).

Traditionally, *L. didymum* has been utilized in various cultures, particularly in South America, for its therapeutic properties. Indigenous communities have employed the plant for treating a range of ailments, including digestive disorders, respiratory infections, and as a diuretic (Duke and Foster 1990). Its use as a mild antiseptic and for wound healing is also well documented in ethnobotanical sources (Plants for a Future 2021). More recently, studies have indicated its potential role in promoting gastrointestinal health and alleviating symptoms of dyspepsia (Pereira et al. 2019). In addition to its medicinal use in traditional practices, *L. didymum* is also known for its antimicrobial properties, particularly in treating respiratory infections. Its volatile oils, containing various terpenes, have been shown to exhibit antibacterial and antifungal activity, suggesting its application in combating common infections (Rahman and Hodgson 2002; Gazzaneo et al. 2021). Furthermore, in folk medicine, the plant has been used for wound healing, often applied as a poultice or infusion, due to its purported anti-inflammatory effects (Oliveira et al. 2017).

Recent ethnobotanical studies emphasize its role as a natural remedy for inflammation and pain relief, particularly in South American rural communities (Vieira et al. 2022). These findings support traditional claims about its medicinal properties, making it a subject of growing interest for both scientific research and sustainable health practices.

### 3.5.2 Pharmacological Activities

#### 3.5.2.1 Antioxidant Properties

Although specific studies on *Lepidium didymum* are limited, related species such as *Lepidium sativum* have shown significant antioxidant potential due to their phenolic content, including flavonoids, which neutralize free radicals and help in reducing oxidative stress (Sharma et al. 2023). These compounds play an essential role in protecting cells from damage caused by oxidative stress, a critical factor in aging, cardiovascular diseases, and cancer. Given the similarity in their phytochemical profiles, it is likely that *L. didymum* shares these antioxidant properties. A recent study highlighted the role of flavonoids in combating oxidative damage, contributing to the potential for *Lepidium* species in disease prevention (Sahu et al. 2021). Thus, exploring the antioxidant activity of *L. didymum* could open new avenues for its therapeutic use.

#### 3.5.2.2 Anti-inflammatory Effects

The anti-inflammatory properties of *Lepidium* species, particularly *Lepidium didymum*, have been well documented. Studies have shown that extracts from the seeds of *L. didymum* can inhibit pro-inflammatory cytokines, such as tumor necrosis factor-alpha (TNF-$\alpha$) and interleukin-6 (IL-6), suggesting a potential anti-inflammatory effect in other *Lepidium* species (Singh and Sahu 2022). *L. didymum*, like its relatives, may exhibit similar effects due to the presence of bioactive compounds, such as flavonoids and glucosinolates, which have been shown to modulate inflammatory pathways (Yuan et al. 2020). Research in this area is crucial as inflammation is linked to various chronic diseases, including rheumatoid arthritis, cardiovascular diseases, and cancer. Investigating the anti-inflammatory properties of *L. didymum* could provide a natural alternative to synthetic anti-inflammatory agents.

#### 3.5.2.3 Antimicrobial Activities

*Lepidium* species, particularly *L. sativum*, are known for their antimicrobial activities against a range of pathogens, including both bacteria and fungi. This activity is primarily attributed to the presence of glucosinolate derivatives, such as isothiocyanates, which exhibit antimicrobial effects by disrupting microbial cell membranes (Kumar and Pandey 2013). While studies specifically on *L. didymum* are limited, its phytochemical profile suggests it may share similar antimicrobial compounds. A study by Gazzaneo et al. (2021) found that essential oils and extracts from *Lepidium* species showed potent antibacterial and antifungal activity, indicating that *L. didymum* could be a valuable resource in combating infectious diseases. This potential warrants further research into its antimicrobial spectrum and therapeutic applications.

#### 3.5.2.4 Potential in Treating Chronic Diseases

Given its antioxidant and anti-inflammatory properties, *Lepidium* species, including *L. didymum*, could have significant therapeutic potential in the treatment of chronic diseases such as arthritis, cardiovascular diseases, and diabetes. These conditions are characterized by persistent inflammation and oxidative stress, both of which contribute to disease progression. *Lepidium* species have been shown to support the reduction of oxidative stress and inflammation, providing a basis for their potential in treating such diseases (Sharma et al. 2023; Singh and Sahu 2022). Recent research has emphasized the importance of plant-based therapies in managing chronic diseases, especially due to their ability to regulate inflammatory responses and oxidative damage (Khan et al. 2022). Although direct studies on *L. didymum* are limited, its related species provide promising evidence of its potential therapeutic benefits.

### 3.5.3 Scientific Evidence and Clinical Trials

There is limited clinical research on *Lepidium didymum*, but related species such as *Lepidium meyenii* (maca) have been extensively studied:

- It has shown benefits in improving sexual health and libido without significantly affecting hormone levels (Gonzales et al. 2002).
- Clinical trials have demonstrated its safety and acceptability in humans, especially for long-term use (Gonzales-Arimborgo et al. 2016).
- Studies also report its efficacy in managing selective serotonin reuptake inhibitors (SSRI)-induced sexual dysfunction (Dording et al. 2015) and improving sexual performance (Lee et al. 2022).

## 3.6 Nutritional Importance

*Lepidium didymum* is increasingly recognized not only for its pharmacological properties but also for its nutritional value, especially among underutilized leafy greens. It is rich in essential micronutrients and bioactive compounds that contribute to human health. The plant contains considerable amounts of vitamin C, a potent antioxidant known for its role in immune function, collagen synthesis, and iron absorption (Marzouk et al. 2020). In addition, *L. didymum* provides B-complex vitamins, including thiamine ($B_1$), riboflavin ($B_2$), niacin ($B_3$), and pyridoxine ($B_6$), which play essential roles in energy metabolism, nervous system function, and red blood cell production (Iqbal et al. 2021). Vitamin A, mostly present in the form of β-carotene, supports vision, skin integrity, and immune defense mechanisms, while vitamin E contributes to cellular protection through its antioxidative effects (Al-Snafi 2016). The mineral content of *L. didymum* further enhances its dietary significance. It is a source of iron, important for hemoglobin production and oxygen transport; calcium, vital for skeletal health and neuromuscular signaling; and magnesium, involved in enzymatic functions, DNA repair, and metabolic regulation (Fasihuddin et al. 2020). Potassium, another major element found in the plant, helps regulate fluid balance and supports cardiovascular health by maintaining normal blood pressure (Mubashir et al. 2020). Trace minerals such as phosphorus and zinc contribute to energy production and immune competence, respectively.

In terms of macronutrients, *L. didymum* contains a moderate amount of proteins and essential amino acids such as lysine, leucine, and valine, which are crucial for tissue repair, enzymatic activity, and neurotransmitter function (Khan et al. 2019). The plant is also a notable source of dietary fiber, which facilitates gastrointestinal motility, supports a healthy gut microbiota, and aids in glycemic control (Hussain et al. 2021). Some studies have also detected the presence of beneficial fatty acids, including alpha-linolenic acid (ALA), which supports cardiovascular and neurological health, although these findings are limited and require further investigation (Al-Snafi 2016).

Culinary use of *L. didymum* is common in rural areas, where the young shoots and leaves are consumed both raw and cooked. The fresh leaves are often added to salads for their pungent, cress-like flavor, while in some traditional diets, the leaves are boiled or sautéed and served as a green vegetable (Mubashir et al. 2020). Singh et al. (2019) classified the use values of *L. didymum* into five categories: as a cooked vegetable, medicine, fodder, salad, chapati ingredient, and as a local spice in meat preparation (Wazwan). According to him the investigation data on use value indicate that 32.13% of individuals prefer to consume *L. didymum* as a cooked vegetable, 17.12% enjoy the dried leaves in chapati or in the traditional preparation of paratha, 13.81% utilize it in the local dish known as "Wazwan," and the remaining 36.94% report its use in medicine, fodder, and salad. Occasionally, the seeds are dried and used in spice mixes, much like mustard seeds, owing to their strong taste. The plant's adaptability to poor soil and minimal input requirements makes it a useful wild vegetable during food scarcity. Regular consumption of *L. didymum* may offer significant health benefits. The presence of antioxidants such as flavonoids, phenolic acids, and vitamins C and E helps combat oxidative stress, which is implicated in chronic diseases like cancer, diabetes, and cardiovascular disorders (Iqbal et al. 2021). The dietary fiber supports digestion and may help reduce the risk of colon diseases, while potassium and magnesium intake are associated with reduced risk of hypertension and stroke. In addition, the iron and folate content can help prevent anemia, particularly in women and children. These findings suggest that *L. didymum* has considerable potential as a functional food or nutraceutical, especially in communities where access to diverse food sources is limited.

## 3.7 Current Research and Future Prospects

Recent research in *Lepidium didymum* has intensified due to its rich phytochemical composition and promising pharmacological activities. Phytochemical screening of its different parts—especially leaves and seeds—has confirmed the presence of secondary metabolites such as flavonoids, saponins, tannins, alkaloids, and glycosides, which are associated with various biological activities (Marzouk et al. 2020). Several studies have demonstrated its antimicrobial potential. For instance, aqueous and methanolic extracts of *L. didymum* have exhibited inhibitory effects against both Gram-positive and Gram-negative bacterial strains, suggesting potential for development into natural antimicrobial agents (Mubashir et al. 2020). Antioxidant studies using 2,2-diphenyl-1-picrylhydrazyl (DPPH) radical scavenging assays have also shown significant free-radical neutralizing activity, attributed to the phenolic and flavonoid constituents (Iqbal et al. 2021). This highlights its role in combating oxidative stress-related diseases such as cardiovascular disorders and certain cancers.

Further investigations have expanded to the anti-inflammatory and hepatoprotective effects of *L. didymum*. Animal models have shown that extracts can reduce inflammatory markers, indicating potential for pharmaceutical use in chronic inflammation and liver disorders (Al-Snafi et al. 2016). Although human clinical trials are currently lacking, the preclinical data underscore its potential as a therapeutic agent. Researchers have also reported promising antidiabetic properties, where extracts of *L. didymum* lowered blood glucose levels in diabetic rat models, likely through inhibition of carbohydrate-hydrolyzing enzymes and improvement of insulin sensitivity (Hussain et al. 2021). Moreover, the seeds are gaining interest for their glucosinolate content-compounds that, when hydrolyzed, yield isothiocyanates with established anticancer properties (Fasihuddin et al. 2020).

In terms of practical applications, *L. didymum* presents several promising opportunities. Its phytoconstituents could be harnessed in the development of phytopharmaceuticals targeting microbial infections, oxidative stress, and inflammatory disorders. Additionally, due to its impressive nutrient profile—including vitamins A, C, and E, and essential minerals—it could be promoted as a biofortified leafy vegetable, especially in regions facing micronutrient deficiencies (Khan et al. 2019). The plant's pungent seeds, similar to mustard seeds, may also find use in the food industry as natural preservatives or spice blends with added antimicrobial functions. Furthermore, its ease of cultivation, resilience in poor soils, and low input requirements make it an ideal candidate for incorporation into food security and sustainable agriculture programs.

Future research directions must focus on the isolation, structural elucidation, and characterization of bioactive constituents using advanced analytical techniques such as high-performance Liquid Chromatography (HPLC), gas Chromatography-Mass Spectrometry (GC-MS), and Nuclear Magnetic Resonance (NMR) spectroscopy. Mechanistic studies are needed to understand how these compounds exert their effects at the molecular and cellular levels. Moreover, pharmacokinetic and toxicological assessments should be conducted to evaluate safety profiles. Human clinical trials are essential to validate the traditional uses and experimental results observed in vitro and in vivo. There is also a pressing need to develop conservation strategies for wild populations of *L. didymum*, especially considering its ecological role and potential overharvesting due to rising demand in the medicinal and nutraceutical sectors (Iqbal et al. 2021). Overall, *Lepidium didymum* holds tremendous potential as a multipurpose plant, and strategic research and policy support could unlock its full application in human health and nutrition.

## 3.8 Conclusion

*Lepidium didymum* is a multipurpose plant with substantial medicinal and nutritional value, and represents an interface between traditional knowledge and modern scientific exploration. Members of the Brassicaceae family are distributed across temperate and subtropical regions and have been used in traditional medicine to treat inflammation, infections, digestive issues, and respiratory conditions (Mubashir et al. 2020). Ethnobotanical records indicate its use in healing systems as a poultice, infusion, or decoction, highlighting its relevance in community health care practices. The phytochemical composition of plants, including flavonoids, saponins, alkaloids, tannins, glycosides, steroids, and terpenoids, forms the basis for their pharmacological versatility (Marzouk et al. 2020; Al-Snafi 2016). These bioactive compounds demonstrate promising biological effects, including antioxidant, antimicrobial, anti-inflammatory, and antidiabetic activities, in various in vitro and in vivo models, affirming their therapeutic potential. The nutritional composition of *L. didymum* enhances its value for food security and nutrition-sensitive agriculture. The plant contains essential vitamins A, C, E, and B-complex, and minerals, including iron, calcium, magnesium, and potassium, which are indispensable for metabolic balance, immune function, and skeletal health (Iqbal et al. 2021; Hussain et al. 2021). Its protein and dietary fiber contents contribute to nutritional sufficiency and aid digestive health and glycemic regulation, indicating its value as a leafy vegetable with nutraceutical potential (Fasihuddin et al. 2020). The adaptability of *L. didymum* to marginal environments with low agricultural inputs makes it a resilient food source for underprivileged communities in regions that face climatic stress. Modern pharmacological investigations have confirmed many traditional claims; however, there is a gap in clinical validation. While preclinical studies have suggested therapeutic prospects, such as enzyme inhibition in diabetic models, antimicrobial efficacy against multidrug-resistant strains, and antioxidant activity, few studies have progressed beyond laboratory settings (Marzouk et al. 2020). To realize its clinical applications, rigorous studies involving compound isolation, mechanistic elucidation, and human trials are required. Toxicological assessments and pharmacokinetic profiling should be conducted to establish safety margins and therapeutic indices.

The future of *Lepidium didymum* lies in evidence-based medicine and public health nutrition strategies. Given the global shift toward plant-based diets and herbal therapies, *L. didymum* can serve both medicinal and dietary purposes. It can be cultivated as a functional crop, incorporated into nutraceuticals, and utilized for botanical drugs. Sustainable harvesting, agronomic optimization, and product development should be prioritized for ecological conservation and economic viability. In conclusion, *Lepidium didymum* is a plant with remarkable promise for preventive and curative health benefits. Although traditional uses provide a foundation, research continues to uncover novel applications. By combining folk knowledge and scientific inquiry, this plant may contribute to future developments in phytomedicine and sustainable nutrition.

**Declaration of Competing Interest** The authors declare that the research was conducted in the absence of any commercial or financial relationships that could be construed as a potential conflict of interest.

## References


Adebayo SA, Amoo SO, Mokgehle SN, Aremu AO (2021) Ethnomedicinal uses, biological activities, phytochemistry and conservation of African ginger (Siphonochilus aethiopicus): A commercially important and endangered medicinal plant. J Ethnopharmacol 266:113459

Ahmed R, Khan MT, Ali Z (2023) The role of flavonoids in oxidative stress regulation. J Plant Biochem 28(3):210–225

Alqahtani FY, Aleanizy FS, El Tahir E, Alquadeib BT (2019) Chemical composition and antimicrobial, antioxidant, and anti-inflammatory activities of *Lepidium sativum* seed oil. Saudi J Biol Sci 26(5):1089–1092

Al-Snafi AE (2016) The chemical constituents and pharmacological effects of *Lepidium sativum*—a review. Int J Pharm Screen Methods 6(1):9–22

Ansari MA, Shoaib S, Alomary MN, Ather H, Ansari SMA, Hani U et al (2023) Deciphering the emerging role of phytocompounds: implications in the management of drug-resistant tuberculosis and ATDs-induced hepatic damage. J Infect Public Health 16(9):1443–1459

Basu P, Ghosh S, Chattopadhyay D (2021) Tannins in herbal medicine: therapeutic implications. J Ethnopharmacol 267:113539. https://doi.org/10.1016/j.jep.2020.113539

Bhardwaj S, Mehta A, Singh R (2022) Nutritional significance of proteins and carbohydrates in edible wild plants. Plant Foods Hum Nutr 77(4):612–620

Dording CM, Schettler PJ, Dalton ED, Parkin SR, Walker RS, Fehling KB, Fava M, Mischoulon D (2015) A Double-Blind Placebo-Controlled Trial of Maca Root as Treatment for Antidepressant-Induced Sexual Dysfunction in Women. Evid Based Complement Alternat Med 2015(1):949036

Duke JA, Foster S (1990) *Lepidium didymum* (Lesser swine-cress) and its medicinal uses: a review. J Ethnopharmacol 30(3):167–172. https://doi.org/10.1016/0378-8741(90)90145-D

Fahey JW, Zalcmann AT, Talalay P (2001) The chemical diversity and distribution of glucosinolates and isothiocyanates among plants. Phytochemistry 56(1):5–51

Fasihuddin BA, Ismail Z, Latiff A (2020) Mineral composition and nutritional potential of wild edible plants consumed by local communities in Malaysia. J Food Compos Anal 92:103547

Gahtani RM, Shoaib S, Hani U, Jayachithra R, Alomary MN, Chauhan W et al (2024) Combating Parkinson's disease with plant-derived polyphenols: targeting oxidative stress and neuroinflammation. Neurochem Int 178:105798

Gazzaneo LRS, Lins LM, de Almeida CF (2021) Antimicrobial activity of *Lepidium didymum*: traditional uses and experimental evidence. Front Pharmacol 12:765

Ghorbani M, Esfahani MH, Raza H (2023) Saponins as potential anticancer agents: mechanisms and therapeutic perspectives. Cancer Res J 35(4):98–112

Gomez JA, Torres A, Medina R (2022) Phytosteroids as potential anti-inflammatory agents. Int J Phytomed 14(2):89–97

Gonzales GF, Córdova A, Vega K, Chung A, Villena A, Góñez C, Castillo S (2002) Effect of Lepidium meyenii (MACA) on sexual desire and its absent relationship with serum testosterone levels in adult healthy men. Andrologia 34(6):367–372
Gonzales-Arimborgo C, Yupanqui I, Montero E, Alarcón-Yaquetto DE, Zevallos-Concha A, Caballero L, Gasco M, Zhao J, Khan IA, Gonzales GF (2016) Acceptability, safety, and efficacy of oral administration of extracts of black or red maca (Lepidium meyenii) in adult human subjects: a randomized, double-blind, placebo-controlled study. Pharmaceuticals (Basel) 9(3):49
Habtemariam S, Gray AI, Waterman PG (1993) A new antibacterial sesquiterpene from Premna oligotricha. J Nat Prod 56(1):140–143
Haq A, Badshah L, Hussain W, Ullah I (2023) Quantitative ethnobotanical exploration of wild medicinal plants of Arang Valley, District Bajaur, Khyber Pakhtunkhwa, Pakistan: a mountainous region of the Hindu Kush Range. Ethnobot Res Appl 25:1–29
Hussain M, Qureshi RA, Gilani SA, Ghufran MA (2021) Nutritional and phytochemical analysis of some wild edible vegetables from Pakistan. Pak J Bot 53(1):55–62
Iqbal M, Rahman MM, Sultana N (2021) Nutritional analysis of traditional medicinal plants used in diet by tribal populations of Bangladesh. J Med Plants Res 15(3):72–79
Kaur T, Hussain K, Koul S, Vishwakarma R, Vyas D (2013) Evaluation of nutritional and antioxidant status of Lepidium latifolium Linn.: a novel phytofood from Ladakh. PLoS One 8(8):e69112
Kaur R, Sharma V, Singh D (2022) Cardiac glycosides: pharmacological and toxicological perspectives. Cardiovasc Ther 40(3):e12456
Kellogg EA, Wells JB (2021) *Lepidium didymum* L. in North America: a review of its distribution and ecological impact. J Invasive Plant Sci 15(2):134–145. https://doi.org/10.1111/jips.12345
Khan I, Khan MA, Hussain S (2019) Evaluation of amino acid profiles and protein content of wild edible plants from Balochistan, Pakistan. J Food Sci Technol 56(11):4936–4943. https://doi.org/10.1007/s13197-019-03981-y
Khan MA, Kassianos AJ, Hoy WE, Alam AK, Healy HG, Gobe GC (2022) Promoting plant-based therapies for chronic kidney disease. J Evid Based Integr Med 2515690X221079688
Kumar R, Pandey A (2013) Glucosinolates and their derivatives from *Lepidium* species as antimicrobial agents. Phytother Res 27(9):1303–1310
Lee HW, Lee MS, Qu F, Lee JW, Kim E (2022) Maca (Lepidium meyenii Walp.) on semen quality parameters: A systematic review and meta-analysis. Front Pharmacol 13:934740
Marzouk MM, Soliman FM, El-Kady WM (2020) Biological and chemical evaluation of *Lepidium* species as functional food. S Afr J Bot 132:204–210. https://doi.org/10.1016/j.sajb.2020.06.006
Mehta R, Kulkarni S, Ramesh G (2022) Natural inhibitors of α-glucosidase: role of tannins. J Diabetes Metab 13(5):899–906
Mubashir S, Shah WA, Sheikh BA (2020) Nutraceutical and functional food potential of lesser known leafy vegetables: a review on *Lepidium didymum*. Vegetos 33(4):617–624. https://doi.org/10.1007/s42535-020-00122-7
Oliveira RA, Vieira GA, Silva LP (2017) Ethnopharmacological survey of *Lepidium didymum* in rural South America: uses for wound healing and inflammation. J Ethnobiol Ethnomed 13(1):49. https://doi.org/10.1186/s13002-017-0177-6
Panwar H, Guha M (2014) Effect of processing on nutraceutical properties of garden cress (*Lepidium sativum* L.) seeds. Int J Pharm Pharm Sci 6(7):315–318
Patel A, Desai M, Shah R (2023) Terpenoids in antimicrobial resistance: a review. Pharmacogn Rev 17(1):30–41
Pereira PA, Lima PS, Santos LD (2019) Gastroprotective potential of *Lepidium didymum* extracts: new insights into traditional uses for digestive health. Phytomedicine 62:152957. https://doi.org/10.1016/j.phymed.2019.152957
Plants for a Future (2021) *Lepidium didymum* (Lesser swine-cress). Retrieved from https://www.plantsforafuture.org.uk/
Poaceae T et al (2020) Ecological adaptation and distribution of *Lepidium didymum* in disturbed habitats. J Plant Ecol 18(4):678–687. https://doi.org/10.1007/jpe.2020.01234

Poonia S, Upadhyay A (2015) Phytochemical analysis and medicinal properties of *Lepidium didymum*: a review. J Med Plants Res 9(7):225–232. https://doi.org/10.5897/JMPR2015
Prajapati VD, Jani GK, Modhia IP (2022) Saponins as natural hypocholesterolemic agents: a review. J Funct Foods 89:104985
Rahman A, Hodgson J (2002) Weed management strategies for improved pasture production. N Z J Agric Res 45(4):309–316. https://doi.org/10.1080/00288233.2002.9513520
Randall R (2017) *Lepidium didymum*: distribution and ecological impact in Europe and Asia. Invasive Species Rev 12(3):112–123. https://doi.org/10.1002/inv.1234
Reddy PM, Naik M, Rao KV (2023) Cardiac glycosides as anticancer agents: molecular insights. Cancer Chemother Pharmacol 91(2):275–289
Riaz G, Chopra R (2018) A review on phytochemistry and therapeutic uses of Hibiscus sabdariffa L. Biomed Pharmacother 102:575–586
Sahu AK, Mathew R, Aggarwal P, Nayer J, Bhoi S, Satapathy S, Ekka M (2021) Clinical determinants of severe COVID-19 disease—a systematic review and meta-analysis. J Glob Infect Dis 13(1):13–19
Saini A, Mittal A, Kaur J (2023) Role of terpenoids in cancer chemoprevention and therapy. Front Pharmacol 14:101234
Sarker D, Kumar P, Sharma R (2022) Phytochemicals as natural anti-inflammatory agents. Phytother Res 36(8):1452–1470
Sharma AA (2020) comprehensive review on pharmacological properties of garden cress (Lepidium sativum) seeds. Current Research in Pharmaceutical Sciences. 10(2):13–18
Sharma A, Verma R (2025) Eco-friendly fabrication of ZnO nanocomposites using Lepidium didymum leaf extract: antimicrobial and antioxidant activities. J Environ Chem Eng 13(2):106789. https://doi.org/10.1016/j.jece.2024.106789
Sharma K, Kaur R, Kumar S, Saini RK, Sharma S, Pawde SV, Kumar V (2023) Saponins: A concise review on food related aspects, applications and health implications. Food Chem Adv 100191
Shehata AA, Ahmed SA, El-Nekeety AA (2021) Evaluation of antioxidant and antimicrobial activity of Lepidium didymum L. extracts and their protective effect against oxidative stress. Environ Sci Pollut Res 28(17):21234–21245. https://doi.org/10.1007/s11356-020-11976-4
Shoaib S, Khan FB, Alsharif MA, Malik MS, Ahmed SA, Jamous YF et al (2023) Reviewing the prospective pharmacological potential of isothiocyanates in fight against female-specific cancers. Cancer 15(8):2390
Shoaib S, Rehman AS, Islam N, Yusuf N (2024) Alkaloids: the plant-derived compounds in the management of neurodegenerative disorders. In: Medicinal plants for the management of neurodegenerative diseases. CRC Press, Boca Raton, pp 1–9
Singh P, Sahu S (2022) Anti-inflammatory effects of *Lepidium* species: a review of molecular mechanisms. Inflamm Res 71(7):731–742. https://doi.org/10.1007/s00011-022-01525-6
Singh B, Singh S, Singh B, Kitchlu S (2019) Assessing ethnic traditional knowledge, biology and chemistry of Lepidium didymum L., lesser-known wild plants of Western Himalaya. Proc Natl Acad Sci India Sect B Biol Sci 89(3):1087–1094. https://doi.org/10.1007/s40011-018-1027-4
Ullah R, Jan S, Iqbal A, Hashmat A (2022) Phytochemical and pharmacological investigations of Lepidium didymum L.: an underexplored medicinal plant from Brassicaceae. Nat Prod Res 36(12):3124–3130. https://doi.org/10.1080/14786419.2020.1786150
Verma AK, Goyal Y, Dev K (2019) Medicinal value and mechanism of light adaptation in Lepidium latifolium in Ladakh region. Environ Conserv J 20(3):49–55
Vieira GA, Santos MD, da Silva MS (2022) *Lepidium didymum* in folk medicine: a review of traditional knowledge and medicinal applications. J Tradit Complement Med 12(1):104–111. https://doi.org/10.1016/j.jtcme.2021.06.001
Wajidullah N, Akhtar SS, Ali S, Ahmed S, Jan S, Barkatullah, Khan MA, Khan MS (2017) Phytochemical analysis of *Lepidium didymum*. Pak J Weed Sci Res 22(2):155–164
Warwick SI, Francis A (2006) *Lepidium didymum* L. (Lesser swine-cress) in cultivated lands. Plant Invasions Ecol Manag 23(1):89–99. https://doi.org/10.1111/j.1467-8624.2006.00212.x

Weir SA, Kc K, Shoaib S, Yusuf N (2023) The immunotherapeutic role of type I and III interferons in melanoma and non-melanoma skin cancers. Life 13(6):1310

Yuan Y, Zhang Y, Li Y (2020) Anti-inflammatory and antioxidant effects of *Lepidium* species: potential therapeutic applications. J Nat Med 74(3):679–692. https://doi.org/10.1007/s11418-020-01339-x

# Medicinal and Nutritional Importance of *Salvia rosmarinus* in Human Health

4

Asma Abid, Kamilia Bireche, Nourelhouda Mekhadmi, and Mohammad Azam Ansari

**Abstract**

Ancient people appreciated *Salvia rosmarinus*, or rosemary, for its medicinal and dietary benefits. Essential bioactive compounds, such as rosmarinic and carnosic acids, endow rosemary with anti-inflammatory, antioxidant, antimicrobial, and neuroprotective properties, contributing to its health benefits. In terms of nutrition, rosemary is an excellent source of key vitamins, minerals, and dietary fiber, which help promote overall well-being. Rosemary serves not only in culinary and medicinal applications, but also in cosmetic and aromatherapy fields, underscoring its multifaceted functions and historical significance. Furthermore, the conditions under which rosemary is cultivated and harvested also influence the preservation of its compounds. This chapter examines the health of rosemary as part of the integration of traditional knowledge and modern scientific research by

A. Abid (✉) · K. Bireche
Laboratory of Valorization and Promotion of Saharan Resources (VPRS), Faculty of Mathematics and Matter Sciences, University of Ouargla, Ouargla, Algeria

Higher School of Saharan Agriculture, El Oued, Algeria

Laboratory of Valorization and Promotion of Saharan Resources (VPRS), Faculty of Mathematics and Matter Sciences, University of Ouargla, Higher School of Saharan Agriculture El , Ouargla, Algeria

N. Mekhadmi
Laboratory of the Development and Technology of Saharan Resources (VTRS), Echahid Hamma Lakhdar El Oued University, El-Oued, Algeria

Faculty of Life and Natural Sciences, Department of Biology, University of El Oued, El-Oued, Algeria

M. A. Ansari
Department of Epidemic Disease Research, Institutes for Research and Medical Consultations (IRMC), Imam Abdulrahman Bin Faisal University, Dammam, Saudi Arabia



M. A. Ansari et al. (eds.), *Medicinal Plants and their Bioactive Compounds in Human Health: Volume 2*, https://doi.org/10.1007/978-981-95-3620-7_4

highlighting the health-protective value of rosemary and gives practical recommendations for its safe and effective use in diverse fields.

**Keywords**

Medicinal properties · Nutritional importance · Phytochemical compounds · Rosemary · *Salvia rosmarinus*

## Abbreviations

| | |
|---|---|
| CAT | Catalase (an enzyme that decomposes hydrogen peroxide) |
| DPP-4 | Dipeptidyl peptidase 4 (an enzyme involved in glucose metabolism, relevant to diabetes research) |
| GPx | Glutathione peroxidase (an enzyme that detoxifies peroxides) |
| MBC | Minimum bactericidal concentration (the lowest concentration of an antimicrobial agent required to kill a particular bacterium) |
| MIC | Minimum inhibitory concentration (the lowest concentration of an antimicrobial agent that inhibits the visible growth of a microorganism) |
| NF-κB | Nuclear factor kappa B (a protein complex involved in inflammation) |
| NO | Nitric oxide (a signaling molecule involved in various biological processes) |
| ROS | Reactive oxygen species (molecules that can damage cells and are involved in aging and diseases) |
| SOD | Superoxide dismutase (an enzyme involved in antioxidant defense) |

## 4.1 Introduction

A recent evolutionary investigation has combined the genus *Rosmarinus* with the genus *Salvia* (*Salvia rosmarinus* syn *Rosmarinus officinalis*). This indicates that the species under study is no longer properly known as *Rosmarinus officinalis. Salvia rosmarinus* is the new specific epithet for this species, as the merger already claimed the term *Salvia officinalis* (Appiah et al. 2022). This small, fragrant shrub, which is native to Europe and the Mediterranean coast, is a member of the Lamiaceae family and is also known as rosemary, sea dew, or antos (Charles 2012). People have used *S. rosmarinus*, one of the oldest Mediterranean plants, as a natural remedy and culinary flavoring for more than a millennium (Han et al. 2023).

*Kingdom*: Plantae
*Phylum*: Spermaphyte
*Class*: Dicotyledons (Magnoliopsida)
*Order*: Lamiales
*Family*: Lamiaceae
*Genus*: *Salvia*
*Species*: *Salvia rosmarinus*

*Salvia rosmarinus* is an aromatic herb native to the Mediterranean region, renowned for its rich historical background and diverse traditional uses. Valued for centuries, rosemary has played a significant role in culinary, medicinal, and cultural practices. In Mediterranean cuisine, it enhances the flavor of meats, soups, and stews. Beyond its culinary applications, rosemary is also utilized in traditional medicine for its purported health benefits, including memory enhancement and the treatment of neurodegenerative disorders (Meziane et al. 2024; Oresanya and Orhan 2024). Historically, the rosemary has been utilized in folk medicine to treat a variety of ailments, such as digestive issues, headaches, and muscle pain, In traditional European medicine, it is well documented, particularly for its neuroprotective properties (Oresanya and Orhan 2024).

## 4.2 Botanical Description of *Salvia rosmarinus*

### 4.2.1 Morphological and Anatomical Characteristics

Rosemary is an aromatic shrub, easily recognized from afar by its distinctive scent. This evergreen plant can reach up to two meters in height and is highly branched. Its leaves are leathery, sessile, linear, entire, and have rolled edges, with a distinctive venation pattern that includes primary, secondary, tertiary, and quaternary veins. They are green and rough on the upper surface, while white and tomentose (hairy) underneath (Mahmoud 2023; Quezel and Santa 1963). The flowers are pale blue or whitish, subsessile, and arranged in small axillary and terminal clusters.

*Flower Morphology*: The calyx, corolla, and stamens of *Salvia rosmarinus* are important for sectional and specific delimitations within the *Salvia* genus (Mahmoud 2023; Quezel and Santa 1963). The calyx is bell-shaped, bilabiate, and powdery, while the corolla is bilabiate with a prominent tube, a bifid upper lip forming a helmet. It contains two stamens with prominent filaments inserted at the throat of the corolla, each equipped with a small tooth near the base, and linear anthers. The stem is twisted, angular, and fragile, while the root is deep and tap rooted. The fruit is an oval, dry, and smooth berry (Fig. 4.1). In addition, the plant exhibits both glandular and non-glandular trichomes, which are significant for taxonomic classification. The mesophyll structure and the presence of a hypodermis are also notable anatomical features (Mahmoud 2023).

### 4.2.2 Geographic Distribution

The species *Salvia rosmarinus*, with a broad geographic distribution, is primarily concentrated in the Mediterranean region but also found in various other parts of the world. Due to its economic and medicinal importance, both natural growth and human cultivation influence this distribution. The Mediterranean coasts and Southwest Asia are where the *Salvia rosmarinus* plants originate. The Jin dynasty brought them to China. These shrubs are now grown all over the world, primarily as

**Fig. 4.1** Morphological aspect of *Salvia rosmarinus* (Quezel and Santa 1963)

an ornament in gardens (Waggas and Balawi 2010). Spain, Tunisia, Morocco, Algeria, Italy, France, and Portugal cultivate rosemary for the extraction of essential oils. In India, the production of rosemary began in the late 1980s and continued to grow in the 1990s (Panda 2005).

### 4.2.3 Specific Regional Studies

Research on rosemary accessions in Ethiopia has found significant phenotypic variety, indicating that local populations have adapted to particular regional environments. For conservation and breeding initiatives that seek to improve desired qualities, this diversity is essential (Zigene et al. 2022). Researchers have recognized the Oriental region of Morocco as a key location for the production of essential oils from *Salvia rosmarinus*. The plant's ability to adapt to various climatic conditions is demonstrated by the seasonal variations in the chemical composition of these oils (Annemer et al. 2022).

## 4.3 Cultivation and Harvesting

Several crucial aspects of *Salvia rosmarinus*'s culture and harvesting influence the production and quality of its essential oils and bioactive substances.

### 4.3.1 Cultivation Techniques

*Salvia rosmarinus* grows in Mediterranean climes, which are ideal. For optimal yield and quality, the plant needs well-drained soil and sunlight (Musolino et al. 2023). Similar species like *Salvia officinalis* show that plant density and weed control can dramatically impact essential oil yield and quality. Low plant densities and weed control via mulching can boost plant growth and oil content (Tibaldi et al. 2022). To ensure consistent and high-quality plant material for growing, *S. rosmarinus* could use micropropagation like *Salvia sclarea* (Grigoriadou et al. 2020).

### 4.3.2 Harvesting Periods

The quantity and chemical content of *S. rosmarinus* essential oils fluctuate with the season. For example, elevated yields and an increased concentration of 1,8-cineole are noted in the summer, whereas winter yields are diminished with a higher α-pinene content. The optimal period for harvesting to get maximum yield of essential oils is from late spring to early summer. This timeframe aligns with the maximum concentration of favorable chemicals such as 1,8-cineole (Annemer et al. 2022).

## 4.4 *Salvia rosmarinus* Phytochemical Compounds and Medicinal Importance

*Salvia rosmarinus* is crucial, according to 2019 research. Medline had 51 entries, Embase 47, BVS Regional Portal 62, Science Direct 567, CAPES Periodicals 174, and Scopus 43 (Borges et al. 2019). The medicinal herb *Salvia rosmarinus* contains polyphenols, flavonoids, terpenoids, and essential oils (Andrade et al. 2018). These chemicals make *Salvia rosmarinus* valuable in traditional and modern medicine. These substances derive from the plant's antioxidant, anti-inflammatory, antibacterial, and neuroprotective capabilities.

### 4.4.1 Essential Oils Composition

According to Borges et al. (2019), *S. rosmarinus* essential oil samples contain approximately 150 chemical components, with 1,8-cineole, α-pinene, and camphor being the most commonly reported molecules. *Salvia rosmarinus* essential oils are mostly monoterpenes and sesquiterpenes (Borges et al. 2019). Sesquiterpenes and monoterpenes essentially make up *Salvia rosmarinus*'s essential oils. The key chemicals founds in *S. rosmarinus* essential oils are camphor, borneol, α-pinene, and 1,8-cineole (Leporini et al. 2020) (Fig. 4.2). Each study finds these components; however, their quantities vary depending on drying and extraction processes (Al-Ghoul et al. 2023; Mahmoud 2023). Different drying procedures alter oil composition. According to Al-Ghoul et al. (2023), shade drying preserves the highest

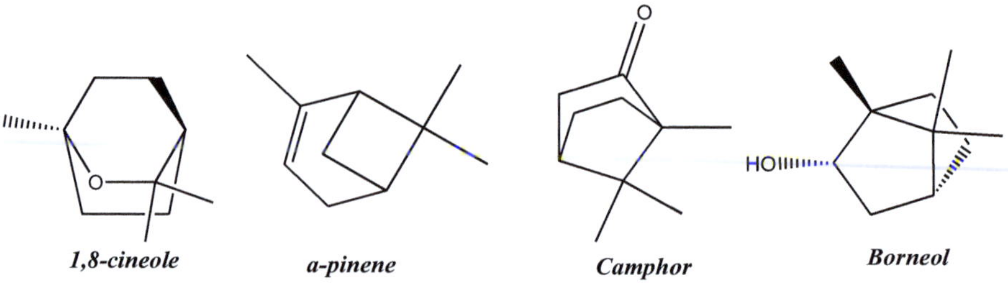


**Fig. 4.2** Chemical structures frequently identified in *S. rosmarinus* essential oils

levels of α-pinene and 1,8-cineole, whereas oven drying enhances borneol and camphor concentration (Al-Ghoul et al. 2023). Different rosemary cultivars have varying essential oil compositions, with some containing more camphor and others α-pinene and myrcene (Pieracci et al. 2021).

The essential oil of *S. rosmarinus* has antibacterial, antioxidant, antifungal, and anti-inflammatory effects (Kačániová et al. 2017). Furthermore, conventional wisdom acknowledges its ability to alleviate muscular discomfort and improve cognitive disorders like Alzheimer's disease (Leporini et al. 2020). Monoterpenes such as 1,8-cineole, borneol, pinene, limonene, camphene, camphor, and myrcene are responsible for the biological activities of *S. rosmarinus* essential oil. Studies have shown that *S. rosmarinus* essential oil reduces inflammation by blocking NF-κB (nuclear factor kappa B) transcription and the arachidonic acid cascade (Takaki et al. 2008; Borges et al. 2018). Antioxidants protect against inflammation-related damage. Also, *S. rosmarinus* essential oil reduces airway inflammation by relaxing smooth muscles (Borges et al. 2019).

### 4.4.2 Terpenoid Compounds

#### 4.4.2.1 Diterpenoids

Diterpene phenols and quinones, such as carnosic acid and carnosol, are found in *Salvia rosmarinus* (Fig. 4.3). There are 62 diterpenoids in *S. rosmarinus*, including 55 abietane, 3 icetexane, and 4 open-rings. These can change into each other in certain situations and have different biological effects. Out of 62 diterpenoids in *S. rosmarinus*, 40 show antioxidant, anti-inflammatory, anti-tumor, and neuroprotective properties (Liang et al. 2024). Diterpene compounds' antioxidant and anti-inflammatory properties are associated with specific chemical functional groups. Some diterpenes have phenolic hydroxyl, carbonyl, epoxy, or methyl groups that can neutralize reactive oxygen species (ROS) by giving up hydrogen atoms or electrons. Carbonyl and carboxylic acid groups may help reduce inflammation by modulating cyclooxygenase and lipoxygenase. Diterpenes with alkylated, hydroxylated, or epoxylated groups can interact with DNA, stopping the production of nucleic acids and killing tumor cells. In *S. rosmarinus*:

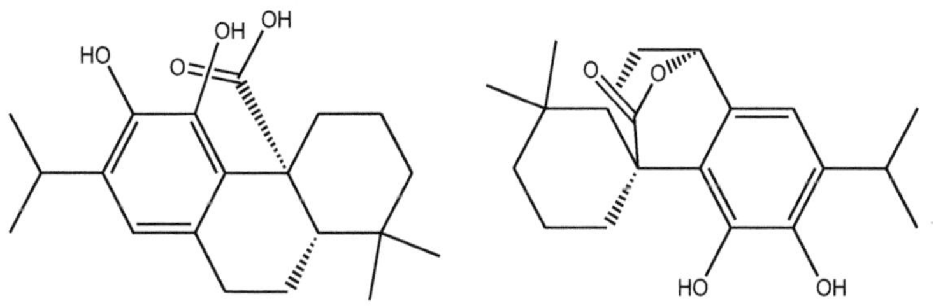


*Carnosic acid* *Carnosol*

**Fig. 4.3** Chemical structures of carnosic acid and carnosol identified in *S. rosmarinus*

- Twenty-seven diterpenoids in *S. rosmarinus* had been identified with antioxidant effect by reducing ROS, nitric oxide (NO), lipid peroxidation, and increasing the level of antioxidant defense (such as the Glutathione(GSH) level), as well as promoting the activity of antioxidant enzymes (such as superoxide dismutase [SOD], catalase [CAT], glutathione peroxidase [GPx]) (Habtemariam 2023).
- Eight diterpenoids had been reported with anti-inflammatory effects including downregulating NF-κB protein expression, inhibiting Mitogen-Activated Protein Kinase(MAPK) phosphorylation, suppressing of STAT1 and STAT3 activation, promoting Nrf2 release into the nucleus for Antioxidant Response Element (ARE) binding, and decreasing NLRP3 inflammasome (Habtemariam 2023).
- Ten diterpenoids reported in *S. rosmarinus* have anti-tumor effects, which could decrease the cytotoxic activity of five tumor cell lines (A-549, SMMC-7721, HL-60, MCF-7, and SW480) and one non-tumor cell line (BEAS-2B) (Zheng et al. 2020).
- Six diterpenoids in *S. rosmarinus* had neuroprotective effects (Liang et al. 2024).

#### 4.4.2.2 Triterpenoids

Previous studies have shown that the most effective active triterpenoids in rosemary extracts are ursolic acid and betulinic acid (Fig. 4.4) (Alavi et al. 2021). Both of these compounds are pentacyclic triterpenoid compounds, and they exhibit a wide range of biological actions, including neuroprotective, anti-inflammatory, and antioxidant effects (Farzan et al. 2024). Zhou et al. (2021) discovered five norursane-type triterpenoids: three new ones, 3β-28-norursa-12,20(30)-dien-3-ol, 3β-28-norursa-12,16,20(30)-triene-3-ol, and 3β-28-norursa-12-ene-3-ol, and two known ones. These phytochemicals greatly reduced NO production in RAW264.7 cells activated by Lipopolysaccharide (LPS). All of these compounds showed strong anti-inflammatory activity (Zhou et al. 2021). In 2022, Zhong et al. studied an ethanolic extract of *S. rosmarinus* and discovered three ursane-type triterpenoids: 3-oxours-12-en-20,28-olide, 3β-hydroxyurs-12-en-20,28-olide, and $3^2$-hydroxy-urs-11, 13(18)-dien-20,28-olide. These compounds inhibited LPS-activated macrophages' NO generation, causing considerable anti-inflammatory effects (Zhong et al. 2022).

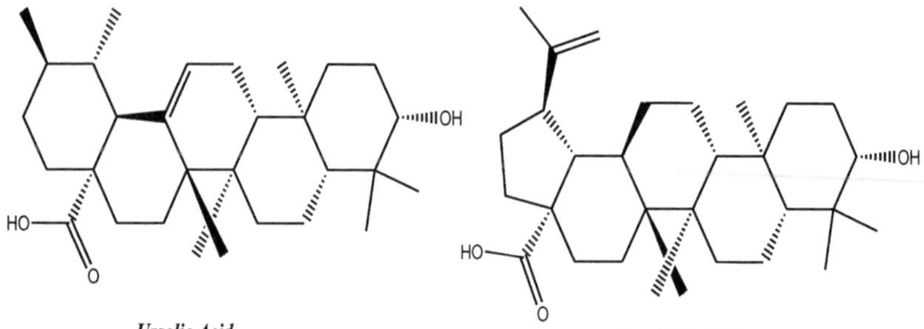

*Ursolic Acid* *Betulinic Acid*

**Fig. 4.4** Chemical structures of most effective active triterpenoids identified in rosemary

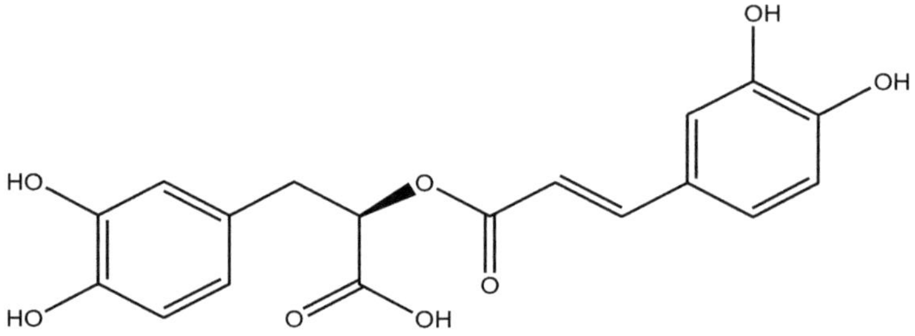

***Rosmarinic acid***

**Fig. 4.5** Chemical structure of most polyphenol (rosmarinic acid) identified in rosemary

### 4.4.3 Polyphenols and Flavonoids

In *Salvia rosmarinus*, several polyphenolic compounds have been identified, among which apigenin, diosmin, luteolin, genkwanin, and a variety of phenolic acids—including rosmarinic, chlorogenic, and caffeic acids—are the most prominent (Andrade et al. 2018). Notably, rosmarinic acid and carnosic acid are considered the principal bioactive molecules present in this species. Rosmarinic acid (Fig. 4.5), in particular, is one of the most abundant caffeic acid derivatives in the plant kingdom. It was originally isolated from *S. rosmarinus* and is chemically characterized as an ester formed between caffeic acid and 3,4-dihydroxyphenyllactic acid. This compound has garnered significant scientific interest due to its broad spectrum of pharmacological and therapeutic activities (Petersen 2013).

## 4.5 Nutritional Benefits of *Salvia rosmarinus*

Essential oils, vitamins, and minerals make rosemary a versatile herb with health advantages. Essential oils like 1,8-cineole, α-pinene, and camphor contribute to its scent and medicinal benefits. It contains vitamins A, B-complex, and E, which boost

**Table 4.1** Summary of the nutritional composition and health benefits of rosemary

| Nutrient category | Key components | Health benefits | References |
|---|---|---|---|
| Essential oils | 1,8-Cineole, α-pinene, camphor, linalool, camphene, borneol, limonene, verbenone, myrtenole, myrcene, α-thujene, B-pinene, trephine, 3-carene, P-cymenene, camphonelal, and iso-pinocamphone | Therapeutic properties; distinctive fragrance | Aljaafari et al. (2021), Oualdi et al. (2023), Ribeiro-Santos et al. (2015) |
| Vitamins | Ascorbic acid, thiamin, riboflavin, niacin, vitamin B6, vitamin E, folate, vitamin B12, vitamin A, Retinol Activity Equivalent (RAE), and vitamin D (D2 + D3) | Immune support; antioxidant defense; metabolism regulation | Meziane et al. (2024), Ribeiro-Santos et al. (2015) |
| Minerals | Ag, Al, As, B, Ba, Bi, Ca, Cd, Co, Cr, Cu, Fe, K, Li, Mg, Mn, Na, Ni, P, Pb, Se, Sr, Ti, V, and Zn | Bone health; enzymatic support; physiological balance | Aziz et al. (2022), Arslan and Özcan (2008) |
| Macronutrients | Proteins, lipids, and carbohydrates | Contributes to overall energy and digestive health | Meziane et al. (2024) |
| Fatty acids | Saturated, monounsaturated, and polyunsaturated fatty acids | Beneficial for healthy cholesterol levels | Meziane et al. (2024), Ribeiro-Santos et al. (2015) |
| Dietary fiber | Present in notable amounts | Promotes digestive health | Cui et al. (2019), Gill et al. (2021) |

immunity, antioxidant defense, and metabolism. Rosemary also contains calcium, magnesium, potassium, and iron, which are essential for bone and enzymatic health. Rosemary has many nutritional uses due to these components (Meziane et al. 2024). Table 4.1 lists the essential nutrients of rosemary and their health advantages.

The bioactive properties of rosemary provide many health benefits. Rosmarinic acid and carnosic acid, potent antioxidants, neutralize free radicals to protect cells from oxidative damage, reducing the risk of chronic diseases and boosting health (El Boukhari and Fatimi 2023; Aziz et al. 2022; Gonzalez-Vallinas et al. 2013). Rosemary may help treat arthritis due to carnosic acid, which reduces inflammation (Gonçalves et al. 2022). Rosemary also lowers cholesterol, improves blood circulation, and boosts High-Density Lipoprotein (HDL) (Akbari et al. 2022; Musolino et al. 2023). People have used rosemary to treat bloating, indigestion, and gas and to improve digestion by increasing bile production (Fornari et al. 2014). The herb's neuroprotective components, especially rosmarinic acid, boost memory and minimize mental tiredness (Hatami et al. 2020). Rosemary also promotes fat metabolism, reduced fat formation, and weight control, preventing obesity-related diseases (Rafie et al. 2017). According to González-Vallinas et al. (2015), rosemary's bioactive components have anti-cancer effects, limiting cancer cell development and lowering tumor formation, making it a helpful herb for overall health (González-Vallinas et al. 2015).

## 4.6 Historical Significance and Traditional Uses

Mediterranean cuisine relies on rosemary to flavor meats, soups, and stews. In addition to its culinary use, it has a long history in traditional medicine for memory augmentation and neurodegenerative illnesses (Meziane et al. 2024; Oresanya and Orhan 2024). Rituals and ceremonies have used rosemary as a symbol of remembering, affection, and loyalty (Han et al. 2023). Folk medicine, particularly in Europe, has utilized it to treat digestive disorders, migraines, and muscle discomfort because of its neuroprotective qualities (Shimira et al. 2022; Oresanya and Orhan 2024).

With its aromatic leaves and essential oils, rosemary flavors meats, vegetables, and breads (Musolino et al. 2023; Veenstra and Johnson 2021). It inhibits cholinesterase and affects neurotransmitter systems, increasing cognitive function and perhaps treating Alzheimer's disease (Oresanya and Orhan 2024; Ayoub et al. 2022). Its antioxidant and antibacterial characteristics make it a significant cosmetic and pharmaceutical element for skin benefits and natural preservatives (Meziane et al. 2024; Reuter et al. 2007). Due to its anti-inflammatory and analgesic properties, people have also used it locally for muscle discomfort and internally for liver and intestinal health (Cedillo-Portillo et al. 2024; Veenstra and Johnson 2021). Rosemary has many traditional purposes, but recent research is expanding its usage. Ultrasound-assisted extraction has made bioactive compound extraction more useful across industries (Al-Ghoul et al. 2023). Genomic investigations are also revealing its biosynthesis routes, which could lead to novel drugs (Selmi et al. 2017).

## 4.7 Medicinal Properties of *Salvia rosmarinus* and Its Mechanism of Action

### 4.7.1 Anti-Inflammatory Effects

The various medicinal properties and mode of action of *S. rosmarinus* have been represented in Fig. 4.6. Components such as carnosol, carnosic acid, ursolic acid, oleanolic acid, and micromeric acid are responsible for the anti-inflammatory properties of rosemary extract (Altinier et al. 2007; Reuter et al. 2007). Carnosic acid is a powerful inhibitor of nitric oxide (NO), a pro-inflammatory mediator (Lo et al. 2002). At concentrations as low as 6.2 μg/mL, it inhibits NO by 72%, with complete inhibition at >12.5 μg/mL (Yimam et al. 2017). Additionally, rosemary extract shows strong antioxidant and antiplatelet activity, with maximum platelet inhibition at 31 μg/mL of carnosic acid (Yimam et al. 2017). In a study on mice with atopic dermatitis, topical treatment with carnosol significantly reduced skin lesions, immunoglobulin E, neutrophils, and inflammatory cytokines (tumor necrosis factor-alpha [TNF-α], interleukin-1 beta [IL-1β]). Histologically, carnosol-treated skin was thinner, with less inflammatory cell infiltration and fewer mast cells, and it reduced the expression of enzymes like Inducible Nitric Oxide Synthase (iNOS) and Cyclooxygenase-2 (COX-2) (de Macedo et al. 2020; Lim et al. 2016). While rosemary extracts did not show significant anti-inflammatory effects on existing

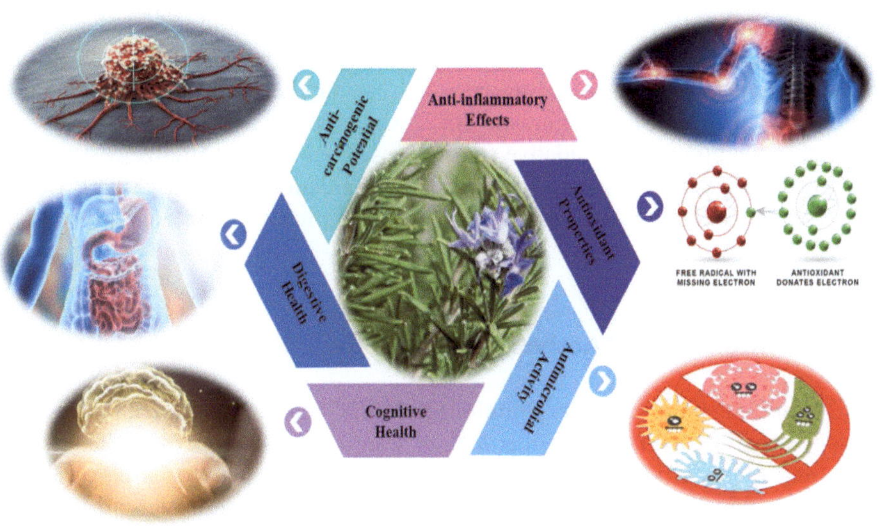


**Fig. 4.6** Medicinal properties of *Salvia rosmarinus*

dermatitis in human volunteers, they reduced inflammation when applied simultaneously with an irritant, indicating a protective effect. In another test using croton oil-induced ear edema, rosemary extracts prepared with *n*-hexane and chloroform demonstrated strong, dose-dependent anti-inflammatory effects comparable to indomethacin, a known anti-inflammatory agent (Lee et al. 2017).

The anti-inflammatory effects of rosemary are mediated through multiple mechanisms involving its bioactive compounds. Primarily, these effects are attributed to the modulation of inflammatory mediators and inhibition of specific signaling pathways. Salicylic acid derivatives present in rosemary inhibit prostaglandin production, reducing inflammation when applied topically (Reller and Kretschmar 1980). Furthermore, rosemary inhibits key inflammatory mediators such as nitric oxide (NO), tumor necrosis factor-alpha (TNF-$\alpha$), interleukin-1 beta (IL-1$\beta$), and interleukin-6 (IL-6) by preventing the activation of nuclear factor kappa B (NF-$\kappa$B), a critical transcription factor in the inflammatory response (Li et al. 2012). Furthermore, flavonoids such as salvigenin enhance its anti-inflammatory effects by diminishing the release of these mediators, as evidenced in animal studies (Roh et al. 2015).

### 4.7.2 Antioxidant Properties

Rosemary contains antioxidants like carnosol, carnosic acid, and rosmanol (Wenkert et al. 1965). These chemicals effectively scavenge ROS such as peroxyl, hydroxyl, and hydrogen peroxide (Wijeratne and Cuppett 2007). Carnosic acid and carnosol stop liposomal and microsomal lipid peroxidation. They are responsible for 90% of rosemary's antioxidant power (Aruoma et al. 1992). This makes rosemary a beneficial antioxidant for biological systems and food preservation by reducing oxidative degradation (Djenane et al. 2002). The Nrf2 transcription factor activates redox-dependent signaling pathways in rosemary's antioxidant mechanism. Carnosic acid and carnosol stimulate Nrf2, which boosts SOD and GPx synthesis, protecting the body from oxidative stress. These chemicals neutralize free radicals to prevent lipid peroxidation and cell and tissue oxidation. Rosemary's antioxidants decrease lipid oxidation in food systems, prolonging meat and oil's shelf life (Nieto et al. 2018).

### 4.7.3 Antimicrobial Activity

MIC (minimum inhibitory concentration) and MBC (minimum bactericidal concentration) experiments have shown rosemary's antibacterial properties. Sienkiewicz et al. and others attributed rosemary's antibacterial activities to its polyphenolic extract, specifically the synergy between rosmarinic acid and carnosic acid. Carnosic acid and carnosol concentrations strongly correlate with antimicrobial action (Sienkiewicz et al. 2013). Zaouali et al. found that α-pinene, a key component of rosemary essential oil, enhances antibacterial activity against *Staphylococcus aureus*. Research suggests that terpenes like α-pinene can disrupt bacterial cell membranes, leading to cell lysis (Zaouali et al. 2010). Studies also show that rosemary essential oil's ability to kill *Escherichia coli* is due to the action of several minor volatile components working together, not just one main component (Micić et al. 2021). In beef meatballs, cooked beef, pig sausage, and lamb, rosemary inhibits *E. coli*, *Bacillus cereus*, and *S. aureus* development (Nieto et al. 2018). Rosemary extracts help protect meat packing from rotting germs, improving food safety (Kaur et al. 2023). Mihajilov-Krstev et al. (2010) used the broth microdilution method to find that rosemary essential oil was very good at killing Gram-negative bacteria like *E. coli*, with MIC values ranging from 0.025 to 0.78 μL/mL (Mihajilov-Krstev et al. 2010). Other investigations show that rosemary oil substantially inhibits *E. coli* at >6.4 mg/L (Prabuseenivasan et al. 2006). The antibacterial properties of rosemary essential oil come from carnosol, rosmarinic acid, rosmaridiphenol, and isorosmanol. These chemicals damage bacterial cell membranes, altering nutrition and electron transport, cellular leakage, and fatty acid synthesis. They can interact with membrane proteins, destroying membrane shape and function (Fung et al. 1977). Carnosic acid is more effective against pathogenic microorganisms than rosmarinic acid (Vegara et al. 2011).

### 4.7.4 Cognitive Health

Rosemary may improve cognitive function and treat forgetfulness, dementia, and delirium. These illnesses impede learning, memory, and cognition. Rosemary and its components are neuroprotective, according to research. Researchers found that combining rosemary lemon oil with lavender oil in senior persons increased cognitive functioning, sleep quality, and reduced daytime tiredness (Yıldırım and Kitiş 2020). Also, Li et al. (2020) found that rosmarinic acid reduced cognitive impairment in hypoxia-ischemia-induced mice. The study showed that the compound not only improved movement and spatial memory but also protected against myelin degradation in neurons (Li et al. 2020). Song et al. (2016) showed that rosemary extract reduced cognitive deficits in rats after recurrent mild traumatic brain injury. Rosemary lowered neuronal degeneration and oxidative stress, improving cognition. These results suggest that rosemary's antioxidant and anti-inflammatory capabilities may treat cognitive problems, particularly those related to oxidative stress and neuroinflammation (Song et al. 2016).

Rosemary's antioxidant, anti-inflammatory, and anti-apoptotic qualities come from carnosic acid and rosmarinic acid (Rahbardar and Hosseinzadeh 2020; de Oliveira 2016; Faridzadeh et al. 2022). These stimulate the Keap1/Nrf2 pathway, increasing antioxidant enzymes and reducing oxidative stress (de Oliveira 2016). Rosemary inhibits pro-inflammatory cytokines and saves neurons by preserving mitochondria (Habtemariam 2023). As a cholinesterase inhibitor, it preserves acetylcholine levels, increasing cognitive function (Kamli et al. 2022). Rosemary may treat neurodegenerative illnesses like Alzheimer's and Parkinson's due to these processes (Kamli et al. 2022).

### 4.7.5 Digestive Health

Rosemary, rich in phenolic compounds like rosmarinic acid, provides notable digestive health benefits, although around 60% of rosmarinic acid degrades during gastrointestinal digestion, limiting its bioactivity post-ingestion (Gonçalves et al. 2019). Rosemary shows potential in antiproliferative applications; while crude and digested extracts lack strong effects, fermented extracts exhibit significant cytotoxicity against HeLa cells, suggesting therapeutic value for digestive health and disease prevention (Amar et al. 2017; Jardak et al. 2017). Protecting rosmarinic acid and similar bioactives through microencapsulation could further optimize rosemary's health benefits as a functional food ingredient (Ribeiro et al. 2016). In dietary applications, adding rosemary extract at 1–3 g/kg to rainbow trout diets improves digestive enzyme activity (amylase, lipase, trypsin) and supports nutrient absorption, while also reducing liver steatosis, aiding lipid metabolism, and enhancing liver health (Karataş et al. 2020). Rosemary extract also boosts antioxidant enzymes like superoxide dismutase and catalase, which help reduce oxidative stress and protect digestive tissues from reactive oxygen species, further underscoring its protective role in digestive health (Yildirim et al. 2017, 2018).

### 4.7.6 Anti-carcinogenic Potential

The anti-carcinogenic potential of rosemary, particularly through compounds like rosmarinic acid and carnosic acid, is demonstrated through a range of mechanisms that target multiple cancer types, including liver, colon, breast, pancreatic, and lung cancers. A primary action of rosmarinic acid and carnosic acid is their antiproliferative effect, which inhibits cancer cell growth and spread by inducing cell cycle arrest, a critical step in controlling unchecked cell division typical in cancer (Jin et al. 2020; Wang et al. 2019). Additionally, rosmarinic acid and carnosic acid promote apoptosis or programmed cell death in cancer cells by upregulating pro-apoptotic proteins, such as Bax, and downregulating anti-apoptotic proteins like Bcl-2. This selective cytotoxicity allows for the targeted destruction of cancer cells without harming healthy ones (Han et al. 2018; Messeha et al. 2020). Beyond this, rosemary-derived compounds exhibit anti-angiogenic properties, effectively inhibiting the development of blood vessels that supply tumors with nutrients, thus restricting resources essential for tumor growth. Furthermore, rosmarinic acid and carnosic acid show a synergistic effect with chemotherapy drugs, enhancing their efficacy and potentially reducing required doses, minimizing side effects, and addressing chemotherapy resistance, a significant challenge in cancer treatment (Zhou et al. 2022). Their strong antioxidant properties also contribute to their therapeutic profile by reducing oxidative stress, a factor associated with cancer development, thereby helping to prevent cellular damage that could lead to cancer. Rosmarinic acid and carnosic acid inhibit key signaling pathways, like the PI3K/Akt/mTOR pathway, essential for cancer cell survival, growth, and proliferation (Sirajudeen et al. 2024).

## 4.8 Applications in Human Health

### 4.8.1 Medicinal Uses

Rosemary has long been used in traditional medicine, particularly for its neuroprotective effects. Historically employed to enhance memory and address neurodegenerative disorders, modern studies confirm that rosemary may support memory improvement and neuroprotection at the molecular level (Oresanya and Orhan 2024). In addition, rosemary possesses strong anti-inflammatory and antioxidant properties, largely attributed to compounds like rosmarinic acid and carnosic acid, which help in managing asthma and reducing oxidative stress (Castañeta et al. 2022; Rajizadeh et al. 2024). Its benefits extend further, with notable antimicrobial and anti-cancer properties. Active compounds in rosemary inhibit the growth of cancer cells and provide natural defenses against infections (Cedillo-Portillo et al. 2024; Shimira et al. 2022). Rosemary's role in liver health also underscores its medicinal use, as it has been traditionally employed for its hepatoprotective effects, helping to protect liver tissue from potential damage (Veenstra and Johnson 2021).

### 4.8.2 Ethnopharmacological Applications

In the realm of ethnopharmacology, rosemary holds relevance for managing metabolic and neurological conditions. For example, it is traditionally used to aid in the management of type 2 diabetes mellitus through its enzyme inhibiting effects, such as dipeptidyl peptidase (DPP-4), which is beneficial for regulating blood sugar and supporting metabolic health (Shimira et al. 2022). Moreover, rosemary extracts are utilized in conductive education, an approach developed to assist children with neurological disorders, due to its phytotherapeutic properties that may aid neurological function and support overall health (Vissi et al. 2021).

### 4.8.3 Cosmetic Applications

Rosemary's extensive benefits make it a valuable ingredient in cosmetics. Due to its potent antioxidant and anti-inflammatory properties, rosemary is commonly incorporated into skincare formulations to treat conditions like acne and eczema. Compounds such as carnosol, carnosic acid, and ursolic acid are credited with reducing oxidative stress and inflammation, making rosemary an effective component in anti-aging and skincare products (de Macedo et al. 2020). Beyond skincare, rosemary's wound-healing properties and ultraviolet (UV)-protective effects make it a popular addition to sunscreens and after-sun care products, while its versatility extends to treatments for both pathological and non-pathological skin conditions like cellulite and alopecia (de Macedo et al. 2020). Additionally, rosemary's natural preservative qualities, owing to its antimicrobial activity, allow for extended shelf life of cosmetic products without synthetic preservatives (Meziane et al. 2024).

### 4.8.4 Aromatherapy and Mental Health

The essential oils of *Salvia rosmarinus* are rich in monoterpenes such as 1,8-cineole, α-pinene, and camphor, contributing to its therapeutic applications in aromatherapy. These essential oils are commonly used to enhance relaxation, mood, and cognitive function. Their ability to improve mental clarity and focus is a noted benefit (Leporini et al. 2020). Rosemary essential oil also exhibits cholinesterase inhibitory activity, which is valuable in managing Alzheimer's disease symptoms, further supporting cognitive health and mental clarity. Additionally, inhalation of rosemary oil has been shown to reduce cortisol levels, making it an effective agent for stress and anxiety relief, promoting a sense of well-being (Leporini et al. 2020).

### 4.8.5 Natural Food Preservative

Beyond health and wellness applications, rosemary is also valued as a natural food preservative. Its significant antioxidant and antimicrobial activities inhibit microbial growth and prevent spoilage through oxidation, making it an effective preservative in the food industry (Veenstra and Johnson 2021).

## 4.9 Safety and Toxicity

### 4.9.1 Potential Toxicity Concerns

While rosemary is generally considered safe, some compounds, such as thujone found in related *Salvia* species, can pose toxicity risks if consumed in excessive amounts. Regulatory bodies have set safe limits for such compounds to prevent adverse effects (Garg and Kumar 2024). Additionally, studies have shown that rosemary extracts may have cytotoxic effects on healthy cell lines, indicating the need for careful dosage and application to avoid harmful impacts (Irtegun Kandemir et al. 2022).

### 4.9.2 Recommended Dosages

The recommended dosages for rosemary are typically small and controlled, particularly when using extracts or essential oils due to their potent bioactive compounds. However, specific dosage details were not provided in the cited studies. Safe limits have been established for certain compounds like thujone to avoid toxicity (Garg and Kumar 2024).

### 4.9.3 Potential Side Effects and Interactions

Compounds such as thujone, found in rosemary, can be toxic when consumed in high doses, potentially leading to adverse effects such as seizures (Garg and Kumar 2024). Additionally, the essential oil of rosemary, which contains monoterpenes and oxygenated monoterpenes, may cause skin irritation or allergic reactions in sensitive individuals (Leporini et al. 2020). Rosemary can also interact with medications, particularly by enhancing the effects of anticoagulants, requiring caution for individuals taking blood-thinning treatments (Shimira et al. 2022).

### 4.9.4 Safety in Various Populations

- *Pregnant women*: The safety of rosemary during pregnancy is not well documented. Due to its potential uterotonic effects, it is generally recommended to avoid high doses during pregnancy (Shimira et al. 2022).
- *Children*: There is limited safety information regarding the use of rosemary in children. Given its potential toxicity, it should be used cautiously in pediatric populations (Garg and Kumar 2024).

- *General population*: Rosemary is considered safe for the general population when used in culinary amounts. However, individuals allergic to plants in the Lamiaceae family should be cautious (Natali et al. 2022).

## 4.10 Conclusion and Future Prospects

Researchers have thoroughly investigated *Salvia rosmarinus*, commonly referred to as rosemary, for its diverse pharmacological and nutritional benefits. Rosmarinic acid and carnosic acid are two bioactive compounds that have strong anti-inflammatory, antioxidant, and antimicrobial effects. This supports the use of rosemary for health benefits. Furthermore, including rosemary in food enhances its important vitamins and minerals, which are advantageous to different sectors, including cosmetics and traditional medicine. Future research on rosemary will focus on its potential in treating neurological and metabolic illnesses to establish new pharmacological targets. The genomics of biosynthetic pathways and microencapsulation are two new ways to extract and store chemicals that may make the medicinal parts of rosemary more available and increase their amount. The sustainable cultivation and harvesting of rosemary are essential for its preservation as a valuable plant material.

## References


Akbari S, Sohouli MH, Ebrahimzadeh S, Ghanaei FM, Hosseini AF, Aryaeian N (2022) Effect of rosemary leaf powder with weight loss diet on lipid profile, glycemic status, and liver enzymes in patients with nonalcoholic fatty liver disease: a randomized, double-blind clinical trial. Phytother Res 36:2186–2196

Alavi MS, Fanoudi S, Ghasemzadeh Rahbardar M, Mehri S, Hosseinzadeh H (2021) An updated review of protective effects of rosemary and its active constituents against natural and chemical toxicities. Phytother Res 35:1313–1328

Al-Ghoul AM, Alwardat S, Abdelhalim AR (2023) Chemical composition of essential oil from fresh *Salvia rosmarinus* from Jordan at different drying methods. Int J Membr Sci Technol 10:4143–4149

Aljaafari MN, AlAli AO, Baqais L, Alqubaisy M, AlAli M, Molouki A, Ong-Abdullah J, Abushelaibi A, Lai K-S, Lim S-HE (2021) An overview of the potential therapeutic applications of essential oils. Molecules 26:628

Altinier G, Sosa S, Aquino RP, Mencherini T, Loggia RD, Tubaro A (2007) Characterization of topical antiinflammatory compounds in *Rosmarinus officinalis* L. J Agric Food Chem 55:1718–1723

Amar Y, Meddah B, Bonacorsi I, Costa G, Pezzino G, Saija A, Cristani M, Boussahel S, Ferlazzo G, Meddah AT (2017) Phytochemicals, antioxidant and antiproliferative properties of *Rosmarinus officinalis* L on U937 and CaCo-2 cells. Iran J Pharm Res: IJPR 16:315

Andrade JM, Faustino C, Garcia C, Ladeiras D, Reis CP, Rijo P (2018) *Rosmarinus officinalis* L.: an update review of its phytochemistry and biological activity. Future Sci OA 4:FSO283

Annemer S, Farah A, Stambouli H, Assouguem A, Almutairi MH, Sayed AA, Peluso I, Bouayoun T, Talaat Nouh NA, El Ouali Lalami A (2022) Chemometric investigation and antimicrobial activity of *Salvia rosmarinus* Spenn essential oils. Molecules 27:2914

Appiah KS, Omari RA, Onwona-Agyeman S, Amoatey CA, Ofosu-Anim J, Smaoui A, Arfa AB, Suzuki Y, Oikawa Y, Okazaki S (2022) Seasonal changes in the plant growth-inhibitory effects of rosemary leaves on lettuce seedlings. Plants 11:673

Arslan D, Özcan MM (2008) Evaluation of drying methods with respect to drying kinetics, mineral content and colour characteristics of rosemary leaves. Energy Convers Manag 49:1258–1264

Aruoma O, Halliwell B, Aeschbach R, Löligers J (1992) Antioxidant and pro-oxidant properties of active rosemary constituents: carnosol and carnosic acid. Xenobiotica 22:257–268

Ayoub IM, George MY, Menze ET, Mahmoud M, Botros M, Essam M, Ashmawy I, Shendi P, Hany A, Galal M (2022) Insights into the neuroprotective effects of *Salvia officinalis* L. and *Salvia microphylla* Kunth in the memory impairment rat model. Food Funct 13:2253–2268

Aziz E, Batool R, Akhtar W, Shahzad T, Malik A, Shah MA, Iqbal S, Rauf A, Zengin G, Bouyahya A (2022) Rosemary species: a review of phytochemicals, bioactivities and industrial applications. S Afr J Bot 151:3–18

Borges RS, Lima ES, Keita H, Ferreira IM, Fernandes CP, Cruz RAS, Duarte JL, Velázquez-Moyado J, Ortiz BLS, Castro AN (2018) Anti-inflammatory and antialgic actions of a nanoemulsion of *Rosmarinus officinalis* L. essential oil and a molecular docking study of its major chemical constituents. Inflammopharmacology 26:183–195

Borges RS, Ortiz BLS, Pereira ACM, Keita H, Carvalho JCT (2019) Rosmarinus officinalis essential oil: a review of its phytochemistry, anti-inflammatory activity, and mechanisms of action involved. J Ethnopharmacol 229:29–45

Castañeta G, Cifuentes N, Sepulveda B, Bárcenas-Pérez D, Cheel J, Areche C (2022) Untargeted metabolomics by using UHPLC–ESI–MS/MS of an extract obtained with ethyl lactate green solvent from *Salvia rosmarinus*. Separations 9:327

Cedillo-Portillo JJ, Villastrigo-López WY, Castañeda-Facio AO, Esparza-González SC, Múzquiz-Ramos EM, Sáenz-Galindo A (2024) *Salvia rosmarinus* Spenn. Main applications and ultrasonic extraction of secondary metabolites: a general review. Rev Mex Ing Biomed 45:35–61

Charles D (2012) Handbook of herbs and spices. Woodhead Publishing series in food science, technology and nutrition. Cambridge, Woodhead Publishing

Cui J, Lian Y, Zhao C, Du H, Han Y, Gao W, Xiao H, Zheng J (2019) Dietary fibers from fruits and vegetables and their health benefits via modulation of gut microbiota. Compr Rev Food Sci Food Saf 18:1514–1532

de Macedo LM, Santos ÉM d, Militão L, Tundisi LL, Ataide JA, Souto EB, Mazzola PG (2020) Rosemary (*Rosmarinus officinalis* L., syn *Salvia rosmarinus* Spenn.) and its topical applications: a review. Plants 9:651

de Oliveira MR (2016) The dietary components carnosic acid and carnosol as neuroprotective agents: a mechanistic view. Mol Neurobiol 53:6155–6168

Djenane D, Sánchez-Escalante A, Beltrán JA, Roncalés P (2002) Ability of α-tocopherol, taurine and rosemary, in combination with vitamin C, to increase the oxidative stability of beef steaks packaged in modified atmosphere. Food Chem 76:407–415

El Boukhari R, Fatimi A (2023) A review of the patentability of rosemary-derived drugs and bioactive compounds. Drugs Drug Candidates 2:172–188

Faridzadeh A, Salimi Y, Ghasemirad H, Kargar M, Rashtchian A, Mahmoudvand G, Karimi MA, Zerangian N, Jahani N, Masoudi A (2022) Neuroprotective potential of aromatic herbs: rosemary, sage, and lavender. Front Neurosci 16:909833

Farzan M, Farzan M, Shahrani M, Navabi SP, Vardanjani HR, Amini-Khoei H, Shabani S (2024) Neuroprotective properties of Betulin, Betulinic acid, and Ursolic acid as triterpenoids derivatives: a comprehensive review of mechanistic studies. Nutr Neurosci 27:223–240

Fornari T, Ramírez de Molina A, & Reglero G (2014) Phenolic diterpenes from rosemary as enhancement agents of usual chemotherapeutic drugs for colorectal cancer therapy. In J. Hu (Ed.), New Developments in Terpenes Research (pp. 47–78). Nova Science Publishers, Inc. ISBN 978-1-62948-760-1

Fung DY, Taylor S, Kahan J (1977) Effects of butylated hydroxyanisole (BHA) and butylated hydroxytoluene (BHT) on growth and aflatoxin production of aspergillus flavus. J Food Saf 1:39–51

Garg D, Kumar V (2024) A review of therapeutic properties and uses of *Salvia officinalis*. J Pharma Insights Res 2:146–154

Gill SK, Rossi M, Bajka B, Whelan K (2021) Dietary fibre in gastrointestinal health and disease. Nat Rev Gastroenterol Hepatol 18:101–116

Gonçalves GA, Corrêa RC, Barros L, Dias MI, Calhelha RC, Correa VG, Bracht A, Peralta RM, Ferreira IC (2019) Effects of in vitro gastrointestinal digestion and colonic fermentation on a rosemary (*Rosmarinus officinalis* L) extract rich in rosmarinic acid. Food Chem 271:393–400

Gonçalves C, Fernandes D, Silva I, Mateus V (2022) Potential anti-inflammatory effect of *Rosmarinus officinalis* in preclinical in vivo models of inflammation. Molecules 27:609

Gonzalez-Vallinas M, Molina S, Vicente G, de la Cueva A, Vargas T, Santoyo S, Garcia-Risco MR, Fornari T, Reglero G, de Molina AR (2013) Antitumor effect of 5-fluorouracil is enhanced by rosemary extract in both drug sensitive and resistant colon cancer cells. Pharmacol Res 72:61–68

González-Vallinas M, Reglero G, Ramírez de Molina A (2015) Rosemary (*Rosmarinus officinalis* L.) extract as a potential complementary agent in anticancer therapy. Nutr Cancer 67:1223–1231

Grigoriadou K, Trikka F, Tsoktouridis G, Krigas N, Sarropoulou V, Papanastasi K, Maloupa E, Makris A (2020) Micropropagation and cultivation of Salvia sclarea for essential oil and sclareol production in northern Greece. In Vitro Cell Dev Biol-Plant 56:51–59

Habtemariam S (2023) Anti-inflammatory therapeutic mechanisms of natural products: insight from rosemary diterpenes, carnosic acid and carnosol. Biomedicine 11:545

Han Y-H, Kee J-Y, Hong S-H (2018) Rosmarinic acid activates AMPK to inhibit metastasis of colorectal cancer. Front Pharmacol 9:68

Han D, Li W, Hou Z, Lin C, Xie Y, Zhou X, Gao Y, Huang J, Lai J, Wang L (2023) The chromosome-scale assembly of the *Salvia rosmarinus* genome provides insight into carnosic acid biosynthesis. Plant J 113:819–832

Hatami SM, Mehrpooya N, Vagharseyyedin SA, Rahimi H (2020) Effect of aromatherapy with rosemary essential oil on occupational fatigue and depression experienced by ambulance technicians. Herb Med J 5:76–84

Irtegun Kandemir S, Fidan HS, Yener I, Mete N, Ertas A, Topcu G, Kolak U (2022) Investigation of cytotoxic and apoptotic effects of 63 compounds obtained from Salvia species: promising anticancer agents. J Food Biochem 46:e14226

Jardak M, Elloumi-Mseddi J, Aifa S, Mnif S (2017) Chemical composition, anti-biofilm activity and potential cytotoxic effect on cancer cells of *Rosmarinus officinalis* L. essential oil from Tunisia. Lipids Health Dis 16:1–10

Jin B, Liu J, Gao D, Xu Y, He L, Zang Y, Li N, Lin D (2020) Detailed studies on the anticancer action of rosmarinic acid in human Hep-G2 liver carcinoma cells: evaluating its effects on cellular apoptosis, caspase activation and suppression of cell migration and invasion. J BUON 25:1383–1389

Kačániová M, Terentjeva M, Kántor A, Tokár M, Puchalski C, Ivanišová E (2017) Antimicrobial effect of sage (L.) and rosemary (L.) essential oils on microbiota of chicken breast. Proc Latvian Acad Sci Sect B 71:461–467

Kamli MR, Sharaf AAM, Sabir JS, Rather IA (2022) Phytochemical screening of *Rosmarinus officinalis* L. as a potential anticholinesterase and antioxidant–medicinal plant for cognitive decline disorders. Plants 11:514

Karataş T, Korkmaz F, Karataş A, Yildirim S (2020) Effects of rosemary (*Rosmarinus officinalis*) extract on growth, blood biochemistry, immunity, antioxidant, digestive enzymes and liver histopathology of rainbow trout, *Oncorhynchus mykiss*. Aquac Nutr 26:1533–1541

Kaur R, Gupta TB, Bronlund J, Kaur L (2023) The potential of rosemary as a functional ingredient for meat products—a review. Food Rev Int 39:2212–2232

Lee DY, Hwang CJ, Choi JY, Park MH, Song MJ, Oh KW, Son DJ, Lee SH, Han SB, Hong JT (2017) Inhibitory effect of carnosol on phthalic anhydride-induced atopic dermatitis via inhibition of STAT3. Biomol Ther 25:535–544

Leporini M, Bonesi M, Loizzo MR, Passalacqua NG, Tundis R (2020) The essential oil of *Salvia rosmarinus* Spenn. from Italy as a source of health-promoting compounds: chemical profile and antioxidant and cholinesterase inhibitory activity. Plants 9:798

Li M, Zhang L, Cai RL, Gao Y, Qi Y (2012) Lipid-soluble extracts from *Salvia miltiorrhiza* inhibit production of LPS-induced inflammatory mediators via NF-κB modulation in RAW 264.7 cells and perform antiinflammatory effects in vivo. Phytother Res 26:1195–1204

Li M, Cui M-M, Kenechukwu NA, Gu Y-W, Chen Y-L, Zhong S-J, Gao Y-T, Cao X-Y, Wang L, Liu F-M (2020) Rosmarinic acid ameliorates hypoxia/ischemia induced cognitive deficits and promotes remyelination. Neural Regen Res 15:894–902

Liang H, Guan M, Li T, Li S, Ma C, Li C (2024) Critical review on biological effect and mechanisms of diterpenoids in *Rosmarinus officinalis*. Food Med Homol 2:9420021

Lim SJ, Kim M, Randy A, Nam EJ, Nho CW (2016) Effects of *Hovenia dulcis* Thunb. extract and methyl vanillate on atopic dermatitis-like skin lesions and TNF-α/IFN-γ-induced chemokines production in HaCaT cells. J Pharm Pharmacol 68:1465–1479

Lo A-H, Liang Y-C, Lin-Shiau S-Y, Ho C-T, Lin J-K (2002) Carnosol, an antioxidant in rosemary, suppresses inducible nitric oxide synthase through down-regulating nuclear factor-κB in mouse macrophages. Carcinogenesis 23:983–991

Mahmoud WA (2023) Morphological, anatomical and chemical studies of *Salvia rosmarinus* Spenn. growing in Al-Marj (Libya). Al-Mukthar J Sci 38:1–11

Messeha SS, Zarmouh NO, Asiri A, Soliman KF (2020) Rosmarinic acid-induced apoptosis and cell cycle arrest in triple-negative breast cancer cells. Eur J Pharmacol 885:173419

Meziane H, Zraibi L, Albusayr R, Bitari A, Oussaid A, Hammouti B, Touzani R (2024) *Rosmarinus officinalis* Linn.: unveiling its multifaceted nature in nutrition, diverse applications, and advanced extraction methods. J Umm Al-Qura Univ Appl Sci 11:1–29

Micić D, Đurović S, Riabov P, Tomić A, Šovljanski O, Filip S, Tosti T, Dojčinović B, Božović R, Jovanović D (2021) Rosemary essential oils as a promising source of bioactive compounds: chemical composition, thermal properties, biological activity, and gastronomical perspectives. Foods 10:2734

Mihajilov-Krstev T, Radnović D, Kitić D, Stojanović-Radić Z, Zlatković B (2010) Antimicrobial activity of *Satureja hortensis* L. essential oil against pathogenic microbial strains. Arch Biol Sci 62:159–166

Musolino V, Macrì R, Cardamone A, Tucci L, Serra M, Lupia C, Maurotti S, Mare R, Nucera S, Guarnieri L (2023) *Salvia rosmarinus* spenn.(Lamiaceae) hydroalcoholic extract: phytochemical analysis, antioxidant activity and in vitro evaluation of fatty acid accumulation. Plants 12:3306

Natali LA-B, Diaz-Morales AP, Vázquez-Núñez E, Peréz-Ríos SR, Cortés-Hernández M, Medina-Peréz G, Cecilia B (2022) Health and medicinal properties of rosemary oleoresin. In: Handbook of oleoresins. CRC Press, Boca Raton, pp 235–259

Nieto G, Ros G, Castillo J (2018) Antioxidant and antimicrobial properties of rosemary (*Rosmarinus officinalis*, L.): a review. Medicine 5:98

Oresanya IO, Orhan IE (2024) Deciphering neuroprotective effect of *Rosmarinus officinalis* L. (syn. *Salvia rosmarinus* Spenn.) through preclinical and clinical studies. Curr Drug Targets 25:330–352

Oualdi I, Diass K, Azizi S-e, Dalli M, Touzani R, Gseyra N, Yousfi EB (2023) *Rosmarinus officinalis* essential oils from Morocco: new advances on extraction, GC/MS analysis, and antioxidant activity. Nat Prod Res 37:2003–2008

Panda H (2005) Aromatic plants cultivation, processing and uses: how to start a successful aromatic plants business, how to start aromatic plants cultivation industry in India, how to start aromatic plants farm? How to start aromatic plants production business, Indian aromatic plant, list of aromatic plants and their uses. Asia Pacific Business Press Inc, New Delhi

Petersen M (2013) Rosmarinic acid: new aspects. Phytochem Rev 12:207–227

Pieracci Y, Ciccarelli D, Giovanelli S, Pistelli L, Flamini G, Cervelli C, Mancianti F, Nardoni S, Bertelloni F, Ebani VV (2021) Antimicrobial activity and composition of five rosmarinus (now salvia spp. and varieties) essential oils. Antibiotics 10:1090

Prabuseenivasan S, Jayakumar M, Ignacimuthu S (2006) In vitro antibacterial activity of some plant essential oils. BMC Complement Altern Med 6:1–8

Quezel P, Santa S (1963) Nouvelle flore de l'Algérie et des régions désertiques méridionales. Eds. du Centre Nat. de la Recherche Scientifique

Rafie H, Soheila H, Grant E (2017) *Rosmarinus officinalis* (rosemary): a novel therapeutic agent for antioxidant, antimicrobial, anticancer, antidiabetic, antidepressant, neuroprotective, anti-inflammatory and anti-obesity treatment. J Herb Med 3:8

Rahbardar MG, Hosseinzadeh H (2020) Therapeutic effects of rosemary (*Rosmarinus officinalis* L.) and its active constituents on nervous system disorders. Iran J Basic Med Sci 23:1100

Rajizadeh MA, Bejeshk MA, Aminizadeh A, Yari A, Rostamabadi F, Bagheri F, Najafipour H, Nematollahi MH, Amirkhosravi A, Mehrabani M (2024) Inhalation of spray-dried extract of *Salvia rosmarinus* spenn alleviates lung inflammatory, oxidative, and remodeling changes in asthmatic rats. Pharmacology 109:10–21

Reller HH, Kretschmar HC (1980) Analgesic and anti-inflammatory compositions for topical application. Google Patents

Reuter J, Jocher A, Hornstein S, Mönting JS, Schempp CM (2007) Sage extract rich in phenolic diterpenes inhibits ultraviolet-induced erythema in vivo. Planta Med 73:1190–1191

Ribeiro A, Caleja C, Barros L, Santos-Buelga C, Barreiro MF, Ferreira IC (2016) Rosemary extracts in functional foods: extraction, chemical characterization and incorporation of free and microencapsulated forms in cottage cheese. Food Funct 7:2185–2196

Ribeiro-Santos R, Carvalho-Costa D, Cavaleiro C, Costa HS, Albuquerque TG, Castilho MC, Ramos F, Melo NR, Sanches-Silva A (2015) A novel insight on an ancient aromatic plant: the rosemary (*Rosmarinus officinalis* L.). Trends Food Sci Technol 45:355–368

Roh JS, Lee H, Woo S, Yoon M, Kim J, Park SD, Shin SS, Yoon M (2015) Herbal composition Gambigyeongsinhwan (4) from *Curcuma longa*, *Alnus japonica*, and Massa Medicata Fermentata inhibits lipid accumulation in 3T3-L1 cells and regulates obesity in Otsuka long-Evans Tokushima fatty rats. J Ethnopharmacol 171:287–294

Selmi S, Rtibi K, Grami D, Sebai H, Marzouki L (2017) Rosemary (*Rosmarinus officinalis*) essential oil components exhibit anti-hyperglycemic, anti-hyperlipidemic and antioxidant effects in experimental diabetes. Pathophysiology 24:297–303

Shimira F, Zahid G, Nyirahabimana F (2022) An overview and renewed emphasis on ethnopharmacology of rosemary (*Salvia rosmarinus*). Curr Perspect Med Aromat Plants (CUPMAP) 5:30–41

Sienkiewicz M, Łysakowska M, Pastuszka M, Bienias W, Kowalczyk E (2013) The potential of use basil and rosemary essential oils as effective antibacterial agents. Molecules 18:9334–9351

Sirajudeen F, Malhab LJB, Bustanji Y, Shahwan M, Alzoubi KH, Semreen MH, Taneera J, El-Huneidi W, Abu-Gharbieh E (2024) Exploring the potential of rosemary derived compounds (rosmarinic and carnosic acids) as cancer therapeutics: current knowledge and future perspectives. Biomol Ther 32:38

Song H, Xu L, Zhang R, Cao Z, Zhang H, Yang L, Guo Z, Qu Y, Yu J (2016) Rosemary extract improves cognitive deficits in a rats model of repetitive mild traumatic brain injury associated with reduction of astrocytosis and neuronal degeneration in hippocampus. Neurosci Lett 622:95–101

Takaki I, Bersani-Amado L, Vendruscolo A, Sartoretto S, Diniz S, Bersani-Amado C, Cuman R (2008) Anti-inflammatory and antinociceptive effects of *Rosmarinus officinalis* L. essential oil in experimental animal models. J Med Food 11:741–746

Tibaldi G, Hazrati S, Hosseini SJ, Ertani A, Bulgari R, Nicola S (2022) Cultivation techniques and drying process can affect the inflorescence essential oil composition of three selections of *Salvia officinalis*. Ind Crop Prod 183:114923

Veenstra JP, Johnson JJ (2021) Rosemary (*Salvia rosmarinus*): health-promoting benefits and food preservative properties. Int J Nutr 6:1

Vegara S, Funes L, Martí N, Saura D, Micol V, Valero M (2011) Bactericidal activities against pathogenic bacteria by selected constituents of plant extracts in carrot broth. Food Chem 128:872–877

Vissi T, Zelkó R, Földesi R, Túri I (2021) Traditional application of sage (salvia) in conductive education and its potential evidence–based background. Heliyon 7:e08114

Waggas AM, Balawi AE (2010) Possible therapeutic role of rosemary (*Rosemarinus officinalis* L.) leave extract against acrylamide-induced neurotoxicity in different brain areas of the male albino rats. Egypt J Exp Biol 6:23–29

Wang L, Yang H, Wang C, Shi X, Li K (2019) Rosmarinic acid inhibits proliferation and invasion of hepatocellular carcinoma cells SMMC 7721 via PI3K/AKT/mTOR signal pathway. Biomed Pharmacother 120:109443

Wenkert E, Fuchs A, McChesney JD (1965) Chemical artifacts from the family Labiatae. J Org Chem 30:2931–2934

Wijeratne SS, Cuppett SL (2007) Potential of rosemary (*Rosemarinus officinalis* L.) diterpenes in preventing lipid hydroperoxide-mediated oxidative stress in Caco-2 cells. J Agric Food Chem 55:1193–1199

Yıldırım TA, Kitiş Y (2020) The effect of aromatherapy application on cognitive functions and daytime sleepiness in older adults living in a nursing home. Holist Nurs Pract 34:83–90

Yildirim S, Oto G, Comba B, Ekin S, Cinar DA (2017) An investigation of the protective effects of resveratrol on some biochemical parameters and histopathological findings in experimentally-induced chronic flurosis in rats. Fluoride 50:365–373

Yildirim S, Ekin S, Huyut Z, Oto G, Comba A, Uyar H, Sengul E, Cinar DA (2018) Effect of chronic exposure to sodium fluoride and 7, 12-dimethylbenz [a] anthracene on some blood parameters and hepatic, renal, and cardiac histopathology in rats. Fluoride 51:278–290

Yimam M, Lee YC, Jiao P, Hong M, Brownell L, Jia Q (2017) A standardized composition comprised of extracts from *Rosmarinus officinalis*, *Annona squamosa* and *Zanthoxylum clava-herculis* for cellulite. Pharm Res 9:319

Zaouali Y, Bouzaine T, Boussaid M (2010) Essential oils composition in two *Rosmarinus officinalis* L. varieties and incidence for antimicrobial and antioxidant activities. Food Chem Toxicol 48:3144–3152

Zheng G, Kadir A, Zheng X, Jin P, Liu J, Maiwulanjiang M, Yao G, Aisa HA (2020) Spirodesertols a and B, two highly modified spirocyclic diterpenoids with an unprecedented 6-isopropyl-3 H-spiro [benzofuran-2, 1′-cyclohexane] motif from Salvia deserta. Organ Chem Front 7:3137–3145

Zhong X-J, Na Z, Xin W, Jin-Jie L, Hui M, Yue J, Jia-Hui X, Peng-Cheng L, Shang X-Y (2022) Three new ursane-type triterpenoids from *Rosmarinus officinalis* and their biological activities. Chin J Nat Med 20:155–160

Zhou N, Wang Z-y, Wu Y, Zhong X-j, Wang X, Li J-j, Shang X-y (2021) Norursane-type triterpenoids from Rosmarinus officinalis and their anti-inflammatory activity evaluation. Fitoterapia 153:104982

Zhou X, Wang W, Li Z, Chen L, Wen C, Ruan Q, Xu Z, Liu R, Xu J, Bai Y (2022) Rosmarinic acid decreases the malignancy of pancreatic cancer through inhibiting Gli1 signaling. Phytomedicine 95:153861

Zigene ZD, Asfaw BT, Bitima TD (2022) Phenotypic diversity of rosemary (*Salvia rosmarinus* Schleid.) accessions for qualitative characters. Heliyon 8:e11895

# Medicinal and Nutritional Importance of *Rhodomyrtus tomentosa* (Aiton) Hassk

# 5

Dwi Marlina Syukri, Tessa Sjahriani, Tosin Ayodeji Oladosu, Sudarshan Singh, and Mohit Agrawal

**Abstract**

*Rhodomyrtus tomentosa*, commonly known as the Rose Myrtle or Downy Rose Myrtle, is a species of flowering plant from the family Myrtaceae. *Rhodomyrtus tomentosa* is native to Southeast Asia and can be found in countries including Thailand, Malaysia, Indonesia, and the Philippines. Moreover, *Rhodomyrtus tomentosa* is an evergreen shrub or small tree that grows up to 3–5 m in height with dark green, glossy leaves that are elliptical in shape and covered in fine, soft hairs. The flowers are usually solitary, white, and have a distinctive star-like shape, whereas dark purple to black berry-like fruit is edible in nature. Literature reported that *Rhodomyrtus tomentosa* is well adaptable to various habitats and can be found in both lowland and montane forests. *Rhodomyrtus tomentosa* contains numerous bioactive phytopharmaceuticals, including rhodomyrtone and its by-products, myricitrin, piceatannol, quercetin, dihydromyricetin, kaempferol, and vitexin. Additionally, antimicrobial assays demonstrated that the plant contains effective antimicrobial complexes with a diversity of mechanisms of action. This chapter provides updated information on the chemical conformation, pharmacology, and toxicology of the herb and its primary bioactive compounds.

D. M. Syukri · T. Sjahriani
Faculty of Medicine, Malahayati University, Bandar Lampung, Lampung, Indonesia

T. A. Oladosu
Department of Chemistry and Biochemistry, University of Minnesota Duluth, Duluth, MN, USA

S. Singh (✉)
Faculty of Pharmacy, Chiang Mai University, Chiang Mai, Thailand
e-mail: sudarshansingh83@hotmail.com

M. Agrawal
Department of Pharmacology, School of Medical and Allied Sciences, K.R. Mangalam University, Gurugram, Haryana, India



M. A. Ansari et al. (eds.), *Medicinal Plants and their Bioactive Compounds in Human Health: Volume 2*, https://doi.org/10.1007/978-981-95-3620-7_5

**Keywords**

Antimicrobial · Antioxidant · Anti-inflammatory · Antidiabetic · Anticancer · *Rhodomyrtus tomentosa*

## 5.1 Introduction

Amid the vastness of our natural environment, some floral jewels are particularly noteworthy for their aesthetic appeal and ecological significance (Singh 2017). *Rhodomyrtus tomentosa* is one such jewel; it is a species that weaves a tapestry of biodiversity, cultural legacy, and medicinal potential that goes beyond mere floral beauty. Originating in Southeast Asia, *Rhodomyrtus tomentosa*, also referred to as Rose Myrtle, emerges as a remarkable herb with a fascinating combination of decorative appeal and significant ecological significance. Its lustrous, dark-green, evergreen foliage sets the stage for an amazing show of colorful berry-like fruits and dainty pink blooms. In addition to drawing the eye, this color symphony alludes to the rich tapestry of life that is being revealed both within and outside of this endearing shrub (Nwabor and Singh 2022).

*Rhodomyrtus tomentosa* has found use in traditional medicine due to its strong antibacterial qualities (Vo and Ngo 2019), such as rhodomyrtone, which makes it more than just an attractive adornment. Beyond folklore, this species' pharmacological potential captivates academics and medical professionals alike, providing a natural defense against illnesses brought on by oxidative stress. *Rhodomyrtus tomentosa* stands out among the more than 450 genera and 3200 species that make up the Myrtaceae family (Anane 2022). It not only adds to the overall richness of biodiversity but also embodies a genetic variety that has captivated experts all over the world. Recent investigations into its genetic composition, for example, have shed light on the complex gene dance that has transformed this species into a botanical wonder.

Stories of biodiversity may be found within the leaves and flowers of *Rhodomyrtus tomentosa* (Zhao et al. 2020). This plant is home to a wide variety of insects, birds, and other plants, fostering an environment that is peaceful and emphasizes the interdependence of all living things. Intrigued by its genetic variety (Hue 2015), scientists have worked hard to decipher the complex patterns that turn this species into more than simply a botanical marvel, rather a living example of the evolutionary dance that sculpts our world.

For indigenous communities, *Rhodomyrtus tomentosa* is more than just a plant; it is a living chapter in cultural histories. Its fruits, leaves, and roots are used in traditional ceremonies, culinary endeavors, and as a symbol of resilience against changing times (Geetha et al. 2010).

*Rhodomyrtus tomentosa* has great cultural value in addition to its place in the complex web of biodiversity. Indigenous people have created myths around the plant's presence and have used its many components for ceremonial, medicinal, and culinary purposes. In-depth research on *Rhodomyrtus tomentosa* is undertaken in this book, which intertwines the species' botanical characteristics, ecological significance, and historical cultural influences. Come explore the intricacies of this remarkable botanical story, where tradition and science come together to honor *Rhodomyrtus tomentosa's* eternal beauty and vitality (Wu et al. 2015).

## 5.2 Plant Description

The evergreen shrub *Rhodomyrtus tomentosa*, sometimes denoted as Rose Myrtle, is a member of the Myrtaceae family (Abd Hamid et al. 2017). This species is indigenous to Southeast Asia and can be found there thriving in a variety of environments, from highland to coastal.

## 5.3 The Morphological Attributes

1. *Size and growth form*: Reaching heights of 1.5–4 m, *Rhodomyrtus tomentosa* normally develops as a compact, bushy shrub. Because of its spreading nature and lush leaves, it is a great ornamental for landscaping (Mitra et al. 2012).
2. *Leaves*: *Rhodomyrtus tomentosa* has opposing, simple, elliptical leaves with a glossy, dark green color (Hmoteh 2018). Fine tomentum, or hairs, cover the underside of the leaves, giving them a unique feel (Hue 2015).
3. *Flowers*: The plant bears pretty pink flowers with five petals that can be found alone, in clusters, or both. During the blossoming season, the intricately beautiful blooms blanket the bush, producing a visually spectacular show.
4. *Fruits*: *Rhodomyrtus tomentosa* produces tiny, berry-like fruits with a diameter of roughly 1 cm after it flowers. These tasty fruits range in hue from pink to purple-black and contribute to the plant's attraction (Wilson 2010).

## 5.4 Chemical Composition

Numerous phytochemicals found in *Rhodomyrtus tomentosa* contribute to its nutritional and therapeutic qualities. Although individual plants may differ in their precise chemical composition, the following components have been identified:

1. *Rhodomyrtone*: *Rhodomyrtus tomentosa* is renowned for containing rhodomyrtone, a natural antibiotic with demonstrated efficacy against various bacterial strains, including *Staphylococcus aureus* (Limsuwan et al. 2009; Mordmuang et al. 2019).
2. *Flavonoids and polyphenols*: The leaves and fruits of *Rhodomyrtus tomentosa* contain flavonoids and polyphenols (Kuntorini et al. 2023), known for their antioxidant properties, contributing to the plant's potential health benefits (Vo and Ngo 2019).

3. *Essential oils*: The leaves and flowers of *Rhodomyrtus tomentosa* yield essential oils with a complex composition that includes compounds like eucalyptol and α-pinene. These oils contribute to the plant's aromatic qualities and may have therapeutic effects (Kuntorini et al. 2023).

Fruit has muffled quantities of protein (2.63%), lipids (1.59%–3.5%), and sugars (5.65%), but elevated quantities of dietary fiber (69.94%–87.43%), α-tocopherol (38.90%–51.87%), manganese (>100%), and copper (44.44%). Fruit samples have a high concentration of linoleic acid (75.36% of total fatty acids) (Lai et al. 2015). According to a previous study, the total phenolic content of *R. tomentosa* berries is 49.21 ± 0.35 mg gallic acid equivalent/g dry weight (Lai et al. 2015). Furthermore, the total flavonoid content of *R. tomentosa* is 5.21 ± 0.20 mg rutin equivalent/g dry weight. An ultraviolet (UV) spectrophotometer revealed that a flavonoid-rich extract purified with AB-8 macroporous resin contained 62.09 ± 2.63% rutin equivalent (Wu et al. 2015). In the meantime, Lai et al. investigated the effects of various maturity stages and environmental factors on the total flavonoid content of *R. tomentosa* in Vietnam. They found that at the fourth maturity stage, the berries of *R. tomentosa* contained approximately 1.60 ± 0.06 mg of cyanidin-3-glucoside equivalent/g dry weight of total anthocyanins and 1.52 ± 0.14 mg of quercetin equivalent/g dry weight of total flavonols.

Potassium (221.76 mg/150 g fruit), calcium (73.65 mg/150 g fruit), manganese (3.23 mg/150 g fruit), iron (1.54 mg/150 g fruit), zinc (0.61 mg/150 g fruit), and copper (0.40 mg/150 g fruit) are all highly concentrated in *R. tomentosa* fruit. *R. tomentosa* fruits contain 5.62 mg of vitamin C and 3.89 mg of vitamin E per 150 g of fruit, respectively.

## 5.5 Biological Source

The Greek words are the source of *Rhodomyrtus*'s name. For red, use rhodon; for myrtle, use myrtose. However, this species is known for its tomentose due to the characteristics of its leaves, which have a glossy green appearance on the upper side and a dense, soft, hairy presence on the underside (Langeland and Burks 2008). *Rhodomyrtus tomentosa* is a beautiful ornamental plant that grows abundantly and is a representative of the Myrtaceae family. This shrub is native to Southeast Asia and is evergreen. In addition, it has edible bell-shaped fruits that are dark purple in color and rose-colored flowers (Winotai et al. 2005).

## 5.6 Description of Plant

The term "small plant or tree species up to 2–3 m tall" can be used to describe this plant. Their leathery leaves are 3(−5) nerved, elliptic to ovate-elliptic, 5–8 cm long, and 1.5–4 cm wide (Fig. 5.1 and Table 5.1). Furthermore, the petioles on leaves are 0.4–1 cm long, with a glossy, glabrate upper surface and a densely tomentose lower

**Fig. 5.1** Photographs of *Rhodomyrtus tomentosa* plant

**Table 5.1** Morphological characteristics of *Rhodomyrtus tomentosa*

| Characteristics | Description |
|---|---|
| Size | 1.5–4 m in height |
| Growth form | Compact, bushy shrub |
| Leaves | Opposite, elliptical, simple, glossy, tomentose |
| Flowers | Pink, five-petaled, solitary, or clustered |
| Fruits | Small, berry-like, pink to purple, black |

surface. Moreover, axillary cymes have one to five flowers, each of which is supported by two tiny bracts; the petals are pink and measure roughly 10–13 mm in length, while the sepals are orbicular and *tomentose*, measuring 3–4 mm (Maskam 2011). The fruits of this plant are edible, and it looks like blueberry fruits. They contain sugar, vitamin, and mineral. In addition, the fruit is an ellipsoid berry with a persistent calyx that has a diameter of 1–1.5 cm. Additionally, green skin and an astringent flavor characterize unripe fruits. The berry turns to a purplish dark when ripe, and the pulp is purplish, soft, and sweet. In addition, the berries have a large number of 1.5 mm diameter deltoid seeds arranged in six pseudo-locules separated by thin false septa. Fruits have 40–45 disk-shaped, 2–3 mm long seeds (Staples and Herbst 2005; Lim 2012).

## 5.7 Overview of *Rhodomyrtus tomentosa* Research

*Rhodomyrtus tomentosa* is a type of flowering plant native to Southeast Asia (Abd Hamid et al. 2017). It is often denoted as the wax apple or downy rose myrtle. The many pharmacological, horticultural, and ecological qualities of this evergreen shrub or small tree have drawn a lot of interest in the scientific community. Highlighting the most important areas of research on *Rhodomyrtus tomentosa*, this overview covers the plant's horticultural importance, therapeutic applications, and potential ecological effects.

*Rhodomyrtus tomentosa* is a member of the Myrtaceae family and is distinguished by its eye-catching white to pink blooms, downy undersides, and glossy leaves. The plant can grow to a height of 2–5 m, and it is well known for its adaptability to different types of soil and climate.

Research has focused on the potential medical benefits of *Rhodomyrtus tomentosa*. Its fruits, bark, and leaves have all been shown to contain bioactive substances, such as flavonoids, polyphenols, and essential oils. Because of these chemicals' antibacterial, antioxidant, and anti-inflammatory properties, traditional medicine and pharmaceutical research are interested in studying this plant. This plant is a small evergreen shrub with edible purple berries that can be eaten raw, cooked into jam and pies, or added to salads. It also has attractive fuzzy leaves and vibrant rose-colored flowers. The fruit of this plant is used as a material to produce wine and jellies in Vietnam. *R. tomentosa* also exhibits potential for use in fire breaks in the Himalayas as a species that resists fire. In some countries including China, Vietnam, Indonesia, Thailand, and Malaysia, for a long time, *R. tomentosa* has been utilized in conventional medicine, such as, the native people in Malaysia use the berries as a medication for dysentery and diarrhea (Ong and Nordiana 1999). Additionally, segments of the roots and trunk are used as traditional medicine for postpartum women and for stomach ailments. The native people of Indonesia have long used the crushed leaves of *R. tomentosa* as a wound remedy. In addition, *R. tomentosa* is a medication used also in Thailand to treat dysentery, diarrhea, and fever (Chuakul 2005). In China, urinary tract infections can be treated using *R. tomentosa*. Furthermore, in Singapore, *R. tomentosa* is used as a conventional medicine to treat pain, heartburn, and snake bites (Limsuwan et al. 2012). In the meantime, Vietnam has been using *R. tomentosa* fruits to strengthen the immune system and treat dysentery and diarrhea (Do 2011). *Rhodomyrtus tomentosa* has become known for its horticulture usefulness in addition to its medicinal benefits (Sarah et al. 2014). The plant is used mostly for ornamental purposes; gardens and landscapes favor it for its lovely flowers and foliage. Its appeal as a multipurpose horticultural species is further enhanced by its adaptability to many soil types and climates. The ecological effects of *Rhodomyrtus tomentosa* have drawn attention in areas where it has been introduced (Yang et al. 2016). The plant has the capacity to become naturally occurring and outcompete surrounding vegetation, altering regional ecosystems. It is essential to do research on the ecological effects of its introduction in order to comprehend and mitigate its invasiveness. *Rhodomyrtus tomentosa* has a long history of traditional applications across a wide range of civilizations. Various plant parts have been used medicinally by indigenous societies to treat respiratory problems, skin conditions, and gastrointestinal illnesses. Studies on ethnobotany have investigated the traditional knowledge and cultural importance of *Rhodomyrtus tomentosa* (Odedina et al. 2015). Research on *Rhodomyrtus tomentosa*'s genetic composition has been made easier by developments in molecular biology (Huang et al. 2019). Understanding its genetic diversity, population structure, and potential for selective breeding has been made possible by genomic research and molecular markers. Comprehending the genetic makeup of plants is essential for conservation initiatives and maximizing desired results. Beyond its usage in decorative and medicinal plants, *Rhodomyrtus tomentosa* has demonstrated potential in agricultural settings. Research has been done on its potential as a biopesticide, crop enhancer, or source of natural chemicals for plant protection. There is curiosity in the plant's allelopathic effects on other species as well as its potential to improve soil health.

## 5.8 Antimicrobial Activities of *Rhodomyrtus tomentosa*

*Rhodomyrtus tomentosa* has the potential to be a valuable medicinal compound with strong antimicrobial activity. Fruit and leaf extracts of *R. tomentosa* with an ethanol as a solvent show action against *Bacillus cereus* and *Candida albicans*, of 0.42 and 0.35, correspondingly. Leaves, stem, twig, and fruit of the shrub have comparative activity with reference to chloramphenicol against *Salmonella typhi* and *Propionibacterium acnes* (activity index of 0.19–0.50) (Kusuma et al. 2016). All *Staphylococcal* bacteria isolated from milk were effectively inhibited by the *R. tomentosa* leaf ethanol extract, with minimum inhibitory and minimum bactericidal concentration values ranging from 16 to 64 µg/mL and 64 to 128 µg/mL, respectively (Mordmuang and Voravuthikunchai 2015). Additionally, it demonstrated antibacterial activity with Minimum Inhibitory Concentration (MIC) values of 31.25, 15.62, and 1000 µg/mL against *S. aureus* ATCC 25923, *Streptococcus mutans*, and *C. albicans* ATCC 90028, respectively (Limsuwan et al. 2014). Furthermore, *Streptococcus agalactiae* and *Streptococcus iniae* isolated from infected tilapia were also effectively hindered by ethanolic extract, with minimum inhibitory concentration values ranging from 7.8 to 62.5 µg/mL (Na-Phatthalung et al. 2018). Additionally, ethanolic extract inhibited clinical isolates of *Streptococcus pyogenes*, with an MIC value range of 3.91–62.5 µg/mL (Limsuwan et al. 2012).

Rosli and coworker reported that strong inhibition qualities were shown by the methanol extract of *R. tomentosa* against *Escherichia coli* and *S. aureus*, with inhibition zones measuring 10 mm for leaves, 16 and 12 mm for fruits, and 10 and 13 mm for stems, respectively. In addition, *R. tomentosa* ethanolic leaf extracts were effective as a biocontrol agent against *Listeria monocytogenes* (Odedina et al. 2015) and *E. coli* O157:H7 (Hmoteh 2018), which are important foodborne pathogens.

The antibacterial properties of the isolated rhodomyrtone bioactive compounds from *R. tomentosa* have also been examined. Rhodomyrtone, a natural antibacterial drug derived from *R. tomentosa*, has been demonstrated to be a promising new candidate. *Staphylococcus epidermidis*, *Bacillus subtilis*, *Clostridium difficile*, *Enterococcus faecalis*, *S. aureus*, *methicillin-resistant S. aureus*, *Streptococcus gordonii*, *S. mutans*, *Streptococcus pneumoniae*, *S. pyogenes*, and *Streptococcus salivarius* are among the Gram-positive bacteria that are significantly inhibited by rhodomyrtone's antibacterial properties (Limsuwan et al. 2009) and pathogens resistant to antibiotics, such as strains of *Enterococcal* bacteria resistant to vancomycin, epidemic *Methicillin-resistant S. aureus*, and *Vancomycin-intermediate S. aureus* (Leejae et al. 2013). Furthermore, rhodomyrtone demonstrated an MIC90 value of 0.5 μg/mL, effectively inhibiting *P. acnes*. After receiving rhodomyrtone treatment, the quantity of bacterial cells was lowered by at least 99% in a matter of hours (Saising and Voravuthikunchai 2012). In particular, it has been noted that rhodomyrtone can inhibit the formation of biofilms and eliminate established biofilms of *S. mutans*, *S. aureus*, *S. epidermidis*, and *P. acnes* (Wunnoo et al. 2017).

## 5.9 Antibacterial Mechanism

Previous study reported that rhodomyrtone was found to inhibit the biosynthesis of staphyloxanthin in bacteria, increasing the pathogen's susceptibility to singlet oxygen killing and $H_2O_2$. Additionally, rhodomyrtone inhibited the activities of enzymes involved in acid production and tolerance, such as the phosphotransferase system and membrane-bound F-ATPase enzymes, as well as the glycolysis enzymes pyruvate kinase and glyceraldehyde phosphate dehydrogenase in the cytoplasm. This resulted in a suppression of acid production from bacteria. Limsuwan and colleagues have found that rhodomyrtone's antibacterial activity is attributed to its interference with metabolic pathways, including gluconeogenesis, amino acid metabolism, and glycolysis. Additionally, it inhibits the expression of streptococcal toxins, including streptococcal pyrogenic exotoxin C and the Cyclic Adenosine Monophosphate (CAMP) factor (Limsuwan et al. 2012).

Antibacterial effect of rhodomyrtone, caused by the disruption of bacterial cell wall biosynthesis and the suppression of staphylococcal antigenic proteins, immunodominant antigen A, and staphylococcal secretory antigen involved in cell wall hydrolysis and cell division, resulted in notable modifications such as cell wall alterations, aberrant septum formation, cellular lysis, and disintegration (Sianglum et al. 2011). Moreover, rhodomyrtone changed enzymes and metabolites implicated in multiple metabolic pathways, such as amino acid biosynthesis, nucleic acid biosynthesis, and glucid and lipid metabolism. In the bacterial cells exposed to rhodomyrtone, there was a noticeable decrease in the amounts of three metabolites (Uridine diphosphate (UDP)-glucose, UDP-glucuronic acid, and UDP-*N*-acetyl-*D*-galactosamine) and two enzymes (glycosyl-transferase and Uridine-5′-triphosphate (UTP)-glucose-1-phosphate uridylyl-transferase) involved in the synthesis of the pneumococcal capsule (Mitsuwan et al. 2017). Additionally, previous study reported that rhodomyrtone is a bacterial cell membrane-damaging agent (Sianglum et al. 2018). Specifically, rhodomyrtone produced large membrane intussusceptions with a sharp rise in fluidity, drawing a variety of membrane and trap proteins. Moreover, rhodomyrtone temporarily binds to phospholipid head groups and distorts lipid packing, as demonstrated by molecular dynamics simulations. These findings explain the induction of membrane curvature and membrane fluidization. With a novel mechanism of action, its capacity to form protein-trapping membrane vesicles and its transient binding mode make it an attractive new antibiotic candidate (Sianglum et al. 2018). Strong antifungal activity was exhibited by *R. tomentosa* against a range of human pathogens. According to Jeenkeawpieam and colleague, endophytic fungi from *R. tomentosa* can be a valuable cradle of possible natural products with antifungal properties (Jeenkeawpieam et al. 2012).

## 5.10 Antioxidant Activities of *Rhodomyrtus tomentosa*

Antioxidant activities of *R. tomentosa* fruit extract were found to be due to phenolic bioactive compounds such as anthocyanin and piceatannol. Geetha and coworker reported ethyl acetate and aqueous alcoholic extract had comparable antioxidant efficacy against rutin and ascorbic acid (Geetha et al. 2010). Moreover, Lavanya et al. reported that lipid peroxide generation was significantly inhibited by an acetone extract of *R. tomentosa* leaves. At 1 mg/mL, the extract exhibited a quick and enhanced tendency to convert $Fe_3$+ to $Fe_2$+, or 10.8 ± 1.12 mM gallic acid and 30.5 ± 5.22 mM ellagic acid, correspondingly. Additionally, the extract showed a 0.93 ± 0.07 mM gallic acid inhibition capacity at 100 µg/mL for lipid peroxidation (Lavanya et al. 2014). Furthermore, *R. tomentosa* extract demonstrated protection against CCl4-induced decreases in Superoxide dismutase (SOD), Catalase (CAT), and Glutathione Peroxidase (GPx) enzyme activities in blood, liver, and kidneys.

Overabundance of innate inflammatory cells, including neutrophils, atherosclerosis, arthritic cells, and macrophages, can harm the body and result in conditions like septic shock, diabetes, cancer, and atherosclerosis (Toltl et al. 2008). Therefore, currently one strategy for preventing or treating such diseases has been conducted, to inactivate the inflammatory responses. Ethno-pharmacological treatments have been researched as potential anti-inflammatory herbal medicine entrants in an attempt to create a medication that effectively treats these conditions. *Rhodomyrtus tomentosa*, commonly recognized as downy rose myrtle or wax apple, is a plant native to Southeast Asia in places such as Vietnam, Indonesia, Sri Lanka, and Thailand. Its leaves and fruits have been used traditionally in folk medicine for various purposes, including their anti-inflammatory activity. One of the compounds isolated from *Rhodomyrtus tomentosa*, called rhodomyrtone, has been found to have strong anti-inflammatory properties. It can suppress the activation of nuclear factor-kappa B (NF-κB), a critical regulator of inflammation, and inhibit the production of pro-inflammatory cytokines like interleukin-6 (IL-6) and tumor necrosis factor-alpha (TNF-α). This extract inhibited inflammatory mediator production at the transcriptional level, with upregulated inducible nitric oxide synthase (iNOS) and cyclooxygenase-2 (COX-2) mRNA levels in RAW264.7 cells. Furthermore, *Rhodomyrtus tomentosa* extract reduced NF-kB and AP-1 promoter activities in HEK293 cells, implying that this extract could target the transcriptional activation of NF-kB and AP-1 (Jeong et al. 2013). Moreover, *Rhodomyrtus tomentosa* extract can block the kinase activities of purified Src, which is involved in the production of nitric oxide (NO) and Prostaglandin E2 ($PGE_2$), and is able to inhibit the activation of AP-1 signaling. In addition, this extract also inhibited the phosphorylation of the upstream signaling enzymes, MEK1/2, c-Raf-1, and MKK4/7, which are responsible for the activation of Extracellular signal-Regulated Kinase (ERK) and c-Jun

N-terminal kinase (JNK) in inflammation (Jeong et al. 2013). Another previous study reported that *Rhodomyrtus tomentosa* extract reduced pro-inflammatory cytokine levels and increased tight junction protein expression to reduce intestinal barrier dysfunction and inflammation (Wang et al. 2022). *Rhodomyrtus tomentosa* extract, with its rich phytopharmaceutical composition, has been conventionally used to address inflammatory conditions. Recent scientific studies are shedding light on the mechanisms through which this shrub exerts its anti-inflammatory effects. *Rhodomyrtus tomentosa* extract is endowed with a plethora of bioactive compounds, many of which contribute to its anti-inflammatory properties. The primary bioactive compounds include phloroglucinols, flavonoids, tannins, and phenolic acids. Each of these compounds plays a unique role in modulating inflammatory pathways, making *Rhodomyrtus tomentosa* extract an encouraging candidate for management of inflammatory syndromes (Idris et al. 2023). These compounds can modulate inflammatory pathways by inhibiting the production of pro-inflammatory cytokines and enzymes. Tannins, as a polyphenolic compound, have the ability to interact with proteins and modulate inflammatory signaling cascades. Tannins may exert anti-inflammatory effects by inhibiting the activity of inflammatory enzymes and reducing the production of pro-inflammatory mediators. In the meantime, flavonoids have the ability to reduce inflammatory markers such as cyclooxygenase-2 (COX-2), inducible nitric oxide synthase (iNOS), and interleukin (IL)-1β. Furthermore, phenolic acids have the ability to reduce inflammation by scavenging free radicals and adjusting immune responses. The ability of phenolic acids to control nuclear factor-kappa B (NF-κB) signaling, a crucial pathway intricate in inflammation, is frequently credited with their anti-inflammatory properties (Fujiati and Haryati 2022).

## 5.11 Potential Toxicity Screening Reports

*Rhodomyrtus tomentosa*, also known as Rose Myrtle or Downy Rose Myrtle, is often cultivated for its edible fruits and medicinal properties. The leaves, fruits, and roots of *Rhodomyrtus tomentosa* have been used traditionally for various purposes including as an antiseptic, anti-inflammatory, and for wound healing. In addition, according to the World Health Organization (WHO), among the thousands of plants in Indonesia that are utilized in traditional medicine is *Rhodomyrtus tomentosa*, also known as haramonting. Because haramonting leaves contain several bioactive compounds, they are used to treat wounds, bleeding, diarrhea, dysentery, colic, and abdominal pain. Nevertheless, at high concentrations, some of these substances can be hazardous, changing and affecting the histology of the heart, kidney, lungs, and brain (Situmorang et al. 2020).

## 5.12 Formulative Application of *Rhodomyrtus tomentosa*

*Rhodomyrtus tomentosa* has various applications due to its medicinal properties and potential health benefits, including:

1. *Traditional medicine*: In traditional medicine practices of Southeast Asia, various parts of *Rhodomyrtus tomentosa*, including the leaves, fruits, and roots, are used to prepare herbal remedies. These remedies are often used for their antimicrobial, anti-inflammatory, wound-healing properties and digestive health. One of the most well-known traditional uses of *Rhodomyrtus tomentosa* is as a natural antimicrobial agent. The leaves and fruits contain bioactive compounds such as flavonoids, tannins, and phenolic acids that exhibit antimicrobial activity against bacteria, fungi, and other pathogens. In traditional medicine, extracts or infusions of *Rhodomyrtus tomentosa* leaves are used to treat various bacterial and fungal infections, including skin infections, respiratory infections, and gastrointestinal infections (Limsuwan et al. 2009; Bach et al. 2020). Moreover, *Rhodomyrtus tomentosa* has been used traditionally to alleviate inflammation and pain. Extracts from the plant are applied topically to reduce swelling, relieve muscle pain, and promote wound healing. Additionally, oral consumption of *Rhodomyrtus tomentosa* extracts may help to reduce inflammation internally, making it useful for conditions like arthritis and inflammatory bowel disease (Zhang et al. 2018). In addition, the astringent and antiseptic properties of *Rhodomyrtus tomentosa* make it valuable in wound care. Traditional remedies often involve applying crushed leaves or extracts directly to cuts, wounds, and burns to promote healing and prevent infection. The plant's ability to stimulate tissue regeneration and inhibit bacterial growth contributes to its efficacy in wound management (Ningsih et al. 2021; Nguyen et al. 2023). In traditional medicine, *Rhodomyrtus tomentosa* is used to support digestive health and treat gastrointestinal disorders. Infusions or decoctions made from the leaves or roots are consumed to alleviate symptoms of diarrhea, dysentery, and stomachaches. The plant's antimicrobial properties may help to combat gastrointestinal infections, while its astringent properties can help to reduce diarrhea and improve bowel function (Wang et al. 2022). *Rhodomyrtus tomentosa* has been used conventionally to manage respiratory situations such as coughs, colds, and bronchitis. Infusions or syrups made from the leaves or fruits are consumed to soothe sore throats, alleviate coughing, and clear congestion. The plant's expectorant and antimicrobial properties may help to relieve symptoms of respiratory infections and promote recovery (Li et al. 2023).
2. *Topical preparations*: Extracts or oils derived from *Rhodomyrtus tomentosa* are used in topical formulations such as ointments, creams, or lotions. These formulations are applied to the skin to treat various skin conditions like cuts, wounds, burns, and infections (Ontong et al. 2020).
3. *Oral supplements*: *Rhodomyrtus tomentosa* extracts may be incorporated into oral supplements such as capsules, tablets, or tinctures. These supplements are

consumed for their potential health benefits, including immune support, anti-inflammatory effects, and gastrointestinal health (Ontong et al. 2020).

4. *Cosmetics*: Extracts from *Rhodomyrtus tomentosa* are sometimes incorporated into cosmetic products like facial creams, serums, and masks. These products may be marketed for their antioxidant properties and potential benefits for skin health and anti-aging (Kusuma et al. 2016).
5. *Food and beverage additives*: In some cultures, *Rhodomyrtus tomentosa* fruits are consumed fresh or it is used to prepare beverages like juices, teas, or syrups. The fruits may also be used as flavorings or natural food additives in various culinary applications.
6. *Agricultural applications*: Some research suggests that extracts from *Rhodomyrtus tomentosa* possess pesticidal properties, which may have potential applications in agriculture for pest control and crop protection.

## 5.13 Anticancer and Cell Protective Effects of *Rhodomyrtus tomentosa*

One of the leading causes of mortality wide-reaching, cancer claimed the lives of over 9.6 million people in 2018 (Pasaribu et al. 2021). Terpenoids, phenols, and lipids are among the diverse phytochemical compositions found in various regions of *R. tomentosa*. From the leaves of *R. tomentosa*, triterpenoids such as botulin, β-amyrin, β-amyrenonol, and lupeol have been discovered. Rhodomyrtone A, B, C, and D are examples of the phenolic compounds (major component) that were found in *R. tomentosa* (Tay et al. 2019). Moreover, *R. tomentosa* has antibacterial, anticancer, anti-inflammatory, and antioxidant properties. The methanol extract showed encouraging activity on the two types of cancer cells when tested for its in vitro cytotoxicity against MCF-7, HeLa, A549, and B16 cancer cell lines. The findings demonstrated that several living cells decreased as the methanol extract's cytotoxic effect against MCF-7, HeLa, A549, and B16 cells rose with extract concentration. When the extract concentration was low, the cells had a needle-like shape with close spacing between them. In the meantime, dead cells with spherical and floating shape are visible at high extract concentrations. At the moment, the primary focus is on research on anticancer medications that are not resistant. When compared to conventional medicines, bioactive chemicals derived from plants with their various chemical structures may be able to combat cancer with fewer adverse effects (Zhang et al. 2018; Tay et al. 2019).

In the diabetic wound condition, *R. tomentosa* elevates the GLUT-2 protein level. Previous research showed that the bioactive chemical found in its leaves, rhodomyrtone, significantly inhibited the capacity of epidermoid carcinoma A431 cells to migrate, adhere, and invade other cells (Tayeh et al. 2017). According to reports, rhodomyrtone derived from *R. tomentosa* leaves is a potentially effective natural anticancer drug that can inhibit the viability of HaCaT keratinocyte cells in vitro (Kamarudin et al. 2021). According to certain research conducted in 2017, *R. tomentosa* and a number of isolated chemicals show anticancer effects on a variety of models (Zhang et al. 2018). Presented results on anticancer activities were restricted to the anticancer activity of the chemical rhodomyrtone, which is only contained in the leaves of *R. tomentosa*.

Furthermore, *R. tomentosa* contains a high concentration of antioxidants, and its phenolic content may contain compounds that are beneficial to health. *R. tomentosa* contains antioxidants such as anthocyanins, acyl phloroglucinol, flavonoids, tannins, and triterpenes that protect from hypoxia and cell death (Vo and Ngo 2019). GLUT-1 expression in cancer may reflect the metabolic and vascular requirements of the tumor, which may have therapeutic implications for existence and management planning (Pragallapati and Manyam 2019). Because of its importance in oncogenesis, GLUT-1 has been the subject of several studies aimed at determining its prognostic value in malignancies (Zambrano et al. 2019; Barbosa and Martel 2020). Increased glucose metabolism in cancer cells could be related to GLUT-1 overexpression (Barbosa and Martel 2020). It has been reported that antioxidants lessen harmful side effects during cancer management. Rich in antioxidants, *R. tomentosa* has been linked to effective cancer treatment with minimal side effects; this led to the creation of the antioxidant intervention (Situmorang et al. 2023).

Antioxidants heal tissue damage from cervical cancer and lower MMP-9, GLUT-1, and cell death in the body's tissues and blood vessels. Numerous diseases can be brought on by biological processes like oxidative stress, which is caused by cells lacking in antioxidants (George and Abrahamse 2020). Furthermore, DNA damage has a role in the development and spread of cancer. Since cancer and antioxidants are strongly related, antioxidants are now a regular therapy method. Antioxidants are not used in chemotherapy or radiation therapy, which permanently harm tissue by eliminating tumor cells through a process. The right antioxidant inhibitors can aid in the effective treatment of cancer (George and Abrahamse 2020; Gulcin 2020).

Additionally, aryl phloroglucinol, flavonoids, tannins, and triterpenes—all of which have substantial advantages for repairing cells—are present in the plant's extracts (Vo and Ngo 2019). The way in which rhodomyrtone suppresses the A431 epidermoid carcinoma cells' ability to proliferate is also covered. Additionally, rhodomyrtone prevented the migration, invasion, and metastasis of SW1353 cells, suggesting that it had anticancer characteristics and could be useful in antitumor therapy. Furthermore, in a variety of scenarios, the mixed extracts of *R. tomentosa* have demonstrated antitumor and anticancer properties (Tayeh et al. 2017).

*Rhodomyrtus tomentosa* lowers the amount of IL-1β—which may stop or slow the development of inflammation (Srisuwan et al. 2018)—the proteins that impact the spread of cancer. This plant's potential as an anticancer agent in cervical cancer cases has been examined in earlier research by measuring the production of Transforming Growth Factor Beta 1 (TGFβ1) and Vascular Endothelial Growth Factor Receptor (VEGFR) (Situmorang et al. 2023). Ursonic acid was the second prediction made by the ethyl acetate subfraction of the ethanolic extract of *R. tomentosa* leaves. These are naturally occurring products present in a variety of plants. Plantago's ursonic acid biologically inhibits cyclooxygenase 2/cyclooxygenase 1 as well as prostaglandin formation. This metabolite also has the ability to inhibit the growth and survival of many types of cancer cells. Ursonic acid has antihyperglycemic, anti-inflammatory, antiviral, and antioxidant properties (Vo et al. 2020), including the naturally occurring polyphenolic piceatannol. It is said to have a number of medicinal qualities and positive health effects, including anti-inflammatory, antitumor, antidiabetic, and antioxidant actions (Sang et al. 2021).

## 5.14 Antidiabetic Activity of *Rhodomyrtus tomentosa*

Diabetes is the primary cause of heart disorder, renal failure, and stroke, a serious disease. Diabetes type 2, also known as adult-onset or non-insulin-dependent diabetes, is brought on by the body's inadequate utilization of insulin. Blood sugar levels rise over normal when the body does not use insulin as well as it should. Diabetes is still commonly treated with traditional remedies that use medicinal plants as a therapeutic agent through several different processes. One typical method involves blocking the enzymes that catalyze the hydrolytic breakdown of carbohydrates, such as α-glucosidase and α-amylase. This slows down the blood's absorption of glucose (Khadayat et al. 2020). Subsequent investigation revealed that the methanol (MeOH) extract of *R. tomentosa* inhibited the α-amylase enzyme more than acarbose. Based on the *t*-test, there were significant differentiations between the extracts and the positive control in the anti-α-glucosidase and anti-α-amylase activity results.

Alternative medications made from medicinal plants can be used to treat diabetes with fewer negative effects. Large and intractable starch molecules are broken down into smaller, more readily absorbed units by pancreatic and salivary α-amylases (Kumar et al. 2011). Delaying the absorption of glucose can be a useful strategy in the interim for managing diabetes mellitus and reducing postprandial hyperglycemia. This occurs when the digestive organs' α-glucosidase and α-amylase activities are blocked, resulting in the hydrolysis of carbohydrates (Wresdiyati et al. 2015). Rat blood glucose levels were examined in relation to the methanol extract of *R. tomentosa* leaves, and the results indicated that the extract decreased blood glucose levels by 15.96%–48.51% (Idris et al. 2022).

The methanol extract from *R. tomentosa* leaves dramatically ($p < 0.05$) reduced the blood glucose levels in diabetic rats. The outcomes of the Analysis of Variance (ANOVA) and Lysergic acid diethylamide (LSD) analyses showed a significant difference between these results and the normal group. Plants exhibit antihyperglycemic effect because they can decrease intestinal glucose absorption, promote insulin-dependent activities, and restore pancreatic tissue function, which increases insulin production (Balamurugan et al. 2014). As a result, β-cell protection and decreased blood glucose fluctuation are two benefits of using herbal medications to treat diabetes. On the other hand, compounds including flavonoids, terpenoids, and glycosides—which are frequently linked to antidiabetic effects—have been found in a variety of plants (Malviya et al. 2010).

One treatment option for type 2 diabetes mellitus is the suppression of α-glucosidase. This enzyme is frequently located near the small intestine's border (Fatmawati et al. 2011). An α-glucosidase inhibitor inhibits the synthesis of sugar by delaying the absorption of complex carbs, hence lowering blood glucose levels and absorption over time (Van de Laar 2008; Standl and Schnell 2012). Based on previous research, the methanol extract was used for in vitro anti-α-amylase evaluation and in vivo antidiabetic investigation because the anti-α-glucosidase assay demonstrated that it had the strongest inhibitory capacity.

Based on the *t*-test, there were significant variances between the extracts and the positive control in the anti-α-glucosidase and anti α amylase activity ($p < 0.01$). One treatment option for type 2 diabetes mellitus is the suppression of α-glucosidase. This enzyme is frequently located near the small intestine's border (Fatmawati et al. 2011). An α-glucosidase inhibitor inhibits the synthesis of sugar by delaying the absorption of complex carbohydrates, meaning that the blood level of glucose is gradually reduced (Van de Laar 2008, Standl and Schnell 2012). Alternative medications made from medicinal plants can be used to treat diabetes with fewer negative effects. Large and intractable starch molecules are broken down into smaller, more readily absorbed units by pancreatic and salivary α-amylases (Kazeem et al. 2013). Delaying the absorption of glucose can be a useful strategy in the interim for managing diabetes mellitus and reducing postprandial hyperglycemia. This occurs when the digestive organs' α-glucosidase and α-amylase activity are blocked, resulting in the hydrolysis of carbohydrates (Wresdiyati et al. 2015).

*Rhodomyrtus tomentosa* slows intestinal glucose absorption, increases insulin production via restoring pancreatic tissue function, and promotes insulin-dependent processes' metabolism. These actions give rise to the plant's antihyperglycemic effect (Balamurugan et al. 2014). Advanced glycation end products (AGEs) are created when proteins, lipids, or nucleic acids combine with glucose to generate different Amadori products in the blood and tissues (Singh et al. 2001; Ahmed et al. 2005). The development and problems of diabetes mellitus are brought on by AGEs, which are lipids and results of protein modification. The formation of AGEs will be decreased by good glucose management, which will prevent AGE accumulation in tissue (Hafizur et al. 2012; Siddiqui et al. 2014). Reduced AGEs will result in less proinflammatory cytokines being produced, which will allow damage to be repaired and cell function to deteriorate (Hasibuan et al. 2015).

## 5.15 Conclusions

In conclusion, *Rhodomyrtus tomentosa*, also known as the rose myrtle, is a versatile and attractive plant that is suitable for a variety of landscaping purposes with multifaceted biomimetic attributes. Its ornamental flowers, attractive foliage, and edible fruits make it a desirable choice not only for gardening but also for several pharmacological uses to modulate the radical scavenging efficacy as well as antibacterial, antiviral, and anticancer activities. These wide range of pharmacological properties transform this plant from ornamental use to an important candidate for the progress of new drugs and natural remedies. Further, research is needed to fully understand its mechanistic insight and safety profile for clinical use.

## References

Abd Hamid H, Mutazah S, Yusoff MM (2017) *Rhodomyrtus tomentosa*: a phytochemical and pharmacological review. Asian J Pharm Clin Res 10:10–16

Ahmed N, Babaei-Jadidi R, Howell SK, Thornalley PJ, Beisswenger PJ (2005) Glycated and oxidized protein degradation products are indicators of fasting and postprandial hyperglycemia in diabetes. Diabetes Care 28:2465–2471

Anane J (2022) Green synthesis of Spiro compounds and the phytochemical investigation of diethyl ether fraction of Camu-Camu (*Myrciaria dubia*) leaves. The University of Texas Rio Grande Valley, Edinburg

Bach QN, Hongthong S, Quach LT, Pham LV, Pham TV, Kuhakarn C, Reutrakul V, Nguyen PT (2020) Antimicrobial activity of rhodomyrtone isolated from *Rhodomyrtus tomentosa* (Aiton) Hassk. Nat Prod Res 34:2518–2523

Balamurugan K, Nishanthini A, Mohan VR (2014) Antidiabetic and antihyperlipidaemic activity of ethanol extract of *Melastoma malabathricum* Linn. leaf in alloxan induced diabetic rats. Asian Pac J Trop Biomed 4:S442–S448

Barbosa A, Martel F (2020) Targeting glucose transporters for breast cancer therapy: the effect of natural and synthetic compounds. Cancers (Basel) 12:154

Chuakul W (2005) Medicinal plants in the Khok Pho district, Pattani province (Thailand)

Do T (2011) Medicine plants and remedies of Vietnam. Thoi Dai Publisher, Hanoi

Fatmawati S, Shimizu K, Kondo R (2011) Ganoderol B: a potent α-glucosidase inhibitor isolated from the fruiting body of *Ganoderma lucidum*. Phytomedicine 18:1053–1055

Fujiati F, Haryati H (2022) Anti-inflammatory and anti-remodelling potential of ethanol extract *Rhodomyrtus tomentosa* in combination of asthma and coal dust models. Rep Biochem Mol Biol 10:686

Geetha K, Sridhar C, Murugan V (2010) Antioxidant and healing effect of aqueous alcoholic extract of *Rhodomyrtus tomentosa* (Ait.) Hassk on chronic gastric ulcers in rats. J Pharm Res 3:2860–2862

George S, Abrahamse Ḥ (2020) Redox potential of antioxidants in cancer progression and prevention. Antioxidants 9:1156

Gulcin İ (2020) Antioxidants and antioxidant methods: An updated overview. Arch Toxicol 94:651–715

Hafizur RM, Kabir N, Chishti S (2012) Asparagus officinalis extract controls blood glucose by improving insulin secretion and β-cell function in streptozotocin-induced type 2 diabetic rats. Br J Nutr 108:1586–1595

Hasibuan R, Ilyas S, Hannum S (2015) Effect of leaf extract Haramonting (*Rhodomyrtus tomentosa*) to lower blood sugar levels in mice induced by alloxan. Int J PharmTech Res 8(6):284–291

Hmoteh J (2018) Effects of *Rhodomyrtus tomentosa* (Aiton) Hassk. Leaf extract on human neutro. Prince of Songkla University, Hat Yai

Huang Y, Yang Z, Huang S, An W, Li J, Zheng X (2019) Comprehensive analysis of *Rhodomyrtus tomentosa* chloroplast genome. Plants 8:89

Hue TS (2015) Propagation, genetic variation and evaluation of kemunting (*Rhodomyrtus tomentosa* (Aiton) Hassk) populations in Malaysia for landscape use. PhD Thesis, Universiti Putura, Malaysia. 1–165

Idris M, Sukandar ER, Purnomo AS, Martak F, Fatmawati S (2022) Antidiabetic, cytotoxic and antioxidant activities of *Rhodomyrtus tomentosa* leaf extracts. RSC Adv 12:25697–25710

Idris M, Purnomo AS, Martak F, Kim Y-U, Fatmawati S (2023) Chemical compounds and pharmaceutical properties of *Rhodomyrtus tomentosa*: a traditional medicinal herb from South Kalimantan, Indonesia. J Hunan Univ Nat Sci 50:99–123

Jeenkeawpieam J, Phongpaichit S, Rukachaisirikul V, Sakayaroj J (2012) Antifungal activity and molecular identification of endophytic fungi from the angiosperm *Rhodomyrtus tomentosa*. Afr J Biotechnol 11:14007–14016

Jeong D, Yang WS, Yang Y, Nam G, Kim JH, Yoon DH, Noh HJ, Lee S, Kim TW, Sung G-H (2013) In vitro and in vivo anti-inflammatory effect of *Rhodomyrtus tomentosa* methanol extract. J Ethnopharmacol 146:205–213

Kamarudin E, Zainol H, Anuar TS, Hussain R (2021) *Rhodomyrtus tomentosa* leaves phytochemistry and biological activities: an update review. Malays J Med Health Sci 17:334–343

Kazeem M, Adamson J, Ogunwande I (2013) Modes of inhibition of α-amylase and α-glucosidase by aqueous extract of *Morinda lucida* Benth leaf. Biomed Res Int 2013:527570

Khadayat K, Marasini BP, Gautam H, Ghaju S, Parajuli N (2020) Evaluation of the alpha-amylase inhibitory activity of Nepalese medicinal plants used in the treatment of diabetes mellitus. Clin Phytoscience 6:1–8

Kumar S, Narwal S, Kumar V, Prakash O (2011) α-glucosidase inhibitors from plants: a natural approach to treat diabetes. Pharmacogn Rev 5:19

Kuntorini EM, Sari SG, Fariani R (2023) The morphoanatomy, histochemistry, and phytochemistry of the leaves and fruits of *Rhodomyrtus tomentosa*. Biodivers J Biol Divers 24:98–105

Kusuma IW, Ainiyati N, Suwinarti W (2016) Search for biological activities from an invasive shrub species rosemyrtle (*Rhodomyrtus tomentosa*). Nus Biosci 8:55–59

Lai TNH, André C, Rogez H, Mignolet E, Nguyen TBT, Larondelle Y (2015) Nutritional composition and antioxidant properties of the Sim fruit (*Rhodomyrtus tomentosa*). Food Chem 168:410–416

Langeland K, Burks K (2008) Identification and biology of non-native plants in Florida's natural areas, SP257. University of Florida, Gainesville

Lavanya G, Reddy LM, Padmavathi V, Padmaja A (2014) Synthesis and antimicrobial activity of (1, 4-phenylene) bis (arylsulfonylpyrazoles and isoxazoles). Eur J Med Chem 73:187–194

Leejae S, Hasap L, Voravuthikunchai SP (2013) Inhibition of staphyloxanthin biosynthesis in Staphylococcus aureus by rhodomyrtone, a novel antibiotic candidate. J Med Microbiol 62:421–428

Li J, Chen W, Liu H, Liu H, Xiang S, You F, Jiang Y, Lin J, Zhang D, Zheng C (2023) Pharmacologic effects approach of essential oils and their components on respiratory diseases. J Ethnopharmacol 304:115962

Lim T (2012) Garcinia cowa. Edible medicinal and non-medicinal plants. Springer, Dordrecht

Limsuwan S, Trip EN, Kouwen TR, Piersma S, Hiranrat A, Mahabusarakam W, Voravuthikunchai SP, van Dijl JM, Kayser O (2009) Rhodomyrtone: a new candidate as natural antibacterial drug from *Rhodomyrtus tomentosa*. Phytomedicine 16:645–651

Limsuwan S, Kayser O, Voravuthikunchai SP (2012) Antibacterial activity of *Rhodomyrtus tomentosa* (Aiton) Hassk. leaf extract against clinical isolates of *Streptococcus pyogenes*. Evid Based Complement Alternat Med 2012:697183

Limsuwan S, Homlaead S, Watcharakul S, Chusri S, Moosigapong K, Saising J, Voravuthikunchai SP (2014) Inhibition of microbial adhesion to plastic surface and human buccal epithelial cells by *Rhodomyrtus tomentosa* leaf extract. Arch Oral Biol 59:1256–1265

Malviya N, Jain S, Malviya S (2010) Antidiabetic potential of medicinal plants. Acta Pol Pharm 67:113–118

Maskam MF (2011) Antioxidant effects of *Rhodomyrtus tomentosa* (kemunting) extract on the development of experimental atherosclerotic-induced New Zealand white rabbit. Master thesis, University of Malaya

Mitra S, Irenaeus T, Gurung M, Pathak P (2012) Taxonomy and importance of Myrtaceae. In: Proceedings of the 3rd international symposium on guava and other Myrtaceae, vol 959. ISHS, pp 23–34

Mitsuwan W, Olaya-Abril A, Calderón-Santiago M, Jiménez-Munguía I, González-Reyes JA, Priego-Capote F, Voravuthikunchai SP, Rodríguez-Ortega MJ (2017) Integrated proteomic and metabolomic analysis reveals that rhodomyrtone reduces the capsule in *Streptococcus pneumoniae*. Sci Rep 7:2715

Mordmuang A, Voravuthikunchai SP (2015) *Rhodomyrtus tomentosa* (Aiton) Hassk. leaf extract: an alternative approach for the treatment of staphylococcal bovine mastitis. Res Vet Sci 102:242–246

Mordmuang A, Brouillette E, Voravuthikunchai SP, Malouin F (2019) Evaluation of a *Rhodomyrtus tomentosa* ethanolic extract for its therapeutic potential on *Staphylococcus aureus* infections using in vitro and in vivo models of mastitis. Vet Res 50:1–11

Na-Phatthalung P, Teles M, Voravuthikunchai SP, Tort L, Fierro-Castro C (2018) Immunomodulatory effects of *Rhodomyrtus tomentosa* leaf extract and its derivative compound, rhodomyrtone, on head kidney macrophages of rainbow trout (*Oncorhynchus mykiss*). Fish Physiol Biochem 44:543–555

Nguyen QT, Nguyen VT, Nguyen DT, Pham NT, Nguyen TK, Nhi TTY, Tran NQ, Le Thi P (2023) Injectable hydrogel combining alginate and *Rhodomyrtus tomentosa* medicine with antibacterial, anti-inflammatory and cellular proliferation properties as potential wound dressing material. Mater Today Commun 35:106243

Ningsih ESP, Rizki MI, Budianto WY (2021) Potential antiseptic of *Rhodomyrtus tomentosa* leaves extract on healing wound in male Wistar rats. Res J Pharm Technol 14:3143–3149

Nwabor OF, Singh S (2022) A systematic review on *Rhodomyrtus tomentosa* (Aiton) Hassk: a potential source of pharmacological relevant bioactive compounds with prospects as alternative remedies in varied medical conditions. Int J Pharm Sci Nanotechnol (IJPSN) 15:5875–5891

Odedina GF, Vongkamjan K, Voravuthikunchai SP (2015) Potential bio-control agent from *Rhodomyrtus tomentosa* against *Listeria monocytogenes*. Nutrients 7:7451–7468

Ong H, Nordiana M (1999) Malay ethno-medico botany in Machang, Kelantan, Malaysia. Fitoterapia 70:502–513

Ontong JC, Singh S, Nwabor OF, Chusri S, Voravuthikunchai SP (2020) Potential of antimicrobial topical gel with synthesized biogenic silver nanoparticle using *Rhodomyrtus tomentosa* leaf extract and silk sericin. Biotechnol Lett 42:2653–2664

Pasaribu YP, Fadlan A, Fatmawati S, Ersam T (2021) Biological activity evaluation and in silico studies of polyprenylated benzophenones from *Garcinia celebica*. Biomedicine 9:1654

Pragallapati S, Manyam R (2019) Glucose transporter 1 in health and disease. J Oral Maxillofac Pathol: JOMFP 23:443

Saising J, Voravuthikunchai SP (2012) Anti Propionibacterium acnes activity of rhodomyrtone, an effective compound from *Rhodomyrtus tomentosa* (Aiton) Hassk. leaves. Anaerobe 18:400–404

Sang VT, Minh NHN, Quang NX, Hoai PN, Giang BL, Hung NH, Hoa NLH, Dai Hung N (2021) Investigation of anti-allergic activity of piceatannol from *Rhodomyrtus tomentosa* fruits. Vietnam J Biotechnol 19:119–128

Sarah B, Thohirah L, Kamal M, Rosenani A (2014) Preference for *Molineria latifolia* var. megacarpa and *Rhodomyrtus tomentosa* as native urban landscape plants. Pertanika J Trop Agric Sci 37:457–473

Sianglum W, Srimanote P, Wonglumsom W, Kittiniyom K, Voravuthikunchai SP (2011) Proteome analyses of cellular proteins in methicillin-resistant Staphylococcus aureus treated with rhodomyrtone, a novel antibiotic candidate. PLoS One 6:e16628

Sianglum W, Saeloh D, Tongtawe P, Wootipoom N, Indrawattana N, Voravuthikunchai SP (2018) Early effects of rhodomyrtone on membrane integrity in methicillin-resistant *Staphylococcus aureus*. Microb Drug Resist 24:882–889

Siddiqui BS, Hasan M, Mairaj F, Mehmood I, Hafizur RM, Hameed A, Shinwari ZK (2014) Two new compounds from the aerial parts of *Bergenia himalaica* Boriss and their anti-hyperglycemic effect in streptozotocin-nicotinamide induced diabetic rats. J Ethnopharmacol 152:561–567

Singh HP (2017) Landscape gardening for ecological and aesthetic gains. In: Malhotra SK, Ram L (eds) Floriculture and landscape gardening. Nagaland, Central Institute of Horticulture, pp 1–10

Singh R, Barden A, Mori T, Beilin L (2001) Advanced glycation end-products: a review. Diabetologia 44:129–146

Situmorang PC, Ilyas S, Hutahaean S, Rosidah R (2020) Components and acute toxicity of nanoherbal haramonting (*Rhodomyrtus tomentosa*). J Herbmed Pharmacol 10:139–148

Situmorang PC, Simanullang RH, Syahputra RA, Hutahaean MM, Sembiring H, Nisfa L, Sari ER (2023) Histological analysis of TGFβ1 and VEGFR expression in cervical carcinoma treated with *Rhodomyrtus tomentosa*. Pharmacia 70:217–223

Srisuwan S, Mackin KE, Hocking D, Lyras D, Bennett-Wood V, Voravuthikunchai SP, Robins-Browne RM (2018) Antibacterial activity of rhodomyrtone on *Clostridium difficile* vegetative cells and spores in vitro. Int J Antimicrob Agents 52:724–729

Standl E, Schnell O (2012) Alpha-glucosidase inhibitors 2012–cardiovascular considerations and trial evaluation. Diab Vasc Dis Res 9:163–169

Staples G, Herbst DR (2005) A tropical garden flora: plants cultivated in the Hawaiian islands and other tropical places. Bishop Museum Press, Honolulu

Tay K-C, Tan LT-H, Chan CK, Hong SL, Chan K-G, Yap WH, Pusparajah P, Lee L-H, Goh B-H (2019) Formononetin: a review of its anticancer potentials and mechanisms. Front Pharmacol 10:820

Tayeh M, Nilwarangoon S, Mahabusarakum W, Watanapokasin R (2017) Anti-metastatic effect of rhodomyrtone from *Rhodomyrtus tomentosa* on human skin cancer cells. Int J Oncol 50:1035–1043

Toltl LJ, Swystun LL, Pepler L, Liaw PC (2008) Protective effects of activated protein C in sepsis. Thromb Haemost 100:582–592

Van de Laar FA (2008) Alpha-glucosidase inhibitors in the early treatment of type 2 diabetes. Vasc Health Risk Manag 4:1189–1195

Vo TS, Ngo DH (2019) The health beneficial properties of *Rhodomyrtus tomentosa* as potential functional food. Biomolecules 9:76

Vo TS, Le TT, Kim SY, Ngo DH (2020) The role of myricetin from *Rhodomyrtus tomentosa* (Aiton) Hassk fruits on downregulation of FcεRI-mediated mast cell activation. J Food Biochem 44:e13143

Wang R, Yao L, Meng T, Li C, Wang L (2022) *Rhodomyrtus tomentosa* (Ait.) Hassk fruit phenolic-rich extract mitigates intestinal barrier dysfunction and inflammation in mice. Food Chem 393:133438

Wilson PG (2010) Myrtaceae. In: Flowering plants eudicots: sapindales, cucurbitales, myrtaceae. Springer, pp 212–271

Winotai A, Wright T, Goolsby JA (2005) Herbivores in Thailand on *Rhodomyrtus tomentosa* (Myrtaceae), an invasive weed in Florida. Fla Entomol 88:104–105

Wresdiyati T, Sa'diah S, Winarto A, Febriyani V (2015) Alpha-glucosidase inhibition and hypoglycemic activities of *Sweitenia mahagoni* seed extract. HAYATI J Biosci 22:73–78

Wu P, Ma G, Li N, Deng Q, Yin Y, Huang R (2015) Investigation of in vitro and in vivo antioxidant activities of flavonoids rich extract from the berries of *Rhodomyrtus tomentosa* (Ait.) Hassk. Food Chem 173:194–202

Wunnoo S, Saising J, Voravuthikunchai SP (2017) Rhodomyrtone inhibits lipase production, biofilm formation, and disorganizes established biofilm in *Propionibacterium acnes*. Anaerobe 43:61–68

Yang L, Chen Y, Huang Y, Li Y, Wen M, Liu N, Wang J (2016) Effects of shrub islands created by *Rhodomyrtus tomentosa* (Aiton) Hassk. on the growth, chlorophyll fluorescence, and chloroplast ultrastructure of pine seedlings in degraded land of South China. Land Degrad Dev 27:729–737

Zambrano A, Molt M, Uribe E, Salas M (2019) Glut 1 in cancer cells and the inhibitory action of resveratrol as a potential therapeutic strategy. Int J Mol Sci 20:3374

Zhang Y-B, Li W, Jiang L, Yang L, Chen N-H, Wu Z-N, Li Y-L, Wang G-C (2018) Cytotoxic and anti-inflammatory active phloroglucinol derivatives from *Rhodomyrtus tomentosa*. Phytochemistry 153:111–119

Zhao Z, Wu L, Xie J, Feng Y, Tian J, He X, Li B, Wang L, Wang X, Zhang Y (2020) *Rhodomyrtus tomentosa* (Aiton.): a review of phytochemistry, pharmacology and industrial applications research progress. Food Chem 309:125715

# Medicinal and Nutritional Importance of *Withania somnifera* in Human Health

6

Ishani Yadav, Neelu Sinha, and Udita Tiwari

**Abstract**

Recent research has discovered that *Withania somnifera* plays a major role in health and nutritional aspects of humans as it has anti-inflammatory, antidiabetic, cardioprotective, and sedative properties, like tranquilizers, as well as antioxidative effects on nervous system to cure neurological disorders such as dementia, obsessive compulsive disorder, cancer, and hypoglycemia. Ashwagandha has phytochemical compounds such as volatile oils, reducing sugars, and steroidal lactones, for example withaferin A, withanoside II, withanoside IV, withanoside D, and sitoindoside, which are present in their roots, which possesses pharmacological activities and helps in curing diseases through a natural method with minimum or no side effects. Ashwagandha also possesses secondary metabolites such as flavonoids, condensed tannins, and alkaloids, for example, isopelletierine, anaferine, cuscohygrine, and anahygrine. In this review article, we will identify the role of ashwagandha in curing stress, depression, neurological disorders, and post-Coronavirus Disease-2019 (COVID-19) effects.

**Keywords**

*Withania somnifera* · Amino acid · Secondary metabolites · Sedatives · Ashwagandha

I. Yadav · N. Sinha · U. Tiwari (✉)
Department of Biochemistry, School of Life Sciences, Swami Vivekanand Campus, Dr. Bhimrao Ambedkar University, Agra, UP, India



M. A. Ansari et al. (eds.), *Medicinal Plants and their Bioactive Compounds in Human Health: Volume 2*, https://doi.org/10.1007/978-981-95-3620-7_6

## Abbreviations

| | |
|---|---|
| ROS | Radioactive oxygen species |
| C | Carbon |
| GABA | Gaba aminobutyric acid |
| SARS-COV-2 | Severe acute respiratory syndrome coronavirus |
| Th | T-helper cells |
| ACE | Angiotensin converting enzyme |
| IL | Interleukin |
| IFN | Interferon gamma |
| TGF | Transforming growth factor |
| CSF | Cerebrospinal fluid |
| TNF | Tumor necrosis factor |
| NK | Natural killer cells |
| Ig E | Immunoglobulin E |
| DNA | Deoxyribonucleic acid |
| RNA | Ribonucleic acid |
| CD | Clusters of cell differentiation |

**Learning Objectives**

- Biochemical profile, phytochemicals with their activity and part in which they are present in *Withania somnifera.*
- Taxonomy, source of information and area of cultivation.
- Biological properties of ashwagandha in various areas such as male infertility and central nervous system.
- Pharmacological characteristics to cure diseases.

## 6.1 Introduction

Ashwagandha or *Withania somnifera* has traditionally been used for medicinal purposes for 3000 years. The main source of information for ashwagandha can be seen in *Charak Samhita* and *Sushrut Samhita* (two Vedic texts in Sanskrit) that are said to be the most authentic source of Ayurveda and ancient medical science. *Withania somnifera* is a dicotyledon plant that belongs to Solanaceae family possessing a wide range of pharmacological effects.

It is an important herb of Ayurvedic medical system, which is classified as adaptogen, and two different varieties of it are also defined in Unani literature, which is asgand nagori and asgand dakani (Kaur et al. 2013).

*Withania somnifera* is widely cultivated in the area of Madhya Pradesh, Gujarat, Uttar Pradesh, Punjab, Rajasthan, and Himalayan areas of India. It is also cultivated in countries like Africa, Canary Island, Cape of Good Hope, and Australia. Extracts of ashwagandha is used for curing insomnia, goiter, asthma, anxiety, ulcer, and neurological disorders. All these pharmacological activities are related to anti-stress, adaptogenic, antioxidative, and immunomodulatory properties.

*Withania somnifera* is a highly studied and researched topic defining active phytochemicals to optimize its usage.

According to recent researches, following results are reported:

- Approximately 70 withanolides also known as group of steroidal lactones are reported from leaves.
- Twenty-seven sitoindosides are derived from aerial part and roots of *Withania somnifera.*
- It also consists of various amino acids, alkaloids, phenols, glycosides, volatile oils, and reducing sugars (Gautam et al. 2004).

In 2019, some people were experiencing symptoms of pneumonia in Wuhan city of China. Afterward, an unknown virus was discovered that was termed as Severe Acute Respiratory Syndrome Coronavirus-2 or SARS-COV-2. Having no cure resulted in a pandemic situation all over the globe (De Giorgio et al. 2020). It has been suggested to consume ashwagandha as it possesses immune-modulatory properties enhancing immune system (Wang et al. 2021).

## 6.2 Taxonomy

Kingdom: Plant.
Division: Angiosperm.
Class: Dicotyledonous.
Order: Tubiflorae.
Family: Solanaceae.
Genus: Withania.
Species: Somnifera.

According to Ayurveda, ashwagandha is one of the most important herbs having various biologically active chemical compounds in different parts of plant such as roots and leaves. It is a sedative possessing rejuvenating and life-prolonging properties. Over the past few years, sale of ashwagandha has increased drastically in international market. Besides, many companies cultivate ashwagandha and manufacture their own roots and leaves extract so that they can market them in the name of ashwagandha supplements. Ashwagandha not only has pharmaco-commercial applications but also is used in food industries as an ingredient of functional foods such as herbal cookies, ghee, and beverages.

Many organizations consider it as a pharmaceutically and neutraceutically important herb, therefore its powdered form is incorporated in various drugs used for treatment. Root extract promotes cell cycle, increases semen, regulates blood pressure, strengthens muscles, and increases vital fluid production in body, and many more such activities (Singh et al. 2021). Plants are vital source of medicines in the world and are also a derivative of various drugs. A survey performed by the World Health Organization explains that 80% of people globally use plant extracts as a primary cure of any disease.

## 6.3 Biochemical Profiling

There are several phytochemicals present in parts of the plant *Withania somnifera*. These phytochemicals are biologically active chemicals present in the treatment of disorders. More than 12 alkaloids are found in roots of the plant, for example, isopelletierine, anaferine, cuscohygrine, and anahygrine, and around 40 steroidal lactones also known as withanolide are found, and it is a 28-carbon compound with C-22 and C-26 containing oxygen that forms a lactone ring. There are three broad classes of which are withaferin, glycosides, and sitoindosides. According to the recent researches, seven latest withanosides are derived from roots of the plant. It is a methanolic extract having steroidal lactone. Withanolide D is structurally similar to withanolide A with an exception of hydroxyl group at C-20 and not on C-27.

Unlike others, withanolide E and withanolide J have hydroxyl group at C-17 and not on C-20. Besides, withanolide E and withanolide F are alpha oriented while withanolide G and withanolide J are beta oriented. In molecular structure of sitoindoside, acyl group and glucose units are substituted at C-27 of saponins. Another methanolic root extract isolated from plant is withasomidienone with three double bonds at first, fourth and 24th carbon, whereas withanolide N, withanolide O, withanolide D, withanolide E, and withanolide P are derived from alcoholic extract (Saleem et al. 2020). Other phytochemicals present are tannins, flavonoids, phenols, and several amino acids such as aspartic acid, glycine, tryptophan, glutamic acid, and cysteine. Molecular structure of various phytochemicals present in ashwagandha is shown in Fig. 6.1.

## 6.4 Biological Activity

Ashwagandha is biologically active and every phytochemical exhibits some specific biological activities such as antimicrobial, anti-inflammatory, anti-stress, psychotropic activity, insecticidal, and antidiabetic.

## 6.5 Antimicrobial Activity

In the past many studies have been carried out on the antimicrobial activities of this plant on various microbial species such as *Escherichia coli*, *Pseudomonas aeruginosa*, *Candida albicans*, *Staphylococcus aureus*, and many more, using disk diffusion assay. Mostly root and leaf extracts are used in the assessment of antimicrobial activity in which flavonoids present in roots have given results against many bacteria. Acetone, benzene, chloroform, water, methanol, toluene are used as solvents to study antifungal effects (Speers et al. 2021).

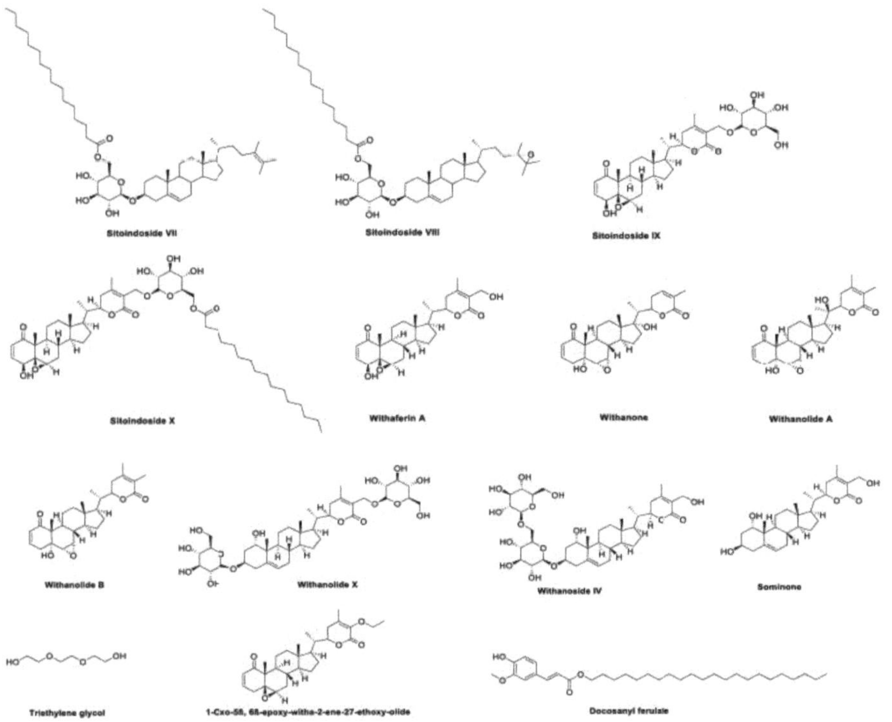


**Fig. 6.1** Types of phytochemicals present in ashwagandha with their well-defined molecular structures. (Source: Speers et al. 2021)

## 6.6 Anti-inflammatory Activity

Ashwagandha exhibits anti-inflammatory activity against aluminum chloride-induced neurological toxicity. Aluminum raises lipid oxidation in cortex and decreases glutathione level. It also exhibits anti-inflammatory keratinocytes by promoting the expression of mRNA specific to inflammatory cytokine whose inhibition results in inflammation. Ashwagandha exhibits its anti-inflammatory activities on interleukin (IL), interferon gamma (IFN), transforming growth factor (TGF), cerebrospinal fluid (CSF), and tumor necrosis factor (TNF). Inflammation can be acute or chronic; therefore, in case of acute inflammation, ashwagandha acts on protease enzyme and lipids (Sikandan et al. 2018).

## 6.7 Cardioprotective Activity

Root powder in aqueous alcoholic solution is used as cardioprotective compound. It consists of alkaloid that lowers the lipid and cholesterol level and increases the activity of 3-hydroxyl-3-methylglutyral coenzyme-A reductase. It also contains flavonoid and phenol, which lowers blood glucose level. It is taken in a dosage of 25 mg per kg, 50 mg per kg, and 100 mg per kg. Regular uptake of *Withania somnifera* has proven to be significantly effective on reducing systolic and diastolic blood pressure (Sandhu et al. 2010).

## 6.8 Anticarcinogenic Activity

Ashwagandha has a significant effect on carcinogens. In many researches *Withania somnifera* is found to be effective in reducing the size of tumor as a result of withaferine A, a steroidal lactone arresting cell cycle at G2/M phase or G0 phase.

Alkaloids present in this plant bind on proteins responsible for tumor, resulting in the disruption of mitotic phase. Major constituents of this plant responsible for anticarcinogenic activity are withaferine A, withanolode VI, withanolide A, withanoside IV, withanone, and withanolide D (Rai et al. 2016).

## 6.9 Antispasmodic Activity

Alkaloids present in ashwagandha extract play a major role in relaxing muscles and show antispasmodic activity against several spasmodic agents such as acetylcholine, leukotriene, endothelia, thrombin, and histamine present in different body parts like intestine, bronchia, trachea, uterus, and vascular muscles (Uddin et al. 2012).

## 6.10 Antidiabetic Activity

*Withania somnifera* has been used for diabetes since ancient time and also demonstrate a pharmaceutical activity in lowering blood glucose level as well as increasing insulin production. Withanolides present in leaf and root extract decrease blood glucose level. Here leaf extract is used for lowering blood glucose level, whereas root extract is used for increasing insulin production in the body.

## 6.11 Male Infertility

Male infertility is a case where males have lower semen quality or abnormal and immature spermatozoa with less mobility. This situation can be a result of either stress, which is related to reactive oxygen species (ROS)/free radicals or smoking. Smoking induces many chemicals that can be harmful for human health. Sitoindoside of steroidal lactone is an effective compound exhibiting anti-stress activity. Taking 5 g of ashwagandha daily for about 3 months regulates spermatogenesis process and results in curing male infertility (Mahadi et al. 2011).

## 6.12 Central Nervous System

*Withania somnifera* has proven to be effective in many diseases related to central nervous system such as anxiety, depression, insomnia, and Alzheimer's disease. Most important phytochemicals that are effective in curing these diseases are withanolide A and withanoside IV (Jain et al. 2012).

## 6.13 Pharmacological Effects

### 6.13.1 Anti-stress

Stress is a biological response by body to external and internal environmental conditions. Stress can lead to multiple disorders depending on its type and frequency.

During stress conditions, adrenal gland present on the top of kidney releases adrenalin and extra cortisol in the blood stream. Anti-stress activity is performed by phytochemicals present in ashwagandha such as glycosides, exclusively sitoindoside VII and sitoindoside VIII. Ashwagandha consists of withanolides in its root. At present, 40 withanolides are present and act as precursor. In case of excess release of hormones, withanolide, which is a precursor, occupies all the cell receptor so that hormones will not be able to attach and exert their function anymore (Chandrasekhar et al. 2012).

### 6.13.2 Immune Modulator

Immune modulators are the compounds that modulate the immune system of any organism. Such substances can either increase or decrease the immune response. Steroidal lactones and withanolide glycosides are majorly responsible for immune response and increase the level of immunoglobulin significantly. Steroidal lactone withaferine A raises innate immunity and T-helper (Th) cells by increasing the production of innate cells such as dendritic cells, macrophages, and basophils. This immune modulation property of ashwagandha plays a major role in boosting immune system of individuals who suffered from Coronavirus during the pandemic as it weakens the immunity. Various immune modulators present in ashwagandha are shown in Table 6.1 with their activities and immune-modulatory activities in Fig. 6.2 (Gambhir et al. 2015).

### 6.13.3 Antioxidative

Oxidative stress plays a major role in depressive disorders. Generally, oxidative stress occurs due to lack of antioxidants. Free oxygen molecules with unpaired free electrons or free radicles are termed as reactive oxygen species (ROS). It is a highly reactive molecule responsible for modification in DNA, RNA, proteins, and lipids.

**Table 6.1** The immune modulators boosted by ashwagandha and their specific activities that they perform in an individual, for example, platelets responsible for inflammation and thrombosis, macrophages responsible for boosting innate immunity as well as homeostasis process and so on

| Immune modulators | Activity |
|---|---|
| Th 1 | Defense against viral and intracellular infections |
| Macrophages | Innate immunity and homeostasis |
| Clusters of cell differentiation (CD)8, CD3, CD45, natural killer (NK) cells | Responsible for viral and humeral immunity |
| Immunoglobulin | Responsible for humeral immune response |
| TNF, immunoglobulin E (IgE) | Allergic responses and T-helper cells 2 mediated activity |
| Platelets | Perform thrombosis and inflammation |
| Corticosterone | Suppression of immune response |

Source: Gambhir et al. (2015)

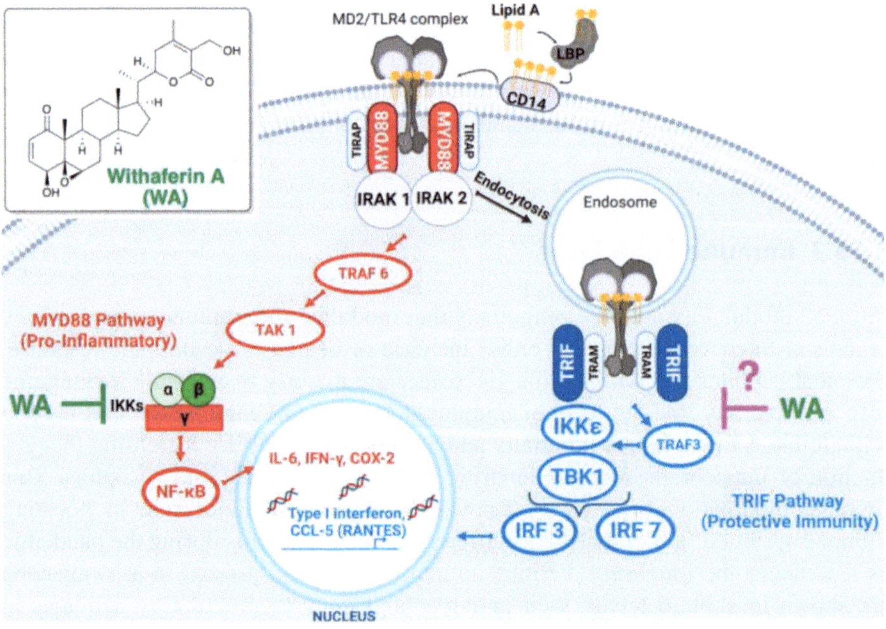


**Fig. 6.2** Immune-modulatory activity of withaferin A providing inhibition against I kappa B kinase (IKK-beta) due to inhibition of myeloid differentiation primary response (MYD-88) signaling. It is a pro-inflammatory pathway modulating the TRIF (Toll/ intraleukin -1 receptor (TIR) domain containing adapter inducing interferon beta) pathway and discussing oxidative stress and immune signaling mechanism. (Source: Kalpana et al. 2023)

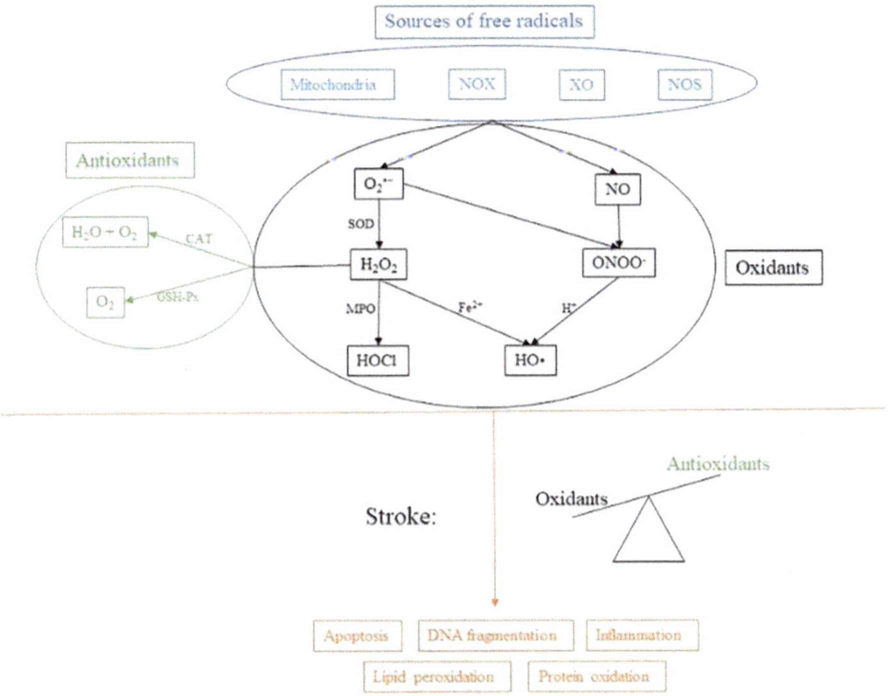


**Fig. 6.3** An oxidative stress pathway to reduce the number of free radicals from our body. Shown are the main sources through which free radicals are produced either as an end product or formed during chemical reactions or mitochondria resulting in DNA fragmentation, inflammation, and many more such conditions. (Source: Jelinek et al. 2021)

Normally, ROS are balanced out by antioxidants but in case of imbalance in antioxidative defense system, major depressive disorder (MDD) occurs (Jelinek et al. 2021). Ashwagandha plant contains flavonoid and phenolic compounds that act as a free radicle scavenger. Flavonoids and phenolic compounds react with free radicles of oxygen resulting in balancing out the oxidative stress. An antioxidative pathway for scavenging free radical is given in Fig. 6.3 (Alam et al. 2012).

### 6.13.4 Gamma-Amino Butyric Acid (GABA) Activity

GABA is a neurotransmitter that plays a key role in neurological disorders. There are two types of GABA receptors, that is, GABA-A, which is an inotropic receptor, and GABA-B, which is a metabotropic receptor. Imbalance in GABA-A activity or lower calcium influx can result in neurological problems. GABA-A is the primary site of action for ashwagandha regulating calcium influx or increasing GABA signaling. GABA and its receptors act as primary inhibitor of neurotransmitters. Its imbalance can cause neurological disorders, anxiety, as well as disturb sleep cycle. Ashwagandha possesses withanolide glycosides that have the ability to cross all the blood-brain

**Table 6.2** Major groups of phytochemicals, their relative activities, and also the part of *Withania somnifera* plant where they can be observed in higher amount

| Phytochemical | Plant part | Activity |
|---|---|---|
| Steroidal lactones | Root | Anti-inflammatory, anti-stress, anticancer, immune modulator |
| Alkaloids | Root | Cardioprotective, anti-inflammatory |
| Phytophenols | Leaves | Antioxidative, anti-inflammatory |
| Flavonoids | Leaves | Antioxidant, anticancer, antiviral |
| Glycosides | Leaves | Anti-stress, neurological disorders |

Source: Saleem et al. (2020)

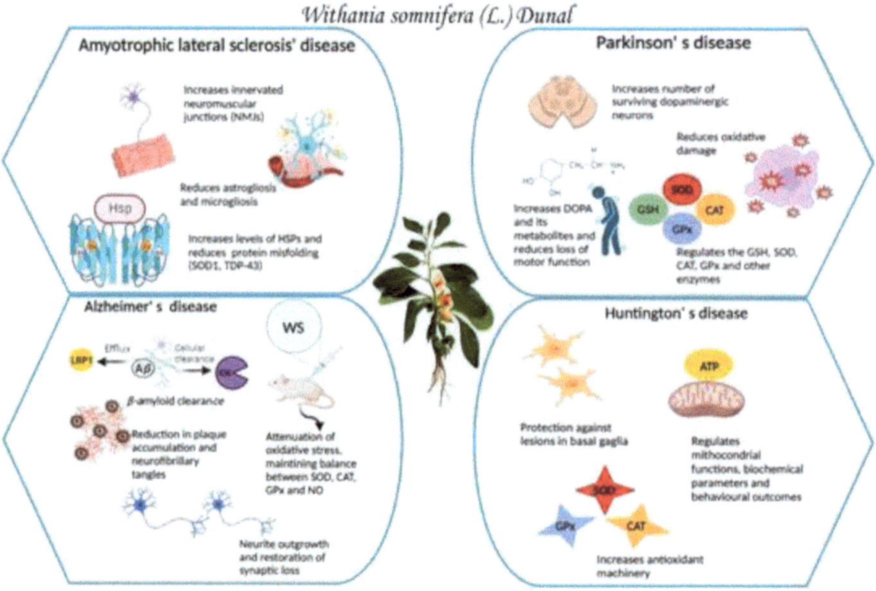


**Fig. 6.4** Role of *Withania somnifera* in neurological disorders that can occur due to brain disorders or any historical background. As ashwagandha consists of phytochemicals, it results in suppression of oxidative stress leading to cure various brain disorders. (Source: Lerose et al. 2024)

barriers and regulate GABA receptor and calcium influx, resulting in proper sleep pattern, low stress level, and prevention from several neurological disorders. Phytochemicals present in *Withania somnifera* are given in Table 6.2 with their location as well as activity, and their effects in neurological disorders are shown in Fig. 6.4.

## 6.14 Post-COVID-19 Effects

In 2019, during December month, many patients with symptoms of pneumonia were reported in Wuhan city of China. Later, this virus was identified as Coronavirus Disease-2019 or COVID-19. This virus spread globally creating a pandemic situation affecting many countries.

To tackle this problem, scientists looked back to Ayurvedic science as herbs are the treasure that can cure or prevent this situation. Later on, ashwagandha was found as a prominent herb for Coronavirus as according to many studies and researches ashwagandha possesses immune modulation and antiviral properties.

According to researches, ashwagandha upgrades the concentration and creation of immunoglobulin, Th1 and Th2 cytokines, B cells, and NK cells. In the second wave of COVID-19, people got affected by a modified form of Coronavirus also known as SARS-COV-2 (Severe Acute Respiratory Syndrome Coronavirus-2), having two areas of action, which is S-protein and Angiotensin converting enzyme 2 (ACE 2) receptor. Due to these actions of ashwagandha, all the patients suffering from COVID-19 developed a strong immune system to tackle Coronavirus (Verma et al. 2020).

## 6.15 Applications in Medical and Food System

*Withania somnifera* is a herb widely known for its traditional uses in medical system for curing various diseases as well as in food such as bakery products and beverages. A detailed aspect of its application in food system and medical system is as follows:

### 6.15.1 Medical System

Ashwagandha, an adaptogen belongs to Solanaceae family, consists of numerous medicinal properties. Mostly roots and leaves are used for making Ayurvedic medicines as well as its extracts are mixed with various drugs for allopathic treatment.

Leaf extracts of *Withania somnifera* are used in eye-related problems, insomnia, aging, and anxiety. Other attributes of ashwagandha include anti-stress, anti-inflammatory, antioxidative, antitumor, and immune-modulating activities. According to recent researches, it plays a major role in neurological disorders, Parkinson's disease, and liver-related disorders (Umadevi et al. 2012).

In females, it regulates menstrual cycle, hormonal imbalance, increases fertility, and maintains sexual health if added in regular diet. On the other hand, it also has some negative effects such as it should be avoided during pregnancy as it might raise the chances of miscarriage.

### 6.15.2 Food System

Nowadays, ashwagandha is used in various food items as an ingredient such as in juice and beverages, dairy products, sweets, as well as bakery items. It is present in the market as functional food. *Withania somnifera* is incorporated into food in powdered form with a range up to 10% to avoid over-ingestion. Recently, herbal cookies are also sold into market as functional food. These cookies are made with leaf powder of *Withania somnifera* so that its taste, color, and shelf-life is maintained. Another example can be taken as addition of *Withania somnifera* into dairy products such as ghee, as it contains antioxidative properties (Sharifirad et al. 2021).

## 6.16 Pharmaco-Commercial Applications

Various parts of ashwagandha are used for the treatment of several diseases such as diabetes, cardiac disorders, liver-related issues, fertility problems, and many more. Different parts of this plant are powdered and incorporated in drugs; for example, root powder is added in drugs for asthma, leukoderma, bronchitis, tuberculosis, liver- and heart related disorders, viral and bacterial diseases, and neurological disorders. If root extract is aqueous, chloroform in nature 3.75 mg is added in 1 mL, in methanolic extract 20 mg per mL is added, and so on. Leaves added in glacial acetic acid and toluene nature extract is added 7.5 mg per mL, and leaves in aqueous methanol nature extract is added 20 mg per mL. Some standard drugs having ashwagandha are gentamycin, chloramphenicol, tibirim, trimethoprim, streptomycin sulfate, bavistin, ketoconazole, and nysatatin (Saleem et al. 2020).

## 6.17 Future Perspective and Conclusion

From the above given data, we can conclude that there are several phytochemicals present in leaves and roots of ashwagandha. Consuming extracts of ashwagandha has been proven to have several pharmacological effects in an organism. Withanolide acts as precursor and performs anti-stress activities; besides, it also maintains the calcium influx and signaling in GABA receptors to avoid any type of neurological disorder and increase calmness, improves sleep, and decreases stress level. Ashwagandha also contains immune-modulatory activity that enhances immune system by increasing innate cells. Withaferin A is the key component responsible for this activity. Flavonoids and phenolic compounds present in ashwagandha act as free radical scavengers and perform antioxidative properties that regulate level of reactive oxygen species (ROS).

*Withania somnifera* is a multipurpose plant with lots of pharmaceutical and nutraceutical properties attracting researchers to gain in-depth knowledge of herbal and traditional medicinal plants. These herbal plants are also mixed in food products for increasing the nutritive value of food as well as added with drugs in a specific amount. Using these herbal plants in our regular diet has proven to be beneficial for human health.

## 6.18 Chapter Summary

Ashwagandha is a plant used in Ayurveda to cure various medical conditions. Its botanical name is *Withania somnifera* that belongs to Solanaceae family and it is a dicotyledon plant. According to research till date, it possesses up to 70 steroidal lactones, phytochemicals, and sitoindosides. Different parts of ashwagandha plant contain different chemicals that are biologically active, such as alkaloids, glycosides, withaferins, and sitoindosides, as well as their structures form lactone ring.

Due to the presence of flavonoids, root and leaf extracts are used to cure antimicrobial effects. Similarly, ashwagandha is also beneficial for inflammation whether acute or chronic. It is also useful in other medical conditions such as controlling stress level or arresting carcinogenic cells at Go, G2, or M phase, so that it cannot transform into a cancer. Ashwagandha also lowers blood glucose, increases insulin, and modulates spermatogenesis process as well as neurological disorders. Ashwagandha also regulates immune-related compounds such as immunoglobulins, macrophages, T-helper cells, and platelets. Ashwagandha has been identified as one of the most effective Ayurvedic plants during second wave of COVID-19 due to its crucial property of immune modulation.

## References


Alam N, Hossain M, Khalil MI, Moniruzzaman M, Sulaiman SA, Gan SH (2012) Recent advances in elucidating the biological properties of *Withania somnifera* and its potential role in health benefits. Phytochem Rev 11(1):97–112

Bhattacharya SK, Keshary PR, Sanyal AK (1978) Immobilization stress-induced antinociception in rats: possible role of serotonin and prostaglandins. Eur J Pharmacol 50:83

Chandrasekhar K, Kapoor J, Anishetty S (2012) A prospective, randomized double-blind, placebo controlled study of safety and efficacy of a high concentration full-spectrum extract of Ashwagandha root in reducing stress and anxiety in adults. Indian J Psychol Med 34:255–262

Cooley K, Szczurko O, Perri D et al (2009) Naturopathic care for anxiety: a randomized controlled trial. PLoS One 4:e6628

De Giorgio MR, Di Nioa S, Morciano C, Conte D (2020) The impact of SARS—CoV -2 on skeletal muscles. Acta Myol 39(4):307–312

Gambhir L, Checker R, Sharma D, Thoh M, Patil A, Degani M et al (2015) Thiol dependent NF-kB suppression and inhibition of T- cell mediated adaptive immuneresponses by a naturally occurring steroidal lactone Withaferin A. Toxicol Appl Pharmacol 289(2):297–312

Gautam M, Diwanay SS, Gairola S, Shinde YS, Jadhav SS, Patwardhan BK (2004) Immune response modulation to DPT vaccine by aqueous extract of *Withania somnifera* in experimental system. Int Immunopharmacol 4(6):841–849

Jain R, Kachhwaha and SL Kothari (2012) Phytochemistry, pharmacology, and biotechnology of Withania somnifera and Withania coagulans:A review. Journal of Medical Plants Research 6(41):5388–5399

Jelinek M, Jurajde M, Duris K (2021) Oxidative stress in brain: basic concepts and treatment strategies in stroke. Antioxidants (Basel) 10(12):1886

Kalpana K, Yap S, Tsuji M, Kawamura A (2023) Molecular mechanism behind the safe immunostimulatory effect of *Withania Somnifera*. Biomolecules 13(5):828

Kaur N, Niazi J, Bains R (2013) A review on pharmacological profile of *Withania somnifera* (Ashwagandha). Res Rev 2:6–14

Lerose V, Ponticelli M, Benedetto N, Carlucci V, Lela L, Tzvetkov NT, Milella L (2024) *Withania somnifera* (L.) Dunal, a potential source of phytochemicals for treating neurodegenerative diseases: a systematic review. Plants 13(6):771

Mahadi AA, Shukla KK, Ahmad MK, Singh R, Shankhwar SN, Singh V, Dalela D (2011) *Withania somnifera* improves semen quality in stress—related male fertility. Evid Based Complement Alternat Med 2011:576962

Rai M, Jogee PS, Agarkar G, Santos CAD (2016) Anticancer activities of *Withania Somnifera*: current research, formulations and future prospectives. Pharm Biol 54:189–197

Sharifirad J, Quispe C, Ayatollahi SA, Kobarfard F, Satniak M, Stepien A, Czopek K, Sen S, Acharya K, Matthews KR, Sener B, Devkota HP, Kirkin C, Ozcelik B, Victoriano M, Martorell M, Suleria HAR, Alshehri MM, Chandran D, Kumar M, Martins NC, Cho WC (2021) Article ID 8985179) Chemical composition, biological activity, and health—promoting effects of *Withania somnifera* for Pharma- Food Industry Applications. J Food Qual 1:14

Saleem S, Muhammad G, Hussain MA, Altaf M, Bukhri SNA (2020) *Withania somnifera* L.: insights into the phytochemical profile, therapeutic potential, clinical trials, and future prospective. Iran J Basic Med Sci 23(12):1501–1526

Sandhu JS, Shah B, Shenoy S, Chauhan S, Lavekar GS, Padhi MM (2010) Effects of *Withania somnifera* (Ashwagandha) and Terminalia arjuna (Arjuna) on physical performance and cardiorespiratory endurance in healthy youth adults. Int J Ayurveda Res 1(3):144–149

Singh R, Goel S, Bourgeade P, Aleya L, Tewari D (2021) Ayurveda Rasayana as antivirals and immunomodulators: potential applications in COVID-19. Environ Sci Pollut Res 28:55925–55951

Speers AB, Cabey KA, Soumyanath A, Wright KM (2021) Effects of withania somnifera (Ashwagandha) on stress and the stress- related neuropsychiatric disorders anxity, depression and insomnia. Curr Neuropharmacol 19(9):1468–1495

Sikandan A, Shinomiya T and Nagahara Y (2018) Ashwagandha root extract exerts anti - inflammatory effects in HaCaT cells by inhibiting the MAPK/ NF- kB pathways and by regulating cytokines. International journal of Molecular Medicine 42:425–434

Uddin Q, Samiulla L, Singh VK, Jamil SS (2012) Phytochemical and pharmacological profile of *Withania somnifera* Dunal: a review. J Appl Pharm Sci 02(01):170–175

Umadevi M, Rajeshwari R, Rahale CS, Selvavenkadesh S, Pushpa R, Kumar KPS, Bhowmik D (2012) Traditionaland medicinal uses of *Withania somnifera*. Pharma Innov 1(9):102–110

Verma V, Parwe S, Nisargandha M (2020) Systematic review of anti-viral and immunomodulatory potential of Ashwagandha (*Withania somnifera* Linn.) in light of covid-19 pandemic. Int J Bot Stud 5(5):494–497

Wang J, Zhang H, Kaul A, Li K, Priyandoko D, Kaul SC and Wadhwa R (2021) Effect of Ashwagandha Withanolides on muscle cell differentiation. Biomolecules 11(10):1454

# Medicinal and Nutritional Importance of *Bacopa monnieri* in Human Health

7

Mohd Hanzala Khan, Sabnam Farhat, Ruhi Firdaus, Mohd Ashraf Ashfaq, Md Salik Noorani, and Asghar Ali

**Abstract**

Popularly known as "Bramhi" in Ayurveda, meaning "knowledge deriving from gods," *Bacopa monnieri* is cultivated in India for its traditional and modern applications in medicine. It is the first choice as a memory enhancer and sharpener in Indian traditional medicine. Jaundice, diabetes, cough, leprosy, swelling, and diabetes are just a few of the many ailments for which *Bacopa* has been prescribed. This chapter is a modest overview of several aspects of the biology and biotechnology of *Bacopa monnieri.* It discusses its origin, distribution, morphology, and agronomy. Further, it outlines the various secondary metabolites found in it along with their biosynthetic pathways and discusses the different approaches that help in increasing its secondary metabolite pool. More importantly, central to the title of the chapter, it presents an unbiased view of its traditional and modern uses in medicine.

**Keywords**

Bramhi · Diabetes · Leprosy · Alzheimer's disease · Ayurveda

---

Mohd Hanzala Khan and Sabnam Farhat contributed equally to this chapter

M. H. Khan · S. Farhat · M. A. Ashfaq · M. S. Noorani · A. Ali (✉)
School of Chemical and Life Sciences, Jamia Hamdard, New Delhi, India
e-mail: asgharali@jamiahamdard.ac.in

R. Firdaus
School of Unani Medicine and Surgery, Jamia Hamdard, New Delhi, India



M. A. Ansari et al. (eds.), *Medicinal Plants and their Bioactive Compounds in Human Health: Volume 2*, https://doi.org/10.1007/978-981-95-3620-7_7

**Classification**

Kingdom—Plantae
Subkingdom—Viridiplantae
Infra kingdom—Streptophyta
Super division—Embryophyta
Division—Tracheophyta
Subdivision—Spermatophytina
Class—Magnoliopsida
Subclass—Asteridae
Superorder—Asteranae
Order—Lamiales
Family—Plantaginaceae
Genus—*Bacopa*
Species—*monnieri*

## 7.1 Introduction

*Bacopa monnieri*, also known as *Bacopa monniera*, *Herpestis monniera*, water hyssop, and commonly referred to as "Brahmi," has a long-standing history of use in Ayurvedic medicine. Traditionally, it has been valued as a cognitive enhancer, promoting memory, learning, and concentration (Mukherjee and Dey 1966), as well as providing therapeutic benefits for individuals suffering from anxiety and epilepsy (Chopra 1958). In addition to its neurological applications, *Bacopa* has been utilized in India and Pakistan as a cardiac tonic, digestive support, and to aid respiratory function, particularly in cases involving bronchoconstriction (Nadkarni 1988). Contemporary research has largely concentrated on its nootropic properties, affirming the traditional claims regarding improvements in memory and cognitive performance. Furthermore, studies have also corroborated *Bacopa*'s efficacy in managing anxiety, epilepsy, bronchitis, asthma, irritable bowel syndrome, and gastric ulcers, thereby supporting its traditional applications. Its notable antioxidant activity suggests potential protective effects against oxidative stress linked to cardiovascular diseases and certain cancers. *Bacopa monnieri* is a small, perennial, herbaceous plant classified under the family Scrophulariaceae. Renowned in Indian traditional medicine, particularly Ayurveda, it is widely recognized for its memory-enhancing properties. The plant's phytochemical profile includes several active constituents such as the alkaloids brahmine, nicotinine, and herpestine; the saponins bacosides A and B, and saponins A, B, and C; as well as triterpenoid saponins, stigmastanol, β-sitosterol, betulinic acid, D-mannitol, stigmasterol, α-alanine, aspartic acid, glutamic acid, serine, and pseudo-jujubogenin glycosides. *Bacopa* is a key ingredient in medhya rasayana, a category of Ayurvedic formulations aimed at enhancing mental function and intellect (Sivarajan and Balachandran 1994). Numerous studies have highlighted the therapeutic potential of *Bacopa*, particularly in the treatment and prevention of neurological disorders and in the enhancement of higher cognitive functions (Singh and Dhawan 1997; Vohora et al. 2000; Das et al. 2002; Russo

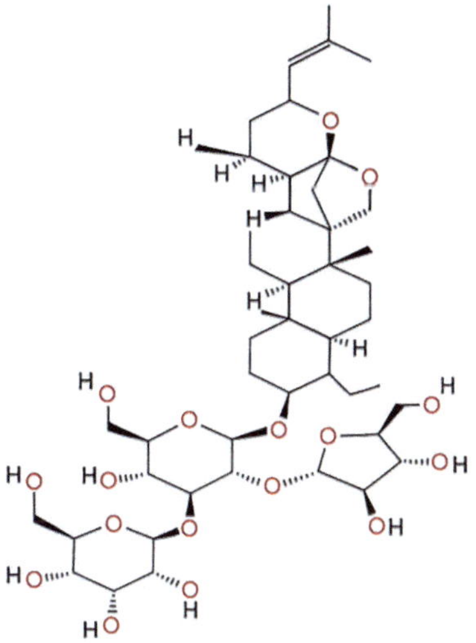

**Fig. 7.1** Chemical structure of *Bacoside* A. Chemical formula-C51H84022, Molecular mass-1111.22 g/mol

and Borrelli 2005). Subsequent investigations revealed that the cognitive-enhancing effects are primarily attributed to the presence of bacoside A (Fig. 7.1) and B, the principal active saponins extracted through ethanol (Singh and Dhawan 1992).

## 7.2 Habit, Habitat, and Morphology

*Bacopa monnieri* is characterized as a small, creeping herb distinguished by its numerous branches, small oblong leaves, and light purple blossoms. It thrives naturally in India and other tropical regions, particularly in moist soils, shallow water, and marshy environments. The plant is adaptable to a wide range of elevations, from sea level up to approximately 4400 feet, and can be readily cultivated under conditions of sufficient water supply. Flowering and fruiting occur during the summer months, and the entire plant is traditionally utilized for medicinal purposes (Stough et al. 2001). The root system consists of thin, wiry, creamish-yellow roots that are small and highly branched. The stems are slender, green or purplish-green in color, measuring approximately 1–2 mm in thickness, with prominent nodes and internodes; they are smooth to the touch and slightly bitter in taste. The flowers are small, solitary, and axillary, displaying five petals in shades of white, purple, pink, or pale violet, with pedicels ranging from 6 to 30 mm in length, and bracteoles shorter than the pedicels. The fruits are glabrous, ovoid capsules, measuring up to 5 mm in length, with a pointed apex (Trivedi et al. 2011). Nutritional analysis reveals that *Bacopa monnieri* contains approximately 5.9 grams of carbohydrates, 1.05 grams of crude fiber, and 1.9 grams of ash per standard measurement. It is also a notable source of minerals and vitamins, providing 202 mg of calcium, 16 mg of phosphorus, 63 mg of ascorbic acid (vitamin C), 0.3 mg of nicotinic acid, and 7.8 mg of iron, contributing an energy value of about 38 calories.

## 7.3 Anatomy

The transverse section of the stem reveals a sequential arrangement of the epidermis, cortex, and endodermis (Mondal et al. 2023). The hypodermis is composed of two to three layers of parenchymatous cells, which contain chlorophyll and tannin. Aerenchyma, developed from modified cortical cells, occupies more than two-thirds of the cross-section. The presence of tannin-containing cells at the junctions of adjacent cells imparts a characteristic Y-shaped appearance to the structure (Pandey 2024). The endodermis consists of a single cell layer, with vascular bundles arranged in a radial pattern. The xylem is well-developed, with the protoxylem oriented toward the pith and the metaxylem directed toward the pericycle. A large pith occupies the central region. In the transverse section of the leaf, the epidermis is prominently visible (Anju et al. 2017). The lamina consists of two types of tissues: palisade and spongy parenchyma. Beneath the epidermis, there are typically three to five layers of palisade cells. Both palisade and spongy tissues are composed of chlorenchymatous cells. A vascular bundle is present within the midrib region (Gubbannavar et al. 2013). Calcium oxalate crystals are observed in both the lamina and midrib. The leaf exhibits two stomatal types: diacytic and anisocytic. Furthermore, Pandey (2022) reported the chromosome number of the species as $2n = 64$.

## 7.4 Cultivation

### 7.4.1 Soil and Climate

Studies have indicated that *Bacopa monnieri* is capable of adapting to varying soil types and climatic conditions (Pankaj et al. 2020). It demonstrates optimal growth in subtropical regions, particularly in waterlogged areas and soils that maintain adequate nutrient levels. Cultivation is most effective during the summer or windy seasons, characterized by elevated temperatures (33–40 °C) and high humidity levels (65–80%) (Seeliger. 1992). Prior to planting, the field must be thoroughly ploughed and cleared of weeds to prepare a suitable environment. The soil should be well-saturated the day before transplantation to ensure proper hydration for the plant cuttings (Sharma et al. 2016). Propagation is typically achieved using cuttings approximately 4–5 cm in length, containing several leaves, stems, and roots, which serve as ideal starting materials (Vijayakumar et al. 2010). The mother plants are divided into smaller sections by carefully detaching the roots. Cuttings are transplanted into moist soil at a spacing of 40 cm × 40 cm, followed by the implementation of a flood irrigation system to maintain appropriate moisture levels (Alamgir 2017). For optimal cultivation, transplantation is recommended around June during the spring, allowing the plants to proliferate through the hot and wet conditions typical of summer until September, at which point harvesting can be initiated. To maintain the plants' long-term viability, it is advisable to harvest twice annually, typically in June and October following seasonal storms (Finn 1996).

### 7.4.2 Irrigation

The survival of transplanted plants is largely influenced by the effectiveness of their irrigation system following relocation (Lastiri-Hernández et al. 2023). Typically, fields are irrigated through periodic flooding, with water being applied approximately every 7–8 days. However, during periods of rainfall, supplementary irrigation is generally unnecessary (Sharma et al. 2012).

### 7.4.3 Harvesting/Post-harvesting

Following the harvest in October and November, both the biomass of the plant and the yield of bacosides begin to decrease (Bafort and Jijakli 2024). Through the practice of ratooning, the plant can be sustainably harvested by trimming approximately 4–5 cm of the stem above the base, allowing the remaining stems to regenerate for future harvests. Traditionally, the drying process involves spreading the harvested plants on the ground under cover, allowing them to air dry naturally at ambient temperature. However, to better preserve the bacoside-A content, a more effective method involves treating the material at 80 °C in an oven for 30 min immediately after harvesting (Topuz et al. 2023). After this initial treatment, further drying can be performed either by leaving the material to air dry at room temperature or by placing it in an oven maintained at 37 °C. Prior to storage, the dried plant material must be cleaned to eliminate any external impurities (Lastiri-Hernández et al. 2023). It is recommended to store the dried product in cool, dry conditions using large, flat-bottomed containers placed away from walls, and to take precautions against fungal growth and pest infestations.

### 7.4.4 Manures, Fertilizers, and Pesticides

The cultivation of medicinal plants should prioritize methods that exclude the use of chemical fertilizers and synthetic pesticides (Husain et al. 2023). Instead, depending on the specific requirements of the plant species, organic fertilizers such as Farmyard Manure (FYM), vermicompost, and green compost can be employed to enhance soil fertility. Additionally, natural substances like neem derivatives (oil, seeds, and leaves), Chitrakmool, Dhatura, and cow urine can serve as effective biopesticides, either individually or in combination, to safeguard plants against various diseases (Naseer et al. 2019).

### 7.4.5 Yield

Following the September harvest, Brahmi is capable of generating approximately 300 quintals per hectare of fresh biomass and around 60 quintals per hectare of dry biomass (Mondal et al. 2023). In terms of bacoside production, yields can reach up to 85 kilograms per hectare. Additionally, after the primary harvest, a secondary yield of about 40 quintals of dry biomass can be obtained in June, resulting in a total annual dry biomass yield of approximately 100 quintals (Lamo Judith Mary and John et al. 2023).

## 7.5 Medicinal Importance

Referred to as "Brahmi" in Ayurveda, meaning "knowledge bestowed by the gods," *Bacopa monnieri* is native to India (Pandey 2022). It holds a significant place in traditional Indian medicine, particularly valued for its role in enhancing and sharpening memory. *Bacopa* has been traditionally employed to treat a wide range of conditions, including jaundice, diabetes, leprosy, cough, swelling, and various skin and blood disorders (Jain et al. 2016). Its longstanding reputation as a memory booster and general tonic is well documented (Russo and Borrelli 2005). Ancient Ayurvedic texts, notably the *Charak Samhita* (circa 2500 B.C.) and the *Sushruta Samhita* (circa 2300 B.C.), attribute multiple therapeutic properties to *Bacopa monnieri*, such as detoxification, cognitive enhancement, anti-inflammatory effects, memory strengthening, and cooling of the body. Additionally, it is recognized for its blood-purifying effects and its ability to alleviate skin diseases and expel phlegm (Suresh et al. 2017). The therapeutic applications of *Bacopa monnieri* are also cited in the *Charak Samhita*, a foundational text of the Indian medical tradition. The herb has been utilized for treating conditions such as pharyngitis, constipation, and blood clot disorders. Furthermore, it offers potential benefits for individuals suffering from anxiety, cognitive impairments, and concentration difficulties (Rai et al. 2017).

### 7.5.1 Phytochemical and Bioactive Compounds

Phytochemical investigations have demonstrated the presence of a wide range of bioactive constituents in *Bacopa monnieri*. Significant quantities of salts, carbohydrates, phenolic compounds, and flavonoids were reported (Pawar et al. 2015). Specific findings included salt content at 2.03 ± 0.11 mg/gm, flavonoids at 10.54 ± 1.71 mg/gm, and both carbohydrate and protein levels measured at 150.63 ± 6.61 mg/gm (Pandey 2022). Further extraction and analysis of plant materials led to the identification of several bioactive compounds such as phlebotomine, terpenoids, saponins, taurine, and cardiac glycosides (Kannan et al. 2011). Twelve triterpenoids were isolated, comprising ten saponins and two glucosides, with jujubogenin and pseudojujubogenin derivatives identified as key components (Sandeep et al. 2020). *Bacopa monnieri* was also found to contain various sterol glycosides, including bacosterol, bacosine, Bacopaside-I, Bacopaside-II, and Bacopa saponin-C. Advanced analytical techniques such as infrared spectroscopy, 1D and 2D nuclear magnetic resonance (HMQC, HMBC, COSY), high-resolution electrospray ionization mass spectrometry (HR-ESI), QTOF-MS, and electron ionization (EI) were employed to elucidate their structures (P. Gupta et al. 2021). Using a specialized reversed-phase high-performance liquid chromatography (RP-HPLC) method, twelve *Bacopa* saponins were quantified in extracts of *Bacopa monnieri* (P. Gupta et al. 2021). Additionally, cucurbitacins A to E were isolated from plant parts by Negi et al. (2011), employing ESI-QTOF-MS/MS coupled with NMR techniques. Kulkarni et al. (2011) confirmed the presence of numerous phytochemicals, identifying approximately 52 constituents such as cinnamic acid, D-mannitol, Bacopa saponins A–D, and bacosides A, E, F, and G. Further studies identified a diverse chemical profile including compounds such as betulinic acid, cucurbitacins (A–E), stregaic acid, ascorbic acid, rosavin, asiatic acid, brahmic acid, oroxindin, llaliolide, wogonin, stigmasterol, β-sitosterol, and bacosine, as well as long-chain hydrocarbons like octacosane, nonacosane, heptacosane, triacontane, hentriacontane, and dotriacontane (H. Kumar et al. 2012). Bioactive flavonoids such as quercetin, apigenin, luteolin, ursolic acid, and asiaticoside were also detected. Saesong et al. (2019) noted that the floral structure, particularly three distinct petals labeled A3, VI, and VII, contributed to the identification of these compounds. Mondal et al. (2023) reported the successful isolation of 26 distinct chemical compounds from *Bacopa monnieri*, including vitamin E, phenols, folic acid, tridecane, octadec-9-enoic acid, and N-hexadecanoic acid. Other notable compounds such as hexadecanoic acid derivatives, 2,6,10-trimethyl, 14-ethylene-14-pentadecene methyl ester, (3-β-24R)-ergot-5-en-ol, and heneicosane were identified, alongside esters of cis-10-nonadecanoic acid.

### 7.5.2 Pharmacognostic and Physiochemical Properties

Powder analysis of *Bacopa monnieri* indicated the presence of starch, cuticle layers, and raphide-type calcium oxalate crystals, although stone cells were notably absent (Alamgir 2017). In a study comparing various extraction techniques, Phrompittayarat et al. (2011) found that macerating *Bacopa monnieri* in 95% methanol for 3 days resulted in the highest yield of Bacopa saponins, as confirmed by HPLC analysis. Furthermore, research by Silpa et al. (2021) demonstrated that drying fresh *Bacopa monnieri* herbage at 50 °C for 12 h in a cabinet dryer preserved the maximum Bacoside content.

### 7.5.3 Mechanisms of Action of Bioactives in *B. monnieri*

As *Bacopa monnieri* is primarily utilized for its cognitive-enhancing effects, the majority of research has concentrated on uncovering the mechanisms that account for these benefits. The triterpenoid saponins, particularly the bacosides, are considered key contributors to *Bacopa*'s ability to promote nerve impulse transmission. Bacosides support the repair of neuronal damage by stimulating kinase activity, encouraging neuronal regeneration, and restoring synaptic function, thereby enhancing (Table 7.1) neural communication (Singh and Dharwan 1997). A major pathological hallmark of Alzheimer's disease is the reduction in cholinergic neuronal

**Table 7.1** Pharmacological attributes of *Bacopa monnieri* and their mechanisms of action

| Sl. no. | Medicinal property | Mechanism of action | Reference |
|---|---|---|---|
| 1. | Memory-enhancing effect | Reduces acetylcholinesterase activity across the brain | Ahirwar et al. (2012) |
| 2. | Anti-inflammatory activity | Inhibits prostaglandin synthesis | Williams et al. (2014) |
| 3. | Antioxidant activity | Suppresses phospho-tau expression | Dubey et al. (2023) |
| 4. | Anticancer activity | Induces cell cycle arrest and apoptosis, inhibits colon cancer cell growth; Bacoside II downregulates aquaporin I expression in HT-29, SW480, SW620, and HCT116 cells | Nakhjavani et al. (2021) |
| 5. | Anticonvulsant activity | Brahmi ghrita showed greater effects during extension and recovery phases compared to control; significant reduction in seizures | Giramkar et al. (2013) |
| 6. | Antifungal activity | Complete inhibition of *Fusarium oxysporum*, *Sclerotium rolfsii*, *Alternaria* spp., and *Rhizoctonia solani* at 50–62.5μg/mL | Jain et al. (2017) |
| 7. | Antibacterial activity | *Bacopa* extract (10 mg/mL) inhibited a range of bacterial strains including *Staphylococcus aureus*, *B. subtilis*, *Pseudomonas aeruginosa*, *Vibrio cholerae*, and *Escherichia coli* | Pawar and Jadhav (2016) |
| 8. | Antiulcer activity | Ethanolic extract (400 mg/kg) demonstrated 82% ulcer protection and significantly reduced ulcer index | Jacob et al. (2021) |
| 9. | Antidepressant activity | Reduces immobility duration via serotonergic and noradrenergic pathways at 80 mg/kg | Girish et al. (2016) |
| 10. | Anti-Alzheimer's activity | *Bacopa* improved escape latency time in the Morris water maze test | Ganjewala et al. (2011) |
| 11. | Hepatoprotective activity | Normalizes elevated liver enzyme levels and improves antioxidant status in liver-injured rats | Kapoor et al. (2009) |
| 12. | Immunostimulatory effects | Increases serum levels of IgA:Immunoglobulin A, IgG:Immunoglobulin G, IgM:Immunoglobulin M, and IgE:Immunoglobulin E compared to control, Echinacea, and Withania groups | Yamada et al. (2011) |
| 13. | Antihyperglycemic activity | *Bacopa* extract lowered blood glucose by up to 49.21% in diabetic rats | Pandey (2022) |
| 14. | Antinociceptive activity | Extract (400 mg/kg) inhibited abdominal constriction more effectively than aspirin | Taznin et al. (2015) |
| 15. | Wound-healing activity | Promotes DNA synthesis and increases hydroxyproline levels in healing tissues | Murthy et al. (2013) |
| 16. | Adaptogenic activity | Significantly reduces plasma ALT levels in rats subjected to acute stress | Banerjee et al. (2014) |

(continued)

**Table 7.1** (continued)

| Sl. no. | Medicinal property | Mechanism of action | Reference |
|---|---|---|---|
| 17. | Antidiarrheal activity | Tannins in *Bacopa* contribute to its antidiarrheal effects | Hossain (2012) |
| 18. | Anti-*helicobacter* activity | 75% inhibition of *helicobacter pylori* at 1000μg/mL; enhances PGE: Prostagladin E and $PGI_2$: Prostagladin I2 levels by 40.8% | Al-Snafi (2019) |
| 19. | Antitoxic activity | Prevented increases in ALT: Alanine Aminotransferase, AST: Aspartate Aminotransferase, and creatinine levels compared to standard treatment in opioid-administered rats | Sumathi and Nongbri (2008) |
| 20. | Antihypertensive and antiallergic activities | Reduces blood pressure via nitric oxide release and stabilizes mast cells similarly to disodium cromoglycate | Kamkaew et al. (2011) and Shende et al. (2020) |
| 21. | Anti-parkinsonian effect | Decreases alpha-synuclein, pro-inflammatory cytokines, and ROS in male Wistar rats | Neto et al. (2024) |
| 22. | Antifertility activity | Reduces sperm motility, viability, and morphology, along with changes in. seminiferous tubule structure in mice | Patel et al. (2017) |
| 23. | Antiaging activity | Protects human astrocytes from benzo[a] pyrene-induced apoptosis and senescence; inhibits ROS production and cell cycle arrest | Saha et al. (2020) |
| 24. | Antiemetic activity | Methanolic and butanolic fractions demonstrated antiemetic effects in *Suncus murinus* models of chemotherapy-induced emesis | Nandy et al. (2019) |

function within the hippocampus (Enz et al. 1993). Evidence from animal studies suggests that bacosides possess antioxidant properties, particularly in the hippocampus, frontal cortex, and striatum (Bhattacharya et al. 2000). Furthermore, experimental research in animals indicates that *Bacopa* extracts can influence the activity of enzymes involved in both the production and elimination of reactive oxygen species (ROS) in the brain (Chowdhuri et al. 2002). In vitro studies have also demonstrated that *Bacopa* provides protective effects against DNA damage in astrocytes (Russo et al. 2003) as well as in human fibroblast cells (Russo et al. 2003). In animal studies, *Bacopa monnieri* has been shown to exert a relaxant effect on the pulmonary arteries, aorta, trachea, as well as ileal and bronchial tissues, potentially through the inhibition of calcium ion influx across cell membranes (Channa et al. 2003). In vitro findings suggest that *Bacopa* may contribute to mast cell stabilization (Samiulla et al. 2001) and exhibit anti-inflammatory properties, likely through the suppression of prostaglandin synthesis and the stabilization of lysosomal membranes (Jain et al. 1994). Additionally, in vitro research indicates that *Bacopa* extracts may possess anticancer activity, possibly by interfering with DNA replication in cancer cell lines (Elangovan et al. 1995).

### 7.5.4 Memory Enhancement

Experimental investigations in animal models have indicated that *Bacopa monnieri* enhances motor learning, supports the acquisition and retention of newly learned behaviors, and delays the extinction of these behaviors (Singh and Dharwan 1997). Furthermore, the methanolic extract and various fractions of *B. monnieri* were examined for their potential antidepressant effects using the forced swimming test (FST) and tail suspension test (TST) in mice. The findings indicated that the methanolic extract, as well as the ethanol and butanol fractions, significantly reduced immobility in both the forced swimming test (FST) and the tail suspension test (TST) in mice following oral administration for five consecutive days. Importantly, at the effective doses used for the FST and TST, none of the tested samples demonstrated any inhibitory effect on locomotor activity (Shen et al. 2009). Additionally, it was observed that bacosides enhanced anterograde memory and mitigated anterograde amnesia induced experimentally by scopolamine and sodium nitrite, likely through mechanisms involving the improvement of acetylcholine levels and alleviation of hypoxic conditions, respectively. Moreover, bacosides were found to reverse BN52021-induced retrograde amnesia, an effect that may be attributed to increased synthesis of platelet-activating factor through the enhancement of cerebral glutamate levels (Choudhuri 1996). *Bacopa* treatment was found to reverse memory impairments caused by scopolamine-induced cholinergic blockade. Furthermore, *Bacopa* enhanced memory performance even in cognitively healthy individuals, while Pycnogenol specifically improved working memory (Ryan et al. 2008). Benzodiazepines, which are known to induce amnesia through the activation of the GABAergic system and disruption of long-term potentiation, were also counteracted by *Bacopa monnieri*, as behavioral studies demonstrated a significant reversal of diazepam-induced amnesia (Saraf et al. 2008). Additionally, *Bacopa* administration alongside phenytoin was shown to mitigate phenytoin-induced cognitive deficits, resulting in notable improvements in both memory acquisition and retention (Vohora et al. 2000). A clinical investigation evaluating the effects of 12 weeks of *Bacopa monnieri* supplementation (300 mg/day) in individuals over 55 years of age reported significant enhancements in memory acquisition and retention (Morgan and Stevens 2010). Consistent cognitive enhancement has also been documented with long-term use of *Bacopa* extracts. In a 12-week, double-blind, placebo-controlled study employing identical selection criteria and dosing (300 mg/day of an extract containing 55% combined bacosides), 46 healthy participants aged 18 to 60 years were randomly assigned to either a treatment or a placebo group. In the study, the same battery of assessments used in the acute dosage trial was administered at baseline, and subsequently at 5 and 12 weeks following the commencement of treatment. By the conclusion of the 12-week period, participants in the treatment group exhibited significant improvements in verbal learning, memory consolidation, and the speed of early information processing compared to the placebo group. These cognitive enhancements were not evident at baseline or at the 5-week evaluation (Stough et al. 2001). The *Bacopa* supplement utilized in the study was commercially available under the name KeenMind™ (Flordis), a formulation derived from the stems, leaves, and roots of *Bacopa monnieri* and extracted using 50%

ethanol. It was standardized to contain 55% ± 5% active bacosides. KeenMind™ is considered beneficial for developing new preventive health strategies and nutritional/pharmacological interventions aimed at promoting cognitive and brain health in older adults. *Bacopa* is believed to exert its cognitive benefits through multiple mechanisms, including (i) pro-cholinergic activity; (ii) antioxidant effects via flavonoids; (iii) metal chelation; (iv) anti-inflammatory properties; (v) enhancement of cerebral blood flow; (vi) adaptogenic activity; and (vii) the clearance of β-amyloid deposits (Stough et al. 2012). However, in a separate double-blind, randomized, placebo-controlled trial involving 76 adults aged 40–65 years, *Bacopa monnieri* supplementation did not affect the rate of learning across various memory tasks. Instead, it was suggested that *Bacopa* primarily reduced the rate of forgetting newly acquired information. No significant effects were observed on attention, verbal and visual short-term memory, retrieval of previously learned information, or self-reported measures of everyday memory function and anxiety levels (Roodenrys et al. 2002).

### 7.5.5 Side Effect and Cytotoxic Activity

The therapeutic efficacy of *Bacopa monnieri* has been enhanced through the development of formulations with enriched phytochemical profiles (Kumar et al. 2016). Sub-chronic toxicity evaluations in Sprague Dawley rats demonstrated no significant changes in neurological status, dietary intake, body weight gain, hematological markers, or biochemical parameters, and no toxic effects were detected (Joshua Allan et al. 2007). The median lethal dose ($LD_{50}$) was established at 2400 mg/kg body weight. Additional studies assessing both acute and chronic toxicity in the same rat model were undertaken (Akbar 2020). In the acute toxicity phase, a single oral dose of 5000 mg/kg was administered to female rats and monitored over 7 days, during which no health abnormalities or behavioral deviations were observed. The chronic toxicity study involved daily oral administration of varying doses (30, 60, 300, and 1500 mg/kg) to both male and female rats over a 270-day period (Sireeratawong et al. 2016). Parameters such as body and organ weights, hematological data, histological structures, and clinical chemistry values were examined, with no evidence of toxicity reported. Collectively, these findings suggest that *Bacopa monnieri* extract is well-tolerated in rodents under both acute and long-term exposure conditions. Furthermore, the extract's cytotoxic potential was assessed using a brine shrimp lethality bioassay, providing additional safety validation (Abbas and Mohani 2011).

## 7.6 Nutritional Importance of *Bacopa monnieri*

Brahmi has several health advantages and is beneficial in all sections. Its leaves contain strong alkaloids and triterpene saponins that stimulate brain activity and improve memory, learning, and mental clarity. Understanding this herb's bioactive components, immune-boosting capabilities, and formulation safety is essential for its optimal use. Research demonstrates that iron, magnesium, calcium, and potassium are among the vital micronutrients found in *Bacopa monnieri* leaves, along with vitamins A, C, and B (Patel et al. 2018). These nutrients are essential for immune system support, energy metabolism enhancement, and general health maintenance. Brahmi extracts can be added to vitamin C-rich herbal drinks. These drinks are healthful and enjoyable without sweets, artificial colors, or flavors (Gouthami 2024). Brahmi and other Ayurvedic herbs can improve health in underdeveloped areas lacking access to good healthcare (Table 7.2). Due to its nutritional properties, *Bacopa monnieri* is often consumed as part of dietary supplements. Its extracts are found in various forms, including capsules, tablets, and powders, contributing to its widespread use for cognitive and general health benefits (Madda et al. 2016).

## 7.7 Secondary Metabolite Biosynthetic Pathway in *Bacopa monnieri*

The synthesis of bioactive secondary metabolites in *Bacopa monnieri,* similar to numerous therapeutic plants, commences with primary metabolism, which is founded on fundamental cellular metabolism. Primary metabolism includes essential metabolic activities like glycolysis, the tricarboxylic acid cycle (TCA) and the pentose phosphate pathway (Bharti et al. 2013). The formation of secondary metabolites begins with photosynthesis, which supplies the primary carbon chains. The fixed carbon integrates into primary carbon metabolism, yielding important intermediates like erythrose-4-phosphate, phosphoenolpyruvate, pyruvate, and 3-phosphoglycerate. Aromatic amino acids are produced by the shikimic acid pathway using erythrose-4-phosphate and phosphoenolpyruvate (Fig. 7.2) (Sharma et al. 2017). These amino acids serve as building blocks for secondary metabolites that contain nitrogen. The tricarboxylic acid cycle (TCA) produces nitrogenous substances by generating aliphatic amino acids. The mevalonic acid (MVA) and methylerythritol phosphate (MEP) routes, which jointly generate terpenoid chemicals, including significant saponins like bacosides, depend on acetyl-CoA, which is produced from pyruvate. Additionally, melonic acid pathways derived from acetyl-CoA promote the synthesis of phenolic compounds (Kumari et al. 2023), another type of secondary metabolite. Following primary carbon metabolism in *Bacopa monnieri*, secondary biosynthetic pathways are initiated, resulting in the synthesis of essential chemicals. The mevalonic acid and MEP pathways employ intermediates such as acetyl-CoA to synthesize terpenoid chemicals, notably bacosides, which are the principal bioactive secondary metabolites of the plant (Vishwakarma et al. 2015).

**Table 7.2** A summary of the several products that are available globally, which have Brahmi as an ingredient (Patel et al. 2018)

| Sl. no | Product | Origin | Constituent compounds | Mechanism of action |
|---|---|---|---|---|
| 1. | Verjinia Wondy Custard Powder | India | Edible starch, edible common salt, permitted food colors, artificial flavors, Brahmi extract | It is well known for improving concentration and memory. The 160 g container of this plant-based food comes in two separate flavor packs, each weighing 80 g. |
| 2. | Paradise Gold Premium Brahmi Almond Syrup | India | Brahmi booti, rose water, giribadam, istakhadus- booti, cardamom, shankpushpi, sugar, skimmed milk, water, citric acid, permitted synthetic food colors (19,140, 42,090), artificial flavor, natural flavor, nature identical flavor, permitted class II | Puddings, cereals, lassi, ice cream, and dessert toppings can all benefit from its natural sweetening properties. There is a 700 mL recyclable bottle available for this plant-based product |
| 3. | Patanjali Brahmi Sharbat | India | Brahmi extract *Bacopa monnieri*, sugar, jala nimbu sat, preservative, sugandhit, dravya color | |
| 4. | Amrapali Mums Breakfast cereals | India | Corn, sugar, salt, Brahmi, malt extract, baking powder, sodium bicarbonate, niacin | The 500 g carton of Amrapali mum's corn flakes has been redesigned. Marketed under the "memory plus" idea, this cereal is entirely vegetarian and enhanced with natural herbs that are believed to improve memory. Iron, vitamin A, vitamin B, and vitamin C are also abundant in these corn flakes. |
| 5. | B-Natural Brahmi Ready to Serve Fruit Beverage | India | Water (69.4%), Brahmi leaf extract (16%), deionized apple concentrate (14.5%), citric acid (1%) | Brahmi has been shown to assist the physiological process underlying calm and improve mental clarity. Its active components help reduce tension and promote mental calm. |
| 6. | Dabur Chyawanprash | India | Dashmool, bala, mudgaparni, mashaparni, karkatshring, tamalaki, draksha, jivanti, pushkara, haritaki, guduchi, karchura, musta, punarnava, utpala, vasa, yashti, kakrasika, varahi, ashwagandha | The product enhances growing children's mental and physical health and boosts immunity. It helps keep young people healthy by protecting them from common illnesses like colds and coughs. It also gives active older folks antioxidants, which boost their immune systems. |

(continued)

**Table 7.2** (continued)

| Sl. no | Product | Origin | Constituent compounds | Mechanism of action |
|---|---|---|---|---|
| 7. | Haritage Almond Syrup | India | Brahmi booti (*Bacopa* monnieri), rose water, giribadam (almond clusters), istakhadus-booti, cardamom, shankphushpi (convolvulus pluricalis), sugar, skimmed milk, water, citric acid, color (19,140, 42,090) | This concentrate can be used to make a beverage with an almond flavor or to make puddings, cereals, lassi, ice creams, and toppings sweeter. |
| 8. | Natura Tammy Raw Mango Flavoured Jelly | India | Aloe vera (10 mg) (aloe barbadensis), Brahmi (5 mg), nimbu satva (32 mg), shankhpushpi (3 mg), stabilizer, preservative (sodium benzoate), color (pea, green supra FCF), added flavor fructooligosaccharide (dietary fiber) | The 4 g pack of the ayurvedic medication is good for vegetarians. |
| 9. | Sandali Brahmi Badam Sharbat | India | Sugar, water, arq (kewra, kashni, anjbar, rose, khus), almonds, shankhpushpi extract, sandalwood oil, Brahmi booti, citric acid (E330), permitted synthetic food color, added flavor, permitted class II preservative (E224) | It is believed to carry the rich scent of the Mughal region into the summer. To prepare, combine one part of the vegetarian food with three parts milk or cooled water. Sugar doesn't need to be added. |
| 10. | Panchwati Health Prash | India | Brahmi, shankpushpi, arjuna, pipali, amla, ashwagandha, honey | It is said to be perfect for grownups and is advised for students. The product comes in a jar weighing one kilogram. |
| 11. | Amrita Drugs Brahmi Powder | India | Brahmi, sankhapushpi, ashwagandha, shatavari, jatamansi, gokshur, sugar | Brahmi powder from amrita drugs is said to improve memory. In addition to being a great tonic, the ayurvedic medication is supposed to enhance memory and focus, alleviate issues related to excessive stress in intellectuals, professionals, students, and the elders. |

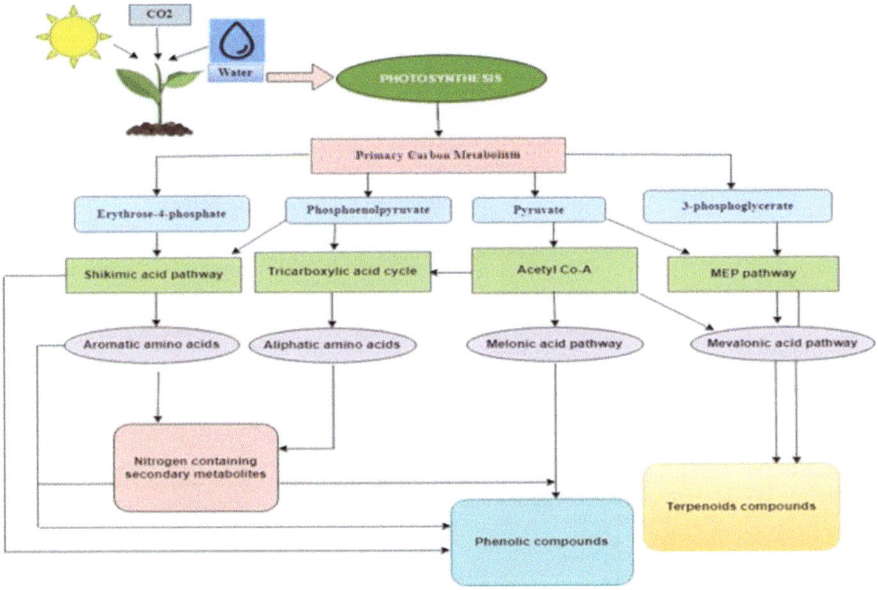


**Fig. 7.2** Photosynthesis initiates primary metabolism, playing a crucial role in the synthesis of secondary metabolites, which encompass nitrogen-containing compounds, phenolic compounds, and terpenoids

### 7.7.1 Biosynthesis of Triterpenoid Saponins via MVA and MEP Pathway

*Bacopa monnieri* contains highly bioactive compounds that are triterpenoid saponins. These are the main compounds that give the plant its neuroprotective and adaptogenic qualities (Kumari et al. 2023). The biosynthesis of secondary metabolites, particularly triterpenoids, involves two major isoprenoid pathways: the mevalonate (MVA) pathway in the cytoplasm and the methylerythritol phosphate (MEP) pathway in the chloroplast. The MVA pathway begins with acetyl-CoA, leading to the formation of mevalonic acid through HMG-CoA and the enzyme HMG-CoA reductase (Kumari et al. 2023). Both pathways produce the universal five-carbon precursors, isopentenyl pyrophosphate (IPP), and dimethylallyl pyrophosphate (DMAPP). This is further converted to isopentenyl pyrophosphate (IPP, C5). Concurrently, the MEP pathway utilizes glyceraldehyde-3-phosphate and pyruvate to yield IPP and dimethylallyl pyrophosphate (DMAPP). These C5 units serve as building blocks for higher isoprenoids. Sequential condensation of these units produces geranyl pyrophosphate (GPP, C10), farnesyl pyrophosphate (FPP, C15), and geranylgeranyl pyrophosphate (GGPP, C20). FPP is a key intermediate for the formation of squalene (C30) via presqualene pyrophosphate (PSPP) (Vishwakarma et al. 2015). Squalene is further oxidized to 2,3-oxidosqualene, which undergoes cyclization by oxidosqualene cyclases (OSCS) to form the triterpenoid skeleton (Fig. 7.3). This skeleton is modified by cytochrome P450 monooxygenases (CYP450s) and glycosyltransferases (GTs) to produce diverse secondary

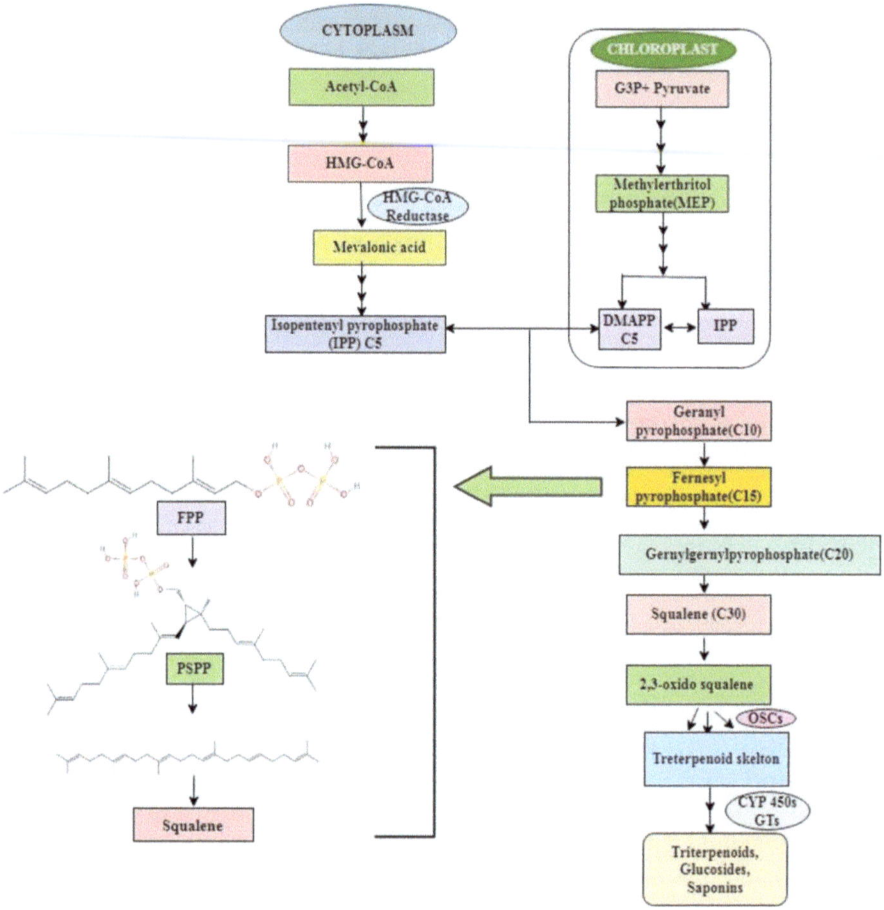


**Fig. 7.3** Illustrative depiction of the triterpenoid saponin biosynthesis through the (MVA) mevalonate and (MEP) methyl-D-erythritol 4-phosphate pathways is presented schematically. Key intermediates in this process include (IPP) isopentenyl diphosphate, dimethylallyl diphosphate (DMAPP) dimethylallyl diphosphate, geranyl pyrophosphate (GPP), and farnesyl pyrophosphate (FPP). Enzymes such as oxidosqualene cyclases (OSCS), cytochrome P450S (CYP450S), and glycosyltransferases (GTS) play pivotal roles in these biochemical transformations. (C5 represents carbon number)

metabolites, including triterpenoids, glucosides, and saponins (Fig. 7.3), which are pharmacologically significant in *Bacopa monnieri*.

## 7.8 Metabolic Pathway Engineering for the Enhancement of Bacoside

The main source of bacosides is *Bacopa monnieri*, which are triterpenoid saponins of the dammarane type. Most recent research has concentrated on enhancing the production of these secondary metabolites through various methods to alter

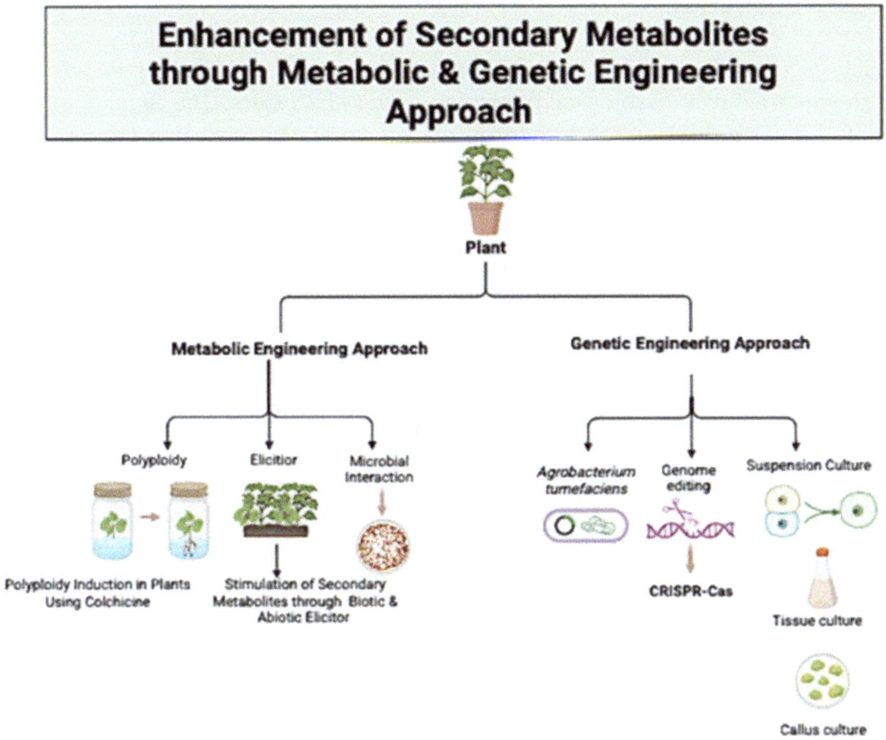


**Fig. 7.4** Metabolic and genetic engineering strategies for enhancing plant secondary metabolites

metabolic pathways and incorporate genes involved in biosynthesis, and through genetic engineering (Jeena et al. 2025). The mevalonate (MVA) and methylerythritol phosphate (MEP) pathways, which provide squalene and thus triterpenoid saponins, are the pathways by which bacosides are made. Key enzymes such as 3-hydroxy-3-methylglutaryl-CoA reductase (HMGR) and squalene synthase (SQS) are significant players in this process. The methods for increasing bacoside production are presented in Fig. 7.4 (Abbassi et al. 2015).

### 7.8.1 Polyploidization

High-bacoside content in *Bacopa monnieri* is rare, limiting its usage in medicine. To increase growth and bacoside levels, researchers doubled the plant's chromosomes with colchicine. A day or two after treating nodal segments with varied colchicine concentrations, they moved them to shoot-inducing medium (Manzoor et al. 2019). Chromosome doubling worked effectively at greater doses and extended exposure. Tetraploid plants (doubled chromosomes) grew and produced more bacoside than diploid ones (Inthima and Sujipuli 2019). They are promising for better pharmacological and medical purposes.

### 7.8.2 Elicitation

Elicitors are utilized to increase *B. monnieri's* synthesis of secondary metabolites with important pharmacological properties, such as bacosides. The elicitation technique mimics stress signals and initiates plant defense pathways; they can be either biological or abiotic. Non-biological elicitors include salts of inorganic chemicals, Cu and Cd ions, Ca2+, and high pH conditions. Biotic elicitors are biological elements that comprise polysaccharides (such as pectin, cellulose, chitin, glucans), glycoproteins, G-proteins, and intracellular proteins (Bhaskar et al. 2022). Yeast extract, salicylic acid, and methyl jasmonate are all effective elicitors for *B. monnieri* (Sanyal et al. 2022). By the in vitro method, they are usually introduced to liquid cultures or agar media; To avoid toxicity and ensure the best possible metabolite synthesis, timing and concentration are crucial (Otari et al. 2023).

### 7.8.3 Microbial Interaction

Plants inoculated with beneficial microorganisms like *Pseudomonas monteilli, Cedecea davisae, Cronobacter dublinensis*, and *Pseudomonas aeruginosa* showed an increase in the expression of genes involved in the bacoside biosynthesis pathway when *Meloidogyne incognita* was present or absent (Gupta et al. 2017). Other microorganisms like *Chitiniphilus* and *Streptomyces* can modulate plant metabolism and immunity and boost defense and secondary metabolism. The study found that these microorganisms increase bacoside-A production and *Meloidogyne incognita* resistance separately and together. Under nematode stress, the combination of microbial therapy upregulated critical bacoside biosynthesis genes (Otari et al. 2023). In addition to increased bacoside content, treated plants had higher chlorophyll, defensive enzymes, phenolics (gallic, syringic, ferulic, and cinnamic acids), and physical barriers like lignin and callose. Through nitrogen fixation and phosphate solubilization, microorganisms enhance nutrient absorption; they also strengthen *Bacopa monnieri* defense system and metabolic pathways; and they produce phytohormones such as auxins that promote root development, plant growth, and increased biomass and metabolite production. (Gupta et al. 2017).

### 7.8.4 Genetic Engineering Approaches for the Enhancement of Bacoside

#### 7.8.4.1 In Vitro Plant Tissue Culture

*Bacopa monnieri*, which has just a small amount of genetic variety, is generally renowned for the memory-enhancing characteristics it possesses. The species is in danger of going extinct as a result of the extreme natural harvesting that occurs in addition to the great commercial demand for it. This can be attributed to the plant's restricted capacity for seed propagation, which can be attributed to low seed viability and limited availability. The majority of this is the result of harvesting practices that are not sustainable (Tiwari et al. 2001). Because the demand for *B. monnieri* is expected to continue to grow on a global scale, it is of the utmost importance to devise methods that are both sustainable and capable of facilitating the widespread proliferation of this plant To undertaking its preservation and fulfill the ever-increasing need for bacoside production, two primary methods can be utilized: (a) the traditional method of propagation, which involves the use of seeds or cuttings, and (b) the in vitro planting tissue culture techniques. There is a viable way for producing elite plants that involves the use of tissue culture techniques ("An Efficient Micropropagation Protocol of Bacopa monnieri (L.) Pennell through Two-Stage Culture of Nodal Segments and Ex Vitro Acclimatization," 2015). These techniques include callus induction, somatic embryogenesis, cell suspension, and organogenesis. Because they enable regulated conditions and consistent yields, these approaches are perfect for increasing the synthesis of valuable secondary metabolites, as they allow for the creation of metabolites. In one study, friable green callus was successfully produced from leaf explants by employing a medium that contained 0.5 mg/L kinetin, 1 mg/L naphthaleneacetic acid (NAA), 1 mg/L casein hydrolysate, and 30 g/L sucrose. Each of these components was present in the medium. During the process of transferring this callus to a liquid media in full darkness and maintaining the same composition, the formation of bacoside-A occurred (Kishore Mohanta and Sahoo 2014).

#### 7.8.4.2 In Vitro Cell Suspension Culture/Callus Culture

Researchers have employed cell suspension and callus culture to increase *Bacopa monnieri* bacoside synthesis in vitro. These approaches are suitable for large-scale plant preservation and propagation (Bhaskar et al. 2022). Genetic transformation has been used to insert biosynthetic genes to increase the metabolic pathway synthesis of important chemicals. Genetic transformation, suspension culture, micropropagation, agrobacterium-mediated transformation, biotransformation, and molecular marker-based approaches have been used to increase bacoside synthesis (Ramesh et al. 2011) (Sharma et al. 2017).

#### 7.8.4.3 Suspension Culture

In recent years, several strategies have been employed for the production and biomass accumulation in *Bacopa monnieri*. In comparison to tissue and organ cultures, cell suspension culture techniques provide a faster and more efficient approach for industrial applications and significant secondary metabolite synthesis. This approach provides a consistent source of natural goods. Numerous efforts have been dedicated to achieving commercial-scale production through suspension cultures (Bansal et al. 2017). The application of plant growth regulators to increase the bacoside content in *Bacopa monnieri*, cell suspension culture, was utilized. In the beginning, callus was induced from leaf explants by administering MS media containing 2,4-D. For the suspension cultures, the MS liquid medium was enhanced with a constant quantity of α-naphthalene acetic acid and casein hydrolysate. Additionally, different concentrations of benzyladenine, kinetin, thidiazuron, and chitosan were added to the media (Jauhari et al. 2019). The formation of bacoside was greatly boosted by decreased quantities of benzyladenine, kinetin, and thidiazuron, among. In particular, thidiazuron was the most successful in encouraging the accumulation of bacosides, although chitosan demonstrated a discernible influence on the levels of bacosides in the culture (Bansal et al. 2017).

#### 7.8.4.4 Tissue Culture Studies

*B. monnieri* cell suspension cultures were successfully created on MS medium that was enhanced with 0.5 mg/L NAA and 0.1 mg/L BAP. Inoculum size, sucrose, and $KH_2PO_4$ concentrations were found to be the main determinants of biomass yield using Plackett–Burman analysis (Leonard et al. 2018). In the absence of exogenous plant growth regulators (PGRs), direct shoot organogenesis has been effectively triggered utilizing leaf and internodal explants. However, the presence of cytokinins like kinetin and BAP (6-benzylaminopurine), frequently in conjunction with auxins like NAA (naphthaleneacetic acid), greatly increased the efficiency of organogenesis and shoot multiplication (M. Sharma et al. 2015).

#### 7.8.4.5 Micropropagation

Distinct phytoconstituents in Brahmi, such as alkaloids (brahmine, nicotinine, and herpestine), as well as saponins (bacosides). These phytoconstituents have several kinds of therapeutic values. In vitro micropropagation is an advantageous method for producing various secondary metabolites, which are often obtained from the apical meristem or other actively dividing tissues and are sterilized to eliminate any contaminants (Antony Ceasar et al. 2010). The sterile explants are then put on a nutritional medium that contains specialized plant growth regulators. Following multiplication, the explants begin to grow and develop into tiny plantlets in the nutrient medium. The growth regulators in the media are modified to encourage fast cell growth. The elongated shoots are transferred to a medium containing rooting hormones to induce the development of roots (Haque et al. 2017). Once the roots are formed, the plantlets can be transferred to soil or another suitable growing medium; they are gradually acclimated to the external environment by exposing them to ambient conditions (Hamdeni et al. 2022).

Micropropagation techniques utilizing axillary meristems and de novo organogenesis have been extensively explored to address propagation limitations. In vitro strategies, including organogenesis, callus induction, and cell suspension cultures, have been employed not only for the propagation of the plant but also for the enhancement of pharmacologically important secondary metabolites (Sanyal et al. 2022).

#### 7.8.4.6 Callus Culture

Callus, an organization of undifferentiated plant cells, plays a vital role in plant tissue culture, with its development significantly influenced by the selection of explant and the hormonal conditions present (Saha et al. 2020). In *Bacopa monnieri*, callus induction was evaluated using both leaf and stem segments on MS medium supplemented with various auxins. Leaf explants exhibited a more pronounced response compared to stem segments, with optimal outcomes observed when 2,4-D was administered, particularly at a concentration of 3 mg/L. A reduction in callus induction was observed with the application of IBA and IAA (Zote et al. 2018). Notably, the combination of 2,4-D and BAP facilitated significant callus proliferation. The results obtained are consistent with previous research indicating that leaf tissues exhibit heightened responsiveness attributable to their physiological and hormonal characteristics (Ahuja et al. 2021).

#### 7.8.4.7 Agrobacterium-Mediated Transformation

Agrobacterium-mediated genetic transformation is an important biotechnological technique that has been effectively used to improve *Bacopa monnieri's* genetic profile, especially in terms of increasing bacoside production and adding new characteristics. Through steady genetic integration, this technique makes use of *Agrobacterium tumefaciens* natural ability to transfer a portion of its DNA (T-DNA) into the genome of host plants. This approach has been improved in the instance of *B. monnieri* to enrich metabolites, increase regeneration efficiency, and improve gene delivery (Yadav et al. 2014). Nodal, internodal, and leaf disc segments with high regeneration potential are commonly used. These explants are pre-cultured on Murashige and Skoog (MS) medium supplemented with cytokinins like 6-benzylaminopurine (BAP) and auxins such as indole-3-acetic acid (IAA) or naphthalene acetic acid (NAA) to promote shoot initiation and enhance transformation efficiency (Nisha et al. 2003). One phase in the transformation process is inoculation with target gene-carrying strains of *Agrobacterium* (EHA105/LBA4404). Co-cultivation on dark media for 2–3 days, to get rid of bacteria, uses antibiotics (cefotaxime/ticarcillin). Following selection and regeneration on kanamycin and hygromycin-containing medium, rooting and acclimation occur. This method sustainably produces high-bacoside lines and makes functional research easier. *B. monnieri* may be further enhanced for use in medicine by combining this with CRISPR and gene editing (Nisha et al. 2003).

#### 7.8.4.8 Biotransformation

Plant tissue cultures from explants like leaves, stems, roots, meristems, cotyledons, embryos, and protoplasts are identified in sterile circumstances for secondary metabolite mining and rapid growth. Strain enhancement, high-producing cell line selection, and medium optimization boost secondary metabolite production. Many plant cell cultures yield poor outcomes (Sudheer et al. 2023). The use of nanomaterials to improve the generation of secondary metabolites has been investigated more and more in *B. monnieri* biotransformation techniques. One noteworthy method involved applying different concentrations of zinc oxide nanoparticles (ZnO NPs, approximately 24 nm) to suspension cultures of *B. monnieri* (0, 0.25, 0.50, 0.75, and 1.0 ppm). The purpose of this treatment was to evaluate their effects on bacoside-A accumulation, cell proliferation, and the expression of important biosynthetic genes, especially HMG-CoA reductase, an enzyme that is essential to the bacoside biosynthesis pathway (Bhardwaj et al. 2018).

#### 7.8.4.9 Molecular Marker Based

Research utilizing molecular markers, specifically RAPD, ISSR (Kareem et al. 2015), and SSR markers, has proven to be highly effective in assessing the level of clonal fidelity between mother plants and tissue-cultured regenerants (Dey et al. 2019). Furthermore, these markers have been utilized to examine the genetic diversity recognized within wild populations of *B. monnieri* (Ramesh et al. 2011).

## 7.9 Application of Gene Editing Techniques for Enhancement of Secondary Metabolites in *Bacopa monnieri*

*Bacopa monnieri* is well-known for its abundance of secondary metabolites, especially bacosides, which are the substances that give it its enhanced cognitive properties. But conventional breeding methods and even in vitro culture techniques can occasionally fall short when it comes to significantly increasing or altering the metabolic pathways that are involved in the creation of metabolites. The secondary metabolites produced by *Bacopa monnieri* can be increased in a specific, efficient, and repeatable manner, recognizing recent advancements in gene editing technologies, especially those based on CRISPR/Cas (Gao et al. 2025). Gene editing facilitates precise changes to the plant genome by either knocking out negative regulatory genes or activating transcription factors and biosynthetic genes involved in secondary metabolite pathways. For instance, genes encoding enzymes such as squalene synthase, cytochrome P450s, and glycosyltransferases are essential for the triterpenoid saponin biosynthesis pathway, which involves the synthesis of bacosides (Singh 2023). By employing CRISPR/Cas9 or CRISPR activation systems (CRISPRa) to modify these genes, bacoside accumulation can be significantly boosted. Furthermore, by using CRISPR interference (CRISPRi) systems to downregulate competing pathways (Bhojiya and Joshi 2024) and channel more precursors toward bacoside production. The metabolic flux can be steered towards the intended synthesis of secondary metabolites. By combining genome-wide transcriptomics and metabolomics with gene editing techniques, significant regulatory networks, transcription factors like MYB (myeloblastosis), bHLH (basic Helix-Loop-Helix), and WRKY (WRKYGQK domain), along with gene clusters linked to bacoside biosynthesis. CRISPR elements can be incorporated into *Bacopa* cells using biolistic methods or agrobacterium-mediated transformation (Yamada and Sato 2021). When coupled with tissue culture techniques for regeneration and selection, modified lines with a greater bacoside content can be produced and maintained. This approach increases output and offers a sustainable alternative to wild gathering, which promotes conservation and economic scalability (Puchta 2017).

## 7.10 Future Prospective

*Bacopa monnieri's* traditional history for enhancing memory and cognition is still upheld in contemporary herbal therapy, as evidenced by its inclusion in twenty-two distinct herbal formulations described by Neto et al. 2024). Given its increasing popularity as a health supplement, *Bacopa* has the potential to expand even further in the nutraceutical industry. Through ethnobotanical research, this plant contains medicinal properties, such as its anti-venomous, anti-paralytic, and anti-sickling properties, which have been identified (Silambarasan and Ayyanar 2015). However, there is currently not enough of experimental scientific research on these widely held views. Researchers are encouraged to further investigate the therapeutic potential of its bioactive components as there is currently no thorough clinical or pharmacological confirmation. From the perspective of nutraceuticals, *Bacopa monnieri* offers a lot of potential. Alkaloids, flavonoids, saponins, and other phytochemicals, which can be used as a functional dietary component to lower anxiety, improve mental clarity, and fend off neurodegenerative diseases (Valotto Neto et al. 2024). It is also a good option for formulations aimed at mood control, stress management, and overall well-being due to its adaptogenic qualities. Standardized Brahmi extracts, capsules, teas, and fortified foods may hold great medical and financial potential in light of the rising consumer desire for natural, plant-based health products (Patel et al. 2018).

## 7.11 Conclusion

Scientific studies on Brahmi have shown that it has medical, nutraceutical, and practical uses. The biosynthesis of bacosides remains a mystery, despite several investigations into their presence, chemical composition, and various medicinal uses. *Bacopa monnieri* is renowned not just for the curative effects it possesses, but also for the substantial nutritional content it contains. The plant has been the subject of a substantial amount of research on previous occasions. Bacosides have been shown to possess outstanding antioxidant effects, as indicated by the findings of this research. When it comes to the health of the brain, the effects of these chemicals are especially beneficial. Several factors contribute to the enhancement of its therapeutic potential. One of these factors is the fact that it possesses significant anti-inflammatory properties, which include the reduction of enzymes and cytokines. To develop high-yield bacoside cultivars, which can be created by genetic engineering, the study of the biosynthesis of *B. monnieri* at the genetic level has a great deal of promise. This plant has made substantial contributions to the fields of health and nutrition, which highlight its significance. The numerous contributions that this plant has made highlight the significance that it plays, even though we are still in the process of uncovering the intricacy of this plant.

## 7.12 Summary

*Bacopa monnieri* is considered an essential herb by Ayurveda due to its neuroprotective qualities and ability to sustain adaptogenic function, which improves cognitive function and reduces stress. By providing vitamins, minerals, and amino acids to meet the body demands, the plant adds value to the diet. According to scientific studies on its cytotoxicity, prolonged or excessive usage of the drug may pose health risks. Consuming large amounts of *Bacopa* may cause adverse effects such as fatigue, vomiting, and gastrointestinal issues. By measuring the amount of ash and extractable components, the medicinal plant *Bacopa monnieri* satisfies pharmacological requirements while displaying distinctive microscopic patterns. *Bacopa monnieri's* varied phytochemical components, which include flavonoids, alkaloids, sterols and bacosides A and B, are the main source of its therapeutic benefits. Biochemical processes of high complexity are essential for *B. monnieri* to create secondary metabolites. The Mevalonate (MVA) pathway, together with methyl-D-erythritol 4 phosphate (MEP), produces triterpenoid saponins through sequential isoprenoid synthesis until bacosides are formed. The developmental phases of the plant, together with environmental conditions, substantially affect the operating biosynthetic pathways. Plant cell cultures involving callus and suspension, along with growth regulators and elicitors, enhance the production of *Bacopa monnieri* metabolites. The production of plant metabolites also increases through the modifications of plant metabolism by endophytes and rhizobacteria. Genetic engineering, alongside the CRISPR/Cas9 technology, allows scientists to enhance bacoside production through pathway refining strategies. *B. monnieri* uses its intrinsic traditional use values to power biotech engineering for medical applications, combined with agricultural benefits and environmental cleanup practices.

**Acknowledgement** The authors acknowledge the support from the Department of Botany and Biochemistry, Jamia Hamdard, New Delhi, for providing the space, library, and Internet connection. AA. would like to acknowledge the Department of Health Research (DHR), Government of India (File No R.12014/61/2022-HR), for the Young Scientist award. We are also thankful to the UGC for providing UGC-SAP DRS-II grant [File No. F.3-17/2018/DRS-II (SAP-II) and DST for providing DST FIST grant (SR/FST/LS-I/2018/149 (C) to the Department of Biochemistry, School of Chemical and Life Sciences, Jamia Hamdard, New Delhi. Flowchart of secondary metabolites and diagrams are made by draw.io and bio-render.

## References

Abbas M, Mohani N (2011) The involvement of opioidergic mechanisms in the activity of *Bacopa monnieri* extract and its toxicological studies. J Ethnopharmacol 5(8):1120–1124

Abbassi SJ, Vishwakarma RK, Patel P, Kumari U, Khan BM (2015) Bacopa monniera recombinant mevalonate diphosphate decarboxylase: biochemical characterization. Int J Biol Macromol 79:661–668. Available from: https://www.sciencedirect.com/science/article/pii/S0141813015003827

Ahirwar S, Tembhre M, Gour S, Namdeo A (2012) Anticholinesterase efficacy of Bacopa monnieri against the brain regions of rat-a novel approach to therapy for Alzheimer's disease. Asian J Exp Sci 26(1):65–70

Ahuja A, Tripathi MK, Tiwari S, Tripathi N, Tiwari G, Mishra N et al (2021) Recent advancements on callus and cell suspension cultures: an effectual reserve for the production of pharmaceutically significant metabolites. In: Current aspects in pharmaceutical Research and Development, Vol 6. Book Publisher International (SCIENCEDOMAIN International), pp 96–111

Akbar S (2020) *Bacopa monnieri* (L.) Wettst. (Plantaginaceae): a comprehensive review of their traditional medical uses and scientific justifications. In: *Handbook of 200 medicinal plants*, pp 401–412. https://doi.org/10.1007/978-3-030-16807-0_41

Alamgir ANM (2017) Cultivation of herbal drugs, biotechnology, and in vitro production of secondary metabolites, high-value medicinal plants, herbal wealth, and herbal trade. In: Therapeutic use of medicinal plants and their extracts: volume 1: pharmacognosy, pp 379–452. https://doi.org/10.1007/978-3-319-63862-1_9

Al-Snafi AE (2019) The pharmacology of Bacopa monniera. Int J Pharma Sci Res 13(4): Available from: https://www.ijpsr.info/docs/IJPSR13-04-12-005.pdf

An efficient micropropagation protocol of Bacopa monnieri (L.) Pennell through two-stage culture of nodal segments and ex vitro acclimatization (2015) J Appl Biol Biotechnol

Anju V, Naresh C, Avinash P (2017) Anatomical markers and phytochemical study of different plant parts of *Bacopa monnieri* (L.) Wettst. Int J Life Sci 5(3) www.ijlsci.in

Antony Ceasar S, Lenin Maxwell S, Bhargav Prasad K, Karthigan M, Ignacimuthu S (2010) Highly efficient shoot regeneration of Bacopa monnieri (L.) using a two-stage culture procedure and assessment of genetic integrity of micropropagated plants by RAPD. Acta Physiol Plant [Internet] 32(3):443–452. https://doi.org/10.1007/s11738-009-0419-8

Bafort F, Jijakli MH (2024) Vertical farming of medicinal plants. In: Priyadarshan SM, Jain PS, Al-Khayri JM (eds) Digital agriculture: a solution for sustainable food and nutritional security, pp 129–177. https://doi.org/10.1007/978-3-031-43548-5_5

Banerjee R, Hazra S, Ghosh AK, Mondal AC (2014) Chronic administration of Bacopa monniera increases BDNF protein and mRNA expressions: a study in chronic unpredictable stress induced animal model of depression. Psychiatry Investig 11(3):297–306

Bansal M, Sudhakara Reddy M, Kumar A (2017) Optimization of cell growth and bacoside-A production in suspension cultures of Bacopa monnieri (L.) Wettst. using response surface methodology. In Vitro Cell Dev Biol- Plant [internet] 53(6):527–537. https://doi.org/10.1007/s11627-017-9847-0

Bharti N, Yadav D, Barnawal D, Maji D, Kalra A (2013) Exiguobacterium oxidotolerans, a halotolerant plant growthpromoting rhizobacterium, improves yield and content of secondary metabolites in Bacopa monnieri (L.) Pennell under primary and secondary salt stress. World J Microbiol Biotechnol 29(3):379–387. https://doi.org/10.1007/s11274-012-1192-1

Bhardwaj P, Goswami N, Narula P, Jain CK, Mathur A (2018) Zinc oxide nanoparticles (ZnO NP) mediated regulation of bacosides biosynthesis and transcriptional correlation of HMG-CoA reductase gene in suspension culture of *Bacopa monnieri*. Plant Physiol Biochem [Internet] 130:148–156. Available from: https://www.sciencedirect.com/science/article/pii/S0981942818303012

Bhaskar R, Xavier LSE, Udayakumaran G, Kumar DS, Venkatesh R, Nagella P (2022) Biotic elicitors: a boon for the in-vitro production of plant secondary metabolites. Plant Cell Tissue Organ Cult (PCTOC). 149(1):7–24. https://doi.org/10.1007/s11240-021-02131-1

Bhattacharya SK, Bhattacharya A, Kumar A, Ghosal S (2000) Antioxidant activity of Bacopa monniera in rat frontal cortex, striatum and hippocampus. Phytother Res 14(3):174–9

Bhojiya AA, Joshi H (2024) CRISPR gene editing for secondary metabolite production: a review. In: Kumar A, Arora S, Ogita S, Yau YY, Mukherjee K (eds) Gene editing in plants: CRISPR-Cas and its applications [internet]. Springer, Singapore, pp 437–475. https://doi.org/10.1007/978-981-99-8529-6_17

Channa S, Dar A, Yaqoob M, Anjum S, Sultani Z (2003) Broncho-vasodilatory activity of fractions and pure constituents isolated from Bacopa monnieri. J Ethnopharmacol 86(1):27–35

Chopra RN (1958) Indigenous drugs of India, 2nd edn. U.N. Dhur and Sons, Calcutta, p 341

Choudhuri RD (1996) Herbal drugs industry, 2nd edn. Eastern Publishers, New Delhi, pp 1–5

Chowdhuri DK, Parmar D, Kakkar P, Shukla R, Seth PK, Srimal RC (2002) Antistress effects of bacosides of Bacopa monnieri: modulation of Hsp70 expression, superoxide dismutase and cytochrome P450 activity in rat brain. Phytother Res 16(7):639–45

Das A, Shankar G, Nath C, Pal R, Singh S, Singh HK (2002) A comparative study in rodents of standardized extracts of *Bacopa monniera* and *Ginkgo biloba*, anticholinesterase and cognitive enhancing activities. Pharmacol Biochem Behav 73:893–900

Dey A, Hazra AK, Nongdam P, Nandy S, Tikendra L, Mukherjee A et al (2019) Enhanced bacoside content in polyamine treated in-vitro raised Bacopa monnieri (L.) Wettst. S Afr J Botany [Internet] 123:259–269. Available from: https://www.sciencedirect.com/science/article/pii/S0254629918315734

Dubey T, Kushwaha P, Thulasiram HV, Chandrashekar M, Chinnathambi S (2023) Bacopa monnieri reduces Tau aggregation and Tau-mediated toxicity in cells. Int J Biol Macromol 234:123171

Elangovan V, Govindasamy S, Ramamoorthy N, Balasubramanian K, Somasundaram ST (1995) In vitro studies on the anticancer activity of Bacopa monniera. Fitoterapia. 66(3):211–215

Enz A, Amstutz R, Boddeke H, Gmelin G, Malanowski J (1993) Brain selective inhibition of acetylcholinesterase: a novel approach to therapy for Alzheimer's disease. Prog Brain Res 98:431–438

Finn M (1996) The mangrove mesocosm of biosphere 2: design, establishment and preliminary results. Ecol Eng:21–56. https://doi.org/10.1016/0925-8574(95)00050-X

Ganjewala D, Srivastava AK (2011) Recent progress on chemical composition and bioactivities of Bacopa monnieri (Linn.) a plant of Ayurveda. Medic Aromat Plant Sci Biotechnol 5(2):102–108

Gao H, Pei X, Song X, Wang S, Yang Z, Zhu J et al (2025) Application and development of CRISPR technology in the secondary metabolic pathway of the active ingredients of phytopharmaceuticals. Front Plant Sci [Internet] 15. Available from: https://www.frontiersin.org/journals/plant-science/articles/10.3389/fpls.2024.1477894

Girish C, Oommen S, Vishnu R (2016) Evidence for the involvement of the monoaminergic system in the antidepressantlike activity of methanolic extract of Bacopa monnieri in albino mice. Int J Basic Clin Pharmacol 5(3):914–922

Giramkar SA, Jain A, Juvekar AR (2013) Evaluation of anticonvulsant activity of Brahmi Ghrita in mice. Ayu. 34(3):318–321. https://doi.org/10.4103/0974-8520.123134

Gouthami NS (2024) Development of process for value-added products from Brahmi leaves [PhD thesis]. Udaipur: Maharana Pratap University of Agriculture and Technology

Gubbannavar J, Chandola H, Harisha C, Khanpara K, Shukla V (2013) A comparative pharmacognostical and preliminary physico-chemical analysis of stem and leaf of *Bacopa monnieri* (L.) Pennel and *Bacopa floribunda* (R.BR.). Wettst Ayu 34(1):95. https://doi.org/10.4103/0974-8520.115441

Gupta R, Singh A, Srivastava M, Singh V, Gupta MM, Pandey R (2017) Microbial modulation of bacoside a biosynthetic pathway and systemic defense mechanism in Bacopa monnieri under Meloidogyne incognita stress. Sci Rep 7(1):41867. https://doi.org/10.1038/srep41867

Gupta P, Verma A, Rai N, Singh AK, Singh SK, Kumar B, Kumar R, Gautam V (2021) Mass spectrometry-based technology and workflows for studying the chemistry of fungal endophyte derived bioactive compounds. ACS Chem Biol 16(11):2068–2086. https://doi.org/10.1021/acschembio.1c00581

Hamdeni I, Louhaichi M, Slim S, Boulila A, Bettaieb T (2022) Incorporation of organic growth additives to enhance in vitro tissue culture for producing genetically stable plants. Plants [Internet] 11(22):3087. Available from: https://www.mdpi.com/2223-7747/11/22/3087

Haque SM, Chakraborty A, Dey D, Mukherjee S, Nayak S, Ghosh B (2017) Improved micropropagation of Bacopa monnieri (L.) Wettst. (Plantaginaceae) and antimicrobial activity of in vitro and ex vitro raised plants against multidrug-resistant clinical isolates of urinary tract infecting (UTI) and respiratory tract infecting (RTI) bacteria. Clin Phytoscience [Internet] 3(1):17. https://doi.org/10.1186/s40816-017-0055-6

Hossain H, Howlader MSI, Dey SK, Hira A, Ahmed A (2012) Evaluation of analgesic, antidiarrhoeal and cytotoxic activities of ethanolic extract of Bacopa monnieri (L). J Pharm Res Int 2(3):188–196

Husain S, Nandi A, Simnani FZ, Saha U, Ghosh A, Sinha A, Sahay A, Samal SK, Panda PK, Verma SK (2023) Emerging trends in advanced translational applications of silver nanoparticles: a progressing dawn of nanotechnology. J Funct Biomater 14(1). https://doi.org/10.3390/jfb14010047

Inthima P, Sujipuli K (2019) Improvement of growth and bacoside production in Bacopa monnieri through induced autotetraploidy with colchicine. Peer J 2019(10)

Jacob J, Cherian R, Thomas S (2021) Evaluation of antiulcer activity of ethanolic extract of Bacopa monnieri in experimental rats. J Drug Deliv Ther 11(2):122–126. https://doi.org/10.22270/jddt.v11i2.4693

Jain P, Khanna NK, Trehan N, Pendse VK, Godhwani JL. Antiinflammatory effects of an Ayurvedic preparation, Brahmi Rasayan, in rodents. Indian J Exp Biol 32(9):633–636. PMID: 7814042

Jain PK, Das D, Jain P, Jain P (2016) Pharmacognostic and pharmacological aspect of Bacopa monnieri – a review. Innovare J Ayurvedic Sci 4(3):[page numbers]

Jain P, Sharma HP, Basri F, Priya K, Singh P (2017) Phytochemical analysis of Bacopa monnieri (L.) Wettst. and their anti-fungal activities. Indian J Tradit Knowl 16(2):310–318

Jauhari N, Bharadwaj R, Sharma N, Bharadvaja N (2019) Assessment of bacoside production, total phenol content and antioxidant potential of elicited and non-elicited shoot cultures of Bacopa monnieri (L.). Environ Sustain [Internet] 2(4):441–453. https://doi.org/10.1007/s42398-019-00071-3

Jeena GS, Kumar S, Bharti S, Singh N, Joshi A, Lahane V et al (2025) Engineering Bacopa monnieri for improved bacoside content and its neurological evaluation. Appl Microbiol Biotechnol 109(1):83. https://doi.org/10.1007/s00253-025-13453-x

Joshua Allan J, Damodaran A, Deshmukh NS, Goudar KS, Amit A (2007) Safety evaluation of a standardized phytochemical composition extracted from *Bacopa monnieri* in Sprague–Dawley rats. Food Chem Toxicol 45(10):1928–1937. https://doi.org/10.1016/j.fct.2007.04.010

Kamkaew N, Scholfield CN, Ingkaninan K, Maneesai P, Parkington HC, Tare M, Chootip K (2011) *Bacopa monnieri* and its constituents is hypotensive in anaesthetized rats and vasodilator in various artery types. J Ethnopharmacol 137(1):790–795. https://doi.org/10.1016/j.jep.2011.06.045

Kannan R, Arumugam S, Kannan RN, Sudha A, Manimaran A, Saravanan D, Natarajan E (2011) Beneficial effect of *Bacopa monnieri* extract on gentamicin-induced nephrotoxicity and oxidative stress in albino rats. https://www.researchgate.net/publication/286019657

Kapoor R, Srivastava S, Kakkar P (2009) Bacopa monnieri modulates antioxidant responses in brain and kidney of diabetic rats. Environ Toxicol Pharmacol 27(1):62–69

Kareem VKA, Rajasekharan PE, Mohan CK (2015) Assessment of genetic fidelity in micropropagated *Bacopa monnieri* using ISSR marker assay. Interne. Available from: https://www.researchgate.net/publication/259432622

Kishore Mohanta Y, Sahoo S (2014) In vitro culture of highly valuable medicinal plant Bacopa Monnieri (L.) Penn. for rapid and mass multiplication. Int J Pharm Sci Invent:3. Available from: www.ijpsi.org

Kulkarni O, Mukherjee S, Bhandare R, Jagtap S, Dugad S, Pawar N, Pawar P (2011) Evaluation of comparative free-radical quenching potential of Brahmi (*Bacopa monnieri*) and Mandookparni (*Centella asiatica*). Ayu 32(2):258. https://doi.org/10.4103/0974-8520.92549

Kumar H, More SV, Han SD, Choi JY, Choi DK (2012) Promising therapeutics with natural bioactive compounds for improving learning and memory — a review of randomized trials. Molecules 17(9):10503–10539. https://doi.org/10.3390/molecules170910503

Kumar N, Abichandani L, Thawani G, Gharpure V, Naidu MU, Venkat Ramana G (2016) Efficacy of standardized extract of *Bacopa monnieri* (Bacognize®) on cognitive functions of medical students: a six-week, randomized placebo-controlled trial. Evid Based Complement Alternat Med. https://doi.org/10.1155/2016/4103423

Kumari A, Kumar A, Mandal S, Roy P, Sircar D (2023) A metabolic reprogramming in Bacopa monnieri plants induced by methyl-jasmonate and enhanced biosynthesis of triterpene saponins. Ind Crop Prod 204:117241. Available from: https://www.sciencedirect.com/science/article/pii/S0926669023010063

Lamo JM, John L, Rao SR (2023) Medicinal plants of north-east india: biodiversity and their ethnomedicinal values. in medicinal plants: biodiversity, biotechnology and conservation (pp. 219–247). Singapore: Springer Nature Singapore

Lastiri-Hernández MA, Álvarez-Bernal D, Cruz-Cárdenas G, Silva-García JT, Conde-Barajas E, Oregel-Zamudio E (2023) Potential of Epipremnum aureum and Bacopa monnieri (L.) Wettst for saline phytoremediation in artificial wetlands. Water 15(1):194. https://doi.org/10.3390/w15010194

Leonard J, Seth B, Sahu BB, Singh VR, Patra N (2018) Statistical optimization for enhanced bacoside a production in plant cell cultures of Bacopa monnieri. Plant Cell, Tissue Organ Cult (PCTOC) [Internet] 133(2):203–214. https://doi.org/10.1007/s11240-017-1373-6

Madda DR, M L, Pola S, Sandeep BV (2016) Assessment of biological properties of two south Indian medicinal plants of Bacopa monnieri and Anacardium occidentale. Int J Bioassays 5(09):4899

Manzoor A, Ahmad T, Bashir MA, Hafiz IA, Silvestri C (2019) Studies on colchicine induced chromosome doubling for enhancement of quality traits in ornamental plants. Plants 8(7):194. Available from: https://www.mdpi.com/2223-7747/8/7/194

Mondal S, Bhar K, Mondal P, Panigrahi N, Sahoo SK, Swetha P, Chakraborty S, Teja NY, Parveen N (2023) In quest of the mysterious holistic Vedic herb Bacopa monnieri (L.). Pennell Pharmacogn Res 15(3):410–454. https://doi.org/10.5530/pres.15.3.045

Morgan A, Stevens J (2010) Does *Bacopa monnieri* improve memory performance in older persons? Results of a randomized, placebo-controlled, double-blind trial. J Altern Complement Med 16(7):753–759

Mukherjee DG, Dey CD (1966) Clinical trial on Brahmi. I J Exper Med Sci 10:5–11

Murthy KRC, Raju VR, Ashok P, Kiranmai M (2013) Evaluation of wound healing activity of Bacopa monnieri (L.) Wettst. in rats. J Ethnopharmacol 150(2):573–580. https://doi.org/10.1016/j.jep.2013.09.019

Nadkarni KM (1988) Indian materia medica: with Ayurvedic, Unani-Tibbi, Siddha, Allopathic, Homeopathic & Naturopathic & home remedies, appendices & indexes. Mumbai: South Asia Books

Nakhjavani M, Hardingham JE, Palethorpe HM, Tomita Y, Smith E, Price TJ, et al. (2021) Bacopa monnieri (L.) exerts anti-cancer effects in colorectal cancer cells by downregulating aquaporin 1 expression. Cancers (Basel). 13(10):2453

Nandy S, Dey A, Mukherjee A (2019) Advances in dammarane-type triterpenoid saponins from Bacopa monnieri: structure, bioactivity, biotechnology and neuroprotection. In: Bioactive natural products. https://doi.org/10.1016/B978-0-12-817901-7.00015-0

Naseer I, Ahmad M, Nadeem SM, Ahmad I, Zahir ZA (2019) Rhizobial inoculants for sustainable agriculture: prospects and applications. In: Biofertilizers for sustainable agriculture and environment, pp 245–283

Negi JS, Singh P, Pant GJN, Rawat MSM (2011) High-performance liquid chromatography analysis of plant saponins: an update (2005–2010). Pharmacogn Rev 5(10):155–158. https://doi.org/10.4103/0973-7847.91109

Neto LJV, Reverete M, Junior RCM, Machado NM, Joshi RK, dos Santos Buglio D, et al. (2024) The neuroprotective and cognitive-enhancing effects of *Bacopa monnieri*: a systematic review [Internet]. Available from: https://www.preprints.org/manuscript/202402.0207/v1

Nisha KK, Seetha K, Rajmohan K, Purushothama MG (2003) *Agrobacterium tumefaciens*-mediated transformation of Brahmi [*Bacopa monniera* (L.) Wettst.], a popular medicinal herb of India. Curr Sci [Internet] 85(1):85–89. Available from: http://www.jstor.org/stable/24107719

Otari SS, Devkar SP, Patel SB, Ghane SG (2023) Micropropagation and elicited production of triterpenoid saponin glycosides and stigmasterol via precursor and elicitor feeding in Bacopa floribunda (R.Br.) Wettst.—a potential nootropic herb. Front Plant Sci 14:1096842. Available from: https://www.frontiersin.org/journals/plant-science/articles/10.3389/fpls.2023.1096842

Pandey AK (2022) Bacopa monnieri (Linn.) Pennell – a possible plant for impossible diseases (a review). Adv Pharmacol Pharm 10(1):35–53. https://doi.org/10.13189/app.2022.100104

Pandey AK (2024) Quantitative estimation of secondary metabolite, in-vitro antioxidant, Anti-Sickling & Anti-Inflammatory Activity by HRBC membrane stabilization of Ethanolic extract of Bacopa monnieri (L.) Pennell. Adv Pharmacol Pharm. 12(1):19–33. https://doi.org/10.13189/app.2024.120103

Pankaj U, Singh DN, Mishra P, Gaur P, Babu CSV, Shanker K, Verma RK (2020) Autochthonous halotolerant plant growth-promoting rhizobacteria promote bacoside A yield of Bacopa monnieri (L.) Nash and phytoextraction of salt-affected soil. Pedosphere 30(5):671–683. https://doi.org/10.1016/S1002-0160(20)60029-7

Patel SK, Singh S, Singh HK, Singh SK (2017) Effect of standardized extract of Bacopa monnieri (CDRI-08) on testicular functions in adult male mice. J Ethnopharmacol 197:101–109. https://doi.org/10.1016/j.jep.2016.07.026

Patel S, Vigyan K, Saran K, Birwal P, Basu S, Pawar D. Brahmi (Bacopa monnieri) as functional food ingredient in food processing industry [internet]. 2018. Available from: https://www.researchgate.net/publication/325012871

Pawar SS, Jadhav MG (2015) Determination and quantification of bacoside A from Bacopa monnieri (L) by highperformance thin-layer chromatography. Int J Pharmacogn Phytochem Res 7(5):1060–1065

Pawar SS, Jadhav MG, Deokar TG (2016) Study of phytochemical screening, physicochemical analysis and antimicrobial activity of Bacopa monnieri (L) extracts. Int J Pharm Clin Res 8(8):1222–1229

Phrompittayarat W, Jetiyanon K, Wittaya-Areekul S, Putalun W, Tanaka H, Khan I, Ingkaninan K (2011) Influence of seasons, different plant parts, and plant growth stages on saponin quantity and distribution in Bacopa monnieri. Songklanakarin. J Sci Technol 33(2)

Puchta H (2017) Applying CRISPR/Cas for genome engineering in plants: the best is yet to come. Curr Opin Plant Biol [Internet] 36:1–8. Available from: https://www.sciencedirect.com/science/article/pii/S136952661630125X

Rai K, Gupta N, Sen Dharamdasani L, Nair P, Bodhankar P (2017) Bacopa monnieri: a wonder drug changing fortune of people. J Drug Deliv Ther 5:127–132

Ramesh M, Karthikeyan A, Vijayakumar K, Joe M, Largia V, Karutha PS (2011) Agrobacterium-mediated transformation of pharmaceutically important Indian medicinal herb Bacopa monnieri (L.). J Med Plants Res [Internet] 5(11):2316–2321. Available from: http://www.academicjournals.org/JMPR

Roodenrys S, Booth D, Bulzomi S, Phipps A, Micallef C, Smoker J (2002) Chronic effects of Brahmi (Bacopa monnieri) on human memory. Neuropsychopharmacology 27(2):279–281

Russo A, Izzo AA, Borrelli F (2003) Free radical scavenging capacity and protective effect of Bacopa monniera L. on DNA damage. Phytother Res 17(7):870–5

Russo A, Borrelli, F (2005) Bacopa monniera, a reputed nootropic plant: an overview. Phytomedicine 12(4):305–317. https://doi.org/10.1016/j.phymed.2003.12.008

Russo A, Borrelli F. (2005) Bacopa monniera, a reputed nootropic plant: an overview. Phytomedicine 12(4):305–317

Ryan J, Croft K, Mori T, Wesnes K, Spong J, Downey L, Kure C, Lloyd J, Stough C (2008) An examination of the effects of the antioxidant Pycnogenol on cognitive performance, serum lipid profile, endocrinological and oxidative stress biomarkers in an elderly population. J Psychopharmacol 22(5):553–562

Saesong T, Temkitthawon P, Nangngam P, Ingkaninan K (2019) Pharmacognostic and physico-chemical investigations of the aerial part of Bacopa monnieri (L.) Wettst. Songklanakarin. J Sci Technol 41:397–404

Saha PS, Sarkar S, Jeyasri R, Muthuramalingam P, Ramesh M, Jha S (2020) In vitro propagation, phytochemical and neuropharmacological profiles of *Bacopa monnieri* (L.) Wettst.: a review. Plants [Internet] 9(4):411. Available from: https://www.mdpi.com/2223-7747/9/4/411

Samiulla DS, Prashanth D, Amit A (2001) Mast cell stabilising activity of Bacopa monnieri. Fitoterapia 72(3):284–285

Sandeep D, Ghosh S, Atta-ur-Rahman. (2020) Triterpenoids: structural diversity, biosynthetic pathway, and bioactivity. Bioactive Natural Products 67:411–461. https://doi.org/10.1016/B978-0-12-819483-6.00012-6

Sanyal R, Nandi S, Pandey S, Chatterjee U, Mishra T, Datta S et al (2022) Biotechnology for propagation and secondary metabolite production in Bacopa monnieri. Appl Microbiol Biotechnol 106(5):1837–1854. https://doi.org/10.1007/s00253-022-11820-6

Saraf MK, Prabhakar S, Pandhi P, Anand A (2008) Bacopa monniera ameliorates amnesic effects of diazepam qualifying behavioural molecular partitioning. Neuroscience 155(2):476–484

Seeliger. (1992) Coastal Foredunes of southern Brazil: physiography, habitats, and vegetation. Academic, pp 367–381. https://doi.org/10.1016/B978-0-08-092567-7.50029-5

Sharma N, Satsangi R, Pandey R, Singh R, Kaushik N, Tyagi RK (2012) In vitro conservation of Bacopa monnieri (L.) using mineral oil. Plant Cell Tissue Organ Cult 111(3):291–301. https://doi.org/10.1007/s11240-012-0194-x

Sharma M, Ahuja A, Gupta R, Mallubhotla S (2015) Enhanced bacoside production in shoot cultures of Bacopa monnieri under the influence of abiotic elicitors. Nat Prod Res [Internet] 29(8):745–749. https://doi.org/10.1080/14786419.2014.986657

Sharma N, Singh R, Jain SM (2016) In vitro propagation and conservation of Bacopa monnieri L. In: Protocols for in vitro cultures and secondary metabolite analysis of aromatic and medicinal plants. https://doi.org/10.1007/978-1-4939-3332-7_11

Sharma A, Verma N, Verma P, Verma RK, Mathur A, Mathur AK (2017) Optimization of a Bacopa monnieri-based genetic transformation model for testing the expression efficiency of pathway gene constructs of medicinal crops. In Vitro Cell Dev Biol Plant [internet] 53(1):22–32. https://doi.org/10.1007/s11627-017-9804-y

Shende S, Upmanyu N, Shende S, Dangi S, Raj R (2020) Treatment of allergic disorder from the traditional era to the modern era. World. J Pharm Pharm Sci. https://doi.org/10.20959/wjpps20209-17175

Shen YH, Zhang LJ, Li XY, Li YL, Wang W (2009) Antidepressant-like effects of Bacopa monnieri extract in mice: behavioral and neurochemical evidence. Phytomedicine 16(12):1046–1052

Silambarasan R, Ayyanar M (2015) An ethnobotanical study of medicinal plants in Palamalai region of eastern Ghats, India. J Ethnopharmacol [Internet] 172:162–178. Available from: https://www.sciencedirect.com/science/article/pii/S0378874115003864

Silpa SG, Smitha GR, Ranjitha K (2021) Drying and packaging methods impact the bacoside profile and microbiological quality of Brahmi herb (Bacopa monnieri L.) during storage. Ind Crop Prod 159:113064. https://doi.org/10.1016/j.indcrop.2020.113064

Singh H (2023) Enhancement of plant secondary metabolites by genetic manipulation. In: Singh R, Kumar N (eds) *Genetic manipulation of secondary metabolites in medicinal plants* [internet]. Springer, Singapore, pp 59–90. https://doi.org/10.1007/978-981-99-4939-7_4

Singh HK, Dharwan BN (1997) Neuro psycho pharmacological effects of the Ayurvedic nootropic *Bacopa monniera* Linn (Brahmi). Indian J Pharmacol 29:359–365

Singh HK, Dhawan B (1992) Drugs affecting learning and memory. In: Tandon PN, Bijiani V, Wadhwa S (eds) Lectures in neurobiology, vol 1. Wiley Eastern, New Delhi, pp 189–207

Singh HK, Dhawan B (1997) Neuropsychopharmacological effects of the Ayurvedic nootropic *Bacopa monnieri* Linn. (Brahmi). Indian J Pharmacol 29:S359–S365

Sireeratawong S, Jaijoy K, Khonsung P, Lertprasertsuk N, Ingkaninan K (2016) Acute and chronic toxicities of Bacopa monnieri extract in Sprague-Dawley rats. BMC Complement Altern Med 16(1):249. https://doi.org/10.1186/s12906-016-1236-4

Sivarajan VV, Balachandran I (1994) Ayurvedic drugs and their plant sources. Oxford and IBH Publishing, New Delhi, pp 97–99

Stough C, Lloyd J, Clarke J (2001) The chronic effects of an extract of *Bacopa monniera* (Brahmi) on cognitive function in healthy human subjects. Psychopharmacology 156:481–484

Stough CK, Pase MP, Cropley V, Myer S (2012) A randomized controlled trial investigating the effect of Pycnogenol and Bacopa CDRI08 herbal medicines on cognitive, cardiovascular, and biochemical functioning in cognitively healthy elderly people: the Australian Research Council longevity intervention (ARCLI) study protocol (ANZCTR12611000487910). Nutr J 11(11):2–9

Sudheer WN, Thiruvengadam M, Nagella P (2023) A comprehensive review on tissue culture studies and secondary metabolite production in *Bacopa monnieri* L. Pennell: a nootropic plant. Crit Rev Biotechnol [Internet] 43(6):956–970. https://doi.org/10.1080/07388551.2022.2085544

Sumathi T, Nongbri A (2008) Hepatoprotective effect of Bacoside-A, a major constituent of Bacopa monniera Linn. Phytomedicine 15(10):901–905. https://doi.org/10.1016/j.phymed.2007.11.020

Suresh S, Mds SN, Mds MD. Evaluation of antibacterial activity of Bacopa monnieri extract on Periodontogenic bacteria-an in-vitro study. Saudi J Oral Dent Res 2017. https://doi.org/10.21276/sjodr.2017.2.11.2

Taznin I, Mukti M, Rahmatullah M (2015) Bacopa monnieri: an evaluation of antihyperglycemic and antinociceptive potential of methanolic extract of whole plants. Pak J Pharm Sci 28(6):2135–2139

Tiwari V, Tiwari KN, Singh BD (2001) Comparative studies of cytokinins on in vitro propagation of Bacopa monniera. Plant Cell Tissue Organ Cult 66(1):9–16. Available from:. https://doi.org/10.1023/A:1010652006417

Topuz A, Dincer C, Kalkan TA, Asik Aygun SM, Jafari M, Malekjani N (2023) Drying technology in food processing. Woodhead Publishing, pp 543–586. https://doi.org/10.1016/B978-0-12-819895-7.00001-8

Trivedi MN, Khemani A, Vachhni UD, Shah CP, Santani DD (2011) Comparative pharmacognostic and phytochemical investigation of two plant species valued as Medhya Rasayanas. Int J Appl Biol Pharm Technol 2:28–36

Valotto Neto LJ, de Araujo M, Moretti Junior RC, Mendes Machado N, Joshi RK, dos Santos BD et al (2024) Investigating the neuroprotective and cognitive-enhancing effects of Bacopa monnieri: a systematic review focused on inflammation, oxidative stress, mitochondrial dysfunction, and apoptosis. Antioxidants [Internet] 13(4):393. Available from: https://www.mdpi.com/2076-3921/13/4/393

Vijayakumar M, Vijayakumar R, Stephen R (2010) In vitro propagation of Bacopa monnieri L. a multipurpose medicinal plant. Indian J Sci Technol 3(7):782–787. https://doi.org/10.17485/ijst/2010/v3i7/29814

Vishwakarma RK, Patel K, Sonawane P, Kumari U, Singh S, Ruby R et al (2015) Squalene synthase gene from medicinal herb Bacopa monniera: molecular characterization, differential expression, comparative modeling, and docking studies. Plant Mol Biol Report 33(6):1675–1685. Available from:. https://doi.org/10.1007/s11105-015-0864-z

Vohora D, Pal SN, Pillai KK (2000) Protection from phenytoin-induced cognitive deficit by Bacopa monnieri, a reputed Indian nootropic plant. J Ethnopharmacol 71:383–390

Williams R, Münch G, Gyengesi E, Bennett L (2014) Bacopa monnieri (L.) exerts anti-inflammatory effects on cells of the innate immune system in vitro. Food Funct 5(3):517–520

Yadav S, Sharma P, Srivastava A, Desai P, Shrivastava N (2014) Strain specific *Agrobacterium*-mediated genetic transformation of *Bacopa monnieri*. J Genet Eng Biotechnol [Internet] 12(2):89–94. Available from: https://www.sciencedirect.com/science/article/pii/S1687157X14000365

Yamada Y, Sato F (2021) Transcription factors in alkaloid engineering. Biomolecules [Internet] 11(11):1719. Available from: https://www.mdpi.com/2218-273X/11/11/1719

Yamada K, Ogita K, Yoneyama M et al (2011) Anti-Parkinsonian effects of Bacopa monnieri: Insights from transgenic and pharmacological Caenorhabditis elegans models of Parkinson's disease. Biochem Biophys Res Commun 413(4):605–610. https://doi.org/10.1016/j.bbrc.2011.09.010

Zote RK, Patil YK, Londhe SS, Thakur VV, Choudhari NB, Thakur V (2018) In vitro regeneration of *Bacopa monnieri* (L.) from leaf and stem explants. Int J Chem Stud 6(2):1577–1580

# Nutritional and Medicinal Importance of *Zingiber officinale* in Human Health

8

Sabry A. H. Zidan and Hamdy K. Assaf

**Abstract**

Nutraceuticals are bioactive compounds found in foods and herbal products with health benefits. Interest in herbal products has increased due to their safety, affordability, and low cost. The Zingiberaceae family, including premium species such as *Elettaria cardamomum*, *Curcuma longa*, and *Zingiber officinale*, is cultivated for spice and medicinal purposes in tropical and subtropical regions. *Zingiber officinale* (ginger), a perennial herbaceous plant, is known for its pharmacological benefits, including antioxidant, anti-inflammatory, antimicrobial, antidiabetic, antiobesity, antitumor, broncho-protective, neuroprotective, cardiovascular protective, and hepatoprotective activities. Ginger is rich in nutrients such as carbohydrates, fiber, proteins, minerals, vitamins, and essential oils. Ginger-derived oil is used as food-flavoring, spices, and as preservatives. The chemical composition of ginger oils depends on factors such as variety, drying, extraction, and storage conditions. This chapter focuses on the therapeutic potential of ginger, examining its chemical components and mechanisms of action, and providing a scientific foundation for its traditional uses. It also highlights the nutritional advantages of ginger in human diets and its applications in the food industry, while addressing different variables affecting its health and nutrient benefits.

**Keywords**

Ginger · *Zingiber officinale* · Nutraceuticals · Health benefits · Nutrition · Herb

S. A. H. Zidan (✉) · H. K. Assaf
Department of Pharmacognosy, Faculty of Pharmacy, Al-Azhar University, Assiut, Egypt
e-mail: sabryzidan@azhar.edu.eg



M. A. Ansari et al. (eds.), *Medicinal Plants and their Bioactive Compounds in Human Health: Volume 2*, https://doi.org/10.1007/978-981-95-3620-7_8

## 8.1 Introduction

*Zingiber officinale* Roscoe, commonly known as ginger (Fig. 8.1), is a well-known herbaceous plant used as food, condiments, and medicine for various health concerns (Shahrajabian et al. 2019). It has been classified as GRAS (generally regarded as safe) by the U.S. Food and Drug Administration (FDA) and ranks among the top 20 herbal supplements sold in the United States (Khan et al. 2016). Ginger has been utilized in traditional Chinese medicine and modern China, and it is listed in the pharmacopoeias of numerous countries for addressing various health issues (Shahrajabian et al. 2019; Khan et al. 2016).

Ginger, a natural herb, is often employed to alleviate symptoms such as nausea and vomiting during early pregnancy and chemotherapy (Zhang et al. 2021). Ginger is usually available as fresh, dried or preserved form (Shahrajabian et al. 2019). It is rich in bioactive phyto-chemical compounds, including gingerol, shogaol, zingerone, and their analogs. These phyto-chemicals provide ginger with a wide range of pharmacological activities, including antioxidant, anti-inflammatory, antimicrobial, antidiabetic, antiobesity, neuroprotective, antitumor, cardiovascular, bronchoprotective, and hepatoprotective properties (Shahrajabian et al. 2019). To date, over 160 constituents have been isolated and identified from ginger (Zhang et al. 2021).

Ginger also contains various nutrients, including carbohydrates, fiber, proteins, minerals, lipids, vitamins, and essential oils (Beristain-Bauza et al. 2019; Khan et al. 2016). Its oil possesses a strong, sour, and pungent flavor, and it is utilized in a wide range of products, such as soft drinks, bakery items, confectionery, pickles, and sauces as well as its preservatives' utilization (Shahrajabian et al. 2019; Otunola et al. 2010). The aromatic properties of ginger are attributed to the volatile compounds present in its essential oil such as, $\beta$-phellandrene, $\alpha$-farnesene, geraniol, and cineole (Beristain-Bauza et al. 2019). The chemical composition of ginger oil is

**Fig. 8.1** Photographs of *Z. officinale* plant (right) and its rhizomes (left)

influenced by factors such as variety, agronomic management, drying, and storage conditions (Beristain-Bauza et al. 2019).

Ginger is a plant native to warm tropical climates and is cultivated in approximately 50 countries worldwide for both spice and medicinal purposes (Zhang et al. 2021; Khan et al. 2016). The planting of ginger typically occurs between April and May, and the harvest takes place from late autumn to winter (Zhang et al. 2021). The main producers of ginger are India, China, and Nigeria, accounting for 35, 19, and 12% of the global production, respectively (Beristain-Bauza et al. 2019; Shahrajabian et al. 2019; Khan et al. 2016). Among these countries, India is the largest producer, with an annual output capacity of 1.07 million tons in 2017. China is the second-largest producer, with 0.58 million tons produced in 2017 (Beristain-Bauza et al. 2019; Khan et al. 2016). In addition to these countries, ginger is also cultivated on a large scale in Bangladesh, Thailand, Nepal, Indonesia, and Nigeria, and it is usually exported to several countries around the globe (Shahrajabian et al. 2019).

*Z. officinale*, a perennial plant belonging to the Zingiberaceae family, has many tuberous roots, aerial shoots, and branched rhizomes. Its height ranges from 0.5 to 1.0 m, and the roots develop from the rhizome, which is a creeping horizontal pseudo-stem that extends 50–100 cm (Zhang et al. 2021). The rhizomes are the primary part of ginger used for therapeutic and nutritional purposes (Pandotra et al. 2013).

Ginger rhizomes are known for their fragrant and spicy flavor, and are typically light yellow or taupe brown in color, with an uneven mass consisting of thick, fleshy, scaly structures with fingers-branching and tuberous joints, and so it is shape is described as "Hand" (Fig. 8.1) (Beristain-Bauza et al. 2019; Zhang et al. 2021).

The polyphenolic compounds found in ginger rhizomes have been shown to possess detoxifying properties against alcohol abuse and bromobenzene intoxication (Stoilova et al. 2007). Additionally, research has demonstrated that ginger rhizomes possess hepatic anticancer, antihyperlipidemic, and antidiabetic properties (Akimoto et al. 2015; Hamed et al. 2012).

Gingerols, the primary compounds found in fresh ginger, are susceptible to dehydration and conversion to shogaols when exposed to heat and/or acidic environments, a process that occurs during the extraction or drying of ginger (Zhang et al. 2021; Beristain-Bauza et al. 2019). As a result, the quantities of gingerols and shogaols in ginger products can vary depending on the heating and drying techniques used (Beristain-Bauza et al. 2019). Dried ginger is typically more pungent than fresh ginger due to the higher concentration of shogaols, which are twice as pungent as gingerols (Zhang et al. 2021).

Additionally, zingerone, does not occur in fresh ginger, can be produced from gingerols through hydrolysis during heating and drying process (Beristain-Bauza et al. 2019). Ginger's essential oil components are also significantly impacted by the drying process, extraction, and storage which can affect the synthesis and breakdown of bioactive substances, as well as their therapeutic and nutritional benefits (Beristain-Bauza et al. 2019). Accordingly, these processes affect the use of ginger products for health and nutritional purposes, and should be taken in consideration during selection the type of ginger products for certain disease.

Thus, the main purpose of this chapter is to delve into the therapeutic potential of ginger, examining the chemical components and mechanisms responsible for its pharmacological properties, and to provide a scientific foundation for its traditional uses. It also sheds light on the nutritional advantages of ginger in human diets and its applications in the food industry, while simultaneously addressing various confounding variables that must be taken into account when assessing the health benefits of ginger in relation to illnesses.

## 8.2 Nutritional Benefits of *Zingiber officinale*

Ginger is widely utilized as an ingredient in both the culinary and pharmaceutical industries in many countries (Shahrajabian et al. 2019). It has a long history of being used as a food flavoring agent in soft drinks, pickles, confections, and baked goods (Beristain-Bauza et al. 2019; Shahrajabian et al. 2019). Additionally, it is used as a preservative in meat and seafood products (Beristain-Bauza et al. 2019; Shahrajabian et al. 2019). Ginger can be employed in various forms, including extracts, powders, essential oils, and additives in edible coatings and films (Beristain-Bauza et al. 2019). The FDA granted GRAS classification to ginger essential oil, oleoresins, and natural extracts in 2017 (Beristain-Bauza et al. 2019). The beverage and fragrance industries made extensive use of ginger oil, with over 300–400 tons produced globally in 2017 (Zhang et al. 2021).

Ginger is rich in vitamins (thiamine, riboflavin, niacin, and vitamin C), minerals (such as magnesium, calcium, iron, phosphorus, and zinc), essential amino acids (leucine and lysine), and unsaturated fatty acids (palmitoleic acid, linoleic acid, and oleic acid) (Khan et al. 2016; Otunola et al. 2010). Fresh ginger consists mainly of water (80.9%), carbohydrates (12.3%), fiber (2.4%), proteins (2.3%), minerals (1.2%), and fats (0.9%) (Zadeh and Kor 2014; Otunola et al. 2010). The powdered rhizome is composed of ash (8%), water (9–10%), protein (9%), fatty oil (3–6%), carbohydrates (60–70%), crude fiber (3–8%), and volatile oil (0.3–3%) (Zhang et al. 2021; Khan et al. 2016; Zadeh and Kor 2014).

So, ginger represents an important source for these valuable nutrients, and dietary intake of ginger has been shown to have beneficial effects on weight loss and certain obesity-associated parameters in overweight and obese adults (Sang et al. 2020).

### 8.2.1 Mineral Contents of *Zingiber officinale*

Ginger rhizome is a significant source of essential minerals in the human diet (Khan et al. 2016). The metals found in ginger rhizome can be categorized into three classes: macronutrients, micronutrients, and heavy metals (Table 8.1). Generally, ginger contains minimal amounts of heavy metals like lead and cadmium, which are far below the recommended limits established by the World Health Organization (WHO) and the European Committee Directives in 2008 (Koch et al. 2017).

**Table 8.1** The average mineral content of fresh and dried *Zingiber officinale* (mg/kg) (Koch et al. 2017)

| | | Fresh ginger | Dry ginger |
|---|---|---|---|
| Macroelements | Ca | 159.0 ± 31.3 | 1587 ± 337 |
| | Mg | 239.8 ± 21.8 | 2572 ± 485 |
| | K | 4052 ± 443 | 40,963 ± 9359 |
| | Na | 115.6 ± 31.4 | 1261 ± 427 |
| Trace elements | Zn | 2.54 ± 0.44 | 25.0 ± 6.90 |
| | Cu | 1.03 ± 0.06 | 11.1 ± 1.76 |
| | Fe | 12.5 ± 2.67 | 140.5 ± 49.4 |
| | Mn | 69.9 ± 10.0 | 758.4 ± 150.3 |
| | Cr | 0.064 ± 0.01 | 0.795 ± 0.054 |
| Heavy metals | Ni | 0.163 ± 0.02 | 1.67 ± 0.28 |
| | Pb | 0.021 ± 0.006 | 0.218 ± 0.032 |
| | Cd | 0.002 ± 0.0008 | 0.02 ± 0.004 |

The mineral composition of ginger varies depending on its geographical origin (Koch et al. 2017; Khan et al. 2016; Wagesho and Chandravanshi 2015), the type of ginger used (fresh or dry), the soil's metal content, and cultivation techniques (Koch et al. 2017). For instance, the metal content of ginger from the ecological plantation on Shikoku Island, Japan, (Koch et al. 2017) is in the following order: Cd < Pb < Ni < Cr < Cu < Zn < Fe < Mn < Na < Ca < Mg < K, while the metal content of ginger root from Ethiopia (Wagesho and Chandravanshi 2015) is in the following order: Cd < Cu < Co < Ni < Cr < Zn < Fe < Mn < Ca < Mg, with no Pb detected. Additionally, analysis of ginger root from Malaysia and Nigeria revealed the following metal levels, in increasing order: Cd < Co < Pb < Ni < Cr < Cu < Ba < Zn < Mn < Na < Ca < Fe < Mg < K. Thus, the location where the plants are collected and the type used can significantly impact the ginger's biological and therapeutic properties, as well as the consumers' health (Koch et al. 2017).

Considering that consuming only 20 g of ginger can provide more than 80 mg of potassium, it can be considered an essential source of this mineral in the human diet due to its high potassium concentration (Koch et al. 2017). However, given that the average daily consumption of fresh ginger root is 20 grams, it cannot be considered a sufficient dietary source of calcium and magnesium, and other additives should be provided (Koch et al. 2017).

Manganese was discovered to be the main microelement present in the rhizomes. Plant-based foods like grains and vegetables are the primary sources of manganese for humans (Koch et al. 2017). Owing to ginger's high manganese concentration, a single part of this plant provides more than 60% of the needed daily intake of this element, which is estimated to be 2.3 mg for males and 1.8 mg for women. Thus, raw ginger rhizome may provide a significant supply of this element in the diet of humans (Koch et al. 2017). Despite being a rich source of iron, copper, zinc, and chromium, ginger should not be regarded as a primary source of these nutrients in the human diet; instead, other additions should be added (Koch et al. 2017).

When compared to products grown in regular cultivars, the elemental composition of ginger rhizomes grown in ecological plantations is better for human health, because it contains higher levels of potassium and manganese, lower levels of toxic heavy metals, and a low sodium content (Koch et al. 2017).

### 8.2.2 Essential Oil Contents of *Zingiber officinale*

Ginger's volatile essential oil (0.3–3%), which may be extracted by steam distillation, is responsible for the spice and scent of ginger (Zhang et al. 2021; Khan et al. 2016; Zadeh and Kor 2014). Analytical tools like Gass Chromatography Mass Spectroscopy (GC-MS), High Perormance Liquid Chromatography (HPLC), and Liquid Chromatography Mass Spectroscopy (LC-MS) can be used to determine the exact constituents of ginger volatile oils (Beristain-Bauza et al. 2019). Ginger oil is mostly utilized in the fragrance and soft drink sectors, as well as a culinary flavoring agent in pickles, sauces, confections, baked goods, and confections (Zhang et al. 2021).

Ginger oil is composed of more than 70 ingredients, the majority of which are monoterpene hydrocarbons, oxygenated monoterpenes, and sesquiterpene hydrocarbons (Fig. 8.2). Ginger oil mostly consists of $\alpha$-zingiberene (30–70%), with lesser levels of $\beta$-bisabolene (10–15%) and $\beta$-sesquiphellandrene (15–20%) (Zhang et al. 2021). A significant amount of other components, including curcumene, $\beta$-phellandrene, $\alpha$-farnesene, geranial, geraniol, geranyl acetate, borneol, cineole, terpineol, limonene, camphene, and linalool, has also been found in ginger oil (Khan et al. 2016; Zadeh and Kor 2014).

Specifically, linalool, geraniol, geranial, neral, and borneol were among the oils with the highest value in flavor factors (An et al. 2016). The food sector safely uses ginger oil, with 300–400 tons produced worldwide in 2017 (Zhang et al. 2021).

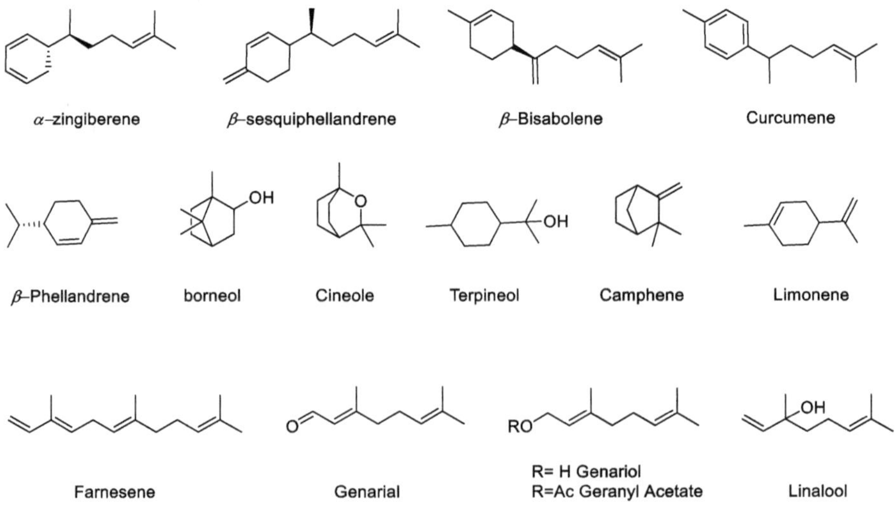


**Fig. 8.2** Chemical structures of the main constituents of *Zingiber officinale* essential oil

While these compounds are characteristics of ginger volatile oil, there is also a significant amount of variation in the quantities and compositions of various sample batches (Zhang et al. 2021). The primary volatile component variations can be ascribed to the ginger's distinct geographical origin, cultivation conditions, drying procedure, storage conditions, extraction techniques, and types of solvents employed (Zhang et al. 2021; An et al. 2016). While convective and intermittent microwave drying would be extremely promising for thermosensitive material, air and infrared drying were the first choice drying strategies to preserve volatile compounds of ginger (An et al. 2016).

Many of these essential oils are employed in a variety of therapeutic formulations in addition to their unique flavor and aroma and application in numerous food items. Furthermore, studies in immuno-suppressed mice have demonstrated that ginger essential oil improved the humoral immune response (Carrasco et al. 2009).

### 8.2.3 Application of *Zingiber officinale* Products in Food Industry

The two main factors that cause food products to deteriorate are microbial growth and lipid oxidation. The usage of ginger in food products has grown recently to prevent food oxidation and microbial growth due to its antibacterial and antioxidant qualities. This has made ginger a good natural preservative in a variety of food products (Fig. 8.3). Ginger extract represents an example of a natural antioxidant

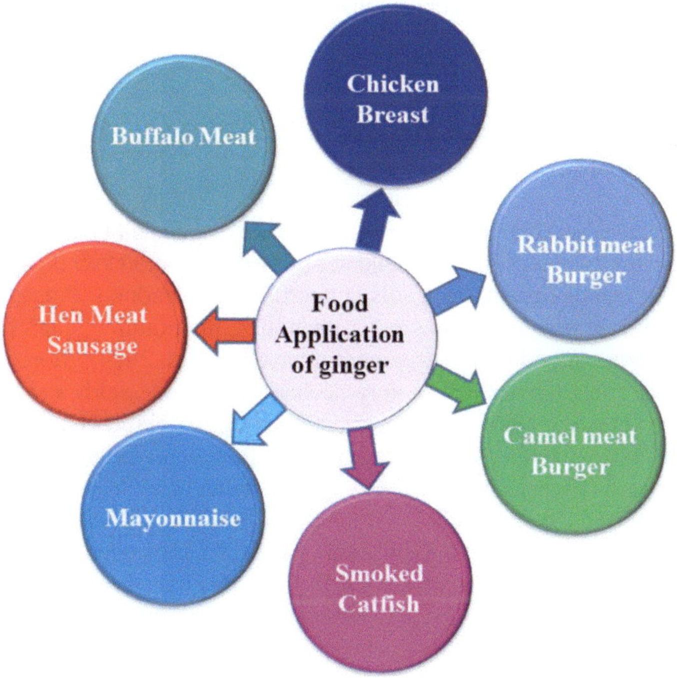


**Fig. 8.3** Food application of *Zingiber officinale*-derived products in some food products

that can be used in processed foods that contain lipids, because it showed the strongest dose-dependent antioxidant effects of lipids, particularly at high temperatures (Si et al. 2018). Additionally, ginger powder and essential oil are added to the formulation of edible films to give them antioxidant and antifungal qualities without changing the film's permeability features (Noori et al. 2018).

By inhibiting the growth of yeasts, molds, and psychrophilic bacteria, an edible coating containing 6% of ginger essential oil extended the shelf life of chicken breast (Noori et al. 2018).

Also, the addition of 1–2% of ginger powder to both raw and cooked rabbit burgers resulted in a significant increase in their polyunsaturated fatty acids and a reduction in bacterial growth. Furthermore, a greater antioxidant capability was noted with an increase in ginger content (Beristain-Bauza et al. 2019). Ginger extract reduced the levels of total coliforms, yeasts, molds, and thiobarbituric acid in hen meat banger during storage (Beristain-Bauza et al. 2019).

Additionally, ginger extract has been utilized to modify the proteins in meat. For instance, after 3 months of freezing, burger patties made of camel meat had better lipid stability, increased volatile chemicals, and fragmented myofibrils (Beristain-Bauza et al. 2019). Additionally, cooked buffalo meat's shear force values were lowered and collagen, sarcoplasmic, and myofibrillar protein solubility was raised by addition of 5% ginger extract (Naveena and Mendiratta 2004). Free fatty acids, trimethylamine, and molds in smoked catfish were all decreased by addition of (1–10%) ginger aqueous solution (Idris et al. 2010), and mayonnaise's lipid peroxidation was also decreased without compromising its sensory or rheological qualities (Beristain-Bauza et al. 2019). In general, the shelf life of many food goods—especially those high in proteins and fats—is extended by the preservation properties of various ginger-derived products.

## 8.3 Health Benefits of *Zingiber officinale* Ingredients and Bioactive Compounds

Ginger contains several phytoconstituents, such as terpene compounds (bisabolene, curcumene, zingiberene, farnesene, and sesquiphellandrene) (Yeh et al. 2014) and phenolic compounds (mainly shogaols, gingerols, and paradols) (Prasad and Tyagi 2015), in addition, ginger contains other phenolics, for example, quercetin, zingerone, gingerenone-A and 6-dehydrogingerdione (Yeh et al. 2014). Moreover, it also contains other constituents such as polysaccharides, organic acids and lipids (Mao et al. 2019). Fresh ginger contains gingerols (e.g., 6-gingerol, 8-gingerol and 10-gingerol), but with long-time storage or heat treatment, gingerols may be converted to shogaols (Schadich et al. 2016). Shogaols can be hydrogenated to produce paradols (Stoner 2013) (Fig. 8.4).

Many bioactive components of ginger are essential for its wide range of pharmacological effects. Ginger's pharmacological properties have been thoroughly studied, including antioxidant activity, anti-inflammatory activity, antimicrobial activity, antidiabetic activity, antiobesity activity, neuroprotective activity, antitumor

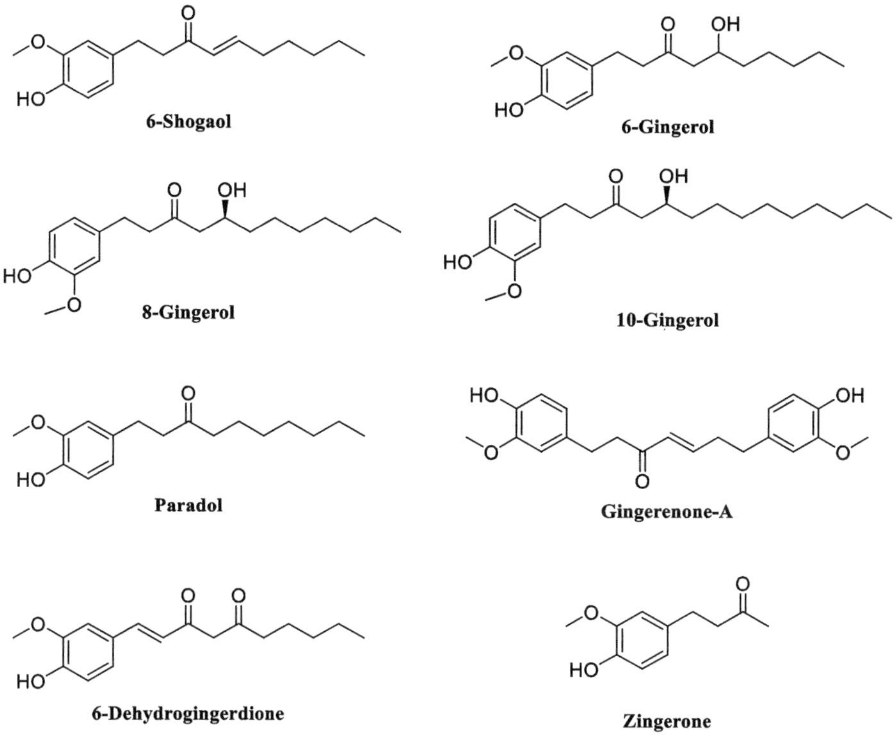


**Fig. 8.4** Chemical structure of bioactive major constituents of *Zingiber officinale*

activity, cardiovascular protective activity, antiemetic activity, broncho-protective effect, and hepatoprotective activity (Ma et al. 2021; Mao et al. 2019) (Table 8.2).

### 8.3.1 Antioxidant Activity

*Zingiber officinale* extract exhibited antioxidant activity by scavenging DPPH and reducing its capacity, comparable with butylated hydroxytoluene (BHT) by inhibiting effect of the hydroxyl radicals, this is because the ginger extract's polyphenols have a greater potential to chelate, which prevents the production of hydroxyl radicals, which are known to cause lipid peroxidation (Stoilova et al. 2007). Moreover, ginger's phenolic components, including 6-gingerol and its derivatives, have strong antioxidant properties (Maizura et al. 2011); due to these compounds having a substituent on their alkyl chain, they may have both radical scavenging and inhibitory effects on the peroxyl radical-induced liposome peroxidation (Ghasemzadeh et al. 2011). Moreover, 6-shogaol has demonstrated the strongest antioxidant activity compared with other gingerols against DPPH radical, superoxide radical, and hydroxyl radical, when the concentrations of these compounds increased the free radical scavenging activity increased (Dugasani et al. 2010).

**Table 8.2** Health benefits of *Zingiber officinale*

| Health benefit | Extract/compounds | Mechanism | Reference |
|---|---|---|---|
| Antioxidant activity | Ginger extract | Chelatoforming capacity resulting in the inhibition of hydroxyl radical production, which is what causes lipid peroxidation. | Stoilova et al. (2007) |
| | 6-Gingerol and its derivatives. | Inhibiting the peroxidation of liposomes caused by peroxyl radicals. | Ghasemzadeh et al. (2011) and Maizura et al. (2011) |
| | 6-Shogaol | Comprising a pro-oxidant iron species that undergoes the Fenton reaction to produce hydroxyl radicals, the encapsulation demonstrated a commensurate inhibitory effect on Fenton-generated hydroxyl radicals. | Dugasani et al. (2010) |
| Anti-inflammatory activity | Ginger extract | Prevents the induction of many genes that code for inflammatory response proteins, including chemokines and cytokines, also reduces leukotriene biosynthesis by blocking 5-lipoxygenase and suppresses prostaglandin synthesis by inhibiting COX-1 and COX-2. | Grzanna et al. (2005) |
| | 6-Shogaol | Inhibiting phosphatidylinositol-3-kinase/Akt signaling and NF-κB signaling, suggesting that intestinal inflammation plays a protective effect against barrier disruption. | Luettig et al. (2016) |
| | 6-Dehydroshogaol | Inhibition of LPS-induced PGE2 and NO and production. | Zhang et al. (2013) |
| Antimicrobial activity | Ginger essential oil | Destroying the cell membrane's ability to function mostly because of macromolecules such proteins and nucleic acids interfering with the metabolism of energy. | Wang et al. (2020) |
| | Crude extract of ginger and methanol fraction | Preventing *Streptococcus mutans* from adhering, forming biofilms, and synthesizing glucan by downregulating its virulence genes. | Hasan et al. (2015) |
| | Essential oil of ginger | Decreasing the synthesis of ergosterol, influencing membrane integrity, and producing less fumonisin B1 and B2. | Yamamoto-Ribeiro et al. (2013) |

(continued)

**Table 8.2** (continued)

| Health benefit | Extract/compounds | Mechanism | Reference |
|---|---|---|---|
| Antidiabetic activity | Aqueous extract of ginger | Causing an increase in the amount of insulin secreted by the pancreas from β cells or the release of bound insulin. | Al-Amin et al. (2006) |
| | 6-Gingerol and 6-shogaol | Preventing the development of diabetes problems, and preventing the AGEs' precursor, MGO, from being produced by trapping it. | Zhu et al. (2015) |
| | 6-Gingerol | Protect pancreatic β-cells from against oxidative damage and reduced hepatic glucose synthesis and consequently, blood glucose concentrations by regulating the hepatic gene expression of glucose metabolism-related enzymes toward increases in glycogenesis and decreases in gluconeogenesis and glycogenolysis. | Son et al. (2015) |
| Antiobesity activity | An aqueous extract of ginger | Caused a significant decrease in the weight of parametrial adipose tissue, which may have been caused via inhibition of intestine dietary fat absorption. | Han et al. (2005) |
| | Gingerenone A | Inhibiting the process of fat formation and adipogenesis in preadipocyte cells, this was followed by an adjustment of the metabolism of fatty acids through the in vivo and in vitro activation of AMPK. Also, decreased inflammatory response in adipose tissue through the suppression of pro-inflammatory cytokines and the inhibition of macrophage recruitment. | Suk et al. (2017) |
| | 6-Shogaol and 6-gingerol | Acts as a particular PPARδ and increases the expression of genes dependent on PPARδ in human skeletal muscle myotubes that are grown with ensuing impacts on energy homeostasis that promoted fat utilization via improved PPARδ signaling, reducing diet-induced obesity and increasing endurance capacity. | Misawa et al. (2015) |

(continued)

**Table 8.2** (continued)

| Health benefit | Extract/compounds | Mechanism | Reference |
|---|---|---|---|
| Neuroprotective activity | Ginger extract | Enhancing neurotrophic factor BDNF and GAP-43 in the brain and reducing IL-1b, GFAP and NF-κB levels. | Sahin et al. (2019) |
| | 6-Gingerol | Increasing the BDNF protein production through the CREB and protein kinase B/Akt signaling pathway. | Kim et al. (2018) |
| | 6-Shogaol | Enhancing Nrf2, scavenging free radicals, and raising the amounts of many phase II antioxidant molecules, including NQO1 and HO-1. | Peng et al. (2015) |
| Antitumor activity | Ginger extract | Lowering the expression of genes associated with the PI3K/Akt and Ras/ERK pathways, such as the v-Ki-ras2 KRAS, Bcl-xL, ERK, and Akt. | Tahir et al. (2015) |
| | 6-Shogaol | Suppresses NF-κB and STAT3, lowers the expression of cyclin D1, survivin, c-Myc, and Bcl-2, and increases Bax expression. | Saha et al. (2014) |
| | 10-Gingerol | Lowering cell division and causing apoptosis and S phase cell cycle arrest. | Bernard et al. (2017) |
| Cardiovascular protective activity | Aqueous extract of ginger | Inhibitory effect on ACE and a few pro-oxidants in rat heart caused by lipid peroxidation. | Akinyemi (2013) |
| | Ginger extract | Decreasing platelet ADA activity and increasing the level of adenosine, which prevented platelet aggregation. | Akinyemi et al. (2016) |
| | Aqueous extract of ginger | A dual inhibitory action facilitated by muscarinic receptor activation and blockade of calcium channels. | Ghayur et al. (2005) |
| | 6-Gingerol, 8- gingerol and 10-gingerol | Atropine-resistant vasodilator action. | Ghayur et al. (2005) |
| Antiemetic activity | Ginger extract | Inhibiting intestinal neurons' activation of 5-HT receptors. | Walstab et al. (2013) |
| | 6-Shogaol, 6-gingerol and zingerone | Decreasing the transmission of emetic signals in vagal afferent neurons by inhibiting the 5-HT receptor. | Jin et al. (2014) |

(continued)

**Table 8.2** (continued)

| Health benefit | Extract/compounds | Mechanism | Reference |
|---|---|---|---|
| Broncho-protective effect | Ethanol and aqueous ginger extracts | Suppressing the Th2-mediated immune response, which was further supported by a decrease in IL-4 and IL-5 mRNA expression. | Khan et al. (2015) |
| | Ginger extract, (6, 8 and 10-gingerols) | Changing the intracellular calcium ion in smooth muscle airways. | Townsend et al. (2013) |
| | 6-Gingerol, 8-gingerol and 6-shogaol | Suppression of phosphodiesterase 4D. | Townsend et al. (2014) |
| | Ginger extract and 6-shogaol | Increasing the cAMP concentrations in CD4 cells and prevents NF-κB from being induced signaling and proinflammatory cytokine release. | Yocum et al. (2020) |
| Hepatoprotective activity | Ginger extract | Decreasing liver marker enzymes and decreasing proinflammatory cytokine levels and the downregulation of the TGF-β1/Smad3 and NF-κB/IκB pathways in rats given $CCl_4$. | Hasan et al. (2016) |
| | Ginger extract | Decreasing increased levels of immune-expression of the proapoptotic protein (Bax), alkaline phosphatase, and serum aminotransferase. | Huang (2019) |
| | 6-Gingerol | Reducing the serum levels of total bilirubin, hepatic marker enzymes (AST, ALT, and ALP), and hepatic malondialdehyde production, as well as the reduction of the liver's antioxidant state. | Sabina et al. (2011) |

### 8.3.2 Anti-inflammatory Activity

Ginger effectively exhibited anti-inflammatory activity via many mechanisms such as suppression of cyclooxygenase (COX) and lipoxygenase (LOX) and control of prostaglandins, cytokines, nitric oxide, and interleukins (Gunathilake and Rupasinghe 2015). Ginger extract modifies biochemical pathways activated by chronic inflammation by inhibiting the activation of numerous genes encoding inflammatory response proteins, including chemokines and cytokines, further inhibiting the production of prostaglandins by blocking COX-1 and COX-2, as well as the production of leukotrienes by blocking 5-lipoxygenase (Grzanna et al. 2005). Also, 6-shogaol prevents barrier disruption caused by the pro-inflammatory cytokine tumor necrosis factor (TNF-α) by inhibiting phosphatidylinositol-3-kinase/Akt signaling and nuclear factor-κB (NF-κB) signaling, suggesting that intestinal inflammation plays a protective effect against barrier disruption (Luettig et al.

2016). Furthermore, 6-dehydroshogaol showed more potent anti-inflammatory activity than 6-shogaol and 6-gingerol, by inhibition of lipopolysaccharide (LPS)-induced nitric oxide (NO) and prostaglandin E2 (PGE2) production in RAW 264.7 cells than 6-gingerol and 6-shogaol (Zhang et al. 2013).

### 8.3.3 Antimicrobial Activity

Many reported data have showed that ginger extract can suppress the growth of many bacteria, fungi, and viruses (Munda et al. 2018; Juwita et al. 2018). Ginger essential oil showed bactericidal action against pathogenic bacteria *Staphylococcus aureus* and *Escherichia coli* by destroying the cell membrane's ability to function mostly as a result of macromolecules including proteins and nucleic acids interfering with the metabolism of energy (Wang et al. 2020). Also, the crude extract and methanolic fraction of *Zingiber officinale* suppress biofilm formation, glucan production, and the adherence of *Streptococcus mutans*; the primary causes of the observed decline in virulence traits are thought to be the downregulation of the virulence genes. Consistent with an in vitro investigation, it was showed that a group of rats treated with *Streptococcus mutans* had less caries formation (Hasan et al. 2015). Moreover, the essential oil of ginger effectively inhibited antifungal activity against the growth of *Fusarium verticillioides* by not only reducing ergosterol biosynthesis and affecting membrane integrity but also decreasing the production of fumonisin B1 and fumonisin B2 (Yamamoto-Ribeiro et al. 2013). Furthermore, the ginger ethanolic extract showed decrease hepatitis C virus (HCV) viral load, liver function parameters, alpha fetoprotein, and improved clinical condition to patients (Abdel-Moneim et al. 2013).

### 8.3.4 Antidiabetic Activity

Studies have shown that ginger has hypoglycemic effects both in vitro and in vivo, supporting the traditional usage of ginger in the treatment of diabetes mellitus (Gunathilake and Rupasinghe 2015). Reversing the altered metabolism of fats and carbohydrates, as well as insulin release and sensitivity, are the processes underpinning the antidiabetic effect (Srinivasan 2017). This is caused by the active constituents in ginger, shogaol, and gingerols. Additionally, ginger has demonstrated notable protective effects against the secondary complications of diabetes that affect the kidney, liver, eyes, and nervous system (Li et al. 2012). Ginger aqueous extract was shown to lower the levels of triacylglycerol, cholesterol, and blood glucose when administered at a dose of 500 mg/kg body weight for 7 weeks to diabetic rats induced with streptozotocin (STZ). Ginger may have a hypoglycemic activity because of its effects on serotonin receptors, an increase in insulin secretion from β cells in the pancreas, or the release of bound insulin (Al-Amin et al. 2006). Moreover, 6-gingerol and 6-shogaol prevented the progression of diabetic complications in vitro, and they also inhibited the production of advanced glycation end-products (AGEs) by trapping methylglyoxal (MGO), which is the precursor of AGEs (Zhu et al. 2015). In addition, 6-gingerol inhibited hypoglycemic effect by the suppression of the increased levels of fasting blood glucose, and increased the obese diabetic mice's sensitivity to glucose. It also protected pancreatic β-cells from oxidative damage and controlled the hepatic gene expression of glucose metabolism-related enzymes. This promoted glycogenesis, and decreased gluconeogenesis and glycogenolysis, which helped to lower hepatic glucose synthesis and, consequently, blood glucose concentrations (Son et al. 2015).

### 8.3.5 Antiobesity Activity

Ginger has been used in managing and preventing obesity, according to several studies (Suk et al. 2017; Mahmoud and Elnour 2013). In vitro experiment in mice exhibited that an aqueous extract of ginger resulted in a significant decrease in parametrial adipose tissue weight, which may have been induced by ginger's active ingredients' suppression of intestinal fat absorption (Han et al. 2005). Moreover, gingerenone A inhibited the process of fat formation and adipogenesis in preadipocyte cells; this was followed by an adjustment of the metabolism of fatty acids through the in vivo and in vitro activation of AMP-activated protein kinase (AMPK). Also, gingerenone A decreased inflammatory response in adipose tissue through the suppression of pro-inflammatory cytokines and the inhibition of macrophage recruitment (Suk et al. 2017). In addition, 6-gingerol and 6-shogaol acted as specific peroxisome proliferator-activated receptor δ (PPARδ) and stimulated PPARδ-dependent gene expression in cultured human skeletal muscle myotubes with ensuing implications for energy homeostasis that reduced obesity brought on by food and greater endurance through elevated PPARδ signaling that stimulates fat utilization; this is an important therapy for obesity treatment (Misawa et al. 2015).

### 8.3.6 Neuroprotective Activity

Neuroprotection refers to protect the central nervous system from acute and long-term neurodegenerative diseases that cause neuronal damage, for example, Parkinson's disease and Alzheimer's disease, especially elderly people (Uddin et al. 2013). Ginger has been shown in many studies to have anti-inflammatory properties and to improve memory, both of which may help manage and prevent neurodegenerative illnesses diseases (Park et al. 2013; Huh et al. 2018). Ginger extract showed neuroprotection in a model of traumatic brain damage by improving brain derived neurotrophic factors (BDNF) and growth-associated protein-43 (GAP-43) in the brain and decreasing the levels of nuclear factor-κB (NF-κB), glial fibrillary acidic protein (GFAP), and interleukin 1-beta (IL-1b); so, it is possible that ginger's anti-oxidative and anti-inflammatory properties contribute to its neuroprotective activity (Sahin et al. 2019). Furthermore, 6-gingerol in mice with scopolamine-induced amnesia showed a notable improvement in cognitive dysfunctions and memory via increasing the BDNF protein production through the cAMP-response element binding protein (CREB) and protein kinase B/Akt signaling pathway (Kim et al. 2018). Moreover, 6-shogaol showed neuroprotective activity by enhancing nuclear factor erythroid 2-related factor 2 (Nrf2), scavenging free radicals, and increasing the concentrations of various phase II antioxidant compounds, including heme oxygenase-1 (HO-1) and NAD(P)H:quinone oxidoreductase 1 (NQO1) in neuron-like rat pheochromocytoma PC12 cell lines (Peng et al. 2015).

### 8.3.7 Antitumor Activity

Ginger is thought to have anticancer properties primarily because it scavenges free radicals, modifies gene expression, and induces apoptosis, all of which work together to prevent or delay the initiation, development, and advancement of tumors (Baliga et al. 2011). *Zingiber officinale* extract induced apoptosis by reducing the expression of genes connected the Ras/extracellular signal-regulated kinase (ERK) and PI3K/Akt pathways, such as the v-Ki-ras2 Kirsten rat sarcoma viral oncogene homolog (KRAS), ERK, Akt, and B-cell lymphoma-extralarge (Bcl-xL). Additionally, it enhanced the production of caspase 9, which induced apoptosis in HT-29 colorectal cancer cells (Tahir et al. 2015). In addition, 6-shogaol may have a greater effect on cell survival and apoptosis in human and mouse prostate cancer cells than 6-gingerol and 6-paradol because it suppresses NF-κB and signal transducer and activator of transcription 3 (STAT3), lowers the expression of cyclin D1, survivin, c-Myc, and B-cell lymphoma 2 (Bcl-2), and increases Bax expression (Saha et al. 2014). Moreover, 10-gingerol inhibited the growth of breast cancer cells in mice and humans. It caused apoptosis, S phase cell cycle arrest, and decreased cell division (Bernard et al. 2017).

### 8.3.8 Cardiovascular Protective Activity

Hypertension and dyslipidemia increase the risk of cardiovascular diseases, such as coronary heart disease and stroke (Akinyemi et al. 2015). Many reported data has shown that ginger decreases the levels of lipids and blood pressure, contributing to protection from cardiovascular diseases (de Las et al. 2017). An in vitro experiment in rats heart represented the aqueous ginger extract showed antihypertensive activity by the inhibition of some pro-oxidants and the angiotensin-converting enzyme (ACE) brought on by lipid peroxidation (Akinyemi 2013). Moreover, ginger extract provided protection in hypertensive rats, against problems resulting from hypertension by lowering the activity of platelet adenosine deaminase (ADA) and elevating adenosine levels. This inhibited platelet aggregation and encouraged vasodilation (Akinyemi et al. 2016). Furthermore, the aqueous extract of ginger as well as its phenolic compounds (6-gingerol, 8-gingerol, and 10-gingerol) exhibited hypotensive and vasodilator effects. The aqueous extract of ginger decreased blood pressure by stimulating muscarinic receptors and by blocking calcium channels. While its phenolic compounds (6-gingerol, 8-gingerol, and 10-gingerol) decreased blood pressure via promising atropine-resistant vasodilator effect (Ghayur et al. 2005).

### 8.3.9 Antiemetic Activity

Traditionally, ginger has been used to alleviate gastrointestinal symptoms, and current studies have shown that ginger can successfully relieve emesis and nausea (Marx et al. 2016; Bossi et al. 2017). Ginger extract reduced nausea and vomiting brought on by chemotherapy by inhibiting enteric neurons' 5-hydroxytryptamine (5-HT) receptor activity (Walstab et al. 2013). In addition, ginger helped patients experience less nausea after antiretroviral therapy and antituberculosis medications, and it decreased how frequently individuals experienced mild, moderate, and severe episodes of nausea (Emrani et al. 2016). Moreover, 6-shogaol, 6-gingerol, and zingerone suppressed the transmission of emetic signals in vagal afferent neurons in an in vitro experiment by inhibiting the 5-HT receptor; also 6-shogaol demonstrated the greatest inhibitory efficiency (Jin et al. 2014). Although the exact mechanism of ginger's antiemetic activity is still unknown, it is thought that ginger's active constituents improve gastric tone and motility by acting on the GIT's antiserotonergic and anticholinergic receptors (Abdel-Aziz et al. 2006).

### 8.3.10 Broncho-Protective Effect

Many studies have shown that ginger and its bioactive components have bronchodilating properties (Mangprayool et al. 2013; Rouhi et al. 2006). The ethanol and aqueous extract of ginger reduced Th2-mediated immune response, which may have been achieved by blocking allergic inflammation; it was further supported by a drop in IL-4 and IL-5 mRNA expression levels (Khan et al. 2015). Furthermore, ginger extract and its phenolic compounds (6-gingerol, 8-gingerol, and 10-gingerol) revealed bronchodilation rapid relaxation in the smooth muscle of the isolated human airway via adjusting the calcium ion levels in airway smooth muscle (Townsend et al. 2013). In addition, 6-gingerol, 8-gingerol, and 6-shogaol enhanced β-agonist-induced relaxation in human airway smooth muscle through the inhibition of phosphodiesterase 4D (Townsend et al. 2014). Moreover, ginger extract and 6-shogaol exhibited antiasthmatic effect by increasing the cyclic adenosine monophosphate (cAMP) in CD4 cells, preventing the activation of NF-κB signaling, and reducing the production of proinflammatory cytokines (Yocum et al. 2020).

### 8.3.11 Hepatoprotective Activity

Ginger has been demonstrated to exhibit hepatoprotective effects through its antioxidant activity (Liu et al. 2013). Ginger extract prevented liver injury by $CCl_4$-induced inflammation via the reduced liver marker enzymes and the decreased proinflammatory cytokine levels and the downregulation of the transforming growth factor-β1 (TGF-β1)/Smad3 and NF-κB/IκB pathways in rats given $CCl_4$ (Hasan et al. 2016). Also, ginger extract exhibited hepatoprotective effect in piroxicam-induced liver toxicity by lowering increased levels of alkaline phosphatase, serum aminotransferase, and immune-mediated proapoptotic protein (Bax) expression (Huang 2019). Moreover, 6-gingerol exhibited a strong hepatoprotective activity by reducing serum levels of total bilirubin and hepatic marker enzymes (aspartate aminotransferase [AST], alanine aminotransferase [ALT], and alkaline phosphatase [ALP]). It also stopped the livers of mice given acetaminophen from becoming overly reactive to malondialdehyde and from losing their antioxidant status (Sabina et al. 2011).

### 8.3.12 Other Pharmacological Activities

In addition to the previously listed bioactivities, ginger provides other health benefits, including the following:

*Antirheumatic effect*: Preclinical research has demonstrated the antirheumatic properties of regularly used ginger and its phytochemicals, such as 6-shogaol, zingerone, and cedrol. They cause pro-inflammatory cytokines to be suppressed and modify signaling pathways that are crucial to the pathogenesis of rheumatoid arthritis and osteoarthritis (Szymczak et al. 2024).

*Antiallergic effect*: Ginger exhibited antiallergic activity and inhibited allergic reactions by decreasing release of β-hexosaminidase from rat basophilic leukemia (RBL-2H3) cells (Chen et al. 2009). Also, ginger lowered serum total IgE, making it an effective treatment for allergic rhinitis after 4 weeks of treatment (Alsamarai et al. 2015).

*Nephroprotective effect*: The ginger aqueous ethanolic extract exhibited nephroprotective effect via increasing glutathione activity and through stopping the reduction in renal antioxidant status by doxorubicin. (Ajith et al. 2008). Moreover, gentamicin-induced nephrotoxicity was deceased by the gingerol-enriched fraction, and this protection was linked to decreases in oxidative stress and NO generation as well as an inhibition of the expression of inflammatory genes (Rodrigues et al. 2014).

*Larvicidal activity*: 10-shogaol, 6-shogaol, 6-gingerol, 10-gingerol, and hexahydrocurcumin have larvicidal activity against larvae of *Angiostrongylus cantonensis* or decreased their spontaneous movements in a manner that was dependent on both time and dose (Lin et al. 2010).

*Antischistosomal activity*: The aqueous extract of ginger exhibited antischistosomal effect of with *Schistosoma mansoni*; the male worms that were collected from mice which were given ginger treatment had abnormal surface architecture, with their surface displaying a partial loss of tubercle spines, significant erosion in the tegumental regions between tubercles, and a large number of minute blebs around tubercles (Mostafa et al. 2011).

*Effect on migraine*: Ginger extract showed increased effectiveness in preventing migraines by reducing headache duration and severity in a clinical trial that was randomized, placebo-controlled (Helli et al. 2022).

*Effect of ginger on eye*: In the rabbit eyes, ginger extract reduces intraocular pressure (IOP), and it may also be a helpful treatment for IOP in people, when IOP increase is the primary risk factor for the development of glaucoma (Akpalaba et al. 2009).

*Radio protective activity*: Ginger extract significantly protected rats against whole-body gamma-rays and helped to mitigate the effects of radiation. The findings showed that ginger extract decreased radiation-induced chromosomal abnormality and micronucleus formation, as well as DNA oxidation by restoring serum levels of 8-hydroxy-2′-deoxyguanosine. Furthermore, pretreatment with ginger extract preserved the integrity of the sperm DNA and decreased the incidence of sperm abnormalities following irradiation. This reduced DNA damage in both bone marrow cells and epididymal sperm (Abd El-Monem and Elwakeel 2020).

## 8.4 Conclusion

Nutraceuticals, bioactive substances with health benefits, are increasingly popular in food and pharmaceutical products. *Zingiber officinale* (ginger) has numerous pharmacological properties, including antioxidant, anti-inflammatory, and antimicrobial effects. It contains highly valued nutrients such as minerals, vitamins, and essential oils. Ginger oil is used as a spice, preservative, and flavoring in food. The chemical composition of ginger is influenced by several production variables. This chapter explores ginger's chemical composition, medicinal and nutritional potentials, emphasizing its health and nutritional benefits in human diets and food sectors.

## References

Abd El-Monem D, Elwakeel S (2020) Radioprotective efficacy of ginger (*Zingiber officinale*) extract against gamma-ray radiation-induced genotoxicity in rats. Int J Radiat Res 18(1):43–55

Abdel-Aziz H, Windeck T, Ploch M, Verspohl EJ (2006) Mode of action of gingerols and shogaols on 5-HT3 receptors: binding studies, cation uptake by the receptor channel and contraction of isolated Guinea-pig ileum. Eur J Pharmacol 530(1–2):136–143

Abdel-Moneim A, Morsy BM, Mahmoud AM, Abo-Seif MA, Zanaty MI (2013) Beneficial therapeutic effects of *Nigella sativa* and/or *Zingiber officinale* in HCV patients in Egypt. EXCLI J 12:943

Ajith T, Aswathy M, Hema U (2008) Protective effect of *Zingiber officinale* roscoe against anticancer drug doxorubicin-induced acute nephrotoxicity. Food Chem Toxicol 46(9):3178–3181

Akimoto M, Iizuka M, Kanematsu R, Yoshida M, Takenaga K (2015) Anticancer effect of ginger extract against pancreatic cancer cells mainly through reactive oxygen species-mediated autotic cell death. PLoS One 10(5):e0126605

Akinyemi A (2013) Ginger varieties (*Zingiber officinale*) inhibit key enzyme linked to hypertension (Angiotensin-I converting enzyme) and some pro-oxidants induced lipid peroxidation in rat heart: in *vitro*. J Clin Exp Cardiolog 4:4

Akinyemi AJ, Thome GR, Morsch VM, Stefanello N, Goularte JF, Belló-Klein A, Oboh G, Schetinger MRC (2015) Effect of dietary supplementation of ginger and turmeric rhizomes on angiotensin-1 converting enzyme (ACE) and arginase activities in L-NAME induced hypertensive rats. J Funct Foods 17:792–801

Akinyemi AJ, Thomé GR, Morsch VM, Bottari NB, Baldissarelli J, de Oliveira LS, Goularte JF, Belló-Klein A, Oboh G, Schetinger MRC (2016) Dietary supplementation of ginger and turmeric rhizomes modulates platelets ectonucleotidase and adenosine deaminase activities in normotensive and hypertensive rats. Phytother Res 30(7):1156–1163

Akpalaba R, Agu C, Adeleke O, Asonye C (2009) Effect of orally administered *Zingiber Officinale* on the intra ocular pressure of experimental rabbits. Int J Health Res 2(3):273–278

Al-Amin ZM, Thomson M, Al-Qattan KK, Peltonen-Shalaby R, Ali M (2006) Anti-diabetic and hypolipidaemic properties of ginger (*Zingiber officinale*) in streptozotocin-induced diabetic rats. Br J Nutr 96(4):660–666

Alsamarai AM, Hamid MA, Alobaidi AHA (2015) Therapeutic and immunologic effects of *Zingiber officiale* in allergic rhinitis. In: Allergic diseases: new insights, p 153

An K, Zhao D, Wang Z, Wu J, Xu Y, Xiao G (2016) Comparison of different drying methods on Chinese ginger (*Zingiber officinale* roscoe): changes in volatiles, chemical profile, antioxidant properties, and microstructure. Food Chem 197:1292–1300

Baliga MS, Haniadka R, Pereira MM, D'Souza JJ, Pallaty PL, Bhat HP, Popuri S (2011) Update on the chemopreventive effects of ginger and its phytochemicals. Crit Rev Food Sci Nutr 51(6):499–523

Beristain-Bauza SDC, Hernández-Carranza P, Cid-Pérez TS, Ávila-Sosa R, Ruiz-López II, Ochoa-Velasco CE (2019) Antimicrobial activity of ginger (*Zingiber officinale*) and its application in food products. Food Rev Intl 35(5):407–426

Bernard MM, McConnery JR, Hoskin DW (2017) [10]-Gingerol, a major phenolic constituent of ginger root, induces cell cycle arrest and apoptosis in triple-negative breast cancer cells. Exp Mol Pathol 102(2):370–376

Bossi P, Cortinovis D, Fatigoni S, Rocca MC, Fabi A, Seminara P, Ripamonti C, Alfieri S, Granata R, Bergamini C (2017) A randomized, double-blind, placebo-controlled, multicenter study of a ginger extract in the management of chemotherapy-induced nausea and vomiting (CINV) in patients receiving high-dose cisplatin. Ann Oncol 28(10):2547–2551

Carrasco FR, Schmidt G, Romero AL, Sartoretto JL, Caparroz-Assef SM, Bersani-Amado CA, Cuman RKN (2009) Immunomodulatory activity of *Zingiber officinale* Roscoe, *Salvia officinalis* L. and *Syzygium aromaticum* L. essential oils: evidence for humor-and cell-mediated responses. J Pharm Pharmacol 61(7):961–967

Chen B-H, Wu P-Y, Chen K-M, Fu T-F, Wang H-M, Chen C-Y (2009) Antiallergic potential on RBL-2H3 cells of some phenolic constituents of Zingiber officinale (ginger). J Nat Prod 72(5):950–953

de Las HN, Valero-Muñoz M, Martín-Fernández B, Ballesteros S, López-Farré A, Ruiz-Roso B, Lahera V (2017) Molecular factors involved in the hypolipidemic-and insulin-sensitizing effects of a ginger (*Zingiber officinale* roscoe) extract in rats fed a high-fat diet. Appl Physiol Nutr Metab 42(2):209–215

Dugasani S, Pichika MR, Nadarajah VD, Balijepalli MK, Tandra S, Korlakunta JN (2010) Comparative antioxidant and anti-inflammatory effects of [6]-gingerol,[8]-gingerol,[10]-gingerol and [6]-shogaol. J Ethnopharmacol 127(2):515–520

Emrani Z, Shojaei E, Khalili H (2016) Ginger for prevention of antituberculosis-induced gastrointestinal adverse reactions including hepatotoxicity: a randomized pilot clinical trial. Phytother Res 30(6):1003–1009

Ghasemzadeh A, Jaafar HZ, Rahmat A (2011) Effects of solvent type on phenolics and flavonoids content and antioxidant activities in two varieties of young ginger (*Zingiber officinale* roscoe) extracts. J Med Plants Res 5(7):1147–1154

Ghayur MN, Gilani AH, Afridi MB, Houghton PJ (2005) Cardiovascular effects of ginger aqueous extract and its phenolic constituents are mediated through multiple pathways. Vasc Pharmacol 43(4):234–241

Grzanna R, Lindmark L, Frondoza CG (2005) Ginger—an herbal medicinal product with broad anti-inflammatory actions. J Med Food 8(2):125–132

Gunathilake K, Rupasinghe HV (2015) Recent perspectives on the medicinal potential of ginger. In: Botanics: targets and therapy, pp 55–63

Hamed MA, Ali SA, El-Rigal NS (2012) Therapeutic potential of ginger against renal injury induced by carbon tetrachloride in rats. Sci World J 2012:840421

Han L-K, Gong X-J, Kawano S, Saito M, Kimura Y, Okuda H (2005) Antiobesity actions of *Zingiber officinale* roscoe. Yakugaku zasshi: J Pharm Soc Japan 125(2):213–217

Hasan S, Danishuddin M, Khan AU (2015) Inhibitory effect of zingiber officinale towards *Streptococcus mutans* virulence and caries development: in *vitro* and in *vivo* studies. BMC Microbiol 15:1–14

Hasan IH, El-Desouky M, Hozayen WG, Abd el Aziz GM (2016) Protective effect of *Zingiber officinale* against CCl4-induced liver fibrosis is mediated through downregulating the TGF-ß1/Smad3 and NF-κB/IκB pathways. Pharmacology 97(1–2):1–9

Helli B, Anjirizadeh F, Mehramiri A, Shalilahmadi D, Latifi SM (2022) The effect of ginger (*Zingiber officinale* Rosc.) consumption in headache prophylaxis in patients with migraine: a randomized placebo-controlled clinical trial. Jundishapur J Nat Pharm Prod 17(3):e120449

Huang Y-S (2019) The hepatoprotective effect of ginger. J Chin Med Assoc 82(11):805–806

Huh E, Lim S, Kim HG, Ha SK, Park H-Y, Huh Y, Oh MS (2018) Ginger fermented with *Schizosaccharomyces pombe* alleviates memory impairment via protecting hippocampal neuronal cells in amyloid beta 1–42 plaque injected mice. Food Funct 9(1):171–178

Idris G, Omojowo FS, Omojasola P, Adetunji C, Ngwu E (2010) The effect of different concentrations of ginger on the quality of smoked dried catfish (*Clarias gariepinus*). Nat Sci 8(4):59–63

Jin Z, Lee G, Kim S, Park C-S, Park YS, Jin Y-H (2014) Ginger and its pungent constituents non-competitively inhibit serotonin currents on visceral afferent neurons. Korean J Physiol Pharmacol 18(2):149

Juwita T, Puspitasari IM, Levita J (2018) Torch ginger (Etlingera elatior): a review on its botanical aspects, phytoconstituents and pharmacological activities. Pak J Biol Sci 21(4):151–165

Khan AM, Shahzad M, Raza Asim M, Imran M, Shabbir A (2015) *Zingiber officinale* ameliorates allergic asthma via suppression of Th2-mediated immune response. Pharm Biol 53(3):359–367

Khan S, Pandotra P, Qazi AK, Lone SA, Muzafar M, Gupta AP, Gupta S (2016) Medicinal and nutritional qualities of *Zingiber officinale*. In: Fruits, vegetables, and herbs. Elsevier, pp 525–550

Kim C-Y, Seo Y, Lee C, Park GH, Jang J-H (2018) Neuroprotective effect and molecular mechanism of [6]-gingerol against scopolamine-induced amnesia in C57BL/6 mice. Evid Based Complement Alternat Med 2018:8941564

Koch W, Kukula-Koch W, Marzec Z, Kasperek E, Wyszogrodzka-Koma L, Szwerc W, Asakawa Y (2017) Application of chromatographic and spectroscopic methods towards the quality assessment of ginger (*Zingiber officinale*) rhizomes from ecological plantations. Int J Mol Sci 18(2):452

Li Y, Tran VH, Duke CC, Roufogalis BD (2012) Preventive and protective properties of *Zingiber officinale* (ginger) in diabetes mellitus, diabetic complications, and associated lipid and other metabolic disorders: a brief review. Evid Based Complement Alternat Med 2012:1

Lin R-J, Chen C-Y, Chung L-Y, Yen C-M (2010) Larvicidal activities of ginger (*Zingiber officinale*) against Angiostrongylus cantonensis. Acta Trop 115(1–2):69–76

Liu C-T, Raghu R, Lin S-H, Wang S-Y, Kuo C-H, Tseng YJ, Sheen L-Y (2013) Metabolomics of ginger essential oil against alcoholic fatty liver in mice. J Agric Food Chem 61(46):11231–11240

Luettig J, Rosenthal R, Lee IFM, Krug SM, Schulzke JD (2016) The ginger component 6-shogaol prevents TNF-α-induced barrier loss via inhibition of PI3K/Akt and NF-κB signaling. Mol Nutr Food Res 60(12):2576–2586

Ma R-H, Ni Z-J, Zhu Y-Y, Thakur K, Zhang F, Zhang Y-Y et al (2021) A recent update on the multifaceted health benefits associated with ginger and its bioactive components. Food Funct 12(2):519–542

Mahmoud R, Elnour W (2013) Comparative evaluation of the efficacy of ginger and orlistat on obesity management, pancreatic lipase and liver peroxisomal catalase enzyme in male albino rats. Eur Rev Med Pharmacol Sci 17(1):75–83

Maizura M, Aminah A, Wan AW (2011) Total phenolic content and antioxidant activity of kesum (*Polygonum minus*), ginger (*Zingiber officinale*) and turmeric (*Curcuma longa*) extract. Int Food Res J 18(2)

Mangprayool T, Kupittayanant S, Chudapongse N (2013) Participation of citral in the bronchodilatory effect of ginger oil and possible mechanism of action. Fitoterapia 89:68–73

Mao Q-Q, Xu X-Y, Cao S-Y, Gan R-Y, Corke H, Beta T et al (2019) Bioactive compounds and bioactivities of ginger (*Zingiber officinale* roscoe). Foods 8(6):185

Marx W, McCarthy A, Ried K, Mckavanagh D, Lohning A, Vitetta L, Sali A, Isenring E (2016) Standardized ginger extract improves quality of life associated with chemotherapy-induced nausea and vomiting. Asia Pac J Clin Oncol 12(S5):87

Misawa K, Hashizume K, Yamamoto M, Minegishi Y, Hase T, Shimotoyodome A (2015) Ginger extract prevents high-fat diet-induced obesity in mice via activation of the peroxisome proliferator-activated receptor δ pathway. J Nutr Biochem 26(10):1058–1067

Mostafa OM, Eid RA, Adly MA (2011) Antischistosomal activity of ginger (*Zingiber officinale*) against Schistosoma mansoni harbored in C57 mice. Parasitol Res 109:395–403

Munda S, Dutta S, Haldar S, Lal M (2018) Chemical analysis and therapeutic uses of ginger (*Zingiber officinale* Rosc.) essential oil: a review. J Essent Oil Bear Plants 21(4):994–1002

Naveena B, Mendiratta S (2004) The tenderization of buffalo meat using ginger extract. J Muscle Foods 15(4):235–244

Noori S, Zeynali F, Almasi H (2018) Antimicrobial and antioxidant efficiency of nanoemulsion-based edible coating containing ginger (*Zingiber officinale*) essential oil and its effect on safety and quality attributes of chicken breast fillets. Food Control 84:312–320

Otunola GA, Oloyede OB, Oladiji AT, Afolayan AJ (2010) Comparative analysis of the chemical composition of three spices–Allium sativum L. *Zingiber officinale* Rosc. and *Capsicum frutescens* L. commonly consumed in Nigeria. Afr J Biotechnol 9(41):6927–6931

Pandotra P, Gupta AP, Husian MK, Gupta S (2013) Genetic and chemo-divergence in eighteen core collection of *Zingiber officinale* from north-West Himalayas. Sci Hortic 160:283–291

Park G, Kim HG, Ju MS, Ha SK, Park Y, Kim SY et al (2013) 6-Shogaol, an active compound of ginger, protects dopaminergic neurons in Parkinson's disease models via anti-neuroinflammation. Acta Pharmacol Sin 34(9):1131–1139

Peng S, Yao J, Liu Y, Duan D, Zhang X, Fang J (2015) Activation of Nrf2 target enzymes conferring protection against oxidative stress in PC12 cells by ginger principal constituent 6-shogaol. Food Funct 6(8):2813–2823

Prasad S, Tyagi AK (2015) Ginger and its constituents: role in prevention and treatment of gastrointestinal cancer. Gastroenterol Res Pract 2015:142979

Rodrigues FA, Prata MM, Oliveira IC, Alves NT, Freitas RE, Monteiro HS et al (2014) Gingerol fraction from *Zingiber officinale* protects against gentamicin-induced nephrotoxicity. Antimicrob Agents Chemother 58(4):1872–1878

Rouhi H, Ganji F, Nasri H (2006) Effects of ginger on the improvement of asthma [the evaluation of its treatmental effects]. Pak J Nutr 5(4):373–376

Sabina EP, Pragasam SJ, Kumar S, Rasool M (2011) 6-gingerol, an active ingredient of ginger, protects acetaminophen-induced hepatotoxicity in mice. Zhong xi yi jie he xue bao= J Chin Integr Med 9(11):1264–1269

Saha A, Blando J, Silver E, Beltran L, Sessler J, DiGiovanni J (2014) 6-Shogaol from dried ginger inhibits growth of prostate cancer cells both in vitro and in *vivo* through inhibition of STAT3 and NF-κB signaling. Cancer Prev Res 7(6):627–638

Sahin K, Kilic E, Balcikanli Z, Ates N, Orhan C, Tuzcu M, Juturu V (2019) Ginger provides neuroprotection in experimental model of traumatic brain injury. FASEB J 33(S1):795.16-.16

Sang S, Snook HD, Tareq FS, Fasina Y (2020) Precision research on ginger: the type of ginger matters. J Agric Food Chem 68(32):8517–8523

Schadich E, Hlaváč J, Volná T, Varanasi L, Hajdúch M, Džubák P (2016) Effects of ginger phenylpropanoids and quercetin on Nrf2-ARE pathway in human BJ fibroblasts and HaCaT keratinocytes. Biomed Res Int 2016:2173275

Shahrajabian MH, Sun W, Cheng Q (2019) Clinical aspects and health benefits of ginger (*Zingiber officinale*) in both traditional Chinese medicine and modern industry. Acta Agric Scand, Sect B – Soil Plant Sci 69(6):546–556

Si W, Chen YP, Zhang J, Chen Z-Y, Chung HY (2018) Antioxidant activities of ginger extract and its constituents toward lipids. Food Chem 239:1117–1125

Son MJ, Miura Y, Yagasaki K (2015) Mechanisms for antidiabetic effect of gingerol in cultured cells and obese diabetic model mice. Cytotechnology 67:641–652

Srinivasan K (2017) Ginger rhizomes (Zingiber officinale): a spice with multiple health beneficial potentials. Pharma Nutr 5(1):18–28

Stoilova I, Krastanov A, Stoyanova A, Denev P, Gargova S (2007) Antioxidant activity of a ginger extract (*Zingiber officinale*). Food Chem 102(3):764–770

Stoner GD (2013) Ginger: is it ready for prime time? Cancer Prev Res 6(4):257–262

Suk S, Kwon GT, Lee E, Jang WJ, Yang H, Kim JH et al (2017) Gingerenone a, a polyphenol present in ginger, suppresses obesity and adipose tissue inflammation in high-fat diet-fed mice. Mol Nutr Food Res 61(10):1700139

Szymczak J, Grygiel-Górniak B, Cielecka-Piontek J (2024) *Zingiber Officinale* roscoe: the Antiarthritic potential of a popular spice—preclinical and clinical evidence. Nutrients 16(5):741

Tahir AA, Sani NFA, Murad NA, Makpol S, Ngah WZW, Yusof YAM (2015) Combined ginger extract & Gelam honey modulate Ras/ERK and PI3K/AKT pathway genes in colon cancer HT29 cells. Nutr J 14:1–10

Townsend EA, Siviski ME, Zhang Y, Xu C, Hoonjan B, Emala CW (2013) Effects of ginger and its constituents on airway smooth muscle relaxation and calcium regulation. Am J Respir Cell Mol Biol 48(2):157–163

Townsend EA, Zhang Y, Xu C, Wakita R, Emala CW (2014) Active components of ginger potentiate β-agonist–induced relaxation of airway smooth muscle by modulating cytoskeletal regulatory proteins. Am J Respir Cell Mol Biol 50(1):115–124

Uddin R, Kim HH, Lee J-H, Park SU. Neuroprotective effects of medicinal plants. 2013

Wagesho Y, Chandravanshi BS (2015) Levels of essential and non-essential metals in ginger (*Zingiber officinale*) cultivated in Ethiopia. Springerplus 4:1–13

Walstab J, Krüger D, Stark T, Hofmann T, Demir I, Ceyhan G et al (2013) Ginger and its pungent constituents non-competitively inhibit activation of human recombinant and native 5-HT3 receptors of enteric neurons. Neurogastroenterol Motil 25(5):439–e302

Wang X, Shen Y, Thakur K, Han J, Zhang J-G, Hu F, Wei Z-J (2020) Antibacterial activity and mechanism of ginger essential oil against *Escherichia coli* and Staphylococcus aureus. Molecules 25(17):3955

Yamamoto-Ribeiro MMG, Grespan R, Kohiyama CY, Ferreira FD, Mossini SAG, Silva EL, Abreu Filho BA, Mikcha JMG, Junior MM (2013) Effect of *Zingiber officinale* essential oil on Fusarium verticillioides and fumonisin production. Food Chem 141(3):3147–3152

Yeh H-y, Chuang C-h, Chen H-c, Wan C-j, Chen T-l, Lin L-y (2014) Bioactive components analysis of two various gingers (*Zingiber officinale* roscoe) and antioxidant effect of ginger extracts. LWT-Food Sci Technol 55(1):329–334

Yocum GT, Hwang JJ, Mikami M, Danielsson J, Kuforiji AS, Emala CW (2020) Ginger and its bioactive component 6-shogaol mitigate lung inflammation in a murine asthma model. Am J Phys Lung Cell Mol Phys 318(2):L296–L303

Zadeh JB, Kor NM (2014) Physiological and pharmaceutical effects of ginger (*Zingiber officinale* roscoe) as a valuable medicinal plant. Eur J Exp Biol 4(1):87–90

Zhang G, Nitteranon V, Chan LY, Parkin KL (2013) Glutathione conjugation attenuates biological activities of 6-dehydroshogaol from ginger. Food Chem 140(1–2):1–8

Zhang M, Zhao R, Wang D, Wang L, Zhang Q, Wei S, Lu F, Peng W, Wu C (2021) Ginger (*Zingiber officinale* Rosc.) and its bioactive components are potential resources for health beneficial agents. Phytother Res 35(2):711–742

Zhu Y, Zhao Y, Wang P, Ahmedna M, Sang S (2015) Bioactive ginger constituents alleviate protein glycation by trapping methylglyoxal. Chem Res Toxicol 28(9):1842–1849

# Medicinal and Nutritional Importance of Garlic (*Allium sativum* L.) Human Health

9

Gulzar Ahmed Rather, Mohd Yasin Bhat, Mudasir Irfan Dar, Mohmmad Ashaq Sofi, Raman Yadav, Humeraha Nazneen, and Asma Abid

**Abstract**

*Allium sativum* L., or garlic, is a widely grown Amaryllidaceae family plant with powerful therapeutic and nutritional benefits. Garlic's therapeutic properties emerge from its bioactive components, especially organosulfur compounds like allicin. Garlic, originally from Middle Asia or West China, was revered by

G. A. Rather (✉)
AMR and Nanotherapeutics Laboratory, Department of Pharmacology, Saveetha Dental College and Hospitals, Saveetha Institutes of Medical and Technical Sciences (SIMATS), Chennai, Tamil Nadu, India

M. Y. Bhat
Government Degree College Baramulla (Autonomous), Higher Education Department, Government of Jammu and Kashmir, Jammu and Kashmir, India

M. I. Dar
Government Degree College Sogam, Higher Education Department, Government of Jammu and Kashmir, Jammu and Kashmir, India

M. A. Sofi
Department of Biomedical Engineering, Sathyabama Institute of Science and Technology, Chennai, Tamil Nadu, India

R. Yadav
Department of Pharmacology School of Pharmacy Noida International University Gautam Budh Nagar, Uttar pradesh, India

H. Nazneen
Nizam College Osmania University, Hyderabad, Telangana, India

A. Abid
Laboratory of Valorization and Promotion of Saharan Resources (VPRS), Faculty of Mathematics and Matter Sciences, University of Ouargla, Ouargla, Algeria

Higher School of Saharan Agriculture, Higher School of Saharan Agriculture, El Oued, Algeria



M. A. Ansari et al. (eds.), *Medicinal Plants and their Bioactive Compounds in Human Health: Volume 2*, https://doi.org/10.1007/978-981-95-3620-7_9

Egyptians, Greeks, Chinese, and Indians. Modern studies of its cardiovascular protection, antioxidant, anti-inflammatory, anticancer, antibacterial, antifungal, and antiviral activities are in compliance with its historical significance. Garlic lowers blood pressure, atherosclerosis, and thrombosis. Its antioxidant and anti-inflammatory abilities fight oxidative damage and chronic inflammation, while its effects on cell proliferation and apoptosis fight cancer, notably digestive tract malignancies. Garlic's antimicrobial qualities also enhance immunity and combat many infections. This chapter presents its medicinal and nutritional significance from historical and modern scientific perspectives, advocating for its continued usage in health promotion and illness prevention.

**Keywords**

*Allium sativum* · Garlic · Inflammation · Anticancer · Cardiovascular

## 9.1 Introduction

Kingdom: *Plantae*
Phylum: *Tracheophyta*
Class: *Liliopsida*
Order: *Asparagales*
Family: *Amaryllidaceae*
Genus: *Allium*
Species: *Allium sativum* L.

*Allium sativum* L. commonly known as garlic (Fig. 9.1) is one of the widely cultivated members of the bulbous family. It is grown as a shallow rooted vegetable crop with strong taste and flavor (Ammarellou et al. 2022). Few more species of the genus Allium which are closely related to this plant are onion, leek, rakkyo, chive,

**Fig. 9.1** *Allium sativum* L (garlic)

**Table 9.1** Some other names of garlic in vernacular and foreign languages

| Language | Name of garlic |
|---|---|
| Assamese | Naharu |
| Bengali | Rahsun |
| Gujrati | Lasan |
| Kannada Tamil | Bellulli Ullipundu, Vellaippundu |
| Kashmiri | Ruhan |
| Malayalam | Velluli |
| Marathi | Lusson |
| Oriya | Rasuna |
| Punjabi | Lassan, Lasun |
| Kannada | Bellull |
| Telugu | Velluri |
| Urdu | Lassun, Lehsan |
| Chinese | Suan |
| French | Ail |
| Russian | Chesnok |
| Dutch | Knoflook |
| Arabic | Thum |
| Swedish | Vitlok |
| German | Knoblanch |
| Japanese | Ninniku |
| Spanish | Ajo |

Source: Spices board of India official website, https://www.indianspices.com/

shallot, etc. (Fenwick et al. 1985). It is an odorous and aromatic herb whose bulb is used as a spice and flavoring agent. Commercially, it is in great demand throughout the world owing to its culinary and ethomedicinal use for a considerable number of diseases (Tudu et al. 2022). Apart from its name, it is known by the names of lahsun, Lasun, and Lessan in Hindi, and Lashuna in Sanskrit languages. The other Indian vernacular and foreign names are mentioned in Table 9.1.

Carolus Linnaeus (1707–1778) carried forward the name used by Caspar Bauhin in his Theatri Botanici (1623) and subsequently described *Allium sativum* in the first edition of Species plantarum (1753). The name may have its background in Roman language where it was called Allium or Allioum.

## 9.2 Origin and History

Garlic has been recognized since ancient times as a valuable herbal remedy, widely used across various cultures for treating ailments and promoting overall health. Due to its similarity in morphology and genetic make-up, it was believed that it originated from wild ancestor *Allium longicuspis*. However, this speculation is marked with a great doubt because of its sterility. As a result, *Allium truncatum*, *Allium tuncelianum*, and *Allium macrochaetum* are regarded potential wild ancestors (Li et al. 2022a, b). The native land of garlic is believed to be middle Asia. However, some historians speculate its origin in west China. The mention of garlic was in calendar of the Hsia written 2000 years before Jesus Christ. Apart from its legends of garlic associated with Chinese Yellow emperor, Huang Di, the therapeutic uses of garlic in traditional Chinese culture were numerous like exhibiting influence on spleen, kidney, stomach, and removal of toxins from the body. Another view is that, Sumerians (2600–2100 BC) brought it to China since they already cultivated it due to its therapeutic roles in their culture. Alphonse De Condolle in his "Origin of cultivated plants" (1883) described that it was indigenous to Europe and West Asia and other regions just cultivated it later giving it different names. Because garlic has so many name variations in different cultures, it is highly possible that it has been under cultivation from antiquity. Irrespective of its origin, the benefits of garlic to human health was a common conclusion in several isolated cultures in the world. In the ancient Egypt, the garlic was an ingredient in the daily diet of laborers involved in the building of gigantic pyramids. In fact, it has been found in many ancient burial sites like the tomb of Tutankhamen. Garlic with oil was used to treat skin ailments and to stimulate milk production in women after giving birth to babies in Coptic medicine. In ancient Greece, garlic was an important component of the diet of soldiers during battles to keep them robust, alert, and productive. There are also evidences of garlic been utilized as a performance enhancing agent before competitive athletics in earliest Olympic Games held in Greece. The Father of Medicine Hippocrates used to advocate the consumption of garlic for pulmonary disorders, and for abdominal growths. Pedanius Dioscorides in his famous pharmacological text De Materia medica (50–70 AD) described that garlic can be used to expel flatulence but with certain side effects like drying of stomach, instigate thirst, and blurring the eyesight. He also recommended it for cardiovascular ailments, as an anthelmintic, cough reliever, leprosy, tooth aches, stimulant to menstrual flow, etc. Likewise in Roman culture, garlic was recommended to both soldiers and sailors. The field of medicine in ancient Rome was significantly impacted by the renowned Greek physician Pliny the Elder (23–79 AD). Pliny attributed a multitude of health benefits to garlic, including its ability to guard against poisons and diseases (Rivlin 2001). Garlic is believed to have been utilized in China for medical purposes, as a component in culinary dishes, and as a means of preserving food, dating back to or before to 2000 BC. The earliest known reference to this medicinal plant appears in the Avesta, a collection of Zoroastrian sacred texts thought to have been compiled in the sixth century BC (Ajami and Vazirijavid 2019). In the ancient Chinese system of medicine, it was prescribed for the treatment of diarrhea, worm infestation, and respiratory complaints. In India, the ancient medical systems like Ayurveda, Unani, and Tibbi emphasized the use of garlic in treatment of several disorders. Charaka-Samhita recommended the use of garlic for treatment of arthritis and heart diseases 2000 years ago. In the ancient sacred book of India, the Vedas (1200–200 BC), there is mention of garlic among other herbs and has been used for healing procedures by the general population and physicians over generations (Ried and Fakler 2014). The ancient Sanskrit name of garlic in Ayurvedic system is Mahaaushada which means best medicine for diseases. Garlic was advocated by Al Qanoon Fil Tib, the renowned author Avicenna, as a beneficial remedy for numerous ailments, including arthritis, parasitic infections, toothache, chronic coughs, and insect bites, infectious (Ajami and Vazirijavid 2019; Bayan et al. 2014). The legacy of garlic use albeit with some repugnant attitude continued through middle ages and in the Renaissance England, garlic was considered a good medicine for protection against any kind of pestilence in the times of plague. It was an ingredient of pharmaceutical preparations and was consumed with honey, wine, vinegar, oil, etc. In the eighteenth century, the recent times, various claimed health benefits of the consumption of garlic have been scientifically validated by a large number of research studies. Although garlic is not a central ingredient in traditional Japanese cuisine, a distinctive preparation called garlic homogenate—created by kneading and grinding crushed garlic with egg yolk—has been used as a traditional food for a long time (Nakasone et al. 2013). Apart from food and medicinal use, it was regarded as a protective plant to ward off evil influences by many cultures like Hindus, Scandinavians, Greeks, and Germans, etc. Likewise, it is bought on Saint John's Day to bring financial prosperity for the rest of the year in some European countries. There are a variety of hypotheses that are discussed in relation to this herb; some of them were so illogical that they vanished over the course of time, while others have persisted up until the present day. Some people even in times the modern times use the previously assigned terms such as "Russian penicillin" "plant talisman," "natural antibiotic," and "rustic's theriac" for garlic (Petrovska and Cekovska 2010). Even in the era of modern medicine, the popularity of garlic has not vanished but has in fact gained validation as a part of complementary and alternative medicine throughout the world.

## 9.3 Botanical Description of Garlic Plant

*Allium sativum* L. known as garlic belongs to family Amaryllidaceae and monocot genus *Allium*. It is a perennial herbaceous plant with a bulb at the base (underground) covered with fibrous roots on all over its lower side growing in all directions that can extend up too few inches. They can almost be 40 to 60 in number. The roots are characteristically hairless, smooth, thin, elongated, delicate and white, red, purple to ivory in color. The roots may also contain little bit pungent taste and odor. These characteristics vary with the variety as well as the soil in which garlic is grown. They help to normally anchor the plant in the soil and also to absorb water along with other requisite mineral nutrients from the soil. The garlic plant at maturity stands as green large erect stem known as scape that can extend from tens of centimeters up to almost 1 m. The underground modified stem in the form of bulb is the primary part used as a spice, condiment in cuisines, and has a tremendous medicinal significance. The pungent bulb is enclosed in a thin, white, or pinkish papery sheath, within which lie the individual bulblets, or cloves. Each garlic bulb usually holds between ten and twenty individual cloves (Tudu et al. 2022) (Fig. 9.2). The cloves are generally asymmetric in shape, with the exception of those located near the center. The overlapping leaf sheaths at the base of the scape and which usually form a pseudo stem are linear, flattened, solid, stout few centimeters long with

**Fig. 9.2** Garlic *Allium sativum* L. (**a**) A field of garlic plants growing in Jammu and Kashmir (India). (**b**) A growing bulb (prematurely removed) with fibrous roots on all over its lower side. (**c**) Mature garlic bulbs dried after harvest. (**d**) Bulblets (cloves) of garlic covered with thin papery sheath

a pointed or acute apex. The garlic plant has pink-to-violet flowers in the form of umbel like inflorescence and interspersed bulbils that blossom in the northern hemisphere during monsoons. The spherical flower cluster is at first surrounded in papery tapered bracts called spathe which split open to bloom them later. It produces hermaphrodite flowers (Sopha et al. 2024). The cross-pollination of garlic depends much on bees, butterflies, moths, and other insects. The sterility observed in many garlic cultivars limits their ability to cross with wild relatives, which makes identifying common garlic varieties more difficult. This makes figuring out tough, the wild progenitor of common garlic (Li et al. 2022a, b).

## 9.4 Therapeutic Properties of Garlic

Garlic's medicinal benefits are well-documented, primarily attributed to its abundance of bioactive compounds, particularly organosulfur compounds (OSCs) like allicin, diallyl sulfide (DAS), diallyl disulfide (DADS), and diallyl trisulfide (DATS). These compounds are believed to contribute significantly to garlic's health-promoting properties. Studies have shown that garlic offers a wide range of medicinal benefits, like cardiovascular support, potential cancer-preventive effects, antimicrobial activity, anti-inflammatory effects, immune modulation, neuroprotection, and wound healing.

### 9.4.1 Cardiovascular Protection

Research on both humans and animals suggests that garlic can help prevent and treat various cardiovascular diseases (CVDs) like high blood pressure, clogged arteries, thrombosis, and atherosclerosis (Varshney and Budoff 2016). Garlic also enhances several cardiovascular features, including postocclusive reactive hyperemia, microcirculation, adipose tissue in the epicardial and perivascular areas, and carotid artery intima-media thickness (Imaizumi et al. 2023). Studies on garlic have shown it can decrease blood pressure at both the systolic and diastolic levels, therefore potentially reducing the chances of cardiovascular illnesses (Varshney and Budoff 2016). Thus, garlic's numerous cardiovascular protection properties are well-established, as discussed below.

#### 9.4.1.1 Antihypertensive Properties

Garlic treatment led to a significant decrease in blood pressure in the majority of patients who were diagnosed with hypertension, according to the findings indicated by Stabler et al. (2012) and Omar (2013), demonstrating its efficacy as a natural remedy. Furthermore, the research by Ried et al. (2013a) suggests that a dosage of aged garlic extract within the range of 240–960 mg, containing approximately 0.6–2.4 mg of S-allylcysteine (SAC), resulted in a significant decrease in blood pressure by approximately 12 mm/Hg over a period of 3 months. This highlights the critical role of both dosage and treatment duration in attaining therapeutic benefits. Some of the organosulfur compounds (OSCs) found in garlic are responsible for the positive effects that garlic has on the cardiovascular system, especially S-allylcysteine (SAC) and allicin. SAC is abundantly found in garlic and aged garlic extract (Cruz et al. 2007), increased vascular stiffness, decreased lipid abnormalities, and inflammation, helping uncontrolled hypertensives maintain blood pressure (Ried et al. 2010). Experimental evidence, as demonstrated by Sobenin et al. (2009), highlights allicin's role as a systemic vasodilator. Allicin induces the dilation of mesenteric circulation through mechanisms independent of prostaglandin secretion and the β-adrenergic pathway, further supporting its potential in managing hypertension.

Garlic-derived organosulfur compounds (OSCs) exert an antihypertensive effect through various mechanisms, including the production of hydrogen sulfide ($H_2S$), scavenging of free radicals, and gene regulation in the mechanisms that regulate cholesterol homeostasis (Tocmo et al. 2015). $H_2S$, considered the third endogenous gasotransmitter after NO and CO, possesses vasodilating properties crucial in regulating blood pressure (Benavides et al. 2007). It acts as a biomarker associated with the prevention of cardiovascular disease (Ghantous et al. 2020). $H_2S$ is produced enzymatically and nonenzymatically in mammalian tissues, while NO is primarily produced via the L-arginine–NO synthase (NOS) pathway (Nagpure and Bian

2016). Both NO and $H_2S$ widen blood vessels. $H_2S$ does this by opening $K_{ATP}$ channels, allowing $K^+$ out, which prevents $Ca^{2+}$ from entering cells. This relaxation of smooth muscle cells leads to vessel dilation (Benavides et al. 2007). Organic polysulfides in garlic, like diallyl disulfide and diallyl trisulfide, are converted by human red blood cells (RBCs) into $H_2S$. This $H_2S$ release relaxes vascular smooth muscle cells via $K_{ATP}$ channel activation (Benavides et al. 2007). These studies suggest that garlic's beneficial effects are primarily attributed to its potential to produce hydrogen sulfide ($H_2S$). It is predicted that around two cloves of garlic are capable of releasing sufficient $H_2S$ to keep the function of blood vessels in a state of equilibrium (Benavides et al 2007; Ried and Fakler 2014).

Another plausible mechanism for garlic's antihypertensive action is linked to angiotensin converting enzyme (ACE), which transforms angiotensin I into some other hormone known as angiotensin II. As a vasoconstrictor, angiotensin II comes into play, causing blood vessels to constrict and blood pressure to rise (Batiha et al. 2020). Organopolysulfides (DAS, DADS, and DATS) isolated from garlic have demonstrated antihypertensive actions by inhibiting ACE, thereby assisting in lowering blood pressure (Al-Malki 2016). The above discussion, supported by experimentation, confirms the antihypertensive properties of garlic and suggests that consuming or administering garlic would likely be safer than traditional antihypertensive medication (Ried et al. 2013b; Emamat et al. 2020).

#### 9.4.1.2 Antiatherosclerotic Properties

Atherosclerosis is among the most prevalent chronic diseases in humans, developing gradually over time in the form of plaque, often unnoticed for extended periods (Fredman and Spite 2013; Kirichenko et al. 2020). It arises from various factors, with hypercholesterolemia being a major contributor (Milind and Sushila 2012). Experimental studies on hypercholesterolemic rats fed a cholesterol-rich diet have shown that garlic administration significantly lowers serum l LDL levels, and triglyceride, contributing to a lowered risk of cardiovascular disease (Maulik 2013). Garlic is believed to lower cholesterol by interfering with the production of 3-hydroxy-3-methylglutaryl-CoA production (Tesfaye 2021). In experimental atherosclerosis caused by a high cholesterol diet, garlic preparations decreased aortic atheromatous lesions by 50% over time (Maulik 2013). Garlic protects atherosclerosis by decreasing the lipids in the artery membrane. Allicin, SAC, and DADS in garlic oil have an antiatherosclerotic action (Singh et al. 2020). Studies have shown that garlic extracts inhibit the blood plasma enzyme sialidase, which is in charge of atherogenic low-density lipoprotein (LDL) production (Sobenin et al. 2019). Aged garlic extract limits the progression of atherosclerosis by suppressing CD36 receptor activity in macrophages, reducing the accumulation of oxidized LDL and foam cells (Morihara et al. 2011).

#### 9.4.1.3 Antithrombotic Properties

When blood vessels are damaged, platelets become activated, aggregate, and adhere to one another, leading to clot formation and the onset of thrombosis (Yun et al. 2016). This impedes the natural flow of blood in vessels and is a significant contributor to cardiovascular diseases worldwide (Hou et al. 2015). Platelets activate the fibrinogen receptor GPIIb/IIIa, generate thromboxane-A2 (TXA2), and temporarily elevate intracellular calcium ions (Rahman 2007). Lipoxygenase metabolites, cyclic adenosine monophosphate (cAMP), cyclic guanosine monophosphate (cGMP), and nitric oxide also contribute to platelet aggregation. Garlic inhibits platelet aggregation, which has been thoroughly investigated in in vitro trials to avoid thrombosis (Rahman 2007). Mice treated with garlic showed delayed platelet aggregation within pial micro vessels and inhibited platelet adhesion to the wall of pial microvessels, indicating a protective effect of garlic on microvessel endothelium (Fahim and El-Sabban 2001). The oral and intraperitoneal treatment of garlic extract (up to 500 mg/kg) significantly reduced rat serum thromboxane B2 (TXB2) (Bordia et al. 1996). Phospholipase A2, Cyclooxygenase -1 (COX-1)/COX-2, and TXAS synthesize thromboxane-A2 (TXA2) from arachidonic acid. TXA2 strongly aggregates platelets and contracts vascular smooth muscle cells. TXA2 binding to its receptor causes $Ca^{2+}$ to enter, contracting vascular smooth muscle cells. TXA2 vasoconstriction brings platelets together, causing clots. Garlic blocks COX activity and suppresses intraplatelet $Ca^{2+}$ mobilization to prevent platelet aggregation in vitro (Rahman 2007; Bayan et al. 2014). Garlic components have a short half-life and are swiftly removed, while its active elements may build up slowly. Interestingly, while higher doses of garlic did not effectively inhibit platelet aggregation, long-term consumption of smaller amounts of fresh garlic was found to significantly reduce serum thromboxane levels—by up to 80%—indicating its potential in preventing thrombosis (Bordia et al. 1996).

Plasma fibrinolytic activity (FA) also plays crucial role in antithrombosis by activating plasminogen and inhibiting thrombin (Hvas and Larsen 2023). Individuals with lower plasma fibrinolytic activity are more susceptible to coronary artery disease due to excessive fibrin deposition, leading to thrombosis in their vessels. This suggests that higher FA activity results in a more favorable antithrombotic effect. Significant enhancement in FA has been observed following the administration of garlic extract, garlic oil, or garlic juice in investigations involving both humans and animals. While conducting research on a rat version, Fukao et al. (2007) reported that garlic administration modulated FA by enhancing plasminogen activation and inhibiting thrombin. In patients with cerebral atherosclerosis, a two-week treatment with garlic powder tablets (Allicor) led to approximately a 22% increase in fractional anisotropy (FA), indicating improved vascular function (Sobenin et al. 2010). However, another study reported a potential increase in the risk of myocardial infarction associated with garlic consumption, observed in both healthy individuals and patients recovering from acute myocardial infarction (Sobenin et al. 2012).

Among the organosulfur compounds found in garlic, allicin stands out for its strong antiplatelet aggregation properties, supporting the herb's broader antithrombotic potential (Cavagnaro et al. 2007). Allicin has been shown in vitro to reduce platelet aggregation in humans without influencing the activity of cyclooxygenase, thromboxane synthase, vascular prostacyclin synthase, or AMP levels. H2S reduces allicin into diallyl polysulfides, therefore avoiding myocardial fibrosis and injury. Antiplatelet actions of methyl allyl trisulfides (MATS) are due to their arachidonic acid cascade inhibition (Choi and Park 2012). Ajoene, another OSC in garlic extract regulates platelet aggregation by limiting arachidonic acid metabolism via cyclooxygenase and lipoxygenase pathways (Teranishi et al. 2003), decreasing thromboxane A2 and 12-Hydroxyeicosatetraenoic acid (12-HETE) formation. Ajoene may directly interact with platelets' fibrinogen receptor (GPIIb/IIIa) to prevent aggregation (Taylor et al. 2006). Aged garlic extract helps stop platelet clumping by blocking GPIIb/IIIa receptors and raising cAMP levels, which stops ADP-activated platelets from sticking to fibrinogen (Allison et al. 2012). Unlike allicin, another garlic compound—sodium 2-propenyl thiosulfate—prevents platelet clumping in dogs by affecting cyclooxygenase activity. (Chang et al. 2005). Thus, garlic seems to be rich in myriad of bioactive compounds that can prevent thrombosis and the development of atherosclerosis by impeding platelet aggregation through a plethora of mechanisms.

### 9.4.2 Antioxidant Properties

Reactive oxygen species (ROS) and nitrogen species, collectively known as free radicals, carry unpaired electrons and play significant roles in biological processes. Nitric oxide synthase and Nicotinamide adenine dinucleotide phosphate (NADPH) oxidase, myeloperoxidase, enzymes govern extracellular RONS synthesis in mammals' innate immune systems to fight germs. However, excessive release of Reactive Oxygen and Nitrogen species (RONS), both intracellularly and extracellularly, can lead to oxidative damage to biomolecules, inducing oxidative stress (Weidinger and Kozlov 2015). To counteract this, cellular systems employ various enzymatic and nonenzymatic antioxidant mechanisms (Dar et al. 2017). Garlic's antioxidant properties come from its high content of organosulfur compounds (OSCs) like allicin, alliin, DADS, DATS, and S-allylmercaptocysteine (SAMC). These chemicals protect lipids, scavenge free radicals, boost antioxidant enzyme activity, and control gene expression (Moosavian et al. 2020). Biochemical evidence suggests that garlic-derived OSCs inhibit TLR4 dimerization, downregulating nuclear factor-kappa B (NF-kappaB) activation, Inducible Nitric Oxide Synthase (iNOS) expression and cyclooxygenase-2, reducing nitric oxide generation and oxidative stress (Ryu et al. 2015).

Allicin, a compound found in garlic, helps the body fight oxidative stress by boosting antioxidant enzymes and substances like glutathione, which neutralize harmful free radicals (Ansari et al. 2017). SIRT1 activation, triggered by garlic, boosts antioxidant defenses by modifying DNA-binding proteins, reducing ROS, and safeguarding blood vessel cells (Pan et al. 2022).

Other garlic components like alliin can lower the production of peroxides and the activity of an enzyme called iNOS, while also preventing the loss of glutathione. This supports the body's natural antioxidant defenses, potentially reducing the risk or progression of atherosclerosis (Kim et al. 2001). Another garlic compound, S-allyl cysteine (SAC), can block harmful effects caused by oxidized LDL (a form of bad cholesterol) and inhibit NF-κB activation, a key player in oxidative stress and inflammation (Li et al. 2022a, b).

In rats fed a high-fat diet, diallyl disulfide (DADS) was found to activate the KEAP1/Nrf2 pathway, leading to higher levels of antioxidant enzymes like NQO1, γ-GCS, and SOD-1, which helped reduce oxidative stress in the liver (Chien et al. 2021). This effect was linked to SAC's ability to neutralize damaging ROS such as superoxide (O2−) (Maldonado et al. 2003). Additionally, diallyl sulfide (DAS), another garlic compound, helped reduce harmful oxygen and nitrogen species by either blocking a liver enzyme (cytochrome P450-2E1) or increasing the activity of protective genes like Nrf2 and Heme oxygenase-1 (Abdel-Daim et al. 2018).

Other active components in garlic, like saponins and phenols, also show antioxidant action. Antioxidant effects of garlic total saponins (TSG) include Malondialdehyde (MDA) inhibition and Superoxide dismutase (SOD) activity restoration (Miao et al. 2020). Garlic saponins act as antioxidants within cells, defending mouse C2C12 myoblasts against the damaging effects of hydrogen peroxide on growth and DNA integrity (Kang et al. 2016).

Interestingly, the antioxidant potency of garlic is significantly influenced by processing. Raw garlic exhibits stronger antioxidant properties compared to cooked garlic, as shown by DPPH, 2,2'- Azino-bis(3-ethylbezothiazoline-6-sulfonic acid) (ABTS), and Ferric reducing antioxidant power (FRAP) tests (Shang et al. 2019). However, tests such as FRAP, DPPH, $H_2O_2$ ABTS and $Fe^{2+}$ chelation indicate that aged garlic actually has even stronger antioxidant strength than fresh garlic. Old garlic's antioxidant effect comes from sulfur compounds like SAC and SAMC as well as flavonoids and phenols (Jang et al. 2018). The Nrf2 pathway, activated by aged garlic extract, boosts the production of key antioxidant enzymes like HO-1 and GCLM, which help protect endothelial cells from oxidative injury (Batiha et al. 2020).. Garlic is a potential antioxidant source because of its many components and mechanisms.

### 9.4.3 Anti-inflammation Properties

For many centuries, garlic has been recognized as an effective natural remedy for preventing and addressing inflammation-associated conditions (Hitchcock et al. 2021). Many studies have confirmed its anti-inflammatory benefits. For example, clinical trials showed that taking aged garlic extract (AGE) for about 4 months helped reduce gum inflammation caused by the bacteria Porphyromonas gingivalis (Ohtani and Nishimura 2020). A separate study showed that garlic extract alleviated liver inflammation caused by *Eimeria papillate* (Ahmad et al. 2016). In addition, a daily dose of 300 mg of aged garlic extract (AGE) over 18 months has been shown to reduce periodontal disease—an inflammatory condition caused by gum detachment from the teeth (Zini et al. 2020). Garlic reduces inflammation by its organosulfur compounds to suppress the NF-κB signaling pathway, resulting in decreased levels of cytokines like interleukin-6 (IL-6) and tumor necrosis factor-α (TNF-α) (Arreola et al. 2015; Abdel-Daim et al. 2020). Additionally, garlic blocks the TLR-4 pathway in macrophages, further suppressing NF-κB activity and inflammation (You et al. 2019).

AGE also activates AMPK, a protein that helps regulate inflammation, by increasing its activity in the tissues of sick animals (Miki et al. 2020). Diallyl trisulfide, a garlic component, has been demonstrated to decrease inflammation in gum disease and neurodegenerative disorders by inhibiting NF-κB and MAPK pathways (Ho and Su 2014; Fu et al. 2015). Another compound in garlic, Z-ajoene, also has strong anti-inflammatory effects. It works by decreasing cytokines like IL-6, and IL-12β, while increasing IL-10, an anti-inflammatory molecule. It does this by suppressing the activity of STAT3 and COX2, which are involved in controlling inflammation (Hitchcock et al. 2021).

### 9.4.4 Anticancer Properties

Numerous studies have demonstrated that garlic formulations exhibit potent anticancer effects (Nicastro et al. 2015; Dorrigiv et al. 2020). In vitro studies have consistently shown that Garlic extract (GE) decrease cell proliferation by approximately 90% in cancer cell lines of the breast, liver, and prostate (Bagul et al. 2015). Particularly noteworthy is garlic's strong anticancer activity, especially concerning digestive tract tumors. Human population research has indicated that consuming garlic significantly reduces the chances of esophagus, stomach, and colon cancer (Wang et al. 2022). Additionally, individuals who consumed more garlic have been observed to have about 54% reduced risk of pancreatic cancer compared to those consuming lesser amounts (González et al. 2006)

Garlic contains several organosulfur compounds (OSCs) with strong anticancer potential, including allicin, alliin, DAS, DADS, DATS, and allyl mercaptan (Iciek et al. 2012; Imaizumi et al. 2023). These compounds work by affecting different biological processes related to cancer, such as preventing DNA damage caused by carcinogens, reducing mutations, eliminating harmful free radicals, helping break down carcinogens, apoptosis, and controlling abnormal cell growth (Xiao et al. 2018).

Garlic and its various physiologically active ingredients work as antagonists in anticancer medications by inhibiting growth and causing death. Garlic extract has been shown through histological analysis to inhibit colorectal cancer progression by decreasing the expression of key regulators of the cell cycle, including cyclin B1 and CDK1 (Jikihara et al. 2015). In the case of gastric cancer, allicin—one of garlic's major bioactive compounds—has been found to disrupt cell division at the G2/M phase, likely through interference with transforming growth factor-α (TGF-α) signaling and its interaction with Epidermal growth factor receptor (EGFR) (Capasso 2013).

Additionally, allicin has been established to promote apoptosis and prevent cell growth in gastrointestinal cancer cells by upregulating p38 expression and cleaving caspase 3, or by altering the BAX/BCL2 ratio via the mitochondrial route (Zhang et al. 2015). Furthermore, allicin has been reported to prevent cell proliferation by targeting microtubules or tubulins, leading to apoptosis (Prager et al. 2007; Rana et al. 2011). Another organic substance that may be obtained from garlic is called ajoene, and it is responsible for inducing a cell death in leukemic cells and other blood cells by activating caspase-3 and caspase-8 (Bayan et al. 2014).

### 9.4.5 Antibacterial Properties

The wide antibacterial capabilities of garlic have been acknowledged for centuries, with studies validating its effectiveness against several bacterial types (Gebreselema and Mebrahtu 2013). Research has demonstrated its efficacy against various pathogens, including *Aeromonas hydrophila*, *Escherichia coli*, *Salmonella*, *Pseudomonas aeruginosa*, *Staphylococcus aureus*, *Streptococcus mutans*, *Klebsiella*, *Micrococcus*, *Bacillus cereus*, and *Helicobacter pylori* (Zardast et al. 2016; Singh et al. 2020). Different concentrations of garlic extract have antibacterial activities effective against the microorganisms present in human dental plaque (Houshmand et al. 2013). Notably, garlic's antibacterial effects are attributed to compounds like allicin, which inhibit sulfhydryl enzymes crucial for bacterial survival and pathogenesis. (Batiha et al. 2020). Allicin, a key organosulfur compound (OSC) in garlic, modifies sulfhydryl enzymes, including alcohol dehydrogenases, thioredoxin reductase, RNA polymerase, and cysteine proteases in bacterial pathogens. This action disrupts bacterial functions vital for their survival and virulence, thus impeding their pathogenicity (Batiha et al. 2020). Bacteria often develop resistance to antibiotics more rapidly than to allicin, underscoring the potential of garlic as an alternative or adjunctive treatment for drug-resistant infections. For instance, while *Escherichia coli* showed resistance to vancomycin, it remained susceptible to the antibacterial effects of allicin and garlic extract (Sasi et al. 2021).

Garlic extract also proved to be very effective against gram negative pathogen *Pseudomonas aeruginosa* which otherwise showed antibiotic resistance to many drugs in clinical trials. Therefore, it is suggested that garlic could be used against various drug resistant infections (Karuppiah and Rajaram 2012). Moreover, garlic demonstrates synergistic effects when combined with antibiotics like vancomycin and ciprofloxacin, enhancing their antibacterial activity. Allicin's interaction with sulfhydryl groups on enzymes encoded by transposons responsible for antibiotic resistance can increase susceptibility to antibiotics, further combating drug-resistant bacteria (Zain al-abdeen et al. 2013; Sahidur et al. 2023).

Furthermore, it has been shown that the thiosulfinates found in garlic inhibit the quorum sensing systems of bacteria. Diallyl disulfide (DADS) inhibits the expression of virulence factor genes regulated by quorum sensing in pathogens such as *Hafnia alvei* and *Pseudomonas aeruginosa* (Hongman et al. 2018; Li et al. 2021). Another sulfur-rich substance found in garlic, ajoene, likewise inhibits *Pseudomonas aeruginosa*'s quorum sensing systems, which lowers the production of harmful factors (Jakobsen et al. 2017). Additionally, garlic extract has antibiofilm qualities that increase the susceptibility of bacteria to antibiotic therapy and immune cell phagocytic assault (Bjarnsholt et al. 2005). Garlic's potential as a natural treatment for bacterial infections, particularly those resistant to conventional antibiotics, is underscored by its intricate mechanisms of action.

### 9.4.6 Antifungal Properties

The antifungal activities of garlic extracts and its bioactive compounds have been extensively researched and documented. Various studies have highlighted the effectiveness of garlic against a wide range of fungal species, including *Trichophyton, Botrytis, Candida, Epidermophyton, Aspergillus, Rhizopus, Microsporum, Alternaria, Neofabraea,* and *Penicillium* (Pârvu et al. 2019; Fufa 2019). Different garlic extracts, such as aqueous, ethanolic, methanolic, and petroleum ether extracts, have been shown to exhibit antifungal actions. These extracts have demonstrated inhibitory effects on fungal germination and growth, sometimes comparable to conventional antifungal drugs like ketoconazole (Shams-Ghahfarokhi et al. 2006). Garlic oil, abundant in bioactive constituents such as diallyl disulfide and diallyl trisulfide, influences fungal spore and cell germination by infiltrating cellular and organelle membranes, instigating permanent alterations in cell ultrastructure, and resulting in the efflux of cellular contents (Li et al. 2016). DADS and DATS have shown antifungal effectiveness against a variety of fungal species, such as *Blastoschizomyces capitatus* and *Candida albicans.* Garlic-derived saponins have antifungal activity against *Trichoderma harzianum* and *Botrytis cinerea* (Zhen et al. 2006). Ajoene, derived from garlic, is acknowledged as an effective topical antifungal drug, suppressing the proliferation of fungus like Aspergillus niger, Candida albicans, and Paracoccidioides (Ledezma and Apitz-Castro 2006). When it comes to the treatment of uncommon fungal infections of the brain that are brought on by *Cryptococcus meningitis*, it has been discovered that the administration of garlic by intravenous administration is more effective and less harmful than the antibiotic Amphotericin-B would be. The potential therapeutic efficacy of garlic in clinical settings is brought to light by this. When taken as a whole, the antifungal capabilities of garlic and the bioactive compounds that it contains present intriguing avenues for the development of novel antifungal medicines and therapeutic treatments, both topical and systemic, for the treatment of fungal infections. It is necessary to conduct additional study in order to ensure that the mechanisms of action of garlic-derived antifungal drugs are well understood and that therapeutic applications are optimized. (Lemar et al. 2007; Alam et al. 2016).

### 9.4.7 Antiviral Properties

Traditionally, garlic has been valued in various medicinal systems for its effectiveness in managing viral infections. The active compounds exhibit antiviral activity by targeting many stages of the virus's life cycle. Certain components of garlic may impede viruses from penetrating human cells by modifying the virus's outer surface, therefore preventing the onset of infection (Rouf et al. 2020). Research has shown that garlic extracts are effective against several viruses, including the Newcastle disease virus (Meléndez-Villanueva et al. 2019; Harazem et al. 2019), influenza A (H1N1) and the measles virus.

The principal antiviral compounds in garlic are diallyl trisulfide, allicin, and ajoene. These medications may impair virus attachment to host cells, inhibit genome translation, and prevent viral replication by interfering with enzymes such as RNA polymerase (Sharma 2019). Garlic extracts have also been shown to reduce the activity of RNA polymerase and nucleoprotein production in viruses such as H1N1 and avian infectious bronchitis virus (Chavan et al. 2016; Shojai et al. 2016). Garlic-derived compounds may inhibit reverse transcriptase, an essential enzyme for viral replication, in retroviruses such as HIV-1 (Rouf et al. 2020). Furthermore, allicin's potential to mitigate immunological damage caused by viruses, including the reticuloendotheliosis virus, is also under research (Wang et al. 2017; Rouf et al. 2020).

The global outbreak of coronavirus disease 2019 (COVID-19), caused by the severe acute respiratory syndrome coronavirus 2 (SARS-CoV-2) virus, refocused attention on garlic's potential for therapeutic use. Studies suggest that certain compounds in garlic may offer protective effects against SARS-CoV-2 and other coronaviruses, including the infectious bronchitis virus (IBV). Quercetin, a notable target for antiviral treatment, is a garlic compound recognized for its ability to block the virus's primary protease (Mpro) (Khubber et al. 2020). The World Health Organization (Tudu et al. 2022) has dismissed claims that garlic consumption may prevent COVID-19. While garlic may alleviate symptoms, it has not been shown to halt the condition.

## 9.5 Future Perspective

Garlic (*Allium sativum*) is well known for its health and nutritional benefits, supported by both traditional use and modern scientific research. We've covered garlic's plant classification, its historical importance, and its powerful active compounds, especially organosulfur compounds such as allicin, which contribute to its medicinal benefits. Evidence suggests garlic is beneficial for heart health, acts as an antioxidant, reduces inflammation, helps prevent cancer, and fights bacteria, fungi, and viruses. These effects contribute to lowering blood pressure, reducing plaque build-up in arteries, preventing blood clots, protecting against oxidative stress, easing long-term inflammation, and strengthening the immune system.

Looking forward, garlic shows significant promise for enhancing health and preventing illness. Continued research is expected to uncover exactly how garlic works in the body, which could lead to new medical uses and more effective supplement forms. Advances in how garlic is processed may help improve the strength and availability of its health-promoting compounds. Additionally, studying how garlic works together with other natural ingredients or conventional medicines might lead to more holistic treatment options. As we learn more about how garlic supports the body, it will likely become an even more important part of modern healthcare and nutrition. This chapter emphasizes the importance of further research and encourages the ongoing use of garlic in both preventive care and disease treatment.

## 9.6 Authors Contribution

*Gulzar Ahmed Rather* helped in study conception, literature review, and chapter writing. *Mohd Yasin Bhat* was involved in data collection and input on nutritional aspects. *Mudasir Irfan Dar* helped in analysis support and feedback on medicinal content. *Raman Yadav* contributed to scientific review, formatting, and references. *Asma Abid* helped in source compilation and editing. *Mohmmad Ashaq Sofi* helped in proofreading and publication compliance. *Humeraha Nazneen* was involved in writing and formatting.

## References


Abdel-Daim MM, Shaheen HM, Abushouk AI et al (2018) Thymoquinone and diallyl sulfide protect against fipronil-induced oxidative injury in rats. Environ Sci Pollut Res 25:23909–23916

Abdel-Daim MM, Abushouk AI, Bungău SG et al (2020) Protective effects of thymoquinone and diallyl sulphide against malathion-induced toxicity in rats. Environ Sci Pollut Res 27:10228–10235

Ahmad TA, Eweida AE, Sheweita SA (2016) B-cell epitope mapping for the design of vaccines and effective diagnostics. Trials Vaccinol 5:71–83

Ajami M, Vazirijavid R (2019) Garlic (*Allium sativum* L.). In: Nonvitamin and nonmineral nutritional supplements. Academic Press, London, pp 227–234

Alam K, Hoq O, Uddin S (2016) Medicinal plant *Allium sativum*. A review. J Med Plant Stud 4(6):72–79

Allison GL, Lowe GM, Rahman K et al (2012) Aged garlic extract inhibits platelet activation by increasing intracellular cAMP and reducing the interaction of GPIIb/IIIa receptor with fibrinogen. Life Sci 91(25–26):1275–1280

Al-Malki AL (2016) Inhibition of α-glucosidase by thiosulfinate as a target for glucose modulation in diabetic rats. Evid Based Complement Alternat Med 2016(1):7687915

Ammarellou A, Yousefi AR, Heydari M et al (2022) Biochemical and botanical aspects of *Allium sativum* L. sowing. BioTech 11(2):16

Ansari IA, Khan I, Khan MS et al (2017) Organosulfur compounds of garlic as potent chemotherapeutic agents against cancer. Anticancer Plants Clin Trials Nanotechnol 3:253–272

Arreola R, Quintero-Fabián S, López-Roa RI et al (2015) Immunomodulation and anti-inflammatory effects of garlic compounds. J Immunol Res 2015(1):401630

Bagul M, Kakumanu S, Wilson TA et al (2015) Crude garlic extract inhibits cell proliferation and induces cell cycle arrest and apoptosis of cancer cells in vitro. J Med Food 18(7):731–737

Batiha GE-S, Beshbishy AM, Wasef LG, Elewa YHA, Al-Sagan AA, Abd El-Hack ME, Taha AE, Abd-Elhakim YM, Devkota HP (2020) Chemical constituents and pharmacological activities of garlic (Allium sativum L.): a review. Nutrients 12(3):872

Bayan L, Koulivand PH, Gorji A (2014) Garlic: a review of potential therapeutic effects. Avicenna J Phytomed 4(1):1

Benavides GA, Squadrito GL, Mills RW et al (2007) Hydrogen sulfide mediates the vasoactivity of garlic. Proc Natl Acad Sci 104(46):17977–17982

Bjarnsholt T, Jensen PØ, Rasmussen TB et al (2005) Garlic blocks quorum sensing and promotes rapid clearing of pulmonary *Pseudomonas aeruginosa* infections. Microbiology 151(12):3873–3880

Bordia T, Mohammed N, Thomson M et al (1996) An evaluation of garlic and onion as antithrombotic agents. Prostaglandins Leukot Essent Fat Acids 54(3):183–186

Capasso A (2013) Antioxidant action and therapeutic efficacy of Allium sativum L. Molecules 18(1):690–700

Cavagnaro PF, Camargo A, Galmarini CR et al (2007) Effect of cooking on garlic (Allium sativum L.) antiplatelet activity and thiosulfinates content. J Agric Food Chem 55(4):1280–1288

Chang HS, Yamato O, Yamasaki M et al (2005) Modulatory influence of sodium 2-propenyl thiosulfate from garlic on cyclooxygenase activity in canine platelets: possible mechanism for the anti-aggregatory effect. Prostaglandins Leukot Essent Fat Acids 72(5):351–355

Chavan RD, Shinde P, Girkar K et al (2016) Assessment of anti-influenza activity and hemagglutination inhibition of *Plumbago indica* and *Allium sativum* extracts. Pharm Res 8(2):105

Chien LH, Wu CT, Deng JS et al (2021) Salvianolic acid c protects against cisplatin-induced acute kidney injury through attenuation of inflammation, oxidative stress and apoptotic effects and activation of the CaMKK–AMPK–sirt1-associated signaling pathway in mouse models. Antioxidants 10(10):1620

Choi YH, Park HS (2012) Apoptosis induction of U937 human leukemia cells by diallyl trisulfide induces through generation of reactive oxygen species. J Biomed Sci 19:1–11

Cruz C, Correa-Rotter R, Sánchez-González DJ et al (2007) Renoprotective and antihypertensive effects of S-allylcysteine in 5/6 nephrectomized rats. Am J Physiol Renal Physiol 293(5):F1691–F1698

Dar MI, Naikoo MI, Khan FA et al (2017) An introduction to reactive oxygen species metabolism under changing climate in plants. In: Reactive oxygen species and antioxidant systems in plants: role and regulation under abiotic stress, pp 25–52. https://doi.org/10.1007/978-981-10-5254-5_2

Dorrigiv M, Zareiyan A, Hosseinzadeh H (2020) Garlic (*Allium sativum*) as an antidote or a protective agent against natural or chemical toxicities: a comprehensive update review. Phytother Res 34(8):1770–1797

Emamat H, Tangestani H, Totmaj AS et al (2020) The effect of garlic on vascular function: a systematic review of randomized clinical trials. Clin Nutr 39(12):3563–3570

Fahim MA, El-Sabban FF (2001) Garlic prevents ultrastructural alterations caused by dehydration in mouse cerebral microvessels. Anat Rec 263(1):85–90

Fenwick GR, Hanley AB, Whitaker JR (1985) The genus allium. Part 2. Crit Rev Food Sci Nutr 22(4):273–377

Fredman G, Spite M (2013) Recent advances in the role of immunity in atherosclerosis. Circ Res 113(12):e111–e114

Fu E, Tsai MC, Chin YT et al (2015) The effects of diallyl sulfide upon *P orphyromonasgingivalis* lipopolysaccharide stimulated proinflammatory cytokine expressions and nuclear factor-kappa B activation in human gingival fibroblasts. J Periodontal Res 50(3):380–388

Fufa BK (2019) Anti-bacterial and anti-fungal properties of garlic extract (*Allium sativum*): a review. Microbiol Res J Int 28(3):1–5

Fukao H, Yoshida H, Tazawa I et al (2007) Antithrombotic effects of odorless garlic powder both in vitro and in vivo. Biosci Biotechnol Biochem 71(1):84–90

Gebreselema G, Mebrahtu G (2013) Medicinal values of Garlic : A review. Int J Med Med Sci 5 (9):401–408

Ghantous CM, Kamareddine L, Farhat R et al (2020) Advances in cardiovascular biomarker discovery. Biomedicine 8(12):552

González CA, Pera G, Agudo A et al (2006) Fruit and vegetable intake and the risk of stomach and oesophagus adenocarcinoma in the European Prospective Investigation into Cancer and Nutrition (EPIC–EURGAST). Int J Cancer 118(10):2559–2566

Harazem R, Rahman SA, El-Kenawy A et al (2019) Evaluation of antiviral activity of Allium cepa and Allium sativum extracts against Newcastle disease virus. Alexandria J Vet Sci 61(1):108

Hitchcock JK, Mkwanazi N, Barnett C et al (2021) The garlic compound Z-Ajoene, S-Thiolates COX2 and STAT3 and dampens the inflammatory response in RAW264. 7 macrophages. Mol Nutr Food Res 65(3):2000854

Ho SC, Su MS (2014) Evaluating the anti-neuroinflammatory capacity of raw and steamed garlic as well as five organosulfur compounds. Molecules 19(11):17697–17714

Hongman H, Yifang W, Gongliang Z et al (2018) Effects of sulfide flavors on AHL-mediated quorum sensing and biofilm formation of Hafnia alvei. J Food Sci 83(10):2550–2559

Hou Y, Carrim N, Wang Y et al (2015) Platelets in hemostasis and thrombosis: novel mechanisms of fibrinogen-independent platelet aggregation and fibronectin-mediated protein wave of hemostasis. J Biomed Res 29(6):437

Houshmand B, Mahjour F, Dianat O (2013) Antibacterial effect of different concentrations of garlic (Allium sativum) extract on dental plaque bacteria. Indian J Dent Res 24(1):71–75

Hvas CL, Larsen JB (2023) The fibrinolytic system and its measurement: history, current uses and future directions for diagnosis and treatment. Int J Mol Sci 24(18):14179

Iciek M, Kwiecień I, Chwatko G et al (2012) The effects of garlic-derived sulfur compounds on cell proliferation, caspase 3 activity, thiol levels and anaerobic sulfur metabolism in human hepatoblastoma HepG2 cells. Cell Biochem Funct 30(3):198–204

Imaizumi VM, Laurindo LF, Manzan B et al (2023) Garlic: a systematic review of the effects on cardiovascular diseases. Crit Rev Food Sci Nutr 63(24):6797–6819

Jakobsen TH, Warming AN, Vejborg RM et al (2017) A broad range quorum sensing inhibitor working through sRNA inhibition. Sci Rep 7(1):9857

Jang HJ, Lee HJ, Yoon DK et al (2018) Antioxidant and antimicrobial activities of fresh garlic and aged garlic by-products extracted with different solvents. Food Sci Biotechnol 27:219–225

Jikihara H, Qi G, Nozoe K et al (2015) Aged garlic extract inhibits 1, 2-dimethylhydrazine-induced colon tumor development by suppressing cell proliferation. Oncol Rep 33(3):1131–1140

Kang JS, Kim SO, Kim GY et al (2016) An exploration of the antioxidant effects of garlic saponins in mouse-derived C2C12 myoblasts. Int J Mol Med 37(1):149–156

Karuppiah P, Rajaram S (2012) Antibacterial effect of *Allium sativum* cloves and *Zingiber officinale* rhizomes against multiple-drug resistant clinical pathogens. Asian Pac J Trop Biomed 2(8):597–601

Khubber S, Hashemifesharaki R, Mohammadi M et al (2020) Garlic (*Allium sativum* L.): a potential unique therapeutic food rich in organosulfur and flavonoid compounds to fight with COVID-19. Nutr J 19:1–3

Kim KM, Chun SB, Koo MS et al (2001) Differential regulation of NO availability from macrophages and endothelial cells by the garlic component S-allyl cysteine. Free Radic Biol Med 30(7):747–756

Kirichenko TV, Sukhorukov VN, Markin AM et al (2020) Medicinal plants as a potential and successful treatment option in the context of atherosclerosis. Front Pharmacol 11:403

Ledezma E, Apitz-Castro R (2006) Ajoene the main active compound of garlic (Allium sativum): a new antifungal agent. Rev Iberoam Micol 23(2):75–80

Lemar KM, Aon MA, Cortassa S et al (2007) Diallyl disulphide depletes glutathione in Candida albicans: oxidative stress-mediated cell death studied by two-photon microscopy. Yeast 24(8):695–706

Li WR, Shi QS, Dai HQ et al (2016) Antifungal activity, kinetics and molecular mechanism of action of garlic oil against *Candida albicans*. Sci Rep 6(1):22805

Li WR, Zeng TH, Yao JW et al (2021) Diallyl sulfide from garlic suppresses quorum-sensing systems of Pseudomonas aeruginosa and enhances biosynthesis of three B vitamins through its thioether group. Microb Biotechnol 14(2):677–691

Li M, Yun W, Wang G et al (2022a) Roles and mechanisms of garlic and its extracts on atherosclerosis: a review. Front Pharmacol 13:954938

Li N, Zhang X, Sun X et al (2022b) Genomic insights into the evolutionary history and diversification of bulb traits in garlic. Genome Biol 23(1):188

Maldonado PD, Barrera D, Rivero I et al (2003) Antioxidant S-allylcysteine prevents gentamicin-induced oxidative stress and renal damage. Free Radic Biol Med 35(3):317–324

Maulik N (2013) Cardiovascular diseases: nutritional and therapeutic interventions. CRC Press, Boca Raton
Meléndez-Villanueva MA, Morán-Santibañez K, Martínez-Sanmiguel JJ et al (2019) Virucidal activity of gold nanoparticles synthesized by green chemistry using garlic extract. Viruses 11(12):1111
Miao Q, Wang R, Bai D (2020) Antiatherosclerosis properties of total saponins of garlic in rats. Evid Based Complement Alternat Med 2020(1):3683659
Miki S, Suzuki JI, Kunimura K et al (2020) Mechanisms underlying the attenuation of chronic inflammatory diseases by aged garlic extract: involvement of the activation of AMP-activated protein kinase. Exp Ther Med 19(2):1462–1467
Milind P, Sushila K (2012) How to live with rheumatoid arthritis??? Int Res J Pharm 3(3):115–121
Moosavian SP, Arab A, Paknahad Z et al (2020) The effects of garlic supplementation on oxidative stress markers: a systematic review and meta-analysis of randomized controlled trials. Complement Ther Med 50:102385
Morihara N, Ide N, Weise N et al (2011) Aged garlic extract inhibits homocysteine-induced scavenger receptor CD36 expression and oxidized low-density lipoprotein cholesterol uptake in human macrophages in vitro. J Ethnopharmacol 134(3):711–716
Nagpure BV, Bian JS (2016) Interaction of hydrogen sulfide with nitric oxide in the cardiovascular system. Oxidative Med Cell Longev 2016(1):6904327
Nakasone Y, Nakamura Y, Yamamoto T et al (2013) Effect of a traditional Japanese garlic preparation on blood pressure in prehypertensive and mildly hypertensive adults. Exp Ther Med 5(2):399–405
Nicastro HL, Ross SA, Milner JA (2015) Garlic and onions: their cancer prevention properties. Cancer Prev Res 8(3):181–189
Ohtani M, Nishimura T (2020) The preventive and therapeutic application of garlic and other plant ingredients in the treatment of periodontal diseases. Exp Ther Med 19(2):1507–1510
Omar S (2013) Garlic and cardiovascular diseases. In: Natural products: phytochemistry, botany and metabolism of alkaloids, pehnolics and terpenes. Springer, Berlin, pp 3661–3696
Pan Z, Dong H, Huang N et al (2022) Oxidative stress and inflammation regulation of sirtuins: new insights into common oral diseases. Front Physiol 13:953078
Pârvu M, Mot CA, Pârvu AE et al (2019) Allium sativum extract chemical composition, antioxidant activity and antifungal effect against Meyerozymaguilliermondii and Rhodotorulamucilaginosa causing onychomycosis. Molecules 24(21):3958
Petrovska BB, Cekovska S (2010) Extracts from the history and medical properties of garlic. Pharmacogn Rev 4(7):106
Prager KM, Goncharov I, Rabinkov A et al (2007) Allicin inhibits cell polarization, migration and division via its direct effect on microtubules. Cell Motil Cytoskeleton 64(5):321–337
Rahman K (2007) Effects of garlic on platelet biochemistry and physiology. Mol Nutr Food Res 51(11):1335–1344
Rana SV, Pal R, Vaiphei K et al (2011) Garlic in health and disease. Nutr Res Rev 24(1):60–71
Ried K, Fakler P (2014) Potential of garlic (*Allium sativum*) in lowering high blood pressure: mechanisms of action and clinical relevance. Integr Blood Press Control 7:71–82
Ried K, Frank OR, Stocks NP (2010) Aged garlic extract lowers blood pressure in patients with treated but uncontrolled hypertension: a randomised controlled trial. Maturitas 67(2):144–150
Ried K, Frank OR, Stocks NP (2013a) Aged garlic extract reduces blood pressure in hypertensives: a dose–response trial. Eur J Clin Nutr 67(1):64–70
Ried K, Toben C, Fakler P (2013b) Effect of garlic on serum lipids: an updated meta-analysis. Nutr Rev 71(5):282–299
Rivlin RS (2001) Historical perspective on the use of garlic. J Nutr 131(3):951S–954S
Rouf R, Uddin SJ, Sarker DK, Islam MT, Ali ES, Shilpi JA, Nahar L, Tiralongo E, Sarker SD (2020) Antiviral potential of garlic (Allium sativum) and its organosulfur compounds: a systematic update of pre-clinical and clinical data. Trends Food Sci Technol 104:219–234

Ryu JH, Park HJ, Jeong YY et al (2015) Aged red garlic extract suppresses nitric oxide production in lipopolysaccharide-treated RAW 264.7 macrophages through inhibition of NF-κ B. J Med Food 18(4):439–445
Sahidur MR, Islam S, Jahurul MH (2023) Garlic (*Allium sativum*) as a natural antidote or a protective agent against diseases and toxicities: a critical review. Food Chem Adv 3:100353
Sasi M, Kumar S, Kumar M et al (2021) Garlic (*Allium sativum* L.) bioactives and its role in alleviating oral pathologies. Antioxidants 10(11):1847
Shams-Ghahfarokhi M, Shokoohamir MR, Amirrajab N et al (2006) In vitro antifungal activities of *Allium cepa*, *Allium sativum* and ketoconazole against some pathogenic yeasts and dermatophytes. Fitoterapia 77(4):321–323
Shang A, Cao SY, Xu XY et al (2019) Bioactive compounds and biological functions of garlic (*Allium sativum* L.). Foods 8(7):246
Sharma N (2019) Efficiency of garlic and onion against virus. Int J Res Pharm Sci 10(4): 3578–3586
Shojai TM, Langeroudi AG, Karimi V et al (2016) The effect of *Allium sativum* (Garlic) extract on infectious bronchitis virus in specific pathogen free embryonic egg. Avicenna J Phytomed 6(4):458
Singh RN, Kumar P, Kumar N et al (2020) Garlic (*Allium sativum*): pharmaceutical uses for human health. Int J Pharm Sci Res 11(9):4214–4228
Sobenin IA, Andrianova IV, Fomchenkov IV et al (2009) Time-released garlic powder tablets lower systolic and diastolic blood pressure in men with mild and moderate arterial hypertension. Hypertens Res 32(6):433–437
Sobenin IA, Pryanishnikov VV, Kunnova LM et al (2010) The effects of time-released garlic powder tablets on multifunctional cardiovascular risk in patients with coronary artery disease. Lipids Health Dis 9:1–6
Sobenin I, Andrianova I, Ionova V et al (2012) Anti-aggregatory and fibrinolytic effects of time-released garlic powder tablets. Med Health Sci J 10:47–52
Sobenin IA, Myasoedova VA, Iltchuk MI et al (2019) Therapeutic effects of garlic in cardiovascular atherosclerotic disease. Chin J Nat Med 17(10):721–728
Sopha GA, Syahid SF, Cartika I (2024) Aerial bulbils as garlic alternative planting materials, a systematic review. Chile J Agric Res 84(1):123–131
Stabler SN, Tejani AM, Huynh F (2012) Garlic for the prevention of cardiovascular morbidity and mortality in hypertensive patients. Cochrane Database Syst Rev 2012(8):CD007653
Taylor P, Noriega R, Farah C et al (2006) Ajoene inhibits both primary tumor growth and metastasis of B16/BL6 melanoma cells in C57BL/6 mice. Cancer Lett 239(2):298–304
Teranishi K, Apitz-Castro R, Robson SC et al (2003) Inhibition of baboon platelet aggregation *in vitro* and *in vivo* by the garlic derivative, ajoene. Xenotransplantation 10(4):374–379
Tesfaye A (2021) Revealing the therapeutic uses of garlic (Allium sativum) and its potential for drug discovery. Sci World J 2021(1):8817288
Tocmo R, Liang D, Lin Y et al (2015) Chemical and biochemical mechanisms underlying the cardioprotective roles of dietary organopolysulfides. Front Nutr 2:1
Tudu CK, Dutta T, Ghorai M et al (2022) Traditional uses, phytochemistry, pharmacology and toxicology of garlic (*Allium sativum*), a storehouse of diverse phytochemicals: a review of research from the last decade focusing on health and nutritional implications. Front Nutr 9:949554
Varshney R, Budoff MJ (2016) Garlic and heart disease. J Nutr 146(2):416S–421S
Wang L, Jiao H, Zhao J et al (2017) Allicin alleviates reticuloendotheliosis virus-induced immunosuppression viaERK/Mitogen – activated protein kinase pathway in specific pathogen-free chickens. Front immunol 8:1856
Wang Y, Huang P, Wu Y et al (2022) Association and mechanism of garlic consumption with gastrointestinal cancer risk: a systematic review and meta-analysis. Oncol Lett 23(4):1–16
Weidinger A, Kozlov AV (2015) Biological activities of reactive oxygen and nitrogen species: oxidative stress versus signal transduction. Biomolecules 5(2):472–484
Xiao J, Xing F, Liu Y et al (2018) Garlic-derived compound S-allylmercaptocysteine inhibits hepatocarcinogenesis through targeting LRP6/Wnt pathway. Acta Pharm Sin B 8(4):575–586

You BR, Yoo JM, Baek SY et al (2019) Anti-inflammatory effect of aged black garlic on 12-O-tetradecanoylphorbol-13-acetate-induced dermatitis in mice. Nutr Res Pract 13(3):189–195

Yun SH, Sim EH, Goh RY et al (2016) Platelet activation: the mechanisms and potential biomarkers. Biomed Res Int 2016(1):9060143

Zain al-abdeen SS, Abdullah IT, Al-Salihi SS (2013) The synergism effect of aqueous garlic extract and ciprofloxacin against some multi-resistant bacteria. J Microbiol Biotech Res 3(3):136–142

Zardast M, Namakin K, Kaho JE et al (2016) Assessment of antibacterial effect of garlic in patients infected with *Helicobacter pylori* using urease breath test. Avicenna J Phytomed 6(5):495

Zhang X, Zhu Y, Duan W et al (2015) Allicin induces apoptosis of the MGC-803 human gastric carcinoma cell line through the p38 mitogen-activated protein kinase/caspase-3 signaling pathway. Mol Med Rep 11(4):2755–2760

Zhen H, Fang F, Ye DY et al (2006) Experimental study on the action of allitridin against human cytomegalovirus in vitro: inhibitory effects on immediate-early genes. Antivir Res 72(1):68–74

Zini A, Mann J, Mazor S et al (2020) Beneficial effect of aged garlic extract on periodontitis: a randomized controlled double-blind clinical study. J Clin Biochem Nutr 67(3):297–301

# Medicinal and Nutritional Importance of *Prunus dulcis* in Human Health

10

Aman Safdar, Furqan Choudhary, Mohtashim Rehan, Abdur Rahmaan Akhtar, and Humaira Farooqi

**Abstract**

Almond trees under the scientific name *Prunus dulcis* received cultivation from ancient times until today with the main plantations located in the United States and Australia and Spain. The research evaluates how almonds support medical and dietary needs for human well-being. Almonds have anti-inflammatory and antioxidant qualities because of their makeup, which included monounsaturated and polyunsaturated fatty acids, proteins, vitamins, phytosterols and polyphenols. Daily almond consumption brings multiple health benefits to the body by improving cholesterol profiles together with blood sugar management and maintaining cardiovascular health and combatting inflammation and tissue-damaging oxidation. Scientific research confirms that using *Prunus dulcis* nuts as food protection treatment helps protect mice liver and kidney tissues from harm. People who consume almonds receive brain defence benefits through bioactive compounds that also enhance skin health protection. Scientists need to perform additional research to fully comprehend how *Prunus dulcis* benefits human health as well as disease treatment functions.

A. Safdar · F. Choudhary · H. Farooqi (✉)
Department of Biotechnology, Jamia Hamdard, New Delhi, India
e-mail: hfarooqi@jamiahamdard.ac.in

M. Rehan
Paramedical College, Faculty of Medicine, Aligarh Muslim University, Aligarh, Uttar Pradesh, India

A. R. Akhtar
Department of Biosciences, Faculty of Life Science, Jamia Millia Islamia, New Delhi, India



M. A. Ansari et al. (eds.), *Medicinal Plants and their Bioactive Compounds in Human Health: Volume 2*, https://doi.org/10.1007/978-981-95-3620-7_10

**Keywords**

*Prunus dulcis* · Antioxidant · Anti-inflammatory · Cardioprotective · Neuroprotective · Skin health

## 10.1 Introduction

Between 2000 and 3000 BC, the almond (*Prunus dulcis* [Mill.] D.A. Webb) was one of the earliest fruit trees to be cultivated. Two mediterranean countries as its birthplace are Syria and Turkey (Ladizinsky 1999). Today, the United States is the country that produces the most almonds, followed by Australia and Spain. Almond varieties that are grown have unique chemical profiles that are influenced by genetic ecological and processing factors. Nuts have been associated with health advantages, especially in preventing cardiometabolic illness when consistently ingested (Becerra-Tomás et al. 2019; De Souza et al. 2017). Historically acknowledged as a source of essential nutrients, Rosaceae family members are now in great demand as a nutrient-dense food that is gaining popularity among gardeners and consumers alike (Barreca et al. 2020). Its nutritional and therapeutic properties have long been recognized in a number of ethnomedical systems, particularly in Persian medicine (PM) (Karimi et al. 2021). High levels of fatty acids (MUFA, PUFA), proteins, mineral (copper, calcium, magnesium, etc.), vitamin profiles (E, B) and other bioactive components (phytosterols, polyphenols etc.) are all found in almonds, in terms of nutrition. The chemical supports the human body's antioxidant defences while also acting as an anti-inflammatory (Barreca et al. 2020). Furthermore, these components promote several health benefits, including obesity, oxidative stress and inflammation, cardiovascular risk, lipid and glucose management, serum uric acid, prebiotics and secondary metabolites of almonds as a functional food (Barreca et al. 2020). According to the particular study, mice are shielded against the liver and renal damage caused by gentamicin by the nut formulation *Prunus dulcis*. Its ability to regulate inflammatory pathways, as well as its anti-inflammatory effects on the liver, kidneys and cytokines, provides protection against *Prunus dulcis*. Research indicates that large dosages of *Prunus dulcis* are the most effective in causing harm to the kidneys and liver (Sharma et al. 2024), indicating a probable health advantage and the need for more study to determine the full extent of *Prunus dulcis* potential health benefits.

This chapter will discuss the medicinal and nutritional benefits of *Prunus dulcis*, exploring the nutritional composition, pharmacological and medicinal properties, and potential uses in improving the health.

## 10.2 Botanical Description and Taxonomy of *Prunus dulcis*

### 10.2.1 Morphological Characteristics of the *Prunus dulcis*

As a little deciduous tree, almond trees typically reach heights of 4–10 m and trunk diameters of 30 cm; young shoots are originally green, become purple when exposed to sunshine, then turn grey in their second year. About 4–13 cm long and 1.2–4 cm wide, the lanceolate leaves have a serrated edge on the 2.5 cm petiole. The blossoms appear either alone or in pairs before the leaves do in the early spring. They have five petals and are white or light pink, with a diameter of 3–5 cm (Mushtaq et al. 2015). From a botanical context, it is a drupe rather than a nut. The plum and cherry, among other *Prunus* species, have fleshy exocarps, while the hull has a downy, leathery, grey–green covering (Kumar et al. 2023). Microscope measurements show that almond shells have a diameter of 40–60 μm for tiny holes and 300–500 μm for big holes (Li et al. 2018). The lignifies endocarp or shell that envelops the almond kernel provides protection against mould and insect infestation when it is tightly sealed (Gradziel 2020). Since most of the kernel sections are smooth or slightly wrinkled and lack of pubescence was observed, the study discovered that some cultivars have unique kernel morphologies that are beneficial for breeding operations (Genç and Güney 2025). Although physical factors have little effect on the organoleptic quality of kernels, food manufacturers consider them to be significant characteristics. Basic qualities and selection parameters that affect manufacturing output and customer needs should include evaluation of shell properties. Producers must assess the physical characteristics of both the kernel and the shell. The multiple layers of the shell prevent cracks, but the hardness of the shell regulates defence mechanisms including fungal contamination, insect penetration and bird feeding, which can also harm kernels. In addition to kernel qualities, which include aspects, size, shape and colour, kernel characteristics also include the thickness of the seed coat (Socias I Company et al. 2009).

### 10.2.2 Taxonomy of *Prunus dulcis*

| | |
|---|---|
| **Kingdom** | Plantae |
| **Order** | *Rosales* |
| **Family** | *Rosaceae* |
| **Sub-family** | *Prunoideae* |
| **Genus** | *Prunus* |
| **Sub-genus** | *Amygdalus* |
| **Species** | *Prunus dulcis* |

### 10.2.3 Geographic Distribution and Cultivation

The genus *Amygdalus L.* is a unique Irano-Turanian one. The majority of its species may be found in southwest Asia. A small number of species spread the genus's native range to south Mongolia, west China, southwest Europe and the central Mediterranean basin (Browicz and Zohary 1996). Altitude separation occasionally takes the role of geographic isolation. The sole notable exception to this rule appears to be *Amygdalus communis.* This fruit trees's domestication and the subsequent expansion of almond farming brought it into touch with a number of its previously distant vicarious relatives (Browicz and Zohary 1996). It is evidenced by regional studies that wild populations that still exist in the remote parts of the Near East and mountainous areas near there have significant genetic properties in the production of robust cultivars (Gharaghani et al. 2017). Although native to the Levantine and some parts of Iran, *Prunus dulcis* has a fast rate of spreading beyond the territory due to trade regimens such as the Silk Road and human migration. Almonds started their journey in the eastern part and then travelled to Central Asia. In the west, they extended to the Mediterranean basin where they were able to find a climatic condition that led to their natural propagation. Then it spread even to California, the world centre of almond production (Covert 2011). Almond production and establishment in arid and semi-arid regions of California, Australia and Spain, as well as growth in emerging markets in China and India, are examples of how the species adaptability in terms of economic and ecological affinities has caused the domestic population of almonds to be relocated outside of the Mediterranean region (Čolić et al. 2019; Homet-Gutiérrez et al. 2015). Besides, the latest ecological niche modelling through MaxEnt approach has been able to identify that the world potential range of *Prunus dulcis* can reach as far south as Xinjiang in China, the southern parts of South America and even southern Africa and parts of Australia which have Mediterranean-type climates (Wang et al. 2021).

The plant's physiological and biochemical activity is significantly regulated by the climate. The phenological stage's success and, hence, yield output is influenced by temperature, precipitation, humidity and brightness. Almonds grow in regions with yearly mean temperatures between 15 and 20 °C. The species is vulnerable to salt-stressed habitats, but generally exhibits tolerance to drought conditions (Freitas et al. 2023). The type of soil also plays a significant role in the almond cultivation where the plant thrives well in deep and well-drained sandy loams although with the appropriate management skills, the plant can succeed in many soil types: loamy soils, clayey soils and even salt-affected soils (Mutlu et al. 2023). Moreover, recent advancements in technologies of remote sensing have made it possible to develop strong tools of managing and monitoring almond orchards. By monitoring satellite and unmanned aerial vehicle (UAV) photos, scientists can now visualize real-time detection of disease outbreaks, canopy heterogeneity and water stress. Besides making the nutrition and irrigation management easy, the technologies enable growers, with ease, to intervene and prevent major losses in terms of yields occasioned by

**Table 10.1** Comparative assessment of major almond producing nations

| Country | Key cultivars | Climate and irrigation | Production | Practices and challenges | References |
|---|---|---|---|---|---|
| California, USA | Nonpareil, Carmel, Butte | Mediterranean; drip irrigation; frost risk | ~490 M lbs | Mechanized; high water/ pollination costs | Guimarães et al. (2024) and Kumar et al. (2023) |
| Spain | Marcona. Desmayo, Tuono, landraces | Mediterranean; ~92% rainfed | ~374 K metric tonnes | Traditional; lower yields under rainfed | Kumar et al. (2023) and Pérez-Sánchez et al. (2021) |
| Turkey | Local Turkish cultivars | Mediterranean/ semi-arid; mixed irrigation | ~190 K tonnes | Intercropping; emerging drip systems | Guimarães et al. (2024) and Lutsiak et al. (2023) |
| China | Local arid-adapted varieties | Arid; cold winters, dry summers | ~140 K tonnes | Cooperative modernization; export-oriented | Guimarães et al. (2024) and Lutsiak et al. (2023) |
| Australia | Nonpareil, Carmel, price, others | Mediterranean-temperature; advanced irrigation | ~360 K tonnes | Mechanized; water management critical | Calle et al. (2024) and Guimarães et al. (2024) |

pests' disease or poor weather patterns (Guimarães et al. 2024). All the most relevant agronomic strategies applied in every place, climatic situation, key information about cultivars and any quantitative measures of production are all complied Table 10.1.

## 10.3 Phytochemical Constituent and Bioactive Compounds

### 10.3.1 Phenolic Acids and Flavonoids

Almonds contain polyphenols, especially flavonoids and phenolic acids (Smeriglio et al. 2016). Scientific investigation reveals that almond hulls are rich in phenolic acids, with neochlorogenic acid, chlorogenic acid and cryptogenic acid being the most notable acid constituents, based on their data, a number of scientific research have determined how these antioxidant compounds function (Takeoka and Dao 2002). Qureshi et al. found that three main flavonoids found in almond hull, catechin, isorhamnetin and kaempferol, increase the antioxidant qualities of almond hulls. (Qureshi et al. 2019). The presence of these substances increases hopes for almonds possible nutritional worth and other health advantages.

### 10.3.2 Tannins and Terpenoids

Almonds contain a lot of hydrolysed tannins. According to the research that examined the almonds composition, among California almond varieties, the almond hydrolysable tannin content was 54.7 ± 2.3 mg ellagic acid and 27.4 ± 7.3 mg gallic acid per 100 g. Thus, a large portion of the polyphenol content in almonds is derived from hydrolysable tannins (Xie et al. 2012). Terpenoids are also abundant in almonds. The primary constituents of Sicilian almond hulls, including oleanolic, ursolic and betulinic acids, the 2-hydroxy counterparts alphitolic, corosolic and maslinic acids and the associated aldehydes, betulinic, oleanolic and ursolic, are identified by a bioassay-guided fractionation process. Furthermore, betulinic acid demonstrated antiproliferative activity against Michigan Cancer Foundation-7 (MCF-7) cells (GI50 = 0.27 μM), surpassing that of the anticancer medication 5-flourouracil, indicating possible health benefits (Amico et al. 2006). These studies suggest that almonds may provide health benefits beyond simple nourishment.

### 10.3.3 Lipophilic Compounds

There are lot of lipophilic compounds in almonds. In the same study by Keser and Yilmaz, the following compounds were determined: myricetin (1831.52 mu g/g), kaempferol (104.52 mu g/g), naringenin (2.51 mu g/g), vanillic acid (91.70 mu g/g), caffeic acid (85.92 mu g/g), ferulic acid (27.11 mu g/g), rosmarinic acid (0.95 mu g/g), hydroxycinnamic acid (1.35 mu g/g), delta-tocopherol (4.95 mg/kg), mu-tocopherol (104.15 mg/kg), vitamin k (42.25 mg/kg), beta-sitosterol (366.95 mg/kg) and stigmasterol (242.65 mg/kg) (Keser and Yilmaz 2014).

Figure 10.1 displays the bioactive compounds found in kernels together with their by-products (Polskiego et al. 2021).

## 10.4 Nutritional Composition

Almonds are composed of a variety of minor bioactive components and a low water content. They also include around 50% lipids, 25% proteins and 20% carbs. Almonds nutritional value includes both major and tiny elements that support their health benefits. Monosaturated fatty acids (MUFA) and polyunsaturated fatty acids (PUFA) make up the majority of the lipid profile, with fibre, vitamins and minerals serving as important components that promote health (Barreira et al. 2012; Oliveira et al. 2019).

### 10.4.1 Macronutrients (e.g. Fats, Proteins and Carbohydrates)

During ripening period, the lipid content increase, making it noticeably high. In one research, the average lipid content across the several cultivars was 248.8 g/kg at the

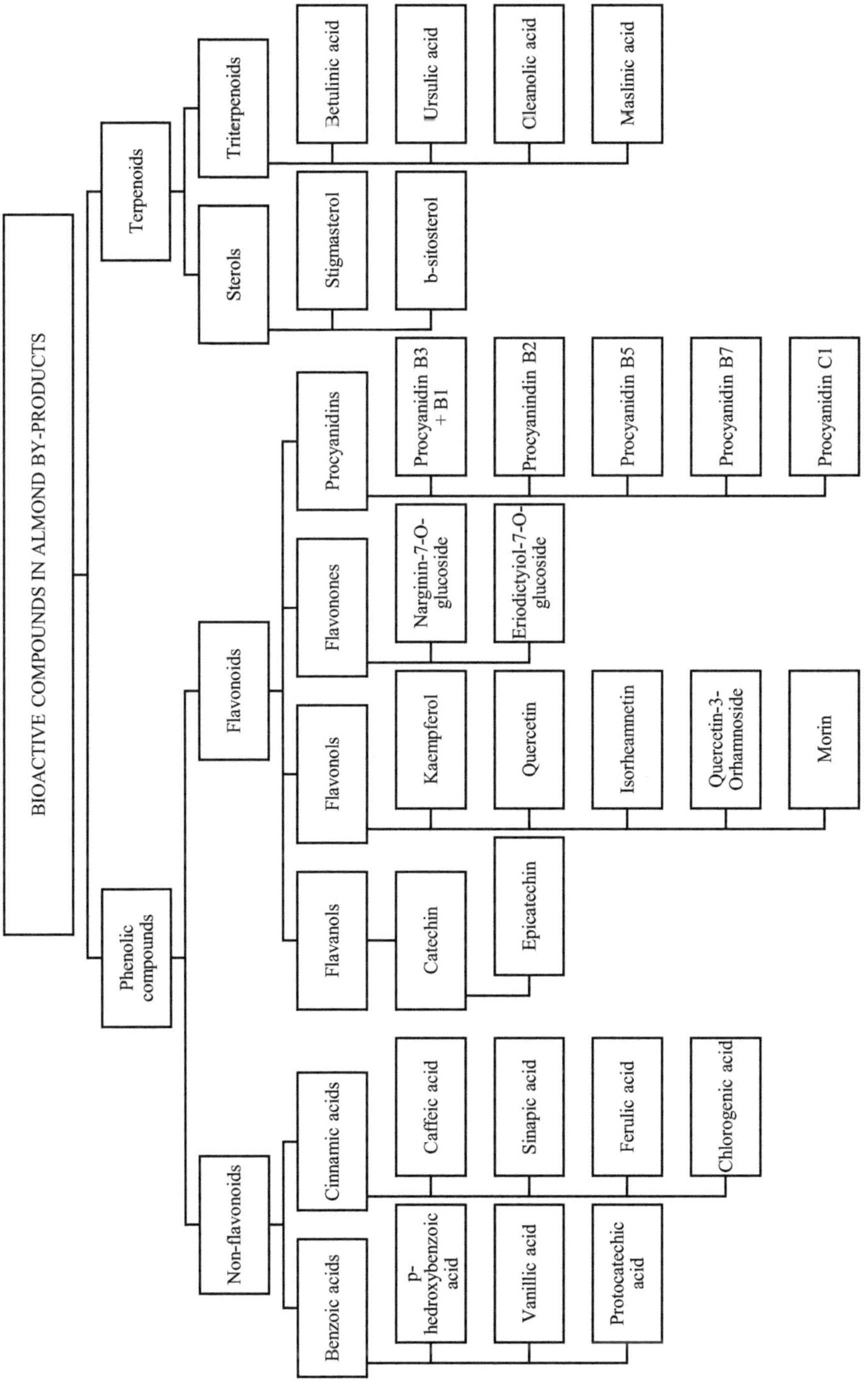


**Fig. 10.1** Bioactive compounds in almond by-products. (Adapted from Prgomet et al. 2017)

early stage (T1) and significantly rose to 487.3 g/kg at full maturity (T2) (Summo et al. 2018). An analysis of the chemical composition of the skin of *Prunus dulcis* L. almonds revealed the presence of fatty acids, including linoleic acid (63.19%), palmitic acid (16.65%), oleic acid (8.41%), stearic acid (4.10%) and palmitoleic acid (3.40%) (Badalamenti et al. 2023). According to a different research, almonds contain fatty acids, including oleic acid (70.61%) and linoleic acid (20.68%). Additionally, the protein level varies from cultivar to cultivar. It averaged 134.8 g/kg in the early (T1) and somewhat rose to 197.9 g/kg at full maturity (T2) (Summo et al. 2018). Lysine was identified as the first-limiting amino acid by amino acid analysis, which produced amino acid scores of 0.53, 0.52, 0.49 and 0.56 for several almond varieties (House et al. 2019). Research also examined the amount of carbohydrates in several cultivars; the average amount was 148.9 g/kg at early maturity (T1) and 226.3 g/kg at full maturity (T2). Humans are only able to digest, absorb and metabolize the sugars, starches and certain sugar alcohols found in almonds to produce energy. The nonstarch polysaccharides have physiological benefits that are good for human health, but they cannot be used as an energy source since they are indigestible (Summo et al. 2018; Yada et al. 2011). These results imply that almonds nutritional value is improved by their notable rise in macronutrient content. Ripe almonds are therefore a nutrient-dense food, indicating potential health advantages.

### 10.4.2 Micronutrients (e.g. Vitamins and Minerals)

Almonds are a good source of vitamins and minerals. The main elements are magnesium (159–404 mg/kg), calcium (160–663 mg/kg), phosphorus (310–938 mg/kg) and potassium (465–1510 mg/kg). With concentrations ranging from 0.463 to 4.76 mg/100 g for copper and 1.31–3.97 mg/100 g for manganese, almonds are regarded as a key source of these minerals. Because of intake from soil, water and fertilizers, the mineral content of almonds is influenced by agronomic and geographical conditions (Piscopo et al. 2010). As a result, depending on geographic location and farming methods, almonds mineral composition might vary (Drogoudi et al. 2013).

Riboflavin (vitamin B2) and vitamin E tocopherols (α, β, γ and δ) are abundant in almonds. Compared to other tree nuts, almonds have higher quantities of tocopherols, which gradually decline over storage (Franklin et al. 2017). Even though β-tocopherol, γ-tocopherol and δ-tocopherol have lesser levels (0.012–0.8 mg/100 g, 0.014–0.084 mg/100 g and 0.002–0.016 mg/100 g, respectively), the main isomer, α-tocopherol, has been shown to vary in kernels from 8.5 to 84.0 mg/100 g (Kodad et al. 2018). Along with their health advantages, tocopherols provide almond oil a substantial antioxidant stability; however, their composition varies greatly amongst the genotypes under study, as well as with temperature and water application. For instance, it was that the quantities of α-tocopherol in seven commercial almond cultivars cultivated in California vary from 18.2 to 32.9 mg/100 g. Recent research has shown that 50 g of Afghan almonds daily, which contain an average of 30 mg/100 g of α-tocopherol, is enough to meet the current recommended daily allowance for this vitamin (Tomishima et al. 2022).

### 10.4.3 Dietary Fibres

One significant modifiable element that has a significant impact on the makeup of the gut microbiota is diet (Rothschild et al. 2018). Only a small number of research examined how eating almonds affected clinical results or gut function (Burns et al. 2016; Ukhanova et al. 2014). Randomized controlled research showed that eating almonds raises butyrate in humans, indicating beneficial changes to microbiome function, but has no effect on microbiota composition. Incorporating almonds into the diet can boost fibre intake without causing digestive issues (Creedon et al. 2022). This reveals *Prunus dulcis* potential as a beneficial dietary fibre source with promising qualities for a range of uses.

A summary of the nutrient composition of almond kernels, hulls and shells of different varieties which include data from USDA 2022 can be found in Table 10.2.

## 10.5 Medicinal Properties

### 10.5.1 Antioxidant Effects

Antioxidant are abundant in almonds (El Bernoussi et al. 2024). Studies have examined the anticancer effects of bioactive compounds from *Prunus dulcis*. It showed that almond extract particularly its catechin component significantly reduced the growth of cancer cells and further increased the production of reactive oxygen species (ROS) and caused apoptosis (Rakshanaa et al. 2023). Additionally, almonds exhibit antimicrobial effects against numerous bacteria and restrict the growth of human breast, colon and liver cancer cell lines (Gomaa 2013). Since vitamin E antioxidants in almonds enable free radicals to attach to particles, they reduce oxidative stress levels, which aid in the development of cancer. Because these nutrients work together to prevent the formation of cancer, humans are benefitted from diets high in antioxidants and oxidative stress components, which are present in almonds (Ejike and Liman 2022). Additionally, another study revealed that the antioxidant compounds in *Prunus dulcis* almond shells combat free radicals, which are associated with the development of cancer and other disease. Antioxidant usage reduces stress, which has anticancer consequences. Cell functions require antioxidant protection to perform at their best. Almond shells from *Prunus dulcis* have been shown to have therapeutic qualities for the development of cancer therapeutic agents that combat oxidative damage linked to the spread of cancer (Khan et al. 2022). Furthermore, studies show that bitter almonds include two active chemicals that serve as antioxidants that serve as antioxidants: oestrone and stigmasterol glycerol. Compounds in bitter almonds boost their therapeutic qualities by combating systems linked to oxidative stress during cancer therapy. The identification of 53 significant protein interactions between NR3C2 and MAOA suggests that bitter almond antioxidants may be used as components of cancer treatment by activating many targets and components (Feng et al. 2023). Therefore, these findings highlight that both sweet and bitter almonds contain physiologically active components that

**Table 10.2** Nutritional composition of almond per 100 g of edible nuts

| Name | Average amount | Unit |
|---|---|---|
| ***Proximates*** | | |
| Water | 4.26 | G |
| Energy (Atwater general factors) | 626 | Kcal |
| Energy (Atwater specific factors) | 584 | Kcal |
| Nitrogen | 4.14 | G |
| Protein | 21.4 | G |
| Total lipid (fat) | 51.1 | g |
| Ash | 3.16 | g |
| ***Carbohydrates*** | | |
| Carbohydrate, by difference | 20 | g |
| Fibre, total dietary | 10.8 | g |
| ***Minerals*** | | |
| Calcium, Ca | 254 | mg |
| Iron, Fe | 3.74 | mg |
| Magnesium, Mg | 258 | mg |
| Phosphorus, P | 503 | mg |
| Potassium, K | 733 | mg |
| Sodium, Na | <2.5 | mg |
| Zinc, Zn | 2.86 | mg |
| Copper, Cu | 0.91 | mg |
| Manganese, Mn | 2.15 | mg |
| Molybdenum, Mo | 45.3 | µg |
| ***Vitamins and other components*** | | |
| Thiamine | 0.16 | mg |
| Niacin | 3.77 | mg |
| Vitamin B-6 | 0.101 | mg |
| Biotin | 57 | µg |
| Vitamin K (phylloquinone) | <0.1 | µg |
| Vitamin K (dihydrophylloquinone) | <0.1 | µg |
| Vitamin K (Menaquinone-4) | <0.1 | µg |
| ***Lipids*** | | |
| Fatty acids, total saturated | 3.78 | g |
| Fatty acids, total monounsaturated | 31.3 | g |
| Fatty acids, total polyunsaturated | 11.2 | g |
| ***Amino acids*** | | |
| Tryptophan | 0.208 | g |
| Threonine | 0.608 | g |
| Isoleucine | 0.774 | g |
| Leucine | 1.49 | g |
| Lysine | 0.618 | g |
| Methionine | 0.154 | g |
| Phenylalanine | 1.2 | g |
| Tyrosine | 0.63 | g |

(continued)

**Table 10.2** (continued)

| Name | Average amount | Unit |
|---|---|---|
| Valine | 0.916 | g |
| Arginine | 2.46 | g |
| Histidine | 0.548 | g |
| Alanine | 0.98 | g |
| Aspartic acid | 2.4 | g |
| Glutamic acid | 4.25 | g |
| Glycine | 1.43 | g |
| Proline | 1.31 | g |
| Serine | 0.99 | g |
| Hydroxyproline | <0.01 | g |
| Cysteine | 0.5 | g |

include antioxidants such as vitamin E, oestrone and stigmasterol glycerol compounds. These chemicals have distinct antibacterial, cancer-fighting and stress-reduction qualities, making it a viable option for more study and advancement in cancer treatment.

### 10.5.2 Anti-inflammatory Effects

The anti-inflammatory qualities of almond skins are mostly attributed to the presence of polyphenols (Arangia et al. 2023). Almonds are effective anti-inflammatory agents for heart health, according to scientific research. Antioxidants and certain molecules in vitamin E work synergistically to effectively reduce physiological stress signals. Because almonds increase the generation of butyrate synthesis and encourage the growth of good bacteria, they enhance the gut microbiota and offer further cardiac protection (Singar et al. 2024). Research shows that almond skin extract (ASE) has been shown to have consistent anti-inflammatory effects. A substantial body of research shows that using almond skin extract (ASE) as a therapy reduces inflammation by inactivating pro-inflammatory molecules, preventing tissue changes and lowering neutrophil counts in live tests. Together with catalase and glutathione antioxidant enzymes, the extract restored superoxide dismutase's normal activity. ASE's anti-inflammatory properties inhibit the activity of the pro-inflammatory nuclear protein nuclear factor kappa B (NF-κB) while also triggering antioxidant pathways via nuclear factor-erythroid 2-related factor 2 (Nrf2) (Arangia et al. 2023). Additionally, almond polysaccharides (AP-1) have anti-inflammatory qualities that make them ideal for treating ulcerative colitis (UC). Almond polysaccharides (AP-1) therapy produced the anti-inflammatory cytokine interleukin-10 (IL-10) and concurrently reduces tumour necrosis factor-α (TNF-α), IL-1 β, IL-6. Since they shield cells and prevent immune cells strikes and antioxidant enzyme activation, almond polysaccharides (AP-1) have anti-inflammatory qualities. By blocking NF-κB signalling pathways, the presence of AP-1 protein reduces the synthesis of both Cyclooxygenase-2 (COX2) and Inducible Nitric Oxide Synthase (iNOS) proteins, improving the regulation of the inflammatory response (Peng et al. 2024). Studies highlight that almonds can suppress the production of inflammatory cytokine mRNA for IL-1β, IL-6 and TNF-α that is triggered by lipopolysaccharide (LPS). Almond extracts had a more noticeable inhibitory effect on IL-6 mRNA than lipopolysaccharide (LPS) treatment alone, resulting in a 61% reduction in IL-1β mRNA expression levels. Furthermore, almonds help patients control inflammatory diseases by lowering inflammatory responses in macrophages (Müller et al. 2019). A study noticed that consuming almonds reduced inflammatory markers in Chinese individuals with type 2 diabetes mellitus. When compared to the standard diet, the almond eating regimen decreased interleukin-6 and C-reactive protein by a median of 103 and 103%, respectively. The TNF-α propensity declined by 157%; therefore, almonds can assist persons with type 2 diabetes lower inflammation when consumed as part of a balanced diet (Liu et al. 2013). As a whole, the data supports and aids in the management of inflammatory diseases such as type 2 diabetes and ulcerative colitis.

### 10.5.3 Cardioprotective Effects

Almond's ability to lower cholesterol improves high-density lipoprotein (HDL) function, control blood sugar and reduce inflammation, all of which enhance lipid profiles and have a heart-protective impact. Research on cardiovascular health often looks at two main areas of antioxidants and how they affect gut flora (Singar et al. 2024). A study found that the primary evaluation criteria for almond cardiovascular protection assessment with overweight Asian Indian participants were insulin sensitivity along with improvements in lipid profiles and decreases in total cholesterol. Over the course of the 12-week study, the participants ingested 43 g of almonds per day in order to get the health benefits (Gayathri et al. 2023). Compared to a nut-free diet, almonds significantly lower triglycerides and systolic blood pressure, suggesting enhanced cardiometabolic health without compromising mood or cognitive function (Coates et al. 2020). Additionally, studies examining the cardioprotective benefits of almonds for individuals with type 2 diabetes by assessing oxidative stress and inflammatory markers found significant decrease in interleukin-6, C-reactive protein, and tumor necrosis factor-α confirming that almonds are good for cardiovascular health in this patient population (Liu et al. 2013). Overall, these findings suggests that almond in the diet improves heart metabolism over time by lowering oxidative stress, inflammation and blood cholesterol levels as well as managing blood sugar levels.

### 10.5.4 Neuroprotective Effects

Almond consumption on a regular basis protects brain cells by protecting neutral tissue with antioxidant properties. Eating almonds on a daily basis enhances brain function and memory (Samuel Sundar Doss et al. 2020). This is because almonds are a good source of phytochemicals, vitamins and macronutrients. Compounds derived from almonds are used to modify the molecular processes associated with Alzheimer's disease. These compounds control oxidative stress and tau phosphorylation and amyloidogenesis pathways. Cholinergic pathways are strongly impacted by almonds anti-inflammatory qualities. Almond's preventive bioactive ingredients serve as a supplement for brain health and a disease-prevention tool for Alzheimer's disease management. To determine all possible uses for these components, further thorough scientific research is required (Vaishnavi and Justin Thenmozhi 2023). Additionally, almonds reduce stress and improve memory, which make it feasible for them to have protective benefits on brain tissue. Studies revealed that when MDA (malondialdehyde) levels dropped and antioxidant enzyme activity rose, almonds eaten before treatment reduced indicators of oxidative stress brought on by scopolamine. Almond antioxidants including polyphenols protect the surrounding cells that sustain damage because these defence mechanisms establish a way to delay aging related memory loss Studies revealed that when MDA (malondialdehyde) levels dropped and antioxidant enzyme activity rose, almonds eaten before treatment reduced indicators of oxidative stress brought on by scopolamine. Almond antioxidants including polyphenols protect the surrounding cells that sustain damage because these defence mechanisms establish a way to delay aging related memory loss because the antioxidants in almonds reach the human body's system. (Batool et al. 2018). Recent randomized control trials have shown the almond ability to have neuroprotective properties that enhance brain function and stabilize heart rate responses in stressful mental situations. According to research, almonds are a food that promotes brain health because of their well-established benefits (Dreher 2021). Furthermore, a study highlights the presence of flavonoids in almonds could have neuroprotective effects on brain. By reducing inflammation in the brain, shielding neurone from harmful chemicals and maybe enhancing cognitive function, exercise performance produces two crucial advantages. Through the creation of superior brain synaptic networks and improved vascular health, services engaged through these mechanisms stop apoptotic processes and may even reduce neural decline and brain degeneration (Mishra 2024). Additionally, research on Muqawwi-i-dimagh (brain tonic) in Unani medicine reveals that substances with high amounts of vitamin E and antioxidants are beneficial for neurological well-being and the treatment of forgetfulness (Siddiqui et al. 2023). Overall, the comprehensive literature of almonds on neuroprotective effects suggests that eating almonds helps protect the brain by improving memory and reducing inflammation and oxidative stress, which may help stave against Alzheimer's disease.

### 10.5.5 Skin Health Effects

Daily consumption of almonds has been linked to improved skin (Rybak et al. 2021). Studies have demonstrated that almonds may slow down the pace at which UVB rays cause face skin to age, which might have positive effects on human skin health (Dreher 2021). Furthermore, a study examined that eating almonds affected the length, depth and severity of wrinkles on the face; they found that wrinkles significantly improved. For 16 weeks, women ($n = 31$) in this control-trial research ate 2.1 oz (340 kcal) of packed almonds. There was no increase in sebum production in the almond group during the intervention period, and there were no additional negative effects associated with its use (Foolad et al. 2019). Another study found that 56 postmenopausal women who also consumed almonds daily saw a substantial reduction in the average severity of wrinkles and the level of facial pigmentation without an increase in sebum production (Rybak et al. 2021). Since vitamin E is incorporated into the cell membrane, it has antioxidant and photoprotective qualities that may help protect against free radicals and the damage they cause. This may be the reason why α-tocopherol helps prevent skin ageing. Through a reduction in matrix metalloproteinase activity, vitamin E may also be able to stop the breakdown of collagen. As a result, eating enough almonds each day may be a viable substitute to stop or slow the development of wrinkles, face pigmentation and ultimately skin ageing (Flores-Balderas et al. 2023). Overall, eating almonds every day may slow down the ageing process and lessen wrinkles and pigmentation because of its antioxidant qualities.

## 10.6 Conclusion

The almond tree, *Prunus dulcis*, is an excellent natural solution for comprehensive medical requirements. Almonds are not only used for their nutritional value but they also contain a number of biological components that have significant health advantages. In order to treat conditions like oxidative stress, chronic inflammation, cardiovascular diseases, neurodegeneration and skin ageing, the functional ingredients in almonds combine various bioactive elements, such as phenolic acids, flavonoids and tannins, with terpenoids, lipophilic compounds, vitamins, minerals and dietary fibres.

Almonds' antioxidant components, such as vitamin E, oestrone and stigmasterol glycerol, work synergistically to protect against oxidative damage and lower the risk of developing a disease that might progress to cancer. Because polyphenols and almond polysaccharides have anti-inflammatory properties, they can help treat ulcerative colitis and type 2 diabetes. Almonds' anti-inflammatory qualities and impacts on blood sugar and cholesterol management make their health-protective qualities clear.

Almonds' anti-inflammatory and antioxidant properties protect the brain, making them a good dietary addition for cognitive growth and potential cognitive ageing prevention. According to studies, almonds shield the skin from ultraviolet B (UVB) rays by reducing the appearance of photoaging and preventing pigmentation disorders and wrinkles. Because almond by-products contain hydrolysable tannins, betulinic acid and other therapeutic bioactive chemicals in addition to the functional substance hydrolysable tannins and betulinic acid, their therapeutic qualities are enhanced.

Almonds' wide range of useful applications is what makes them unique. By choosing alternative growing regions and almond varietals, farmers may optimize their cultivation methods for almond health qualities. Advanced scientific study is necessary to fully understand the nutritional and therapeutic benefits of almonds. Scientific research must examine the internal mechanisms of action of lesser-known almond components as well as create the best possible formulations for clinical testing. These well-established bases can be used to produce new medicinal molecules and nutraceutical goods.

The performance features of *Prunus dulcis* are improved by both modern scientific techniques and established traditions. Consuming *Prunus dulcis* on a regular basis provides a practical way for people to fight oxidative stress and prevent inflammatory reactions, which delays the onset of chronic diseases and enhances overall health. For the sake of human health, a number of studies and ongoing research on almonds will reinterpret the relationships between food consumption and medical systems.

## References

Amico V, Barresi V, Condorelli D, Spatafora C, Tringali C (2006) Antiproliferative terpenoids from almond hulls (Prunus dulcis): identification and structure-activity relationships. J Agric Food Chem 54(3):810–814

Arangia A, Ragno A, Cordaro M, D'Amico R, Siracusa R, Fusco R et al (2023) Antioxidant activity of a Sicilian almond skin extract using in vitro and in vivo models. Int J Mol Sci 24(15):12115

Badalamenti N, Bruno M, Loizzo MR, Puccio V, Gaglio R, Francesca N et al (2023) Antibacterial activity and chemical characterization of almond (Prunus dulcis L.) peel extract. Nat Prod Res 37(10):1680–1686

Barreca D, Nabavi SM, Sureda A, Rasekhian M, Raciti R, Silva AS et al (2020) Almonds (Prunus Dulcis Mill. D. A. Webb): a source of nutrients and health-promoting compounds. Nutrients 12(3):672

Barreira JCM, Casal S, Ferreira ICFR, Peres AM, Pereira JA, Oliveira MBPP (2012) Supervised chemical pattern recognition in almond (Prunus dulcis) Portuguese PDO cultivars: PCA- and LDA-based triennial study. J Agric Food Chem 60(38):9697–9704

Batool Z, Tabassum S, Siddiqui RA, Haider S (2018) Dietary supplementation of almond prevents oxidative stress by advocating antioxidants and attenuates impaired aversive memory in male rats. Plant Foods Hum Nutr 73(1):7–12

Becerra-Tomás N, Paz-Graniel I, Kendall C, Kahleova H, Rahelić D, Sievenpiper JL et al (2019) Nut consumption and incidence of cardiovascular diseases and cardiovascular disease mortality: a meta-analysis of prospective cohort studies. Nutr Rev 77(10):691–709

Browicz K, Zohary D (1996) The genus Amygdalus L. (Rosaceae): species relationships, distribution and evolution under domestication. Genet Resour Crop Evol 43(3):229–247

Burns AM, Zitt MA, Rowe CC, Langkamp-Henken B, Mai V, Nieves C et al (2016) Diet quality improves for parents and children when almonds are incorporated into their daily diet: a randomized, crossover study. Nutr Res. Elsevier 36(1):80–89

Calle A, Aparicio-Durán L, Batlle I, Eduardo I, Miarnau X (2024) Review of agronomic and kernel quality traits of 273 almond cultivars. Genet Resour Crop Evol 72(4):3783–3828

Coates AM, Morgillo S, Yandell C, Scholey A, Buckley JD, Dyer KA et al (2020) Effect of a 12-week almond-enriched diet on biomarkers of cognitive performance, mood, and cardiometabolic health in older overweight adults. Nutrients 12(4):1180

Čolić S, Zec G, Natić M, Fotirić-Akšić M (2019) Almond (Prunus dulcis) oil. Fruit oils: chemistry and functionality. Springer, pp 149–180

Covert MM (2011) The influence of chilling and heat accumulation on bloom timing, bloom length and crop yield in Almonds (Prunus dulcis (Mill.)). Master's theses, San Luis Obispo

Creedon AC, Dimidi E, Hung ES, Rossi M, Probert C, Grassby T et al (2022) The impact of almonds and almond processing on gastrointestinal physiology, luminal microbiology, and gastrointestinal symptoms: a randomized controlled trial and mastication study. Am J Clin Nutr 116(6):1790–1804

De Souza RGM, Schincaglia RM, Pimente GD, Mota JF (2017) Nuts and human health outcomes: a systematic review. Nutrients 9(12):1311

Dreher ML (2021) A comprehensive review of almond clinical trials on weight measures, metabolic health biomarkers and outcomes, and the gut microbiota. Nutrients 13(6):1968

Drogoudi PD, Pantelidis G, Bacchetta L, De Giorgio D, Duval H, Metzidakis I et al (2013) Protein and mineral nutrient contents in kernels from 72 sweet almond cultivars and accessions grown in France, Greece and Italy. Int J Food Sci Nutr 64(2):202–209

Ejike UDI, Liman ML (2022) Role of dietary antioxidants in chemoprevention of nitrosamines-induced carcinogenesis. Handbook of oxidative stress in cancer: therapeutic aspects: volume 1. Spring 1:253–275

El Bernoussi S, Boujemaa I, El Guezzane C, Bou-Ouzoukni Y, Nounah I, Bouyahya A et al (2024) Comparative analysis of nutritional value and antioxidant activity in sweet and bitter almonds. LWT. Academic 206:116587

Feng X, Niu J, Song Z, Tao A, Zong R, Pei W et al (2023) Exploring the mechanism of antioxidant action of bitter almond based on network pharmacology and molecular docking techniques. J Food Qual 1:9187692

Flores-Balderas X, Peña-Peña M, Rada KM, Alvarez-Alvarez YQ, Guzmán-Martín CA, Sánchez-Gloria JL et al (2023) Beneficial effects of plant-based diets on skin health and inflammatory skin diseases. Nutrients 15(13):2842

Foolad N, Vaughn AR, Rybak I, Burney WA, Chodur GM, Newman JW et al (2019) Prospective randomized controlled pilot study on the effects of almond consumption on skin lipids and wrinkles. Phytother Res 33(12):3212–3217

Franklin LM, Chapman DM, King ES, Mau M, Huang G, Mitchell AE (2017) Chemical and sensory characterization of oxidative changes in roasted almonds undergoing accelerated shelf life. J Agric Food Chem 65(12):2549–2563

Freitas TR, Santos JA, Silva AP, Fraga H (2023) Reviewing the adverse climate change impacts and adaptation measures on almond trees (Prunus dulcis). Agriculture 13(7):1423

Gayathri R, Abirami K, Kalpana N, Manasa VS, Sudha V, Shobana S et al (2023) Effect of almond consumption on insulin sensitivity and serum lipids among Asian Indian adults with overweight and obesity – a randomized controlled trial. Front Nutr 9:1055923

Genç B, Güney M (2025) Evaluation of pomological traits and fatty acid profile of almond (Prunus dulcis L. [Mill.] D.A. Webb) genotypes grown in Yozgat province. Eur Food Res Technol:1–11

Gharaghani A, Solhjoo S, Oraguzie N (2017) A review of genetic resources of almonds and stone fruits (Prunus spp.) in Iran. Genet Resour Crop Evol 64(3):611–640

Gomaa EZ (2013) In vitro antioxidant, antimicrobial, and antitumor activities of bitter almond and sweet apricot (Prunus armeniaca L.) kernels. Food Sci Biotechnol 22(2):455–463

Gradziel TM (2020) Redomesticating almond to meet emerging food safety needs. Front Plant Sci 11:526480

Guimarães N, Sousa JJ, Pádua L, Bento A, Couto P (2024) Remote sensing applications in almond orchards: a comprehensive systematic review of current insights, research gaps, and future prospects. Appl Sci 14(5):1749

Homet-Gutiérrez P, Schupp EW, Gómez JM (2015) Naturalization of almond trees (Prunus dulcis) in semi-arid regions of the Western Mediterranean. J Arid Environ 113:108–113

House JD, Hill K, Neufeld J, Franczyk A, Nosworthy MG (2019) Determination of the protein quality of almonds (Prunus dulcis L.) as assessed by in vitro and in vivo methodologies. Food Sci Nutr 7(9):2932–2938

Karimi Z, Firouzi M, Dadmehr M, Javad-Mousavi SA, Bagheriani N, Sadeghpour O (2021) Almond as a nutraceutical and therapeutic agent in Persian medicine and modern phytotherapy: a narrative review. Phytother Res 35(6):2997–3012

Keser S, Yilmaz O (2014) Some bioactive compounds and antioxidant activities of the bitter almond kernel (Prunus dulcis var. amara). J Chem Soc Pak:36

Khan SA, Riaz A, Asad M, Tahir M, Mahmood M, Khan SR et al (2022) Exploration of antioxidant and antimicrobial potential of prunus dulcis and arachis hypogaea shells extracts. Int J Health Sci (Qassim) 6(S7):6875–6892

Kodad O, Socias i Company R, Alonso JM (2018) Genotypic and environmental effects on tocopherol content in almond. Antioxidants 7(1):6

Kumar A, Bhuj BD, Dhar S, Kumar A (2023) Journal of agriculture and aquaculture research article new approaches for Almonds (Prunus amygdalus. Batsch) production in Mediterranean climates: a review. 5(1):1–20

Ladizinsky G (1999) On the origin of almond. Genet Resour Crop Evol 46(2):143–147

Li X, Liu Y, Hao J, Wang W (2018) Study of almond Shell characteristics. Materials 11(9):1782

Liu JF, Liu YH, Chen CM, Chang WH, Chen CYO (2013) The effect of almonds on inflammation and oxidative stress in Chinese patients with type 2 diabetes mellitus: a randomized crossover controlled feeding trial. Eur J Nutr 52(3):927–935

Lutsiak V, Hutsol T, Kovtun O, Koberniuk O, Malaga-Toboła U, Lis A et al (2023) Innovative almond-growing strategy in Ukraine: enterprise level. Sustainability 15(11):8546

Mishra S (2024) Neuroprotective effect of naturally occurring flavonoids. Cent Nerv Syst Agents Med Chem 25:1–16

Müller AK, Schmölz L, Wallert M, Schubert M, Schlörmann W, Glei M et al (2019) In vitro digested nut oils attenuate the lipopolysaccharide-induced inflammatory response in macrophages. Nutrients 11(3):503

Mushtaq A, Khaliq M, Saeed A, Azeem MW, Ben GJ (2015) Almond (Purunus amygdalus L.): a review on health benefits, nutritional value and therapeutic applications. IJCBS 8:103–106

Mutlu B, Alp Ş, Yıl VY (2023) Aesthetic and ecological characteristics of Prunus Taxa in landscape design: a study in eastern and southeastern provinces of Türkiye. Eurasian J For Sci 11(3):116–136

Oliveira I, Meyer AS, Afonso S, Aires A, Goufo P, Trindade H et al (2019) Phenolic and fatty acid profiles, α-tocopherol and sucrose contents, and antioxidant capacities of understudied Portuguese almond cultivars. J Food Biochem 43(7):e12887

Peng Y, Zhu J, Li Y, Yue X, Peng Y (2024) Almond polysaccharides inhibit DSS-induced inflammatory response in ulcerative colitis mice through NF-κB pathway. Int J Biol Macromol 281:136206

Pérez-Sánchez R, Remedios Morales-Corts M, Barba-Espín G, Ramón Acosta-Motos J, Ait Barka E (2021) Agromorphological characterization and nutritional value of traditional almond cultivars grown in the Central-Western Iberian Peninsula. Agronomy 11(6):1238

Piscopo A, Romeo FV, Petrovicova B, Poiana M (2010) Effect of the harvest time on kernel quality of several almond varieties (Prunus dulcis (Mill.) D.A. Webb). Sci Hortic 125(1):41–46

Polskiego K, Anatomicznego T, Towarzystwa P, Komórki Oraz Fundacji B, Komórki B, Molekularnej B (2021) POSTĘPY BIOLOGII KOMÓRKI TOM. 48(2):81–182

Prgomet I, Goncalves B, Domínguez-Perles R, Pascual-Seva N, Barros AIRNA (2017) Valorization challenges to almond residues: phytochemical composition and functional application. Molecules 22(10)

Qureshi MN, Numonov S, Aisa HA (2019) Chemical and pharmacological evaluation of hulls of Prunus dulcis nuts. Int J Anal Chem:5861692

Rakshanaa R, Aashmi JPL, Sheikh RA, Vishnudas D, Kingsley DJ (2023) Anti-cancer effects of the bioactive compounds of Prunus dulcis against ovarian cancer cell lines (PA-1). Res J Biotechnol 18(10):41–47

Rothschild D, Weissbrod O, Barkan E, Kurilshikov A, Korem T, Zeevi D et al (2018) Environment dominates over host genetics in shaping human gut microbiota. Nature 555(7695):210–215

Rybak I, Carrington AE, Dhaliwal S, Hasan A, Wu H, Burney W et al (2021) Prospective randomized controlled trial on the effects of Almonds on facial wrinkles and pigmentation. Nutrients 13(3):785

Samuel Sundar Doss D, Vijayakumar J, Rekha K, Parveen A (2020) A review on neuroprotective effects of almonds. Int J Res Pharmaceut Sci 11(3):4807–4814

Sharma RK, Jha N, Mishra A, Singh G, Sanwelka M, Gupta A (2024) Therapeutic effectiveness of Prunus dulcis for nephroprotective and hepatoprotective in Albino mice. Curr Indian Sci 2(1):1–10

Siddiqui M, Begum W, Scholar P (2023) Almond (Prunus amygdalus L.): a source of revitalizing health and its therapeutic application. J Drug Deliv Ther 13(11):176–182

Singar S, Kadyan S, Patoine C, Park G, Arjmandi B, Nagpal R (2024) The effects of almond consumption on cardiovascular health and gut microbiome: a comprehensive review. Nutrients 16(12):1964

Smeriglio A, Mandalari G, Bisignano C, Filocamo A, Barreca D, Bellocco E et al (2016) Polyphenolic content and biological properties of Avola almond (Prunus dulcis Mill. D.A. Webb) skin and its industrial byproducts. Ind Crops PRO 83:283–293

Socias i Company R, Alonso JM, Kodad O (2009) Fruit quality in almond: physical aspects for breeding strategies. Acta Hortic 814:475–480

Summo C, Palasciano M, De Angelis D, Paradiso VM, Caponio F, Pasqualone A (2018) Evaluation of the chemical and nutritional characteristics of almonds (Prunus dulcis (Mill). D.A. Webb) as influenced by harvest time and cultivar. J Sci Food Agric 98(15):5647

Takeoka GR, Dao LT (2002) Antioxidant constituents of almond [Prunus dulcis (Mill.) D.A. Webb] Hulls. J Agric Food Chem 51(2):496–501

Tomishima H, Luo K, Mitchell AE (2022) The almond (Prunus dulcis): chemical properties, utilization, and valorization of coproducts. Annu Rev Food Sci Technol 13:145–166

Ukhanova M, Wang X, Baer DJ, Novotny JA, Fredborg M, Mai V (2014) Effects of almond and pistachio consumption on gut microbiota composition in a randomised cross-over human feeding study. Br J Nutr 111(12):2146–2152

Vaishnavi G, Justin Thenmozhi A (2023) Nuts and their potential role in Alzheimer's disease. Spring:125–149

Wang W, Li ZJ, Zhang YL, Xu XQ (2021) Current situation, global potential distribution and evolution of six almond species in China. Front Plant Sci 12:619883

Xie L, Roto AV, Bolling BW (2012) Characterization of ellagitannins, gallotannins, and bound proanthocyanidins from California almond (Prunus dulcis) varieties. J Agric Food Chem 60(49):12151–12156

Yada S, Lapsley K, Huang G (2011) A review of composition studies of cultivated almonds: macronutrients and micronutrients. J Food Compos Anal 24(4–5):469–480

# Medicinal and Nutritional Importance of *Cymbopogon citratus* in Human Health

11

Abdelmonem Siddiq, Manisha Das, Muhammed Shabil, Shubham Shrestha, Swarali Yatin Chodnekar, and Nityanand Jain

**Abstract**

*Cymbopogon citratus*, or lemongrass, is renowned for its aromatic properties and its extensive use in traditional medicine, cosmetics, and culinary preparations. A rich source of essential electrolytes, minerals, vitamins, macronutrients, and phytochemicals, lemongrass offers a wide range of biological and therapeutic advantages. These effects are mediated by bioactive compounds, including geranial, nerol, and myrcene, which have been demonstrated to exert anti-inflammatory, antifungal, antineoplastic, and antibacterial properties. Additionally, the high phenolic content in lemongrass contributes to its antioxidant capacity, neutralizing free radicals, chelating metal ions, and inhibiting radical-producing enzymes. Lemongrass has also demonstrated hepatoprotective

A. Siddiq (✉)
Faculty of Pharmacy, Mansoura University, Mansoura, Egypt
e-mail: abdelmonemalsaid555@std.mans.edu.eg

M. Das
Faculty of Medicine, Dhaka Medical College, Dhaka University, Dhaka, Bangladesh

M. Shabil
Department of Pharmacy Practice, Faculty of Pharmacy, M.S. Ramaiah University of Applied Sciences, Bangalore, Karnataka, India

Global Center for Evidence Synthesis (GCES), Chandigarh, India

S. Shrestha
Patan Academy of Health Sciences, Patan Hospital, Lalitpur, Nepal

S. Y. Chodnekar
Faculty of Medicine, Teaching University Geomedi LLC, Tbilisi, Georgia
e-mail: s.chodnekar@geomedi.edu.ge

N. Jain
Independent Consultant, Chandigarh, India



M. A. Ansari et al. (eds.), *Medicinal Plants and their Bioactive Compounds in Human Health: Volume 2*, https://doi.org/10.1007/978-981-95-3620-7_11

effects and the capacity to enhance cardiovascular health. The extracts have been observed to possess analgesic properties, which are primarily attributable to the presence of citral. This provides pain relief by inhibiting inflammation and interacting with opioid receptors, thereby offering natural alternatives for pain management. Furthermore, lemongrass facilitates digestion by stimulating enzyme production, alleviating gastrointestinal discomfort, and preventing gastric ulcers through mucin production and reduced acid secretion. While lemongrass has garnered scientific interest due to its efficacy, health benefits, and minimal side effects, caution is warranted in the use of lemongrass extracts, infusions, and essential oils, particularly in direct skin application and in combination with certain medications due to potential interactions. Proper dosage and moderation are essential to avoid adverse effects. Further research is required to determine the characteristics and potential adverse effects associated with the use of lemongrass. The combination of traditional practices with modern evidence-based scientific knowledge serves to highlight the potential of lemongrass in the development of sustainable and holistic health solutions.

**Keywords**

Cymbopogon · Lemongrass · Plant extracts · Essential oils · Nutritional value · Antioxidants · Complementary therapies

## 11.1 Introduction

*Cymbopogon citratus*, more commonly known as lemongrass, is a tropical–subtropical plant that has a long history of use in complementary medicine and culinary traditions. The plant is visually distinctive, exhibiting thin, long leaves and a pronounced aroma (Saddiq and Khayyat 2010; Akhila 2009; Oladeji et al. 2019). Extracts and essential oils (EO) derived from the plant are known to possess anti-inflammatory, anti-fungal, anti-cancer, and anti-bacterial properties (Fig. 11.1) (Mouta et al. 2022; Haque et al. 2018). The use of herbal products as anti-microbial and antioxidant food preservatives has been gaining prominence among researchers over the past few decades due to several factors, including their accessibility, minimal adverse effects, reduced toxicological profile, and enhanced systemic bioavailability in comparison to other anti-microbial agents that are currently available (Marina et al. 2013; Miguel 2010).

Lemongrass essential oil is extracted from the dried or fresh leaves using steam distillation, a process that also yields hydrosols or aromatic fluids, which are frequently utilized in the treatment of inflammatory illnesses and microbiological infections (Abe et al. 2003b; Tiwari et al. 2010; Alitonou et al. 2006). The plant's therapeutic properties are attributed to the presence of chemical compounds, including geranial/citral A, nerol/citral B, and myrcene, among others (Bayala et al. 2018; Miean and Mohamed 2001; Ekpenyong and Akpan 2017; Figueirinha et al. 2008; De Oliveira et al. 2017). A study conducted by Onawunmi et al., demonstrated that

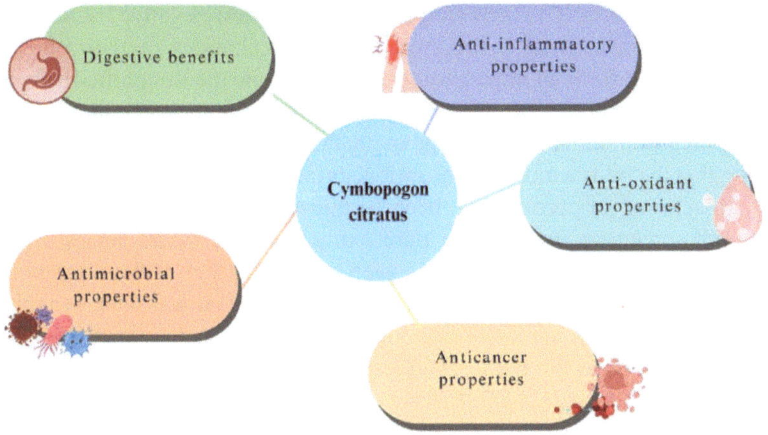


**Fig. 11.1** Overview of the proposed medicinal domains of *Cymbopogon citratus* extracts

the citral compounds act as highly potent terpenoids, while myrcene enhances their actions when combined (Onawunmi et al. 1984). Terpenoids are known to inhibit the production of pro-inflammatory cytokines, including tumor necrosis factor-alpha (TNF-α), and to regulate downstream neutrophil adherence responses (Abe et al. 2003b; Pérez et al. 2011).

## 11.2 Therapeutic Properties of *Cymbopogon citratus*

### 11.2.1 Anti-microbial Properties

Extensive research has been conducted on the anti-bacterial properties of lemongrass extracts against a range of clinically significant bacteria (Hussain et al. 2023). Its essential oil has demonstrated anti-microbial activity against a range of microorganisms, including *Staphylococcus aureus*, *Staphylococcus epidermidis*, *Streptococcus mutans*, and *Candida* spp., both in single and mixed biofilms (Almeida et al. 2013). The antibacterial action of EO targets the bacterial cell membrane and cytoplasmic metabolism. This includes induction of morphological changes such as increased cell membrane fluidity and permeability, alterations in ion transport and channel function, inhibition of membrane formation, increase in membrane protein destruction, spheroplast formation, the formation of blisters or mesosomes, the development of abnormally shaped cells, inhibition of ATPase activities, inhibition of cell respiration, and the induction of lysis (Takaisi-Kikuni et al. 1996; Nguefack et al. 2004; Cowan 1999; Burt and Reinders 2003).

The essential oil has demonstrated efficacy at minimum inhibitory concentrations ranging from 12.5 to 25 μL/mL against the tested pathogens (Mohammed et al. 2020). Furthermore, many studies have indicated that lemongrass essential oils exert a more pronounced effect on gram-positive bacteria than on gram-negative bacteria. The increased resistance of gram-negative bacteria may be attributed to the presence of hydrophilic lipopolysaccharides on their outer membrane, which could form an impenetrable barrier despite the lipophilic-hydrophobic characteristics of essential oils. Nonetheless, studies have shown lemongrass essential oils to be more efficacious than the methanolic extract against these pathogens, indicating that the oil is a more potent antibacterial agent (Subramaniam et al. 2020).

Its antibacterial properties have been further substantiated through synergistic antibacterial screening studies. The pure extract of *C. citratus* demonstrated the most pronounced antibacterial efficacy, even exceeding that of the reference drug. The ethanolic extract exhibited higher concentrations of caffeic acid, synapic acid, and benzoic acid, which are likely responsible for its antibacterial activity (Hussain et al. 2023). Moreover, the liquid residues of ethanolic and aqueous extracts exert antimicrobial effects against *Staphylococcus aureus*, *Staphylococcus epidermidis*, *Pseudomonas aeruginosa*, and *Candida albicans*. These antibacterial properties have significant implications for the prevention and treatment of bacterial infections, particularly in the context of skin infections such as pitted keratolysis, rendering it a potential natural remedy for the condition (Hussain et al. 2023; Naik et al. 2010; Schweitzer et al. 2022).

### 11.2.2 Anti-fungal Properties

In addition to its well-documented anti-bacterial activities, recent research has also underscored the potent anti-fungal properties of *C. citratus*, thereby establishing it as a valuable natural resource in the fight against fungal infections. Its anti-fungal properties can be primarily attributed to the presence of bioactive compounds such as citral, geraniol, and limonene, within its essential oil, which have been demonstrated to exhibit anti-fungal activity against a wide range of pathogenic fungi (Khan and Ahmad 2012; Irkin and Korukluoglu 2007). One of the principal mechanisms through which the essential oil exerts its anti-fungal effects is by disrupting the fungal cell membrane.

The bioactive compounds present in the oil have been observed to disrupt the structural integrity of the fungal cell wall and membrane, resulting in increased permeability and ultimately, cell death (Onawunmi 1989; Cimanga et al. 2002). This mode of action has been demonstrated to be particularly effective against *Candida* spp., a group of opportunistic fungi that are responsible for a number of fungal infections in humans (Galvez et al. 1993; Puatanachokchai et al. 2002). As reported by Abe et al., lemongrass essential oils markedly impeded the mycelial proliferation of *Candida albicans*, suggesting that the oil may prove beneficial in the management of cutaneous candidiasis. Furthermore, the vapors of these oils can result in a reduction in the capacity of fungal cells to adhere to the skin, which subsequently diminishes their virulence and infectiousness (Abe et al. 2003a). The effects of linalool on fungal cell biology have been assessed using electron microscopy. The results indicate that linalool resulted in a reduction in cell size and aberrant germination, thereby validating the extract's suspected antifungal action (Alviano et al. 2005).

Apart from its direct anti-fungal properties, the essential oil also exhibits synergistic effects when combined with conventional anti-fungal drugs. Several studies have demonstrated that the combination of lemongrass essential oil and fluconazole, a commonly used anti-fungal medication, can result in enhanced anti-fungal activity and a reduction in the minimum inhibitory concentration of fluconazole (Fernández et al. 2010; Santoro et al. 2007). This indicates that the essential oil may be employed as an adjuvant therapy, thereby enhancing the efficacy of conventional anti-fungal treatments, and potentially overcoming drug resistance in certain fungal strains. The anti-fungal potential of lemongrass extracts is not limited to its effects on *Candida* spp.

The essential oil has shown efficacy against other clinically relevant fungal pathogens, including *Aspergillus* spp. and *Cryptococcus* spp., which are responsible for a range of invasive fungal infections (Sonker et al. 2015; Bayala et al. 2014). The broad-spectrum anti-fungal activity of lemongrass essential oil renders it a promising candidate for the development of natural, plant-based antifungal therapies. *C. citratus* has also been shown to possess the capacity to impede the development of fungal biofilms. Biofilms are complex communities of microorganisms that can adhere to surfaces and exhibit increased resistance to conventional anti-microbial agents (Akono Ntonga et al. 2014; Blanco et al. 2005).

### 11.2.3 Anti-inflammatory Properties

The presence of polyphenolic compounds, including citral, geraniol, and limonene, results in the potent anti-inflammatory properties of lemongrass essential oils. These compounds function by inhibiting the synthesis of pro-inflammatory mediators and modulating the activity of inflammatory signaling pathways, including prostaglandins and leukotrienes, which play a key role in the inflammatory response (Cheel et al. 2005; Tapia et al. 2007).

A study found that the infusion of dried leaves from *C. citratus* significantly inhibited lipopolysaccharide (LPS)-induced nitric oxide (NO) production and inducible NO synthase (iNOS) protein expression in a skin-derived dendritic cell line (Cheel et al. 2005). This inhibition was attributed to the presence of chlorogenic acid, a polyphenol known to inhibit the activity of the proteasome and the nuclear factor-κB (NF-κB) pathway (Francisco et al. 2013). This mechanism is also favored by the flavonoid-rich fractions such as luteolin glycosides that have been known to diminish NO production and iNOS protein levels (Francisco et al. 2013). Similarly, tannin-rich fractions in *C. citratus* possess anti-inflammatory properties, although the specific compounds responsible have yet to be fully identified (Francisco et al. 2013).

Several studies confirm these findings and demonstrate the attenuation of inflammatory processes in a range of in vitro and in vivo models. For example, Figueirinha et al. showed that an aqueous extract of *C. citratus* effectively inhibited the production of NO and prostaglandin E2, two key inflammatory mediators, in macrophage cell lines (Figueirinha et al. 2010). Similarly, Adedapo et al. demonstrated that the essential oil from *C. citratus* had anti-inflammatory activity in a rat model of carrageenan-induced paw edema (Adedapo et al. 2008). A recent study examined the impact of the essential oil on oxidative stress and inflammation in zebrafish (Tapia et al. 2007). The findings indicate that the essential oil markedly diminished neutrophil migration and caudal fin regeneration, thereby substantiating its anti-inflammatory efficacy. Furthermore, the essential oil demonstrated the capacity to diminish the activity of antioxidant enzymes, including superoxide dismutase (SOD) and catalase (CAT), which may indicate the existence of antioxidant properties (Tapia et al. 2007).

The anti-inflammatory properties of the plant have led researchers to investigate its potential applications in various inflammation-related conditions. We next discuss and present a brief overview of the inflammatory conditions where lemongrass and its extracts have been investigated for potential clinical use. It is important to note that further research and clinical trials are essential to fully validate its therapeutic benefits and to facilitate its integration into mainstream medical practice.

***Arthritis*** Rheumatoid arthritis (RA) and osteoarthritis (OA) are chronic inflammatory diseases of the joints that cause pain and joint degradation. Conventional therapeutic modalities prioritize the management of symptoms over the modification of the underlying disease process. This approach frequently results in the persistence of pain and disability. Recent studies have indicated the potential of lemongrass as a natural therapeutic agent for these conditions. Apart from mediating anti-inflammatory responses (Venzon et al. 2018), it has been observed that the plant extract could significantly reduce cartilage degradation in an in vitro model of osteoarthritis by preserving the extracellular matrix and inhibiting the activity of matrix metalloproteinases (Figueirinha et al. 2010). In another interesting study, the anti-inflammatory and analgesic effects of *C. citratus* essential oil were evaluated in a rat model of arthritis. The essential oil demonstrated efficacy in reducing paw edema and lowering levels of pro-inflammatory cytokines, including TNF-$\alpha$ and Interleukin IL-1$\beta$, within affected tissues (Adedapo et al. 2008). These findings suggest that the essential oil provides not only symptomatic relief but also has the potential to modify the underlying disease processes in arthritis.

***Inflammatory Bowel Diseases (IBD)*** Encompassing Crohn's disease and ulcerative colitis, this group of diseases is characterized by chronic inflammation of the gastrointestinal tract, resulting in severe gastrointestinal symptoms and a notable reduction in quality of life. The current therapeutic approaches are designed to control inflammation and maintain remission, yet they frequently result in considerable adverse effects. A study investigating the effects of *C. citratus* on inflammatory bowel diseases (IBD) found that an aqueous extract could inhibit the production of pro-inflammatory cytokines like IL-6, IL-1$\beta$, and TNF-$\alpha$, which are key in the pathogenesis of IBD (Cheel et al. 2005). The study also demonstrated that the extract reduced intestinal inflammation in a mouse model of colitis, thereby providing a basis for considering the oil as a potential complementary therapy in the management of IBD. Further studies have provided additional support for these findings. Another group reported that the essential oil of *C. citratus* reduced inflammation and oxidative stress in a rat model of colitis, indicating its protective effects on the gastrointestinal lining (Venzon et al. 2018). Moreover, others have demonstrated that citral exerts anti-inflammatory effects by inhibiting the NF-κB signaling pathway and reducing the expression of inflammatory markers in a murine model of IBD (Martins et al. 2017; Boukhatem et al. 2014).

***Cutaneous Conditions*** Inflammatory skin conditions, including contact dermatitis, psoriasis, and atopic dermatitis, are typified by redness, itching, and chronic inflammation (Milani-Nejad et al. 2017). Conventional treatments frequently entail the use of corticosteroids and immunosuppressants, which can elicit adverse effects when used over an extended period (Privitera et al. 2021). *C. citratus* has demonstrated potential as a therapeutic agent for these conditions due to its anti-inflammatory and anti-microbial properties. A study found the essential oil to inhibit inflammatory mediators such as NO and prostaglandins, thereby reducing skin inflammation in a mouse model of contact dermatitis (Rocha et al. 2023). Additionally, the study highlighted the essential oil's anti-microbial activity, which is advantageous in preventing secondary infections in inflamed skin regions. The essential oil has also shown significant anti-inflammatory and wound-healing properties in rat models of skin injury (Adedapo et al. 2008; Lulekal et al. 2019). The essential oil accelerated the healing process by enhancing collagen synthesis and promoting the re-epithelialization of the wound area (Tapia et al. 2007). Research has demonstrated that topical formulations comprising *C. citratus* essential oil can alleviate symptoms of atopic dermatitis by attenuating inflammation and pruritus, offering a natural alternative to steroid-based therapies (Rocha et al. 2023; Mohammed et al. 2022).

### 11.2.4 Analgesic Properties

Recent research studies have demonstrated that lemongrass essential oil has the capacity to significantly diminish pain responses in mice during the acetic acid-induced writhing test (Gawade 2012; De Sousa 2011; Garcia et al. 2022). The essential oil demonstrated pronounced analgesic effects in rats subjected to the tail flick test, a methodology employed to assess pain thresholds (Lorenzetti et al. 1991; Viana et al. 2000). These effects are primarily attributed to the oil's ability to inhibit the production of certain chemicals in the body that are responsible for triggering inflammatory and pain signals. For example, citral compounds are renowned for their anti-inflammatory and analgesic properties (Kiani et al. 2022a; Bachiega and Sforcin 2011). It is postulated that these compounds inhibit enzymes that contribute to inflammation and pain, such as cyclooxygenase-2 (COX-2) (Gonçalves et al. 2020; Lee et al. 2008). Additionally, lemongrass may interact with opioid receptors in the brain, modulating pain perception in a manner analogous to pain-relief medications, yet without the adverse effects (Handbook of Essential Oils: Science, Technology, and Applications 2009; Buckle 2014).

Notable is the ability of the essential oil to provide pain relief from headaches and musculoskeletal discomfort. Topical application of the diluted oil, following the dilution with a carrier oil, has been demonstrated to provide relief from these conditions by reducing inflammation and promoting relaxation. Furthermore, its analgesic properties render it an efficacious natural remedy for menstrual cramps, offering a safer and more natural alternative to conventional painkillers. Inhalation of lemongrass essential oil vapor has been demonstrated to provide relief from respiratory conditions such as bronchitis and asthma. This is achieved by soothing the airways and reducing inflammation (Majewska et al. 2019). This is of particular significance for individuals afflicted with chronic respiratory conditions, as it provides a natural and non-invasive method of managing symptoms. Moreover, research indicates that lemongrass essential oil has a beneficial effect on mental health. The aroma of lemongrass has been demonstrated to have an anxiolytic effect, promoting a sense of calm and relaxation and reducing stress and anxiety levels (Figueirinha et al. 2010). This makes it an excellent addition to any self-care routine, particularly for individuals who experience symptoms of anxiety or depression.

### 11.2.5 Antioxidant Activity

Antioxidants are chemical compounds that neutralize free radicals, which are unstable molecules that can cause oxidative damage to cells and contribute to the development of various diseases. *C. citratus* has been demonstrated to exhibit substantial antioxidant activity, which is ascribed to the presence of a range of bioactive compounds, including flavonoids, phenolic acids, and terpenoids (Oladeji et al. 2019; Sousa et al. 2021; Shah et al. 2011).

Lemongrass displays considerable free radical scavenging activity because of its elevated phenolic compound content. These compounds neutralize unstable molecules that can cause cellular damage, thereby preventing the development of chronic diseases. Lemongrass can chelate metal ions, such as iron and copper, which serve as catalysts for the production of detrimental free radicals. By forming complexes with these metals, lemongrass inhibits the generation of radicals. Lemongrass has been demonstrated to inhibit the activity of enzymes such as xanthine oxidase and NADPH oxidase, which are involved in the generation of free radicals. The inhibition of these enzymes consequently reduces the rate of radical formation, thereby providing additional protection against oxidative stress. It has been demonstrated in studies that the aqueous extract of *C. citratus* is an effective protective agent against hydrogen peroxide (H2O2)-induced oxidative stress and injury in a variety of tissues (Rahim et al. 2013).

A study conducted on rats revealed that the extract markedly diminished malonaldehyde (MDA) levels and augmented glutathione (GSH) levels within the testes and epididymis, thereby safeguarding against hydrogen peroxide-induced testicular atrophy and sperm damage (Blanco et al. 2005). The extract demonstrated protective effects against $H_2O_2$-induced liver injury in male rats, as evidenced by reduced oxidative stress and inflammation (Rahim et al. 2013). Its antioxidant properties are attributed to its polyphenol content, particularly chlorogenic acid, isoorientin, and swertiajaponin. These compounds have been demonstrated to impede the generation of reactive oxygen species (ROS) in human umbilical vein endothelial cells (HUVECs) subjected to high glucose, hydrogen peroxide, or oxidized low-density lipoprotein (LDL) (Campos et al. 2014). Additionally, the polyphenols demonstrated a nitric oxide-independent vasodilatory effect on blood vessels, indicating a potential therapeutic avenue for cardiovascular disease prevention (Campos et al. 2014).

The formulation of extracts has a marked impact on its antioxidant efficacy (Sousa et al. 2021). It has been observed that lemongrass extracts encapsulated in gelatin capsules demonstrate superior antioxidant activity in comparison to other forms, due to the protection of active compounds from degradation and enhanced absorption in the gastrointestinal tract (Sousa et al. 2021). The antioxidant potential of lemongrass essential oil, which is concentrated with active compounds, may vary depending on the volatility of its components. In contrast, lemongrass tea, another popular form of consumption, may demonstrate reduced antioxidant potential due to the variability in preparation methods and extraction efficiency.

The hepatoprotective effects of *C. citratus* have been substantiated by a substantial body of research. For example, a study demonstrated that the extract markedly diminished liver damage and inflammation in rats induced by paracetamol, a prevalent cause of liver toxicity. The antioxidant properties of the extract helped to mitigate oxidative stress and protect against liver damage (Saenthaweesuk et al. 2017). Oxidative stress is a pivotal factor in the pathogenesis of cardiovascular diseases, including atherosclerosis and hypertension. The antioxidants present in lemongrass, particularly flavonoids and phenolic compounds, have been demonstrated to counteract lipid peroxidation and enhance endothelial function (Silva et al. 2022). This may result in enhanced arterial flexibility and a reduction in blood pressure, which could subsequently diminish the overall risk associated with cardiovascular conditions (Gayathri et al. 2010).

The dietary incorporation of lemongrass may prove advantageous in the modulation of cardiovascular disease risk factors, largely due to its anti-inflammatory and anti-hyperlipidemic effects, as evidenced by previous research (Sousa et al. 2021; Gayathri et al. 2010). These antioxidants assist in the reduction of DNA damage, a precursor to cancer development, and the inhibition of cancer cell proliferation pathways (Koh et al. 2012). The anti-cancer properties of citral have been demonstrated in the inhibition of the growth of specific cancer cell lines, including those of the breast and colon (Kapur et al. 2016; Balusamy et al. 2020). The results of the study indicate that citral induces apoptosis (programmed cell death) and disrupts the cell cycle in malignant cells, thereby reducing the likelihood of cancer progression (Balusamy et al. 2020; Cortes-Torres et al. 2023).

The relationship between oxidative stress and neurodegenerative diseases such as Alzheimer's and Parkinson's disease is well documented in the scientific literature (Barnham et al. 2004). The antioxidants present in lemongrass may confer neuroprotective benefits by protecting neurons from oxidative damage (Temitayo et al. 2020). Lemongrass extracts have been demonstrated to mitigate the effects of oxidative stress in brain cells, thereby potentially slowing the progression of neurodegenerative diseases (Siddique et al. 2021). These protective effects may be mediated through the upregulation of endogenous antioxidant defenses in neuronal cells, which contributes to a reduction in the accumulation of toxic proteins and a delay in neurodegeneration (Siddique et al. 2021; da Silva et al. 2024).

### 11.2.6 Anti-cancer Properties

Many studies have indicated the potential anti-cancer effects of lemongrass. A study demonstrated that lemongrass extract induces apoptosis in colon cancer cells in vitro in a time- and dose-dependent manner, while exhibiting no effect on healthy cells (Ruvinov et al. 2019; Alwaili 2022). The oral administration of this extract was also observed to effectively inhibit colon cancer xenograft growth in mice, enhance the efficacy of the chemotherapy regimen folinic acid, fluorouracil and oxaliplatin (FOLFOX), and reduce its adverse effects (Ruvinov et al. 2019). Another study demonstrated anticarcinogenic action against human colon cancer cell proliferation in vitro and significant increases in HO-1 mRNA levels in treated cells, indicating the potential to induce antioxidant responses (Alwaili 2022). The administration of lemongrass extracts has been observed to elevate levels of reactive oxygen species and induce apoptosis in cancer cells (Mukarram et al. 2021). It was observed that citral induces endoplasmic reticulum stress and exerts an antiproliferative effect on cancer cells. Furthermore, lemongrass tea has been observed to possess anti-inflammatory, antimicrobial, and anticancer properties, which may assist in the prevention of cellular dysfunction and cancerous growth (Mukarram et al. 2021). These findings collectively indicate that lemongrass may serve as a significant complementary treatment or preventative measure for colon cancer.

Although the evidence of anti-cancer properties from the studies is encouraging, the evidence remains preliminary, and further rigorous research is required to definitively ascertain its efficacy. The current findings are primarily derived from in vitro studies and animal models, which may not always directly translate to human health outcomes. Further clinical trials with human participants are essential to establish the potential therapeutic benefits of lemongrass, elucidate its mechanisms of action, and evaluate its safety and efficacy as a cancer treatment. Such studies would provide a more robust basis for recommending lemongrass as a component of cancer prevention or treatment regimens. These uses differ between *Cymbopogon* species as shown in Table 11.1 (Tibenda et al. 2022).

### 11.2.7 Gastro-stimulatory and Gastro-protective Functions

Lemongrass has a long history of use as a digestive aid due to its ability to stimulate the production of digestive enzymes and gastric juices, which are essential for the

**Table 11.1** Different *Cymbopogon* species and their medicinal uses

| Species | Essential components | Therapeutic effects | References |
|---|---|---|---|
| *C. caesius* (Hook and Arn.) Stapf | Carvone, limonene, perillyl alcohol, citronellol, and citronellal | Treating morning sickness in pregnant women | Leistner (2000) and Russell et al. (1990) |
| *Cymbopogon citratus* (DC.) Stapf | Camphene, limonene, nonan-4-ol, citronellal, citronellol, neral, geraniol, and citral | Antidiarrheal, antibacterial, antifilarial, antifungal, anti-inflammatory, antimalarial, hypoglycemic, and hypolipidemic effects | Shah et al. (2011) and Adeneye and Agbaje (2007) |
| *Cymbopogon densiflorus* (Steud.) Stapf | Trans-p-mentha-2,8-dien-1-ol, cis-p-mentha-2,8-dien-1-ol, trans-p-mentha-1(7),8-dien-2-ol, cis-pmentha-1(7),8-dien-2-ol and cis-piperitol | Antitumor and antibacterial effects | Takaisi-Kikuni et al. (2000) and Pereira et al. (2021) |
| *Cymbopogon flexuosus* (Nees ex Steud.) | Citral, geranyl acetate, myrcene, and citronellol | Anti-inflammatory effect and cancer chemoprevention | Sharma et al. (2009) and Han and Parker (2017) |
| *Cymbopogon giganteus* Chiov. | Trans-p-1(7),8-menthadien-2-ol, cis-p-1(7),8-menthadien-2-ol, trans-p-2,8-menthadien-1-ol, cis-p-2,8-menthadien-1-ol | Antimicrobial, antitrypanosomal, antiplasmodial, and anti-inflammatory agent | Kpoviessi et al. (2014), Jirovetz et al. (2007) and Alitonou et al. (2006) |
| *Cymbopogon jwarancusa* subsp. olivieri (Boiss.) S. Soenarko | Piperitone-carene, citronellal, p-cymene, geraniol, β-pinene, and γ-terpinene | Antipyretic, antirheumatic, antitussive, aromatic, blood purifier, treatment of fever, cough, rheumatism, gout, dyspepsia, cholera, and anti-fungi | Bhuyan et al. (2015) and Heinrich (2003) |
| *Cymbopogon proximus* (Hochst. ex A. Rich.) Chiov. | Piperitone, carene, limonene, and elemol | Anticonvulsant, anxiolytic, hypotensive, anticonvulsant, and antiemetic | El-Tahir and Abdel-Kader (2008) |
| *Cymbopogon procerus* (R.Br.) | Elemicin and pinene | Medicinal wash | Thompson and Forster (2021) |
| *Cymbopogon schoenanthus* (L.) Spreng | Piperitone, limonene, dihydrocarveol, δ-terpinene, α-terpineol, and elemol | Anti-acetylcholinesterase effect | Khadri et al. (2008) and Ketoh et al. (2006) |
| *Cymbopogon nardus* (L.) Rendle | Citronellal, 2,6-octadienal, caryophyllene, citronellol, limonene, and eugenol | Antibacterial | Piéron (1928) |

(continued)

**Table 11.1** (continued)

| Species | Essential components | Therapeutic effects | References |
|---|---|---|---|
| *Cymbopogon olivieri* (Boiss.) Bor | Piperitone, β-caryophyllene, delta-3-carene, α-eudesmol, α-terpinene, and elemol | Antimicrobial effect | Mahboubi and Kazempour (2012) |
| *Cymbopogon winterianus* Jowitt ex Bor | geraniol, citronellol, citronellal, caryophyllene, citronellyl acetate, elemol, geranyl acetate, linalyl acetate, methyl-iso-eugenol, and nerol | Antifungal agent | Andila et al. (2018) and Deletre et al. (2015) |
| *Cymbopogon validus* (Stapf) Stapf ex Burtt Davy | artemisia ketone, linalool, northujane, verbenone, naphthalene, δ-cadinene, hedycaryol, and α-eudesmol | Anti-inflammatory, emetic, anti-infective, anti-plasmodic, and morning sickness | Chagonda et al. (2000), Essential oil-bearing grasses: the genus cymbopogon (2009) and Rungqu et al. (2016) |
| *Cymbopogon martinii* Stapf | geraniol, citral, citronellol, linalool, and geranyl acetate | Anthelmintic and antimicrobial effects | Prasad et al. (2010), Katiki et al. (2011) and Mishra et al. (2015) |
| *Cymbopogon pendulus* Stapf | citral-a, citral-b, geranyl acetate, β-caryophyllene, elemol, geraniol, and linalool | Antifungal effect | Pandey et al. (1996) |
| *Cymbopogon distans* (Nees ex Steud.) | Terpineol, piperitone, geraniol, limonene, and methyl eugenol | Anti-inflammatory, cough, common cold, asthma, chronic bronchitis, antibacterial, and antifungal effects | Zhang et al. (2011) |

breakdown of food and the subsequent improvement of nutrient absorption (Nambiar and Matela 2012). The compounds citral and geraniol have been demonstrated to possess carminative properties, which facilitate the alleviation of symptoms associated with bloating, flatulence, and discomfort. This is achieved by the expulsion of gas and the soothing of the digestive tract (Lei et al. 2019). Moreover, these compounds may facilitate the digestive process, thereby reducing symptoms of indigestion and discomfort associated with slow digestive processes (Lei et al. 2019).

Lemongrass is frequently used in traditional medicine to alleviate stomach pain, cramps, and discomfort (Nambiar and Matela 2012). Its antispasmodic effects can relax the smooth muscles of the gastrointestinal tract, which are beneficial for relieving symptoms of irritable bowel syndrome (IBS) and other spasmodic gastrointestinal disorders. Research suggests that lemongrass can significantly reduce abdominal pain and improve the quality of life for individuals suffering from digestive issues by mitigating the intensity and frequency of gastrointestinal spasms (Babelghaith et al. 2024; Matera et al. 2023).

The gastroprotective properties of lemongrass are valuable in preventing and healing gastric ulcers. For example, a study found that the essential oil of *C. citratus* reduced the area of gastric lesions in mice by 96.57% at a dose of 100 mg/kg (Sagradas et al. 2015). Another study found that the plant infusion leaves significantly reduced gastric mucosal injury in rats (Boeira et al. 2020). One key mechanism is the inhibition of gastric acid secretion, which can help in reducing the risk of gastric ulcers. Studies have shown that the essential oil inhibits gastric acid secretion by blocking the activity of the enzyme H+/K + -ATPase, which is responsible for acid production in the stomach (Venzon et al. 2018). Another mechanism of gastroprotection is the enhancement of mucin production, which helps to protect the stomach lining from damage. Studies have shown that the infusion of leaves increases mucin production in the stomach, which can help to heal gastric ulcers (Wu et al. 2021).

Additionally, it promotes better blood circulation to the stomach, enhancing the healing process of damaged tissues (Fernandes et al. 2012). This can prevent the development of peptic ulcers and offer therapeutic benefits to individuals suffering from chronic gastritis, potentially reducing their reliance on conventional medications (Fernandes et al. 2012; Sagradas et al. 2015). Lemongrass exhibits strong antimicrobial properties that are effective against various pathogenic bacteria, including *Helicobacter pylori*, which is often implicated in the development of stomach ulcers and chronic gastritis (Kiani et al. 2022b). Its anti-inflammatory properties are particularly beneficial in reducing gastrointestinal inflammation, which is crucial in the management of IBD, such as Crohn's disease and ulcerative colitis (Venzon et al. 2018). The essential oils containing citral and geraniol can modulate inflammatory pathways, leading to reduced inflammation and associated symptoms.

Studies indicate that lemongrass may enhance liver function by protecting the liver from oxidative stress and damage caused by toxins (Rahim et al. 2014). This hepatoprotective effect is particularly beneficial for individuals with liver disorders or those exposed to factors that can impair liver function (69). By reducing oxidative damage and enhancing the liver's detoxifying capacities, lemongrass can contribute to better overall health and resilience against liver diseases (Uchida et al. 2017).

## 11.3 Nutritional Profile of *Cymbopogon citratus*

*C. citratus* is a rich source of electrolytes and minerals, including sodium, potassium, calcium, copper, magnesium, manganese, selenium, phosphorus, iron, and zinc. Additionally, it contains vitamins such as folate, niacin, pyridoxine, riboflavin, and vitamins A, C, and E, as well as macronutrients like carbohydrates and proteins, with a minor quantity of fat (Kandhro et al. 2011). The available evidence increasingly suggests that these phytochemical components contribute to the diverse biological and therapeutic effects of *C. citratus*. As indicated by the USDA National Nutrient Database 2011, the plant contains essential vitamins and minerals that are vital for maintaining optimal bodily function (T. Mohammed 2019). Based on the analysis of the U.S. Department of Agriculture, the main phytochemicals of *C. citratus* and their composition are detailed in Table 11.2 (FoodData Central 2024).

### 11.3.1 Vitamins

Vitamins play crucial physiological roles in the body's normal functioning. The concentration of vitamins in *C. citratus* follows the order: vitamin C > vitamin B1 > vitamin B2 > vitamin A > vitamin E > vitamin B9 (Uraku et al. 2015). Despite the relatively low vitamin content of the leaves, ingestion of this plant can assist in fulfilling the daily vitamin requirements for healthy adults (The National Academies

**Table 11.2** Nutritional composition of *C. citratus* (Portion: 100 g)

| Principle | Nutrient value |
|---|---|
| Energy | 99 kcal |
| Carbohydrates | 25.31 g |
| Protein | 1.82 g |
| Total fat | 0.49 g |
| Cholesterol | 0 mg |
| **Vitamins** | |
| Folates | 75 μg |
| Niacin | 1.101 mg |
| Pyridoxine | 0.080 mg |
| Riboflavin | 0.135 mg |
| Thiamin | 0.065 mg |
| Vitamin A | 6 mg |
| Vitamin C | 2.6 mg |
| **Electrolytes** | |
| Sodium | 6 mg |
| Potassium | 723 mg |
| **Minerals** | |
| Calcium | 65 mg |
| Copper | 0.266 mg |
| Iron | 8.17 mg |
| Magnesium | 60 mg |
| Manganese | 5.244 mg |
| Selenium | 0.7 μg |
| Zinc | 2.23 mg |

Source: USDA National Nutrient Database

Press 2024). Lemongrass contains citral, an essential oil with a range of medicinal and practical applications. Following distillation, citral can be employed in the synthesis of ionones and vitamins A and E, in addition to a range of citral acetals, which find utility in perfumery and the reduction of antibacterial activity. Citral plays a pivotal role in the synthesis of flavoring and fragrance components, including ionones and methyl ionones, as well as vitamins A and E (Shahzadi 2017).

In the pharmaceutical industry, plant extracts are of significant value for their active ingredients, which are utilized in the development of medicinal products, as well as for their potential as raw material sources. For example, lemongrass is employed in the production of vitamin A and in perfumery, and it is beneficial when prepared as a medicinal tea (Kiani et al. 2022a). Lemongrass is widely used in Ayurvedic medicine for its healing properties in treating cough and cold. Its vitamin C content, along with other beneficial components, may help relieve nasal blockages, flu, and respiratory disorders such as bronchial asthma (Mohammed 2019).

Other flavonoids extracted from *C. citratus*, such as licochalcone A and B, have potent antioxidant effects comparable to vitamin E. These phytochemicals, along with vitamins A, C, and E found in the extract, contribute to its powerful antioxidant properties. The presence of antioxidants like flavonoids, tannins, saponins, alkaloids, and vitamins A, C, and E in *C. citratus* may also contribute to its hemodynamic effects, considering the role of oxidative stress in hypertension pathogenesis (Ekpenyong et al. 2015). Additionally, the folate and potassium content in the stem and leaves of lemongrass support DNA synthesis and cell division (Mohammed 2019). The control of neuronal growth and signaling in animals has been related to zinc and folates. These trace components contribute to the health advantages of *C. citratus* tea by enhancing memory, focus, and the ability to process information (Akchiche et al. 2012).

### 11.3.2 Minerals

Lemongrass is a valuable source of essential minerals including potassium, sodium, calcium, magnesium, manganese, iron, zinc, copper, and phosphorus. These minerals play critical roles in various physiological functions, from maintaining cardiovascular and bone health to supporting immune and metabolic processes. Regular consumption of lemongrass can help meet daily mineral requirements and promote overall health (Sambo et al. 2024).

In a study on *C. citratus*, the primary minerals identified included potassium (K), sodium (Na), calcium (Ca), magnesium (Mg), manganese (Mn), iron (Fe), zinc (Zn), phytate, and phosphorus (P) (Boukhatem et al. 2014). Additional minerals detected in lemongrass include chromium (Cr), nickel (Ni), copper (Cu), arsenic (As), cadmium (Cd), and lead (Pb) (Anal 2014). Among the minerals analyzed, calcium was found to be the most abundant. Calcium is vital for bone and teeth health, cardiac muscle function, nerve impulses, and blood and milk clotting (Nimenibo-Uadia and Nwosu 2020).

Lemongrass is also a good source of potassium, magnesium, and phosphorus, though it has relatively low levels of iron and copper (Nimenibo-Uadia and Nwosu 2020). Potassium is crucial for cell and body fluid balance, helping control heart rate and blood pressure [147]. Magnesium is required by enzymes involved in glycolysis and is found predominantly in bones and muscle tissues. Manganese acts as a cofactor for the antioxidant enzyme superoxide dismutase, aiding in protecting cells from oxidative damage.

Iron is an essential trace element that is a component of hemoglobin and is involved in the metabolism of lipids, carbohydrates, and proteins. A deficiency of this element can result in the development of anemia. Zinc is a vital component in the secretion of insulin, the process of wound healing, the attainment of normal growth and development, and the formation of behavioral patterns. Copper is another essential microelement that is necessary for the absorption and incorporation of iron into hemoglobin. It acts as a cofactor for enzymes such as superoxide dismutase and cytochrome c oxidase. Selenium plays a pivotal role in the prevention of liver necrosis and functions as a cofactor for the antioxidant enzyme glutathione peroxidase (Parvez 2013).

Moreover, lead, a heavy metal, has been linked to the development of cancerous and genetically altered cells in experimental animals, as well as to disorders affecting the central nervous system. Cadmium and mercury are toxic metals that have been demonstrated to have a deleterious impact on the urinary and respiratory systems. However, the levels of cadmium (0.006 ppm), mercury (0.033 ppm), and lead (0.097 ppm) in lemongrass were found to be within the permissible limits, indicating that their concentrations are negligible with regard to potential health concerns (Sambo et al. 2024).

### 11.3.3 Fiber Content

Trowell et al. defined dietary fiber as plant cell remnants resistant to digestion by human enzymes, including components like hemicellulose, cellulose, lignin, oligosaccharides, pectins, gums, and waxes (Trowell et al. 1985). Lemongrass is known to be composed of about 27% crude fiber on a dry weight basis. Dietary fiber offers several health benefits, such as enhancing intestinal motility, impacting post-meal glucose and insulin levels, lowering cholesterol, and providing a prebiotic effect (Barber et al. 2020).

Dietary fiber, derived from carbohydrates, is non-digestible in the human gut but can be fermented in the colon by beneficial microorganisms. This includes cellulose, pectin, β-glucans, fructans, galactooligosaccharides, and resistant starch (Gropper 2005). Studies have shown that dietary fiber is linked to various health benefits, including enhanced digestive health (Kaczmarczyk et al. 2012). Fiber fermentation in the colon results in the production of short-chain fatty acids (SCFA) like acetate, propionate, and butyrate, which improve gut health by promoting the growth of beneficial microorganisms (Kaczmarczyk et al. 2012). The SCFAs also enhance blood flow, oxygen uptake, mucus production, and cell differentiation in the colon, which helps prevent colon cancer and colitis (den Besten et al. 2013; Scheppach 1994).

Dietary fiber intake is strongly associated with weight loss. This is due to its ability to modulate key enzymes and hormones related to carbohydrate and lipid metabolism, enhancing insulin sensitivity (Kaczmarczyk et al. 2012). Fiber creates a feeling of fullness, which reduces overall calorie intake. Additionally, the fermentation of dietary fiber in the colon produces short chain fatty acids (SCFAs) that play a role in regulating fatty acid, glucose, and cholesterol metabolism (Anderson and Bridges 1984).

Dietary fiber can help reduce cholesterol levels by various mechanisms. The SCFAs produced from fiber fermentation in the colon, particularly propionate, inhibit cholesterol synthesis in the liver by blocking the enzyme HMG-CoA reductase (Lin et al. 1995). This enzyme is also targeted by statin drugs used to treat high cholesterol (Pichandi et al. 2011). Preventing the reabsorption of bile acids in the liver also prevents cholesterol formation through the enterohepatic circulation of bile acids. This results in an increased excretion of cholesterol and a subsequent reduction in blood cholesterol levels (Caballero et al. 2005).

Bioactive compounds in dietary fiber slow down glucose absorption in the intestines. This effect is achieved either by delaying the digestion of carbohydrates or reducing the glycemic index of carbohydrate-rich foods. Additionally, dietary fiber can moderate postprandial blood glucose levels by inhibiting pancreatic α-amylase and α-glucosidase enzymes, which are involved in carbohydrate digestion. These mechanisms offer potential benefits in managing hyperglycemia and preventing type 2 diabetes (Etxeberria et al. 2012).

By suppressing digestive enzymes and regulating lipid metabolism, dietary fiber derived from *C. citratus* exhibited anti-obesity properties. An in vitro investigation conducted by Villalobos et al. demonstrated that propionate from *C. citratus* total dietary fiber fermentation decreased α-amylase activity (an enzyme that breaks down starch molecules to release glucose subunits). Dietary fiber also inhibited the rise in fasting blood sugar levels, lowered total cholesterol, and minimized the increase in low-density lipoprotein (LDL) + very low-density lipoprotein (VLDL) levels (Villalobos et al. 2021).

A study on Filipino confectioneries demonstrates the nutritional advantages of combining lemongrass, providing a variety of options from high-fiber and protein-rich snacks to decadent desserts. These show even further that adding lemongrass to confections broadens their nutritional profiles and creates new opportunities for consumer education, market segmentation, and product launching. This can result in the creation of confections that suit contemporary dietary trends and tastes while maintaining desirable taste and offering several health benefits (Rosete and Torres 2024).

### 11.3.4 Phytochemicals

The therapeutic potential of medicinal herbs is often correlated with their diverse array of phytochemicals. A variety of chemical constituents have been identified in Cymbopogon species, including hydrocarbons, alcohols, ketones, esters, phenols (including flavonoids and tannins), volatile and non-volatile terpenoids, acids, carotenoids, and other miscellaneous compounds. Notable phytoconstituents, including essential oils, flavonoids, terpenoids, phenols, and tannins, are notably present (Tibenda et al. 2022). Essential oils are primarily composed of monoterpenes, monoterpenoids, sesquiterpenes, and sesquiterpenoids, with the presence of minor fatty alcohols such as 1-octanol and 4-nonanol (Piaru et al. 2012).

The antioxidant properties of flavonoids are well-documented. Isoorientin and tricin were isolated from both a dichloromethane extract and the entire plant of *C. parkeri,* demonstrating muscular relaxation activity (Rizk et al. 1995). In addition, flavonoid compounds like cynaroside, luteolin, isoscoparin, and 20-O-rhamnosyl isoorientin were identified in the leaves and rhizomes of *C. citratus*. The aerial parts of *C. citratus* yielded apigenin, kaempferol, caffeic acid, catechol, quercetin, chlorogenic acid, hydroquinone, and elemicin (Asif and Khodadadi 2013).

A study by Ressurreicao et al. indicated that *C. citratus* extracts and their polyphenols can inhibit cholesterol absorption in the intestine by disrupting micelles. The phenolic compounds show promise as alternatives to traditional treatments like statins, which are the primary treatment for hypercholesterolemia but come with various side effects such as fatigue, muscle pain, and gastrointestinal issues (Da Ressurreição et al. 2022).

Lemongrass essential oils find extensive use in the fragrance, cosmetic, food, and flavor industries. These oils are associated with biological effects such as pain relief and blood sugar regulation (Dikshit and Husain 1984). Essential oils are mixtures of lipophilic and volatile compounds (often terpenoids) present in plants. Essential oil components, such as those found in *C. citratus*, have a lot in common in terms of absorption, metabolism, and excretion. They are quickly absorbed following oral, pulmonary, and cutaneous application. Most are metabolized and eliminated by the kidney in the form of glucuronides or exhaled as $CO_2$. Because of their fast clearance and short biological half-lives, they are unlikely to accumulate in the body.

Lemongrass oil is used medicinally to enhance circulation, relieve coughs and colds, boost immunity, stabilize menstrual cycles, and treat rheumatism, bladder, and digestive disorders. It has significant commercial value in flavors, cosmetics, perfumes, detergents, and pharmaceuticals. Its chemical compounds provide antioxidant, anti-inflammatory, antibacterial, antifungal, antiparasitic, insect-repellent, analgesic, and antipyretic effects. It prevents gout by inhibiting xanthine oxidase, shows anticancer properties, lowers cholesterol, protects the liver, and benefits male reproductive health by mitigating oxidative stress and improving sperm quality (Abdelkader et al. 2023).

Previous research demonstrated how oxidants increase aflatoxin synthesis, whereas antioxidants (such as citral, cinnamaldehyde, and eugenol) can

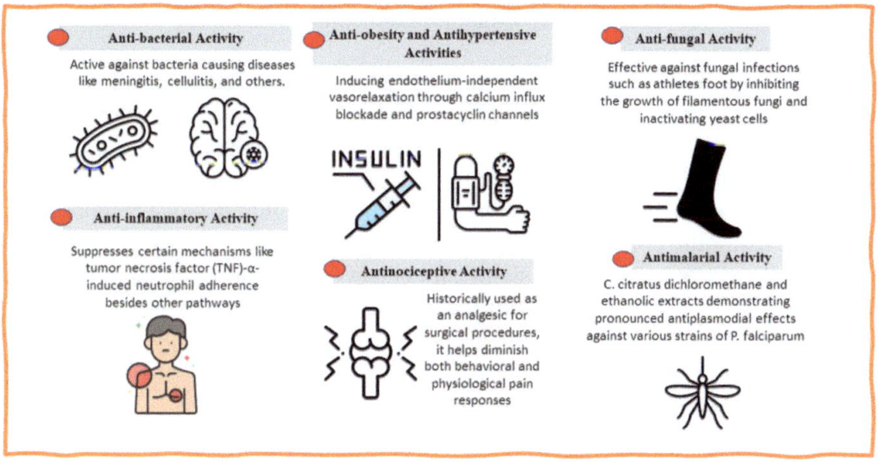


**Fig. 11.2** Health benefits of using *C. citratus*

dramatically lower aflatoxin production (Sun et al. 2016), (Fountain et al. 2016). According to Soares et al., essential oils exhibit antioxidant activity that is higher than that of the methanolic extracts obtained and comparable to that of butylated hydroxytoluene (BHT), a synthetic antioxidant. Due to its anti-oxidant qualities, essential oils have been employed in experiments to preserve a variety of food substrates, such as cooked and raw pork patties (Soares et al. 2013). Since eugenol, present in a trace amount in essential oil, has the ability to scavenge free radicals, Nathchnok et al. evaluated the use of essential oil in raw food packaging in place of the commercial antioxidant BHT (Petchsoongsakul and Pechyen 2012). These nutritional effects are correlated with health benefits as shown in Fig. 11.2.

## 11.4 Safety Considerations and Precautions

### 11.4.1 Allergies and Sensitivities

Although lemongrass essential oil is generally regarded as safe for most individuals, it should be used with caution. Direct application of the essential oil to the skin without adequate dilution may result in irritation or an allergic reaction (Bampidis et al. 2022; Kusuma et al. 2024). It is similarly recommended to refrain from excessive or prolonged use, as this may result in gastrointestinal disturbances or interactions with certain medications. This underscores the necessity for moderation and professional guidance in its utilization (Kusuma et al. 2024). As the potential of natural remedies such as lemongrass essential oil is further elucidated, it becomes imperative to integrate traditional wisdom with modern scientific practices. This synergy has the potential to facilitate the development of more sustainable, accessible, and effective health solutions, thereby reducing our reliance on synthetic drugs and their associated risks, while simultaneously embracing a more holistic approach to health and wellness.

Lemongrass oil has been documented to act as a potent sensitizer and/or a significant irritant. There is sufficient evidence to establish a causal link between lemongrass oil and the clinical course of the eruption. This includes the eruption's appearance following appropriate exposure, its rapid disappearance following exposure cessation while conducting routine tasks, and its highly positive response to a patch test using the contaminated wood. Lemongrass oil has a variety of applications and should be considered a significant cutaneous irritant when identifying the causative agents of contact dermatitis (Mendelsohn 1944).

### 11.4.2 Interactions with Medications

In their study, Silva et al. examined the impact of essential oils on the sleep cycle of male Swiss mice exposed to barbiturates as well as to three agents, namely pentylenetetrazol (PTZ), pilocarpine, and strychnine. The findings indicated that *C. citratus* was more efficacious than the other essential oils in lengthening the latency to the first convulsion and the latency to death in the pentylenetetrazol-induced convulsion paradigm. The higher doses of essential oils (200 mg/kg) yielded superior outcomes, although this effect was less pronounced in the pilocarpine-induced convulsion model. A comparable outcome was observed in the model of convulsions caused by strychnine. The sedative effect of barbiturates was not intensified by *C. citratus* as rapidly as by the other essential oils. Considering the evidence that the essential oils delayed the onset of convulsions induced by pentylenetetrazol, it seems plausible that they are interfering with the function of GABAergic neurons (Adelman et al. 1979).

Even higher doses, 0.5 and 1 g/kg of essential oils, were observed to increase the pentobarbital-induced sleeping time, delay clonic types of seizures induced by PTZ, and block tonic extensions induced by maximal electroshock (Blanco et al. 2009). Higher doses of the essential oils extended the death latency, although the onset of convulsions was lessened when pilocarpine, an agonist for cholinergic receptors, was used as the convulsing drug (Smolders et al. 1997). Following the essential oil treatment, pentobarbital was seen to elicit a reduction in sleeping latency and an increase in sleeping duration. These are standard measures associated with a sedative/hypnotic characteristic (Fujimoto et al. 1960). The only beneficial outcome observed was obtained with an intraperitoneal dosage of 100 mg/kg of citral. The barbiturate effect could not be enhanced with the same dosage of citral administered orally (Carlini et al. 1986).

To reduce the possibility of these medications' adverse effects being worse, it is also advised to avoid taking them with medications that are substrates for cytochrome p450 and glutathione-S-transferase. Additionally, in order to mitigate the effects of citronella oil and its related compounds on the development of unborn children, mothers who are pregnant or nursing should abstain from ingesting lemongrass tea (Lemongrass | Memorial Sloan Kettering Cancer Center 2024).

### 11.4.3 Proper Dosage and Moderation

The conventional diclofenac, a well-known anti-inflammatory non-steroidal medication, was compared to lemongrass oil's anti-inflammatory effect. Diclofenac showed comparable edema reduction efficacy (86.2%) obtained from a 50 mg/kg dosage (Majewska et al. 2019). The antifungal potential of *C. citratus* essential oil on Malassezia yeasts, as observed in preclinical testing, and the safety of formulations at the applied concentration of 1.25 μL/mL, have been confirmed by the mycological cure rate for patients treated with this oil (Rodriguez-Tudela et al. 2008).

Gentamicin and *C. citratus* were administered concurrently to the animals, and the result was normal glomeruli and tubules with a hazy appearance and regeneration activity. In addition, animals given *C. citratus* showed intact parenchyma and no signs of tubular necrosis or notable glomerular abnormalities. Renal damage associated with aminoglycosides was successfully avoided by co-administration of a daily dose of 200 mg/kg of *C. citratus* for three weeks, as determined by renal functional parameters and histological analysis (Ullah et al. 2013).

Regarding cancer treatment, the overall number of tumors, the frequency of tumors per group (which was 44% lower with the 200 mg/kg/day dose), the average tumor volume, the cumulative tumor volume, necrosis, and mitosis were all reduced by the essential oil at the 50, 100, and 200 mg/kg/day body weight (Rojas-Armas et al. 2020). According to a study, the antiapoptotic proteins Bcl-xL and Mcl-1 were expressed less frequently when essential oil and citral were added, which inhibited the growth and survival of small lung cancer cells (SCLC cells) (Ranitha et al. 2014). Additionally, reduced cell division and raised intracellular reactive oxygen species changed the potential of the mitochondrial membrane and started cervical cancer cell lines' death. ME-180 and HeLa (Najar et al. 2020). Related research indicates that while *C. citratus* essential oil's genotoxicity and toxicological biomarkers have a low toxicity level, they are appropriate and safe for long-term use at levels up to 100 mg/kg (Bidinotto et al. 2011).

## 11.5 Conclusions

*C. citratus* is an ideal candidate for further scientific investigation across a range of disciplines, including agronomy, food production, cosmetics, and pharmaceuticals. This is due to the plant's extensive traditional use, rich phytochemical profile, and potential for pharmacological applications. Consequently, further research is necessary to ascertain the precise therapeutic efficacy and organ toxicity of specific species or active components. Although certain fundamental pharmacokinetic characteristics of the isolated pharmaco-active compounds have been documented, it is not possible to positively associate or contrast them with corresponding information on the complex blends of essential oils. Consequently, a substantial amount of information regarding this intriguing herb remains to be discovered. This may be achieved by identifying and clarifying the actions of its additional bioactive components, particularly regarding their effects on humans.

**Authors' Contributions** Abdelmonem Siddiq (AS), considered the first author and corresponding author, proposed the project, contributed to the conception, formulation, and drafting of the article, and reviewed and revised the manuscript. MD, MS, SS, and SYC participated in writing and revising the final manuscript. NJ participated in writing, reviewing, and editing of the final manuscript. All authors approved the final manuscript as submitted and agreed to be accountable for all aspects of the work.

## References

Abdelkader S, Elkhishin I, Mesallam D, Abdelwahab M (2023) Lemongrass (Cymbopogon Citratus) essential oil: health beneficial perspective lemongrass (Cymbopogon Citratus) essential oil: health beneficial perspective. Eur Chem News 12:3422–3426. https://www.researchgate.net/publication/376857629_Lemongrass_Cymbopogon_Citratus_Essential_Oil_Health_Beneficial_Perspective_Lemongrass_Cymbopogon_Citratus_Essential_Oil_Health_Beneficial_Perspective

Abe S, Sato Y et al (2003a) Anti-Candida Albicans activity of essential oils including lemongrass (Cymbopogon Citratus) oil and its component, Citral. Nippon Ishinkin Gakkai Zasshi 44(4):285–291. http://www.jstage.jst.go.jp/article/jjmm1990/44/4/44_4_285/_article/-char/ja/

Abe S, Maruyama N et al (2003b) Suppression of tumor necrosis factor-alpha-induced neutrophil adherence responses by essential oils. Mediat Inflamm 12(6):323–328. https://onlinelibrary.wiley.com/doi/10.1080/09629350310001633342

Adedapo AA, Jimoh FO, Afolayan AJ, Masika PJ (2008) Antioxidant activities and phenolic contents of the methanol extracts of the stems of Acokanthera Oppositifolia and Adenia Gummifera. BMC Complement Altern Med 8(1):1–7. https://link.springer.com/articles/10.1186/1472-6882-8-54 (June 30, 2024)

Adelman RD, Conzelman G, Spangler W, Ishizaki G (1979) Comparative nephrotoxicity of gentamicin and Netilmicin: functional and morphological correlations with urinary enzyme activities. Curr Probl Clin Biochem 9:166–182. http://www.ncbi.nlm.nih.gov/pubmed/446071

Adeneye AA, Agbaje EO (2007) Hypoglycemic and hypolipidemic effects of fresh leaf aqueous extract of Cymbopogon Citratus Stapf. in rats. J Ethnopharmacol 112(3):440–444

Akchiche N et al (2012) Homocysteinylation of neuronal proteins contributes to folate deficiency-associated alterations of differentiation, vesicular transport, and plasticity in hippocampal neuronal cells. FASEB J 26(10):3980–3992. https://onlinelibrary.wiley.com/doi/abs/10.1096/fj.12-205757

Akhila A (2009) Essential oil-bearing grasses: the genus Cymbopogon, 1–247. https://www.taylorfrancis.com/books/mono/10.1201/9780849378584/essential-oil-bearing-grasses-anand-akhila (July 3, 2024)

Akono Ntonga P et al (2014) Activity of Ocimum Basilicum, Ocimum Canum, and Cymbopogon Citratus essential oils against Plasmodium Falciparum and mature-stage larvae of Anopheles Funestus. Parasite 21. /pmc/articles/PMC4082313/ (June 30, 2024)

Alitonou G et al (2006) Investigations on the essential oil of Cymbopogon Giganteus from Benin for its potential use as an anti-inflammatory agent. Int J Aromather 16(1):37–41. https://linkinghub.elsevier.com/retrieve/pii/S0962456206000038

Almeida RBA et al (2013) Antimicrobial Activity of the Essential Oil of Cymbopogon Citratus (DC) Stapf. on Staphylococcus Spp., Streptococcus Mutans and Candida Spp. Revista Brasileira de Plantas Medicinais 15(4):474–482. https://www.scielo.br/j/rbpm/a/pBr7yrjcc8TNzKpXdBcDwTS/ (June 30, 2024)

Alviano WS et al (2005) Antimicrobial activity of *croton Cajucara* Benth linalool-rich essential oil on artificial biofilms and planktonic microorganisms. Oral Microbiol Immunol 20(2):101–105. https://onlinelibrary.wiley.com/doi/10.1111/j.1399-302X.2004.00201.x

Alwaili MA (2022) Protective effects of lemongrass (*Cymbopogon Citratus* STAPF) extract mediated mitochondrial fission and glucose uptake inhibition in SW1417. Food Sci Technol

43:e94522. https://www.scielo.br/j/cta/a/YwZrtspBFZZMFxrPrMK9Lvs/?lang=en (July 1, 2024)

Anal JMH (2014) Trace and essential elements analysis in Cymbopogon Citratus (DC.) Stapf samples by graphite furnace-atomic absorption spectroscopy and its health concern. J Toxicol 2014:690758. http://www.ncbi.nlm.nih.gov/pubmed/25525430

Anderson JW, Bridges SR (1984) Short-chain fatty acid fermentation products of plant fiber affect glucose metabolism of isolated rat hepatocytes. Proc Soc Exp Biol Med (New York, NY) 177(2):372–376. http://www.ncbi.nlm.nih.gov/pubmed/6091151

Andila PS et al (2018) The phytochemistry of Cymbopogon Winterianus essential oil from Lombok Island, Indonesia and its antifungal activity against phytopathogenic fungi. Nusantara Biosci 10(4):232–239

Asif M, Khodadadi E (2013) Medicinal uses and chemistry of flavonoid contents of some common edible tropical plants. Arch Adv Biosci 4(3)

Babelghaith SD, Sales I, Syed W, Al-Arifi MN (2024) The use of complementary and alternative medicine for functional gastrointestinal disorders among the Saudi population. Saudi Pharmaceut J 32(6):102084

Bachiega TF, Sforcin JM (2011) Lemongrass and Citral effect on cytokines production by murine macrophages. J Ethnopharmacol 137(1):909–913

Balusamy SR et al (2020) Citral induced apoptosis through modulation of key genes involved in fatty acid biosynthesis in human prostate cancer cells: in silico and in vitro study. BioMed Res Int 2020(1):6040727. https://onlinelibrary.wiley.com/doi/full/10.1155/2020/6040727 (July 1, 2024)

Bampidis V, et al (2022) Safety and efficacy of a feed additive consisting of an essential oil from the aerial parts of Anethum Graveolens L. (dill herb oil) for use in dogs and cats (FEFANA Asbl). EFSA J 20(12):e07689. https://onlinelibrary.wiley.com/doi/full/10.2903/j.efsa.2022.7689 (June 30, 2024)

Barber TM, Kabisch S, Pfeiffer AFH, Weickert MO (2020) The health benefits of dietary fibre. Nutrients 12(10):3209. http://www.ncbi.nlm.nih.gov/pubmed/33096647

Barnham KJ, Masters CL, Bush AI (2004) Neurodegenerative diseases and oxidative stress. Nat Rev Drug Discov 3(3):205–214. https://www.nature.com/articles/nrd1330 (July 1, 2024)

Bayala B et al (2014) Chemical composition, antioxidant, anti-inflammatory and anti-proliferative activities of essential oils of plants from Burkina Faso. PLoS One 9(3):e92122. https://journals.plos.org/plosone/article?id=10.1371/journal.pone.0092122 (June 30, 2024)

Bayala B et al (2018) Cymbopogon Citratus and Cymbopogon Giganteus essential oils have cytotoxic effects on tumor cell cultures. Identification of Citral as a new putative anti-proliferative molecule. Biochimie 153:162–170. https://linkinghub.elsevier.com/retrieve/pii/S0300908418300567

Bhuyan PD, Tamuli P, Boruah P (2015) In-vitro efficacy of certain essential oils and plant extracts against three major pathogens of Jatropha Curcas L. Am J Plant Sci 6(2):362–365

Bidinotto LT et al (2011) Protective effects of lemongrass (*Cymbopogon Citratus* STAPF) essential oil on DNA damage and carcinogenesis in female Balb/C mice. J Appl Toxicol 31(6):536–544. https://analyticalsciencejournals.onlinelibrary.wiley.com/doi/10.1002/jat.1593

Blanco AR et al (2005) Epigallocatechin Gallate inhibits biofilm formation by ocular staphylococcal isolates. Antimicrob Agents Chemother 49(10):4339–4343. https://journals.asm.org/doi/10.1128/aac.49.10.4339-4343.2005 (June 30, 2024)

Blanco MM et al (2009) Neurobehavioral effect of essential oil of Cymbopogon Citratus in mice. Phytomedicine 16(2–3):265–270. https://linkinghub.elsevier.com/retrieve/pii/S0944711307000712

Boeira CP et al (2020) Phytochemical characterization and antimicrobial activity of Cymbopogon Citratus extract for application as natural antioxidant in fresh sausage. Food Chem 319:126553

Boukhatem MN et al (2014) Lemon grass (Cymbopogon Citratus) essential oil as a potent anti-inflammatory and antifungal drugs. Libyan J Med 9(1) https://www.ajol.info/index.php/ljm/article/view/108546 (June 30, 2024)

Buckle J (2014) Clinical aromatherapy-e-book: essential oils in practice. Elsevier Health Sciences

Burt SA, Reinders RD (2003) Antibacterial activity of selected plant essential oils against Escherichia Coli O157:H7. Lett Appl Microbiol 36(3):162–167. https://academic.oup.com/lambio/article/36/3/162/6704508

Caballero B, Allen L, Prentice A (2005) Encyclopedia of human nutrition. Elsevier Science. https://books.google.com.np/books?id=1GdqAAAAMAAJ

Campos J et al (2014) Lemon grass (Cymbopogon Citratus (D.C) Stapf) polyphenols protect human umbilical vein endothelial cell (HUVECs) from oxidative damage induced by high glucose, hydrogen peroxide and oxidised low-density lipoprotein. Food Chem 151:175–181

Carlini EA et al (1986) Pharmacology of lemongrass (Cymbopogon Citratus Stapf). I. Effects of teas prepared from the leaves on laboratory animals. J Ethnopharmacol 17(1):37–64. https://linkinghub.elsevier.com/retrieve/pii/0378874186900723

Chagonda LS, Makanda C, Chalchat J-C (2000) The essential oils of wild and cultivated Cymbopogon Validus (Stapf) Stapf Ex Burtt Davy and Elionurus Muticus (Spreng.) Kunth from Zimbabwe. Flavour Fragr J 15(2):100–104

Cheel J, Theoduloz C, Rodríguez J, Schmeda-Hirschmann G (2005) Free radical scavengers and antioxidants from lemongrass (Cymbopogon Citratus (DC.) Stapf.). J Agric Food Chem 53(7):2511–2517. https://pubs.acs.org/doi/abs/10.1021/jf0479766 (June 30, 2024)

Cimanga K et al (2002) Correlation between chemical composition and antibacterial activity of essential oils of some aromatic medicinal plants growing in the Democratic Republic of Congo. J Ethnopharmacol 79(2):213–220

Cortes-Torres AG et al (2023) Cymbopogon Citratus essential oil: extraction, GC–MS, phytochemical analysis, antioxidant activity, and in silico molecular docking for protein targets related to CNS. Curr Issues Mol Biol 45(6):5164–5179. https://www.mdpi.com/1467-3045/45/6/328/htm (July 1, 2024)

Cowan MM (1999) Plant products as antimicrobial agents. Clin Microbiol Rev 12(4):564–582. https://journals.asm.org/doi/10.1128/CMR.12.4.564

da Silva D, Cândido K et al (2024) Evaluation of the anti-inflammatory and antioxidant potential of Cymbopogon citratus essential oil in zebrafish. Animals 14(4):581. https://www.mdpi.com/2076-2615/14/4/581/htm (July 1, 2024)

De Oliveira MAC et al (2017) Cymbopogon Citratus essential oil: effect on polymicrobial caries-related biofilm with low cytotoxicity. Braz Oral Res 31(0) http://www.scielo.br/scielo.php?script=sci_arttext&pid=S1806-83242017000100273&lng=en&tlng=en

De Sousa DP (2011) Analgesic-like activity of essential oils constituents. Molecules 16(3):2233–2252. https://www.mdpi.com/1420-3049/16/3/2233/htm (June 30, 2024)

Deletre E et al (2015) Electrophysiological and behavioral characterization of bioactive compounds of the thymus vulgaris, Cymbopogon Winterianus, Cuminum Cyminum and Cinnamomum Zeylanicum essential oils against anopheles Gambiae and prospects for their use as Bednet treatments. Parasit Vectors 8:1–14

den Besten G et al (2013) The role of short-chain fatty acids in the interplay between diet, gut microbiota, and host energy metabolism. J Lipid Res 54(9):2325–2340. http://www.ncbi.nlm.nih.gov/pubmed/23821742

Dikshit A, Husain A (1984) Antifungal action of some essential oils against animal pathogens

Ekpenyong CE, Akpan EE (2017) Use of Cymbopogon Citratus essential oil in food preservation: recent advances and future perspectives. Crit Rev Food Sci Nutr 57(12):2541–2559. https://doi.org/10.1080/10408398.2015.1016140

Ekpenyong CE, Akpan E, Nyoh A (2015) Ethnopharmacology, phytochemistry, and biological activities of Cymbopogon Citratus (DC.) Stapf extracts. Chin J Nat Med 13(5):321–337. https://linkinghub.elsevier.com/retrieve/pii/S1875536415300236

El-Tahir KEH, Abdel-Kader MS (2008) Chemical and pharmacological study of Cymbopogon Proximus volatile oil. In: Essential oil-bearing grasses: the genus Cymbopogon. CRC Press, Boca Raton

Etxeberria U et al (2012) Antidiabetic effects of natural plant extracts via inhibition of carbohydrate hydrolysis enzymes with emphasis on pancreatic alpha amylase. Expert Opin Ther Targets 16(3):269–297. http://www.ncbi.nlm.nih.gov/pubmed/22360606

Fernandes C et al (2012) Investigation of the mechanisms underlying the gastroprotective effect of Cymbopogon Citratus essential oil. J Young Pharm 4(1):28–32
Fernández MA et al (2010) Anti-inflammatory activity of abietic acid, a Diterpene isolated from Pimenta Racemosa Var. Grissea. J Pharm Pharmacol 53(6):867–872. https://doi.org/10.1211/0022357011776027
Figueirinha A et al (2008) Cymbopogon Citratus leaves: characterization of flavonoids by HPLC–PDA–ESI/MS/MS and an approach to their potential as a source of bioactive polyphenols. Food Chem 110(3):718–728. https://linkinghub.elsevier.com/retrieve/pii/S0308814608002306
Figueirinha A et al (2010) Anti-inflammatory activity of Cymbopogon Citratus leaf infusion in lipopolysaccharide-stimulated dendritic cells: contribution of the polyphenols. 13(3):681–690. https://home.liebertpub.com/jmf. https://www.liebertpub.com/doi/10.1089/jmf.2009.0115 (July 1, 2024)
FoodData Central (2024). https://fdc.nal.usda.gov/fdc-app.html#/food-details/168573/nutrients
Fountain JC et al (2016) Oxidative stress and carbon metabolism influence Aspergillus Flavus transcriptome composition and secondary metabolite production. Sci Rep 6(1):38747. https://www.nature.com/articles/srep38747
Francisco V et al (2013) Anti-inflammatory activity of *Cymbopogon Citratus* leaves infusion via proteasome and nuclear factor-KB pathway inhibition: contribution of chlorogenic acid. J Ethnopharmacol 148(1):126–134. https://www.sciencedirect.com/science/article/pii/S037887411300247X
Fujimoto JM, Pearce KB, Plaa GL (1960) Barbiturate metabolism as affected by certain agents acting on the liver. J Pharmacol Exp Ther 129:139–143. http://www.ncbi.nlm.nih.gov/pubmed/13825580
Galvez J et al (1993) Antidiarrhoeic activity of euphorbia hirta extract and isolation of an active flavonoid constituent. Planta Medica 59(4):333–336. http://www.thieme-connect.de/DOI/DOI?10.1055/s-2006-959694 (June 30, 2024)
Garcia R et al (2022) Evaluation of anti-inflammatory and analgesic activities of Cymbopogon Citratus in vivo-polyphenols contribution. Res J Med Plant 9(1):1–13. https://estudogeral.uc.pt/handle/10316/101140 (June 30, 2024)
Gawade SP (2012) Acetic acid induced painful endogenous infliction in writhing test on mice. J Pharmacol Pharmacother 3(4):348
Gayathri K, Jayachandran KS, Vasanthi HR, Rajamanickam GV (2010) Cardioprotective effect of lemon grass as evidenced by biochemical and histopathological changes in experimentally induced cardiotoxicity. 30(8):1073–1082. https://doi.org/10.1177/0960327110386391. https://journals.sagepub.com/doi/abs/10.1177/0960327110386391 (June 30, 2024)
Gonçalves ECD et al (2020) Citral inhibits the inflammatory response and hyperalgesia in mice: the role of TLR4, TLR2/Dectin-1, and CB2 cannabinoid receptor/ATP-sensitive K+ channel pathways. J Nat Prod 83(4):1190–1200. https://pubs.acs.org/doi/abs/10.1021/acs.jnatprod.9b01134 (June 30, 2024)
Gropper SAS (2005) Advanced nutrition and human metabolism. Thomson/Wadsworth, Belmont. http://archive.org/details/advancednutritio0004grop
Han X, Parker TL (2017) Lemongrass (Cymbopogon Flexuosus) essential oil demonstrated anti-inflammatory effect in pre-inflamed human dermal fibroblasts. Biochimie open 4:107–111
Handbook of Essential Oils: Science, Technology, and Applications (2009) Handbook of essential oils. https://www.taylorfrancis.com/books/mono/10.1201/9781420063165/handbook-essential-oils-husnu-baser-gerhard-buchbauer (June 30, 2024)
Haque AN, Ahsanul M, Remadevi R, Naebe M (2018) Lemongrass (Cymbopogon): a review on its structure, properties, applications and recent developments. Cellulose 25:5455–5477
Heinrich M (2003) Plants and people of Nepal-Narayan P. Manandhar with the assistance of Sanjay Manandhar. Timber Press, Portland, p 599. US $69.95 (Hardcover), ISBN 0-88192-527-6; Scientific and Common Botanical Name Index, Bibliography, Numerous Black An. J Ethnopharmacol 2(88): 300–301

Hussain S et al (2023) Synergistic antibacterial screening of Cymbopogon Citratus and Azadirachta Indica: phytochemical profiling and antioxidant and hemolytic activities. ACS Omega 8(19):16600–16611. https://pubs.acs.org/doi/full/10.1021/acsomega.2c06785 (June 30, 2024)

Irkin R, Korukluoglu M (2007) Control of Aspergillus niger with garlic, onion and leek extracts. Afr J Biotechnol 6(4):384–387

Jirovetz L et al (2007) Composition and antimicrobial activity of Cymbopogon Giganteus (Hochst.) Chiov. Essential flower, leaf and stem oils from Cameroon. J Essent Oil Res 19(5):485–489

Kaczmarczyk MM, Miller MJ, Freund GG (2012) The health benefits of dietary fiber: beyond the usual suspects of type 2 diabetes mellitus, cardiovascular disease and colon cancer. Metabolism 61(8):1058–1066. http://www.ncbi.nlm.nih.gov/pubmed/22401879

Kandhro A et al (2011) Determination of different trace and essential element in Lemon grass samples by X-ray fluorescence spectroscopy technique. Int Food Res J 18:265–270. https://www.researchgate.net/publication/261025462_Determination_of_different_trace_and_essential_element_in_lemon_grass_samples_by_X-ray_fluorescence_spectroscopy_technique

Kapur A et al (2016) Modulation of oxidative stress and subsequent induction of apoptosis and endoplasmic reticulum stress allows citral to decrease cancer cell proliferation. Sci Rep 6(1):1–14. https://www.nature.com/articles/srep27530 (July 1, 2024)

Katiki LM et al (2011) Anthelmintic activity of Cymbopogon Martinii, Cymbopogon Schoenanthus and Mentha Piperita essential oils evaluated in four different in vitro tests. Vet Parasitol 183(1–2):103–108

Ketoh GK, Koumaglo HK, Glitho IA, Huignard J (2006) Comparative effects of Cymbopogon Schoenanthus essential oil and Piperitone on Callosobruchus Maculatus development. Fitoterapia 77(7–8):506–510

Khadri A et al (2008) Antioxidant and antiacetylcholinesterase activities of essential oils from Cymbopogon Schoenanthus L. Spreng. Determination of chemical composition by GC–mass spectrometry and 13C NMR. Food Chem 109(3):630–637

Khan MSA, Ahmad I (2012) Antibiofilm activity of certain phytocompounds and their synergy with fluconazole against Candida Albicans biofilms. J Antimicrob Chemother 67(3):618–621. https://doi.org/10.1093/jac/dkr512

Kiani HS et al (2022a) Phytochemical composition and pharmacological potential of lemongrass (Cymbopogon) and impact on gut microbiota. AppliedChem 2(4):229–246. https://www.mdpi.com/2673-9623/2/4/16

Kiani HS et al (2022b) Phytochemical composition and pharmacological potential of lemongrass (Cymbopogon) and impact on gut microbiota. Appl Chem 2(4):229–246. https://www.mdpi.com/2673-9623/2/4/16/htm (June 30, 2024)

Koh PH, Mokhtar RAM, Iqbal M (2012) Antioxidant potential of Cymbopogon Citratus extract: alleviation of carbon tetrachloride-induced hepatic oxidative stress and toxicity. Hum Exp Toxicol 31(1):81–91. https://journals.sagepub.com/doi/full/10.1177/0960327111407226 (June 30, 2024)

Kpoviessi S et al (2014) Chemical composition, cytotoxicity and in vitro antitrypanosomal and antiplasmodial activity of the essential oils of four Cymbopogon species from Benin. J Ethnopharmacol 151:652

Kusuma IY et al (2024) Exploring the clinical applications of lemongrass essential oil: a scoping review. Pharmaceuticals 17(2):159. https://www.mdpi.com/1424-8247/17/2/159/htm (June 30, 2024)

Lee HJ et al (2008) Inhibitory effect of citral on NO production by suppression of INOS expression and NF-KB activation in RAW264.7 cells. Arch Pharm Res 31(3):342–349. https://link.springer.com/article/10.1007/s12272-001-1162-0 (June 30, 2024)

Lei Y, Fu P, Jun X, Cheng P (2019) Pharmacological properties of geraniol – a review. Planta Medica 85(1):48–55. http://www.thieme-connect.com/products/ejournals/html/10.1055/a-0750-6907 (July 1, 2024)

Leistner OA (2000) Seed plants of Southern Africa: families and genera

Lemongrass | Memorial Sloan Kettering Cancer Center (2024). https://www.mskcc.org/cancer-care/integrative-medicine/herbs/lemongrass

Lin Y et al (1995) Differences in propionate-induced inhibition of cholesterol and triacylglycerol synthesis between human and rat hepatocytes in primary culture. Br J Nutr 74(2):197–207. http://www.ncbi.nlm.nih.gov/pubmed/7547837

Lorenzetti BB et al (1991) Myrcene mimics the peripheral analgesic activity of lemongrass tea. J Ethnopharmacol 34(1):43–48

Lulekal E et al (2019) Phytochemical analysis and evaluation of skin irritation, acute and sub-acute toxicity of Cymbopogon Citratus essential oil in mice and rabbits. Toxicol Rep 6:1289–1294

Mahboubi M, Kazempour N (2012) Biochemical activities of Iranian Cymbopogon Olivieri (Boiss) Bor. essential oil. Indian J Pharm Sci 74(4):356

Majewska E et al (2019) Lemongrass (Cymbopogon Citratus) essential oil: extraction, composition, bioactivity and uses for food preservation – a review. Pol J Food Nutr Sci 69(4):327–341. http://journal.pan.olsztyn.pl (June 30, 2024)

Marina D, Jasmina M, Ana D (2013) Natural products from plants and fungi as fungicides. In: Nita M (ed) Fungicides – showcases of integrated plant disease management from around the world. INTECH. http://www.intechopen.com/books/fungicides-showcases-of-integrated-plant-disease-management-from-around-the-world/natural-products-from-plants-and-fungi-as-fungicides

Martins HB et al (2017) Anti-inflammatory activity of the essential oil citral in experimental infection with staphylococcus aureus in a model air pouch. Evid Based Complement Altern Med 2017(1):2505610. https://onlinelibrary.wiley.com/doi/full/10.1155/2017/2505610 (June 30, 2024)

Matera R, Lucchi E, Valgimigli L (2023) Plant essential oils as healthy functional ingredients of nutraceuticals and diet supplements: a review. Molecules 28(2):901. https://www.mdpi.com/1420-3049/28/2/901/htm (July 1, 2024)

Mendelsohn HV (1944) Dermatitis from lemon grass oil (Cymbopogon Citratus or Andropogon Citratus). Arch Dermatol 50(1):34. http://archderm.jamanetwork.com/article.aspx?doi=10.1001/archderm.1944.01510130037010

Miean KH, Mohamed S (2001) Flavonoid (Myricetin, Quercetin, Kaempferol, Luteolin, and Apigenin) content of edible tropical plants. J Agric Food Chem 49(6):3106–3112. https://pubs.acs.org/doi/10.1021/jf000892m

Miguel MG (2010) Antioxidant and anti-inflammatory activities of essential oils: a short review. Molecules 15(12):9252–9287. https://www.mdpi.com/1420-3049/15/12/9252

Milani-Nejad N, Zhang M, Kaffenberger BH (2017) Association of dermatology consultations with patient care outcomes in hospitalized patients with inflammatory skin diseases. JAMA Dermatol 153(6):523–528

Mishra PK et al (2015) Efficacy of Caesulia Axillaris, Cymbopogon Khasans and Cymbopogon Martinii essential oils as plant based preservatives against degradation of raw materials of Andrographis Paniculata by fungal spoilage. Int Biodeterior Biodegradation 104:244–250

Mohammed T (2019) Lemon grass (Cymbopogon, L Spreng) valuable grass but underutilized in Northern Nigeria. https://www.researchgate.net/profile/Toungos-Mohammed/publication/334164105_Lemon_Grass_Cymbopogon_L_spreng_Valuable_Grass_But_Underutilized_In_Northern_Nigeria/links/5d1b259d458515c11c096dea/Lemon-Grass-Cymbopogon-L-spreng-Valuable-Grass-But-Underutilize

Mohammed TK, Aqel N, Al-Dujaili EAS (2020) Antimicrobial activity of liquid residues of Cymbopogon Citratus oil extracts. J Phys Conf Ser 1660(1):012006. https://iopscience.iop.org/article/10.1088/1742-6596/1660/1/012006 (June 30, 2024)

Mohammed M et al (2022) Antifungal, antioxidant and antibiofilm activities of essential oils of Cymbopogon Spp. Antibiotics 11(6):829. https://www.mdpi.com/2079-6382/11/6/829/htm (July 1, 2024)

Mouta LFGL et al (2022) Cymbopogon Citratus essential oil increases the effect of Digluconate chlorhexidine on microcosm biofilms. Pathogens 11(10):1067. https://www.mdpi.com/2076-0817/11/10/1067

Mukarram M et al (2021) Lemongrass essential oil components with antimicrobial and anticancer activities. Antioxidants 11(1):20. https://www.mdpi.com/2076-3921/11/1/20/htm (July 1, 2024)

Naik MI, Fomda BA, Jaykumar E, Bhat JA (2010) Antibacterial activity of lemongrass (Cymbopogon Citratus) oil against some selected pathogenic bacterias. Asian Pac J Trop Med 3(7):535–538

Najar B, Shortrede JE, Pistelli L, Buhagiar J (2020) Chemical composition and *in vitro* cytotoxic screening of sixteen commercial essential oils on five cancer cell lines. Chem Biodivers 17(1):e1900478. https://onlinelibrary.wiley.com/doi/10.1002/cbdv.201900478

Nambiar VS, Matela H (2012) Potential functions of lemon grass (Cymbopogon Citratus) in health and disease. Int J Pharm Biol Arch 3(5):1035–1043. www.floridata.com (July 1, 2024)

Nguefack J, Leth V, Amvam Zollo PH, Mathur SB (2004) Evaluation of five essential oils from aromatic plants of Cameroon for controlling food spoilage and mycotoxin producing fungi. Int J Food Microbiol 94(3):329–334. https://linkinghub.elsevier.com/retrieve/pii/S0168160504001205

Nimenibo-Uadia R, Nwosu EO (2020) Phytochemical, proximate and mineral elements composition of lemongrass (Cymbopogon Citratus (DC) Stapf) grown in Ekosodin, Benin City, Nigeria. Nigerian J Pharm Appl Sci Res 9(2):52–56. https://www.nijophasr.net/index.php/nijophasr/article/view/332

Oladeji OS, Adelowo FE, Ayodele DT, Odelade KA (2019) Phytochemistry and pharmacological activities of Cymbopogon Citratus: a review. Sci Afr 6:e00137

Onawunmi GO (1989) Evaluation of the antimicrobial activity of citral. Lett Appl Microbiol 9(3):105–108. https://doi.org/10.1111/j.1472-765X.1989.tb00301.x

Onawunmi GO, Yisak W-A, Ogunlana EO (1984) Antibacterial constituents in the essential oil of Cymbopogon Citratus (DC.) Stapf. J Ethnopharmacol 12(3):279–286. https://linkinghub.elsevier.com/retrieve/pii/0378874184900576

Pandey MC, Sharma JR, Dikshit A (1996) Antifungal evaluation of the essential oil of Cymbopogon Pendulus (Nees Ex Steud.) Wats. Cv. Praman. Flavour Fragr J 11(4):257–260

Parvez M (2013) Euphorbia Granulata Forssk as a source of mineral supplement. Am Eurasian J Agric Environ Sci 13. https://www.researchgate.net/publication/299362560_Euphorbia_granulata_Forssk_as_a_Source_of_Mineral_Supplement

Pereira GL d C et al (2021) Antitumor effect of Cymbopogon Densiflorus (Linneu) essential oil in bladder cancer cells. Nat Prod Res 35(23):5238–5242

Pérez G, Salud MZS, Lucina Arias G, Miguel Ramos L (2011) Anti-inflammatory activity of some essential oils. J Essent Oil Res 23(5):38–44. http://www.tandfonline.com/doi/abs/10.1080/10412905.2011.9700480

Petchsoongsakul N, Pechyen C (2012) Characterization of eugenol extracted from lemongrass (*Cymbopogon Citratus* (DC.) Stapf) for food packaging materials. Adv Mater Res 506:603–606. https://www.scientific.net/AMR.506.603

Piaru SP et al (2012) Chemical composition, anti-angiogenic and cytotoxicity activities of the essential oils of Cymbopogan Citratus (Lemon Grass) against colorectal and breast carcinoma cell lines. J Essent Oil Res 24(5):453–459. https://doi.org/10.1080/10412905.2012.703496

Pichandi S et al (2011) The role of statin drugs in combating cardiovascular diseases – a review. Int J Cur Sci Res 1(2):47–56

Piéron H (1928) Gravitational sensitivity and geotropic reactions in slugs: an analysis of the laws of excitation and of the internal factors in reaction. Pedagog Semin J Genet Psychol 35(1):3–17

Prasad CS, Shukla R, Kumar A, Dubey NK (2010) In vitro and in vivo antifungal activity of essential oils of Cymbopogon Martini and Chenopodium Ambrosioides and their synergism against dermatophytes. Mycoses 53(2):123–129

Privitera G et al (2021) Combination therapy in inflammatory bowel disease – from traditional Immunosuppressors towards the new paradigm of dual targeted therapy. Autoimmun Rev 20(6):102832

Puatanachokchai R et al (2002) Inhibitory effects of Lemon Grass (Cymbopogon Citratus, Stapf) extract on the early phase of hepatocarcinogenesis after initiation with diethylnitrosamine in male Fischer 344 rats. Cancer Lett 183(1):9–15

Rahim SM et al (2013) Protective effect of Cymbopogon Citratus on hydrogen peroxide-induced oxidative stress in the reproductive system of male rats. Syst Biol Reprod Med 59(6):329–336. https://www.tandfonline.com/doi/abs/10.3109/19396368.2013.827268 (June 30, 2024)

Rahim SM et al (2014) Hepatoprotective effect of Cymbopogon Citratus aqueous extract against hydrogen peroxide-induced liver injury in male rats. Afr J Tradit Complement Altern Med 11(2):447–451. https://www.ajol.info/index.php/ajtcam/article/view/105070 (July 1, 2024)

Ranitha M et al (2014) Comparison of chemical composition and toxicity of essential oils from lemongrass (Cymbopogon Citratus) extracted with microwave-assisted hydrodistillation (MAH) and conventional hydrodistillation (HD) methods. J Neurosurg 1:10–15. https://api.semanticscholar.org/CorpusID:94687653

Ressurreição D, Sandrine SP, Batista MT, Figueirinha A (2022) Effect of phenolic compounds from Cymbopogon Citratus (DC) Stapf. leaves on micellar solubility of cholesterol. Molecules 27(21):7338. https://www.ncbi.nlm.nih.gov/pmc/articles/PMC9655935/

Rizk et al (1995) Constituents of plants growing in Qatar Part Xxvii: flavonoids of Cymbopogon Parkeri

Rocha SF et al (2023) Revealing the protective effect of topically applied cymbopogon citratus essential oil in human skin through a contact model. Cosmetics 10(1):29. https://www.mdpi.com/2079-9284/10/1/29/htm (June 30, 2024)

Rodriguez-Tudela JL et al (2008) Clinical relevance of resistance to antifungals. Int J Antimicrob Agents 32:S111–S113. https://linkinghub.elsevier.com/retrieve/pii/S0924857908700104

Rojas-Armas JP et al (2020) The essential oil of Cymbopogon Citratus Stapt and Carvacrol: an approach of the antitumor effect on 7,12-dimethylbenz-[α]-anthracene (DMBA)-induced breast cancer in female rats. Molecules 25(14):3284. https://www.mdpi.com/1420-3049/25/14/3284

Rosete SC, Torres RG (2024) Sensory description and consumer hedonic perception of lemongrass (Cymbopogon Citratus) And Wiregrass (Eleusine Indica) as applied to Filipino confections. https://zenodo.org/doi/10.5281/zenodo.12242018

Rungqu P et al (2016) Anti-inflammatory activity of the essential oils of Cymbopogon Validus (Stapf) Stapf Ex Burtt Davy from Eastern Cape, South Africa. Asian Pac J Trop Med 9(5):426–431

Russell GEG et al (1990 Grasses of Southern Africa

Ruvinov I et al (2019) Lemongrass extract possesses potent anticancer activity against human colon cancers, inhibits tumorigenesis, enhances efficacy of FOLFOX, and reduces its adverse effects. Integr Cancer Ther 18. https://journals.sagepub.com/doi/full/10.1177/1534735419889150 (July 1, 2024)

Saddiq AA, Khayyat SA (2010) Chemical and antimicrobial studies of monoterpene: Citral. Pestic Biochem Physiol 98(1):89–93. https://linkinghub.elsevier.com/retrieve/pii/S0048357510000829

Saenthaweesuk S et al (2017) Hepatoprotective and antioxidant effects of Cymbopogon Citratus stapf (lemon grass) extract in paracetamolinduced hepatotoxicity in rats. Trop J Pharmaceut Res 16(1):101–107. https://www.ajol.info/index.php/tjpr/article/view/150951 (June 30, 2024)

Sagradas J et al (2015) Gastroprotective effect of Cymbopogon Citratus infusion on acute ethanol-induced gastric lesions in rats. J Ethnopharmacol 173:134–138

Sambo M, Jakwa AG, Garba IH (2024) Antioxidant potential, phytochemicals, minerals and amino acids content of lemongrass (Cymbopogan Citratus). J Pharmacogn Phytochem 13(3):103. https://doi.org/10.22271/phyto.2024.v13.i3b.14950 (July 1, 2024)

Santoro GF et al (2007) Anti-proliferative effect of the essential oil of Cymbopogon Citratus (DC) Stapf (Lemongrass) on intracellular amastigotes, bloodstream trypomastigotes and culture epimastigotes of Trypanosoma Cruzi (Protozoa: Kinetoplastida). Parasitology 134(11):1649–1656.

https://www.cambridge.org/core/journals/parasitology/article/abs/antiproliferative-effect-of-the-essential-oil-of-cymbopogon-citratus-dc-stapf-lemongrass-on-intracellular-amastigotes-bloodstream trypomastigotes-and-culture-epimastigotes-of-trypanosoma-cruzi-protozoa-kinetoplastida/3DF8FFD32DBA77DB49076336AF77700F (June 30, 2024)

Scheppach W (1994) Effects of short chain fatty acids on gut morphology and function. Gut 35(1 Suppl):S35–S38. http://www.ncbi.nlm.nih.gov/pubmed/8125387

Schweitzer B et al (2022) Antibacterial effect of lemongrass (Cymbopogon Citratus) against the aetiological agents of pitted keratolyis. Molecules 27(4):1423. https://www.mdpi.com/1420-3049/27/4/1423/htm (June 30, 2024)

Shah G et al (2011) Scientific basis for the therapeutic use of Cymbopogon Citratus, Stapf (Lemon Grass). J Adv Pharmaceut Technol Res 2(1):3–8. https://journals.lww.com/japtr/fulltext/2011/02010/scientific_basis_for_the_therapeutic_use_of.3.aspx (June 30, 2024)

Shahzadi MP (2017) Lemon grass (Cymbopogon Citratus). In: Grasses – benefits, diversities and functional roles. IntechOpen. https://www.intechopen.com/chapters/55939

Sharma PR et al (2009) Anticancer activity of an essential oil from Cymbopogon Flexuosus. Chem Biol Interact 179(2–3):160–168

Siddique YH, Naz F, Mantasha I, Shahid M (2021) Lemongrass extract alleviates oxidative stress and delayed the loss of climbing ability in transgenic Drosophila model of Parkinson's disease. Lett Drug Des Discov 18(10):987–997

Silva H, Bárbara R, Silva H, Bárbara R (2022) Exploring the anti-hypertensive potential of lemongrass – a comprehensive review. Biology 11(10):1382. https://www.mdpi.com/2079-7737/11/10/1382/htm (June 30, 2024)

Smolders I et al (1997) Effectiveness of Vigabatrin against focally evoked pilocarpine-induced seizures and concomitant changes in extracellular hippocampal and cerebellar glutamate, gamma-aminobutyric acid and dopamine levels, a microdialysis-electrocorticography study in Freel. J Pharmacol Exp Ther 283(3):1239–1248. http://www.ncbi.nlm.nih.gov/pubmed/9399999

Soares MO et al (2013) Evaluation of antioxidant and antimicrobial properties of the Angolan Cymbopogon Citratus essential oil with a view to its utilization as food biopreservative. J Agric Sci 5(7):p36. http://www.ccsenet.org/journal/index.php/jas/article/view/26302

Sonker N, Pandey AK, Singh P (2015) Efficiency of Artemisia Nilagirica (Clarke) Pamp. Essential oil as a mycotoxicant against postharvest mycobiota of table grapes. J Sci Food Agric 95(9):1932–1939. https://onlinelibrary.wiley.com/doi/full/10.1002/jsfa.6901 (June 30, 2024)

Sousa R, Figueirinha A, Batista MT, Pina ME (2021) Formulation effects in the antioxidant activity of extract from the leaves of Cymbopogon Citratus (DC) Stapf. Molecules 26(15):4518. https://www.mdpi.com/1420-3049/26/15/4518/htm (June 30, 2024)

Subramaniam G, Yew XY, Sivasamugham LA (2020) Antibacterial activity of Cymbopogon Citratus against clinically important bacteria. S Afr J Chem Eng 34:26–30

Sun Q et al (2016) Cinnamaldehyde inhibits fungal growth and aflatoxin B1 biosynthesis by modulating the oxidative stress response of Aspergillus Flavus. Appl Microbiol Biotechnol 100(3):1355–1364. http://link.springer.com/10.1007/s00253-015-7159-z

Takaisi-Kikuni NB, Krüger D, Gnann W, Wecke J (1996) Microcalorimetric and electron microscopic investigation on the effects of essential oil from Cymbopogon Densiflorus on Staphylococcus Aureus. Microbios 88(354):55–62. http://www.ncbi.nlm.nih.gov/pubmed/9121380

Takaisi-Kikuni NB, Tshilanda D, Babady B (2000) Antibacterial activity of the essential oil of Cymbopogon Densiflorus. Fitoterapia 71(1):69–71

Tapia A et al (2007) Free radical scavengers from Cymbopogon Citratus (DC.) Stapf plants cultivated in bioreactors by the temporary immersion (TIS) principle. Zeitschrift fur Naturforschung Sect C J Biosci 62(5–6):447–457. https://www.degruyter.com/document/doi/10.1515/znc-2007-5-620/html (July 1, 2024)

Temitayo GI et al (2020) Inhibitory potentials of Cymbopogon Citratus oil against aluminium-induced behavioral deficits and neuropathology in rats. Anat Cell Biol 53(3):342–354. https://synapse.koreamed.org/articles/1145495 (July 1, 2024)

The National Academies Press (2024). https://nap.nationalacademies.org/

Thompson EJ, Forster PI (2021) Cymbopogon Procerus (R. Br.) Domin, the correct name for Schizachyrium Mitchelliana BK Simon (Poacaeae: Andropogoneae), and Lectotypification of Andropogon Exaltatus R. Br. Austrobaileya 11:41–44

Tibenda JJ, Yi Q, Wang X, Zhao Q (2022) Review of phytomedicine, phytochemistry, ethnopharmacology, toxicology, and pharmacological activities of Cymbopogon genus. Front Pharmacol 13:997918. https://www.ncbi.nlm.nih.gov/pmc/articles/PMC9465289/

Tiwari M, Dwivedi UN, Kakkar P (2010) Suppression of oxidative stress and pro-inflammatory mediators by Cymbopogon Citratus D. Stapf extract in lipopolysaccharide stimulated murine alveolar macrophages. Food Chem Toxicol 48(10):2913–2919. https://linkinghub.elsevier.com/retrieve/pii/S0278691510004850

Trowell H, Burkitt D, Heaton K (1985) Dietary fibre, fibre-depleted foods and disease. https://www.cabidigitallibrary.org/doi/full/10.5555/19862030935

Uchida NS et al (2017) Hepatoprotective effect of citral on acetaminophen-induced liver toxicity in mice. Evid Based Complemen Altern Med 2017(1):1796209. https://onlinelibrary.wiley.com/doi/full/10.1155/2017/1796209 (July 1, 2024)

Ullah N, Khan MA, Khan T, Ahmad W (2013) Cymbopogon Citratus protects against the renal injury induced by toxic doses of aminoglycosides in rabbits. Indian J Pharm Sci 75(2):241–246

Uraku AJ et al (2015) Nutritional and anti-nutritional quantification assessment of Cymbopopgon Citratus leaf. Pharmacol Pharm 6(8):401–410

Venzon L et al (2018) Essential oil of Cymbopogon Citratus (lemongrass) and geraniol, but not Citral, promote gastric healing activity in mice. Biomed Pharmacother 98:118–124

Viana GSB, Vale TG, Pinho RSN, Matos FJA (2000) Antinociceptive effect of the essential oil from Cymbopogon Citratus in mice. J Ethnopharmacol 70(3):323–327

Villalobos MC, Nicolas MG, Trinidad TP (2021) Antihyperglycemic and cholesterol-lowering potential of dietary fibre from lemongrass (Cymbopogon Citratus Stapf.). Mediterr J Nutr Metab 14(4):453–467. https://www.medra.org/servlet/aliasResolver?alias=iospress&doi=10.3233/MNM-210568

Wu Y et al (2021) Preventive effect of Gonggan (Citrus Reticulata Blanco Var. Gonggan) peel extract on ethanol/HCl-induced gastric injury in mice via an anti-oxidative mechanism. Front Pharmacol 12:715306. www.frontiersin.org (July 1, 2024)

Zhang JS et al (2011) Repellent constituents of essential oil of Cymbopogon Distans aerial parts against two stored-product insects. J Agric Food Chem 59(18):9910–9915

# Medicinal and Nutritional Importance of *Ziziphus mauritiana* in Human Health

12

Md. Khokon Miah Akanda, Sanzia Mehjabin, Nazia Hoque, A. H. M. Nazmul Hasan, and G. M. Masud Parvez

**Abstract**

Medicinal plants are precious gifts from nature that may provide humanity with both food and medicine. *Ziziphus mauritiana* (Rhamnaceae) is a seasonal fruit popularly known as "BAU Kul" in Bangladesh or Indian jujube. In traditional medicine, it treats anxiety, fever, asthma, inflammation, and ulcers. *Z. mauritiana* fruits and leaves have been used to cure gonorrhea, wounds, abscesses, and diarrhea. The dried, candied, pickled, juice forms of these fruits are popular because of their nutritional, nutraceutical qualities. Vitamin C, A, antioxidants, polyphenols, and other nutrients are abundant in this plant. These substances prevent chronic diseases and represent analgesic, anti-diabetic, anti-aging, anti-inflammatory, and anti-cancer effects. This chapter reviews new research on its nutritional, therapeutic effects. Memory improvement, neurological, cardiac, and snake venom detoxification are also addressed. Clinical experiment, toxicological investigation, dose and safety profile, and plant future orientation are also covered. In conclusion, *Ziziphus mauritiana* is a therapeutic fruit worth investigating to improve healthcare.

Md. K. M. Akanda (✉) · A. H. M. Nazmul Hasan
Department of Pharmacy, University of Asia Pacific, Dhaka, Bangladesh
https://scholar.google.com/citations?user=q7lgBRsAAAAJ&hl=en

S. Mehjabin
Department of Pharmacy, Varendra University, Rajshahi, Bangladesh

N. Hoque
Department of Pharmacy, East West University, Dhaka, Bangladesh

G. M. M. Parvez
Department of Pharmaceutical and Biomedical Sciences, College of Pharmacy, University of Georgia, Athens, GA, USA



M. A. Ansari et al. (eds.), *Medicinal Plants and their Bioactive Compounds in Human Health: Volume 2*, https://doi.org/10.1007/978-981-95-3620-7_12

**Keywords**

*Ziziphus mauritiana* · Traditional medicine · Nutritional benefits · Medicinal properties · Pharmacological activities

## Abbreviations

| | |
|---|---|
| TSS | Total suspended solids |
| MAF | Most active fraction |
| COX | Cyclooxygenase |
| LOX | Lipoxygenase |
| HCT | Hematocrit |
| RBC | Red blood cell |
| IC50 | Inhibitory concentration 50 |
| SOD | Superoxide dismutase |
| MDA | Malondialdehyde |
| DPPH | 2,2-diphenyl-1-picrylhydrazyl |
| ALT | Alanine aminotransferase |
| ALP | Alkaline phosphatase |
| AST | Aspartate transaminase |
| TB | Tuberculosis |
| Th-1 | T helper |
| IFN | Interferon |

**Learning Objective**

- Identify the key medicinal and nutritional benefits of *Ziziphus mauritiana* for human health.
- Understand the traditional uses of different parts of the plant, including the fruits, leaves, and stems.
- Analyze the pharmacological activities of *Z. mauritiana*, including its anticancer, anti-inflammatory, and antidiabetic effects.
- Examine the chemical constituents of *Z. mauritiana* and their role in its medicinal properties.
- Evaluate the safety profile of *Z. mauritiana* through toxicological studies and its potential for future clinical use.

## 12.1 Introduction

Nature has endowed humans with unique medicinal herbs to build a disease-free and healthy lifestyle. In the past, most plants were altered for medicinal uses; ancient medical schools used extracts and decoctions from these plants (Parvez et al. 2016). Plants have historically been applied to cure a wide range of human ailments and conditions. Based on fossil evidence, people have been using plants as medicines for about 60,000 years (Solecki and Shanidar 1975; Afroja et al. 2017). Numerous natural products have helped to preserve and enhance human health, and one such product that has attracted attention in pharmaceutical research is the modulation of immune responses for the purpose of alleviating sickness (Anjum et al. 2024; Akanda et al. 2024a). Traditional medicine has always made use of plants, and the pharmaceutical firms have placed more emphasis on this area (Afzal et al. 2017). However, the popularity of herbal medicine worldwide is due to several reasons, especially as an alternative medication used prominently in low-income countries because of a lack of access to appropriate medical treatment and inexpensive conventional pharmaceuticals (Uzzaman et al. 2018). Furthermore, fewer adverse effects were presented with natural phytochemicals isolated from medicinal plants compared to synthetic drugs (Owolarafe et al. 2020). It is estimated that around 80% of the populations in developing nations depend upon plant-based natural products for their primary health care (Akanda et al. 2025a).

Mostly trees, shrubs, and some vines, Rhamnaceae, the Buckthorn family, is a broad family of flowering plants with about 870–900 species in approximately 50–60 genera. Rhamnaceae can be found all over the world, with the highest prevalence in subtropical and tropical regions (Afzal et al. 2017). *Ziziphus mauritiana*, a member of the Rhamnaceae family, is widely known as Ber, BAU Kul, or Indian jujube. It is a plant found in many climatic locations worldwide, utilized for nutritional and medicinal purposes (Akhtar et al. 2016, Akanda and Hsasan 2021). *Z. mauritiana* has a long history of use in traditional medicine for a wide variety of conditions, including gastrointestinal issues, parasites, respiratory issues, immune system deficiencies, and wound healing. The tree's leaves, fruits, bark, and seeds are all part of this history (Outlaw et al. 2002, Huang et al. 2008).

The extensive literature on Jujube fruit and its components' nutritional and nutraceutical properties makes it a promising target for discovering new potent bioactive compounds. Dried BAU Kul fruits have long been consumed as food, such as in confections and beverages, due to their excellent nutraceutical properties (Chiou et al. 2020). Not only it is a tasty fruit, but it also has great medicinal properties. This tropical tree species is economically important in part for its nutritious and medicinal qualities. This fruit is very healthy and likewise filled with vitamin C. A lot of the time this will certainly be pickled, raw, or eaten in juice form as well. BAU Kul is regarded as the second most nutritious fruit following guava and boasts a higher vitamin C content than any citrus fruit. Gently unripe fruits may be candied by piercing and plunging in a saline arrangement; ready natural products might be sun-dried and ground for use outside the season. Apart from being eaten raw, the fruit is also consumed candied, pickled, dried, as juice, or even as Indian jujube butter (Jha et al. 2024). In addition, a multitude of botanical components have demonstrated chemo preventive, antibacterial, antioxidant, anti-diabetic, hypolipidemic, wound healing, and hepatoprotective attributes (Akanda and Hasan 2021; Jha et al. 2024). In conclusion, the chapter comprehensively intends to discuss recent advancements regarding *Z. mauritiana* traditional medicinal use, nutritional aspects, pharmacological activities, and safety concerns.

## 12.2 Plant Description

With short spines and an evergreen habit, *Z. mauritiana* shows several drooping branches on every plant, usually ornamented with pointed stipules (Prakash et al. 2021; Rathore et al. 1992).

### 12.2.1 Stem

Often reaching heights of up to 15 m, BAU Kul is well-known for its fast development, spiky look, and remarkable lifetime. It has a spreading crown and relaxed twigs, which give it an unusual aspect (Prakash et al. 2021).

### 12.2.2 Leaves

BAU Kul has alternate ovate-oblong-elliptic leaves. Typically, they are 1.8–3.8 cm wide and 2.5–4.0 cm long. The top is rounded with three basal veins. Complete leaves have also been found (Prakash et al. 2021).

### 12.2.3 Flowers

*Z. mauritiana* bears three to five petals with white or yellow blooms. Male structures bloom before female structures in these plants (Prakash et al. 2021).

### 12.2.4 Fruits

Soft, sugary, meaty, and crisp *Z. mauritiana* fruits are edible. Their 2–5 cm round or oval fruits are red, golden brown, and white. Their apple-like smell is lovely (Prakash et al. 2021; Rathore et al. 1992).

## 12.3 Geographical Source

*Z. mauritiana* grows across tropical Africa, including Tunisia, Algeria, Kenya, Egypt, Libya, and Uganda. It is also distributed in Bangladesh, Afghanistan, Nepal, Pakistan, and India. Australia, China, and Southeast Asia, particularly Indonesia and Malaysia, have the species (Mirheidari et al. 2022). It grows throughout the

Indian subcontinent and the Paleotropics (tropical, subtropical, and warm-temperate zones) (Chiou et al. 2020).

## 12.4 Traditional Uses

Fresh *Z. mauritiana* fruit and juice are believed to improve digestion and liver function. BAU Kul juice and Chalta (*Dillenia indica*) cure fever. The leaf juice is said to improve liver function and treat jaundice, while dried mature fruit is used for wound healing. For wound healing, boils, diarrhea, dysentery, and gingivitis, bark paste is applied. Traditional pastes from the roots treat chronic fever, old wounds, ulcers, and headaches. However, the leaves treat asthma, diarrhea, abscesses, gonorrhea, swelling, and liver diseases. Overall, the fruits serve as wound healers, sedatives, anodynes, and tonics with potential anticancer activity, highlighting the diverse traditional medicinal applications of BAU Kul and its value as a natural remedy (Akanda and Hasan 2021; Prakash et al. 2021; Naaz et al. 2020; Muhammad et al. 2012). Traditional medicine uses the fruit as an analgesic, sedative, wound healer, tonic, and topical treatment for ulcers, wounds, and asthma. Various investigations have shown promising anticancer benefits from the fruit (Yahia et al. 2020; Owolarafe et al. 2020).

## 12.5 Chemical Constituents

The most major phytochemical study of jujube found carbohydrates, flavonoids, alkaloids, saponins, cardioglycosides, and steroids (Yahia et al. 2020; Akanda and Hasan 2021). It contains abundant with diverse chemicals, the main chemical ingredients of different parts of *Z. mauritiana* are summarized in Table 12.1, and the principal constituents are illustrated in Fig. 12.1.

## 12.6 Nutritional Value

The *Z. mauritiana* fruit is enriched with nutrients. In comparison to apples and mangoes, it has higher quantities of vitamin C, protein, and minerals, and it has more iron and phosphorus than oranges (Khera and Singh 1976). Its fruits may contain more iron than apples, and iron is essential for oxygen transport (Butt et al. 2021). Having an abundant supply of numerous vital vitamins, minerals, fats, proteins, and carbohydrates, including calcium, phosphorus, iron, magnesium, potassium, vitamin C, and the B-complex. The fruit is inexpensive and has a good nutritional value. The organic acids citric, succinic, malic, and oxalic contribute to the taste of Jujube. To meet the daily need for vitamin B complex, vitamin C in the diet, 1–3 *Z. mauritiana* fruits should be consumed. Good quantities of fiber found in *Z. mauritiana* contribute to the body's ability to maintain appropriate plasma cholesterol and glycemic response. Jujube pulp is a fantastic option for a functional diet

**Table 12.1** Chemical constituents of *Z. mauritiana*

| Plant parts | Phytoconstituents | Category | References |
|---|---|---|---|
| Aerial parts | 1-Hexacosanol (ceryl alcohol), protopine, berberine | Fatty alcohol<br>Alkaloid | Delfanian (2016), Prakash (2021) and Thanatcha and Pranee (2011) |
| | Quercetin, kaempferol | Flavonoid antioxidant | |
| | Sitosterol, stigmasterol | Sterol | |
| | Lanosterol | Tetracyclic triterpenoid | |
| | Diosgenin | Steroid saponin | |
| Stem | 14-membered cyclopeptide alkaloid | Cyclopeptide alkaloids | Prakash (2021) |
| Leaves | – | Cardiac glycosides, cyclopeptide alkaloids, flavonoids, glycosides, eucoanthocyanidins, mucilage, sugar, polyphenolic hydrolysable tannins, saponins, sterols, triterpenes, and triterpenoids | Ghasham (2017), Jossang (1996) and Kundu (1989) |
| Fruit | Carotene, vitamin A, vitamin C | Vitamin | Cheng (2000) and Prakash (2021) |
| | – | Fatty acids | |
| | Quercetin | Antioxidant flavonoid (Flavanol) | |
| | Calcium, phosphorus, iron, fluoride | Inorganic compound | |
| | Citric acid, maleic acid, oxalic acid | Organic compound | |
| | Carotenoids, thiamine, riboflavin, niacin, ascorbic acid | Vitamin | |
| | Pectin (fiber) | Carbohydrates, dietary fiber, fats, protein, and sugar (reducing and nonreducing) | |

| | | | |
|---|---|---|---|
| Entire plant | The p- or m-ansa 13-, 14-, or 15-membered ring structures comprise a styrylamine moiety along with two or three amino acid residues | Alkaloids | Afzal (2017), Delfanian (2016), Elaloui (2016), Kang (2015), Panseeta (2011), Prakash (2021), Suksamrarn (2006), Tan and Zhou (2006), Thanatcha and Pranee (2018) and Yahia (2020) |
| | – | Ceanothane triterpenes | |
| | – | Cardiac glycoside | |
| | – | Dietary fiber | |
| | – | Fats | |
| | – | Flavonoid | |
| | Sodium, calcium, potassium, magnesium, phosphorus, chlorine, and sulfur | Inorganic compound | |
| | – | Phenolic compound | |
| | – | Protein | |
| | Lupane | Saponins/triterpenoid | |
| | – | Tannins | |
| | Oleanolic acid, maslinic acid, alphitolic acid, and colubrinic acid | Terpenoids | |

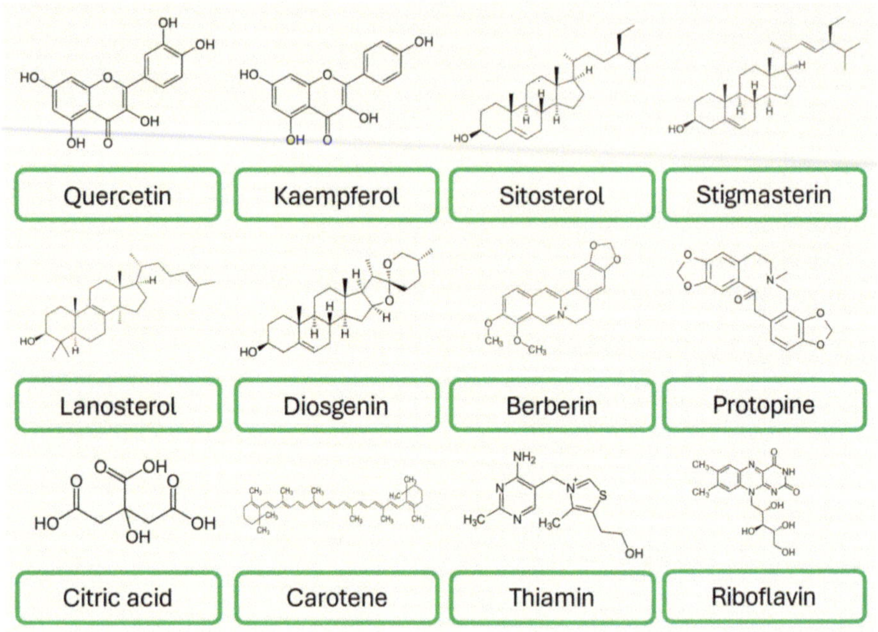


**Fig. 12.1** Principal chemical constituents of *Z. mauritiana*

because it is also high in phytonutrients. Fruit has an acidity of 0.12–1.5% and a total suspended solids (TSS) concentration of 12–20% (Muchuweti et al. 2005). Figure 12.2 shows the nutritional value of *Z. mauritiana* (Hussain et al. 2021).

## 12.7 Medicinal Importance

### 12.7.1 Antiallergic Activity

The most active fraction (MAF) derived from BAU Kul has shown impressive anti-inflammatory and allergy properties. MAF is a natural substance with no negative effects on the body (Talmale et al. 2015). The complement system has been significantly inhibited by the MAF and has also been proven to have inhibitory effects on Cyclooxygenase-1 (COX-1), COX-2, and 5-lipoxygenase (5-LOX) enzymes (Parvez and Akanda 2019). Its more potent suppression of COX-2 over COX-1 is especially notable, and this is advantageous because medications that specifically block COX-2 are linked to a lower risk of stomach injuries. Important components linked to allergy and inflammatory reactions are potently inhibited by the MAF (Hong et al. 2002).

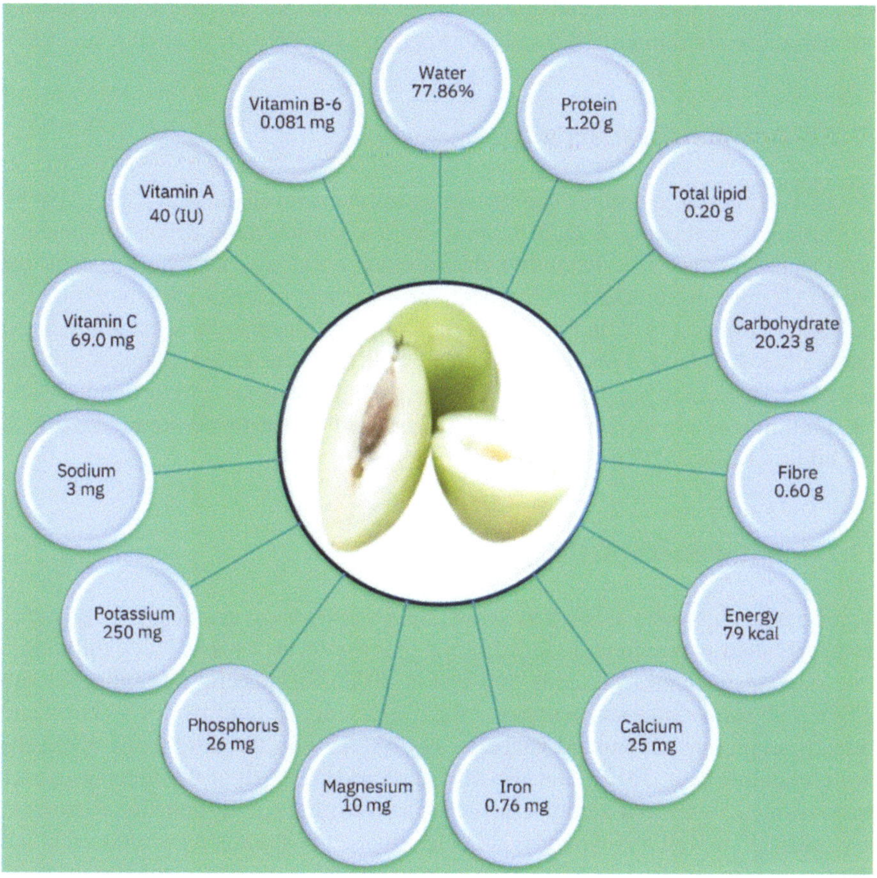


**Fig. 12.2** Nutritional value of *Z. mauritiana*

## 12.7.2 Anticancer and Antitumor Activity

*Z. mauritiana* leaf chloroform extract was the most effective anticancer drug against HCT-116 (IC50 = 18.32 μg/mL) and U937 (IC50 = 27.78 μg/mL) cells. Hexane extract had IC50 values of 48.91 μg/mL and 55.09 μg/mL against HCT-116 and U937 cell growth, respectively (Ashraf et al. 2015). The aqueous-ethanolic seed extract led to a notable decrease in tumor volume and viable cells in Swiss albino mice, enhancing hemoglobin levels, red blood cell (RBC) count, survival duration, and tumor suppression and proportion of Ehrlich ascites carcinoma-infected mice. Reduced lipid peroxidation and elevated levels of glutathione, catalase, and superoxide dismutase indicated enhanced antioxidant status in mice treated with the extract (Mishra et al. 2011). BAU Kul leaf extracts in methanol, chloroform, and hexane were tested for their ability to suppress the plant pathogen *Agrobacterium tumefaciens* using a bench-top potato disc assay. The study investigated the maximum tumor inhibition and tumor number in methanol, chloroform, and hexane extracts at concentrations of 0.8 and 0.025 mg/mL, respectively (Ashraf et al. 2015).

### 12.7.3 Antidiarrheal Activity

The isolated rabbit jejunum's spontaneous pendular movement was suppressed by the methanolic extract of *Z. mauritiana* roots in a concentration-dependent manner, and the rat ileum's acetylcholine-induced contraction was also hindered by the extract (Dahiru et al. 2006). Significant anti-diarrheal efficacy was shown by the methanolic extracts of *Z. mauritiana*; at 400 mg/kg, the stem bark extract reduced the number of defective pellets by 75.68%, while the seed extract showed a reduction of 62.16% at the same dose. The presence of advantageous phytochemicals and the inhibition of prostaglandin synthesis generated by castor oil are credited with this effect (Akanda and Hasan 2021).

### 12.7.4 Antidiabetic Activity

Significant antidiabetic action was shown by *Z. mauritiana*; at a dose of 400 mg/kg b.wt., the stem bark extract reduced blood glucose levels by 44.27%, exhibiting a dose-dependent impact (Akanda and Hasan 2021). In diabetic rats induced with alloxan, *Z. mauritiana* fruit extracts effectively reduced both hyperlipidemia and hypoglycemia that accompany diabetes. 400 mg/kg was the dose that worked the best (Abubakar et al. 2018). The non-polysaccharide fraction of the aqueous extract demonstrated significant antihyperglycemic and hypoglycemic benefits. Petroleum ether extract showed antihyperglycemic activity, while seed extracts reduced blood glucose in alloxan diabetic mice after sub-acute and acute treatment (Jarald et al. 2009; Bhatia and Mishra 2010; Mehjabin et al. 2025).

### 12.7.5 Anti-inflammatory and Analgesic Activity

*Ziziphus mauritiana* methanolic extracts significantly reduced mice's writhing in the acetic acid-induced writhing test, suggesting anti-inflammatory action. When compared to the seed extract, the stem bark extract had superior analgesic effects, and both extracts had higher writhing inhibition than the prescribed dosage of diclofenac sodium (Akanda and Hasan 2021). Furthermore, the anti-inflammatory qualities of *Ziziphus mauritiana* roots have been studied; results indicate that these roots contain substances that can block inflammatory pathways, reducing pain and inflammation (Talmale et al. 2015).

### 12.7.6 Antimicrobial Activity

When *Z. mauritiana* was tested for antimicrobial activity, it was found that only the stem bark extract had bactericidal activity against gram-positive and gram-negative bacteria, including *S. typhi* and *Escherichia coli*, as well as *Staphylococcus aureus and Bacillus megaterium* (Mosaddik et al. 2024). However, when tested against the identified bacteria, the seed extract exhibited no antibiotic action (Akanda and Hasan 2021; Akanda et al. 2024b). The agar well-diffusion procedure can be used to test the possible antibacterial effect of BAU Kul's methanolic extract. *Pseudomonas aeruginosa* is the organism that is most poisonous to it, with *S. paratyphi*, *S. aureus*, *E. coli*, and *C. violaceum* following (in that order) (Tanvir et al. 2015). The ethanol extracts showed antibacterial activity, inhibiting Trichophyton spp., *S. dysenteriae*, and *S. aureus*, with a strong minimum inhibitory concentration (15.625 μg/mL) using the disc diffusion method (Karon et al. 2011; Mm et al. 2018). Six pathogenic bacteria were tested against extracts of BAU Kul leaves, *Pasteurella multocida*, *S. aureus*, *Aspergillus niger*, *Fusarium solani*, *B. subtilis*. An analysis of variance revealed that the extracts had considerably (Ashraf et al. 2015; Mbahi et al. 2018).

### 12.7.7 Antioxidant Activity

*Ziziphus mauritiana* exhibited strong antioxidant activity, with the stem bark extract showing higher total phenol (69.591 mg/g) and flavonoid content (33.000 mg/g) than the seed extract. Both extracts demonstrated remarkable free radical scavenging activity, with IC50 values of 4.13 μg/mL for the stem bark and 4.53 μg/mL for the seed extract (Akanda and Hasan 2021). A different investigation showed that rats exposed to dichloromethane exhibited increased levels of superoxide dismutase (SOD) and T-AOC, alongside decreased levels of malondialdehyde (MDA) in their plasma (Afzal et al. 2017). Methanol, chloroform, and hexane were tested for 2,2-diphenyl-1-picrylhydrazyl (DPPH) free radical scavenging in BAU Kul leaf extract. A dose-dependent increase in extract DPPH free radical scavenging activity from 0.025 to 0.8 mg/mL was observed (Ashraf et al. 2015).

### 12.7.8 Antiulcer Activity

According to one study, 100% ethanol caused linear and distributed dot-like hemorrhagic lesions in the stomach glands. Over time, these lesions fused to produce enormous hemorrhagic areas (Mehjabin et al. 2024a). But when *Z. mauritiana* stem bark methanolic extract was applied in different amounts, it dramatically decreased the quantity and duration of these ethanol-induced stomach ulcers. This finding suggests that the extract may have anti-ulcer properties. Subsequent analysis found that the methanolic extract significantly decreases the incidence of ulcers in mice. It reduced ulcerated area by 8, 66.67, and 82.67% at a dose of 100, 250, and 500 mg/

kg (Akanda and Mehjabin 2024). Furthermore, the extract had strong anti-ulcer effects in an aspirin-induced stomach ulcer model, especially at 250 and 500 mg/kg. Control mice' stomach tissue showed necrotic penetrations, congestion, and altered mucosal surface integrity under the microscope. Nonetheless, mice administered at 100, 250, and 500 mg/kg of the methanolic extract had intact epithelium devoid of necrosis damage or oedema. The study found that the methanolic extract of *Z. mauritiana* stem bark contains phytochemicals that may be beneficial in treating ulcers and lesions by maintaining stomach epithelium (Siddharth et al. 2010; Akter et al. 2020; Khushtar et al. 2016; Tarique et al. 2016).

### 12.7.9 Cytotoxicity Activity

In the brine shrimp lethality test, *Z. mauritiana* seed extract displayed high cytotoxicity (LC50 = 1.46 μg/mL), but stem bark extract had lower lethality comparing to vincristine sulfate (Akanda and Hasan 2021). The lethality test on brine shrimps was used to assess the root portion. At 1000 μg/mL, the dichloromethane extracts demonstrated 70% mortality (Afzal et al. 2017). Comparing the crude extractives to the conventional vincristine sulphate, an ex-vivo brine shrimp lethality test indicated modest to moderate cytotoxic effect (Karon et al. 2011; Akanda et al. 2025b).

### 12.7.10 Dengue Fever Treatment

People contract dengue fever by mosquito bites from infected Aedes species, such as *Aedes aegypti*. The study revealed that the mortality rate of Aedes aegypti larvae increased with the concentration of *Z. mauritiana* leaf extract and the duration of exposure, indicating its potential in controlling the mosquito population. While *Z. mauritiana* leaf extract has shown promise as a larvicide, its direct effectiveness in treating dengue fever in humans (Askur et al. 2021). Another study showed that five different fractions of *Z. mauritiana* showed inhibitory activities against dengue virus (Batool et al. 2018).

### 12.7.11 Hepatoprotective Activity

After receiving a 100 mg/kg dose of aqueous *Z. mauritiana* extract, rats had considerably lower pre-treatment alanine aminotransferase (ALT), alkaline phosphatase (ALP), aspartate transaminase (AST), and tuberculosis (TB) than those given alcohol alone. Before the extract was added, glutathione peroxidase, catalase, reductase, and superoxide dismutase activities increased. These adjustments may reduce alcohol-induced liver damage. The BAU Kul aqueous extract also reduced alcohol-induced morphological alterations. The phytoconstituents in the extract, especially phenolic compounds, tannins, and saponins, may reduce signs of lipid peroxidation and tissue injury. The findings imply that *Z. mauritiana* aqueous leaf extract may reduce alcohol induced liver damage due to its phytoconstituents (Dahiru et al. 2005; Prakash et al. 2021; Dahiru and Obidoa 2007).

### 12.7.12 Immunomodulatory Activity

*Z. mauritiana* seed extract increased humoral (sheep red blood cell) immunological response and Th-1-mediated IFN-gamma but decreased Th-2-mediated IL-4 (Mishra and Bhatia 2010). Dichloromethane extract from the root had a 50% inhibitory concentration of 55.43, demonstrating a substantial immunomodulatory effect (Afzal et al. 2017).

### 12.7.13 Wound Healing Activity

The ethanolic extract from BAU Kul leaves was tested for wound healing in Wistar albino rats using excision, incision, and dead space wound models for ten days. Hydroxyproline was measured using the dead space wound model. The ethanolic extract promoted wound contraction, shortened epithelization, strengthened skin, and increased dry granulation tissue weight. The extract also raised hydroxyproline, indicating improved collagen deposition in the wound. Better collagen synthesis, wound closure, and tissue regeneration showed that *Z. mauritiana* leaf ethanolic extract improves wound healing (Sumanth and Bhargavi 2014; Tahir et al. 2022; Saidurrahman et al. 2022).

## 12.8 Toxicological Studies

In mice, *Z. mauritiana* extracts had no adverse effects or mortality at 4000 mg/kg body weight, suggesting they are safe. Acute oral toxicity testing showed no toxins in seed and stem bark extracts (Akanda and Hasan 2021). Additional acute toxicity testing shows that the extract has an LD50 above 5000 mg/kg body weight, indicating non-toxicity (Mehjabin et al. 2024b). A subchronic toxicological investigation showed liver and renal function abnormalities but no hematological negative effects, save for white blood cell count (Owolarafe et al. 2022). Sub-acute liver toxicity research shows leaf extract toxicity in mice's livers, altering cell shape at 600 and 700 mg/kg BW, but no adverse reactions were observed in allergy testing (Ma'ruf et al. 2023).

## 12.9 Future Perspective and Challenges

*Z. mauritiana*'s ethnomedicinal use stresses its significance in traditional medicine and possible therapeutic characteristics, including anti-inflammatory, antidiabetic, antioxidant, and antibacterial. Phytochemical analysis discovered phenolic chemicals, alkaloids, and flavonoids that support its therapeutic properties. Despite the abundance of *Z. mauritiana* information, many topics remain unexplored. Future research should focus on rigorous scientific inquiries to support its traditional uses and explain its operation. It is also crucial to identify and isolate its bioactive components that cause pharmacological actions. Explore *Z. mauritiana*'s pharmacological research and development potential to create standardized herbal treatments or novel medications. Traditional knowledge combined with bioassay-guided fractionation and pharmacological screens may identify new therapeutic compounds and promote evidence-based herbal therapy. The fruit's numerous vitamins, minerals, and antioxidants suggest usefulness in dietary supplements and functional foods, although clinical studies are lacking. Scientifically studying its benefits may uncover new approaches to improve diet and health.

## 12.10 Conclusion

The chapter offers significant insights into the ethnomedicinal use of *Z. mauritiana* and underscores its potential as a source of natural treatments. These findings highlight the necessity of safeguarding traditional knowledge and pursuing further study to fully realize this extraordinary plant's therapeutic potential. BAU Kul has the potential to significantly enhance unique and effective healthcare solutions by integrating traditional medicine with modern science. By including it in one's diet or herbal medicine, *Z. mauritiana* can become an excellent addition to general health. Further investigation is necessary to thoroughly investigate this compound's therapeutic potential and tailor it for various clinical conditions.

## 12.11 Summary

*Z. mauritiana* offers significant therapeutic benefits and is rich in vitamins, minerals, and antioxidants. The plant is used to treat various ailments like wounds, digestive issues, and inflammation in traditional medicine. The plant contains important phytochemicals such as flavonoids, alkaloids, and saponins. It demonstrates different pharmacological activities like anticancer, antidiabetic, anti-inflammatory, and antioxidant properties. *Z. mauritiana* has shown a good safety profile with no major toxicity observed in animal studies.

## References

Abubakar SM, Umar SA, Alexander I, Abubakar N, Abdulazeez MA, Sule MS (2018) Evaluation of hypoglycaemic, hypolipidaemic and non-toxic effect of hydro-methanolic extracts of *Ziziphus mauritiana, ziziphus spina christi* fruit and glibenclamide on alloxan induced diabetic rats. J Drug Deliv Ther 8(3):82–92

Afroja S, Falgunee FN, Jahan MM, Akanda KM, Mehjabin S, Parvez GMM (2017) Antibacterial activity of different citrus fruits. Spec J Med Res Health Sci 2(1):25–32

Afzal S, Batool M, Ch BA, Ahmad A, Uzair M, Afzal K (2017) Immunomodulatory, cytotoxicity, and antioxidant activities of roots of *Ziziphus mauritiana*. Pharmacogn Mag 13(Suppl 2):S262–S265

Akanda MKM, Hasan AHMN (2021) Characterization of pharmacological properties of methanolic seed and stem bark extracts of *Ziziphus mauritiana* (BAU Kul) using in-vitro and in-vivo animal (Swiss albino male mice) model. Clin Phytoscience 7(1):1–13

Akanda MKM, Mehjabin S (2024) Prebiotics and synbiotics in management of metabolic syndrome and chronic metabolic disease. In: Applications of synbiotics in metabolic disorder management. CRC Press, Taylor & Francis, pp 39–54

Akanda MKM, Mehjabin S, Parvez GMM (2024a) Honey for nutrition and health benefits: an overview. In: Kumar R, Hajam YA, Bala Dhull S, Giri A (eds) Honey in food science and physiology. Springer Nature Singapore, Singapore, pp 33–56

Akanda MKM, Sultana R, Rana MM, Hossain MA, Barik SMA, Islam MR, Islam MN, Parvez GMM (2024b) Prevalence, causes, and management strategies of fungal diseases in northern regions of Bangladesh. Sciphar 3(1):24–34

Akanda MKM, Hasan AHMN, Mehjabin S, Parvez GMM, Yasmin S, Akhtar MS et al (2025a) *Carica papaya* in health and disease: a review of its bioactive compounds for treating various disease conditions, including anti-inflammatory and anti-arthritic activities. Inflammopharmacology 33(5):1–34

Akanda MKM, Mehjabin S, Mosaddik A (2025b) Quinones in the treatment of cardiovascular diseases. In: Dar UA, Shahnawaz M, Hakeem KR (eds) Quinone-based compounds in drug discovery. Elsevier, pp 103–117

Akhtar N, Ijaz S, Khan HMS, Uzair B, Reich A, Khan BA (2016) *Ziziphus mauritiana* leaf extract emulsion for skin rejuvenation. Trop J Pharm Res 15(5):929–936

Akter MR, Akanda MKM, Islam MS, Rana MM (2020) Assessment of drug utilization pattern among outpatients in the orthopedic department of several private clinics in Rajshahi city, Bangladesh. Int J Pharm Sci Res 5(4):13–16

Anjum R, Hasan AHMN, Al Asma Hawa M, Miah Akanda MK, Binte Wahed T, Abdus Salam FB, Rabiul Islam MRI, Shahriar M (2024) Characterization of pharmacological properties and isolation of two bioactive compounds (ursolic acid and palmitoleic acid) from the stem bark extract of *Lannea coromandelica*. Trends Phytochem Res 2(2):106

Ashraf A, Sarfaraz RA, Sarfaraz RA, Anwar F, Shahid SA (2015) Chemical composition and biological activities of leaves of *Ziziphus mauritiana* l native to Pakistan. Pak J Bot 47(1):367–376

Askur A, Adiningsih R, Ganning A (2021) Utilization of bidara leaf (*Ziziphus mauritiana* l.) extract as a natural larvacide. Urban Health 3(1)

Batool R, Aziz E, Mahmood T, Tan BK, Chow VT (2018) Inhibitory activities of extracts of *Rumex dentatus*, *Commelina benghalensis*, *Ajuga bracteosa*, *Ziziphus mauritiana* as well as their compounds of gallic acid and emodin against dengue virus. Asian Pac J Trop Med 11(4):265

Bhatia A, Mishra T (2010) Hypoglycemic activity of *Ziziphus mauritiana* aqueous ethanol seed extract in alloxan-induced diabetic mice. Pharm Biol 48(6):604–610

Butt S, Hussain S, Munawar K (2021) Phytochemistry of *Ziziphus mauritiana*: an overview of its nutritional and pharmaceutical potential. Sci Inq Rev 5(2):1–15

Cheng G, Bai Y, Zhao Y, Tao J, Liu Y, Tu G, Ma L, Liao N, Xu X (2000) Flavonoids from *Ziziphus jujuba Mill var. spinosa*. Tetrahedron 56(45):8915–8920

Chiou C-Y, Shih H-C, Tsai C-C, Jin X-L, Ko Y-Z, Mantiquilla JA, Weng I-S, Chiang Y-C (2020) The genetic relationships of Indian jujube (*Ziziphus mauritiana* lam.) cultivars using SSR markers. Heliyon 6(10):e05078

Dahiru D, Obidoa O (2007) Pretreatment of albino rats with aqueous leaf extract of *Ziziphus mauritiana* protects against alcohol-induced liver damage. Trop J Pharm Res 6(2):705–710

Dahiru D, William ET, Nadro MS (2005) Protective effect of *Ziziphus mauritiana* leaf extract on carbon tetrachloride-induced liver injury. Afr J Biotechnol 4(10):1177–1179

Dahiru D, Sini J, John-Africa L (2006) Antidiarrhoeal activity of *Ziziphus mauritiana* root extract in rodents. Afr J Biotechnol 5(10):941–945

Delfanian M, Esmaeilzadeh Kenari R, Sahari MA (2016) Utilization of jujube fruit (*Ziziphus mauritiana* Lam.) extracts as natural antioxidants in stability of frying oil. Int J Food Prop 19(4):789–801

Elaloui M, Laamouri A, Ennajah A, Cerny M, Mathieu C, Vilarem G, Chaar H, Hasnaoui B (2016) *Phytoconstituents* of leaf extracts of *Ziziphus jujuba* Mill. plants harvested in Tunisia. Ind Crop Prod 83:133–139

Ghasham AA, Muzaini MA, Qureshi KA, Osman G, Khan RA, Farhana SA, Hashmi S, El-Agamy E, Abdallah WE (2017) Phytochemical screening, antioxidant and antimicrobial activities of methanolic extract of *Ziziphus mauritiana* lam. Leaves collected from Unaizah, Saudi Arabia. Int J Pharmaceut Res Allied Sci 6(3):33–46

Hong CH, Hur SK, Oh O-J, Kim SS, Nam KA, Lee SK (2002) Evaluation of natural products on inhibition of inducible cyclooxygenase (COX-2) and nitric oxide synthase (iNOS) in cultured mouse macrophage cells. J Ethnopharmacol 83(1):153–159

Huang Y-L, Yen G-C, Sheu F, Chau C-F (2008) Effects of water-soluble carbohydrate concentrate from Chinese jujube on different intestinal and fecal indices. J Agric Food Chem 56(5):1734–1739

Hussain SZ, Naseer B, Qadri T, Fatima T, Bhat TA (2021) Ber/jujube (*Ziziphus mauritiana*): morphology, taxonomy, composition and health benefits. In: Hussain SZ, Naseer B, Qadri T, Fatima T, Bhat TA (eds) Fruits grown in highland regions of the Himalayas: nutritional and health benefits. Springer, Cham, pp 157–168

Jarald EE, Joshi SB, Jain DC (2009) Antidiabetic activity of extracts and fraction of *Zizyphus mauritiana*. Pharm Biol 47(4):328–334

Jha D, Hangargekar P, Akbar M, Parihar AS, Kashyap S, Joshi A, Rahman MA (2024) *Ziziphus mauritiana*: an in-depth review of its medicinal attributes and pharmacological activities. Intell Pharm 2(2):274–283

Jossang A, Zahir A, Diakite D (1996) Mauritine J, a cyclopeptide alkaloid from *Zizyphus mauritiana*. Phytochemistry 42(2):565–567

Kang KB, Ming G, Kim GJ, Ha T-K-Q, Choi H, Oh WK, Sung SH (2015) Jubanines F–J, cyclopeptide alkaloids from the roots of *Ziziphus jujuba*. Phytochemistry 119:90–95

Karon B, Ibrahim I, Mahmood A, Huq M, Chowdhury M, Rashid M (2011) Preliminary antimicrobial, cytotoxic and chemical investigations of *Averrhoa bilimbi Linn.* and *Zizyphus mauritiana* Lam. Bangladesh. Pharm J 14(2):127–131

Khera AP, Singh JP (1976) Chemical composition of some ber cultivars (*Ziziphus mauritiana* L.). Haryana J Hortic Sci 5(1):21–24

Khushtar M, Siddiqui HH, Dixit RK, Khan MS, Iqbal D, Rahman MA (2016) Amelioration of gastric ulcers using a hydro-alcoholic extract of Triphala in indomethacin-induced Wistar rats. Eur J Integr Med 8(4):546–551

Kundu AB, Barik BR, Mondal DN, Dey AK, Banerji A (1989) Zizyberanalic acid, a pentacyclic triterpenoid of Zizyphus jujuba. Phytochemistry 28(11):3155–3158

Ma'ruf MT, Dewi PS, Poernomo H (2023) Sub-acute toxicity and allergy studies of Bidara leaf (*Ziziphus mauritiana*) extract as a wound healing material. Odonto Dent J 10(2):194–201

Mbahi MA, Mbahi MA, Amos PI (2018) Phytochemical screening and anti-diarrheal activity of the pulp and seed extracts of *Ziziphus mauritiana* in rats. Int J Biochem Res Rev:1–9

Mehjabin S, Akanda MK, Hasan AHMN, Parvez GMM (2024a) Present status and future prospectives to control metabolic disorders. In: Applications of synbiotics in metabolic disorder management. CRC Press, Taylor & Francis, pp 245–256

Mehjabin S, Akanda MKM, Hoque N, Hasan AHMN, Parvez GMM, Mosaddik A (2024b) Flavonoid intake and risk of toxicity in chronic metabolic diseases. In: Role of flavonoids in chronic metabolic diseases: from bench to clinic. Wiley, pp 511–534

Mehjabin S, Akanda MKM, Hasan AHMN, Parvez GMM (2025) Role of inulin in the management of metabolic disorders. In: Akram W, Mishra N, Haider T (eds) Inulin for pharmaceutical applications. Springer, Singapore, pp 189–214

Mirheidari F, Khadivi A, Saeidifar A, Moradi Y (2022) Selection of superior genotypes of Indian jujube (*Ziziphus mauritiana* Lamk.) as revealed by fruit-related traits. Food Sci Nutr 10(3):903–913

Mishra T, Bhatia A (2010) Augmentation of expression of immunocytes' functions by seed extract of *Ziziphus mauritiana* (Lamk.). J Ethnopharmacol 127(2):341–345

Mishra T, Khullar M, Bhatia A (2011) Anticancer potential of aqueous ethanol seed extract of *Ziziphus mauritiana* against cancer cell lines and Ehrlich ascites carcinoma. Evid Based Complement Alternat Med 1:765029

Mm A, Am M, Ia U, Da A, Pi A (2018) Phytochemical screening and antimicrobial activity of the pulp extract and fractions of *Ziziphus mauritiana*. Biochem Anal Biochem 7(2):1–6

Mosaddik A, Akanda MKM, Pal TR, Elaissari A (2024) Advances in nano-biomaterials for effective antimicrobial therapy. In: Pradhan M, Yadav K, Singh Chauhan N (eds) Biomaterial-inspired nanomedicines for targeted therapies. Springer, Singapore, pp 425–450

Muchuweti M, Zenda G, Ndhlala AR, Kasiyamhuru A (2005) Sugars, organic acid and phenolic compounds of *Ziziphus mauritiana* fruit. Eur Food Res Technol 221(3):570–574

Muhammad N, Saeed M, Khan H (2012) Antipyretic, analgesic and anti-inflammatory activity of Viola betonicifolia whole plant. BMC Complement Altern Med 12(1):59

Naaz F, Agari N, Singh A (2020) Medicinal properties of *Ziziphus mauritiana*: a review article. Int J Pharm Res 19(1):1–13

Outlaw WH, Zhang S, Riddle KA, Womble AK, Anderson LC, Outlaw WM, Outlaw NN, Outlaw EC, Thistle AB (2002) The jujube (*Ziziphus jujuba* Mill.), a multipurpose plant. Ebot 56(2):198–200

Owolarafe TA, Salawu K, Ihegboro GO, Ononamadu CJ, Alhassan AJ, Wudil AM (2020) Investigation of cytotoxicity potential of different extracts of *Ziziphus mauritiana* (Lam) leaf *Allium cepa* model. Toxicol Rep 7:816–821

Owolarafe T, Ihegboro G, Salawu K, Ononamadu C, Fadilu M, Musa B (2022) Toxicological investigation of aqueous extract of *Ziziphus mauritiana* leaves on Wistar rats. Int J Tradit Complement Med Res 3(2):91–100

Panseeta P, Lomchoey K, Prabpai S, Kongsaeree P, Suksamrarn A, Ruchirawat S, Suksamrarn S (2011) Antiplasmodial and antimycobacterial cyclopeptide alkaloids from the root of *Ziziphus mauritiana*. Phytochemistry 72(9):909–915

Parvez GMM, Akanda KM (2019) Foods and arthritis: an overview. In: Bioactive food as dietary interventions for arthritis and related inflammatory diseases. Elsevier, pp 3–22

Parvez GMM, Akanda KM, Karim R, Mou SM, Zahan P, Mosaddik DA (2016) Comparrative phytochemical screening and antimicrobial evaluation of different varieties of banana (*Musa sapientum*). Int J Innov Pharmaceut Sci Res 4(4):372–383

Prakash O, Usmani S, Singh R, Singh N, Gupta A, Ved A (2021) A panoramic view on phytochemical, nutritional, and therapeutic attributes of lam.: a comprehensive review. Phytother Res 35(1):63–77

Rathore TS, Singh RP, Deora NS, Shekhawat NS (1992) Clonal propagation of *Zizyphus* species through tissue culture. Sci Hortic 51(1):165–168

Saidurrahman M, Mujahid M, Siddiqui MA, Alsuwayt B, Rahman MA (2022) Evaluation of hepatoprotective activity of ethanolic extract of Pterocarpus marsupium Roxb. leaves against paracetamol-induced liver damage via reduction of oxidative stress. Phytomed Plus 2(3):100311

Siddharth P, Kailash P, Niraj V, Karuna M, Vimal P, Bharadia P, Pundarikakshudu K (2010) Antiulcer activity of methanolic extract of *Ziziphus mauritiana* stem bark. Int J Pharm Phytochem Res 2:6–11

Solecki RS, Shanidar IV (1975) A Neanderthal flower burial in Northern Iraq. Science 190(4217):880–881

Suksamrarn S, Panseeta P, Kunchanawatta S, Distaporn T, Ruktasing S, Suksamrarn A (2006) Ceanothane- and Lupane-type triterpenes with antiplasmodial and antimycobacterial activities from *Ziziphus cambodiana*. Chem Pharm Bull (Tokyo) 54(4):535–537

Sumanth M, Bhargavi YR (2014) Evaluation of wound healing effect of *Ziziphus mauritiana* L. leaf extract in rats. Int J Green Pharm (IJGP) 8(4):263–266

Tahir M, Rahman MA, Khushtar M (2022) Gastroprotective effect of Hyssopus officinalis L. leaves via reduction of oxidative stress in indomethacin-induced gastric ulcer in experimental rats. Drug Chem Toxicol 45(1):291–300

Talmale S, Bhujade A, Patil M (2015) Anti-allergic and anti-inflammatory properties of Zizyphus mauritiana root bark. Food Funct 6(9):2975–2983

Tan N-H, Zhou J (2006) Plant Cyclopeptides. Chem Rev 106(3):840–895

Tanvir EM, Afroz R, Karim N, Mottalib MA, Hossain MI, Islam MA, Gan SH, Khalil MI (2015) Antioxidant and antibacterial activities of methanolic extract of BAU Kul (*Ziziphus mauritiana*), an improved variety of fruit from Bangladesh. J Food Biochem 39(2):139–147

Tarique M, Siddiqui HH, Mohd K, Rahman MA (2016) Protective effect of hydro-alcoholic extract of *Ruta graveolens* Linn. leaves on indomethacin and pylorus ligation-induced gastric ulcer in rats. J Ayurveda Integr Med 7(1):38–43

Thanatcha PA, Pranee A (2018) Extraction and characterization of mucilage in *Ziziphus mauritiana* Lam. Int Food Res J 18:201–2012

Uzzaman S, Akanda KM, Mehjabin S, Parvez GMM (2018) A short review on a nutritional fruit: guava. Open Access Toxicol Res 1(1):1–8

Yahia Y, Benabderrahim MA, Tlili N, Bagues M, Nagaz K (2020) Bioactive compounds, antioxidant and antimicrobial activities of extracts from different plant parts of two *Ziziphus mill.* species. PLoS One 15(5):e0232599

# Medicinal and Nutritional Importance of *Hypericum perforatum* in Human Health

13

Pranjal Gujarathi, Meghraj Suryawanshi,
and Piyush Gujarathi

**Abstract**

*Hypericum perforatum*, commonly known as St. John's Wort, is a perennial herbaceous plant renowned for its significant medicinal and nutritional properties. This chapter delves into the diverse bioactive compounds present in *Hypericum perforatum*, such as hypericin, hyperforin, and flavonoids, and their mechanisms of action. The herb has been extensively studied for its antidepressant, anti-inflammatory, antimicrobial, and antioxidant activities, which contribute to its therapeutic potential in managing mental health disorders, infections, and inflammatory conditions. Additionally, *Hypericum perforatum's* nutritional profile, including essential vitamins and minerals, underscores its role in promoting overall health and well-being. This chapter provides a comprehensive review of current scientific literature on the pharmacological effects, clinical applications, and safety considerations of *Hypericum perforatum*, offering insights into its integration into modern healthcare practices.

**Keywords**

*Hypericum perforatum* · St. John's Wort · Medicinal importance · Clinical trials · Applications

P. Gujarathi
Department of Pharmacology, Vidhyadeep Institute of Pharmacy, Vidhyadeep University, Anita, Surat, Gujarat, India

M. Suryawanshi (✉)
Department of Pharmaceutics, Sandip Institute of Pharmaceutical Sciences (SIPS), Affiliated To Savitribai Phule Pune University (SPPU, Pune), Nashik, Maharashtra, India

P. Gujarathi
Department of Community Medicine, Vidhyadeep Homeopathic Medical College and Research Centre, Vidhyadeep University, Anita, Surat, Gujarat, India



M. A. Ansari et al. (eds.), *Medicinal Plants and their Bioactive Compounds in Human Health: Volume 2*, https://doi.org/10.1007/978-981-95-3620-7_13

## 13.1 Introduction

### 13.1.1 Overview of *Hypericum perforatum*

The *Hypericum* Tourn, ex L. is a multiracial genre comprising more than 1000 species that disseminate globally (Barnes et al. 2001; Nobakht et al. 2022; Suryawanshi et al. 2024; Zhang et al. 2020). *Hypericum perforatum* (HP) L. alias St. John's Wort (SJW) is an eminent species of this genus that belongs to the family Hypericaceae. HP is a nonwoody perennial Indigenous herb of western Asia, Europe, and northern Africa (Alzahrani et al. 2023; Barnes et al. 2001; Robson 2016). The name hypericum was obtained from Greek terms hyper and eikon which signify over and image, also the termed SJW is coined as it blooms in midsummer nearly to John that Baptist day in June. Similarly, the herb is also known by the name Klamath weed, Tipton's weed, Rosin-rose, or Goat weed (Alzahrani et al. 2023; Barnes et al. 2001; Velingkar et al. 2017). The Iranian name of the SJW is Alaf-e-Chai or Gol-e-Raee (Suryawanshi et al. 2024).

The herb *Hypericum perforatum* grows in sunny habitats, in well-drained, limy soil present at the edges of roads, slopes, wood borders, and stones. The herb grows up to a height of 50–100 cm, with bright yellow, star-shaped flowers. The flowers are present in a bunch of three flowers having five petals. Tiny oil glands are present in the leaves. The black dots cause the fingers to turn red when rubbed between them. The herb is harvested before or during their blooming period in European countries. To prevent degradation of its active principles, especially hyperforin, the plant must be dried immediately in warm weather in a drying room after being collected (Galeotti 2017).

### 13.1.2 Historical Uses in Traditional Medicine

Over millennia, *Hypericum* has been considered an herb for healing by various systems of medicine, such as the Chinese, Islamic, Greek, and Homeopathic systems. *Hypericum perforatum* is commonly called Guan Ye Lian Quao, Xiao Xhong Huang, and Xiao Dui Yeu Cao (Guixhou), Guo Lu Huang Gan, Shan Bian, Quian Ceng Lou, Shang Tian Ti (Sichuan), Shan Han Lin Cao (Jiangsu), Da Dui Ye Cao (Hubei), Xiao ye jin si tao (Henan) as well as Xiao Liu Ji nu (Shaanxi). Traditional Chinese Medicine (TCM) uses HP in the treatment of hematemesis, irregular menstrual bleeding, traumatic hemorrhage, jaundice, acute mastitis, sore throat, urinary tract infection, swelling and eye pain, sore furuncle, burns rheumatic arthritis, wounds, and bruises. The herb is bitter and astringent in taste and possesses a neutral nature according to traditional literature (Clarke 2010; Nobakht et al. 2022).

Greek stalwarts—Galen, Dioscorides, Pliny, and Hippocratess—stated various therapeutic advantages of HP as a diuretic, wound healer, and remedy for menstrual cramps, intestinal worms, snake or reptile bites, gastrointestinal diseases, depression, sciatica, ulcers, superficial wounds, and burns. The oil of HP prepared from the flowering top of it was used as a disinfectant for wounds and to heal bruises (Nobakht et al. 2022; Suryawanshi et al. 2024).

Avicena, Razi, Antaki, Herawi, Ansari, Shirazi, Ghassani, and Ibn Beytar abd Aqili, physicians of Islamic Medicine, stated various therapeutic applications of HP. The poultice of this herb has been used in curing infectious wounds, burns, and bruises according to reference books of Islamic traditional medicine (Nobakht et al. 2022; Suryawanshi et al. 2024; Velingkar et al. 2017). The poultice of HP with olive oil was used topically to treat paralysis. The plant was also used as a diuretic, anti-gout, jaundice, frequent urination, chronic selling, kidney stones and to induce abortion (Galeotti 2017; Nobakht et al. 2022; Suryawanshi et al. 2024).

A great homeopath Sammule Hannhmann mentioned the use of homeopathic dilutions of HP in the treatment of nerve-related pains, wound healing, headaches, pain related to epidural or spinal tap, and pain caused by spinal injury (Lennihan 2017). In Europe, HP is traditionally used to cure the inflammation of the bronchi and urogenital tract, biliary diseases, bladder irritation, common cold, burns, skin diseases, diabetes mellitus, dyspepsia, neuralgia, and neurasthenia. Europeans also used the herb SJW to get benefits from numerous pain conditions like migraine, headache, myalgia, sciatica, rheumatism, lumbago, and pain of dental extraction (Galeotti 2017).

### 13.1.3 Importance in Modern Healthcare

*Hypericum perforatum* possesses numerous ethnomedicinal uses but nowadays, it is widely known for its utilization in alleviating mild-to-moderately severe depression, as a neuroprotective agent in various ailments such as Alzheimer's disease (AD), and Parkinson's disease (PD) (Suryawanshi et al. 2024). Recent studies stated that HP can be used in the treatment of other pathological conditions like trigeminal neuralgia (Assiri et al. 2017), and various pain conditions (Galeotti 2017). Along with these recently various biological actions of HP were reported such as antibacterial, antiviral, antioxidant, analgesic, anti-inflammatory, hypotensive, and spasmolytic (Barnes et al. 2001; Marrelli et al. 2020; Zirak et al. 2019). Various animal trials demonstrated the use of HP as an anxiolytic, sedative, nootropic, antischizophrenic, and anticonvulsant. Not only in this but also HP was found to be more preclinically efficient in the management of addiction related to alcohol, nicotine, and caffeine (Can et al. 2011; Can and Özkay 2012; Galeotti 2017).

## 13.2 Phytochemistry of *Hypericum perforatum*

*Hypericum perforatum* enjoys an extensive range of secondary metabolites, for instance, phenolic acids, proanthocyanidins, flavonoids, naphthodianthrones, acylphloroglucinols, xanthones, and volatile oils; among these phloroglucinol derivatives are the prime components (Greeson et al. 2001; Nahrstedt and Butterweck 1997; Suryawanshi et al. 2024; Zhang et al. 2020). Hydroalcoholic decoction of the aerial parts of herbs is generally used that contains ten various secondary metabolites: flavonoids (rutin, hyperoside, quercetin, quercitrin), naphthodianthrones (hypericin and pseudohypericin), acylphloroglucinols (hyperforin and adhyperforin), proanthocyanidins, procyanidins, tannins, volatile oils, amino acids, phenylpropanes, xanthones as well as water-soluble compounds such as organic acids, peptides, and polysaccharides (Galeotti 2017).

## 13.2.1 Phloroglucinols

Phloroglucinol derivatives, polycyclic polyprenylated acylphloroglucinols (PPAPs), in particular, are an array of hybrid natural substances that exhibit substantial structural and functional variety (Yang et al. 2018). Hyperforin, adhyperforin, and oxygenated analogs of hyperforin are most commonly found prenylated phloroglucinols in HP (Barnes et al. 2001; Yang et al. 2018).

### 13.2.1.1 Phenolics

The hypericum was found to contain the phenolic compounds such as caffeic acid, chlorogenic acid, p-coumaric acid, ferulic acid, p-hydroxy-benzoic acid, and vanillic acids. Phenols in HP are responsible for the nootropic, anti-inflammatory, antiviral, and antimicrobial action (Hu et al. 2016; Tanaka et al. 2016; Yan et al. 2018, 2019; Zeng et al. 2018; Zhu et al. 2016).

## 13.2.2 Anthraquinones

The fresh herb of HP contains hypericin, pseudohypericin and isohypericin, prototype hypericin, prototype psuedohypericin, and cyclopseudohypericin. Both hypericin and pseudohypericin are combinedly termed as total hypericin (Barnes et al. 2001; Vanhaelen and Vanhaelen-Fastré 1984).

## 13.2.3 Flavanoids

HP was found to contain flavonols like kaempferol, quercetin, flavones—luteolin, glycosides such as hyperosides, isoquercitrin, quercitrin and rutin, bioflavonoids inclusive of biapigenin and amentoflavone, and catechins. Among these, rutin and hyperoside were found to be present in higher concentrations than the others (Barnes et al. 2001).

## 13.2.4 Essential Oils

The major compound in the volatile oil of HP is the saturated hydrocarbon methyl-2-octane. Other constituents include n-nonane, methyl-2-decane, and n-undecane, as well as α- and β-pinene, α-terpineol, geraniol, myrcene, limonene, caryophyllene, and humulene (Barnes et al. 2001; Suryawanshi et al. 2024).

### 13.2.5 Other

Other components include tannins particularly proanthocyanidins, acids such as isovalerianic, nicotinic, myristic, palmitic, steric acids, carotenoids, choline, nicotinamide, pectin, beta-sitosterols, straight chain hydrocarbons, and alcohols (Barnes et al. 2001; Suryawanshi et al. 2024). The air-dried and pulverized arial parts of the plant contained a two new β-diketones, 2,6,9-trimethyl-8-decene-3,5-dione and 3.7.10-trimethyl-9-undecebe-4,6-dione. Also, the roots of plant were found to contain two novel biphenyls hyperbiphenyls A and B following extraction using polar solvents, cyclohexane, dichloromethane, and fractionation by flash chromatography and semi-preparative HPLC (Bréard et al. 2018; Radulović et al. 2018).

## 13.3 Pharmacological Actions of *Hypericum perforatum*/SJW

Various preclinical and clinical trials have demonstrated the effectiveness of HP in weak/medium depressive disorders. Despite the primary focus of pharmacological studies on HP antidepressant action, multiple studies have reported biological actions of it. Preclinical trials on HP decoction not only underline the antianxiety, sedative, nootropic, antipsychotic, and anticonvulsant action but also state the importance of HP in alcohol, nicotine, and caffeine addiction. Besides this, the antibacterial, antiviral, anti-inflammatory, antinociceptive, antitumor, antioxidant, cardioprotective, and wound healing properties were also introduced through in vitro and in vivo studies. Figure 13.1 illustrates numerous pharmacological actions of *H. perforatum*.

### 13.3.1 Antidepressant and Neuroprotective Action

According to the World Health Organization (WHO), depression is an ordinary mental disorder, appertaining to constant nervousness and apathy in various activities. Depression can be treated with various antidepressant drugs with or without combining with *H. perforatum*. HP is also registered as a traditional medicine in various European countries; HP-containing products were categorized as dietary supplements, while these products are also registered as licensed medicines in Germany (Booker et al. 2018). Recently in the review, Suryawanshi and co-workers spotlighted the effectiveness of HP in various neurodegenerative diseases such as depression, Alzheimer's disease, and Parkinson's disease (Suryawanshi et al. 2024).

Various classes of secondary metabolites such as hypericin, hyperforin, hypericin, and hyperoxide were important chemical constituents in HP and were found to give antidepressant action (Forsdike and Pirotta 2019; Oliveira et al. 2016;

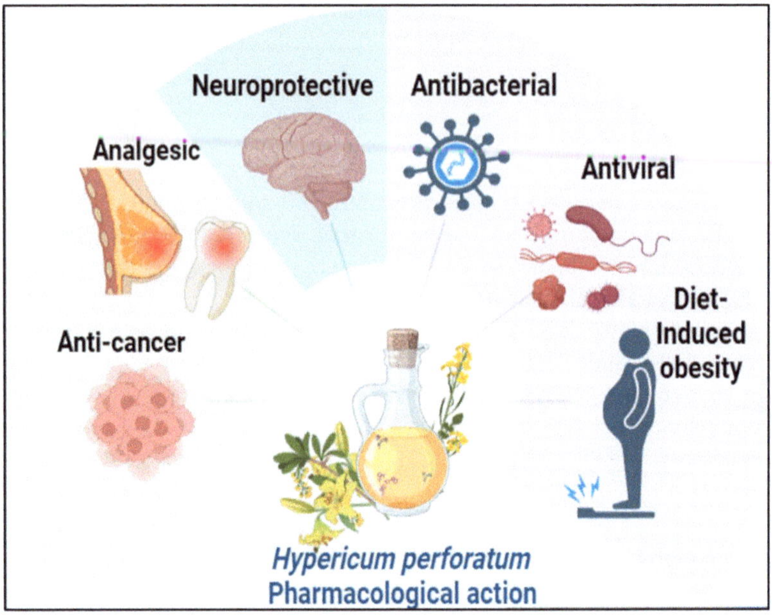


**Fig. 13.1** Various pharmacological actions of *H. perforatum*

Soleymani et al. 2017). The herb produced an antidepressant action, probably by inhibiting the monoamine oxidase (MAO) enzyme. HP was also found to inhibit the reabsorption of dopamine (DA), serotonin (5-HT), noradrenaline (NA), l-glutamate, and gamma-aminobutyric acid axon terminals (Oliveira et al. 2016). Besides, flavonoids like quercetin, luteolin, and kaempferol were responsible for antidepressant effect (Greeson et al. 2001).

The systematic review by Apaydin and co-workers spotlighted that HP extracts are fruitful in treating patients having mild and moderate depression in comparison with placebo and antidepressant drugs. Additionally, there is a paucity of research on the utilization of HP in the treatment of severe depression (Apaydin et al. 2016). The HP showed a comparable response to selective serotonin reuptake inhibitors (SSRIs) in the meta-analysis of 27 clinical studies having 3808 patients from countries like China, Germany, Denmark, Brazil, and the United States. These findings strongly support the effectiveness of St. John's Wort in treating depression; however, further long-term or follow-up research is necessary to determine if the improvement in clinical outcomes with St. John's Wort is sustained over time (Ng et al. 2017). Recently HP extract was evaluated for the stimulation of neurogenesis, the safeguarding action against glutamate or N-methyl-D-aspartate receptor-induced excitotoxicity, and anti-inflammatory action in LPS-activated human macrophages. This study revealed that HP extract possesses protective, neurotrophic, and anti-inflammatory action that can be utilized for curing depressive disorders (Bonaterra et al. 2018). Table 13.1 displays data about in vitro and in vivo antidepressant activity of *H. perforatum*.

**Table 13.1** In vitro and in vivo antidepressant activity of *H. perforatum*

| Plant part/ extract used | Methodology | Result/ probable mechanism | Ref. |
|---|---|---|---|
| Standardized decoction of HP | NMRI mice and brain homogenate of rats' brain MAO A and B assay, synaptosomal uptake analysis, receptor binding assay | 5-HT and NA synaptosomal uptake inhibition, dopamine (DA) inhibitor, β-adrenoreceptor downregulation, and 5-$HT_2$ receptor density downregulation<br>HP extract as found to act similarly to tricyclic antidepressants | Müller et al. (1997) |
| Hyperforin | Mice were used<br>Behavioral studies—open field test (OFT), elevated plus maze test (EPM), forced swim test (FST), sucrose preference test (SPT)<br>Electroencephalogram (EEG) recording | This study provided a reference that the transient receptor potential canonical 6 (TRPC6) channel can be a druggable candidate to treat disorders like anxiety and depression | El Hamdaoui et al. (2022) |
| Active constituents were extracted from hypericum extract | MAO-A inhibition was studied by in-vitro and ex-vivo studies<br>In-vitro studies—Rat brain homogenate<br>Ex-vivo studies—intraperitoneal administration of constituents to rats | Total extract showed a great inhibition of MAO-A and other fractions show inhibition only at a concentration of $10^{-3}$ Mol/L; flavonoid-rich fraction showed 39% inhibition | Bladt and Wagner (1994) |
| Lyophilized alcoholic extract and hypericin | Hypericin and procyanidin B2 were administered locomotor action by OPT was determined<br>Immobility measured by FST in rats | No alterations were observed between animals administered pure hypericin and hypericin combined with procyanidin B12<br>A reduction in HPA axis activity was found | Butterweck et al. (2001) |
| Compounds and extracted fractions for *H. perforatum* | Isolated compounds 1 and 2 in combination with subfraction $IIIc_1$ (mainly contains procyanidins) were administered in male rats<br>FST and OFT were performed | The inverted U-shaped DRC was obtained in FST<br>The in-vivo action was found to be increased<br>Data is suggestive that naphthodianthrones act as antidepressant compounds in HP | Butterweck et al. (1998) |

(continued)

**Table 13.1** (continued)

| Plant part/ extract used | Methodology | Result/ probable mechanism | Ref. |
|---|---|---|---|
| Hyperfoliatin extracted from *H. perforatum* | In-vivo—Mice were subjected to FST and locomotor activity<br>In-vitro—Homogenized solution of the frontal cortex and hypothalamus of rats<br>Synaptosomal uptake of DA, 5-HT, and NA was assessed<br>The binding assay of the DA transporter, NA transporter, and 5-HT transporter was assessed | Hyperfoliatin decreases the immobility time in FST<br>In a concentration-dependent manner Hyperfoliatin inhibited synaptosomal uptake of DA, 5-HT, and NA<br>The result states that the antidepressant effect of HP is may be due to monoamine uptake inhibition | do Rego et al. (2007) |
| Ethanolic mother tincture of *H. perforatum* | Rats were subjected to constant dark and low constant light free running movement to assess the circadian hypotheses<br>*H. perforatum* ethanolic extract was administered in drinking water | The least shortening of the meantime was observed after the stoppage of extract treatment in low-dose group animals in low low-constant light experiment<br>The study showed that the antidepressant action of HP is not caused by circadian mechanism | Francis (2005) |
| A mixture of hyperforin and adhyperforin | Rat cortical brain cells were used to assess the action of chronic hyperforin administration<br>Effect on the brain-derived neurotrophic factor (BDNF) receptor and TrkB and adult hippocampal neurogenesis was evaluated<br>Expression of cyclic adenosine monophosphate response element-binding protein (CERB), phosphor-CERB, TrkB and phosphor-TrkB was evaluated by western blot assay | The drug was found to stimulate the expression of TRPC6 channels and TrkB<br>Hyperforin increased the expression of TrkB in the cortex<br>Hippocampal neurogenesis has not changed<br>Extract was found to act on BDNF/TrkB pathway in brain | Gibon et al. (2013) |

(continued)

**Table 13.1** (continued)

| Plant part/ extract used | Methodology | Result/ probable mechanism | Ref. |
|---|---|---|---|
| Dried flowers, top arial portion, roots of *H. perforatum*—water/ methanol (1:1) extract | Reverse phase-high performance liquid chromatography (RP-HPLC) of plant extract<br>MAO-A inhibition assay and antioxidant action measurement | Human MAO-A was inhibited by HP; flower extract was found to give the greatest inhibition<br>Plant was found to contain pseudohypericin, hypericin, hyperforin, adhyperforin, hyperforin, and flavonoids in HPLC-DAD-MS | Herraiz and Guillén (2018) |
| Extract of *H. perforatum* | In-vitro—Whole brain homogenate was used for paroxetine binding or 5-HT uptake assay<br>Behavioral test-FST, marble-burying test<br>Locomotor action measurement | The synaptosomal 5-HT uptake was inhibited by HP extract with the least paroxetine binding assay<br>HP extract is found to inhibit the marble-burying behavior in mice<br>The immobility time of mice was reduced in FST<br>The evidence reveals that the antidepressant action of HP is not similar to SSRIs | Hirano et al. (2004) |
| Hyperforin from *H. perforatum* ethanolic extract | Alterations in voltage and ligand-gated channels were evaluated by isolated patch-clamped rat central and peripheral neurons | In the p-type calcium channel, hyperforin showed action through its interaction with calmodulin<br>Ethanolic extract was found to inactivate the sodium current<br>Study revealed that alterations in ligand-gated ion channels have a prime role in antidepressant action | Krishtal et al. (2001) |

(continued)

**Table 13.1** (continued)

| Plant part/ extract used | Methodology | Result/ probable mechanism | Ref. |
|---|---|---|---|
| Extract of *H. perforatum* | Mice were used to evaluate the antidepressant action by FST | The positive modulation in behavioral and neurochemical components was observed | Kulkarni et al. (2012) |
| Ethanolic extract of *H. perforatum* | The effect of hypericum, hyperforin, and hyperforin-free hypericum was compared with conventional antidepressant drugs by passive avoidance test in rats.<br>All three components were also evaluated for their ability to increase NA transmission in vivo | The study revealed that a high concentration of hypericum affects the 5-HT reuptake similarly to TCAs than SSRIs<br>Hypericum increases 5-HT and NA transmission in the forebrain limbic brain circuit | Misane and Ögren (2001) |
| HP extract possessing hypericin, hyperforin, and flavonoids | IL_6 knockout mice and wildtype mice were used to evaluate the effect by FST | Study reveals that HP showed its effect by activation of the 5-HT system which is indicative of IL-6 is important for antidepressant action | Calapai et al. (2001) |
| Hydroethanolic extract and hyperforin dicyclohexylammonium (DCHA) | CD-COBS rats were subjected to FST, OFT | The study revealed that HP extract inhibits MAO uptake | Cervo et al. (2002) |
| *H. perforatum* water extract | Sprague-Dawley (SD) rats and C57BL/6 mice were used<br>Chronic unpredictable mild stress model involving fasting and no water,<br>Hot and cold FST, tail pinch, tail suspension, wet cage, cage shaking, and sucrose preference test were used | Hyperoside was found as an antidepressive agent via NLRP1 inflammasome, CXCL/ CXCR2/ BDNF signaling pathway-related effect | Song et al. (2022) |

#### 13.3.1.1 Neuroprotective Action

Hypericin, hyperoside, hyperforin, bipigenin, and kaempferol are the main secondary metabolites in *Hypericum perforatum* responsible for neuroprotection (Suryawanshi et al. 2024). The hypericum was found to produce its action by several mechanisms such as by acting on direct and indirect antioxidant pathways, anti-inflammatory potential, inhibition of cholinesterase enzymes; decreasing the excitations of neurons initiated by GABA; and reducing the binding of antagonist to dopamine -D2 receptor (Suryawanshi et al. 2024). Table 13.2 explained numerous in-vitro and in-vivo neuroprotective activity of *H. perforatum.*

#### 13.3.1.2 Mode of Action as an Antidepressant and Neuroprotective

##### 13.3.1.2.1 Beta Adrenergic and Serotonergic Action of HP

The synaptic transmission of noradrenaline and serotonin inside the central nervous system (CNS) is affected by all antidepressant drugs. Almost all groups of antidepressants were found to enhance the availability of noradrenaline (NA) and serotonin (5-HT) in their synapse at the initiation of their action. Also, the antidepressants and neuroprotective show their action by several mechanisms such as inhibiting neuronal reuptake of NA and 5-HT and inhibiting presynaptic alpha-2 receptors. As a result, the degradation of NA and 5-HT is decelerated and their amount at the receptor site is elevated. Besides, the MAO inhibitors class of antidepressant drugs exerts action by suppressing intra and extraneuronal uptake of MAO enzymes (Müller 2003). Various studies have confirmed that *Hypericum perforatum* occludes both transport systems and acts as an MAO inhibitor responsible for alteration in β-, %-$HT_{1A}$, and $HT_{2A}$ receptors. The antidepressant action of this means may be due to secondary metabolite (Schmidt and Butterweck 2015; Suryawanshi et al. 2024).

The extract void of hyperforin decreased the number of β-adrenoreceptors in rat C6 cell culture to a similar extent as desipramine (De Marchis et al. 2006). Hyperforin and hyperoside preincubated rat C6 cells cause considerable depletion of β-adrenoreceptor binding followed by depletion in cAMP formation. These results state that hyperoside and hyperforin can be used therapeutically for depressive symptoms (Prenner et al. 2007).

##### 13.3.1.2.2 GABA and Glutaminergic Response

The HP decoction exerts a broad spectrum of action along with dopamine serotonin and MAO was found to show its inhibitory action on GABA and L-glutamate synaptosomal uptake. The results obtained from studies revealed that no other antidepressant and neuroprotective agent possesses such a broad-spectrum action as that demonstrated by *Hypericum perforatum* which can inhibit all five neurotransmitters. No other antidepressant drug equally inhibits the activity of all five neurotransmitters on the synaptosomal uptake but *Hypericum perforatum* has the potential to inhibit the same. In preclinical studies, hypericin was found to affect the release of glutamate elicited by 4-aminopyridine in a dose-dependent manner. Hypericin was found to subdue voltage-gated calcium channels and mitogen-activated protein kinase action which lead to the inhibition of glutamate release (Chang and Wang 2010).

##### 13.3.1.2.3 Action on Ion Channels

Hyperforin, a prenylated phloroglucinol derivate, was found to act on voltage-dependent and ligand-gated ion channels by inhibiting conductance generation. The

**Table 13.2** In-vitro and in-vivo neuroprotective activity of H. perforatum

| Neuroprotective effect | Methodology | Result | Ref. |
|---|---|---|---|
| Nootropic effect | Ethanolic extract of HP in rats<br>Animal models used-elevated plus maze test (EPM), passive avoidance test, high-affinity binding assay, electric shock induced memory loss (ECS) | The extract was found to reduce the binding of dopamine D2 receptors and improve the binding of 5-$HT_{2A}$ receptor antagonists and benzodiazepine receptors. | Kumar et al. (2002) |
| Alzheimer's and Parkinson's disease | Standard extract of HP<br>In-vitro pheochromocytoma cell line (PC12) was used and disease was induced by hydrogen peroxide | The extract was found to inhibit the fractionation of DNA and avert the cell shrinking and antioxidant action | Benedí et al. (2004); Lu et al. (2004) |
| Alzheimer's and Parkinson's disease and dementia | Standard extract of HP<br>Rotenone-induced oxidative stress model in rats | The extract was found to show antioxidant property | Sánchez-Reus et al. (2007) |
| Alzheimer's disease | Hydroalcoholic decoction of flowers of HP<br>The action was assessed in nematodes-*Globodera pallida*<br>Anticholinesterase (AChE), Butyrylcholinesterase (BuChE) inhibitory action<br>Beta-secretase 1 site amyloid precursor protein (APP) breaking enzyme 1 (BACE 1) inhibitory action<br>Antioxidant action | AChE and BuChE action inhibitor, BACE1 action inhibitor<br>Antioxidant property | Althobaiti et al. (2022) |
| Alzheimer's disease | Secondary metabolite hyperforin extracted from HP<br>Morris water maze test was performed in rats following the injection of aβ E22G peptide in the hippocampus | The amyloid plaque accumulation and aβ induced neurotoxicity were reduced | Dinamarca et al. (2006) |
| Alzheimer's disease | Hydroalcoholic extract of HP was tested in-vitro in microglial cell line BV2<br>The lactate dehydrogenase assay, reactive oxygen species production, and phagocytosis of amyloid beta were assessed | Fluidity of cell member was changed by extract | Kraus et al. (2007) |

(continued)

**Table 13.2** (continued)

| Neuroprotective effect | Methodology | Result | Ref. |
|---|---|---|---|
| Alzheimer's disease | Hydroethanolic extract of HP was used<br>Mice expressing mutated human APP and mutated presenilin (PS1) and *an* in-vitro cell line of microglia were used<br>Diseased was assessed by Morris water test, aβ-phagocytosis assay and ABC-transporter action assay | Reduction in soluble Aβ42 species | Hofrichter et al. (2013) |
| Alzheimer's disease | 80% methanolic decoction was evaluated in rats<br>The disease was induced by aluminum chloride | The monoamine uptake was inhibited by extract | Cao et al. (2017) |
| Alzheimer's disease | H\P decoction containing 5% hyperforin was evaluated in mice brain homogenate<br>A double transgenic mouse model was used to induce disease | The expression of p-glycoprotein was enhanced, parenchymal beta-amyloid deposition was reduced | Brenn et al. (2014) |
| Diabetes-associated learning and memory disabilities | Water extract of HP was evaluated in rats<br>Streptozotocin was used for diabetes induction, and passive avoidance learning (PAL) test and retention test were used for assessing learning and memory alterations | The extract showed antioxidant and cholinergic facilitatory action | Hasanein and Shahidi (2011) |
| Stress-induced memory impairment | The brown powder of HP containing 0.2 mg hypericin was administered in rats<br>Memory impairment was induced by light/dark and corticosterone | Synaptic spasticity was changed | Trofimiuk et al. (2011) |
| Action on spatial learning and memory | The extract containing 0.3% hypericin was used in rats. The Morris water maze test was used | The 5-HT amount was elevated in the cortex, monoamine amount was altered in the brain | Widy-Tyszkiewicz et al. (2002) |

(continued)

**Table 13.2** (continued)

| Neuroprotective effect | Methodology | Result | Ref. |
|---|---|---|---|
| Antianxiety action | HP extract consisting of hypericin, proto hypericin, and pseudo hypericin was used in rats and cell cultures of cerebral granule cells<br>Light/dark model, open field test | The extract showed an effect on GABA, NMDA, and BZD receptors | Vandenbogaerde et al. (2000) |
| Neuroprotective action in Parkinson's disease | Decoction of HP along with bromocriptine was administered in mice<br>The disease was induced by 1-Methyl-4-phenyl1–1,2,3,6-tetrahydropoyridine (MPTP)<br>Rats were subjected to Rota rod, forepaw stride length, and hand test | HP decoction was found to show antioxidant property | Mohanasundari et al. (2006) |
| Neuroprotective action | Hydroalcoholic decoction of HP was used in-vitro in the hippocampus of rats embryo<br>Beta-amyloid induced neurotoxicity | The decoction was found to show antioxidant action | Silva et al. (2004) |
| Neuroprotective action | Secondary metabolites quercetin, kaempferol, and biapigenin were extracted from HP decoction and used in the neuronal culture of the hippocampus<br>Toxicity was induced by neuronal excitotoxicity and mitochondrial dysfunction | Antioxidant action, calcium dysregulation, and mitochondrial dysfunction properties were shown by sec. Metabolites | Silva et al. (2008) |
| Neuroprotective action | Flavonoid flavonoid-containing faction of HP decoction was used in the PC12 cell culture<br>Cell culture was subjected to hydrogen peroxide-induced apoptosis | The flavonoid fraction of decoction was responsible for antioxidant property | Zou et al. (2010) |
| Neurobehavioral alterations related to acute stress | Decoction of HP was administered in mice and animals were subjected to restrain stress | Detection was found to show antioxidant action | Kumar et al. (2010) |

(continued)

**Table 13.2** (continued)

| Neuroprotective effect | Methodology | Result | Ref. |
|---|---|---|---|
| Anticonvulsant action | 70% ethanolic decoction of HP and butanol and ethereal decoction was administered in rabbits<br>Animals were subjected to a kindling model of seizure | The excitation of neurons was decreased, affecting the amount of catecholamines and acting as pro-epilepticus | Ivetic et al. (2011) |
| Neuroinflammation inhibition | Hyperoside was used in HT22 murine neuronal cell culture<br>The neuroinflammation was induced by lipopolysaccharide (LPS) | Hyperoside was found to show inhibitory action on apoptosis, inflammation<br>And antioxidant action | Huang et al. (2021) |

hyperforin did not affect the C-type transient receptor potential channels isoforms TRPC1, TRPC3, TRPC4, TRPC5, and TRPC7. The study revealed that hyperforin, a constituent of St. John's Wort (SJW), triggered intracellular calcium transients and depolarizing inward currents in hippocampal pyramidal neurons. These effects were sensitive to lanthanum (La3+), an inhibitor of transient receptor potential canonical (TRPC) channels. Notably, these actions resembled those elicited by brain-derived neurotrophic factor (BDNF), a neurotrophin. The findings suggest that the antidepressant effects of SJW may be mediated through a mechanism similar to that engaged by BDNF (Leuner et al. 2013).

Gibon et al. elevated the effect of chronic hyperforin treatment on cortical neuron culture. Study results revealed that hyperforin stimulated the expression of the TRPC6 channel and TrkB (Gibon et al. 2013). The cascade triggered by hyperforin, a key component in St. John's Wort, involves the transcription factor CREB and the neurotrophin BDNF. Calcium ions regulate this process and manifest specifically in the brain's cortical region. Interestingly, this effect is not observed in the hippocampus, a region where the impact of conventional antidepressants is typically expected (Bouron and Lorrain 2014). Figure 13.2 showed mechanisms of action of *Hypericum perforatum* as a neuroprotective agent.

#### 13.3.1.2.4 Action of Hypothalamic–Pituitary–Adrenal (HPA) Axis

Currently, research suggests that HP is associated with the regulation of genes that dominate HPA axis function and action, at least in part, stress-induced effects on neuroplasticity and neurogenesis. A considerable number of the findings from preclinical investigations have stemmed from the administration of elevated doses of purified compounds and extracts, which do not accurately reflect the concentrations of individual bioactive constituents achieved following oral consumption in human subjects (Crupi et al. 2013).

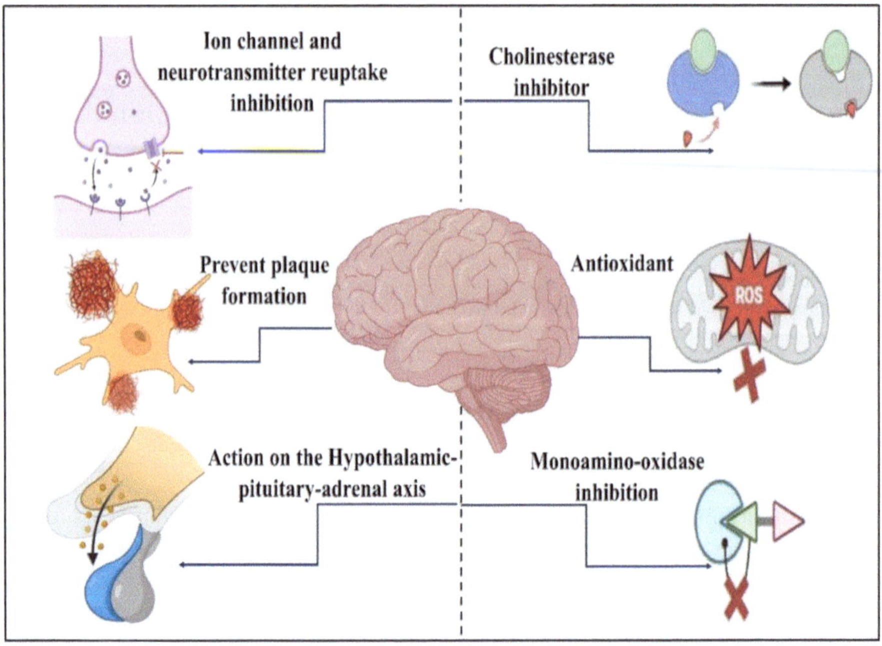


**Fig. 13.2** Various mechanisms of action of *Hypericum perforatum* as a neuroprotective agent

#### 13.3.1.2.5 Inhibitors of Cholinesterase

The cholinergic hypothesis proposes that the cognitive decline observed in neurodegeneration more particularly in Alzheimer's disease (AD) stems from deficits in central cholinergic neurotransmission, resulting from a loss of the neurotransmitter acetylcholine (ACh). The active components found in HP decoction exert their effects through the mechanism of Cholinesterase (ChE) inhibition, acting on the enzymes acetylcholinesterase and butyrylcholinesterase. By inhibiting these enzymes responsible for degrading ACh, the active components in SJW enhance central cholinergic function by increasing the availability of ACh to stimulate nicotinic and muscarinic receptors within the brain (Suryawanshi et al. 2024).

Hypericin, hyperforin, and hyperoside, the active compounds found in St. John's Wort, have been demonstrated to exhibit several neuroprotective properties. These compounds have been shown to inhibit the formation of beta-amyloid plaques, a hallmark of Alzheimer's disease. Additionally, they possess the ability to mitigate neuroinflammatory responses, act as antiapoptotic agents, and function as potent antioxidants. Collectively, these properties contribute to their potential to provide neuroprotection against various neurodegenerative disorders. Figure 13.3 illustrated *H. perforatum* as a cholinesterase inhibitor.

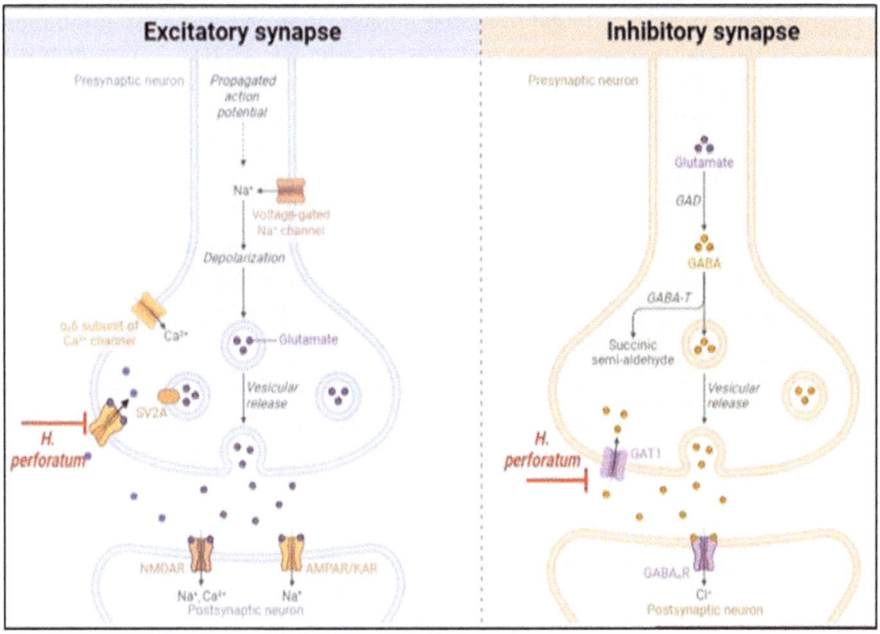


**Fig. 13.3** *H. perforatum* as a cholinesterase inhibitor

## 13.3.2 Analgesic Action

The hypericum species are generally used in China to treat two prime diseases furuncle and mastitis. Thirty-six active components were isolated from various species of *Hypericum* (Zhang et al. 2020). The review study on the therapeutic action of HP stated that minimum doses of dry decoction of herbs were found to show analgesic action and reinforce the action of opioids in acute and chronic preclinical pain models. In-vitro and in-vivo studies demonstrated that hypericin and hyperforin are accountable for their anti-inflammatory action (Galeotti et al. 2010).

Various scientific studies described the action of *H. perforatum* in regulating pain perception. The HP extract and its isolated secondary metabolite were found to relieve the inflammation and produced analgesia and antihyperalgesic action. The hydroalcoholic extract of HP was evaluated for alleviating neuropathic pain in oxaliplatin-induced neurotoxicity. The action of *H. perforatum* decoction on oxaliplatin-induced apoptosis was assessed in rat astrocytes. The extract was found to deplete the elevated caspase-3 action at all tested doses of extract(Cinci et al. 2017). Novelli and colleagues' study examined how *H. perforatum* extract and hyperforin inhibit the fundamental signaling pathways that cytokines stimulate in pancreatic β cells. They also examined the extent to which both of these substances hindered modifications in the level of expression of key target genes involved in β-cell function and differentiation, as well as in the induction of apoptosis and inflammation (Novelli et al. 2016).

Hypericum extract was found to inhibit inflammation and leucocyte infiltration in animals induced by carrageenan and prostaglandin E1 (PGE1). Tumor necrosis factor (TNF) initiates the activation of NF-κB, which is inhibited by hypericin in vitro studies. (Barnes et al. 2001; Brockmöller et al. 1997). It was also found to inhibit the arachidonic acid and leukotriene $B_4$ (Panossian et al. 1996). The intravitreal injection of hypericin subdued the advancement of proliferative vitreoretinopathy (PVR) in the rabbit model (Tahara et al. 1999). According to reports, 12-lipoxygenase activity is impeded by hypericin and pseudohypericin; lipoxygenase products stimulate leukotrienes and other responses could be involved within inflammatory responses (Bezáková et al. 1999). HP is generally used as an herbal remedy in dental practice to get rid of pain after extraction of tooth (Galeotti 2017). Table 13.3 explained and reported many In vivo analgesic activity of *H. perforatum*.

**Table 13.3** In vivo analgesic activity of *H. perforatum*

| Plant part or extract used | Methodology | Results | Ref. |
|---|---|---|---|
| Dry ethanolic extract | Hot-plate method in rats and mouse | The extract was found to induce antinociceptive action | Abdel-Salam (2005) |
| Dry ethanolic extract | Tail flick method in mouse | Antinociceptive action | Galeotti et al. (2010) |
| Extract of HP | Tail electric stimulation method in rats | Antinociceptive action | Abdel-Salam (2005) |
| Ethanolic extract of HP | Acetic acid-induced abdominal constrictions in mouse | Antinociceptive action | Bukhari et al. (2004); Kumar et al. (2001) |
| Dry ethanolic extract of HP | Capsaicin-induced paw licking in rats | Antinociceptive action | Abdel-Salam (2005) |
| Ethanolic extract of HP | Formalin assay in mouse | Antinociceptive action | Hatanaka et al. (2011); Uchida et al. (2008) |
| Dry ethanolic extract | Nitric oxide donor-induced hyperalgesia in mouse | Reclamation of thermal hyperalgesia | Galeotti and Ghelardini (2013a) |
| Dry ethanolic extract | Meningeal nociception in mouse | Reclamation of thermal hyperalgesia | Galeotti and Ghelardini (2013b) |
| Dry ethanolic extract | Chronic constriction hurt in rats<br>Oxaliplatin-induced neuropathic pain in rats and mouse | Reclamation of physical and thermal hyperalgesia | Galeotti et al. (2010) |
| Ethanolic extract of HP | Antiretroviral induced neuropathy in mouse | Reclamation of physical and thermal hyperalgesia | Sanna et al. (2016) |

### 13.3.3 Antimicrobial and Antiviral Action

The antimicrobial constituent isolated from Hp was phloroglucinol derivatives, xanthones, flavonoids, and other phenolic constituents (Zhang et al. 2020). The HP was also found to show the antimicrobial effect against *Staphylococcus aureus*, MRSA, the hyperforin is mainly responsible for its effect (Reichling et al. 2001; Saddiqe et al. 2010). The extract was also possessed to have antifungal and antiyeast potential against pathogenic fungi like *Trichophyton rubrum, Fusarium oxysporum, Microsporum canis, Pichia fermentans, Exophiala dermatitidis, Kluyveromyces marxianus, Candida albicans, and Saccharomyces cerevisiae*. Cao et al. (2017). Besides hypericin and hyperforin were found to show effective against malarial and leishmaniosis parasites (Montoya et al. 2015). The antibacterial activity of several extracts of *H. perforatum* aerial parts has been assessed by Lyles and colleagues. The antibacterial activity of several extracts of *H. perforatum* aerial parts has been assessed by Lyles and colleagues. Each sample was examined for S. aureus MIC values. Recently Lyles and co-workers stated the antibacterial nature of *S. aureus* of various extracts of arial parts of HP (Lyles et al. 2017).

Flavonoids and catechin-rich decoction of HP were responsible for acting against the influenza virus (Minshen). Herpes simplex type 1 and 2 as well as other encapsulated viruses was found to be inhibited by hypericin and pseudohypericin in in-vitro studies (Weber et al. 1994; Wood et al. 1990). Besides human immunodeficiency virus (HIV)-1 was inhibited by both of these constituents (Lavie et al. 1989; Lopez-Bazzocchi et al. 1991). Additionally, hypericin fights against viral protease, which is well-known for its capacity to combat a number of viruses, including bronchitis, herpes simplex, and influenza A (Montoya et al. 2015; Pu et al. 2009).

#### 13.3.3.1 Mechanism of Action

##### 13.3.3.1.1 Antibacterial Action

The active constituents in SJW were found to show its antibacterial by membrane disruption, inhibition of bacterial growth, generation of reactive oxygen species and oxidative stress, and anti-biofilm activity (Lopez-Bazzocchi et al. 1991; Saddiqe et al. 2010).

St. John's Wort (SJW), a perennial herb with yellow flowers, contains a diverse array of active compounds that exhibit potent antibacterial properties. These compounds exert their antimicrobial effects through multiple mechanisms, making them effective against a wide range of bacterial pathogens (Saddiqe et al. 2010; Weber et al. 1994).

One of the primary mechanisms by which SJW compounds combat bacteria is through membrane disruption. The lipophilic nature of these compounds allows them to readily integrate into the bacterial cell membrane, disrupting its structural integrity and permeability. This compromises the membrane's essential functions, such as maintaining electrochemical gradients and regulating the influx and efflux of essential nutrients and waste products, ultimately leading to cell death (Saddiqe et al. 2010; Sherif et al. 2023).

Additionally, SJW compounds have been shown to inhibit bacterial growth by interfering with various cellular processes crucial for proliferation. They can inhibit enzymes involved in DNA replication, protein synthesis, and metabolic pathways, effectively stalling the growth and replication of bacterial cells.

Furthermore, SJW compounds can generate reactive oxygen species (ROS) within bacterial cells, leading to oxidative stress. ROS can damage cellular components, including proteins, lipids, and nucleic acids, ultimately causing cell death or growth inhibition. This mechanism is particularly effective against bacteria with compromised antioxidant defense systems (Saddiqe et al. 2010; Sherif et al. 2023).

Notably, SJW compounds also possess the ability to prevent the formation of biofilms, which are complex communities of bacteria embedded in a self-produced extracellular matrix. Biofilms are notoriously difficult to eradicate and contribute to the persistence of bacterial infections. By inhibiting biofilm formation, SJW compounds can enhance the efficacy of antimicrobial treatments and prevent the establishment of chronic infections (Saddiqe et al. 2010; Sherif et al. 2023).

#### 13.3.3.1.2 Antiviral Action

HP was found to show its antiviral action by inhibiting viral entry at an early stage of infection, inhibiting the enzyme reverse transcriptase, and immunomodulatory effect (Miskovsky 2002).

The antiviral action of HP is believed to occur through multiple mechanisms. First, HP inhibits viral entry at an early stage of the infection cycle, preventing the virus from gaining access to host cells. This is achieved by interfering with the binding of viral particles to cell surface receptors or by disrupting the fusion of viral and cellular membranes (Miskovsky 2002).

Second, HP has been shown to inhibit the activity of the enzyme reverse transcriptase, which is crucial for the replication of retroviruses such as HIV. Reverse transcriptase is responsible for converting the viral RNA genome into DNA, a necessary step for integration into the host cell's genome and subsequent viral replication. By inhibiting this enzyme, HP effectively blocks the replication of retroviruses (Chen et al. 2019).

Furthermore, HP exerts an immunomodulatory effect, which contributes to its antiviral properties. HP has been demonstrated to enhance the activity of various immune cells, including natural killer cells, macrophages, and lymphocytes. These immune cells play a crucial role in the body's defense against viral infections by recognizing and eliminating infected cells. By stimulating the immune response, HP aids in the clearance of viral pathogens from the host organism (Chen et al. 2019).

### 13.3.4 Anticancer Action

Hypericin was identified as the most reliable anticancer agent among the other compounds isolated from HP extracts. During the cancer treatment, hypericin was used in light therapy to induce apoptosis of mitochondrial cells and Golgi complex (Davids and Maduray 2012; Theodossiou et al. 2009). Hypericin was also effective against melanoma and breast cancer cell lines with possible mechanisms of cell shrinkage, membrane integrity loss, and independent apoptosis (Kleemann et al. 2014). The photodynamic therapy (PDT) uses hypericin as a powerful photosensitizer in cancer treatment, the organization of HUVEC endothelial cells in capillary-like endothelial cells is inhibited by it, and hypericin also restrains the angiogenesis in Matrigel sponge assay. Hyperoside induced apoptosis and cell proliferation in cancer cell lines (Liu et al. 2016). The antineoplastic potential of HP extracts was also assessed on human melanoma cells. The extract was found to show comparable phototoxicity on melanoma cells with bergapten (Menichini et al. 2013). The hypericin isolated from HP extract was assessed for anaplastic thyroid cancer combined with laser. The study revealed that hypericin efficaciously acts on anaplastic thyroid cells in hypericin-medicated photodynamic therapy and promotes cell death (Kim et al. 2018). The extract of flowers inhibited the growth of human erythroleukemic cell lines and it was found to cause apoptosis of K562 cells. These results indicate that the action was caused due to nonhypericin constituents in extracts (Valletta et al. 2018). HP extract was evaluated for its metabolic profiling and cytotoxic action against human adenocarcinoma colon Caco-2 cell lines. The study revealed that crude extract was only effective at high concentrations (Sarrou et al. 2018). The anti-proliferative and pro-apoptotic action of ethanolic extract of HP was evaluated by the researcher. The study suggests that extract inhibits the growth and promotes apoptosis of MCF-7 cells via inhibition of AMPK/mTOR and mitochondrial activation pathway in a dose-dependent manner (You et al. 2018).

#### 13.3.4.1 Mechanism of Action

The main compound hypericin was responsible for anticancer and antineoplastic action. Hypericin, a naturally occurring compound, exhibits potent anticancer properties through various mechanisms. It suppresses the production of pro-inflammatory mediators, inhibits endothelial and fibroblast growth factors, disrupts cell adhesion processes, and impedes angiogenesis. Additionally, hypericin targets mitochondrial thioredoxin, a critical regulator of redox homeostasis. Notably, this compound induces apoptosis in cancer cells by elevating the levels of caspase-3 and caspase-4, key executioners of the apoptotic pathway. Furthermore, hypericin arrests the cell cycle at the metaphase stage, leading to programmed cell death. Its multifaceted mode of action also involves modulating the expression patterns of various proteins and genes implicated in cancer progression (Alavi Dana et al. 2023; Choudhary et al. 2022; Menegazzi et al. 2020).

### 13.3.5 Wound Healing Action

Volatile oil and tincture of HP exert a wound-healing action on mild wounds and it can be useful for heat burns fire and muscle pain. The wound-healing action could be a synergistic effect of hypericin, isoquercitrin, rutin, hyperoside, and epicatechin. Another study on wound healing action stated that this effect is caused due to enhanced production and activation of fibroblast collagen cells by the active constituents of HP (Belwal et al. 2019; Süntar et al. 2010; Wölfle et al. 2013).

### 13.3.6 Other Actions

The extract of HP was found to improve the metabolic syndrome. Researchers have studied the effect of HP on diet-induced obesity and metabolic syndrome in animals. The study revealed that the extract ameliorates glucose and fat metabolism and insulin resistance. The extract was further studied for its action on hyperlipidemia and it was found to decrease the Low-density lipoprotein cholesterol (LDL-CH) and total cholesterol without interfering with triglyceride and high-density lipoprotein cholesterol (HDL-CH), improve liver profile as well as reduce oxidative damage (Ghosian Moghaddam et al. 2016; Tian et al. 2015). The study by Novelli et al. explored how *H. perforatum* extracts and hyperforin inhibits key signaling pathways activated by cytokines in pancreatic β cells, revealing their capacity to prevent changes in critical target gene expression related to β-cell functionality, differentiation, inflammatory response, and apoptosis, primarily via the inhibition of multiple phosphorylation steps in the Signal transducer and activator of transcription1 (STAT-1), nuclear factor kappa-light-chain-enhancer of activated B cells (NF-κB), and mitogen-activated protein kinase (MAPK) pathways, suggesting a potential role for *H. perforatum* extract in the prevention or treatment of immune-mediated β-cell dysfunction associated with type 1 diabetes(Novelli et al. 2016). The seed extract of *H. perforatum* was studied for antioxidant potential by Heinrich and co-workers which revealed that the xanthones are responsible for 2,2-diphenyl-1-picrylhydrazyl (DPPH) free radical scavenging action (Heinrich et al. 2017). The extract of HP is considered an effective natural remedy for the prevention and treatment of pressure sores and other dermatological diseases in hospitalized patients (https://doi.org/10.1016/j.jep.2016.12.030). Oral administration of a unit dose of HP extract was found to reduce alcohol intake in preclinical studies; this study is suggestive of its use in drug addiction and dependence (Barnes et al. 2001).

## 13.4 Adverse Effects

According to the data published by national and internal drug safety monitoring bodies, the *Hypericum perforatum* is well tolerated. It was found to produce mild adverse effects such as gastrointestinal symptoms, dizziness, confusion, tiredness, and sedation (Barnes et al. 2001). The HP has been commercially used in European countries and the United States of America as an over-the-counter medicine. While HP may offer potential benefits, there are concerns regarding its unsupervised use for self-medication of potentially serious mental health conditions, as there is no evidence supporting its efficacy or safety for such applications. Additionally, certain compounds found in St. John's Wort are potent enzyme inducers, which could lead to pharmacokinetic interactions when taken concurrently with other medications.

These factors underscore the importance of exercising caution and consulting with healthcare professionals before incorporating St. John's Wort into one's treatment regimen(Marrelli et al. 2020).

St. John's Wort (SJW), a popular herbal supplement globally, has garnered widespread acclaim for its perceived favorable risk–benefit ratio. While generally well-tolerated, mild adverse effects have been reported in certain instances. These effects are primarily attributed to the presence of naphthodianthrones (hypericin and pseudohypericin) and tend to manifest at doses exceeding the recommended maximum of 11.25 mg of total hypericin. The minimum dosages required to achieve pain-relieving effects are unlikely to cause adverse reactions or light sensitivity associated with hypericin (Henderson et al. 2002; Rodríguez-Landa and Contreras 2003; Schulz et al. 2011).

The research conducted by Kahyaoğlu and colleagues evaluated the embryotoxic and teratogenic effects of *Hypericum perforatum* (St. John's Wort) exposure in pregnant rats. Their findings revealed that pre-conception exposure to *H. perforatum* reduced pregnancy rates and decreased fetal numbers. Additionally, exposure during pregnancy extended gestation duration and impaired fetal development, causing liver and kidney tissue damage. Consequently, the authors strongly caution against recommending *H. perforatum* to pregnant women due to its potential adverse impacts on both fatal and maternal health (Kahyaoğlu et al. 2018).

The isolated compounds from HP decoction were found to show acute neuropathy and induced mania. Hypericin was found to show phototoxicity due to photodermatitis when it was administered in high doses. Excessive cutaneous phototoxicity was noticed with a high dose of hypericin related to AIDS treatment. Various in-vitro studies suggest that the HP extract also has a phototoxic effect on the eye. The hypericin and light of UV-visible range damage the human lens epithelial cells and human retinal pigment epithelial cell (Klemow et al. 2011).

## 13.5 Drug Interactions

The major drug interactions of Hypericum perforatum observed were induction of p-glycoprotein expression and hepatic cytochrome P450 enzyme (CYP2A4, CYP2C19, CYP2C9, CYP2D6, and CYP1A2). This effect is produced by hyperforin via the activation of the pregnane X receptor (PXR). Induction of CYP450 causes a decrease in drug concentration metabolized by these cytochromes or enhancement in drug action; for example, clopidogrel converts to its active form by the aid of these enzymes. Other drugs that are overwhelmed by HP are bupropion, oral contraceptives, docetaxel, rifampicin, rivaroxaban, oxycodone, oral s-ketamine, cyclosporine, tacrolimus, atorvastatin, zolpidem, metformin, and simvastatin (Andrén et al. 2007; Suryawanshi et al. 2024). Studies have demonstrated that the concurrent administration of ritonavir and ketoconazole, in combination with St. John's Wort (SJW), results in an elevated concentration of midazolam and a corresponding decrease in its clearance rate. This observation suggests that the inhibitory effects of ritonavir and ketoconazole outweigh the inductive properties of

**Table 13.4** Drug interactions of *H. perforatum*

| Name of drug | Interaction, effect, or advice | Ref. |
|---|---|---|
| Irinotecan | HP effects on metabolizing enzymes of irinotecan<br>Use irinotecan with a delay and after 2 weeks | Mathijssen (2002) |
| Indinavir, ritonavir, and saquinavir | HP reduced the plasma concentration and effectiveness by acting on CYP450 enzymes | Cvetkovic and Goa (2003); Eagling et al. (2002); Piscitelli et al. (2000) |
| Docetaxel, imatinib | HP reduces the plasma concentration and effectiveness by acting on CYP450 enzymes and leads to treatment failure in cancer patients | Frye et al. (2004); Goey et al. (2014) |
| Cyclosporin, tacrolimus | HP reduces the plasma concentration of drugs which can cause organ transplant rejection | Karliova et al. (2000); Mai et al. (2004); Ruschitzka et al. (2000) |
| Warfarin, digoxin | HP decreased the therapeutic concentration of warfarin by enhancing its clearance and the risk of clot formation increased.<br>HP decoction possesses a high amount of hyperforin alters the bioavailable dose of digoxin | Jiang et al. (2004); Mueller (2004) |
| Norethindrone, ethinylestradiol | HP decreases the half-life of the drugs by induction of CYP3A which leads to an increased chance of evaluation and bleeding, unwanted pregnancy | Hall (2003); Murphy et al. (2005) |
| Atorvastatin, simvastatin, rosuvastatin | The total cholesterol and LDL cholesterol increased in combination with HP | (Andrén et al. (2007); Hebert et al. (2004) |
| Oxycodone | HP causes increase in the metabolism of this dis and decreases its analgesic action | Nieminen et al. (2010) |
| Rifampicin | Photosensitivity | Hohmann et al. (2016) |
| Zolpidem | Decreased plasma concentration of this drug, concomitant administration of HP and zolpidem needs dose adjustment | Hojo et al. (2011) |

SJW. However, it is crucial to emphasize that these findings are contingent upon the hyperforin content present in the SJW extracts employed in the research (Fuchs et al. 2013; Hafner et al. 2010). Table 13.4 displayed data about different drug interactions of *H. perforatum*.

## 13.6 Clinical Trials on *Hypericum perforatum*

Table 13.5 summarizes the clinical trial conducted between 2022 and 2024 on *Hypericum perforatum*. This summary examines the efficacy and safety of Shugan Jieyu capsule and St. John's Wort for treating major depressive disorder (MDD). A randomized controlled trial found Shugan Jieyu capsule comparable to St. John's

**Table 13.5** Clinical trials on *H. perforatum*

| Sl. no. | Title of trail | Type of trail | Status | Reference |
|---|---|---|---|---|
| 1. | Efficacy and safety of Shugan Jieyu capsule and St. John's Wort for major depressive disorder with somatic complaints | A randomized, double-blind, positive-controlled, multicenter clinical trial | Active | Xiang et al. (2023) |

Wort in treating MDD with somatic complaints, with potentially greater benefits for male patients.

**Main Findings** Both the Shugan Jieyu capsule and St. John's Wort groups showed significant improvements in the 17-item Hamilton Depression Rating Scale (HDRS-17), but there was no significant difference between the two groups. The Shugan Jieyu capsule was more effective than St. John's Wort in reducing HDRS-17 scores in male patients, but not in female patients. The Shugan Jieyu capsule was as effective as St. John's Wort in treating major depressive disorder (MDD) patients with somatic complaints, particularly in male patients.

**Limitations** Lack of prior evidence directly comparing the two treatments (Shugan Jieyu capsule and St. John's Wort) for depression with somatic complaints. No significant difference in efficacy between the two treatments, despite both showing significant improvements. The Shugan Jieyu capsule may only be more effective than St. John's Wort for male patients, and not for female patients.

## 13.7 Future Perspectives

Recent research highlights its potential in both food and pharmaceutical applications due to its rich phytochemical profile, including naphthodianthrones, flavonoids, and essential oils (Kapoor et al. 2023). The plant demonstrates a wide range of therapeutic effects, including antidepressant, antioxidant, antimicrobial, and anti-inflammatory properties (2020). While *H. perforatum* is well-known for its efficacy in treating mild-to-moderate depression, ongoing studies explore its neuroprotective, anxiolytic, and antitumor activities (Budantsev et al. 2021). The genus *Hypericum* contains over 450 species; many of which have been used in traditional medicine for various ailments(Marrelli et al. 2016). Current research focuses on elucidating the molecular mechanisms of *H. perforatum's* actions, particularly in nervous system disorders, and investigating its potential in treating metabolic and genitourinary disorders(Budantsev et al. 2021).

Recent studies have shown promising results in pain management, with preclinical trials indicating antinociceptive and analgesic properties at low doses(Galeotti 2017). The plant contains various bioactive compounds, including hypericin and hyperforin, which contribute to its medicinal effects (Marrelli et al. 2016). Research has revealed antimicrobial, neuroprotective, antioxidant, anti-inflammatory, and wound healing properties (2020). Hypericin, a key component of *H. perforatum*, has shown antineoplastic, antitumor, and antiviral activities, as well as potential in enzyme inhibition and photodynamic therapy for cancer treatment (Kubin et al. 2005). While these findings are promising, further clinical research is needed to validate the efficacy and safety of *H. perforatum* in various therapeutic applications, particularly in pain management and other emerging areas of interest.

*Hypericum perforatum* (St. John's Wort) has been used for centuries in traditional medicine across Europe, West Asia, and North Africa (Saller and Melzer 2013). Its flowers contain numerous bioactive compounds, including naphthodianthrones, acylphloroglucinols, and flavonoids. These components contribute to its diverse therapeutic properties, including antidepressant, antioxidant, and anti-inflammatory effects (Kapoor et al. 2023; Saller and Melzer 2013). *H. perforatum's* plurivalent nature makes it valuable for treating various conditions, such as gastrointestinal complaints, skin diseases, and depression (Saller and Melzer 2013). Its potential extends to complementary cancer treatment and cardiovascular health (Saller and Melzer 2013). In the Carpathian Mountains, *H. perforatum* is among the most preferred botanical remedies used by local ethnic groups (Kozlowska et al. 2018). The plant's versatility and therapeutic potential make it a promising candidate for integration into both conventional and complementary medicine, as well as for use in food and pharmaceutical applications (Kapoor et al. 2023; Kozlowska et al. 2018).

## References


Abdel-Salam OME (2005) Anti-inflammatory, antinociceptive, and gastric effects of *Hypericum perforatum* in rats. Sci World J 5:586–595. https://doi.org/10.1100/tsw.2005.78

Alavi Dana SMM, Khoshnazar M, Banyarani SP, Mokhtari FS, Mohammadie ZE, Shamsehkohan A, Valizadeh M, Ahangar-Sirous R, Rahimi Z, Poudineh M, Deravi N (2023) Anti-cancer potential of Hypericum spp. with focus on Hypericum Perforatum: a review of the literature. Curr Tradit Med 9. https://doi.org/10.2174/2215083808666220822141221

Althobaiti NA, Menaa F, Dalzell JJ, Albalawi AE, Ismail H, Alghuthaymi MA, Aldawsari RD, Iqbal H, McAlinney C, Green BD (2022) Ethnomedicinal plants with protective effects against Beta-amyloid peptide (aβ)1-42 indicate therapeutic potential in a new in vivo model of Alzheimer's disease. Antioxidants. 11:1865. https://doi.org/10.3390/antiox11101865

Alzahrani MA, Ofisan SB, Alshumaymiri NI, Alghuwainem M, Altamimi M, Alali AY, Rabie M, AboSkena AK, Almaymuni K, Almannie R, Binsaleh S (2023) Effect of St. John's Wort (Hypericum perforatum L) on male sexual and reproductive health: a narrative review. Biomedicines 11:2800. https://doi.org/10.3390/biomedicines11102800

Andrén L, Andreasson Å, Eggertsen R (2007) Interaction between a commercially available St. John's wort product (Movina) and atorvastatin in patients with hypercholesterolemia. Eur J Clin Pharmacol 63:913–916. https://doi.org/10.1007/s00228-007-0345-x

Apaydin EA, Maher AR, Shanman R, Booth MS, Miles JNV, Sorbero ME, Hempel S (2016) A systematic review of St. John's wort for major depressive disorder. Syst Rev 5:148. https://doi.org/10.1186/s13643-016-0325-2

Assiri K, Alyami Y, Uyanik JM, Romero-Reyes M (2017) Hypericum perforatum (St. John's Wort) as a possible therapeutic alternative for the management of trigeminal neuralgia (TN – a case report). Complement Ther Med 30:36–39. https://doi.org/10.1016/j.ctim.2016.10.014

Barnes J, Anderson LA, Phillipson JD (2001) St John's wort (*Hypericum perforatum* L.): a review of its chemistry, pharmacology and clinical properties. J Pharm Pharmacol 53:583–600. https://doi.org/10.1211/0022357011775910

Belwal T, Devkota HP, Singh MK, Sharma R, Upadhayay S, Joshi C, Bisht K, Gour JK, Bhatt ID, Rawal RS, Pande V (2019) St. John's Wort (Hypericum perforatum). In: Nonvitamin and nonmineral nutritional supplements. Elsevier, pp 415–432

Benedí J, Arroyo R, Romero C, Martín-Aragón S, Villar AM. Antioxidant properties and protective effects of a standardized extract of Hypericum perforatum on hydrogen peroxide-induced oxidative damage in PC12 cells. Life Sci 75, 1263–1276 (2004). https://doi.org/10.1016/j.lfs.2004.05.001

Bezáková L, Psenák M, Kartnig T (1999) effect of dianthrones and their precursors from Hypericum perforatum L. on lipoxygenase activity. Pharmazie 54:711

Bladt S, Wagner H (1994) Inhibition of MAO by fractions and constituents of Hypericum extract. J Geriatr Psychiatry Neurol 7:57–59. https://doi.org/10.1177/089198879400700115

Bonaterra GA, Schwendler A, Hüther J, Schwarzbach H, Schwarz A, Kolb C, Abdel-Aziz H, Kinscherf R (2018) Neurotrophic, Cytoprotective, and anti-inflammatory effects of St. John's Wort extract on differentiated mouse hippocampal HT-22 neurons. Front Pharmacol 8. https://doi.org/10.3389/fphar.2017.00955

Booker A, Agapouda A, Frommenwiler DA, Scotti F, Reich E, Heinrich M (2018) St John's wort ( Hypericum perforatum ) products – an assessment of their authenticity and quality. Phytomedicine 40:158–164. https://doi.org/10.1016/j.phymed.2017.12.012

Bouron A, Lorrain E (2014) Effets cellulaires et moléculaires de l'hyperforine, un antidépresseur végétal : revue de la littérature. Encéphale 40:108–113. https://doi.org/10.1016/j.encep.2013.03.004

Bréard D, Viault G, Mezier M-C, Pagie S, Bruguière A, Richomme P, Charreau B, Derbré S (2018) Additional insights into *Hypericum perforatum* content: isolation, Total synthesis, and absolute configuration of Hyperbiphenyls A and B from immunomodulatory root extracts. J Nat Prod 81:1850–1859. https://doi.org/10.1021/acs.jnatprod.8b00325

Brenn A, Grube M, Jedlitschky G, Fischer A, Strohmeier B, Eiden M, Keller M, Groschup MH, Vogelgesang S (2014) St. John's Wort reduces beta-amyloid accumulation in a double transgenic Alzheimer's disease mouse model-role of P-glycoprotein. Brain Pathol 24:18–24. https://doi.org/10.1111/bpa.12069

Brockmöller J, Reum T, Bauer S, Kerb R, Hübner W-D, Roots I (1997) Hypericin and Pseudohypericin: pharmacokinetics and effects on photosensitivity in humans. Pharmacopsychiatry 30:94–101. https://doi.org/10.1055/s-2007-979527

Budantsev AL, Prikhodko VA, Varganova IV, Okovityi SV (2021) Biological activity of hypericum Perforatum L. (hypericaceae): a review. Pharm Pharmacol 9:17–31. https://doi.org/10.19163/2307-9266-2021-9-1-17-31

Bukhari IA, Dar A, Khan RA (2004) Antinociceptive activity of methanolic extracts of St. John's Wort (Hypericum perforatum) preparation. Pak J Pharm Sci 17:13–19

Butterweck V, Petereit F, Winterhoff H, Nahrstedt A (1998) Solubilized Hypericin and Pseudohypericin from *Hypericum perforatum* exert antidepressant activity in the forced swimming test [3]. Planta Med 64:291–294. https://doi.org/10.1055/s-2006-957437

Butterweck V, Korte B, Winterhoff H (2001) Pharmacological and endocrine effects of Hypericum Perforatum and Hypericin after repeated treatment. Pharmacopsychiatry 34:2–7. https://doi.org/10.1055/s-2001-15508

Calapai G, Crupi A, Firenzuoli F, Inferrera G, Ciliberto G, Parisi A, De Sarro G, Caputi AP (2001) Interleukin-6 involvement in antidepressant action of Hypericum Perforatum. Pharmacopsychiatry 34:8–10. https://doi.org/10.1055/s-2001-15507

Can ÖD, Özkay ÜD (2012) Effects of *Hypericum montbretti* extract on the central nervous system and involvement of GABA (A)/benzodiazepine receptors in its pharmacological activity. Phytother Res 26:1695–1700. https://doi.org/10.1002/ptr.4629

Can ÖD, Öztürk Y, Öztürk N, Sagratini G, Ricciutelli M, Vittori S, Maggi F (2011) Effects of treatment with St. John's Wort on blood glucose levels and pain perceptions of streptozotocin-diabetic rats. Fitoterapia 82:576–584. https://doi.org/10.1016/j.fitote.2011.01.008

Cao Z, Wang F, Xiu C, Zhang J, Li Y (2017) Hypericum perforatum extract attenuates behavioral, biochemical, and neurochemical abnormalities in aluminum chloride-induced Alzheimer's disease rats. Biomed Pharmacother 91:931–937. https://doi.org/10.1016/j.biopha.2017.05.022

Cervo L, Rozio M, C. E-S, Guiso G, Morazzoni P, Caccia S (2002) Role of hyperforin in the antidepressant-like activity of Hypericum perforatum extracts. Psychopharmacology 164:423–428. https://doi.org/10.1007/s00213-002-1229-5

Chang Y, Wang S-J (2010) Hypericin, the active component of St. John's wort, inhibits glutamate release in the rat cerebrocortical synaptosomes via a mitogen-activated protein kinase-dependent pathway. Eur J Pharmacol 634:53–61. https://doi.org/10.1016/j.ejphar.2010.02.035

Chen H, Muhammad I, Zhang Y, Ren Y, Zhang R, Huang X, Diao L, Liu H, Li X, Sun X, Abbas G, Li G (2019) Antiviral activity against infectious bronchitis virus and bioactive components of Hypericum perforatum L. Front Pharmacol 10. https://doi.org/10.3389/fphar.2019.01272

Choudhary N, Collignon TE, Tewari D, Bishayee A (2022) Hypericin and its anticancer effects: from mechanism of action to potential therapeutic application. Phytomedicine 105:154356. https://doi.org/10.1016/j.phymed.2022.154356

Cinci L, Di Cesare Mannelli L, Maidecchi A, Mattoli L, Ghelardini C (2017) Effects of *Hypericum perforatum* extract on oxaliplatin-induced neurotoxicity: in vitro evaluations. Z Naturforsch C 72:219–226. https://doi.org/10.1515/znc-2016-0194

Clarke R (2010) Are modern preparation methods compromising the efficacy of Chinese herbal medicines? An investigation using Guan Ye Lian Qiao (Hypericum Perforatum L.)

Crupi R, Abusamra Y, Spina E, Calapai G (2013) Preclinical data supporting/refuting the use of Hypericum perforatum in the treatment of depression. CNS Neurol Disord Drug Targets 12:474–486. https://doi.org/10.2174/1871527311312040006

Cvetkovic RS, Goa KL (2003) Lopinavir/Ritonavir. Drugs 63:769–802. https://doi.org/10.2165/00003495-200363080-00004

Davids LM, Maduray M (2012) The anticancer activity of Hypericin in photodynamic therapy. J Bioanal Biomed s6. https://doi.org/10.4172/1948-593X.S6-004

De Marchis G, Bürgi S, Kientsch U, Honegger U (2006) vitamin E reduces antidepressant-related β-Adrenoceptor Down-regulation in cultured cells. Comparable effects on St. John's Wort and tricyclic antidepressant treatment. Planta Med 72:1436–1437. https://doi.org/10.1055/s-2006-951727

Dinamarca MC, Cerpa W, Garrido J, Hancke JL, Inestrosa NC (2006) Hyperforin prevents β-amyloid neurotoxicity and spatial memory impairments by disaggregation of Alzheimer's amyloid-β-deposits. Mol Psychiatry 11:1032–1048. https://doi.org/10.1038/sj.mp.4001866

do Rego J-C, Benkiki N, Chosson E, Kabouche Z, Seguin E, Costentin J (2007) Antidepressant-like effect of hyperfoliatin, a polyisoprenylated phloroglucinol derivative from Hypericum perfoliatum (Clusiaceae) is associated with an inhibition of neuronal monoamines uptake. Eur J Pharmacol 569:197–203. https://doi.org/10.1016/j.ejphar.2007.05.008

Eagling VA, Wiltshire H, Whitcombe IWA, Back DJ (2002) CYP3A4-mediated hepatic metabolism of the HIV-1 protease inhibitor saquinavir *in vitro*. Xenobiotica 32:1–17. https://doi.org/10.1080/00498250110085845

El Hamdaoui Y, Zheng F, Fritz N, Ye L, Tran MA, Schwickert K, Schirmeister T, Braeuning A, Lichtenstein D, Hellmich UA, Weikert D, Heinrich M, Treccani G, Schäfer MKE, Nowak G, Nürnberg B, Alzheimer C, Müller CP, Friedland K (2022) Analysis of hyperforin (St. John's wort) action at TRPC6 channel leads to the development of a new class of antidepressant drugs. Mol Psychiatry 27:5070–5085. https://doi.org/10.1038/s41380-022-01804-3

Forsdike K, Pirotta M (2019) St John's wort for depression: scoping review about perceptions and use by general practitioners in clinical practice. J Pharm Pharmacol 71:117–128. https://doi.org/10.1111/jphp.12775

Francis AJP (2005) Antidepressant action of St. John's Wort, Hypericum perforatum: a test of the circadian hypotheses. Phytomedicine 12:167–172. https://doi.org/10.1016/j.phymed.2004.04.004

Frye R, Fitzgerald S, Lagattuta T, Hruska M, Egorin M (2004) Effect of St John's wort on imatinib mesylate pharmacokinetics. Clin Pharmacol Ther 76:323–329. https://doi.org/10.1016/j.clpt.2004.06.007

Fuchs I, Hafner-Blumenstiel V, Markert C, Burhenne J, Weiss J, Haefeli WE, Mikus G (2013) Effect of the CYP3A inhibitor ketoconazole on the PXR-mediated induction of CYP3A activity. Eur J Clin Pharmacol 69:507–513. https://doi.org/10.1007/s00228-012-1388-1

Galeotti N (2017) Hypericum perforatum (St John's wort) beyond depression: A therapeutic perspective for pain conditions. J Ethnopharmacol 200:136–146. https://doi.org/10.1016/j.jep.2017.02.016

Galeotti N, Ghelardini C (2013a) Reversal of NO-induced nociceptive hypersensitivity by St. John's wort and hypericin: NF-κB, CREB and STAT1 as molecular targets. Psychopharmacology 227:149–163. https://doi.org/10.1007/s00213-012-2950-3

Galeotti N, Ghelardini C (2013b) St. John's wort relieves pain in an animal model of migraine. Eur J Pain 17:369–381. https://doi.org/10.1002/j.1532-2149.2012.00196.x

Galeotti N, Vivoli E, Bilia AR, Bergonzi MC, Bartolini A, Ghelardini C (2010a) A prolonged protein kinase C-mediated, opioid-related Antinociceptive effect of St John's Wort in mice. J Pain 11:149–159. https://doi.org/10.1016/j.jpain.2009.06.013

Galeotti N, Vivoli E, Bilia AR, Vincieri FF, Ghelardini C (2010) St. John's Wort reduces neuropathic pain through a hypericin-mediated inhibition of the protein kinase C γ and ε activity. Biochem Pharmacol 79:1327–1336. https://doi.org/10.1016/j.bcp.2009.12.016

Ghosian Moghaddam M, Roghani M, Maleki M (2016) Effect of Hypericum perforatum aqueous extracts on serum lipids, aminotransferases, and lipid peroxidation in hyperlipidemic rats. Res Cardiovasc Med 5:5. https://doi.org/10.5812/cardiovascmed.31326

Gibon J, Deloulme J-C, Chevallier T, Ladevèze E, Abrous DN, Bouron A (2013) The antidepressant hyperforin increases the phosphorylation of CREB and the expression of TrkB in a tissue-specific manner. Int J Neuropsychopharmacol 16:189–198. https://doi.org/10.1017/S146114571100188X

Goey AKL, Meijerman I, Rosing H, Marchetti S, Mergui-Roelvink M, Keessen M, Burgers JA, Beijnen JH, Schellens JHM (2014) The effect of St John's Wort on the pharmacokinetics of docetaxel. Clin Pharmacokinet 53:103–110. https://doi.org/10.1007/s40262-013-0102-5

Greeson JM, Sanford B, Monti DA (2001) St. John's wort (Hypericum perforatum): a review of the current pharmacological, toxicological, and clinical literature. Psychopharmacology 153:402–414. https://doi.org/10.1007/s002130000625

Hafner V, Jäger M, Matthée A-K, Ding R, Burhenne J, Haefeli WE, Mikus G (2010) Effect of simultaneous induction and inhibition of CYP3A by St John's Wort and ritonavir on CYP3A activity. Clin Pharmacol Ther 87:191–196. https://doi.org/10.1038/clpt.2009.206

Hall S (2003) The interaction between St John's wort and an oral contraceptive. Clin Pharmacol Ther 74:525–535. https://doi.org/10.1016/j.clpt.2003.08.009

Hasanein P, Shahidi S (2011) Effects of *Hypericum perforatum* extract on diabetes-induced learning and memory impairment in rats. Phytother Res 25:544–549. https://doi.org/10.1002/ptr.3298

Hatanaka J, Shinme Y, Kuriyama K, Uchida A, Kou K, Uchida S, Yamada S, Onoue S (2011) In vitro and in vivo characterization of new formulations of St. John's Wort extract with improved pharmacokinetics and anti-nociceptive effect. Drug Metab Pharmacokinet 26:551–558. https://doi.org/10.2133/dmpk.DMPK-11-RG-041

Hebert MF, Park JM, Chen Y, Akhtar S, Larson AM (2004) effects of St. John's Wort ( *Hypericum perforatum* ) on tacrolimus pharmacokinetics in healthy volunteers. J Clin Pharmacol 44:89–94. https://doi.org/10.1177/0091270003261078

Heinrich M, Lorenz P, Daniels R, Stintzing FC, Kammerer DR (2017) Lipid and phenolic constituents from seeds of *Hypericum perforatum* L. and *Hypericum tetrapterum* Fr. and their antioxidant activity. Chem Biodivers 14. https://doi.org/10.1002/cbdv.201700100

Henderson L, Yue QY, Bergquist C, Gerden B, Arlett P (2002) St John's wort ( *Hypericum perforatum* ): drug interactions and clinical outcomes. Br J Clin Pharmacol 54:349–356. https://doi.org/10.1046/j.1365-2125.2002.01683.x

Herraiz T, Guillén H (2018) Monoamine oxidase-A inhibition and associated antioxidant activity in plant extracts with potential antidepressant actions. Biomed Res Int 2018:1–10. https://doi.org/10.1155/2018/4810394

Hirano K, Kato Y, Uchida S, Sugimoto Y, Yamada J, Umegaki K, Yamada S (2004) Effects of oral administration of extracts of *Hypericum perforatum* (St John's wort) on brain serotonin transporter, serotonin uptake and behaviour in mice. J Pharm Pharmacol 56:1589–1595. https://doi.org/10.1211/0022357045039

Hofrichter J, Krohn M, Schumacher T, Lange C, Feistel B, Walbroel B, Heinze H-J, Crockett S, F.Sharbel T, Pahnke J (2013) Reduced Alzheimer's disease pathology by St. John's Wort treatment is independent of Hyperforin and facilitated by ABCC1 and microglia activation in mice. Curr Alzheimer Res 10:1057–1069. https://doi.org/10.2174/15672050113106660171

Hohmann N, Maus A, Carls A, Haefeli WE, Mikus G (2016) St. John's wort treatment in women bears risks beyond pharmacokinetic drug interactions. Arch Toxicol 90:1013–1015. https://doi.org/10.1007/s00204-015-1532-7

Hojo Y, Echizenya M, Ohkubo T, Shimizu T (2011) Drug interaction between St John's wort and zolpidem in healthy subjects. J Clin Pharm Ther 36:711–715. https://doi.org/10.1111/j.1365-2710.2010.01223.x

Hu L, Xue Y, Zhang J, Zhu H, Chen C, Li X-N, Liu J, Wang Z, Zhang Y, Zhang Y (2016) (±)-Japonicols A–D, Acylphloroglucinol-based meroterpenoid enantiomers with Anti-KSHV activities from *Hypericum japonicum*. J Nat Prod 79:1322–1328. https://doi.org/10.1021/acs.jnatprod.5b01119

Huang J, Zhou L, Chen J, Chen T, Lei B, Zheng N, Wan X, Xu J, Wang T (2021) Hyperoside attenuate inflammation in HT22 cells via upregulating SIRT1 to activities Wnt/β-catenin and sonic hedgehog pathways. Neural Plast 2021:1–10. https://doi.org/10.1155/2021/8706400

Ivetic V, Trivic S, Pogancev MK, Popovic M, Zlinská J (2011) Effects of St John's Wort (Hypericum perforatum L.) extracts on Epileptogenesis. Molecules 16:8062–8075. https://doi.org/10.3390/molecules16098062

Jiang X, Williams KM, Liauw WS, Ammit AJ, Roufogalis BD, Duke CC, Day RO, McLachlan AJ (2004) Effect of St John's wort and ginseng on the pharmacokinetics and pharmacodynamics of warfarin in healthy subjects. Br J Clin Pharmacol 57:592–599. https://doi.org/10.1111/j.1365-2125.2003.02051.x

Kahyaoğlu F, Gökçimen A, Demirci B (2018) Investigation of the embryotoxic and teratogenic effect of Hypericum perforatum in pregnant rats. J Turkish Soc Obstet Gynecol 15:87–90. https://doi.org/10.4274/tjod.84429

Kapoor S, Chandel R, Kaur R, Kumar S, Kumar R, Janghu S, Kaur A, Kumar V (2023) The flower of Hypericum perforatum L.: a traditional source of bioactives for new food and pharmaceutical applications. Biochem Syst Ecol 110:104702. https://doi.org/10.1016/j.bse.2023.104702

Karliova M, Treichel U, Malagò M, Frilling A, Gerken G, Broelsch CE (2000) interaction of Hypericum perforatum (St. John's wort) with cyclosporin A metabolism in a patient after liver transplantation. J Hepatol 33:853–855. https://doi.org/10.1016/S0168-8278(00)80321-9

Kim H, Kim SW, Seok KH, Hwang CW, Ahn J-C, Jin J-O, Kang HW (2018) Hypericin-assisted photodynamic therapy against anaplastic thyroid cancer. Photodiagn Photodyn Ther 24:15–21. https://doi.org/10.1016/j.pdpdt.2018.08.008

Kleemann B, Loos B, Scriba TJ, Lang D, Davids LM (2014) St John's Wort (Hypericum perforatum L.) Photomedicine: Hypericin-photodynamic therapy induces metastatic melanoma cell death. PLoS One 9:e103762. https://doi.org/10.1371/journal.pone.0103762

Klemow KM, Bartlow A, Crawford J, Kocher N, Shah J, Ritsick M (2011) Medical attributes of St. John's Wort (Hypericum perforatum)

Kozlowska W, Wagner C, Moore EM, Matkowski A, Komarnytsky S (2018) Botanical provenance of traditional medicines from Carpathian Mountains at the Ukrainian-polish border. Front Pharmacol 9. https://doi.org/10.3389/fphar.2018.00295

Kraus B, Wolff H, Heilmann J, Elstner EF (2007) Influence of Hypericum perforatum extract and its single compounds on amyloid-β mediated toxicity in microglial cells. Life Sci 81:884–894. https://doi.org/10.1016/j.lfs.2007.07.020

Krishtal O, Lozovaya N, Fisunov A, Tsintsadze T, Pankratov Y, Kopanitsa M, Chatterjee SS (2001) Modulation of ion channels in rat neurons by the constituents of Hypericum Perforatum. Pharmacopsychiatry 34:74–82. https://doi.org/10.1055/s-2001-15510

Kubin A, Wierrani F, Burner U, Alth G, Grunberger W (2005) Hypericin – the facts about a controversial agent. Curr Pharm Des 11:233–253. https://doi.org/10.2174/1381612053382287

Kulkarni SK, Akula KK, Deshpande J (2012) Evaluation of antidepressant-like activity of novel water-soluble curcumin formulations and St. John's Wort in behavioral paradigms of despair. Pharmacology 89:83–90. https://doi.org/10.1159/000335660

Kumar V, Singh PN, Bhattacharya SK (2001) Anti-inflammatory and analgesic activity of Indian Hypericum perforatum L. Indian J Exp Biol 39:339–343

Kumar V, Khanna VK, Seth PK, Singh PN, Bhattacharya SK (2002) Brain neurotransmitter receptor binding and nootropic studies on Indian *Hypericum perforatum* Linn. Phytother Res 16:210–216. https://doi.org/10.1002/ptr.1101

Kumar A, Garg R, Prakash AK (2010) effect of St. John's Wort (Hypericum perforatum) treatment on restraint stress-induced behavioral and biochemical alteration in mice. BMC Complement Altern Med 10:18. https://doi.org/10.1186/1472-6882-10-18

Lavie G, Valentine F, Levin B, Mazur Y, Gallo G, Lavie D, Weiner D, Meruelo D (1989) Studies of the mechanisms of action of the antiretroviral agents hypericin and pseudohypericin. Proc Natl Acad Sci 86:5963–5967. https://doi.org/10.1073/pnas.86.15.5963

Lennihan B (2017) Homeopathy for pain management. Alter Complemen Therap 23:176–183. https://doi.org/10.1089/act.2017.29129.ble

Leuner K, Li W, Amaral MD, Rudolph S, Calfa G, Schuwald AM, Harteneck C, Inoue T, Pozzo-Miller L (2013) Hyperforin modulates dendritic spine morphology in hippocampal pyramidal neurons by activating Ca $^{2+}$ -permeable TRPC6 channels. Hippocampus 23:40–52. https://doi.org/10.1002/hipo.22052

Liu Y, Liu G, Mei J, Wang J (2016) The preventive effects of hyperoside on lung cancer in vitro by inducing apoptosis and inhibiting proliferation through Caspase-3 and P53 signaling pathway. Biomed Pharmacother 83:381–391. https://doi.org/10.1016/j.biopha.2016.06.035

Lopez-Bazzocchi I, Hudson JB, Towers GHN (1991) Antiviral activity of the photoactive plant pigment hypericin. Photochem Photobiol 54:95–98. https://doi.org/10.1111/j.1751-1097.1991.tb01990.x

Lu Y-H, Du C-B, Liu J-W, Hong W, Wei D-Z (2004) Neuroprotective effects of *Hypericum perforatum* on trauma induced by hydrogen peroxide in PC12 cells. Am J Chin Med (Gard City N Y) 32:397–405. https://doi.org/10.1142/S0192415X04002053

Lyles JT, Kim A, Nelson K, Bullard-Roberts AL, Hajdari A, Mustafa B, Quave CL (2017) The chemical and antibacterial evaluation of St. John's Wort oil macerates used in Kosovar traditional medicine. Front Microbiol 8. https://doi.org/10.3389/fmicb.2017.01639

Mai I, Bauer S, Perloff E, Johne A, Uehleke B, Frank B, Budde K, Roots I (2004) Hyperforin content determines the magnitude of the St John's wort–cyclosporine drug interaction. Clin Pharmacol Ther 76:330–340. https://doi.org/10.1016/j.clpt.2004.07.004

Marrelli M, Statti G, Conforti F, Menichini F (2016) New potential pharmaceutical applications of Hypericum species. Mini-Rev Med Chem 16:710–720. https://doi.org/10.2174/1389557515666150709105844

Marrelli M, Statti G, Conforti F (2020) *Hypericum* spp.: An update on the biological activities and metabolic profiles. Mini-reviews. Med Chem 20:66–87. https://doi.org/10.2174/1389557519666190926120211

Mathijssen RHJ (2002) Effects of St. John's Wort on Irinotecan metabolism. CancerSpectrum knowledge Environment 94:1247–1249. https://doi.org/10.1093/jnci/94.16.1247

Menegazzi M, Masiello P, Novelli M (2020) Anti-tumor activity of Hypericum perforatum L. and Hyperforin through modulation of inflammatory signaling, ROS generation and proton dynamics. Antioxidants 10:18. https://doi.org/10.3390/antiox10010018

Menichini G, Alfano C, Marrelli M, Toniolo C, Provenzano E, Statti GA, Nicoletti M, Menichini F, Conforti F (2013) Hypericum perforatum L. subsp. *perforatum* induces inhibition of free radicals and enhanced phototoxicity in human melanoma cells under ultraviolet light. Cell Prolif 46:193–202. https://doi.org/10.1111/cpr.12020

Misane I, Ögren SO (2001) effects of Hypericum perforatum (St. John's wort) on passive avoidance in the rat: evaluation of potential neurochemical mechanisms underlying its antidepressant activity. Pharmacopsychiatry 34:89–97. https://doi.org/10.1055/s-2001-15449

Miskovsky P (2002) Hypericin – a new antiviral and antitumor photosensitizer: mechanism of action and interaction with biological macromolecules. Curr Drug Targets 3. https://doi.org/10.2174/1389450023348091

Mohanasundari M, Srinivasan MS, Sethupathy S, Sabesan M (2006) Enhanced neuroprotective effect by combination of bromocriptine and Hypericum perforatum extract against MPTP-induced neurotoxicity in mice. J Neurol Sci 249:140–144. https://doi.org/10.1016/j.jns.2006.06.018

Montoya A, Daza A, Muñoz D, Ríos K, Taylor V, Cedeño D, Vélez ID, Echeverri F, Robledo SM (2015) Development of a novel formulation with Hypericin to treat cutaneous Leishmaniasis based on photodynamic therapy in *in vitro* and *in vivo* studies. Antimicrob Agents Chemother 59:5804–5813. https://doi.org/10.1128/AAC.00545-15

Mueller S (2004) Effect of St John's wort dose and preparations on the pharmacokinetics of digoxin*1. Clin Pharmacol Ther 75:546–557. https://doi.org/10.1016/j.clpt.2004.01.014

Müller W (2003) Current St. John's wort research from mode of action to clinical efficacy. Pharmacol Res 47:101–109. https://doi.org/10.1016/S1043-6618(02)00266-9

Müller W, Rolli M, Schäfer C, Hafner U (1997) Effects of Hypericum extract (LI 160) in biochemical models of antidepressant activity. Pharmacopsychiatry 30:102–107. https://doi.org/10.1055/s-2007-979528

Murphy PA, Kern SE, Stanczyk FZ, Westhoff CL (2005) Interaction of St. John's Wort with oral contraceptives: effects on the pharmacokinetics of norethindrone and ethinyl estradiol, ovarian activity and breakthrough bleeding. Contraception 71:402–408. https://doi.org/10.1016/j.contraception.2004.11.004

Nahrstedt A, Butterweck V (1997) Biologically active and other chemical constituents of the herb of *Hypericum perforatum* L. Pharmacopsychiatry 30:129–134. https://doi.org/10.1055/s-2007-979533

Ng QX, Venkatanarayanan N, Ho CYX (2017) Clinical use of Hypericum perforatum (St John's wort) in depression: a meta-analysis. J Affect Disord 210:211–221. https://doi.org/10.1016/j.jad.2016.12.048

Nieminen TH, Hagelberg NM, Saari TI, Neuvonen M, Laine K, Neuvonen PJ, Olkkola KT (2010) St John's wort greatly reduces the concentrations of oral oxycodone. Eur J Pain 14:854–859. https://doi.org/10.1016/j.ejpain.2009.12.007

Nobakht SZ, Akaberi M, Mohammadpour AH, Tafazoli Moghadam A, Emami SA (2022) Hypericum perforatum: traditional uses, clinical trials, and drug interactions. Iran J Basic Med Sci 25:1045–1058. https://doi.org/10.22038/IJBMS.2022.65112.14338

Novelli M, Menegazzi M, Beffy P, Porozov S, Gregorelli A, Giacopelli D, De Tata V, Masiello P (2016) St. John's wort extract and hyperforin inhibit multiple phosphorylation steps of cytokine signaling and prevent inflammatory and apoptotic gene induction in pancreatic β cells. Int J Biochem Cell Biol 81:92–104. https://doi.org/10.1016/j.biocel.2016.10.017

Oliveira AI, Pinho C, Sarmento B, Dias ACP (2016) Neuroprotective activity of Hypericum perforatum and its major components. Front Plant Sci 7. https://doi.org/10.3389/fpls.2016.01004

Panossian AG, Gabrielian E, Manvelian V, Jurcic K, Wagner H (1996) Immunosuppressive effects of hypericin on stimulated human leukocytes: inhibition of the arachidonic acid release, leukotriene B4 and interleukin-Iα production, and activation of nitric oxide formation. Phytomedicine 3:19–28. https://doi.org/10.1016/S0944-7113(96)80005-5

Piscitelli SC, Burstein AH, Chaitt D, Alfaro RM, Falloon J (2000) Indinavir concentrations and St John's wort. Lancet 355:547–548. https://doi.org/10.1016/S0140-6736(99)05712-8

Prenner L, Sieben A, Zeller K, Weiser D, Häberlein H (2007) Reduction of high-affinity $\beta_2$-adrenergic receptor binding by Hyperforin and Hyperoside on rat C6 glioblastoma cells measured by fluorescence correlation spectroscopy. Biochemistry 46:5106–5113. https://doi.org/10.1021/bi6025819

Pu X, Liang J, Wang X, Xu T, Hua L, Shang R, Liu Y, Xing Y (2009) Anti-influenza A virus effect of Hypericum perforatum L. extract. Virol Sin 24:19–27. https://doi.org/10.1007/s12250-009-2983-x

Radulović NS, Genčić MS, Stojanović NM, Randjelović PJ, Baldovini N, Kurteva V (2018) Prenylated β-diketones, two new additions to the family of biologically active Hypericum perforatum L. (Hypericaceae) secondary metabolites. Food Chem Toxicol 118:505–513. https://doi.org/10.1016/j.fct.2018.05.009

Reichling J, Weseler A, Saller R (2001) A Current review of the antimicrobial activity of Hypericum perforatum L. Pharmacopsychiatry 34:116–118. https://doi.org/10.1055/s-2001-15514

Robson NKB (2016) And then came molecular phylogenetics—reactions to a monographic study of Hypericum (Hypericaceae). Phytotaxa 255:181. https://doi.org/10.11646/phytotaxa.255.3.1

Rodríguez-Landa JF, Contreras CM (2003) A review of clinical and experimental observations about antidepressant actions and side effects produced by Hypericum perforatum extracts. Phytomedicine 10:688–699. https://doi.org/10.1078/0944-7113-00340

Ruschitzka F, Meier PJ, Turina M, Lüscher TF, Noll G (2000) Acute heart transplant rejection due to Saint John's wort. Lancet 355:548–549. https://doi.org/10.1016/S0140-6736(99)05467-7

Saddiqe Z, Naeem I, Maimoona A (2010) A review of the antibacterial activity of Hypericum perforatum L. J Ethnopharmacol 131:511–521. https://doi.org/10.1016/j.jep.2010.07.034

Saller R, Melzer J (2013) Multimorbidität und Multi-Target-Therapie in der Phytotherapie. Complement Med Res 20:1–1. https://doi.org/10.1159/000351724

Sánchez-Reus MI, Gómez del Rio MA, Iglesias I, Elorza M, Slowing K, Benedí J (2007) Standardized Hypericum perforatum reduces oxidative stress and increases gene expression of antioxidant enzymes on rotenone-exposed rats. Neuropharmacology 52:606–616. https://doi.org/10.1016/j.neuropharm.2006.09.003

Sanna MD, Ghelardini C, Galeotti N (2016) St. John's Wort potentiates anti-nociceptive effects of morphine in mice models of neuropathic pain. Pain Med:pnw241. https://doi.org/10.1093/pm/pnw241

Sarrou E, Giassafaki L-P, Masuero D, Perenzoni D, Vizirianakis IS, Irakli M, Chatzopoulou P, Martens S (2018) Metabolomics assisted fingerprint of Hypericum perforatum chemotypes and assessment of their cytotoxic activity. Food Chem Toxicol 114:325–333. https://doi.org/10.1016/j.fct.2018.02.057

Schmidt M, Butterweck V (2015) The mechanisms of action of St. John's wort: an update. Wien Med Wochenschr 165:229–235. https://doi.org/10.1007/s10354-015-0372-7

Schulz H-U, Schürer M, Bässler D, Weiser D (2011) Investigation of the effect on photosensitivity following multiple Oral dosing of two different Hypericum extracts in healthy men. Arzneimittelforschung 56:212–221. https://doi.org/10.1055/s-0031-1296713

Sherif MM, Elshikh HH, Abdel-Aziz MM, Elaasser MM, Yosri M (2023) *In vitro* antibacterial and phytochemical screening of *Hypericum perforatum* extract as potential antimicrobial agents against multi-drug-resistant (MDR) strains of clinical origin. Biomed Res Int 2023. https://doi.org/10.1155/2023/6934398

Silva BA, Dias ACP, Ferreres F, Malva JO, Oliveira CR (2004) neuroprotective effect ofH. perforatum extracts on β-amyloid-induced neurotoxicity. Neurotox Res 6:119–130. https://doi.org/10.1007/BF03033214

Silva B, Oliveira PJ, Dias A, Malva JO (2008) Quercetin, kaempferol and biapigenin fromhypericum perforatum are neuroprotective against excitotoxic insults. Neurotox Res 13:265–279. https://doi.org/10.1007/BF03033510

Soleymani S, Bahramsoltani R, Rahimi R, Abdollahi M (2017) Clinical risks of St John's Wort ( *Hypericum perforatum* ) co-administration. Expert Opin Drug Metab Toxicol 13:1047–1062. https://doi.org/10.1080/17425255.2017.1378342

Song A, Wu Z, Zhao W, Shi W, Cheng R, Jiang J, Ni Z, Qu H, Qiaolongbatu X, Fan G, Lou Y (2022) The role and mechanism of Hyperoside against depression-like behavior in mice via the NLRP1 Inflammasome. Medicina (B Aires) 58:1749. https://doi.org/10.3390/medicina58121749

Süntar IP, Akkol EK, Yılmazer D, Baykal T, Kırmızıbekmez H, Alper M, Yeşilada E (2010) Investigations on the in vivo wound healing potential of Hypericum perforatum L. J Ethnopharmacol 127:468–477. https://doi.org/10.1016/j.jep.2009.10.011

Suryawanshi MV, Gujarathi PP, Mulla T, Bagban I (2024) Hypericum perforatum: a comprehensive review on pharmacognosy, preclinical studies, putative molecular mechanism, and clinical studies in neurodegenerative diseases. Naunyn Schmiedeberg's Arch Pharmacol 397:3803–3818. https://doi.org/10.1007/s00210-023-02915-6

Tahara Y, Sakamoto T, Oshima Y, Ishibashi T, Inomata H, Murata T, Hinton DR, Ryan SJ (1999) The antidepressant hypericin inhibits progression of experimental proliferative vitreoretinopathy. Curr Eye Res 19:323–329. https://doi.org/10.1076/ceyr.19.4.323.5302

Tanaka N, Yano Y, Tatano Y, Kashiwada Y (2016) Hypatulins A and B, Meroterpenes from *Hypericum patulum*. Org Lett 18:5360–5363. https://doi.org/10.1021/acs.orglett.6b02725

Theodossiou TA, Hothersall JS, De Witte PA, Pantos A, Agostinis P (2009) The multifaceted Photocytotoxic profile of Hypericin. Mol Pharm 6:1775–1789. https://doi.org/10.1021/mp900166q

Tian J, Tao R, Zhang X, Liu Q, He Y, Su Y, Ji T, Ye F (2015) Effect of *Hypericum perforatum L.* Extract on insulin resistance and lipid metabolic disorder in high-fat-diet induced obese mice. Phytother Res 29:86–92. https://doi.org/10.1002/ptr.5230

Trofimiuk E, Holownia A, Braszko JJ (2011) St. John's wort may relieve negative effects of stress on spatial working memory by changing synaptic plasticity. Naunyn Schmiedebergs Arch Pharmacol 383:415–422. https://doi.org/10.1007/s00210-011-0604-3

Uchida S, Hirai K, Hatanaka J, Hanato J, Umegaki K, Yamada S (2008) Antinociceptive effects of St. John's Wort, Harpagophytum Procumbens extract and grape seed Proanthocyanidins extract in mice. Biol Pharm Bull 31:240–245. https://doi.org/10.1248/bpb.31.240

Valletta E, Rinaldi A, Marini M, Franzese O, Roscetti G (2018) Distinct *Hypericum perforatum* L. total extracts exert different antitumour activity on erythroleukemic $K_{562}$ cells. Phytother Res 32:1803–1811. https://doi.org/10.1002/ptr.6114

Vandenbogaerde A, Zanoli P, Puia G, Truzzi C, Kamuhabwa A, De Witte P, Merlevede W, Baraldi M (2000) Evidence that total extract of Hypericum perforatum affects exploratory behavior and exerts anxiolytic effects in rats. Pharmacol Biochem Behav 65:627–633. https://doi.org/10.1016/S0091-3057(99)00233-6

Vanhaelen M, Vanhaelen-Fastré R (1984) Quantitative determination of biologically active constituents in crude extracts of medicinal plants by thin-layer chromatography-densitometry. J Chromatogr A 312:497–503. https://doi.org/10.1016/S0021-9673(01)92806-7

Velingkar VS, Gupta GL, Hegde NB (2017) A current update on phytochemistry, pharmacology and herb–drug interactions of Hypericum perforatum. Phytochem Rev 16:725–744. https://doi.org/10.1007/s11101-017-9503-7

Weber ND, Murray BK, North JA, Wood SG (1994) The antiviral agent Hypericin has *in vitro* activity against HSV-1 through non-specific association with viral and cellular membranes. Antivir Chem Chemother 5:83–90. https://doi.org/10.1177/095632029400500204

Widy-Tyszkiewicz E, Piechal A, Joniec I, Blecharz-Klin K (2002) Long term administration of Hypericum perforatum improves spatial learning and memory in the water maze. Biol Pharm Bull 25:1289–1294. https://doi.org/10.1248/bpb.25.1289

Wölfle U, Seelinger G, Schempp C (2013) Topical application of St. John's Wort (Hypericum perforatum). Planta Med 80:109–120. https://doi.org/10.1055/s-0033-1351019

Wood S, Huffman J, Weber N, Andersen D, North J, Murray B, Sidwell R, Hughes B (1990) Antiviral activity of naturally occurring Anthraquinones and Anthraquinone derivatives. Planta Med 56:651–652. https://doi.org/10.1055/s-2006-961304

Xiang Y, Wang L, Gu P, Wang C, Tian Y, Shi W, Deng F, Zhang Y, Gao L, Wang K, Wang Y, He J, Zhao W, Bi X, Hu J, Zhong L, Guo Y, Zhou X, Wang H, Xie P (2023) A randomized, double-blind, positive-controlled, multicenter clinical trial on the efficacy and safety of ShuganJieyu Capsule and St John's Wort for major depressive disorder with somatic complaints. Int J Ment Health Addict. https://doi.org/10.1007/s11469-023-01222-8

Yan X-T, An Z, Tang D, Peng G-R, Cao C-Y, Xu Y-Z, Li C H, Liu P L, Jiang Z-M, Gao J-M (2018) Hyperelatosides A–E, biphenyl ether glycosides from *Hypericum elatoides* , with neurotrophic activity. RSC Adv 8:26646–26655. https://doi.org/10.1039/C8RA05322G

Yan X-T, An Z, Huangfu Y, Zhang Y-T, Li C-H, Chen X, Liu P-L, Gao J-M (2019) Polycyclic polyprenylated acylphloroglucinol and phenolic metabolites from the aerial parts of Hypericum elatoides and their neuroprotective and anti-neuroinflammatory activities. Phytochemistry 159:65–74. https://doi.org/10.1016/j.phytochem.2018.12.011

Yang X-W, Grossman RB, Xu G (2018) Research Progress of polycyclic Polyprenylated Acylphloroglucinols. Chem Rev 118:3508–3558. https://doi.org/10.1021/acs.chemrev.7b00551

You M-K, Kim H-J, Kook J, Kim H-A (2018) St. John's Wort regulates proliferation and apoptosis in MCF-7 human breast cancer cells by inhibiting AMPK/mTOR and activating the mitochondrial pathway. Int J Mol Sci 19:966. https://doi.org/10.3390/ijms19040966

Zeng Y-R, Yi P, Gu W, Xiao C-X, Huang L-J, Tian D-S, Yan H, Chen D-Z, Yuan C-M, Hao X-J (2018) Hypermonins A and B, two 6-norpolyprenylated acylphloroglucinols with unprecedented skeletons from *Hypericum monogynum*. Org Biomol Chem 16:4195–4198. https://doi.org/10.1039/C8OB00650D

Zhang R, Ji Y, Zhang X, Kennelly EJ, Long C (2020) Ethnopharmacology of Hypericum species in China: A comprehensive review on ethnobotany, phytochemistry and pharmacology. J Ethnopharmacol 254:112686. https://doi.org/10.1016/j.jep.2020.112686

Zhu H, Chen C, Tan D, Li D, Guo Y, Wei G, Zhang J, Wang J, Luo Z, Xue Y, Zhang Y (2016) Sampbenzophenones A–G, prenylated benzoylphloroglucinol derivatives from Hypericum sampsonii. RSC Adv 6:86710–86716. https://doi.org/10.1039/C6RA17885E

Zirak N, Shafiee M, Soltani G, Mirzaei M, Sahebkar A (2019) *Hypericum perforatum* in the treatment of psychiatric and neurodegenerative disorders: Current evidence and potential mechanisms of action. J Cell Physiol 234:8496–8508. https://doi.org/10.1002/jcp.27781

Zou Y, Lu Y, Wei D (2010) Protective effects of a flavonoid-rich extract of *Hypericum perforatum* L. against hydrogen peroxide-induced apoptosis in PC12 cells. Phytother Res 24. https://doi.org/10.1002/ptr.2852

# Medicinal and Nutritional Importance of *Cannabis sativa* in Human Health

14

Roy Dinata, Vanlalhruaii Famhawite, Puranjoy Mipun, Sritapa Basu Mallick, and Rinku Baishya

**Abstract**

*Cannabis sativa* (*C. sativa*) possesses therapeutic potential, contributing to its historical use in ethnomedicine and pharmacology. *C. sativa* generates a distinctive category of terpenophenolic chemicals referred to as cannabinoids, and the therapeutic use of cannabinoids from *C. sativa* includes rheumatism, epilepsy, asthma, skin burns, pain, control of sexually transmitted infections, complications during childbirth, postpartum hemorrhage, and gastrointestinal function. The use of *C. sativa* remains constrained due to its illegality in several countries. This chapter aims to elucidate the biological potential of *C. sativa* phytocompounds in the management of pain, neurological disorders, and psychiatric disorders. This chapter additionally addressed the therapeutic significance and applications of cannabis in the prevention and treatment of life-threatening disorders. Furthermore, psychotropic cannabinoids derived from *C. sativa* were investigated, and the corresponding clinical findings were thoroughly discussed.

R. Dinata
Animal Science Research Center, Division of Animal Sciences, University of Missouri, Columbia, MO, USA

V. Famhawite · R. Baishya (✉)
Centre for Preclinical Studies (CPS), CSIR-North East Institute of Science and Technology (NEIST), Pulibor, Jorhat, Assam, India

P. Mipun
Department of Botany, Bhattadev University, Bajali, Assam, India

S. B. Mallick
Indian Institute of Science Education and Research Kolkata (IISER Kolkata), Mohanpur, West Bengal, India
e-mail: sbm20rs125@iiserkol.ac.in



M. A. Ansari et al. (eds.), *Medicinal Plants and their Bioactive Compounds in Human Health: Volume 2*, https://doi.org/10.1007/978-981-95-3620-7_14

Furthermore, the nutritional profile of *C. sativa* seeds, including proteins, essential fatty acids (omega-3 and omega-6), vitamins, and minerals, was reviewed in details.

**Keywords**

*Cannabis sativa* · Bioactive compounds · Pharmacological activity · Human health

## 14.1 Introduction

### 14.1.1 Historical Perspective on the Use of *Cannabis sativa* in Medicine and Nutrition

*Cannabis*, a member of the *Cannabaceae* family originally from central Asia, is an annual plant pollinated by the wind, known for its elongated flowers and sharp leaves (Li 1974, Mukherjee et al. 2008, McPartland and Guy 2017). Since ancient times, it has been a valuable source of food, fiber, medicine, and psychoactive substances. The two primary species are *C. sativa L.* and *Cannabis indica L.* Distinguishing between them has been a longstanding issue among botanists. In the past, China utilized *Cannabis* for making cloth and fiber (Friedman and Sirven 2017). Subsequently, *Cannabis* started being utilized for medicinal purposes in China to treat human illnesses. Ancient Egyptians utilized *Cannabis* for a variety of issues such as infections, pain relief, and reducing inflammation. The calming properties of *Cannabis* were recorded in ancient Indian Vedic texts around 1500 BCE (Russo 2007). Since that time, the medicinal advantages have been recognized in Asia and North Africa for numerous conditions. In children with epilepsy, the use of the tetrahydrocannabinol (THC) isomer from *Cannabis* was reported for treatment in the late nineteenth century (Davis and Ramsey 1949). *Cannabis* (*C. sativa L.*) is a commonly occurring plant species, present in a variety of environments ranging from coastal areas to the cooler, elevated regions of the Himalayas, thought to have originated around 10,000 years back. The emperors of China were also the first to document the plant's medicinal use. *Cannabis* plant is referenced in Hindu texts and ancient literature found in Iran dating back to 600–700 BC (Asadi et al. 2019). Evidence of growing cannabis for food or medicinal purposes has been discovered in ancient Egyptian mummies. Hence, it is thought that this plant originated in Central Asia (Hazekamp et al. 2010). Historical proof indicates that the initial paper crafted from hemp fibers was manufactured in China, a century prior to the birth of Christ. Hemp cultivation for fiber became popular in Western Asia, Egypt, and eventually Europe from 1000 to 2000 BC. It gained popularity in Europe after the year 500 AD as well. The introduction of this product to South America (Chile) occurred in 1545 (Thomas and ElSohly 2016). The fibrous and industrial hemp types do not produce any narcotic effects and are suitable for large-scale industrial cultivation and usage without any issues (Mead 2017). Presently, in numerous Thai and foreign cuisines, *Cannabis* leaves sections are utilized in beverages like tea, coffee, fruit juices, and cannabis-infused sodas also incorporate cannabis leaves

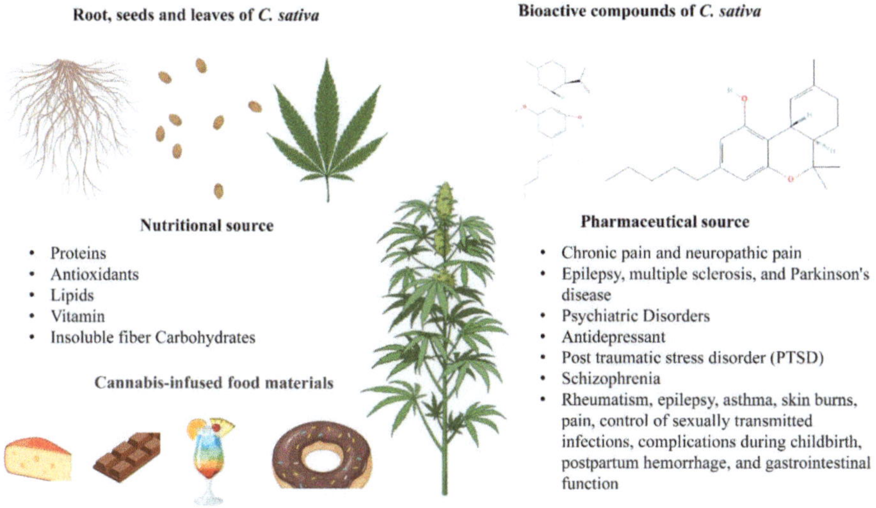


**Fig. 14.1** Overview of *C. sativa* as nutritional and pharmaceutical sources

(Friedman and Sirven 2017; Rasera et al. 2021; Saravanan et al. 2022). Nevertheless, there is limited information in Thailand regarding the nutritional components and bioactive substances found in *Cannabis* leaves to enhance the appropriate utilization of cannabis (Maniin et al. 2022). Further, *Cannabis sativa*, commonly known as cannabis or marijuana, has been studied extensively for its medicinal properties. Its therapeutic potential is attributed to various compounds found in the plant, primarily cannabinoids and terpenes. Ongoing research continues to explore the full spectrum of medicinal benefits and potential risks associated with cannabis. This includes studying long-term effects, optimal dosing, and interactions with other medications (Nabissi et al. 2015). Therefore, future research on cannabis treatments to individual needs and conditions, optimizing efficacy, and safety in human health must be needed. In this chapter, we discuss the diversity, bioactive compounds as well as the therapeutic uses of *C. sativa* in pain management, neurological and psychiatric disorders, cancer treatment, digestive health, and sleep disorders (Fig. 14.1). Further, we discussed the nutritional status and toxicity profiles of *C. sativa* bioactive compounds, and their limitations based on clinical and preclinical studies (Fig. 14.1).

### 14.1.2 Morphological Characteristics of *Cannabis sativa*

*C. sativa L.* is an herb that is usually annual and dioecious, but sometimes monoecious, with male and female flowers growing on different plants and pollinated by the wind. It spreads from seeds, thrives in sunny open areas with well-drained soil, and requires ample nutrients and water. During a growing season lasting four to six months, it has the potential to grow as tall as 5 m (16 ft). Nevertheless, in contemporary breeding and cultivation of recreational marijuana, the favored method for reproducing the plants is through cloning, utilizing cuttings from the designated mother plant. As implied by the term, female plants are preferred because they yield more psychoactive compounds compared to male plants. Before transitioning into the flowering stage, it is impossible to differentiate between the genders of *Cannabis* plants based on anatomy, but once this stage begins, the development of male and female plants diverges significantly. Decreased daylight triggers flowering in plants. The female plant forms many dense clusters of flowers, with a main one at the top of the stem and smaller ones on each branch. Male flowers, on the other hand, hang in loose groups on leafless branches. The male plants complete their pollen release and perish before the seeds in the female plants mature, which occurs four to eight weeks after fertilization. One kilogram of seed can be produced by a big female. If the seed remains intact, it has the potential to sprout in the following spring. Based on the latest botanical classification, *Cannabis* is grouped with *Humulus* in the Cannabinaceae family. However, cannabinoids are specifically present in *C. sativa* despite this connection. Even when grafting *Cannabis* and *Humulus* in the *Humulus* genus, cannabinoids were not found, but instead bitter acids like humulone, adhumulone, and cohumulone were discovered (Hourfane et al. 2023). The evident connection between the two plant types is demonstrated by the fact that cannabinoids and bitter acids are both produced from comparable biosynthetic pathways. The current organized categorization of *Cannabis* is division Angiosperms, class Dicotyledon, subclass Archichlamydeae, order Urticales, family Cannabinaceae, genus Cannabis, species *Sativa L.* (Hazekamp et al. 2010).

### 14.1.3 Cannabinoids (THS, CBD, CBG) and Terpenes as Bioactive Compounds

The *Cannabis* genus is responsible for the cultivation of various plants, including hemp and marijuana. There are 100 different cannabinoids found in the cannabis plant. The main cannabinoids are THC (tetrahydrocannabinol) and CBD (cannabidiol). Tetrahydrocannabinol triggers the body's natural cannabinoid receptors, CB1 and CB2. CB1 activation is the reason why THC has psychoactive effects. In clinical trials, the doses of Cannabidiol being studied do not directly activate those receptors; therefore, it is believed to be non-psychoactive (Pertwee 2008). Further, cannabinoids, present in significant amounts in *Cannabis*, induce distinct physiological effects in humans, with more than a hundred varieties. THC is the psychoactive element behind the marijuana "high," while CBD is a nonaddictive and nonhallucinogenic compound with medicinal benefits (Iseger and Bossong 2015). Products like CBD-infused bud, oil, and tinctures are available for sale to help with lowering inflammation and stress (Turner et al. 1980; Rupasinghe et al. 2020). THC is banned in numerous countries due to its mind-altering properties, despite being commonly used for medical and recreational purposes. Therefore, it is crucial to distinguish between the various categories of *C. sativa* L., such as marijuana and industrial hemp, to participate in legal farming activities. Setting a THC concentration threshold on a dry weight basis is typically used as the main factor in differentiating between the two distinct types. In industrial hemp, THC levels are generally under 1% when measured by dry weight, whereas in marijuana they can vary from 3 to 15% (Raihan and Bijoy 2023; Farinon et al. 2020). Terpenoids or isoterpenoids are also a major category of plant compounds. The isoprenoid pathway creates both primary and secondary metabolites. Isoprenoids play various roles in primary metabolism, such as serving as phytohormones (gibberellic acid, abscisic acid, and cytokinin) and membrane stabilizers (sterols), as well as being utilized in respiration (ubiquinone) and photosynthesis. The hemp plant has multiple uses in different industrial fields, with every part being utilized (Ferber et al. 2020). Fatemeh Shakib and colleagues found that chlorophyll and plastocyanin play a role as secondary metabolites in defense mechanisms such as phytoalexins. *Cannabis* has the potential to produce terpenes which are believed to enhance the beneficial effects of cannabis through the "entourage effect" when combined with cannabinoids (Kumar and Goel 2019). *Cannabis* seeds and roots generate 11 phenolic amides and lignanamides, known as lignans. N-trans-coumaryl-tyramine, N-trans-feruloyltyramine, and N trans-caffeoyltyramine are all examples of phenolic amides. Phenolic amides are stored during plant chemical defense reactions. Phenolic acids also have a significant impact on the process of flowering, sexual organs, and resistance (Andre et al. 2016).

*C. sativa* has a complex phytochemical content that includes numerous compounds from various chemical classes. The majority of the chemical compounds found in hemp are produce as a result of secondary metabolism. *C. sativa* contains a range of chemical substances, such as cannabinoids (cannabigerol, cannabichromene, cannabidiol, cannabisykosol, cannabielsoin, cannabinol, cannabinodial, cannabitriol), enzymes, amino acids, proteins, glycoproteins, sugars, hydrocarbons, nitrogenous components, simple esters, fatty acids, lactones, steroids, terpenes, phenols, flavonoids, vitamins, pigments, and elements. Amino acids, fatty acids, and steroids are part of primary metabolism in the plant, whereas phytocannabinoids, terpenes, and phenolic compounds are considered secondary metabolism metabolites. THC, CBD, cannabidiolic acid (CBDA), CBG, cannabigerolic acid (CBGA), CBGV, and CBGVA are the primary cannabinoids found in cannabis (Turner et al. 1980).

### 14.1.4 Cannabinoids: THC (Tetrahydrocannabinol), CBD (Cannabidiol), CBG (Cannabigerol) and Their Biosynthesis Pathway

*Cannabis* is extremely intricate when it comes to phytochemical compounds with over 480 distinct chemical compounds being discovered in the hemp plant. Some compounds in cannabis are classified as primary metabolites like amino acids, fatty acids, and steroids, whereas cannabinoids, flavonoids, stilbenoids, terpenoids, lignans, and alkaloids are considered secondary metabolites. Thus, cannabis is viewed as an authentic botanical source of secondary metabolites, which holds significance in the pharmaceutical sector (Iffland and Grotenhermen 2017). A group of chemical compounds found in *C. sativa* that interact with the human body's endocannabinoid system (ECS) are known as cannabinoids. It can influence various physiological processes in humans. The most common cannabinoids are tetrahydrocannabinol (THC), cannabidiol (CBD), and cannabigerol (CBG) (Grotenhermen 2003; Tahir et al. 2021). The primary psychoactive component of *Cannabis* is tetrahydrocannabinol (THC). It binds to the CB1 receptors in the brain, which leads to the characteristic "high" associated with marijuana use. THC is associated analgesic, anti-inflammatory, and appetite-stimulating properties (Grotenhermen 2003; Tahir et al. 2021; Jastrząb et al. 2022). Cannabidiol (CBD) is nonpsychoactive. CBD has potential therapeutic effects which include anxiolytic, antiepileptic, and neuroprotective properties. It interacts with various receptors in the ECS and other neurotransmitter systems, contributing to its broad range of effects without significant psychoactivity (Grotenhermen 2003; Gülck and Møller 2020). The precursor of THC, CBD, and other cannabinoids is cannabigerol (CBG). It is non-psychoactive. It is believed to have potential therapeutic applications. For example, it shows anti-inflammatory, antibacterial, and neuroprotective effects (McPartland and Russo 2001) (Table 14.1). Until 1964, all known pathways for cannabinoid biosynthesis involved either cannabidiol (CBD) or cannabidiolic acid (CBDA) as the main intermediate, believed to be derived from a monoterpene and either oleotol or oleoleic acid. Nevertheless, studies have demonstrated that cannabigerol (CBG), which acts as a basic building block for cannabinoids, is created through the merging of geranyl diphosphate with either oleotol or oleoleic acid (OA) (Gülck and Møller 2020). Following this, they determined that

**Table 14.1** Terpenoids known in *Cannabis* spp. and their biological activity

| Terpenoid | Known properties | References |
|---|---|---|
| Beta-myrcene | Analgesic, anti-inflammatory, antibiotic, antimutagenic | McPartland and Russo (2001), Hazekamp et al. (2010) |
| Beta-caryophyllene | Anti-inflammatory, cytoprotective, antimalarial, CB2 agonist | |
| d-limonene | Immune potentiator, antidepressant, antimutagenic | |
| Linalool | Sedative, antidepressant, anxiolytic, immune potentiator | |
| Pulegone | Acetylcholinesterase (AChE) inhibitor, sedative, antipyretic | |
| 1, 8-cineol | AChE inhibitor, stimulant, antibiotic, antiviral, anti-inflammatory, antinociceptive | |
| α-Pinene | Anti-inflammatory, bronchodilator, stimulant,<br>antibiotic, antineoplastic, AChE inhibitor | |
| α-Terpineol | Sedative, antibiotic, AChE inhibitor, antioxidant, antimalarial | |
| Terpincol-4-ol | AChE inhibitor, antibiotic | |
| p-cymene | Antibiotic, anticandidal, AChE inhibitor | |

cannabidiol (CBD), delta-9-tetrahydrocannabinol (THC), and cannabinol (CBN) originate from cannabigerol (CBG) and vary primarily in the process by which this precursor is transformed. A deeper insight into the biosynthesis of cannabinoids was gained when Shoyama and colleagues (1975) found that cannabinoids and cannabinolic acid (CBNA) are not produced by the living plant's free phenolics (non-carboxylic acid). After analyzing various African cannabinoids, Turner and Hadley (Turner et al. 1973) achieved comparable findings. It is currently understood that plant metabolism produces cannabinoids in the form of carboxylic acids, with the substituent group in the carboxyl (-COOH) located at the 2 positions. Research using C13-labeled glucose has verified that geranyl diphosphate (GPP) and OA play specific roles in the biosynthesis of cannabinoids (Livingston et al. 2020). The production of cannabinoids through biological processes involves combining important stages in polyketide and isoprenoid metabolism. Hexanoic acid likely originates from C18 saturated fatty acids and serves as the starting molecule for polyketide synthesis, where it is continually oxidized and cleaved into C6 segments. The production of crucial cannabinoids requires two immediate precursors. The oleoleic acid (OA) is produced from hexanoyl-CoA by the polyketide pathway, while geranyl diphosphate is supplied by the MEP pathway (Mahlberg et al. 1980; Zysk 2018). OA is synthesized from hexanoic acid within the cell's cytosol. The plastid non-mevalonate pathway (MEP) synthesizes GPP, with the initial step being the conversion of geranyl diphosphate and oleic acid to CBGA by the enzyme Geranylpyrophosphate:olivetolate geranyltransferase (GOT). CBGA functions as an aromatic prenyltransferase located on the cell membrane. Continued oxidation of THCA results in the creation of CBNA (McPartland and Russo 2001; Mead 2017).

Cannabinoids are produced in the glandular trichomes, also known as the "marijuana bud," of female flowers, with male flowers lacking trichomes usually having minimal cannabinoids (Mahlberg et al. 1980). Trichomes can also be found on bracts, leaves, and on the underside of the anther lobes of male flowers (Zysk 2018). Trichōma, which means "hair" in Greek (Hammond and Mahlberg 1977), is categorized as stalked, sessile, or bulbous. Small bulbous trichomes produce minimal cannabinoids, while the other two forms are primarily responsible for the majority of cannabinoid production. Short-stalked sessile trichomes consist of a rounded head with multiple secretory cells and a storage cavity for metabolites under the cuticle (Russo 2011). Stalked trichomes are characterized by a slightly bigger spherical tip, protruding a few hundred microns from the surface of the epidermis. It is still uncertain how much sessile and stalked trichome individually contributes to cannabinoid production (Mahlberg et al. 1980). The molecular-level understanding of cannabinoid biosynthesis is still incomplete. In short, cannabinoids all start with the same process: tetraketide synthase (TKS) which is a type III polyketide synthase (PKS), combines hexanoyl-CoA with three malonyl-CoA molecules to produce 3,5,7-triox ododecaneoyl-CoA. Olivetolic acid is cyclized and aromatized by olivetolic acid cyclase (OAC), resulting in the formation of olivetolic acid (OLA), with coenzyme A being lost in the process. The prenyl group is added by aromatic prenyltransferase to the highly nucleophilic 2-resorcinol position, producing cannabigerolic acid (CBGA). This central intermediate then splits off to create the cannabinolic acids

(THCA, CBDA, and CBCA), which are converted to THC, CBD, and CBC through nonenzymatic decarboxylation (Jastrząb et al. 2022).

### 14.1.5 Role of Terpenes, Flavonoids, and Other Phytochemicals of Cannabis

*Cannabis* has a substantial quantity of terpenoids along with numerous numbers. The terpenoids found in marijuana have various biological effects that could be responsible for modulating the impact of THC and creating distinct pharmacological effects of their own (Niesink and van Laar 2013). Terpenoid compounds can help reduce or change the unwanted effects of THC. Numerous terpenoids found in *Cannabis* exhibit acetylcholine esterase inhibition, such as pulegone, limonene, limonene oxide, and a-terpinene. Therefore, terpenoids are being studied for potential use in Alzheimer's disease treatment (Hazekamp et al. 2010) (Table 14.1). THC is linked to adverse psychological effects like anxiety and depersonalization (Niesink and van Laar 2013). These effects might be reduced by terpenoids in *Cannabis* due to their calming and antidepressant qualities. Beta myrcene, a terpenoid found in cannabis, demonstrates both analgesic and anti-inflammatory properties. *Cannabis* displays permeability of the blood–brain barrier because of the existence of terpenoids. Terpenoids found in *Cannabis* could have a significant impact on the plant's chemical ecology (Livingston et al. 2020). The cat was eagerly awaiting its dinner, meowing loudly to get the attention of its owner. Terpenoids found in male *Cannabis* plant pollen may be significant in attracting pollinators or deterring pests (Hazekamp et al. 2010).

A total of 23 flavonoids known from *Cannabis* belong to the foliage, blooms, pollen, and branches of the plant. Flavonoids play multiple functions in plant physiology which offer color and taste to plants for further growth and development of plants. They offer protection against predators and pathogens and have the ability to shield plants from the impacts of damaging UV radiation. Moreover, flavonoids have the ability to function as metal chelators and also as scavengers of free radicals. Two flavonoids cannflavin A and B are prenylated flavonoids found in *Cannabis*. A total of 27 nitrogenous compounds have been found in *Cannabis*, with 10 being confirmed as alkaloids. Despite being considered a fascinating group of compounds for their biological activity, there is currently a lack of information on the pharmacological profile of *Cannabis alkaloids*.

Nineteen different stilbenoids have been detected in *Cannabis*. They have been separated from stem, leaves, and resin. Research has shown that some *Cannabis stilbenoids* have antibacterial properties (Hazekamp et al. 2010).

## 14.2 Medicinal Importance of *Cannabis sativa* in Human Health

### 14.2.1 Efficacy of Cannabinoids in Chronic Pain Conditions and Neuropathic Pain

#### 14.2.1.1 Chronic Pain and Neuropathic Pain

Chronic pain is typically defined as pain lasting more than three months. It can be triggered by events like tissue injury or surgery or result from nerve damage, with a variable onset and fluctuating course. Chronic pain involves maladaptive plasticity in the nociceptive circuits of the peripheral and central nervous systems (CNS). Further, this leads to heightened sensitization and a shift toward increased excitation in the spinal dorsal horn, resulting in exaggerated pain responses. Additionally, chronic pain is linked to changes in the endocannabinoid system, including increased cannabinoid receptor levels, altered receptor functionality, changes in endocannabinoid production or release, and interactions with other pain mediators (van den Hoogen et al. 2022; Barakji et al. 2023; Carrascosa et al. 2024). It is well known that one-fifth of the global population experiences chronic pain, characterized by spontaneous pain episodes, allodynia (pain from non-painful stimuli), and hyperalgesia (amplified pain response to painful stimuli). Moreover within the United States alone, in 2016, approximately 20.4% (50 million) of adult Americans experienced chronic pain, with 8% of them facing significant negative impacts on their quality of life (Shehata et al. 2022). Further, a cross-sectional study (2011–2015) identified 10.3 million chronic pain patients in the United States, with 247,949 of them using cannabis for pain relief (Safi et al. 2024). However, the complexity and difficulties associated with managing chronic pain have led to the idea of modulating or even interrupting pain pathways through surgical methods and sophisticated interventional pain procedures (Shehata et al. 2022). Effective management of chronic pain remains difficult, as it frequently shows resistance to standard treatments such as opioids (van den Hoogen et al. 2022). Further, neuropathic pain, a severe and persistent form of chronic pain, occurs due to damage to the nervous system and can result from peripheral nerve injuries, toxic substances, or disease processes. Managing neuropathic pain remains a significant challenge in clinical practice, as it frequently responds inadequately to current treatment options. Furthermore, adverse side effects associated with these treatments often limit their effectiveness, contributing to suboptimal pain control (Rahn and Hohmann 2009). Moreover, an estimated 7–10% of adults globally suffer from chronic neuropathic pain, though its prevalence varies by region. Symptoms typically include spontaneous burning pain, numbness, hyperalgesia (increased sensitivity to painful stimuli), and allodynia (pain in response to non-painful stimuli). The persistent pain can make daily activities challenging, often leading to social and economic difficulties. This physical and social burden can contribute to the onset of depression in some patients. Although opioids are the leading therapeutic option for mitigating chronic neuropathic pain, their use is linked to a high risk of addiction and increased mortality. To address these challenges and improve treatment success, new molecular targets for neuropathic pain management are continuously explored (Campos et al. 2021). CADTH, the Canadian Agency for Drug and Health Technology Assessment, reports that cannabis-based therapies may show potential advantages in the management of neuropathic pain (Barakji et al. 2023).

#### 14.2.1.2 Cannabinoids in Pain Treatment

The first known instance of cannabis being used for pain control traces back to 2737 B.C., where it was employed in traditional Chinese medicine (O'Brien and McDougall 2018). *Cannabis* use has long been framed within a culturally biased paradigm (Mallick-Searle and Marie 2019). Currently, medical cannabinoid use is legally approved in 40 countries. *Cannabis* has been predominantly used as a substitute for opioids, as well as for replacing other medications such as antidepressants, anti-anxiety drugs, and nonsteroidal anti-inflammatory drugs (NSAIDs) (Safi et al. 2024). Further, phytocannabinoids, sourced from cannabis plants, form the basis of most medical cannabis products. Of the more than 60 active compounds found in cannabis, 9-tetrahydrocannabinol (THC) and cannabidiol (CBD) have been the primary focus of scientific research. These compounds primarily interact with endogenous cannabinoid receptors, type 1 (CB1R) and type 2 (CB2R) (van den Hoogen et al. 2022). In terms of biological activity, 9-THC, cannabinol (CBN), and cannabinodiol (CBND) are recognized as psychoactive cannabinoids, whereas CBD and other cannabinoids lack psychoactive effects. THC is the most prominent psychoactive compound, and its toxicity has been the subject of extensive research (Odieka et al. 2022). According to responses from 2969 participants in a self-administered survey, the use of medicinal cannabis was reported for arthritis (26%), chronic pain (25%), multiple sclerosis (MS) and depression (22% each), and neuropathy (19%) (Paland et al. 2023). In contrast to opioid analgesics, cannabis has a reduced overdose potential and is less likely to lead to addiction. However, critics of phytocannabinoid argue that cannabinoids can produce a range of effects, including opposing actions, depending on various factors. Research on cannabinoid agonists in preclinical pain models, encompassing nociceptive, visceral, and neuropathic pain, suggests that their antinociceptive efficacy is comparable to that of opioids. Despite cannabis's long history of use for pain relief, clinical trials have found it to be only minimally effective for acute pain, and in some cases, it may even increase certain pain responses (Carrascosa et al. 2024).

#### 14.2.1.3 Factors Affecting Cannabinoid Bioavailability and Efficacy

Due to its high lipophilicity, THC easily crosses the blood–brain barrier, accumulating in tissues with a large volume of distribution and exhibiting a slow elimination profile (Safi et al. 2024). THC, due to its fat sequestration, has a tissue elimination half-life of around 7 days, with full clearance of a single dose possibly extending to 30 days. This accumulation pattern implies that repeated dosing can lead to prolonged presence and continued effects in the brain. Following absorption, THC and other cannabinoids are rapidly distributed throughout various tissues, with the rate of distribution influenced by blood flow. The bioavailability of THC and CBD formulations via the oral route is relatively low, around 6%, due to their lipophilic properties, variable absorption in the gut, and extensive hepatic first-pass metabolism. Plasma levels sufficient for therapeutic effects are maintained for two to six hours, but blood concentrations are only 25–30% of those achieved via smoking the same dose. This lower level is due to first-pass metabolism by the liver, although the metabolite 11-hydroxy delta-9-THC retains some psychoactive effects. (Carrascosa et al. 2024). Further, it is well known that THC's pain-relieving effects in chronic pain may operate through two central cognitive–emotional modulation areas and their neural connections to somatosensory systems. The analgesic action of THC at the cellular level is due to its complex multimodal mechanisms that involve cannabinoid receptors CB1 and CB2, as well as its modulation of other pathways, such as delta and kappa opioid receptors, the GABAergic and glutamatergic systems, and the noradrenergic system (Safi et al. 2024). When combined with CBD, THC's beneficial effects are enhanced, and its side effects are reduced. CBD also alters THC metabolism, extending its effects. Studies have found that CBD penetrates the skin ten times more effectively than THC, suggesting that CBD is more polar in structure compared to THC. Cannabinoids delivered through the skin can potentially offer a steadier and more regulated release into the body by avoiding first-pass metabolism, unlike edibles (Carrascosa et al. 2024). Additionally, inhaled THC reaches its maximum concentration in the brain within 15 minutes, leading to both somatic and psychological effects that persist for two to four hours before tapering off. Further, orally administered THC peaks between 30 and 120 minutes, with effects lasting 5 to 12 hours. The brain's THC concentration is much higher after inhalation. The variability in bioavailability (10–85%) is influenced by factors such as inhalation technique (puff frequency, duration, and depth), the device used, particle size, vaporizer temperature, and the location of deposition in the respiratory system. Emerging technologies, such as nano or ionized particles and the use of omega fats in carriers, offer potential for improved absorption. Moreover, the development of new water-soluble cannabinoid agonists presents opportunities for enhanced bioavailability (Carrascosa et al. 2024). Additionally, the perimedullary route of cannabinoid administration requires lower doses and leads to more enduring effects relative to other administration routes. Cannabinoids are better absorbed when taken with fats, oils, or solvents like ethanol. Intravenous delivery of cannabinoids is complicated by their low solubility in water, yet it remains the most effective method for administering synthetic cannabinoids. Effects onset is delayed (0.5–2 hours) and may be extended by slow absorption from the gastrointestinal tract (Carrascosa et al. 2024).

#### 14.2.1.4 Pain Pathways and Cannabinoid Receptor Involvement

The process of pain transduction depends on transforming nociceptive stimuli into neuroelectric signals, which are subsequently transmitted to the somatosensory cortex and brain stem through ascending and descending pathways. The ascending pathway facilitates the transmission of pain signals, whereas the descending pathway modulates and inhibits this signal. This descending pain modulation circuit involves neurons connecting the periaqueductal gray matter in the midbrain, the nucleus raphe magnus in the medulla, and the dorsal horn of the spinal cord (Safi et al. 2024). CB1 receptors are located at critical points along the pain pathway and influence both the sensory and emotional aspects of pain. CB2 receptors are expressed on immune cells, though usually at lower levels compared to CB1Rs (van den Hoogen et al. 2022).

#### 14.2.1.5 Efficacy of Cannabinoids in Chronic Pain Conditions

Global estimates of generalized chronic pain are reported to range between 10.6 and 11.8%, according to a recent review and the management of various chronic pains using cannabinoids has a long history. Yet, it is important to note that not all chronic pain react similarly to cannabinoids (Berger et al. 2020).

#### 14.2.1.6 Arthritis-Related Pain

Chronic pain is a trademark of rheumatoid disease, profoundly diminishing the quality of life for those affected. Despite advances in treatment, effective pain management remains elusive, emphasizing the need for alternative approaches. Randomized, double-blind, placebo-controlled trials have assessed the potential of medicinal cannabis extracts in managing chronic pain. In one study, THC-based cannabis extracts, delivered as sublingual sprays, were the most effective for pain control in patients with rheumatoid disease, with mild side effects reported. Another investigation examined nabiximols (Sativex®), a balanced combination of CBD and THC, in 58 patients with rheumatoid arthritis. Over five weeks of oromucosal administration, Sativex® significantly improved pain levels during movement and rest, as well as sleep quality, although morning stiffness was not notably reduced. Importantly, no severe side effects or withdrawal symptoms were observed, highlighting the safety of Sativex® as a potential treatment for pain in rheumatoid arthritis (Paland et al. 2023). In the study on arthritis by Frane et al. 2022, participants who used CBD reported a 44% reduction in pain, reflected by a significant 2.58-point decrease in numerical pain scores. Both increased frequency and duration of CBD use were significantly associated with improvements in pain intensity, physical function, and sleep quality, along with greater reductions on the numerical rating scale (NRS). Among those using CBD, 60.5% either reduced or discontinued the use of other medications, including anti-inflammatories, acetaminophen, and opioids. However, 41% of participants reported experiencing side effects, with the majority (84%) classified as mild, followed by moderate (14%) and severe (2%) (Paland et al. 2023). Also, Dronabinol, a synthetic THC drug, demonstrated a significant reduction in pain scores among patients taking opioids for chronic pain. This suggests that cannabinoids may offer useful supplementary benefits for osteoarthritis patients who rely on opioids for managing their chronic pain (O'Brien and McDougall 2018). Despite various studies and trials conducted by different groups, in a systemic review and meta-analysis conducted Barakji et al. (2023), they concluded that although cannabinoids are proved effective in treating chronic pain when compared to placebo, the clinical relevance of this effect is debatable.

#### 14.2.1.7 Cancer-Related Pain

Medical cannabis use is becoming more common in the treatment histories of chronic pain patients. Globally, this trend is evidenced by 40% of cancer patients in regions like Canada, Germany, and Israel, where medical cannabis is legal, using it for pain management. This raise concerns that cancer patients may be at greater risk of experiencing side effects and developing a dependence on cannabis (Paland et al. 2023). Chronic pain in cancer patients can arise either directly from the tumor or as a consequence of cancer treatments. A review of multiple studies indicates that cannabis use can significantly alleviate cancer-related pain, with four out of five studies demonstrating this benefit. Early research on THC extract administered in sesame oil capsules found that higher doses of THC provided increased pain relief. Specifically, a 20 mg dose of THC was significantly more effective in reducing pain compared to placebo. Further investigations into THC extracts and nabiximols (Sativex®), both administered as oromucosal sprays, showed that while THC spray did not yield significant results compared to placebo, nabiximols demonstrated significant pain relief and reduced the need for additional pain medications. Nabiximols were also superior to placebo in terms of reducing average daily pain. In studies comparing THC doses of 10 and 20 mg against placebo and 60 mg codeine, the 20 mg THC dose was significantly more effective in reducing pain compared to placebo. However, the 10 mg THC dose showed comparable effectiveness to 60 mg of codeine, suggesting that 10 mg of THC might be as effective as standard opioid treatments. An evaluation of THC oromucosal spray using a self-titration approach found a greater reduction in average NRS pain scores with THC compared to placebo (THC −1.01 vs. placebo −0.69), although this difference was not statistically significant. The recommended dosing for nabiximols includes up to 8 sprays every 3 hours or 11–16 sprays per day. Effective pain relief was achieved with doses of 8.75 sprays/day, 1–4 sprays/day, and 6–10 sprays/day, while higher doses (11–16 sprays/day) did not provide additional benefits over placebo (Blake et al. 2017).

#### 14.2.1.8 Fibromyalgia-Related Pain

Fibromyalgia, a multisymptomatic disorder commonly associated with chronic pain, often drives patients to seek relief through medical cannabis or synthetic cannabinoids. The first randomized, double-blind, placebo-controlled trial investigating cannabinoids for fibromyalgia treatment, conducted in 2008, demonstrated that nabilone (Cesamet) led to significant reductions in pain, fibromyalgia impact (FIQ), and anxiety. More recently, a controlled crossover study evaluated vaporized cannabis containing THC and CBD in fibromyalgia patients, finding that significant pain relief occurred only with THC-containing products in evoked pain models, while spontaneous pain remained unaffected. Cannabis use for pain management is widespread among fibromyalgia patients, though this may be influenced by the disorder's comorbidities. Retrospective studies suggest improvements in pain and quality of life, though the overall evidence is limited. Despite these limitations and potential biases in randomized controlled trials (RCTs), cannabinoids have demonstrated a significant impact on reducing pain and associated symptoms in fibromyalgia and other chronic pain conditions (Berger et al. 2020).

#### 14.2.1.9 Efficacy of Cannabinoids in Neuropathic Pain

When central nervous system (CNS) injury or disease occurs, microglia—normally characterized by a highly branched morphology with small cell bodies and long processes that monitor the CNS microenvironment—transition into a reactive state. This reactive state is marked by the secretion of proinflammatory cytokines, chemokines, and other immune mediators. The reactive microglial response is crucial for the development of chronic pain, as it promotes central sensitization. Several signaling pathways contribute to microglial reactivity in neuropathic and inflammatory pain states, including the activation of CSF1, P2X4, and P2X7 receptors. For example, nerve injury-induced mechanical allodynia leads to the upregulation of P2X4 receptors on spinal microglia. Activation of these receptors by ATP results in the release of brain-derived neurotrophic factor (BDNF), which disrupts chloride balance by downregulating potassium-chloride cotransporter 2 (KCC2) in spinal neurons. This disruption enhances excitation and diminishes inhibition of nociceptive signaling. Reactive microglia also exerts their effects through P2X7 receptors, nitric oxide (NO) production, and the release of proinflammatory cytokines such as tumor necrosis factor-alpha (TNFα), interleukin-1β (IL-1β), and cathepsin S (CTSS). The involvement of microglia through multiple signaling mechanisms and the secretion of proinflammatory mediators are central to the development of central sensitization, which leads to dysregulated nociceptive processing in chronic pain (van den Hoogen et al. 2022). Cannabinoids have demonstrated enhanced antinociceptive effects in models of inflammatory and neuropathic pain. Specifically, the central cannabinoid receptor CB2 is essential for eliciting an antiallodynic response. Cannabinoids affect neuropathic pain by activating and subsequently desensitizing TRPV1, a crucial receptor, and by reducing the activation of microglia and astrocytes involved in the onset and maintenance of hypersensitivity associated with neuropathic pain (Carrascosa et al. 2024). Animal studies on chronic neuropathic pain show that morphine remains effective for up to 22 days before tolerance develops, while rats receiving THC and CBD did not exhibit this tolerance (Safi et al. 2024). Research on THC and CBD indicated that THC at 2% did not affect pain, THC at 4% resulted in significant pain reduction, and THC at 6% increased pain, with CBD levels <1%. These findings suggest that the observed analgesic effects were primarily due to THC, indicating a potential therapeutic window where specific doses of THC are effective for pain relief, but higher doses may have the opposite effect (Halawa et al. 2018). A study conducted in Canada with 21 chronic pain patients, who sequentially smoked 25 mg of cannabis with varying THC concentrations (0, 2.5, 6.0, and 9.5%) three times daily for five days, found no significant changes in acute neuropathic pain scores. Only a minimal number of subjects achieved 30% pain relief by the end of the trial. While cannabis vaporizers reduce the amount of potentially carcinogenic polyaromatic hydrocarbons, they do not eliminate them completely (Russo 2008). Oral dronabinol, with typical doses ranging from 15 to 16.6 mg THC, failed to alleviate chronic neuropathic pain and was associated with adverse events that often-required discontinuation (Russo 2008).

#### 14.2.1.10 Chronic Neuropathic Pain

Three trials assessed inhaled or vaporized cannabis for chronic neuropathic pain in single-dose experiments. Two trials (231 patients) showed cannabis reduced pain intensity by at least 30%, with a clinically significant effect (NNTB = 3). In a number of studies that evaluated the potency of CBM, there was no convincing evidence of an analgesic effect for CBM in cancer or neuropathic pain, either in short-term or longer-term studies. The evidence was generally of low to very low quality, with concerns about study design, heterogeneity, and small sample sizes affecting the reliability of the findings. In summary, CBM showed limited to no efficacy in reducing pain in cancer or neuropathic pain, and the quality of evidence was low, making it difficult to draw strong conclusions about its effectiveness. In case of ajulemic acid, a synthetic analogue in the dimethylheptyl class, with efficacy in relieving peripheral neuropathic pain, peak plasma concentrations were observed within one to two hours, though in certain subjects this may extend to four to five hours. Its analgesic effects may be linked to binding with intracellular peroxisome proliferator-activated receptor gamma (Russo 2008; Fisher et al. 2021).

#### 14.2.1.11 MS-Related Neuropathic Pain

Cannabinoids are emerging as a promising option for treating chronic neuropathic pain, particularly in cases of multiple sclerosis (MS)-related pain and spasticity, where evidence indicates notable analgesic efficacy. Nabiximols (Sativex) have been studied across several neuropathic pain types, including brachial plexus injury and unilateral neuropathy, with findings indicating significant pain relief, better management of allodynia, and enhanced sleep quality. Additionally, with respect to THC and CBD in MS-Related Pain and Spasticity, oral Δ9-THC and a THC/CBD cannabis extract treatments improved subjective assessments of pain and spasticity, but did not lead to objective improvements in spasticity on the Ashworth Spasticity Scale. Cannabinoids are recommended as a third-line treatment for neuropathic pain by the Canadian Pain Society, surpassing selective serotonin reuptake inhibitors (SSRIs), methadone, and topical lidocaine in their recommendation. A systematic review has confirmed that oral cannabis extracts are effective for pain related to multiple sclerosis, with nabiximols and dronabinol likely providing beneficial effects (Halawa et al. 2018).

#### 14.2.1.12 HIV-Associated Peripheral Neuropathy

Significant pain relief has been observed in studies on inhaled cannabis for HIV-associated peripheral neuropathy. Additionally, higher doses of smoked cannabis (7–9.4%) were effective for treating neuropathic pain of mixed origins (Halawa et al. 2018). Over a five-day trial, 50 individuals with HIV-associated neuropathy who smoked cannabis with 3.56% THC three times daily reported a 30% reduction in pain, while the placebo group showed a 15% decrease (Russo 2008).

#### 14.2.1.13 Diabetic Neuropathy

In a randomized double-blind crossover trial conducted on the impact of inhaled cannabis on painful diabetic neuropathy, subjects were given aerosolized cannabis at low (1% THC), medium (4% THC), and high (7% THC) doses. The findings showed a significant reduction in pain that was dependent on the dose, but cognitive function was adversely affected at the highest THC dose (O'Brien and McDougall 2018).

## 14.2.2 Use of CBD in Epilepsy, Multiple Sclerosis, and Parkinson's Disease

### 14.2.2.1 Epilepsy

Epilepsy, marked by recurrent seizures, affects approximately 2.9 million people in the United States and 70 million worldwide (Rosenberg et al. 2015; Espinosa-Jovel 2023). This condition can be triggered by genetic syndromes, strokes, infections, and traumatic brain injuries. Beyond seizures, epilepsy is linked to various cognitive, psychological, and social challenges, contributing to increased mortality and a reduced quality of life. The financial burden of epilepsy in the United States is estimated at $15.5 billion annually (Rosenberg et al. 2015). Antiepileptic drugs (AEDs) are commonly used to manage seizures but do not address the underlying causes of epilepsy or prevent disease progression. Approximately 70%–80% of newly diagnosed epilepsy patients achieve complete seizure control with AEDs, including valproate and carbamazepine. However, 20–30% of patients do not respond to these treatments, a condition known as pharmacoresistant epilepsy. These patients are at higher risk of cognitive, behavioral, and psychosocial impairments, as well as increased mortality (Stockings et al. 2018; Espinosa-Jovel 2023). Managing treatment-resistant epilepsy often requires complex drug regimens at higher dosages, leading to more significant side effects. These challenges underscore the need for innovative, targeted, and safer treatment options tailored to different disease stages and patient populations (Rosenberg et al. 2015).

### 14.2.2.2 The Potential of Medical Cannabis in Epilepsy Treatment

Recent anecdotal reports and growing media coverage have highlighted medical cannabis as a potential alternative or supplement to existing AEDs. The cannabis plant contains over 100 phytocannabinoids, including Δ9-THC and cannabidiol (CBD), both of which have shown promise in reducing seizures in animal models with minimal toxicity and good tolerability (Rosenberg et al. 2015). However, systematic reviews by the American Academy of Neurology and the Cochrane Database have determined that the efficacy of medical cannabis for epilepsy remains inconclusive. CBD, although structurally similar to Δ9-THC, has a different pharmacological profile. Unlike Δ9-THC, which binds to CB1 and CB2 receptors, CBD interacts with various molecular pathways, including TRP channels, 5-hydroxytryptamine1α receptors, glycine receptors, and GPR55. CBD's ability to inhibit glutamate release and reduce epileptiform activity in hippocampal models suggests its potential for treating seizures (Rosenberg et al. 2015).

#### 14.2.2.3 Clinical Evidence for CBD in Epilepsy

In clinical studies, CBD has shown promise in reducing seizures in patients with treatment-resistant epilepsy. For instance, a study involving Epidiolex (a 99% CBD product) was conducted on 213 patients, mostly children, with treatment-resistant epilepsy. After 12 weeks, seizures were reduced by about 50%, and some Dravet syndrome (DS) patients became seizure-free. The study also found that CBD was generally well-tolerated, with only 4% of patients discontinuing treatment due to adverse effects such as drowsiness, diarrhea, and fatigue (Rosenberg et al. 2015). In a review by Stockings et al. (2018), randomized controlled trials (RCTs) consistently used CBD in conjunction with placebo controls, showing that CBD could reduce seizure frequency in treatment-resistant children and adolescents. Some studies suggested that the benefits of CBD might be more pronounced when used with clobazam, although further investigation into their interaction is necessary.

#### 14.2.2.4 Use of CBD in Specific Epilepsy Syndromes

Preclinical and clinical evidence supports CBD's potential in treating severe forms of epilepsy, such as DS and Lennox-Gastaut syndrome (LGS). CBD has also shown efficacy in reducing seizures in various genetic and developmental epilepsies that are resistant to standard therapies (Zhou et al. 2021). Aicardi syndrome: 59.2% reduction in seizures, CDKL5 deficiency disorder: 59.7% reduction in seizures, Doose syndrome: 28.8% reduction in seizures, Dup15q syndrome: 38.8% reduction in seizures. In patients with SYNGAP1 Epileptic Encephalopathy, high-dose CBD achieved an 85% reduction in seizures by month 9. Similarly, pediatric patients with refractory Sturge–Weber syndrome experienced a 65% reduction in seizures after 14 weeks of CBD treatment (Zhou et al. 2021). While epilepsy remains a global challenge, particularly in low-income regions where 80% of cases occur, the persistence of refractory epilepsy in 25–35% of patients highlights the urgent need for new therapeutic approaches. The promising results from preclinical and clinical studies suggest that CBD could be a viable option for treating severe forms of epilepsy, though further research is needed to fully understand its efficacy and safety (Espinosa-Jovel 2023).

#### 14.2.2.5 Multiple Sclerosis

Multiple sclerosis (MS) is an autoimmune neurological disorder that causes damage to the central nervous system, affecting approximately 2.3 million people around the world. This condition is characterized by the loss of myelin, which impairs nerve signal transmission between the brain and spinal cord. Common symptoms of MS include spasticity, muscle spasms, tremors, bladder dysfunction, neuropathic pain, dysarthria, and cognitive issues such as memory disturbances. Cannabidiol (CBD) has demonstrated potential in alleviating MS symptoms. Research has consistently shown that CBD can reduce spasticity, pain, inflammation, fatigue, and depressive symptoms in MS patients, thereby improving their overall mobility (Haddad et al. 2022).

#### 14.2.2.6 Spasticity in MS Patients

As multiple sclerosis progresses, spasticity affects 60–84% of patients, caused by upper motor neuron damage and abnormal spinal reflex control, which further aggravates other MS-related symptoms and severely impacts quality of life. MS-related spasticity manifests as muscle stiffness resulting from hyperactive stretch reflexes and is frequently accompanied by symptoms like fatigue, pain, and bladder dysfunction, leading to patient dissatisfaction. Treatment for MS spasticity commonly involves a combination of non-pharmacological and pharmacological approaches. Ongoing research into cannabis-based therapies has shown that topical cannabis may be beneficial due to its pharmacokinetic advantages, though smoking or vaping cannabis is still widely used (Haddad et al. 2022).

#### 14.2.2.7 Cannabinoids for the Treatment of Spasticity in MS Patients

##### 14.2.2.7.1 Sativex

Sativex demonstrated consistent spasticity reduction over a 15-week period in an open-label extension trial with 137 MS patients, as assessed by a visual analog scale. Similarly, a 6-week, double-blind, placebo-controlled study with 189 MS patients found a significant decrease in spasticity in the Sativex group compared to placebo, based on numerical rating scale (NRS) scores ($p = 0.048$). However, no significant benefits were observed on other spasticity-related measures, such as the Ashworth scores (Haddad et al. 2022).

##### 14.2.2.7.2 Nabiximols

A comprehensive phase III trial with two stages demonstrated that nabiximols, delivered as an oromucosal spray, resulted in a 51% reduction in spasticity and a 20% reduction in NRS scores. The highest recommended dose is 12 sprays per day. Based on evidence from various randomized controlled trials, nabiximols has been approved for its efficacy in alleviating spasticity symptoms in MS patients. It is sanctioned as an add-on treatment for moderate-to-severe spasticity in several Western nations (Haddad et al. 2022).

#### 14.2.2.8 Pain and Quality of Life in MS Patients

Pain is a prevalent problem among multiple sclerosis (MS) patients, greatly influencing their overall health-related quality of life (HRQoL). The diverse types of pain associated with MS are often inadequately managed by available treatments. Due to the impact of MS on daily functioning, patients with the disease generally have a lower health-related quality of life (HRQoL) compared to the general population. The effect of cannabis on HRQoL in MS patients has been reported inconsistently in research studies (Haddad et al. 2022).

#### 14.2.2.9 Nabiximols for Treatment of MS-Related Pain and Overall Health-Related Quality of Life (HRQoL)

A randomized, placebo-controlled study with 66 participants demonstrated that nabiximols (oromucosal spray) effectively reduced central pain and alleviated pain-related sleep disturbances in MS patients. In a 2013 double-blind study of 339 patients with central neuropathic pain, nabiximols (administered to 167 patients) led to significant improvements in pain relief compared to placebo (administered to 172 patients) by week 10. Patients with neuropathic pain for less than four years were particularly likely to experience benefit from the treatment. Additionally, a recent meta-analysis of nine randomized controlled trials, involving 1289 participants, found that nabiximols provided greater relief from chronic neuropathic pain than placebo, though the effect size was small (Haddad et al. 2022). Nabiximols was found to positively impact physical and mental health-related quality of life (HRQoL) in MS patients during a three- to four-month period as reported by the MOVE2 research. Despite these initial benefits, follow-up assessments over a year did not reveal significant changes in HRQoL. Furthermore, a 15-week trial with 337 MS patients and a 19-week trial with 241 patients both showed limited and statistically insignificant improvements in HRQoL with nabiximols treatment (Haddad et al. 2022).

### 14.2.3 Parkinsons Disease

#### 14.2.3.1 Parkinson's Disease (PD)

Parkinson's disease (PD) is the second most common neurodegenerative disorder after Alzheimer's disease, affecting 4.5 million people worldwide. The disease presents significant costs and disability challenges, with no available treatments that modify the disease progression. Current symptomatic treatments, such as levodopa, offer partial relief but lead to complications like response fluctuations and levodopa-induced dyskinesia (LID) (Figura et al. 2022; Bougea et al. 2020).

#### 14.2.3.2 CBD's Neuroprotective Potential

CBD, first isolated in 1940, has a structure that was identified in 1963. Its concentration in cannabis varies based on plant type, growing conditions, and extraction methods. CBD has demonstrated calming, seizure-reducing, and anti-inflammatory properties. Unlike THC, CBD does not rely heavily on cannabinoid receptors to exert its effects. Instead, it influences these receptors indirectly and can modulate their activity in various ways. Additionally, CBD may increase levels of natural compounds such as AEA and 2-AG in the body, improve cell membrane fluidity, and affect receptor expression (Ferreira-Junior et al. 2019). CBD's neuroprotective effects in PD may be linked to its ability to reduce harmful processes like inflammation and cell death without directly activating cannabinoid receptors (Ferreira-Junior et al. 2019).

#### 14.2.3.3 Cannabis and the Endocannabinoid System in PD

The interest in cannabis for PD stems from its interaction with the endocannabinoid system, which is often disrupted in neurological disorders. Research suggests that cannabidiol (CBD) may help protect neurons from damage and reduce neurodegeneration through its anti-inflammatory and antioxidative effects. However, the role of cannabinoid receptors, such as CB-2R, in CBD's effects is still debated. Phytocannabinoids might influence brain activity by modulating neurotransmission and providing neuroprotective benefits (Figura et al. 2022).

#### 14.2.3.4 Effect of Cannabis on Motor Symptoms in PD

Research on the effects of cannabis on motor symptoms in PD remains limited. Some studies suggest that smoking cannabis can improve motor symptoms like tremor and rigidity, with reductions in Unified Parkinson's Disease Rating Scale (UPDRS) part III scores. Conversely, clinical trials using CBD have produced mixed results. For instance, a study administering 75 mg and 300 mg of CBD found no significant improvement in motor symptoms compared to placebo. However, the 300 mg dose did improve overall quality of life scores (PDQ-39) and daily activities (Figura et al. 2022). A survey of Czech PD patients revealed that 25% used cannabis as adjunct therapy, with 46% reporting improvements in symptoms like resting tremor, bradykinesia, muscle rigidity, and LID. However, only 4.7% experienced worsened symptoms. Despite these positive results, randomized controlled trials have not consistently validated the benefits of cannabis for motor symptoms in PD (Bougea et al. 2020).

#### 14.2.3.5 Effect of Cannabis on Non-motor Symptoms in PD

##### 14.2.3.5.1 Sleep Disorders

Sleep disorders affect about 80% of patients with advanced PD, and treatment remains challenging due to their multifactorial nature. Initial case studies suggested that CBD might be effective for managing REM Sleep Behavior Disorder (RBD), but a subsequent randomized trial found no significant differences between CBD and placebo in alleviating RBD. However, CBD was associated with transient improvements in sleep satisfaction at four and eight weeks. Another study indicated that CBD improved sleep quality based on SCOPA-sleep scores, though no specific impact on RBD was observed (Figura et al. 2022).

##### 14.2.3.5.2 Pain Management

Pain is a common symptom in PD, and dopaminergic therapies offer only partial relief. Studies have shown that cannabis can reduce pain in PD patients, although these findings are limited by the open-label nature of the trials and the lack of blinding. Patients reported pain relief shortly after cannabis use, and long-term use reduced heat pain thresholds. However, detailed information on dosage and disease severity was not consistently provided (Figura et al. 2022).

##### 14.2.3.5.3 Neuropsychiatric Symptoms

CBD has shown promise in managing neuropsychiatric symptoms in PD. A double-blind, placebo-controlled study found that a 300 mg dose of CBD effectively reduced anxiety and tremors induced by anxiety in PD patients. Additionally, CBD doses between 150 mg and 300 mg improved psychosis symptoms, with significant reductions in both Parkinson's Psychosis Questionnaire and Brief Psychiatric

Rating Scale scores. Adverse effects, such as sedation, hallucinations, and dizziness, were noted in some studies (Figura et al. 2022).

A study on the use of nabilone, a synthetic THC analogue, found significant reductions in anxious mood and sleep disturbances in PD patients. However, the improvements in pain observed during the open-label phase did not carry over into the randomized phase (Figura et al. 2022).

##### 14.2.3.5.4 Efficacy of Cannabis in PD: Mixed Findings

Research into the efficacy of cannabis for both motor and non-motor symptoms in PD has produced mixed results. Early studies on cannabis smoking showed no improvement in resting tremor compared to the benefits of levodopa and apomorphine. However, other studies have found that cannabis use improved both motor and non-motor symptoms, such as pain and sleep quality (Figura et al. 2022). A survey conducted by the Michael J. Fox Foundation revealed that 76.3% of respondents were PD patients, with 36.6% using cannabis primarily for medicinal purposes. These patients reported improvements in mood, memory, and fatigue, and many reduced their prescription medication use after starting cannabis treatment (Bougea et al. 2020).

### 14.2.4 Psychiatric Disorders

Potential benefits in anxiety, depression, posttraumatic stress disorder (PTSD), and schizophrenia (SCZ). Mental health conditions, including mood disorders, anxiety disorders, posttraumatic stress disorder (PTSD), and schizophrenia, are a significant public health concern, affecting between 9.6 and 27.8% of the global population. The mechanisms driving these disorders are complex and not yet fully understood, though a mix of genetic, biological, and environmental factors likely contributes to their development and progression. Antidepressants, benzodiazepines, and antipsychotics are the mainstay treatments, but they do not offer complete relief for 40–60% of patients. This indicates a critical need for alternative or additional treatments. There is growing evidence that drugs targeting the endocannabinoid system, which appears to be dysfunctional in psychiatric disorders, may hold promise as new therapeutic options. The therapeutic use of cannabis dates back centuries, with numerous anecdotal reports highlighting its use by individuals seeking to self-medicate and alleviate psychiatric symptom (Scherma et al. 2020).

#### 14.2.4.1 Anxiety Disorder

Anxiety and mood disorders affect over 260 million people globally, with 25% of Europeans experiencing these issues. These disorders greatly reduce quality of life, making everyday activities, work, and social interactions difficult. They are also a major cause of disability, leading to significant economic costs, with Europe spending around EUR 170 billion annually (García-Gutiérrez et al. 2020).

Anxiety disorders, such as generalized anxiety disorder, panic disorder, and social anxiety disorder, cause symptoms like fear, uneasiness, and trouble sleeping.

Anxiety and depression often occur together, making treatment harder and increasing the risk of suicide. Existing treatments, like antianxiety and antidepressant medications, are not always effective, especially in treating cognitive issues or preventing relapse. These treatments also come with side effects like weight gain and reduced sexual desire. Therefore, there is a need for better treatment options (García-Gutiérrez et al. 2020).

#### 14.2.4.2 Cannabidiol (CBD) as a Potent Treatment for Anxiety Disorder

Cannabidiol (CBD), a compound from cannabis, is showing potential as a new treatment for anxiety and depression, with fewer side effects compared to current medications (García-Gutiérrez et al. 2020). CBD is mostly known for reducing anxiety, but its effects can vary depending on the species studied and factors like previous stress and brain regions involved. Unlike other drugs that affect the same system, such as THC and CB1 receptor agonists, CBD does not cause anxiety at higher doses (Blessing et al. 2015).

#### 14.2.4.3 Mechanism and Challenges of CBD Efficacy

More research is needed to understand why CBD works this way and to see how effective it is when used over a longer period, as current studies show mixed results—sometimes reducing anxiety, but other times increasing it. In studies using animals to model general anxiety, CBD has shown mixed results. Lower doses of CBD helped reduce anxiety, but higher doses were not effective. The anxiety-reducing effects of CBD follow a pattern where moderate doses work best, but larger doses do not. No studies have shown that CBD causes anxiety when given in these models. Research on long-term CBD use for anxiety is limited, with some studies showing benefits and others showing no effect. CBD's ability to reduce anxiety in specific brain areas is partly linked to certain receptors, like the 5-HT1A receptor, but not the CB1 receptor. In some brain regions, CBD can actually increase anxiety in unstressed animals, but it reduces anxiety after stress. The anxiety-reducing effects of CBD also seem to involve another receptor, the $GABA_A$ receptor. However, at higher doses, CBD interacts with TRPV1 receptors, which may reduce its ability to ease anxiety, explaining why higher doses are less effective (Blessing et al. 2015).

#### 14.2.4.4 Clinical and Neuroimaging Findings

Initial trials from the 1970s and 1980s suggested that CBD might reduce anxiety triggered by THC in healthy male subjects. Subsequent research showed that CBD administration orally could lower anxiety levels during public speaking and affect brain activity related to anxiety, particularly by reducing activity in specific brain regions. In double-blind studies, CBD did not lead to significant behavioral changes or brain activation compared to THC and did not cause psychotic symptoms or cognitive impairments (García-Gutiérrez et al. 2020). In patients with Social Anxiety Disorder (SAD), CBD was effective in reducing subjective anxiety and modifying brain activity. A large retrospective case series noted that CBD reduced anxiety and improved sleep in psychiatric patients, though sleep improvements varied. Conversely, in non-clinical volunteers with high levels of paranoia, CBD increased anxiety and did not affect persecutory ideation (García-Gutiérrez et al. 2020). Nabiximols, containing both THC and CBD, was reported to alleviate anxiety and cravings in individuals with cannabis use disorder. CBD vaporization improved emotional processing and prevented THC-related impairments in face recognition. Additionally, CBD was found to decrease anxiety and cravings in heroin users, with sustained effects for up to a week (García-Gutiérrez et al. 2020). Neuroimaging studies have shown that CBD affects brain connectivity related to fear, suggesting possible mechanisms for its anxiolytic effects. Current and upcoming trials are evaluating CBD's efficacy for anxiety reduction, including open-label trials with sublingual CBD and phase 2 and 3 trials using various CBD formulations for anxiety disorders and sleep disturbances (García-Gutiérrez et al. 2020).

#### 14.2.4.5 Depression

Depressive disorders, including major depressive disorder (MDD) and persistent depressive disorder, involve sadness, guilt, and social withdrawal (García-Gutiérrez et al. 2020). Globally, depression is a primary cause of disability, with about 15% of people experiencing it at some point in their lives (Langlois et al. 2021).

#### 14.2.4.6 CBD Antidepressant as a Potential in Humans

Research into the antidepressant effects of CBD in humans is currently limited and inconclusive. Evidence from high doses of nabiximols, a combination of CBD and THC, suggests significant mood enhancement in chronic pain patients. Oral CBD has been effective in reducing depressive and psychotic symptoms associated with cannabis use and in repairing specific brain regions damaged by cannabis. In frequent cannabis users, oral CBD has alleviated depressive and psychotic symptoms and improved cognitive abilities such as attention, learning, and memory (García-Gutiérrez et al. 2020). Nabiximols has also been effective in decreasing depression during cannabis withdrawal. An online survey of 2409 CBD users reported mood improvements across various conditions, but did not differentiate between pure CBD and CBD from marijuana. Overall, while current research shows promise, further large-scale clinical trials are necessary to validate the antidepressant efficacy and safety of CBD (García-Gutiérrez et al. 2020).

#### 14.2.4.7 Anhedonia and Cannabis Use

A key symptom of depression, anhedonia, is marked by reduced brain activity in response to rewarding stimuli, particularly within the ventral striatum and orbitofrontal cortex. *Cannabis* use may intensify anhedonia by further diminishing activity in these brain regions associated with reward processing. The relationship between cannabis use and symptoms of apathy and anhedonia shows variability, with some studies identifying these effects while others do not. Cannabis may contribute to anhedonia and depressive states by impairing brain reward response. Chronic use of cannabis is linked to lower dopamine activity, which might contribute to anhedonia. Conversely, moderate cannabis use might increase dopamine production, potentially alleviating symptoms of anhedonia (Langlois et al. 2021).

#### 14.2.4.8 Cannabinoids and Antidepressants

Preclinical research indicates that compounds targeting CB1 receptors could influence and potentially boost the efficacy of antidepressants. Notably, CB1 receptor activation can exhibit both depressant and antidepressant effects, which might help explain the inconsistent findings reported in studies on the relationship between cannabis and depression (Langlois et al. 2021).

### 14.2.5 Post-Traumatic Stress Disorder (PTSD)

#### 14.2.5.1 Prevalence and Symptoms of PTSD

Approximately 1% of the population is affected by PTSD, with military veterans being disproportionately impacted. The core symptoms of PTSD include reliving the traumatic event through intrusive thoughts, flashbacks, or nightmares, avoiding trauma-related cues, and experiencing hyperarousal. PTSD is increasingly understood as a disorder involving memory processing. The primary treatment modalities for PTSD involve trauma-focused psychotherapies that address the reprocessing and reappraisal of trauma memories, alongside pharmacological interventions. Despite these available treatments, numerous patients struggle to access specialized care, and some do not respond sufficiently to the existing pharmacological options. These barriers emphasize the urgent need for new, more effective interventions (Hindocha et al. 2020).

#### 14.2.5.2 The Role of Amygdala and Hippocampus in PTSD

PTSD often involves overactivity in the amygdala, a brain area that makes people overly alert and unable to process traumatic memories properly. THC, a compound in cannabis, targets CB1 receptors in the amygdala, which may help reduce this overactivity and improve brain connections involved in emotional processing. However, high doses of THC can sometimes make things worse, increasing anxiety and amygdala reactivity to negative stimuli (Hindocha et al. 2020). The hippocampus, another brain area involved in learning and memory, also plays a role in PTSD. Targeting the endocannabinoid system, which includes CB1 receptors in both the amygdala and hippocampus, might help alleviate PTSD symptoms. However, current research isn't strong enough to fully support cannabinoids as a treatment for PTSD due to limitations in study quality and design (Hindocha et al. 2020).

#### 14.2.5.3 Cannabinoids and the Endocannabinoid System in PTSD

Cannabis is utilized for its effects on pain, anxiety, and depression, with the endocannabinoid system, particularly CB1 and CB2 receptors, playing a role in emotional learning and memory. THC acts on mood and perception through CB1 and

CB2 receptors but may also cause cognitive deficits and paranoia. In contrast, CBD provides potential therapeutic benefits without psychoactive effects, possibly by interacting with the 5HT1A receptor. Synthetic cannabinoids like nabilone and dronabinol are used for conditions such as nausea and sleep disorders but pose risks, including tachycardia and seizures. CB1 receptors in the amygdala and prefrontal cortex are crucial for fear memory extinction. PTSD patients exhibit reduced endocannabinoid levels, suggesting potential abnormalities in CB1 receptor signaling. Although synthetic cannabinoids may assist in fear extinction and memory consolidation, their long-term use could lead to tolerance, reducing their efficacy. Animal studies support the role of CB1 receptor activation in fear extinction, and genetic variations in CB1 receptors may explain individual differences in anxiety and fear extinction. Enhancing endocannabinoid neurotransmission, especially when combined with exposure therapies, may represent a promising approach to treating PTSD (Forsythe and Boileau 2022).

#### 14.2.5.4 Potential of CBD in PTSD Treatment

CBD, another compound in cannabis, seems to help with emotional and social processing and might be useful in therapies aimed at reducing trauma responses. But long-term use of cannabis can lead to negative effects like addiction, psychosis, and problems with thinking and decision-making (Hindocha et al. 2020). Despite this, some US states allow cannabinoids for PTSD, largely based on veterans' experiences that cannabinoids may be more effective and have fewer side effects than other treatments. Caution is needed, especially for people with a history of psychosis, as cannabinoids might worsen certain symptoms. There is also interest in using cannabinoids to treat sleep problems associated with PTSD, potentially offering a safer alternative to other medications (Hindocha et al. 2020).

#### 14.2.5.5 Clinical Trial

Clinical research into CBD's potential for treating PTSD is expanding. An open-label study has shown that combining CBD with psychiatric treatments and therapy can significantly reduce PTSD symptoms after eight weeks. Additionally, a placebo-controlled trial (NCT02759185) is evaluating the effects of smoked CBD marijuana (up to 1.8 g/day for 3 weeks) on PTSD symptoms, including anxiety and depression, in 76 veterans (García-Gutiérrez et al. 2020).

### 14.2.6 Schizophrenia

Schizophrenia, a chronic and incapacitating brain disorder, impacts around 1% of the population over their lifetime, presenting with positive symptoms like delusions, hallucinations, and disorganized thinking, as well as negative symptoms including social withdrawal, reduced motivation, and emotional blunting. Cognitive impairments, such as deficits in memory, executive functioning, and processing speed, also contribute to the disorder's impact on social and occupational functioning (Manseau and Goff 2015).

#### 14.2.6.1 Cannabis and Schizophrenia

Cannabis, notably its psychoactive ingredient Δ9-tetrahydrocannabinol (THC), can induce psychotic symptoms and cognitive deficits akin to those found in schizophrenia. Cannabis is also the most commonly misused substance among individuals diagnosed with schizophrenia, and ongoing use is associated with worsening psychosis, increased relapse rates, and deteriorating overall functioning. Heavy cannabis consumption has been linked to a substantially increased risk of developing schizophrenia, with a study by Andreasson et al. (1987) revealing that heavy users had a sixfold greater likelihood of developing the disorder over a 15-year follow-up (Manseau and Goff 2015).

#### 14.2.6.2 The Cannabis-Psychosis Relationship

Research shows that people who use cannabis are more likely to develop psychosis, especially if they use it frequently or start at a young age. However, understanding this relationship is complicated by factors like other substance use, gender differences, and socioeconomic background. It's not clear whether cannabis directly causes psychosis or if people with early symptoms of psychosis use cannabis as a way to cope. Although cannabis use is linked to worse outcomes for people with schizophrenia, not everyone who uses cannabis develops psychosis. Additionally, there hasn't been a noticeable increase in psychosis cases despite more people using cannabis. So, while cannabis use and psychosis are connected, it's uncertain whether cannabis use directly causes psychosis. More research, especially after cannabis legalization, may help clarify this relationship (Manseau and Goff 2015).

#### 14.2.6.3 Mechanism of THC in Psychosis

THC, the primary psychoactive agent in cannabis, exerts its effects through CB1 receptors, leading to increased dopamine and glutamate levels and reduced GABA activity in the brain. THC is known to induce psychosis-like symptoms, including paranoia, disorganization, and altered sensory experiences. These effects are more pronounced with high-THC strains, such as "skunk" or synthetic cannabinoids, which are associated with a higher risk of schizophrenia (Manseau and Goff 2015).

#### 14.2.6.4 Potential of CBD in Psychosis Treatment

The potential of cannabinoids as antipsychotic treatments is gaining attention. Rimonabant, a CB1 receptor antagonist, has shown some effectiveness in reducing anxiety and depressive symptoms, though it failed to improve psychosis in smaller trials. Cannabidiol (CBD), another cannabis-derived compound, shows promise as an antipsychotic. CBD appears to counteract THC-induced psychosis, reduce anxiety, and enhance markers of brain health, such as hippocampal volume. Preclinical studies suggest that CBD also has anti-inflammatory and neuroprotective effects, normalizing schizophrenia-like behaviors in animal models. Clinical studies on CBD's antipsychotic effects in humans have produced mixed results, with some evidence suggesting that CBD reduces psychotic symptoms and improves cognitive performance while having fewer side effects than conventional antipsychotics. However, larger clinical trials are necessary to confirm CBD's effectiveness as an antipsychotic. CBD's favorable side effect profile, particularly the reduction of weight gain and movement disorders, makes it an attractive alternative to existing treatments (Manseau and Goff 2015).

#### 14.2.6.5 Clinical Trial

The positive outcomes from animal research have prompted clinical trials to evaluate cannabidiol (CBD) for managing schizophrenia (SCZ). In trials involving chronic CBD administration to patients already on antipsychotic therapy, CBD did not significantly alter positive or negative symptoms or cognitive impairments. However, it did not induce movement disorders, though sedation was more frequent compared to placebo. In a multicenter randomized controlled trial, CBD was found to significantly alleviate positive psychotic symptoms and showed trends towards improved cognitive function and overall performance. Unlike traditional antipsychotics, CBD did not impact common side effects such as prolactin levels or weight gain, though gastrointestinal symptoms were more common with CBD. CBD's efficacy was comparable to amisulpride, with fewer side effects such as extrapyramidal symptoms and weight gain. It also did not significantly affect liver or cardiac functions and increased anandamide plasma levels. Functional MRI studies indicated that CBD improved brain activity in areas related to verbal learning and normalized hippocampal-striatal connectivity in SCZ patients. Improvements in brain activation were also noted in antipsychotic-naïve individuals at high risk for psychosis. CBD reduced hyperactivation in the left insula/parietal operculum and enhanced activation in the right caudate and parahippocampal gyrus. Ongoing trials (NCT03883360, NCT02926859, NCT04411225) are assessing CBD's effectiveness compared to placebo or olanzapine for psychosis and SCZ, with results expected in the near future. Although results are mixed, CBD shows potential as a treatment for schizophrenia, either alone or as an adjunct. More comprehensive double-blind, placebo-controlled trials are necessary to understand better its efficacy and safety profile (García-Gutiérrez et al. 2020).

#### 14.2.6.6 Nutritional Importance of *C. sativa*

*C. sativa's* stem, seeds, roots, and flowers have all been utilized for food, feed, and medicine due to their high nutritional value. Since ancient times, *C. sativa* seeds have been used to make edible oil, as is widely known (Iftikhar et al. 2021). In addition to 10–15% insoluble fiber, the seed has 30% oil and 25% protein, both of which are high in nutrients. The oil contains an abundance of polyunsaturated fatty acids (PUFA), with the ideal ratio for nutritional research being 2.5–3:1 for linolenic acid (ω-6) to linoleic acid (ω-3) (Iftikhar et al. 2021).

Saturated fatty acids make up just around 10% of the oil's total content, which is high in linoleic acid, oleic acid, stearidonic acid, and α-linolenic acid. Strong antioxidants like polyphenols found in *C. sativa* seeds can help treat a variety of maladies, including anxiety, oxidative stress, and the risk of developing chronic conditions, including cancer, neurological disorders, digestive issues, and skin diseases. *Cannabis* has a more developed root system than other flowering plants, which makes it perfect for phytoremediation of soil contaminated with heavy metals. Fiber residue is gathered from the cannabis stem. The fiber of *C. sativa* is an extremely valuable raw material used to make durable fabrics and specialty paper. Chemicals including linoleic acid, alpha-linolenic acid, tocopherol, cannabidiol, cannabisin A, and caffeoyltyramine make up *C. sativa* (Iftikhar et al. 2021). Alpha-linolenic acid, or ALA, is an n-3 (−3) fatty acid that is mostly present in plant foods like walnuts and flaxseed as well as vegetable oils like soybean and canola. Clinical studies have shown that linoleic acid (ALA) has a cardioprotective impact by lowering total and LDL cholesterol when saturated fat is replaced. A possible treatment option for both conditions is cannabidiol, a non-psychoactive cannabis compound (Iftikhar et al. 2021). The pathophysiology of substance-abusing disorders (SUDs) and psychotic disorders is influenced by cannabinoids. It is well known that caffeoyltyramine and its phenolic amides, such as trans-N-caffeoyltyramine and cis-N-caffeoyltyramine, have antioxidant and antifungal properties (Iftikhar et al. 2021).

## 14.3 Conclusion

Despite the illegality of cannabis growing in numerous countries, the recent assessment, acceptability, and legalization of cannabis products for therapeutic purposes have been progressively increasing worldwide. This chapter encapsulates the therapeutic properties of *Cannabis sativa* and its nutraceutical significance. Moreover, clinical and preclinical studies on *Cannabis* are well-established; however, comprehensive research is essential in future investigations to elucidate the chemical compositions of *C. sativa* extracts, the physical properties of their active constituents, toxicity profiles, and susceptibility to photochemical oxidation.

**Authors' Contribution** RD, VF, PM, and SBM wrote the original draft. RD and RB edited and proofread the final chapter.

## References

Andre CM, Hausman JF, Guerriero G (2016) Cannabis sativa: the plant of the thousand and one molecules. Front Plant Sci 4:7–19

Andréasson S, Allebeck P, Engström A & Rydberg U (1987) Cannabis and schizophrenia. A longitudinal study of Swedish conscripts. Lancet (London, England), 2(8574):1483–1486. https://doi.org/10.1016/s0140-6736(87)92620-1

Asadi S, Moghadam H, Naghdi Badi H, Naghavi MR, Salami SAR (2019) A review on agronomic, phytochemical and pharmacological aspects of cannabis (Cannabis sativa L.). J Med Plants 18(70):1–20

Barakji J, Korang SK, Feinberg J, Maagaard M, Mathiesen O, Gluud C, Jakobsen JC (2023) Cannabinoids versus placebo for pain: a systematic review with meta-analysis and Trial sequential analysis. PLoS One 18(1):e0267420

Berger AA, Keefe J, Winnick A, Gilbert E, Eskander JP, Yazdi C, Kaye AD, Viswanath O, Urits I (2020) Cannabis and cannabidiol (CBD) for the treatment of fibromyalgia. Best Pract Res Clin Anaesthesiol 34(3):617–631

Blake A, Wan BA, Malek L, DeAngelis C, Diaz P, Lao N, Chow E, O'Hearn S (2017) A selective review of medical cannabis in cancer pain management. Ann Palliat Med 6:S215–S222
Blessing EM, Steenkamp MM, Manzanares J, Marmar CR (2015) Cannabidiol as a potential treatment for anxiety disorders. Neurotherapeutics 12(4):825–836
Bougea A, Koros C, Simitsi AM, Chrysovitsanou C, Leonardos A, Stefanis L (2020) Medical cannabis as an alternative therapeutics for Parkinsons' disease: systematic review. Complement Ther Clin Pract 39:101154
Campos RMP, Aguiar AFL, Paes-Colli Y, Trindade PMP, Ferreira BK, de Melo Reis RA, Sampaio LS (2021) Cannabinoid therapeutics in chronic neuropathic pain: from animal research to human treatment. Front Physiol 12:785176
Carrascosa AJ, Navarrete F, Saldaña R, García-Gutiérrez MS, Montalbán B, Navarro D, Gómez-Guijarro FM, Gasparyan A, Murcia-Sánchez E, Torregrosa AB, Pérez-Doblado P, Gutiérrez L, Manzanares J (2024) Cannabinoid analgesia in postoperative pain management: from molecular mechanisms to clinical reality. Int J Mol Sci 25(11):6268
Davis JP, Ramsey HH (1949) Antiepileptic action of marihuana active substances. Federat Proc 8:284–285
Espinosa-Jovel C (2023) Cannabinoids in epilepsy: clinical efficacy and pharmacological considerations. Neurologia 38(1):47–53
Farinon B, Molinari R, Costantini L, Merendino N (2020) The seed of industrial hemp (Cannabis sativa L.): nutritional quality and potential functionality for human health and nutrition. Nutrients 12(7):1935
Ferber SG, Namdar D, Hen-Shoval D, Eger G, Koltai H, Shoval G, Weller A (2020) The "entourage effect": terpenes coupled with cannabinoids for the treatment of mood disorders and anxiety disorders. Curr Neuropharmacol 18(2):87–96
Ferreira-Junior NC, Campos AC, Guimaraes FS, Del-Bel E, Zimmermann PMDR, Brum Junior L, Zuardi AW (2019) Biological bases for a possible effect of cannabidiol in Parkinson's disease. Braz J Psychiatry 42:218–224
Figura M, Koziorowski D, Sławek J (2022) Cannabis in Parkinson's disease—the patient's perspective versus clinical trials: a systematic literature review. Neurol Neurochir Pol 56(1):21–27
Fisher E, Moore RA, Fogarty AE, Finn DP, Finnerup NB, Gilron I, Haroutounian S, Krane E, Rice AS, Rowbotham M, Wallace M (2021) Cannabinoids, cannabis, and cannabis-based medicine for pain management: a systematic review of randomised controlled trials. Pain 162:S45–S66
Forsythe ML, Boileau AJ (2022) Use of cannabinoids for the treatment of patients with post-traumatic stress disorder. J Basic Clin Physiol Pharmacol 33(2):121–132
Frane N, Stapleton E, Iturriaga C, Ganz M, Rasquinha V, Duarte R (2022) Cannabidiol as a treatment for arthritis and joint pain: an exploratory cross-sectional study. J Cannabis Res 4(1):47
Friedman D, Sirven JI (2017) Historical perspective on the medical use of cannabis for epilepsy: ancient times to the 1980s. Epilepsy Behav 70:298–301
García-Gutiérrez MS, Navarrete F, Gasparyan A, Austrich-Olivares A, Sala F, Manzanares J (2020) Cannabidiol: a potential new alternative for the treatment of anxiety, depression, and psychotic disorders. Biomolecules 10(11):1575
Grotenhermen F (2003) Pharmacokinetics and pharmacodynamics of cannabinoids. Clin Pharmacokinet 42:327–360
Gülck T, Møller BL (2020) Phytocannabinoids: origins and biosynthesis. Trends Plant Sci 25(10):985–1004
Haddad F, Dokmak G, Karaman R (2022) The efficacy of cannabis on multiple sclerosis-related symptoms. Life 12(5):682
Halawa OI, Furnish TJ, Wallace MS (2018) Role of cannabinoids in pain management. In: Essentials of pain medicine, pp 509–520
Hammond CT, Mahlberg PG (1977) Morphogenesis of capitate glandular hairs of Cannabis sativa (Cannabaceae). Am J Bot 64(8):1023–1031
Hazekamp A, Fischedick JT, Díez ML, Lubbe A, Ruhaak RL (2010) 3.24—chemistry of cannabis. Comprehensive natural products II. 3:1033–1084

Hindocha C, Cousijn J, Rall M, Bloomfield MA (2020) The effectiveness of cannabinoids in the treatment of posttraumatic stress disorder (PTSD): a systematic review. J Dual Diagn 16(1):120–139

Hourfane S, Mechqoq H, Bekkali AY, Rocha JM, El Aouad N (2023) A comprehensive review on Cannabis sativa ethnobotany, phytochemistry, molecular docking and biological activities. Plants 12(6):1245

Iffland K, Grotenhermen F (2017) An update on safety and side effects of cannabidiol: a review of clinical data and relevant animal studies. Cannabis Cannabinoid Res 2(1):139–154

Iftikhar A, Zafar U, Ahmed W, Shabbir MA, Sameen A, Sahar A, Bhat ZF, Kowalczewski PŁ, Jarzębski M, Aadil RM (2021) Applications of *Cannabis Sativa* L. in Food and its therapeutic potential: from a prohibited drug to a nutritional supplement. Molecules. 26(24):7699

Iseger TA, Bossong MG (2015) A systematic review of the antipsychotic properties of cannabidiol in humans. Schizophr Res 162(1–3):153–161

Jastrząb A, Jarocka-Karpowicz I, Skrzydlewska E (2022) The origin and biomedical relevance of cannabigerol. Int J Mol Sci 23(14):7929

Kumar N, Goel N (2019) Phenolic acids: natural versatile molecules with promising therapeutic applications. Biotechnol Rep 24:e00370

Langlois C, Potvin S, Khullar A, Tourjman SV (2021) Down and high: reflections regarding depression and cannabis. Front Psychiatry 12:625158

Li HL (1974) The origin and use of cannabis in eastern Asia linguistic-cultural implications. Econ Bot 28(3):293–301

Livingston SJ, Quilichini TD, Booth JK, Wong DC, Rensing KH, Laflamme-Yonkman J, Castellarin SD, Bohlmann J, Page JE, Samuels AL (2020) Cannabis glandular trichomes alter morphology and metabolite content during flower maturation. Plant J 101(1):37–56

Mahlberg PG, Hammond CT, Turner JC, Hemphill JK (1980) Structure, development and composition of glandular trichomes of Cannabis sativa L. Vancouver: symposium on biology and chemistry of plant Trichomes, Vancouver, BC

Mallick-Searle T, Marie BS (2019) Cannabinoids in pain treatment: an overview. Pain Manag Nurs 20(2):107–112

Maniin P, Labua S, Onsri K, Tongsak Y (2022) The effects of cannabis breeding lines on nutritional content and bioactive compounds in cannabis leaves (cannabis sativa L.). Pak J Life Soc Sci 20(2)

Manseau MW, Goff DC (2015) Cannabinoids and schizophrenia: risks and therapeutic potential. Neurotherapeutics 12(4):816–824

McPartland JM, Guy GW (2017) Models of cannabis taxonomy, cultural bias, and conflicts between scientific and vernacular names. Bot Rev 83:327–381

McPartland JM, Russo EB (2001) Cannabis and cannabis extracts: greater than the sum of their parts? J Cannabis Ther 1(3–4):103–132

Mead A (2017) The legal status of cannabis (marijuana) and cannabidiol (CBD) under US law. Epilepsy Behav 70:288–291

Mukherjee A, Roy SC, De Bera S, Jiang HE, Li X, Li CS, Bera S (2008) Results of molecular analysis of an archaeological hemp (Cannabis sativa L.) DNA sample from north West China. Genet Resour Crop Evol 55:481–485

Nabissi M, Morelli MB, Amantini C, Liberati S, Santoni M, Ricci-Vitiani L, Pallini R, Santoni G (2015) Cannabidiol stimulates A ml-1a-dependent glial differentiation and inhibits glioma stem-like cells proliferation by inducing autophagy in a TRPV 2-dependent manner. Int J Cancer 137(8):1855–1869

Niesink RJ, van Laar MW (2013) Does cannabidiol protect against adverse psychological effects of THC? Front Psych 4:130

O'Brien M, McDougall JJ (2018) Cannabis and joints: scientific evidence for the alleviation of osteoarthritis pain by cannabinoids. Curr Opin Pharmacol 40:104–109

Odieka AE, Obuzor GU, Oyedeji OO, Gondwe M, Hosu YS, Oyedeji AO (2022) The medicinal natural products of Cannabis sativa Linn.: a review. Molecules 27(5):1689

Paland N, Hamza H, Pechkovsky A, Aswad M, Shagidov D, Louria-Hayon I (2023) Cannabis and rheumatoid arthritis: a scoping review evaluating the benefits, risks, and future research directions. Rambam Maimonides Med J 14(4)
Pertwee R (2008) The diverse CB1 and CB2 receptor pharmacology of three plant cannabinoids: Δ9-tetrahydrocannabinol, cannabidiol and Δ9-tetrahydrocannabivarin. Br J Pharmacol 153(2):199–215
Rahn EJ, Hohmann AG (2009) Cannabinoids as pharmacotherapies for neuropathic pain: from the bench to the bedside. Neurotherapeutics 6(4):713–737
Raihan A, Bijoy TR (2023) A review of the industrial use and global sustainability of Cannabis sativa. Global Sustainab Res 2(4):1–29
Rasera GB, Ohara A, de Castro RJ (2021) Innovative and emerging applications of cannabis in food and beverage products: from an illicit drug to a potential ingredient for health promotion. Trends Food Sci Technol 115:31–41
Rosenberg EC, Tsien RW, Whalley BJ, Devinsky O (2015) Cannabinoids and epilepsy. Neurotherapeutics 12(4):747–768
Rupasinghe HV, Davis A, Kumar SK, Murray B, Zheljazkov VD (2020) Industrial hemp (Cannabis sativa subsp. sativa) as an emerging source for value-added functional food ingredients and nutraceuticals. Molecules 25(18):4078
Russo EB (2007) History of cannabis and its preparations in saga, science, and sobriquet. Chem Biodivers 4(8):1614–1648
Russo EB (2008) Cannabinoids in the management of difficult to treat pain. Ther Clin Risk Manag 4(1):245–259
Russo EB (2011) Taming THC: potential cannabis synergy and phytocannabinoid-terpenoid entourage effects. Br J Pharmacol 163(7):1344–1364
Safi K, Sobieraj J, Błaszkiewicz M, Żyła J, Salata B, Dzierżanowski T (2024) Tetrahydrocannabinol and Cannabidiol for pain treatment—an update on the evidence. Biomedicine 12(2):307
Saravanan AM, Al Hashmi RA, Jesil A, Achuthan M, Walke S, Al Rashdi S, Al Balushi N, Jahan S (2022) Experimental scrutinization on the treatment of synthetic wastewater using neem bark and date palm fiber as an adsorbent. Rasayan J Chem Special:204–215
Scherma M, Muntoni AL, Riedel G, Fratta W, Fadda P (2020) Cannabinoids and their therapeutic applications in mental disorders. Dialogues Clin Neurosci 22(3):271–279
Shehata I, Hashim A, Elsaeidy A, Nair A, Urits I, Viswanath O, Kaye A, Habib M (2022) Cannabinoids and their role in chronic pain treatment: current concepts and a comprehensive review. Health Psychol Res 10(4)
Shoyama Y, Yagi M, Nishioka I & Yamauchi T (1975) Biosynthesis of cannabinoid acids. Phytochemistry, 14(10), 2189-2192. https://doi.org/10.1016/S0031-9422(00)91096-3
Stockings E, Zagic D, Campbell G, Weier M, Hall WD, Nielsen S, Herkes GK, Farrell M, Degenhardt L (2018) Evidence for cannabis and cannabinoids for epilepsy: a systematic review of controlled and observational evidence. J Neurol Neurosurg Psychiatry 89(7):741–753
Tahir MN, Shahbazi F, Rondeau-Gagné S, Trant JF (2021) The biosynthesis of the cannabinoids. J Cannabis Res 3:1–12
Thomas BF, ElSohly MA (2016) The botany of Cannabis sativa L. The analytical chemistry of cannabis. pp 1–26
Turner CE, Hadley KW, Fetterman PS, Doorenbos NJ, Quimby MW, Waller C (1973) Constituents of Cannabis sativa L. IV: stability of cannabinoids in stored plant material. J Pharm Sci 62(10):1601–1605
Turner CE, Elsohly MA, Boeren EG (1980) Constituents of Cannabis sativa L. XVII. A review of the natural constituents. J Nat Prod 43(2):169–234
van den Hoogen NJ, Harding EK, Davidson CE, Trang T (2022) Cannabinoids in chronic pain: therapeutic potential through microglia modulation. Front Neural Circuits 15:816747
Zhou D, Dennis E, Snehal I, Swaminathan A (2021) Cannabinoids in the treatment of epilepsy: a review. Eur Med J
Zysk K (2018) Greek and Indian physiognomics. J Am Orient Soc 138(2):313–325

# Medicinal and Nutritional Importance of *Prunus persica* in Human Health

# 15

Shafia Arshad, Farah Zafar, and Faiza Irshad

**Abstract**

Peach (*Prunus persica* L.), a nutrient-dense fruit of the Rosaceae family, is among the world's most widely consumed stone fruits due to its pleasant flavor, tender texture, and rich nutritional profile. Originating in China around 1100 B.C., it later spread globally, reaching Mexico in 1600 and becoming commercially cultivated in the United States by the nineteenth century. Peaches are known for their high water content (88.87 g/100 g), low fat, and a diverse range of macro- and micronutrients including carbohydrates, organic acids (citric, malic, folic), calcium, magnesium, phosphorus, and potassium. B-complex vitamins dominate their vitamin profile. Peaches also contain phytochemicals like phenolic acids, carotenoids, flavonoids, and volatile compounds. The peel has a higher concentration of antioxidants compared to the flesh. Seeds and kernels are rich in minerals, fatty acids (palmitic, oleic, linoleic), and phytosterols such as β-sitosterol. Bioactive compounds like amygdalin, hydroxycinnamic acid, and catechins exhibit strong therapeutic potential, including anticancer and antiaging effects. Various parts of the plant offer pharmacological benefits—seeds, kernels, and leaves exhibit antioxidant, anti-inflammatory, antidiabetic, antimicrobial, analgesic, and cytotoxic properties. Traditionally, *P. persica* has been used to treat ailments like rashes, eczema, earache, piles, and women's health issues. Its leaves also possess laxative, insecticidal, and anthelmintic effects. Despite its rich bioactive content and traditional uses, more scientific exploration is needed

S. Arshad (✉)
Office of Research Innovation and Commercialization, The Government Sadiq College Women University, Bahawalpur, Pakistan
e-mail: shafia.arshad@iub.edu.pk; shafia.arshad@gscwu.edu.pk

F. Zafar · F. Irshad
University College of Conventional Medicine, Faculty of Medicine & Allied Health Sciences, The Islamia University of Bahawalpur, Bahawalpur, Pakistan



M. A. Ansari et al. (eds.), *Medicinal Plants and their Bioactive Compounds in Human Health: Volume 2*, https://doi.org/10.1007/978-981-95-3620-7_15

to fully validate the medicinal potential of peach. Understanding its biological properties and safety profile is essential for informed therapeutic application. This chapter focuses on the nutritional and medicinal aspects of *Prunus persica*, highlighting its bioactive compounds and their roles in promoting health and preventing disease.

**Keywords**

Nutrition · Phytochemistry · Pharmacology · Therapeutic effects · *P. persica* · Phytochemistry

**Learning Objectives**
By the end of this chapter, the reader will be able to:

- Understand the botanical, taxonomical, and historical background of *Prunus persica* (peach).
- Describe the traditional uses of various parts of the peach plant in folk medicine across different cultures.
- Identify and categorize the macronutrients and micronutrients present in peach fruit and their health benefits.
- Discuss the phytochemical profile of peach including phenolic compounds, carotenoids, flavonoids, and volatile compounds.
- Analyze the pharmacological effects of *P. persica*, including its gastrointestinal, antioxidant, anti-inflammatory, antidiabetic, cytotoxic, and neuroprotective properties.
- Evaluate the therapeutic potential and safety considerations of peach as a functional food and natural remedy.

## 15.1 Introduction/Background

A rising number of patients globally rely on medicinal plants and herbs for their health purposes. As a result, the extensive scientific evaluation of their biological effects, therapeutic potential, and safety profile would be greatly helpful in informed decisions related to their uses (Shahrajabian et al. 2023). Several epidemiological studies have revealed that diets rich in fruits and vegetables offer a plethora of benefits including reduced risk of cardiovascular diseases, cancer, neurodegenerative diseases, and other health issues. According to the World Health Organization (WHO), fruits and vegetables are edible plants with high nutritional value and are recognized as functional foods (Kennedy et al. 2021). They are low in calories and high in fiber, water, and other phytochemicals such as carotenoids, flavonoids, vitamins, and minerals. Based on this information, a joint initiative was launched in 2003 by the Food and Agricultural Organization (FAO) and WHO to encourage the production and consumption of fruits and vegetables (Azzolina et al. 2020).

The peach (*Prunus persica* (L.)), a part of the Rosaceae family, originated in 1100 BC in China. Peach ranks just behind the apple and pear among the commercially important fruits in this family. It is one of the most diversified species of fruit along with commercial cultivars with varying shapes, sizes, and colors, and seed properties reflecting a diverse international germplasm (Bento ct al. 2022). The various parts of peach plants have been employed to treat a variety of illnesses. Peaches are divided into two main types freestone and clingstone based on the adherence of flesh to the stone. Each type can possess either yellow or white flesh based on the mesocarp color. China is the largest producer of peaches in the world followed by Italy, Spain, and the United States (Luo et al. 2022). Peach contains carbohydrates, fibers, carotenoids, alkaloids, flavonoids, and polyphenols with potential health-promoting abilities (Bento et al. 2022). These compounds in peaches are responsible for the broad-spectrum pharmacological properties including anti-inflammatory, antioxidant, antipyretic, analgesic, antiparasitic, antidiabetic, immunological, antitussive, and antihistamine properties (Liu et al. 2024). Despite having a high nutrient profile and being widely consumed globally, the peach remains still thoroughly studied. Peach varieties differ depending on the phenolic profile and influence their potential benefits. Phenolic compounds have been the point of focus owing to their antioxidant, anti-inflammatory, and antibacterial properties. This chapter aims to highlight the nutritional and chemical composition including both micronutrients and macronutrients, pharmacological properties, and therapeutic benefits of *Prunus persica* (*P. persica*).

## 15.2 Plant Profile

### 15.2.1 Synonyms

Amygdalus persica, Amygdalus persica var., Amygdalus persica var. aganonucipersica, compressa, scleronucipersica, Amygdalus persica var. scleropersica, Amygdalus persica var., Persica platycarpa, Persica vulgaris var. compressa, Persica vulgaris, *P. persica*, Prunus daemonifuga, compressa, *P. persica* f. aganonucipersica, lasiocalyx, platycarpa, *P. persica* subsp. *P. persica* f. scleropersica (Al-Snafi 2022).

### 15.2.2 Taxonomic Classification

Kingdom: *Plantae*
Subkingdom: *Viridiplantae*
Infrakingdom: *Streptophyta*
Superdivision: *Embryophyta*
Division: *Tracheophyta*
Subdivision: *Spermatophytina*
Class: *Magnoliopsida*

Superorder: *Rosanae*
Order: *Rosales*
Family: *Rosaceae*
Genus: *Prunus*
Species: *P. persica* (Al-Snafi 2022).

### 15.2.3 Common Names

Arabic: Khokh
English: Peach
Chinese: Tao
Japanese: momo
Korean: bogsunganamu
Persian: Aru
Hindi: Aru, Shaftalu
Portuguese: pêssego
Swedish: persika (Hussain et al. 2021).

### 15.2.4 Botanical Description

The trees which do not produce suckers are devoid of thorns and are 30–100 dm tall. The twigs possess buds and are smooth. The leaves are deciduous, petioles measuring 5–10 mm, smooth, without wings, and bearing glands near the tip, with 1–4 discoid glands. The leaf blades are folded along the midrib or sometimes curved, and are oblong to lanceolate measuring (5–)7–15 × 2–4.5 cm possessing a cuneate to obtuse base (Hans et al. 2020). The margins are blunt, and crenulate, teethes are glandular, pointed apex, and surfaces remain smooth. Inflorescences consist of solitary flowers that sometimes appear in two flowered clusters. Pedicles are smooth and are 0–3 mm long. The flowers bloom before leaf emergence, and the hypanthium cup shape measures 4–5 mm and is smooth externally. The sepals are 3.5–5 mm in length and are oblong-ovate with entire ciliate margins with an abaxial hairy surface along with margins; however, the adaxial side is smooth. The sepals are dark pink, obovate to nearly round measuring 10–17 mm in size. The hairs cover the ovaries, the drupes are yellow to orange with reg ting, globose with a size of 40–80 mm and have a velvety appearance. The mesocarp is fleshy, and stones are ellipsoid, flattened strongly, and deeply pitted (Balkrishan et al. 2021). Different parts of peach plant are exhibited in Fig. 15.1.

### 15.2.5 Distribution

Peach is likely originated in China and gradually spread westward through Asia to Mediterranean countries before reaching Europe. It was introduced in the new

**Fig. 15.1** Different part of *P. persica* plant: (**a**) Fruit, (**b**) leaves, (**c**) flowering buds, (**d**) pericarp, (**e**) Kernal and (**f**) whole plant

world by Spanish explorers by 1600; the fruit was already in Mexico. However, large-scale commercial peach cultivation in the United States did not start until the nineteenth century (Manganaris et al. 2023).

## 15.3 Traditional Uses

*P. persica* has been used for the treatment of a variety of illnesses and for improving well-being. Being a widely used fruit, peach has been a part of human diet for thousands of years. It has traditionally been employed for blood stasis due to its blood-activating properties, which was first reported in Shennong during Hang Dynasty. The consumption of peach for blood stasis, constipation, stomach pain, and rheumatic pain has been found in several medical writings often used in combination with *Carthamus tinctorius, Anglica sinensis,* and *Cinnamomum cassia. P. persica* with other plants such as *Cinnamomi ramulus* and *Rheum palmatum* has been used for uterine fibroid, and postpartum abdominal pain (Liu et al. 2024). Peach has also been used traditionally as a therapeutic agent in Japan and South Korea. In Japan, it is often used with other medicinal plants to enjoy therapeutic benefits, and various characteristics are outlined in Japanese pharmacopeia (Bailly 2020). For instance, the formulation prepared with prunus, Junchoto, and Machinegun is employed for the activation of fluid secretion in the colonic epithelium, hence improving bowel movements (Bento et al. 2022). Moreover, formulations with other medicinal plants are used to cure constipation related to blood stasis by preventing clot formation and endorsing vasodilation. The plant leaves were used as sedative, laxative, vermicidal, anthelmintic, and insecticidal and for the treatment of leukoderma, piles, and whooping cough (Al-Snafi 2022). The fruit of the peach plant possesses therapeutic properties and has been used as demulcent, antipyretic, aphrodisiac, antiscorbutic, brain tonic, and aperient. The seed oil possesses abortifacient properties and also has been used for improving various health conditions including earache, piles, deafness, and stomach upset in children. Fruits are rich sources of fibers, alkaloids, polyphenols, oils, carotenoids, carbohydrates, and minerals with amazing health-promoting effects. Different parts of plants show a variety of pharmacological effects comprising anti-inflammatory, antiparasitic, cytotoxic, antipyretic, analgesic, gastrointestinal, antioxidant, antimicrobial, antihistamine, immunological, anti-diabetic, antitussive, and acetylcholine esterase inhibitory properties (Marilena et al. 2018).

## 15.4 Nutritional Value

*P. persica* plant is packed with essential nutrients, offering a wide range of biological compounds contributing to their significant nutritional and health benefits. They are particularly valued owing to their rich source of vitamins, and dietary fibers, which promote digestive, antioxidant, and cholesterol-lowering effects (Bento et al. 2022; Coultate 2023). Moreover, minerals including potassium, iron, and phosphorus play an integral role in cardiovascular health. Additionally, the presence of phytochemicals such as phenolic acids and flavonoids exhibit anti-inflammatory and antioxidant properties potentially lowering the risk of chronic diseases including cancer and cardiac illnesses. The low-calorie and high-water content make them an ideal choice for managing weight and improving hydration. This diverse composition of the peach plant reflects its role in promoting overall health and preventing nutritional deficiencies (Maatallah et al. 2020).

### 15.4.1 Chemical Composition

Nutrients are the food derivatives utilized by the body to improve growth, maintenance, repair, nourish the body, and help in optimal physiological functions. Nutrients may be classified into macronutrients (energy-yielding substances) or micronutrients (essential for body functions) required in smaller quantities. Macronutrients include carbohydrates, organic acids, water, proteins, fats, and fibers; all of these nutrients help the body maintain energy balance and maintain structural integrity. Micronutrients, on the other hand, comprise vitamins and minerals, which are required for cellular functions and metabolic functions (Coultate 2023). Bioactive compounds or secondary metabolites are another important group in plants that are present in small quantities and have profound effects on physiological and cellular activities. They have also been classified as extra-nutritional substances, though not essential for survival but have remarkable effects on human health (Nowicka et al. 2023). Extensive research has been conducted to assess the chemical constituents of the *P. persica* plant to further explore its nutritional value and pharmacological properties.

The peach is a nutrient-rich fruit comprising both macro- and micronutrients. It possesses low fat content and high water accounting for 88.87 g per 100 g of fruit (Dhariwala 2023). Sugar and organic acids such as malic acid, folic acid, and citric acid are relatively low. The primary macronutrient is carbohydrates as organic acid constitutes only 4% of peach nutrient content (Reig et al. 2023). As far as the micronutrient is concerned, peach contains phosphorus, calcium, potassium, magnesium, and vitamin B-complex. Moreover, phytochemical compounds including volatile compounds, phenolic compounds, flavonoids, and carotenoids have also been observed in peach chemical composition (Bento et al. 2022). The detailed chemical composition is described below.

#### 15.4.1.1 Macronutrient

##### 15.4.1.1.1 Carbohydrates

Carbohydrates represent the most abundant macronutrient in fruits and vegetables constituting about 50–80% of their dry weight (Meena et al. 2021). Being a major energy provider, they are important to human nutrition and contribute to the regulation of gut microflora. Three main groups of carbohydrates including monosaccharides (arabinose, ribose, xylose, glucose, galactose, and fructose), oligosaccharides (sucrose, lactose, maltose, and raffinose) and polysaccharides (starch and glycogen) are included in carbohydrates (Ashraf et al. 2011). Organoleptic properties such as taste and aroma are due to mono- and oligosaccharides. Fructose, sucrose, and glucose are the main sugars present in peaches with a total sugar content of 89.16–184.49 g/L from different cultivars. Sucrose represents the highest percentage 55.74–72.96% followed by glucose 6.65–15.4%, fructose 6.67–16.82%, and sorbitol 1.07–10.75% (Reig et al. 2013). The sugar composition and content are credited for the sweetness intensity of fruit, as glucose and fructose contribute to increased sweetness. The fruit maturation changes the sugar levels, as sucrose surpasses the fructose and glucose in ripe peaches; however, sorbitol is higher during growth and declines as ripening progresses (Xu et al. 2022). Moreover, trace amounts of other sugars such as xylose, maltose, trehalose, glycerol, polyols, and myo-inositol are found in mature peaches. The sugar composition not only influences the taste but also affects the aroma, as high sucrose levels tend to have increased aroma (Wu et al. 2022). During the maturation, the organic acid content decreases with an increase in sugar contributing to fruits' unique taste and sensory profile. The presence of fructose in peaches offers gastrointestinal benefits due to its property to promote the growth of beneficial bacteria including lactobacilli and bifidobacteria in the gastrointestinal tract. In addition to its role as a sweetener, sucrose also serves as a primary energy source (Qi and Tester 2020). Sorbitol, which is a sugar alcohol, promotes the transfer of osmotic water to the gut due to its laxative properties and is used in gut health applications. Not only does the sugar content decide the fruit aroma, but the good balance between sugar and organic acid plays a role in enhancing the sensory experience including both flavor and fruit quality (Mihaylova et al. 2021; Wu et al. 2022; Zhang et al. 2023).

#### 15.4.1.1.2 Fibers

Dietary fibers are often termed as indigestible carbohydrates and have been challenging to define their composition that varies among different foods. Basically, fibers are comprised of plant-based polysaccharides and lignans that cannot be hydrolyzed by human gastrointestinal enzymes (Mihaylova et al. 2021). Both soluble and insoluble fibers play a crucial role in the regulation of human physiology. They protect the body against gastrointestinal, cardiovascular, and metabolic disorders and help in the maintenance of body health. Unpeeled peaches are a rich source of fibers both soluble and insoluble, counting for 1.5 g per 100 g of fruit (Bento et al. 2022). Their regulatory effect on gastrointestinal and circulatory systems helps in weight control and regulates cholesterol levels. When coupled with essential vitamins, dietary fibers strengthen the role of fruit in promoting optimal health and balanced nutrition (Massot-Cladera et al. 2020).

#### 15.4.1.1.3 Fatty Acid

Fatty acid, the integral part of triglycerides and the chief form of fat storage, constitutes a small portion of fruit composition. Fat constitutes 0.25 g per 100 g of fruit—a small proportion (Reig et al. 2023). Despite their low concentration, fatty acids are important and serve as a carrier of lipophilic vitamins and bioactive compounds. They also play a crucial role in the prevention of cardiovascular diseases. Moreover, peach oil contains some fatty acids that are significantly important for cellular functions. Polyunsaturated fatty acids including linoleic acid and palmitic acid have also been discovered in peaches. Linoleic acids are not synthesized in the body, though essential fatty acids are required for the repair of cell membranes, maintenance of nervous tissues, and synthesis of prostaglandins. Polyunsaturated fatty acids possess anti-inflammatory properties and help in the regulation of cholesterol levels, promote high-density lipoprotein, and offer additional protection from chronic diseases (Kapoor et al. 2021).

#### 15.4.1.1.4 Amino Acids

Long-chain amino acids fold into secondary and tertiary structures yielding the proteins. Tertiary structures are the active proteins and serve as subunits in quaternary structures or are sometimes associated with nonprotein component such as the heme group. Dietary proteins are required to provide amino acids that cannot be synthesized by the body and play a crucial role in tissue repair, cell structure, and synthesis of enzymes (Lopez and Mohiuddin 2024). The taste of peach is greatly influenced by its amino acid profile. Amino acids are present in both free form or combined with proteins categorized as tasteless (arginine and lysine), bitter (lucein and tyrosine), and sweet (proline and alanine). Essential amino acids found in peach are lysine, phenylalanine, lysine, and arginine (Sun et al. 2022). Studies revealed that peach juice concentrate contains asparagine to be the most abundant amino acid (2491 ± 25 mg/L) followed by aspartic acid (616.4 ± 5.0 mg/L), serine

(370.2 ± 4.1 mg/L), and glutamic acid (283.0 ± 2.8 mg/L). These findings align with the USDA database which features methionine (0.010 g), histidine (0.013 g), tryptophan (0.010 g), cysteine (0.012 g), and tyrosine (0.014 g) amino acids present in 100 g of peach sample (Bento et al. 2022). Amino acid content varies according to cultivar differences and method of analysis. In addition to these common amino acids, peaches' nutritional profile is also enhanced by γ-aminobutyric acid (GABA), lucein, isoleucine, lysine, threonine, valine, alanine, and glycine (Mihaylova et al. 2021).

#### 15.4.1.1.5 Organic Acid

Sugar content in fruit primarily determines the sweetness and organic acids impart sourness to the fruits. Organic acids, low in molecular weight, are composed of carbon, hydrogen, and oxygen with one or more carboxyl groups. They are abundant in plant roots and help in nutrient uptake and detoxification processes according to their carboxyl group dissociation abilities. The fruit's organoleptic properties are determined by the sugar to organic acid ratio and influence the perception of sweetness and sourness. Different organic acids have varying sweetness-masking impacts; for instance, citric acid causes more sourness than similar concentrations of malic acid. Fumaric acid, shikimic acid, and quinic acid are also found in small quantities (Zheng et al. 2021). According to a study, peach contains total acid content of 5.59–18.50 g/L with malic acid contributing 42.92–84.30%, citric acid 3.72–31.61%, and quinic acid 14.56–57.54%, abundantly found in the peel (Bae et al. 2014). Sucrose, a predominant sugar found in peach, also contributes to the inhibition of oxidation reactions in addition to sweetness. Sorbitol, another sugar alcohol, improves the texture of fruit and plays an essential role in sensory quality. Taste is greatly affected by the balance between malic and citric acid; however, L-ascorbic acid content is quietly low in peaches (Saidani et al. 2017).

### 15.4.1.2 Micronutrients

#### 15.4.1.2.1 Minerals

Based on the distinct roles in plant metabolism, minerals may be macroelement and microelements. Magnesium, potassium, and calcium are essential minerals found in peaches and are responsible for metabolic processes such as the synthesis of amino acids and proteins. Moreover, the elements including zinc, iron, and copper are also part of peach composition (Saidani et al. 2017). In addition to being vital for plant functions, minerals are also important for human health. The peel and pulp of peach contain a variety of minerals including nitrogen, manganese, phosphorus, copper, cobalt, chromium, lead, fluoride, selenium, and potassium. According to the USDA, the most abundant mineral found in peaches is potassium, and predominant in pulp and peel (Mitic et al. 2019).

Being rich in potassium, peaches play a crucial role in maintaining electrolyte balance and homeostasis. Phosphorus also helps in nucleic acid synthesis, energy storage, and cellular maintenance. Magnesium helps stabilize the nervous system and activates alkaline phosphatase. Calcium is primarily involved in skeletal development, serves as a co-enzyme in metabolic regulation, and improves physiological processes including muscle contraction, blood clotting, and regulation of sodium ion transport. Manganese, though required in trace amounts, improves carbohydrate metabolism, and prevents oxidative stress by activation of superoxide dismutase. Iron is important for hemoglobin synthesis, while zinc works as a coenzyme for more than 200 enzymes responsible for immune system regulation, cellular growth, and acid–base balance (Godswill et al. 2020).

##### 15.4.1.2.2 Vitamins

Vitamins are the essential nutrients for human physiology and help in specific biochemical processes for metabolic functions, growth, and vitality. They are present in small quantities in food and are vital for normal healthy functions. Vitamins are categorized into water-soluble (vitamin C, vitamin B, and folic acid) and fat-soluble (vitamins A, D, E, and K) and their deficiencies lead to several pathologies (Tyśkiewicz et al. 2018). Vitamin B-complex, vitamin C, and E are predominantly found in peaches. Peaches contain a higher concentration of vitamin C (10.4–33.3 mg/100 g) than 4.15–14.2 mg/100 g found in pulp alone and protect against cardiovascular diseases (Bento et al. 2022). Vitamin E inhibits the oxidation of low-density lipoprotein and lowers the progression of atherosclerosis. Vitamin C increases nitric oxide production, protects vascular endothelium, and plays an important role in wound healing, immune system strengthening, and prevention of cancer (Morelli et al. 2020).

#### 15.4.1.3 Phytochemical Compounds

Phytochemical compounds are secondary metabolites present in plants, though not essential for survival but possess therapeutic effects. They offer important health benefits beyond basic nutrition and lower the risk of chronic diseases. Secondary metabolites or phytochemical classes of compounds present in peaches include carotenoids, volatile compounds, and phenolic compounds (Kumari et al. 2023).

##### 15.4.1.3.1 Volatile Compounds

Volatile compounds are credited for their ability to impart flavor and aroma to the fruits. These characteristics show dynamic development, developing in response to the quality and quantity of volatile compounds which varies in the ripening stage. Aroma is a complex mixture of several volatile compounds with each fruit showing a distinct aroma that determines the concentration and intensity of these volatiles. Depending on the origin, volatiles are categorized into primary and secondary volatiles; primary volatiles present in intact fruit tissue and secondary are produced on tissue disruption due to intracellular binding with sugar as glycoside or glucosinolate. Secondary volatile compounds are higher in concentration than the primary ones and significantly influence flavor development (Zhang et al. 2022). Flavor volatile compounds are sourced from phytonutrients including amino acids, phenol, carotenoids, fatty acids, and terpenoids. According to literature, about 100 volatile compounds are identified in peach and nectars with C6 compounds such as alcohols, esters, ketones, terpenoids, lactones, aldehydes, and C13 nor-isoprenoids.

Among these, γ- and δ-decalactones and γ- and δ-dodecalactones help define peach aroma and work synergistically with alcohol, terpenoids, and aldehydes which yield spicy, floral, and fruity notes to the peach (Zhang et al. 2022). C6 compounds in unripe peaches offer a grassy flavor to fruit. Moreover, other volatile compounds include hexane, hexanal, 2-hexenal, (Z)-3-hexenol, (E)-2-hexenal, (E)-2-hexenol, and 1-hexanol. These compounds continue to diminish as the fruit matures; however, due to their origin from degradation of enzyme catalysis of fatty acid, they may still be present in ripe peaches. Concurrently, C13 nor isoprenoids and lactones present as a dominant volatile compound in ripe peaches (Rey-Serra et al. 2022).

#### 15.4.1.3.2 Carotenoids

Carotenoids are the pigments present in all photosynthetic plants and provide red and yellow color to the flowers and fruits. They are the most prevalent group of pigments in nature. Carotenoids are classified into oxygenated (xanthophylls: lutein and β-cryptoxanthin) and nonoxygenated hydrocarbons (carotenes: α-carotene and β-carotene, or linear, such as lycopene) (Meléndez-Martínez et al. 2022). Carotenoids are beneficial for eye health, especially lutein and zeaxanthin protect against age-related macular degeneration, and protect against reactive oxygen species. Owing to their pigmenting properties, carotenoids are widely employed in food industry and have greater stability than anthocyanins (Olatunde et al. 2020). The predominantly found carotenes in fresh peaches are β-cryptoxanthin and β-carotene, with concentrations varying from 60 to 360 μg/kg in the peel and 60–160 μg/kg in the flesh for β-cryptoxanthin, and from 110 to 3790 μg/kg in the peel and 40 to 1680 μg/kg in the flesh for β-carotene. Total carotenoid content in peaches is reported to be 0.8–3.7 mg and 0.1–1.9 mg per 100 g of tissues in yellow, white, and red-fleshed peaches. Carotenoid content in nectarines ranges from 126 to 197 μg/100 and β-carotene constitutes 95–97% of total carotenoids (Petruccelli et al. 2023), as shown in Fig. 15.2.

#### 15.4.1.3.3 Phenolic Compounds

Phenolic compounds are a group of phytocompounds derived from plant metabolism and possess significant biological activities such as antioxidant, antimicrobial, and pigmentation. These compounds are responsible for the antioxidant activities of peaches. Moreover, these compounds protect against atherosclerosis, cancer, and brain diseases (Rahaman et al. 2023). Structurally, they are composed of an aromatic ring with one or more hydroxyl groups and are categorized into single (with one aromatic ring) and polyphenols (multiple rings). Based on the structural features, phenolic compounds are classified into flavonoids (flavones, flavanones, flavan-o3-ols, isoflavones, and anthocyanins) and nonflavonoids (stilbenes and phenolic acids). They may be present in free form, conjugate (glycoside), or insoluble forms typically attached to plant cell walls (Ding et al. 2020). According to the study, 19 peach cultivars both red and light fleshed exhibited that red-fleshed peaches have a higher phenolic content than light varieties ranging from 228 to 1260 mg of chlorogenic acid per 100 g of tissue compared to 137–371 mg in light-fleshed peaches. An inverse proportion was seen in phenolic compounds and

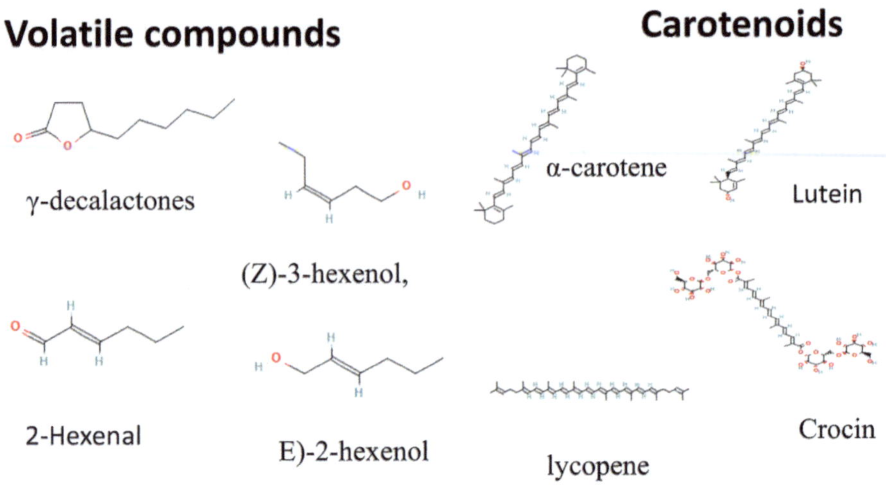


**Fig. 15.2** Volatile compounds and carotenoids present in *P. persica*

anthocyanin compounds in peel and flesh. Depending on the weight percentage of peel and flesh, 70% of compounds were found in the flesh, while 30% in the peel, stating that 20% of total phytochemical concentrations are lost when peels are discarded (Petruccelli et al. 2023). Nonflavonoids, on the other hand, are composed of a single aromatic ring and occur in soluble-free, soluble conjugated, and insoluble bound forms. Structurally, they are in C6-C1 or C6-C3 form, insoluble forms (20% of total phenols) with higher antioxidant activity compared to free or conjugated. The cell walls are not easily digestible and the bound phenols can only be absorbed in the colon. The microflora in the colon break down these molecules and protect the colon from cancer due to their chemoprotective properties. Other phenols can easily be absorbed and after digestion are easily distributed throughout the body (Petruccelli et al. 2023). Table 15.1 shows phytochemical present in *P. persica* and Fig. 15.3 shows structures of phenolic and flavonoids are given.

## 15.5 Pharmacological Effect

### 15.5.1 Gastrointestinal Effect

*P. persica* possesses beneficial effects on gastrointestinal health. It has been reported that ethyl acetate extract from peach showed a stimulatory response in rat colonic tissue in a concentration-dependent manner at the doses $10^{-8}$ to $10^{-5}$ g/mL. Ketotifen, cimetidine, and pyrilamine were used to inhibit the contraction caused by the extract. An increased number of mast cells was observed in rats subjected to peach extract that increased the release of histamine from colonic tissues. It suggests that the prokinetic activity of extract is triggered by mast cell degranulation (Han et al. 2015). Moreover, aqueous extract of peach leaves revealed gut stimulatory activity

**Table 15.1** Phytochemicals and flavonoids present in *P. persica* plant

| Class | Phytochemicals |
|---|---|
| Phenolic acids | Protocatechuic acid, syringic acid, salicylic acid, gallic acid, caffeoyl glucose, neochlorogenic acid, p-Hydroxybenzoic acid, chlorogenic acid, 4-O-β-D-(6-O-gentisoylglucopyranosyl)-vanillic acid, 4-O-caffeoylquinic acid |
| Flavonols | Quercetin-3-O-rutinoside, quercetin-3-O-galactoside, quercetin-3-O-glucoside, isorhamnetin-3-O-rutinoside, luteolin-7-glucoside, isorhamnetin-3-O-glucoside, keampferol-3-O-glucoside, Keampferol-7-neohesperiodise, hesperidin-7-rutinoside |
| Flavan-3-ols | Procyanidin B1, procyanidin B2, (-)-epicatechin, procyanidin A2 |
| Anthocyanins | Cyanidin-3-O-glucoside, cyanidin-3-O-rutinoside |
| Carotenoids | Crocin, cis-violaxanthin, β-cryptoxanthin, All-trans-β-carotene, 13 or 13′Cis-β-carotene, 9 or 9′Cis-β-carotene, β-cryptoxanthin-myristate, -cryptoxanthin-palmitate |

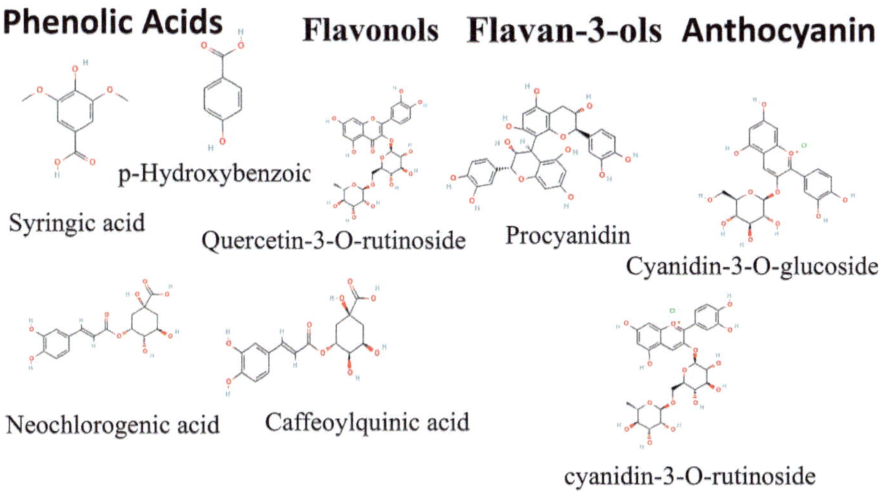


**Fig. 15.3** Flavonoids and phenolic compounds present in *P. persica* plant

in the colon of animals. At a dose of 1–10 mg/mL, the extract showed a moderate spasmogenic effect; however, pretreatment with atropine impeded the contractile activity of the extract. In the rabbit colon, the extract administration results in stimulation of gut, and subsequent relaxation of contraction at high doses. Fractionation assay showed that the spasmogenic activity was more concentrated in ethyl acetate fraction than in aqueous extract. Both aqueous and ethyl acetate extract result in dose-dependent inhibition of potassium-induced contraction (Fu et al. 2023).

### 15.5.2 Analgesic, Anti-inflammatory, and Antipyretic Effect

The anti-inflammatory activity of methanolic extract of aerial part of peaches showed significant anti-inflammatory properties. The extract was shown to reduce the production of proinflammatory mediators and cytokines in lipopolysaccharides-mediated BV2 microglial cells (Elshamy et al. 2019). The extract also can modulate nitrogen-activated protein kinase signaling pathways, as evidenced by increased production of nitric oxide and translocation of NF-kB in primary astrocytes in culture medium. Persicaside, an alkaloid derived from *P. persica* seed (methanol extract) exhibited concentration-dependent suppression of NO and prostaglandin E2 production in osteoblast sarcoma cells through downregulation of expression of inducible nitric oxide synthase and cyclooxygenase. Moreover, *P. persica* extract administration (50 and 100 mg/kg) resulted in a reduction in carrageenan-induced paw edema by 11 and 47% after 1 h, 24 and 33% after 2 h, and 23 and 32% after 4 h, respectively. At a higher dose of 100 mg/kg, the extract significantly decreased rectal temperature in pyretic rats (Elshamy et al. 2019).

### 15.5.3 Cytotoxic Effect

The anticancer activity of biological compounds such as prunasin, mandelic acid β-D-glucopyranoside, amygdalin, amygdalinic acid, and benzyl β-D-glucopyranoside derived from *P. persica* seeds was investigated in both laboratory and animal studies. These compounds inhibit antigen activation of Epstein-Barr virus that is triggered by tumor promoters. These compounds also delay the progression of two-stage carcinogenesis, and this effect is relative to the epigallocatechin gallate sourced from green tea (Esraa Fadl et al. 2023).

In another study, kaempferol 3-O-β-4C1-(6-O-3,4-dihydroxyphenylacetyl glucopyranoside) sourced from the leaves, reduced the viability of human keratinocytes after 24 h with survival rates of 89%. Polyphenols from *P. persica* with a dose of 0.8–1.6 mg/dL per day suppress the tumor growth and lung metastasis of MDA-MB-435 breast cancer cells by downregulating gene expression of metalloproteinase. Moreover, the peach extract administration also inhibits the HEPG2 cell growth, reduces cell migration, and inhibits the expression of extracellular matrix metalloproteinase (MMP) and MMP9. The protective effect of *P. persica* ethanolic extract of leaves also reduces the progression of skin cancer. The topical application of extract before ultraviolet B (UVB) exposure delayed the development of the tumor relative to the control group. The extract inhibits the cancer progression in dose-dependent manner, such as at low and higher concentrations, the inhibition was shown to be 25.9% and 53.9% (Heo et al. 2001).

### 15.5.4 Antiparasitic Effect

The ethyl acetate and ethanol extract of *P. persica* possess anthelmintic activity and kill both *Ascaridia galli* and *Pheretima posthuma*. Extract at the dose of 60 mg/mL demonstrates an anthelmintic effect against *Ascaridia galli* comparable to piperazine citrate. The methanolic extract showed moderate insecticidal activity against

*Tribolium castaneum, Rhyzopertha dominica,* and *Callosobruchus analis.* Dichloromethane fraction exhibited a moderate effect against Rhyzopertha dominica, and chloromethane fraction showed a moderate anthelmintic effect against *Rhyzopertha dominica* and *Callosobruchus analis* with no activity against *Tribolium castaneum* (Kumar and Chaudhary 2015).

### 15.5.5 Antioxidant Effect

The *P. persica* fruit extract showed potent radical scavenging activity with an 86.14 ± 1.33% inhibition at a low IC50 value of 17.5 ± 0.89 μg/mL suggesting strong antioxidant activity. Studies revealed that kaempferol 3-O- β-4C1-(6-O-3,4-dihydroxyphenylacetyl glucopyranoside) derived from leaves shows antioxidant activity with an IC50 of 6.35 ± 3.40 μg/mL compared to standard with an IC50 of 2 ± 0.01 μg/mL (Mostafa et al. 2021). Further analysis of *P. persica* fruit fraction (ethyl acetate, aqueous, n-butanol, and hexane) employing several antioxidant assays-hydrogen peroxide scavenging, iron chelation, superoxide scavenging assay revealed that ethyl acetate and n-butanol fractions of *P. persica* extract have the highest antioxidant effect which is associated with total phenolic and flavonoid content. Moreover, the antiphotoaging and antioxidant effects of MMPP derived from *P. persica* in UV-radiated human skin fibroblast showed radical scavenging activities. The compound suppressed the expression of MMP-1 and upregulated the type-1 procollagen expression highlighting its potential as an antiphotoaging agent (Mousavi et al. 2019).

### 15.5.6 Antiobesity Effect

The fruit extract of *P. persica* plant possesses remarkable antilipase activity and oleanolic acid is considered as the primary active compound for antiobesity activity with an IC50 of 12.64 ± 0.65 μg/mL. It helps speed up hepatic lipid metabolism and shows antiobesity properties. The aqueous extract of *P. persica* flower has the ability to reduce body weight, serum glucose, body fat mass, liver enzyme, and spleen weight (Song et al. 2019). It downregulated the expression of lipogenic gene (stearoyl-CoA desaturase-1 and -2 and fatty acid synthase) and upregulated the genes involved in fatty acid β-oxidation and carnitine palmitoyltransferase-1 in hepatic cells. An herbal extract of *P. persica* combined with another herbal extract also revealed antiobesity effects by regulating the gene expression involved in fatty acid oxidation. Compounds derived from peach flower extract (including methyl caffeate, chlorogenic acid, mandelamide, and ferulic acid) significantly suppressed adipogenesis. Mandelamide showed a significant inhibitory effect with IC50 of 36.04 ± 1.82 μM by downregulating the expression of adipogenic markers such as c-Jun N-terminal kinase, Peroxisome Proliferator-Activated Receptor

Gamma(PPARγ), p38, extracellular signal-regulated kinase, and glucocorticoid receptors (Lee et al. 2021).

### 15.5.7 Acetylcholinesterase Inhibitory Effect

*P. persica* extract oral administration in rats showed acetylcholinesterase activity in the brain cells. Different fractions of the peach methanolic extract showed that chloroform fractions showed the highest inhibition (75%) with an IC50 5.6 μg/mL. Aqueous extract of *P. persica* extract showed the dose-dependent suppression of plasma cholinesterase activity with ID50 at 8.9 mg/kg. The ID50 ratio of brain to plasma was 3:1 and 6:0, showing that the extract successfully crossed the blood brain barrier (Suh et al. 2006). Moreover, the antidementia activities of various parts of *P. persica* were also assessed by employing an ACHE (Acetylcholinesterase) inhibitory assay. The primary active compounds with significant AChE suppression were (+)-4′-O-methyl catechin (IC50: 70.19 ± 1.79 μg/mL), 4,2′,4′-trihydroxy-6′-methoxychalcone 4′-O-β-D-glucopyranoside (IC50: 85.21 ± 0.94 μg/mL), and ferulic acid (IC50: 97.36 ± 2.85 μg/mL). These results showed a strong potential for antidementia activity (Nakagawa et al. 2018).

### 15.5.8 Antimicrobial Effect

The antimicrobial activity of crude methanolic extract of peach and its fraction shows potent antimicrobial properties against several strains of bacteria. They were seen to be effective against *Escherichia coli, Pseudomonas aeruginosa, Salmonella typhi, Enterococcus faecalis, Klebsiella pneumoniae, Bacillus subtilis, Staphylococcus aureus*, and *Shigella flexneri*. The petroleum ether fraction shows low activity against *K. pneumoniae* and *E. faecalis*, highest activity against *E. coli*, moderate activity against *S. aureus*, and no effect against *P. aeruginosa, B. subtilis, S. typhi*, or *S. flexneri* (Oviya et al. 2019). The dichloromethane fraction exhibits an antibacterial effect against *E. coli, K. pneumoniae*, and *S. aureus*, and the ethyl acetate fraction shows a strong antibacterial effect against *K. pneumoniae* and *E. faecalis*. The methanolic extract of the peach plant also exhibits inhibitory activity against several strains of fungi. The methanol, dichloromethane, and chloroform fractions seemed to be highly effective against *Aspergillus flavus, Microsporum canis*, and *Fusarium solani* (Belhadj et al. 2016).

### 15.5.9 Antidiabetic Effect

The ethyl acetate fraction of *P. persica* leaves extract (100 and 200 mg/kg) rich in quercetin showed antidiabetic, antiadipogenic, and radical scavenging activities. In a study of STZ-induced diabetic rats, the administration of 200 mg/kg extract fraction resulted in a hypoglycemic effect and improved body weight. The extract also inhibited the adipogenesis evaluated by preadipocyte differentiation assay. The insulin-secreting effect was also observed when analyzed by glucose-stimulated insulin secretion assay (Lee et al. 2021). The compounds derived from the extract including methyl caffeate, naringenin, astragalin, chlorogenic acid, caffeic acid, and ferulic acid promoted insulin secretion without causing cytotoxic effects. Methyl caffeate upregulated the expression of signaling protein associated with receptors involved in β-cell function and insulin secretion. In addition, leaf extract (1000 mg/kg) lowers the increase of postprandial blood glucose rise by inhibiting glucose absorption in the small intestine. The inhibitory effects were due to acetylated kaempferol glycoside multiforin A derived from methanolic leaf extract of peach (Sharma et al. 2018).

### 15.5.10 Antihistamine and Immunologic Effect

The antihistamine effect of *P. persica* seeds was observed in human mast cells. From the methanol extract of peach, the isolated cyanogenic glycoside and phenolic compounds inhibited histamine release and regulated the expression of proinflammatory cytokines such as IL-6 and TNF-α suggesting the strong antiallergic and anti-inflammatory activities of *P. persica* (Shin et al. 2010). Moreover, the ethanolic extract of peach fruit suppressed the release of 48/8-induced systemic anaphylaxis and immunoglobulin-mediated allergic reactions. The extract reduced histamine release by regulating the calcium levels and downregulating the expression of inflammatory cytokines. The suppression of the nuclear factor kappa B signaling pathway was attributed to the inhibitory effect on cytokine production (Kim et al. 2013).

### 15.5.11 Protective Effect

The protective effects of *P. persica* pericarp extract evaluated in a mouse model (7 days) revealed the protective effect. The administration of extract for a week significantly reduced cisplatin-induced decrease in kidney and liver weight. The rise in serum urea and creatinine, kidney damage markers, and an increase in liver enzymes (ALT, AST) also seemed to be inhibited by the peach extract. It also reverts the cisplatin-induced alteration in serum nitric oxide, glutathione depletion, and lipid peroxidation indicating its ability to reduce oxidative stress and tissue damage (Lee et al. 2008). The bark extract of peach was effective against testosterone-induced benign hyperplasia (BPH) and decreased the testosterone-induced pathological alteration. Moreover, the antioxidant and anti-inflammatory activities of peach extract reflect its potential role in impeding the pathological pathways associated with BPH progression (Jena et al. 2016).

### 15.5.12 Vascular Effect

The ethyl acetate extract of *P. persica* showed beneficial effects against angiotensin II-induced calcium mobilization, reactive oxygen species (ROS) production, and signal transduction pathway in vascular smooth muscle cells (VSMCs). Treatment with ethyl acetate extract reduces the intracellular Ca2+ rise in VSMCs and the production of ROS, which are important for subsequent signaling pathways (Kim et al. 2019). It also shows the ability to reduce the angiotensin-induced phosphorylation of (EGFR) and myosin phosphate target subunit 1 (MYPTI), each of which is involved in the pathogenicity of atherosclerosis and hypertension. Additionally, the vasorelaxant properties of *P. persica* were endothelium-dependent and regulated through NO soluble guanylyl cyclase-cyclic guanosine monophosphate signaling pathway and transduction of muscarinic receptors. The extract also decreased serotonin and angiotensin-induced vasoconstriction through modulation of Ca2+ release by inositol 1,4,5-triphosphate receptors on the sarcoplasmic reticulum membrane (Kono et al. 2013).

### 15.5.13 Dermatological Effect

In an RNA quantification assay in UVB irradiation human skin model, the beneficial effect of fresh unripe peach extract was observed in a clinical study. The application of peach extract caused an upregulation of mRNA levels and restoration of collagen type XVIII suggesting its role in mitigating epidermal junction damage caused by UVB irradiation. The ethanolic extract (50–200 μg/mL) exhibited protective effects against DNA damage resulting from each UVB and UVC radiation in NIH/3T3 skin fibroblast cells (Santin et al. 2019).

### 15.5.14 Antitussive Effect

Polysaccharides sourced from *P. persica* were composed of xylose 20.2%, galactose 33.9%, mannose 2.7%, and arabinose 43.2%. The cough-suppressing effect of peaches' polysaccharides was investigated in cough-induced cats. The finding suggested that these polysaccharides possess significant cough suppression activity surpassing that of antinarcotic standard drugs (Saraswathy et al. 2014).

## 15.6 Conclusion

Peach is a nutrient-dense fruit with a variety of phenolic compounds depending on the cultivar and variety of plant providing the fruit a notable biochemical potential. The nutrition content comprising vitamins, phenolic acid, carotenoids, and flavonoids is crucial in maintaining fruit integrity and commercial value, as these levels vary with ripening and serve as a marker for different cultivars. Being a natural source of plant compounds, peaches greatly contribute to promoting health and well-being. Studies have demonstrated its effect in combating neurodegenerative diseases including Alzheimer's and Parkinson's diseases and providing benefits against chronic diseases such as inflammation, diabetes, cardiovascular diseases, obesity, and certain cancers. There is an increasing interest in natural medicine, particularly herbal plants that are abundant in phytonutrients. Pure peach extract should be further explored for nutritional and pharmaceutical applications. Moreover, any potential adverse effects associated with peach phytochemicals should be considered carefully to optimize its health benefits.

## 15.7 Chapter Summary

This chapter provides a comprehensive overview of *Prunus persica* (peach), a nutritionally rich and globally consumed fruit with deep historical and medicinal roots. Originating in China around 1100 B.C., peach has since spread worldwide and become a staple in both diets and traditional remedies. The fruit is packed with essential nutrients, carbohydrates, fiber, vitamins, and minerals, and is particularly notable for its rich phytochemical content including phenolic acids, flavonoids, carotenoids, and volatile compounds. These constituents contribute to its potent antioxidant, anti-inflammatory, and anticancer activities. Beyond the fruit, the seeds, kernels, and leaves exhibit a broad spectrum of pharmacological effects such as antidiabetic, antimicrobial, and neuroprotective actions. Despite its well-documented traditional uses, further scientific validation is needed to fully harness its medicinal potential. This chapter emphasizes the importance of peach as both a nutritional powerhouse and a therapeutic agent, encouraging its inclusion in functional food and pharmaceutical research.

## References

Al-Snafi AE (2022) Constituents, nutritional and pharmacological importance of Prunus persica-a review. World J Adv Pharm Med Res 3(1):019–029

Ashraf C, Iqbal S, Ahmed D (2011) Nutritional and physicochemical studies on fruit pulp, seed and shell of indigenous Prunus persica. J Med Plants Res 5(16):3917–3921

Azzolina D, Vedovelli L, Gallipoli S, French M, Ghidina M, Lamprecht M et al (2020) Nutrients and caloric intake associated with fruits, vegetables, and legumes in the elderly European population. Nutrients 12(9):2746

Bae H, Yun SK, Yoon IK, Nam EY, Kwon JH, Jun JH (2014) Assessment of organic acid and sugar composition in apricot, plumcot, plum, and peach during fruit development. J Appl Bot Food Qual 87:24–29

Bailly C (2020) Anticancer properties of Prunus mume extracts (Chinese plum, Japanese apricot). J Ethnopharmacol 246:112215

Balkrishan A, Tanwar S, Prajapati UB (2021) Medicinal and nutritional aspect of genus Prunus L. with phytoetymology. Int J Unani Integr Med 5:24–27

Belhadj F, Somrani I, Aissaoui N, Messaoud C, Boussaid M, Marzouki MN (2016) Bioactive compounds contents, antioxidant and antimicrobial activities during ripening of Prunus persica L. varieties from the North West of Tunisia. Food Chem 204:29–36

Bento C, Goncalves AC, Silva B, Silva LR (2022) Peach (Prunus persica): phytochemicals and health benefits. Food Rev Intl 38(8):1703–1734

Coultate T (2023) Food: the chemistry of its components. Royal Society of Chemistry, Cambridge

Dhariwala H (2023) Thermal processing and quality retention of pineapple and peach in jars and pouches: effect of reciprocation agitation and ultrasonic water bath processing. McGill University, Canada

Ding T, Cao K, Fang W, Zhu G, Chen C, Wang X, Wang L (2020) Evaluation of phenolic components (anthocyanins, flavanols, phenolic acids, and flavonols) and their antioxidant properties of peach fruits. Sci Hortic 268:109365

Elshamy AI, Abdallah HMI, El Gendy AE-NG, El-Kashak W, Muscatello B, De Leo M, Pistelli L (2019) Evaluation of anti-inflammatory, antinociceptive, and antipyretic activities of Prunus persica var. nucipersica (nectarine) kernel. Planta Med 85(11/12):1016–1023

Esraa Fadl M, Hany H, El-Mehiry HF, El-Sayed A, Al Deen RIT (2023) Effect of antioxidants and cytotoxic activities of apricot (Prunus Armeniaca L.) and peach (Prunus Persica L.) kernels extracts on experimental rats. Res J Spec Educ 2023(75):387–411

Fu Y, Zhang X, Song Z, Wang L, Zhao P, Chen S (2023) Exploring the active constitutions of peach blossom for amelioration of loperamide-induced mice constipation by ultra-performance liquid chromatography-mass spectrometry combined with network pharmacology. Nat Prod Commun 18(3):1934578X231161184

Godswill AG, Somtochukwu IV, Ikechukwu AO, Kate EC (2020) Health benefits of micronutrients (vitamins and minerals) and their associated deficiency diseases: a systematic review. Int J Food Sci 3(1):1–32

Han W, Xu JD, Wei FX, Zheng YD, Ma JZ, Xu XD et al (2015) Prokinetic activity of Prunus persica (L.) Batsch flowers extract and its possible mechanism of action in rats. Biomed Res Int 2015(1):569853

Hans M, Shah MA, Bansal R (2020) Peach. In: Antioxidants in fruits: properties and health benefits. Springer, Singapore, pp 449–459

Heo MY, Kim SH, Yang HE, Lee SH, Jo BK, Kim HP (2001) Protection against ultraviolet B-and C-induced DNA damage and skin carcinogenesis by the flowers of Prunus persica extract. Mutat Res Genet Toxicol Environ Mutagen 496(1–2):47–59

Hussain SZ, Naseer B, Qadri T, Fatima T, Bhat TA (2021) Peach (Prunus Persica)—morphology, taxonomy, composition and health benefits. In: Fruits grown in highland regions of the Himalayas: nutritional and health benefits. Springer, pp 207–217. https://doi.org/10.1007/978-3-030-75502-7_16

Jena AK, Vasisht K, Sharma N, Kaur R, Dhingra MS, Karan M (2016) Amelioration of testosterone induced benign prostatic hyperplasia by Prunus species. J Ethnopharmacol 190:33–45

Kapoor B, Kapoor D, Gautam S, Singh R, Bhardwaj S (2021) Dietary polyunsaturated fatty acids (PUFAs): uses and potential health benefits. Curr Nutr Rep 10:232–242

Kennedy G, Kanter R, Chotiboriboon S, Covic N, Delormier T, Longvah T et al (2021) Traditional and indigenous fruits and vegetables for food system transformation. Curr Dev Nutr 5(8):nzab092

Kim GJ, Choi HG, Kim JH, Kim SH, Kim JA, Lee SH (2013) Anti-allergic inflammatory effects of cyanogenic and phenolic glycosides from the seed of Prunus persica. Nat Prod Commun 8(12):1934578X1300801221

Kim B, Kim K-W, Lee S, Jo C, Lee K, Ham I, Choi H-Y (2019) Endothelium-dependent vasorelaxant effect of prunus persica branch on isolated rat thoracic aorta. Nutrients 11(8):1816

Kono R, Okuno Y, Nakamura M, Inada K-I, Tokuda A, Yamashita M et al (2013) Peach (Prunus persica) extract inhibits angiotensin II-induced signal transduction in vascular smooth muscle cells. Food Chem 139(1–4):371–376

Kumar N, Chaudhary A (2015) Evaluation of anthelmintic activity of prunus persica (L.). Asian J Pharm Clin Res 8(5):170–172

Kumari N, Kumar M, Puri S, Zhang B, Rais N, Pundir A et al (2023) Peach (Prunus persica (L.) Batsch) seeds and kernels as potential plant-based functional food ingredients: a review of bioactive compounds and health-promoting activities. Food Biosci 54:102914

Lee CK, Park K-K, Hwang J-K, Lee SK, Chung W-Y (2008) The pericarp extract of Prunus persica attenuates chemotherapy-induced acute nephrotoxicity and hepatotoxicity in mice. J Med Food 11(2):302–306

Lee D, Kim J-Y, Qi Y, Park S, Lee HL, Yamabe N et al (2021) Phytochemicals from the flowers of Prunus persica (L.) Batsch: anti-adipogenic effect of mandelamide on 3T3-L1 preadipocytes. Bioorg Med Chem Lett 49:128326

Liu Y-Q, Wu H-L, Zhang Z-Q, Wang W-L, Han G-Q, Zhang C-H et al (2024) Traditional use, phytochemistry, pharmacology, toxicology and clinical applications of persicae semen: a review. Chin J Integr Med:1–11

Lopez MJ, Mohiuddin SS (2024) Biochemistry, essential amino acids. In: StatPearls [Internet]. StatPearls Publishing

Luo C-X, Schnabel G, Hu M, De Cal A (2022) Global distribution and management of peach diseases. Phytopathol Res 4(1):30

Maatallah S, Dabbou S, Castagna A, Guizani M, Hajlaoui H, Ranieri AM, Flamini G (2020) Prunus persica by-products: a source of minerals, phenols and volatile compounds. Sci Hortic 261:109016

Manganaris GA, Sansavini S, Gradziel TM, Bassi D, Crisosto CH (2023) Peach: an introduction. In: Peach. CABI GB, pp 1–16. https://doi.org/10.1079/9781789248456.0001

Marilena H, Oprea E, Ancuceanu R, Dinu M, Anghel AI, Morosan E, Duţu LE (2018) Contributions to the pharmacognostical and phytobiological study of Prunus persica (L.) batsch flowers. Farmacia 66(1):78–82

Massot-Cladera M, Azagra-Boronat I, Franch À, Castell M, Rodríguez-Lagunas MJ, Pérez-Cano FJ (2020) Gut health-promoting benefits of a dietary supplement of vitamins with inulin and acacia fibers in rats. Nutrients 12(8):2196

Meena NK, Choudhary K, Negi N, Meena VS, Gupta V (2021) Nutritional composition of stone fruits. In: Production technology of stone fruits. Springer, Singapore, pp 227–251

Meléndez-Martínez AJ, Mandić AI, Bantis F, Böhm V, Borge GIA, Brnčić M et al (2022) A comprehensive review on carotenoids in foods and feeds: status quo, applications, patents, and research needs. Crit Rev Food Sci Nutr 62(8):1999–2049

Mihaylova D, Popova A, Desseva I, Petkova N, Stoyanova M, Vrancheva R et al (2021) Comparative study of early-and mid-ripening peach (Prunus persica L.) varieties: biological activity, macro-, and micro-nutrient profile. Foods 10(1):164

Mitic S, Stojanovic B, Tosic S, Pavlovic A, Kostic D, Mitic M (2019) Comparative study on minerals in peel and pulp of peach (Prunus persica L.) fruit. Rev Chim Buchar Original Ed 7(6):2281–2285

Morelli MB, Gambardella J, Castellanos V, Trimarco V, Santulli G (2020) Vitamin C and cardiovascular disease: an update. Antioxidants 9(12):1227

Mostafa ES, Maher A, Mostafa DA, Gad SS, Nawwar MA, Swilam N (2021) A unique acylated flavonol glycoside from Prunus persica (L.) var. Florida prince: a new solid lipid nanoparticle cosmeceutical formulation for skincare. Antioxidants 10(3):436

Mousavi M, Zaiter A, Modarressi A, Baudelaire E, Dicko A (2019) The positive impact of a new parting process on antioxidant activity, malic acid and phenolic content of Prunus avium L., Prunus persica L. and Prunus domestica subsp. Insititia L. powders. Microchem J 149:103962

Nakagawa T, Allam AE, Ohnuki K, Shimizu K (2018) Biological activities of extracts from different parts of two cultivars of Prunus persica 'Akatsuki' and 'Fastigiata'. Nat Prod Commun 13(10):1934578X1801301015

Nowicka P, Wojdyło A, Tkacz K, Turkiewicz IP (2023) Quantitative and qualitative determination of carotenoids and polyphenolics compounds in selected cultivars of Prunus persica L. and their ability to in vitro inhibit lipoxygenase, cholinoesterase, α-amylase, α-glucosidase and pancreatic lipase. Food Chem X 17:100619

Olatunde A, Tijjani H, Ishola AA, Egbuna C, Hassan S, Akram M (2020) Carotenoids as functional bioactive compounds. In: Functional foods and nutraceuticals: bioactive components, formulations and innovations. Springer, Cham, pp 415–444

Oviya M, Geetha R, Ganapathy D (2019) Antimicrobial effects of Salvadora persica and Prunus persica on oral pathogens. Drug Invent Today 11(9), p2059

Petruccelli R, Bonetti A, Ciaccheri L, Ieri F, Ganino T, Faraloni C (2023) Evaluation of the fruit quality and phytochemical compounds in peach and nectarine cultivars. Plants 12(8):1618

Qi X, Tester RF (2020) Lactose, maltose, and sucrose in health and disease. Mol Nutr Food Res 64(8):1901082

Rahaman MM, Hossain R, Herrera-Bravo J, Islam MT, Atolani O, Adeyemi OS et al (2023) Natural antioxidants from some fruits, seeds, foods, natural products, and associated health benefits: an update. Food Sci Nutr 11(4):1657–1670

Reig G, Iglesias I, Gatius F, Alegre S (2013) Antioxidant capacity, quality, and anthocyanin and nutrient contents of several peach cultivars [Prunus persica (L.) Batsch] grown in Spain. J Agric Food Chem 61(26):6344–6357

Reig G, Cisneros-Zevallos L, Costa G, Crisosto CH (2023) Peach fruit quality: components, composition, and nutritional and health benefits. In: Peach. CABI GB, pp 226–260. https://doi.org/10.1079/9781789248456.011

Rey-Serra P, Mnejja M, Monfort A (2022) Inheritance of esters and other volatile compounds responsible for the fruity aroma in strawberry. Front Plant Sci 13:959155

Saidani F, Giménez R, Aubert C, Chalot G, Betrán JA, Gogorcena Y (2017) Phenolic, sugar and acid profiles and the antioxidant composition in the peel and pulp of peach fruits. J Food Compos Anal 62:126–133

Santin M, Lucini L, Castagna A, Rocchetti G, Hauser M-T, Ranieri A (2019) Comparative "phenol-omics" and gene expression analyses in peach (Prunus persica) skin in response to different postharvest UV-B treatments. Plant Physiol Biochem 135:511–519

Saraswathy G, Sathiya R, Anbu J, Maheswari E (2014) Antitussive medicinal herbs-an update review. Int J Pharm Sci Drug Res:12–19, Vol.6

Shahrajabian MH, Cheng Q, Sun W (2023) The most important medicinal herbs and plants in traditional Chinese and Iranian medicinal sciences with antioxidant activities. Lett Drug Des Discov 20(9):1171–1184

Sharma G, Kumar S, Sharma M, Upadhyay N, Ahmed Z, Mahindroo N (2018) Anti-diabetic, antioxidant and anti-adipogenic potential of quercetin rich ethyl acetate fraction of Prunus persica. Pharm J 10(3):463–469

Shin T-Y, Park S-B, Yoo J-S, Kim IK, Lee H-S, Kwon TK et al (2010) Anti-allergic inflammatory activity of the fruit of Prunus persica: role of calcium and NF-κB. Food Chem Toxicol 48(10):2797–2802

Song J, Kim Y-S, Kim L, Park HJ, Lee D, Kim H (2019) Anti-obesity effects of the flower of Prunus persica in high-fat diet-induced obese mice. Nutrients 11(9):2176

Suh S-J, Koo B-S, Jin U-H, Hwang M-J, Lee I-S, Kim C-H (2006) Pharmacological characterization of orally active cholinesterase inhibitory activity of Prunus persica L. Batsch in rats. J Mol Neurosci 29:101–107

Sun M, Zhao B, Cai Z, Yan J, Ma R, Yu M (2022) Amino acid profiles in peach (Prunus persica L.) fruit. Foods 11(12):1718

Tyśkiewicz K, Dębczak A, Gieysztor R, Szymczak T, Rój E (2018) Determination of fat-and water-soluble vitamins by supercritical fluid chromatography: a review. J Sep Sci 41(1):336–350

Wu H, Xu Y, Wang H, Miao Y, Li C, Zhao R et al (2022) Physicochemical characteristics, antioxidant activities, and aroma compound analysis of seven peach cultivars (Prunus persica L. Batsch) in Shihezi, Xinjiang. Foods 11(19):2944

Xu G, Li C, Qin S, Xiao W, Fu X, Chen X et al (2022) Changes in sucrose and sorbitol metabolism cause differences in the intrinsic quality of peach fruits cultivated in field and greenhouse environments. Agronomy 12(11):2877

Zhang Y, Liu W, Zhang B, Zhang Y, Cai Z, Song H et al (2022) Analysis of volatile compounds and their potential regulators in four high-quality peach (Prunus persica L.) cultivars with unique aromas. Lwt 160:113195

Zhang X, Su M, Zhou H, Leng F, Du J, Li X et al (2023) Effect of 1-methylcyclopropene on flat peach fruit quality based on electronic senses, LC-MS, and HS-SPME-GC-MS during shelf storage. Lwt 173:114388

Zheng B, Zhao L, Jiang X, Cherono S, Liu J, Ogutu C et al (2021) Assessment of organic acid accumulation and its related genes in peach. Food Chem 334:127567

# Medicinal and Nutritional Importance of *Catharanthus roseus* L. in Human Health

16

Priyanka Tolambiya, Bhagawatilal Jagetiya, and Sujata Mathur

**Abstract**

Medicinal plants are important sources for both newly developed and current medications. Most cultures in the globe, including Tibetan, Chinese, and Indian medicine, use plants directly as remedies. Researchers and medical scientists are working day and night to discover drugs for serious illnesses like diabetes, thyroid, cancer, and neurological disorders. Due to the unparalleled availability of chemical variety, natural products derived from medicinal plants, whether in the form of pure compounds or standardized extracts, provide countless prospects for novel therapeutic leads. Global interest in edible plants has increased due to the growing need for chemical diversity in screening programs and the search for therapeutic medicines from natural sources. Medicinal herbal and botanical preparations contain a variety of bioactive components. The World Health Organization (WHO) reports that 91 countries, including 12 megabiological countries, are home to almost 20,000 medicinal plants. The first steps in using the biologically active chemical found in plant resources are extraction, followed by pharmacological screening, bioactive compound isolation and characterization, toxicological assessment, and clinical assessment.

P. Tolambiya (✉)
Department of Botany, Government PG College, Kekri, Rajasthan, India

B. Jagetiya
Phytotechnology Research Laboratory, Department of Botany, M.L.V. Government College, Bhilwara, Rajasthan, India

S. Mathur
Department of Botany, B.B.D. Government PG College, Jaipur, Rajasthan, India

 
M. A. Ansari et al. (eds.), *Medicinal Plants and their Bioactive Compounds in Human Health: Volume 2*, https://doi.org/10.1007/978-981-95-3620-7_16

**Keywords**

Natural products · Bioactive components · Extraction · Isolation · Characterization · Pharmacological screening

**Learning Objectives**

- Extensive field survey and collection of germplasm
- Physico-chemical analysis of soil samples
- Quantitative estimation of various primary metabolites including cellular metabolites and enzymes
- Extraction and characterization of various secondary metabolites including alkaloids, flavonoids, and phytosterols
- Antimicrobial activity of isolated compound
- In silico study of novel compound.

## 16.1 Introduction

Medicinal plants, medicinal herbs, or simply herbs have been identified and used from prehistoric times. Plants make many chemical compounds for drug making, biological functions, including defense against insects, fungi, and herbivorous mammals. Medicinal plants are widely used to treat disease in nonindustrialized societies, not least because they are far economical than modern medicines. Due to side effect of chemical drug, nowadays medicinal plants have been used as medicine in three different ways. First, their natural chemical constituents can be used directly as extracted forms. Second, they could be employed as agents in a drug combination. The third is that the organic compounds in plants can be shaped into synthetic medications. Alternative methods of managing and treating diabetes have emerged as a result of treatment failure and adverse effects linked to oral hypoglycemic medications. Diabetes is a metabolic disorder characterized by hyperglycemia due to lack of insulin; defective insulin secretion and receptors are insensitive to insulin action. Body fails to utilize the ingested glucose properly; so, excess of sugar is found in blood and urine that leads to serious long-term complications including cardiovascular disease, kidney failure, stroke, and damage to the eyes. Hence, an antidiabetic compound with antioxidant properties would be more beneficial. *Catharanthus roseus* L. is now used as an alternative therapy for diabetes. It is a member of Apocynaceae family. The leaf juice or water decoction of *Catharanthus roseus* L. *var.* alba is used as a folk medicine for the treatment of diabetes all over the world specially in Malaysia, Philippines, West indies, Nigeria, and India. The plant contains more than 400 alkaloids including 130 monoterpenoids indole alkaloids (Heijden et al. 2004; Rasool et al. 2011). Total 65 alkaloids are only found in leaf. The plant has been called a miracle in the prevention of childhood leukemia and cancer treatment. Majorly important protocol used for almost all types of cancer, made up from combination of vinblastine and vincristine (*C. roseus*), carboplatin, epoxide, and methotrexate toxic medicine. Vinblastine and vincristine are the dimers, formed by the coupling of monoindole alkaloids such as catharanthine and vindoline (Meenakshi et al. 2013) mainly present in aerial part of plants used to treat tumor and cancer. Vincristine and vinblastine require both aerial and root parts of a plant to be synthesized (Luca and Laflamme 2001). The aqueous extract of the leaf and stem of *Catharanthus roseus* contained 2,3-butanediol, quinic acids, vindoline, chlorogenic acid, vindoline, secologanin, and quercetin (Jamal and Ahmad 2024). Alkaloid vincristine, vinblastine, vinflunine, vinorelbine, and vindesine are clinically approved for treatment regimen. Millions of tons of Oncovin and Velban, anticancer drugs are sold worldwide (Chaturvedi et al. 2022). These drugs worked as miracle for treatment of deadly disease. To protect Madagascar periwinkle is to protect the future of our children. It has been suggested that, instead of using the side effects causing chemical drugs, people should use *Catharanthus roseus* extraction. The plant is also known for its antihypertensive and antispasmodic properties due to the presence of alkaloids like ajmalicine (vasodilating), serpentine (hypertensive), and reserpine (Mishra et al. 2001). The root bark contains the alkaloids alstonine which has been used for its calming effect and its ability to reduce blood pressure. Vincamine has blood thinning and memory-enhancing properties and is effective in the treatment of vascular dementia. *C. roseus* also shows the presence of Yohimbine along with another flavonoid hirsutidin (Piovan and Fillipini 2007; Millon et al. 2002). Also *Catharanthus roseus* has antiulcer, antidiarrheal, anthelminthic, antiviral, wound-healing, antimalarial, antiallergic, antisterility, and biofungicide property and it bio-accumulates heavy metal like Cd and Cr; so, it is used in phytoremediation. Extracts from these leaves contain 1α, 25-dihydroxy vitamin D3, which inhibits pro-inflammatory signaling (Borah et al. 2024), improves glucose homeostasis in muscles, and contains vitamin C (ascorbic acid), hydroxyproline content which has been linked to improved wound healing and disease resistance, respectively, so that the body's cells, skin, and eyesight can grow normally (Bisla et al. 2013). Other secondary metabolites from *C. roseus* have also been identified, such as steroid (castasterone, brassinolides), flavonoids, anthocyanins, and monoterpenoid glucosides: loganin, secologanin, sweroside, deoxyloganin, and dehydrologanin (Mohammad et al. 2021). Vindolinine and chlorogenic acid demonstrated a strong interaction with the bacterial penicillin-binding protein through molecular docking, suggesting that they could be employed as food preservatives (Jamal and Ahmad 2024). This study has introduced biochemical properties of *Catharanthus roseus* L. by discovering the responsible compound for diabetes through in silico approach along with its antimicrobial, antioxidant properties (Tolambiya and Mathur 2016a, b, 2017) which have potential inhibition capability against various micro-organism and critical diseases. Primary and secondary metabolites along with enzymes are also present in plants which make it as an important medicinal herbs category; further, it is available easily around the world and suitable for almost all weather conditions.

## 16.2 Plant Description and Geographical Distribution

Apocynaceae family member *Catharanthus roseus* L., popularly called the Madagascar periwinkle, is well-known in Madagascar. *Vinca rosea*, *Lochnera rosea*, *Ammocallis rosea*, *Cape periwinkle*, rose periwinkle, rosy periwinkle, and "old-maid" are some additional English names that are infrequently used. Linnaeus published the original name of this species, *Vinca rosea* (1759). Eight species are recognized. Seven of the eight are highly recognized in Madagascar; they are *C. coriaceus*, *C. lanceus*, *C. longifolius*, *C. ovalis*, *C. roseus*, *C. scitulus*, *C. trichophyllus*, and *C. pusillus; the eighth species is indigenous to Sri Lanka and India* (Brun et al. 2001; Almagro et al. 2015). Global adaptation is exhibited by *C. roseus* plants (Pandey et al. 2006; Rahmatzadeh and Kazemitabar 2013). The Greek word for "pure flower" is *Catharanthus*, and the word "roseus" means "red, rose, rosy." It thrives in sun, rain, or seaside conditions, on any kind of soil, and it frequently grows wild. It is also referred to as "Sadabahar," which translates to "always in bloom." Worship is just one of its many uses. These are little shrubs that are perennial herbs with leaves arranged in an opposite or nearly opposite way. Usually, flowers grow alone in the leaf axils. Each has a corolla with five lobes and a tubular throat, and a calyx with five long, narrow lobes. It can reach a height of 20–80 cm and bears pink, purple, or white flowers in bloom (Swanberg and Dai 2008). More than 100 cultivars of *C. roseus* are identified (Ku et al. 2013).

## 16.3 Application of *Catharanthus roseus* L.

| | |
|---|---|
| Drug making | Insecticides |
| Bioremediation | Flavour in food |
| Preservatives | Cosmetic |
| Texturizing agents | Agriculture |
| Forensic | Plastic industry |
| Ornamental | Biogas production |
| Oil and gas industry | Plant growth hormone |

## 16.4 Biochemical Profile of *Catharanthus roseus* L.

| Plant parts | Alkaloids/compound | Activity/application | References |
|---|---|---|---|
| Leaf | 1. Catharanthine<br>2. Vindoline<br>3. Vindolidine<br>4. Vindolicine<br>5. Vindolinine<br>6. Ibogaine<br>7. Yohimbine<br>8. Raubasine<br>9. Vinblastine<br>10. Vincristine<br>11. Leurosine<br>12. Lochnerine<br>13. Vinorelbine<br>14. Vinflunine<br>15. Vindesine<br>16. Vinpocetine<br>17. Citronellol<br>18. Geraniol<br>(Z, E)—pentadecanal<br>19. (E,E)-2,4-Hexagonol<br>20. Phytol<br>21. Palmitic acid<br>22. Methyl myristic acid | 1. Anticancerous<br>2. Antineoplastic alkaloids<br>3. Antidiabetic<br>4. Antitumor<br>5. Menorrhagia<br>6. Rheumatism<br>7. Treat wasp sting<br>8. Stomach cramps<br>9. Vomitive<br>10. Indigestion and dyspepsia<br>11. Gripping pain<br>12. Alzheimer's disease | Luca and Laflamme (2001), Waltz (2004), Muralidharan (2014), Almagro et al. (2015), Aftab et al. (2023) |
| Stem | 1. Leurosine<br>2. Lochnerine<br>3. Catharanthine<br>4. Vindoline<br>5. Isobutyl-3-methoxypyrazine and 2-Phenylethanol<br>6. 1-Phenylethanol volatile oils | 1. Diabetes<br>2. Hypertension<br>3. Insomnia and cancer<br>4. Soar throats<br>5. Chest ailments and laryngitis<br>6. Arrest bleeding | Muralidharan (2014), Aftab et al. (2023) |
| Root | 1. Ajmalicine<br>2. Serpentine<br>3. Catharanthine<br>4. Vindoline<br>5. Leurosine<br>6. Lochnerine<br>7. Reserpine<br>8. Alstonine<br>9. Tabersonine<br>10. Horhammericine<br>11. Lochnericine, Echitovenine | 1. Antihypertensive alkaloids<br>2. Intestinal parasitism<br>3. Vermifungal<br>4. Depurative 5. Hemostatic<br>6. Toothache remedies<br>7. dysentery<br>8. Fever<br>9. Wound healing | Shanks et al. (1998), Muralidharan (2014) |

| Plant parts | Alkaloids/compound | Activity/application | References |
|---|---|---|---|
| Flower | 1. Catharanthine<br>2. Vindoline<br>3. Leurosine<br>4. Lochnerine<br>5. Tricin (flavones)<br>6. Tetracosane<br>7. Heneicoane and tricosane | 1. Eye wash in infants<br>2. Asthma<br>3. Tuberculosis<br>4. Flatulence | Luca and Laflamme (2001), Muralidharan (2014), Aftab et al. (2023) |
| Seed | Vingramine, Methylvingramine | Nasopharynx carcinoma | Jossang et al. (1998) |

### 16.4.1 Anticancer Activity

Ethanolic, methanolic, and aqueous extracts of *Catharanthus roseus* L. leaves, stem, roots, and aerial part were tested on MCF (breast cancer cell lines), HCT116 (colorectal carcinoma cell), Jurkat, 9KB, and P38 cell lines in doses of 50 and 100 mg/kg to demonstrate this activity (Mukhapadhyay and Cordell 1981; Ahmad et al. 2010; Siddiqui et al. 2010; Widowati et al. 2013; Ruskin and Aruna 2014; Harshini et al. 2020).

### 16.4.2 Antidiabetic Activity

500 mg/kg dose of Dichloromethane: methanol extract (1:1) of leaves and twigs of *Catharanthus roseus* L. were tested on streptozotocin (STZ)-induced diabetic rat model to demonstrate this activity (Singh et al. 2001; Bisla et al. 2013).

### 16.4.3 Antioxidant Activity

β-TC6 cells were exposed to 12.5 and 25.0 μg/mL doses of methanolic extract, dichloromethane extract (DE), and ethanolic extracts of *Catharanthus roseus* L. roots (Rahmatzadeh and Kazemitabar 2013).

### 16.4.4 Hypolipidemic Activity

To demonstrate this activity, aqueous extract of *Catharanthus roseus* L. leaves was tested on guinea pigs, alloxan-induced diabetic rats (Jayanthi et al. 2010; Muralidharan 2014).

### 16.4.5 Wound Healing Activity

On streptozotocin-induced mice, doses of 200 and 400 mg/kg of *Catharanthus roseus* L. methanolic extract were tested (Singh et al. 2014; Favali et al. 2014).

### 16.4.6 Antifungal Activity and Antibacterial Activity

Different leaf extracts (acetone, methanol, ethanol) *Catharanthus roseus* L. were tested on *Aspergillus niger*, *Candida albicans*, *Penicillium chrysogenum*, *Staphylococcus aureus*, *Escherichia coli*, *Klebsiella pneumonia*, *Klebsiella oxytoca*, and *Pseudomonas aeruginosa* (Hassan et al. 2011; Khalil 2012; Balaabirami and Patharajan 2012).

### 16.4.7 Antidiarrheal Activity

200 and 500 mg/kg ethanolic extracts of *Catharanthus roseus* L. were tested on Wistar rat (Hassan et al. 2011; Gajalakshmi et al. 2013).

### 16.4.8 Antihelminthic Activity

Dose of 250 mg/mL ethanol extract (250 mg/mL) of *Catharanthus roseus* L. (whole plant) was tested on *Pheretima posthuma* (Kabesh et al. 2015; Agarwal et al. 2011).

### 16.4.9 Antiviral Activity

Selected Catharanthus alkaloids were tested on Vaccinia and polio type III viruses (Farnsworth et al. 1968).

### 16.4.10 Antiparasitic Activity

Due to its distinct peroxide bridge, *C. roseus* has been found to be a possible inhibitor of the malaria-causing protozoan parasite falcipain-2. In a similar vein, antiparasitic effects against the *Trypanosoma cruzi* that causes trypanosomiasis in humans were demonstrated by vinblastine and vincristine, which disrupted the parasite's mitosis and altered its morphology (Grellier et al. 1999). Ag nanoparticles made with stearic acid from *Catharanthus roseus* leaf extract have larvicidal effects on *Aedes aegypti* (yellow fever mosquito) and *Culex quinquefasciatus* (house mosquito) and are effective against *Earias vittella* (Pavunraj et al. 2017).

## 16.5 Materials and Methods

### 16.5.1 Study Area

This study was done in University of Rajasthan Campus, Jaipur, Rajasthan, India. Plant samples were collected to initiate the various experiment of metabolite extraction.

### 16.5.2 Sample Collection and Analysis

In the shade, plant materials were left to dry at room temperature (28 °C). For the further experiments in this study, the leaves, stems, roots, and flowers of *Catharanthus roseus* L. were ground into a fine powder. We quantitatively estimated a number of primary metabolites, including cellular metabolites, and enzymes, including proteins, lipids, carbohydrates, peroxidase, polyphenol oxidase, and indole acetic acid (IAA) oxidase (Tolambiya et al. 2024). Additionally, secondary metabolites were extracted and described, including flavonoids, phytosterols, alkaloids, total phenolic content, and their antioxidant activity (Tolambiya et al. 2023a, b). Organic solvents with different dipole moments and polarities were used to extract the bioactive compound. The main metabolites, proteins, and carbohydrates were measured using the ultra-violet/visible 5300 spectrophotometer (UV-VIS). The 0.1 g dried sample of each plant part was homogenized in pestle and mortar with 20 mL of 80% ethanol separately and left overnight. After centrifuging each sample for 15 min at 1200 rpm, the supernatants were separated and concentrated using Loomis and Shull's (1937) method on a water bath. The phenol sulfuric acid method of Dubois et al. (1951) was used to estimate the amount of carbohydrates in 1 mL of aliquot of each sample. For the standard sugar (glucose) stock solution, a standard regression curve was created. Distilled water was used to create a stock solution of glucose (100 μg/mL). From this solution, 0.1–0.8 was pipetted out into different test tubes, and the volume was increased to 1 mL using distilled water. These tubes were kept on ice. One milliliter of 5% phenol was added in each tube and shaken gently. Five milliliter of conc. sulfuric acid added was rapidly poured. The characteristics yellow orange color developed. Using a ultraviolet/visible spectrophotometer, the optical density was measured at 490 nm after being set for 100% transmission against a blank (80% ethanol). Using Beer's Lambert Law, a standard regression curve was calculated between the known glucose concentration and the corresponding optical density. Quantitative estimation of carbohydrates was reported for maximum in flower (second highest in leaf). The human body uses carbohydrates for energy, helps control blood sugar and insulin metabolism, participates in the metabolism of cholesterol and triglycerides, and promotes fermentation (Holesh et al. 2023). Proteins are of great nutritional value and are directly involved in the chemical processes essential for life. The test samples (50 mg each) were homogenized separately for 30 min in 10 mL of cold 10% trichloroacetic acid (TCA) and then stored for 24 h at 4.0 °C. Separate centrifugations of these mixtures were performed, and the supernatants were disposed of. Every residue was once more suspended in 10 mL of 5% TCA and heated for 30 min at 800° on a water bath; the residue was dissolved in 10 mL of 1 N NaOH and allowed to sit at room temperature. Each of the above samples (1 mL) was taken and the total protein content was estimated using the spectrophotometer through Lowry method (1951). A regression curve was prepared. A stock solution of BSA was prepared in 1 N NaOH (1 $mgL^{-1}$). Eight concentrations (ranging from 0.1 to 0.8 $mgL^{-1}$) were separately measured in test tube and volume of each sample was made to 1 mL by adding distilled water. Five milliliter of a freshly made alkaline solution was added to each. The solution was made by combining 50 mL of 2% Na2CO3 in 0.1 N NaOH with 1 mL of 0.5% CuSO4, 5% H2O in 1% sodium potassium tartrate. Each sample had 0.5 mL of Folin-Ciocalteu reagent added quickly and mixed right away. The optical density of each sample was then measured using a spectrophotometer at 750 nm 30 min later against a blank solution of 10% TCA. To calculate the regression curve, three replicates of each optical density were obtained, and the average value was plotted against their respective concentrations. It was estimated that maximum amount of protein found in leaf. Quantitative estimation of primary metabolite like lipid in all plant sample extract through Jayaram's method (1958). After drying and powdering the test sample, 100 mg was macerated with 10 mL of distilled water and put into a conical flask with 20 mL of methanol and chloroform (2/1: v/v; Jayaram (1958). The colored aqueous layer of methanol, which contained all the water soluble substances and thick interface layer, was discarded in each test sample. The lower layer of chloroform, which contained all the lipids, was carefully collected in the pre-weighed glass vials. The chloroform layers dried in vacuo and weighed. It was estimated that utmost amount of lipids found in leaf part. The food and cosmetics industries, as well as nanotechnology, use lipids (Mashaghi et al. 2013). Peroxidase, polyphenol oxidase, and indole acetic acid oxidase (IAA) are the three types of enzymes that were also measured in *Catharanthus roseus* L. plant parts extract. Laboratory results show that peroxidase inhibits the free radicals and control the reactive oxygen species. Chance and Maehly (1955) method was considered for the oxidation of pyrogallol to purpurogallin in the presence of peroxidase catalyst at 420 nm and 20 °C. Shinshi and Noguchi (1975) method was considered for the oxidation of ortho-diphenolic compounds to their respective quinines in the presence of polyphenol oxidase catalyst at 495 nm. Tannic acid was also used in solution. Enzyme uses the molecular oxygen to oxidize the compound. Saenger (1984) method was used to estimate indole acetic acid (IAA) oxidase activity. Salkowski's reagent was used. IAA participates in biochemical reaction for plant growth and differentiation. The tryptophan is a precursor of IAA. Oxidation of IAA at 530 nm to measure optical density to get the highest antioxidant *Catharanthus roseus* extract. The highest antioxidant activity values were reported for peroxidase and polyphenol oxidase enzymes in leaves and indole acetic acid (IAA) in stem samples. Usually the IAA concentration is found in shoot tip, so result is in favor and responsible for growth and differentiation. The higher oxidase activity inhibits the activity of free radical ions. The free radical ion is present in the body and associated with disease. All traditional medicine and another medicine worked through cell signaling pathways and inhibit the activity of free radical ions. The secondary metabolites flavonoids, phytosterols, and alkaloids extraction and their characterization research work had been conducted on *Catharanthus roseus* L. plant parts by various methods. Flavonoids are a group of polyphenolic compound significantly support in health benefits through cell signaling pathways and antioxidant effects. *Catharanthus roseus* plant samples put into 80% methanol with water bath and concentrated in vacuo with sequential extraction of petroleum ether (Fraction I), diethyl ether (Fraction II) and ethyl acetate (Fraction III) separately. Among all Fraction Fr I had been discarded, while Fr II considered as free flavonoids and Fr III transformed to bound flavonoids by refluxing with 7% sulfuric acid and concentrated in vacuo. To isolate the types of flavonoids in plant sample extracts, thin-layer chromatography is used post adding up the small volume of ethanol in Fr II and Fr III and Silica gel G laminated on thin glass plate, activated on 100 °C for 30 min. Each of the extraction was co-chromatographed with authentic flavonoids as a marker (quercetin) in the presence of solvent mixture (benzene:acetic acid:water: 125:72:3). The developed colored spots had been noted and RF value (retardation factor in chromatography) was calculated by distance traveled by a given component was divided by the distance travelled by the solvent front for a given system at a known temperature, this gives a characteristic and to identify components. The highest total flavonoids had been observed and recorded in leaves and flowers plat samples and separation was done by TLC. Rutin (0.24), Quercetin (0.78), Luteolin (0.56), and Kaempferol (0.86) varieties of flavonoids were found. Leaves and flowers parts of *Catharanthus roseus* have all types of flavonoids. Further infrared spectroscopy, mass, one-dimensional nuclear magnetic resonance spectroscopy (1D NMR), and gas chromatography mass spectroscopy (GCMS) are carried out for the identification and determination of the structure of flavonoids (Rani and Kapoor 2019). People use quercetin as a medicine. It is used for treating conditions of the heart and blood vessels including "hardening of the arteries" (atherosclerosis), high cholesterol, heart disease, and circulation problems. It is also used for diabetes, cataracts, hay fever, peptic ulcer, schizophrenia, inflammation, asthma, gout, viral infections, chronic fatigue syndrome (CFS), preventing cancer, and for treating chronic infections of the prostate. The Kaempferol flavonoids are slightly soluble in water and highly soluble in hot ethanol, ethers, and DMSO (dimethyl sulfoxide). It acts as an antioxidant by reducing oxidative stress. Many studies suggest that consuming kaempferol may reduce the risk of various cancers. A correlation was found between increased level of kaempferol in the diet and a reduced relative risk of type 2 diabetes. Luteolin is common flavonoid that exists in many types of plants including fruits, vegetables, and medicinal herbs. Plants rich in luteolin have been used in Chinese traditional medicine for treating various diseases such as hypertension, inflammatory disorders, and cancer. Rutin is a bioflavonoid; it has powerful antioxidant properties. It also helps our body produce collagen and use vitamin C. It can help lower density lipoprotein (LDL) cholesterol. People with diabetes who had a history of hypertension were given 500 mg of rutin once a day. Phytosterols are plant sterols belonging to triterpene family structurally analogous to cholesterol that act in the intestine to lower cholesterol absorption and play a significant role for human life. Equal amount of sample was put in incubator separately with 30% HCL and obtained liquid discarded post sample filtration. Distilled water is mixed for neutralization and pH 6 taken. The extraction of phytosterols were obtained by incubating in Benzene sovent. Phytosterols isolated through TLC, same as flavonoids and phytosterols RF value were calculated (hexane: acetone: 120 mL: 30 mL). The developed plates were air dried and visualized under UV. Each of the extract was co-chromatographed with authentic phytosterol (Stigmasterols) as a marker. Stigmasterols and β-sitosterols types of phytosterols were isolated through TLC and identification of structure is carried out by infrared spectroscopy, mass, 1D NMR, and GCMS. Therefore, leaf and flower both are having a good amount of phytosterols and can be used for human health. β-Sitosterols, a main dietary phytosterol found in plants, may have the potential for prevention and therapy for human cancer. It is used to make medicine. Research has indicated that stigmasterol may be useful in the prevention of certain cancers, including ovarian, prostate, breast, and colon cancers. Studies have also indicated that a diet high in phytosterols may inhibit the absorption of cholesterol and lower serum cholesterol levels by competing for intestinal absorption. The alkaloids are a large family of nitrogen containing secondary metabolites and alkaline in nature. Equal quantity (5 g of dried powdered) of sample was taken in Erlenmeyer flasks containing 250 mL distilled water and 0.1 mL sulfuric acid mixed and macerated for 3–4 h. Magnesium oxide (approximately 12.5 g) was added in mixture and boiled for 20 min gently. Alcohol (10 mL) was added to remove the mucus and filtered with Whatman filter paper and reconstituted with distilled water after evaporated to dryness in vacuo. Agar well diffusion method was used for characterization and GCMS was done for chemical compound identification. Highest alkaloids, secondary metabolites were found in leaf(19%) and flower(26%) and its characterization were done by antimicrobial activity through agar well diffusion method. Highest Inhibition zone in stem 14 mm had been found. It can be concluded that stem parts having highest inhibition activity may inhibit the free radical ions in the body. After stem alkaloids content retrieved by GCMS, these compounds are required for finding out the actual accountable component in *Catharanthus roseus* through in silico approach for cost-effective drug discovery. Phenolics are chemical compounds that have an aromatic hydrocarbon group directly bonded to an OH. These are widely found in plants and are especially prevalent at higher concentrations as soluble or cell wall-bound compounds as a result of interactions between the plant and its surroundings. Total phenolic content of the extracts was determined by the Folin-Ciocalteu method, considering tannic acid as a stock solution. The antioxidant characteristics of *C. roseus* plant samples verified by DPPH (2,2-diphenyl-1-picrylhydrazyl) radical scavenging method. Further verification had been done by the antimicrobial attributes in vivo grown of *Catharanthus roseus* L., in the presence of polar and nonpolar solvents against four different bacteria and fungi. Methanol is used for arranging *Catharanthus roseus* extract solution due to its polarity and low boiling point property, easily evaporated and it dissolves a larger portion of polar compounds than other polar solvent. Same experiment is done considering nonpolar solvent chloroform ($CHCl_3$) and observed root sample having higher amount of extract post weight measurement. These extracts need to be passing through DMSO for antimicrobial activity due to its important characteristics that it dissolves both the polar and nonpolar compounds and do not participate in antimicrobial activity. The inhibition capability were examined of extract against bacteria and fungi's various sample. *Catharanthus roseus* has successfully inhibited the growth of microorganisms. Agar well diffusion methods (nutrient agar medium for bacteria and potato dextrose agar medium for fungi) were being used for the measurement of inhibition zone. Sterilized distilled water and other solvents used in preparation of extracts were used as negative control. Streptomycin was used as positive control to determine the efficacy of extracts against selected bacteria and Ketoconazole was used as positive control to determine the efficacy of extracts against selected fungi. The experiment is shown that the root extract under methanol (polar) and leaf extract under chloroform (nonpolar) solvent have highest inhibition capability against *Escherchia coli* and *Staphylococcus aureus* bacteria. Also *Penicillium funiculosum* fungi are highly inhibited in the presence of polar and nonpolar solvent by stem extract. No leaf extract inhibits any of the fungi. The plant extract can be used in skin problems, dermatitis, eczema, and acne and autoimmune disease such as HIV/AIDS. A large number of in vivo studies have been conducted on animals to test the claimed activity have demonstrated the hypoglycemic property of many plants. But in silico method also can be used to demonstrate the anti-diabetic property of *Catharanthus roseus* for avoiding post-clinical experiment time and cost. Target identification and molecular docking against inhibitors were done for this purpose. Alkaloids stem extract is used for GCMS. The endoplasmic reticulum contains the nontransmembrane enzyme PTP1B (protein-tyrosine phosphates 1B). Its importance comes from the fact that it negatively regulates both leptin and insulin signaling. It was utilized as a receptor for docking and a crucial target for an anti-diabetic proposed drug. More recently, it has been directly connected to breast cancer and discovered to play a role in the development of tumors. Therefore, PTP1B plays a role in the suppression of leptin and insulin signaling (Villamar-Cruz et al. 2021). It is also thought to be a potential target for therapy. *Catharanthus roseus* contains alkaloids, flavonoids, and phenols, which are natural PTP1B inhibitors. Version 3.5 of the Accelrys Discovery Studio Visualizer was used to view the Auto dock tools' output. Lipinski's rule is applied to every docked structure. Compounds that satisfy the following requirements are more likely to be successful drug candidates and have good absorption and penetration in biological systems, according to the Lipinski "*Rule of Five*":

1. Five or fewer hydrogen-bond donors
2. Ten or fewer hydrogen-bond acceptors
3. Molecular weight less than or equal to 500
4. Calculated log P less than or equal to 5

Drug development is facilitated by Lipinski's rules and molecular docking studies, which eliminate the need for costly post-clinical testing. By using scoring functions to predict the strongest binders, docking enables one to virtually screen the number of compounds. It determines how well the ligand and receptor fit and dock with one another. Among ligands 3-Nitrophthalic Acid compound may be an anti-diabetic agent as it satisfying most of the conditions, follow the all criteria of Lipinski's rules, it is small, hydrophilic compound, having good binding affinity with receptor enzyme protein and having least energy among all others and availability of nitrogen in ring for antihyperglycemic activity and availability of oxygen element for binding to the WPD loop of PTP1B. 3-Nitrophthalic Acid is a precursor of 3-amino benzoic acid which is similar to 2-oxalylaminobenzoic acids, general classes of inhibitors of PTP1B hence the above studies supports the identification of least energy molecule compound which inhibited the PTP1B and it could improve the insulin action.

## 16.6 Conclusion

According to the study mentioned above, *C. roseus* has the potential to be a valuable phytopharmaceutical. Dried plant leaves extract of flavonoids and phytosterols is dissolved in 10 mg/mL ethanol and take 3–3.5 mL sample for GCMS analysis. According to GCMS analysis of plant sample, we have found some novel compound, namely Hentriacontane, 1-Heptacosanol, ethyl 9 hexadecenoate, 2,6 octadien, trimethylsilyl ester, ethyl oleate, ethyl nonadecanonoate, 9,10 anthraquinone having anti-inflammatory, antimicrobial, and antidiarrheal property which are useful in drug making for various disease. Another avenue for the medicinal application of *C. roseus* is made possible by the aqueous extracts of its leaves and stems, which inhibit acetyl cholinesterase (Pereira et al. 2009). However, more investigation is needed to determine what exactly causes this, so that its primary use can be in the treatment of various diseases.

## 16.7 Future Scope

Using data mining and statistical techniques, more ground-breaking research on the statistical relationships between a molecular structure and a particular chemical or biological property can be started for drug discovery. All of these elements are therefore very beneficial for classifying and identifying new compounds as well as predicting their toxicity in the human body. These studies have no negative environmental effects and cut down on the time and expenses associated with the drug discovery process. The primary objective of this study is to investigate the biochemical characterization of *Catharanthus roseus* L. as a medicinal herb in order to open up new avenues for investigation into serious illnesses. *Catharanthus roseus* L. is the subject of several ongoing studies. Many drugs have been developed to treat them effectively and humans are taking advantage of this, but more research is needed to determine which compounds are actually causing the cure in a short amount of time. The scope of the aforementioned research project may be expanded in order to identify specific active ingredients for the treatment of serious illness. For effective toxicity management, the effects of cancer, diabetes, neurological disorders, and deadly microorganism contamination could be lessened. Additionally, the identification of antidiabetic compounds from the remaining plant parts (i.e., leaves and flowers) can be accomplished by using *Catharanthus roseus's* extracted flavonoids and phytosterols in in silico and additional laboratory testing for drug discovery. In conclusion, *Catharanthus roseus* L. has the steadfast determination to combat a variety of diseases and contribute significantly to the field of medicine.

## 16.8 Summary

Medicinal plants are widely used to treat disease in nonindustrialized societies, not least because they are far economical than modern medicines. Due to side effect of chemical drug, nowadays medicinal plants have been used as medicine in different ways. The plant contains more than 400 alkaloids including 130 monoterpenoids indole alkaloids. *Catharanthus roseus* has anticancerous, antidiabetic, antiulcer, antidiarrheal, anthelmintic, antiviral, wound-healing, antimalarial, antiallergic, antisterility, and biofungicide and bio-accumulation properties. It may be noteworthy that the water deficit areas may be well used for the cultivation of medicinal plants like *Catharanthus roseus* (Tolambiya and Jagetiya 2025) so it is also known as "Sadabahar" and the economically important alkaloid production can be enhanced in such plants. This plant also has a defense mechanism against oxidative stress. Phenolics improve the quality and nutritional value of food by preventing lipids from degrading oxidatively. Further described the plant sample of *Catharanthus roseus*, i.e., leaves, stems, roots and flowers extraction and quantitative estimation of Primary, Secondary metabolites and Enzymes in detailed. Organic solvents with different polarity and dipole moment were used to extract the bioactive compound. Primary metabolites carbohydrates and proteins were measured through UV-VIS 5300 spectrophotometer. The secondary metabolites flavonoids, phytosterols, and alkaloids extraction and their characterization research work had been conducted on *Catharanthus roseus* L. Further verification was done by the antimicrobial attributes invivo grown of Catharanthus roseus L., in the presence of polar and nonpolar. *Catharanthus roseus* is successfully inhibited the growth of microorganisms. All traditional medicine and another medicine worked through cell signaling pathways and inhibited the activity of free radical ions. To demonstrate the antidiabetic property of *Catharanthus roseus*, in silico method can be used for avoiding post-clinical experiment, time, and cost. The molecular docking studies and Lipinski's rules facilitate drug development avoiding expensive post-clinical experiments. Small, hydrophilic compounds with low energy, good binding affinities to receptor enzyme proteins, and these characteristics are all very useful for classifying and identifying new compounds as well as predicting their toxicity. In summary, *Catharanthus roseus* L. has the unwavering will to fight a wide range of illnesses and play a revolutionary role in the field of medicine.

**Acknowledgements** The authors are thankful to Rajasthan Agriculture Research Institute, Durgapura, Jaipur, University of Rajasthan, Jaipur, BBD Govt PG College, Chimanpura, Jaipur, MLV Govt PG College, Bhilwara for providing sophisticated lab facility and relevant literature to conduct the experiment.

**Conflict of Interest** There is no conflict of interest disclosed by the writer.

## References

Aftab A, Akram M, Laila U, Zainab R, Iftikhar M, Abdelhak M, Sołowski G, Ozdemir FA, Alinia-Ahandani E, Sfera A (2023) Phytochemistry and phytochemical potential of *Catharanthus roseus*: a narrative review. Int J Med Sci Clin Invent 10:6670–6676

Agarwal S, Jacob S, Chettri N, Bisoyi S, Tazeen A, Vedamurthy AB, Krishna V, Hoskeri HJ (2011) Evaluation of in-vitro anthelmintic activity of *Catharanthus roseus* extract. Int J Pharm Sci Drug Res 3:211–213

Ahmad NH, Rahim RA, Mat I (2010) *Catharanthus roseus* aqueous extract is cytotoxic to Jurkat leukemic T-cells but induces the proliferation of normal peripheral blood mononuclear cells. Trop Life Sci Res 21:101–113

Almagro L, Fernandez-Perez F, Pedreno MA (2015) Indole alkaloids from *Catharanthus roseus*: bioproductionio and their effect on human health. Molecules 20:2973–3000

Balaabirami S, Patharajan S (2012) In vitro antimicrobial and antifungal activity of *Catharanthus roseus* leaves extract against important pathogenic organisms. Int J Pharm Pharm Sci 4:487–490

Bisla G, Choudhary S, Singh E, Chaudhary V (2013) Hyperglycemia and hyperlipidemia mitigating impact of *Catharanthus roseus* (Sadabahar) leaves aqueous extract on type2 diabetes mellitus subjects. Asian J Plant Sci Res 3:170–174

Borah A, Singh A, Saha S (2024) 1α, 25-dihydroxy vitamin D3 containing leaf aqueous extract of *Catharanthus roseus* (L.) G. Don inhibits pro-inflammatory signaling in adipocytes-macrophage co-culture and improves glucose homeostasis in adipocytes and muscle cells. S Afr J Bot 167:519–526

Brun G, Bessière JM, Dijoux-Franca MG, David AM, Mariotte B (2001) Volatile components of Catharanthus roseus (L.) G. Don (Apocynaceae). Flavour Fragr J 16:116–119

Chance B, Maehly AC (1955) Assay of catalase and peroxidases. Methods Enzymol 2:764–775

Chaturvedi V, Saloni Goyal S, Mukim M, Meghani M, Patwekar F, Patwekar M, Khan SK, Sharma GN (2022) A comprehensive review on *Catharanthus roseus* L. (G.) Don: clinical pharmacology, ethnopharmacology and phytochemistry. J Pharmacol Res Dev 94:17–36

Dubois M, Gilles K, Hamilton JK, Rebers PA, Smith F (1951) A colorimetric method for the determination of sugar. Nature 168:167

Farnsworth NR, Svoboda GH, Blomster RN (1968) Antiviral activity of selected *Catharanthus* alkaloids. J Pharmacol Sci 57:2174–2175

Favali M, Muestti R, Benvenuti S, Bianchi A, Pressacco L (2014) *Catharanthus roseus* L. plants and explants infected with phytoplasmas: alkaloid production and structural observations. Protoplasma 223:45–51

Gajalakshmi S, Vijayalakshmi S, Devi RV (2013) Pharmacological activities of *Catharanthus roseus*: a perspective review. Int J Pharm Sci 4:431–439

Grellier P, Sinou V, Garrea de Loubresse N, Bylen E, Boulard Y, Schrevel J (1999) Selective and reversible effects of vinca alkaloids on *Trypanosoma cruzi* epimastigote forms blockage of cytokinesis without inhibition of the organelle duplication. Cell Motil Cytoskeleton 42:36–47

Harshini M, Sheeba L, Selvanayaki M (2020) Anticancer activity of *Catharanthus roseus* and *Murraya koenigii*. J Crit Rev 7:2394–5125
Hassan K, Brenda A, Patrick V, Patrick O (2011) In vivo antidiarrheal activity of the ethanolic leaf extract of *Catharanthus roseus Linn.* (Apocyanaceae) in Wistar rats. Afr J Pharm Pharmacol 5:1797–1800
Heijden RVD, Jacobs DI, Snoeijer W, Hallard D, Verpoorte R (2004) The Catharanthus alkaloids: pharmacognosy and biotechnology. Curr Med Chem 11:607–628
Holesh JE, Aslam S, Martin A (2023) Physiology, carbohydrates. Stat Pearls Publishing, Treasure Island
Jamal QMS, Ahmad V (2024) Identification of metabolites from *Catharanthus roseus* leaves and stem extract, and in vitro and in silico antibacterial activity against food pathogens. Pharmaceuticals (Basel) 17:450
Jayanthi M, Sowbala N, Rajalakshmi G, Kanagavalli U, Sivakumar V (2010) Study of anti-hyperglycemic effect of *Catharanthus roseus* in alloxan induced diabetic rats. Int J Pharm Pharm Sci 2:114–116
Jayaraman J (1958) Laboratory manual in biochemistry. Wiley Eastern Limited, New Delhi, pp 96–97
Jayaram J (1958) Laboratory Manual in Biochemistry. Wiley Eastern Limited, New Delhi:96–97
Jossang A, Fodor P, Bodo B (1998) A new structural class of bisindole alkaloids from the seeds of *Catharanthus roseus*: vingramine and methylvingramine. J Org Chem 63:7162–7167
Kabesh K, Senthilkumar P, Ragunathan R, Raj Kumar R (2015) Phytochemical analysis of *Catharanthus roseus* plant extract and its antimicrobial activity. Int J Pure Appl Biosci 3:162–172
Khalil A (2012) Antimicrobial activity of ethanol leaf extracts of *Catharanthus roseus* from Saudi Arabia. In: Proceedings of 2nd international conference on environment science and biotechnology, vol 48. IACSIT Press, Singapore
Ku C, Chung WC, Chen LL, Kuo CH (2013) The complete plastid genome sequence of Madagascar periwinkle *Catharanthus roseus* (L.) G. Don: plastid genome evolution, molecular marker identification and phylogenetic implications in asterids. PLoS One 8:1–11
Loomis WE, Shull CA (1937) Methods in plant physiology, McGraw-Hill publication in botanical science. McGraw-Hill, p 290
Lowry OH, Rosebrough NJ, Farr AL, Randll RJ (1951) Protein measurement with the Folin-phenol reagent. J Biol Chem 193:265–275
Luca VD, Laflamme P (2001) The expanding universe of alkaloid biosynthesis. Curr Opin Plant Biol 4:225–233
Mashaghi S, Jadidi T, Koenderink G, Mashaghi A (2013) Lipid nanotechnology. Int J Mol Sci 14:4242–4282
Meenakshi K, Neha SL, Amesh C, Sheela C (2013) *Catharanthus roseus* and prospects of its endophytes: a new avenue for production of bioactive metabolites. Int J Pharm Sci Res 4:2705–2716
Millon MJ, Newman TA, Audinot V, Cussac D, Lejeune F, Nicolas JP, Cogé F, Galizzi JP, Boutin JA (2002) Agonist and antagonist actions of Yohimbine as compared to fluparoxan at alpha (2)-adrenergic receptors (AR). Synapse 35:79–95
Mishra P, Uniyal GC, Sharma S (2001) Pattern of diversity for morphological and alkaloid yield related trades among the periwinkle *Catharathus roseus* accessions collected from in and around Indian subcontinent. Genet Res Crop Evol 48:273–286
Mohammad FS, Patwekar MF, Patwekar FI, Basawaraja HSP (2021) Are plant-derived flavonoids the emerging anti-coronavirus agents? INNOSC Theranostics Pharmacol Sci 4:11–16
Mukhapadhyay S, Cordell GA (1981) Catharanthus alkaloid XXXVI isolation of vinca leukoblastine (VLB) and periformyline from Catharanthus trichophyllus and pericyclivine from *Catharanthus roseus*. J Nat Prod 44:335–339
Muralidharan L (2014) *Catharanthus roseus* leaves as an antidiabetic and hypolipidemic agents in alloxan-induced diabetic rats. Am J Phytomed Clin Ther 2:1393–1396

Pandey SR, Mallavarapu GR, Naqvi AA, Yadav A, Rai SK, Srivastava S, Singh D, Mishra R, Kumar S (2006) Volatile components of leaves and flowers of periwinkle *Catharanthus roseus* (L.) G. Don from New Delhi. Flavour Fragr J 21:427–430

Pavunraj M, Baskar K, Duraipandiyan V, Al-Dhabi AN, Benelli G, Rajendhiran V (2017) Toxicity of Ag nanoparticles synthesized using stearic acid from *Catharanthus roseus* leaf extract against *Earias vittella* and mosquito vectors (*Culex quinquefasciatus* and *Aedes aegypti*). J Clust Sci 28:2477–2492

Pereira DM, Valentão P, Sottomayor M, Ferreres F, Andrad PB (2009) Phenolic compounds in *Catharanthus roseus*. Nat Prod 47:1349–1354

Piovan A, Fillipini R (2007) Anthocyanins in *Catharanthus roseus* in vivo and in vitro: a review. Phytochem Rev 6:235–242

Rahmatzadeh S, Kazemitabar SK (2013) Biochemical and antioxidant changes in regenerated periwinkle plantlets due to mycorrhizal colonization during acclimatization. Indian J Adv Chem Sci 5:1535–1540

Rani J, Kapoor M (2019) Gas chromatography-mass spectrometric analysis and identification of bioactive constituents of *Catharanthus roseus* and its antioxidant activity. Asian J Pharm Clin Res 12:461–465

Rasool N, Rizwan K, Zubair M, Naveed K, Ahmed V (2011) Antioxidant potential of different extracts and fractions of *Catharanthus roseus* shoots. Int J Phytomed 3:108–114

Ruskin RS, Aruna SR (2014) In-vitro and in-vivo antitumor activity of *Catharanthus roseus*. Int Res J Pharm Appl Sci 6:1–4

Saenger (1984) Hydrogen bonding in phytohormone-auxin (IAA) and its derivatives

Shanks JV, Bhadra R, Morgan RJ, Rijhwani S, Vani S (1998) Quantification of metabolites in the indole alkaloid pathways of *Catharanthus roseus*: implications for metabolic engineering. Biotechnol Bioeng 58:333–338

Shinshi H, Noguchi M (1975) Relationship between peroxidases, IAA oxidases and polyphenol oxidases. Phytochemistry 14:1255–1258

Siddiqui M, Ismail Z, Aisha A, Majid A (2010) Cytotoxic activity of *Catharanthus roseus* (Apocynaceae) crude extracts and pure compounds against human colorectal carcinoma cell line. Int J Pharmacol 6:43–47

Singh SN, Vats P, Suri S (2001) Effect of an antidiabetic extract of *Catharanthus roseus* on enzymic activities in streptozotocin induced diabetic rats. J Ethnopharmacol 76:269–277

Singh A, Singh PK, Singh RK (2014) Antidiabetic and wound healing activity of *Catharanthus roseus* L. in streptozotocin-induced diabetic mice. Am J Phytomed Clin Ther 2:686–692

Swanberg A, Dai W (2008) Plant regeneration of periwinkle (*Catharanthus roseus*) via organogenesis. Hortic Sci 43:832–836

Tolambiya P, Jagetiya B (2025) Performance of *Catharanthus roseus* L. in different locations of Jaipur city, Rajasthan to find the suitability of soils for cultivation. Plant Arch 25:737–744

Tolambiya P, Mathur S (2016a) A study on potential phytopharmaceuticals assets in *Catharanthus roseus* L. (Alba). Int J Life Sci Biotechnol Pharma Res 5:1–6

Tolambiya P, Mathur S (2016b) Antimicrobial attributes *in-vivo* grown *Catharanthus roseus* L. Eur J Biomed Pharm Sci 3:217–223

Tolambiya P, Mathur S (2017) Identification of cost-effective drug through *in-silico* approach of *Catharanthus roseus* plant ligand-receptor docking for type 2 diabetes NIDDM. World J Pharm Pharm Sci 6:1029–1049

Tolambiya P, Jagetiya B, Mathur S (2023a) Estimation of total phenolic and flavonoid content and antioxidant activity in *Catharanthus roseus* L. (Alba) plant extract. Biol Forum 15:1101–1106

Tolambiya P, Jagetiya B, Mathur S (2023b) Extraction and identification of various phytopharmaceutical constituent of *Catharanthus roseus* L. (Alba) leaves. Plant Arch 23:58–165

Tolambiya P, Jagetiya B, Mathur S (2024) Quantitative estimation of various primary metabolites and enzymes involved in the biochemical defence mechanism of *Catharanthus roseus* L. J Phytol Res 37:163–171

Villamar-Cruz O, Loza-Mejía MA, Arias-Romero LE, Camacho-Arroyo I (2021) Recent advances in PTP1B signaling in metabolism and cancer. Biosci Rep 41:BSR20211994

Waltz, LR (2004) The herbal encyclopedia, A practical guide to the many uses of herbs. I Universe, Inc Lincoln. pp 394

Widowati W, Mozef T, Risdian C, Yellinaty Y (2013) Anticancer and free radical scavenging potency of *Catharanthus roseus*, *Dentrophthoe petandra*, *Piper betle* and *Curcuma mangga* extracts in breast cancer cell lines. Oxid Antioxid Med Sci 2:137–140

# Medicinal and Nutritional Importance of *Sphaeranthus indicus* in Human Health

17

Murtaza Vandeliwala, Ruchi Pathak, and Prinsa R. Nagar

**Abstract**

*Sphaeranthus indicus* Linn., also known as Gorakhmundi, is a key plant in ancient traditions, particularly Ayurveda, and is recognized for its wide spectrum of medicinal and nutritional benefits. This chapter focuses on the numerous therapeutic applications of *S. indicus*, presenting evidence for its anti-inflammatory, antioxidant, antibacterial, and antihepatotoxic properties. The plant is typically found in tropical and subtropical locations and has been used to cure various illnesses, including respiratory, digestive, and skin problems. It also highlights the botanical traits and morphology of *S. indicus*, as well as the regions where it grows. It then discusses phytochemical ingredients and explains the presence of physiologically active constituents like flavonoids, alkaloids, sterols, amino acids, terpenes, and sugars. Many breakthroughs are highlighted, including the plant's effectiveness in blood glucose metabolism and its use in diabetes control, as well as its wound healing and neuroprotective properties. In particular, the manuscript discusses potential future investigations that might validate conventional applications and propose new therapeutic uses for *S. indicus* in modern medicine. Based on existing research, this chapter emphasizes *S. indicus*' value as a beneficial tool in improving human health in terms of medicine and nutrition.

**Keywords**

*Sphaeranthus indicus* · Traditional uses · Pharmacological activities · Medicinal benefits · Phytoconstituents

M. Vandeliwala · R. Pathak · P. R. Nagar (✉)
Department of Pharmaceutical Chemistry, L. J. Institute of Pharmacy, L J University, Ahmedabad, India



M. A. Ansari et al. (eds.), *Medicinal Plants and their Bioactive Compounds in Human Health: Volume 2*, https://doi.org/10.1007/978-981-95-3620-7_17

**Learning Objectives**

- Understanding the ethnopharmacological relevance and Ayurvedic significance of *S. indicus*
- Recognizing the botanical characteristics, morphology, and geographical distribution of the plant
- Insights of the phytochemical profile of *S. indicus*, including key constituents such as alkaloids, flavonoids, sterols, terpenes, and sugars
- The pharmacological properties of *S. indicus* such as inflammatory, antioxidant, antibacterial, and antihepatotoxic effects

## 17.1 Introduction

*Sphaeranthus indicus Linn.*, also known as Gorakhmundi or East Indian globe-thistle, is one the most commonly found Indian herbs in the Asteraceae family in tropical and subtropical areas. Most species of the genus *Sphaeranthus* are either annual or perennial herbs (Basumatary and Bhimrao 2024). It is distributed throughout East India, especially in damp areas of the Himalayas, North and Central India, Bhutan, Bangladesh, and certain regions of Africa and Australia (Ramachandran 2013). In Ayurveda, it is considered one of the most significant traditional medicines due to its outstanding therapeutic advantages as an anti-inflammatory, antioxidant, and antibacterial agent.

### 17.1.1 Botanical Characterization

Gorakhmundi is an annual herbaceous plant with a usual height of 15–60 cm. It showcases branched morphology which is characterized by multiple stems growing from the ground, that are of green color with reddish hue, somewhat angular and are covered with fine hairs. The leaves are arranged alternatively, and possess a lanceolate to oval shape, showcasing a starred margin and rough feel due to the hairy structure of the leaves. The plant bears small, globular flower heads about 1 cm in diameter and attached to long stalks. The florets in the flower head are bisexual, pinkish-purple and possess tubular as well as ligulate components. The fruit is tiny, dry and one seeded which remain closed during maturation. The fruits also possess pappus hairs that help in wind dispersal. The fibrous root structure of the root aids in its survival in adverse conditions (Galani et al. 2010).

### 17.1.2 Taxonomy

| | |
|---|---|
| *Kingdom* | Plantae |
| *Division* | Phanerogamae |
| *Sub-division* | Angiospermae |
| *Class* | Dicotyledonae |

| | |
|---|---|
| *Sub-class* | Gamopetalae |
| *Order* | Asterales |
| *Family* | Asteraceae |
| *Species* | *Indicus* |

## 17.1.3 Vernacular Names

| | |
|---|---|
| *English* | *East Indian globe-thistle* |
| *Hindi* | Gorakhmundi, Mundi |
| *Sanskrit* | Mundi, Sravani, Bhumikadamba |
| *Gujarati* | Gorakhmundi |
| *Marathi* | Gorakhmundi, Mundi |
| *Kannada* | Karande, Mudukattanagida |
| *Malayalam* | Manni |
| *Punjabi* | Gorakhmunda |
| *Bengali* | Mudmudiya, Surmuriya |

## 17.1.4 Characteristics from Microscopic Study

### Root

| | |
|---|---|
| *Epidermis* | Single-layer rectangular cells |
| *Secondary cortex* | A thin wall of oval or tangentially elongated parenchymatous cells, with significant presence of scattered lignified phloem fibers |
| *Middle region* | Lignified secondary xylem with pitted arteries, starch grains, and a prominent hilum located in the secondary cortex |

### Stem

| | |
|---|---|
| *Epidermis* | A single layer of cell enveloped by a thick cuticle |
| *Cortex* | Around four to six layers of oval–polygonal parenchymatous cells with thin wall |
| *Endodermis* | Single layer of barrel-shaped cells. Sometimes surrounded with lignified pericyclic fibers forming a ring |
| *Secondary phloem* | Consist of cellulosic fibers scattered across the region |
| *Secondary xylem* | Consist of a vessel with spiral thickening or simple pitting |

**Leaf**

| | |
|---|---|
| *Epidermis* | Single layered followed by four to six layers of collenchyma and three to four layers of parenchyma |
| *Glandular trichomes* | Have a height of two to three cells and have either uniseriate or biseriate stalk |
| *Mesophyll* | Oval or polygonal, thin-walled parenchymatous cells |
| *Anisocytic stomata* | Present on both sides of surfaces. |

## 17.1.5 Medicinal Properties

*S. indicus* holds a significant place in traditional Ayurvedic medicine due to its numerous therapeutic properties (as discussed in the Fig. 17.1). Topically, the herb is used to treat disorders such as eczema and psoriasis due to its anti-inflammatory and healing properties. It is also used in asthma and bronchitis as it reduces airway inflammation and aids in breathing. The plant is also used to ease the problems of indigestion and flatulence. Due to its hepatoprotective effects, it is also used to treat jaundice. It is also useful for stress and anxiety due to its calming properties. Apart from this, it has been reported that the plant also possesses antibacterial and antifungal properties, making it suitable for treatment.

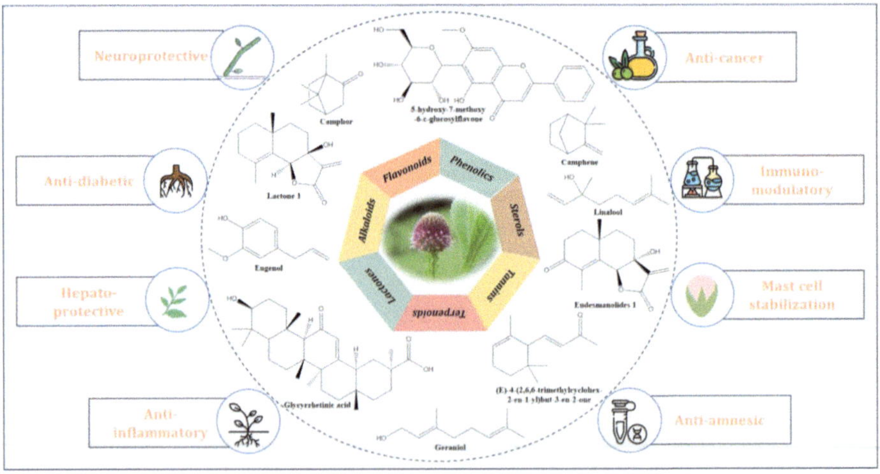


**Fig. 17.1** Some chemically diversified phytoconstituents of *S. indicus* exhibiting significant pharmacological effects

## 17.2 Classification Based on Plant Parts

### 17.2.1 Root

#### 17.2.1.1 Potential Medicinal Effects Exhibited by the Constituents

**Antidiabetic Effects**

Administration of *Sphaeranthus indica linn* methanolic extract (SEM) remarkably reduced STZ-induced increased blood glucose levels, hemoglobin A1c levels, serum lipid parameters, and oxidative stress (Jadhav and Vaghela 2024). The anti-hyperglycemic effects of this plant could be possibly associated with the level of flavonoids present in them, which might be responsible for an increase in the secretion of insulin from pancreatic cells. This effect is most likely due to an improvement in β cell regeneration, which improves the transportation of glucose into the tissues (Hassan et al. 2022). The method of ethanolic extraction of phytoconstituents present in root, essential for antidiabetic effect, is shown in Table 17.1.

**Table 17.1** The antidiabetic, antihyperlipidemic, and antioxidant effects of ethanolic extracts of *Sphaeranthus indicus* roots at different doses (100 and 200 mg/kg) and its effect observed on the animal model

| Name of activity | Extract used | Observation | Reference |
|---|---|---|---|
| Antidiabetic | Ethanolic | In a study involving STZ-induced diabetic rats, administering a dose of 100 and 200 mg/kg ethanolic extract of *S. indicus* root at resulted in a significant decrease in blood sugar level as compared to the diabetic control group. | Ramachandran et al. (2011) |
| Antihyperlipidemic | Ethanolic | In STZ-induced diabetic rats, treatment with ethanolic extract of *Sphaeranthus indicus* roots at doses of 100 and 200 mg/kg resulted in significant ($p < 0.01$) modification of elevated lipid profile as compared to the diabetic control group. | Ramachandran et al. (2011) |
| Antioxidant | Ethanolic | In STZ-induced diabetic rats, administering an ethanolic extract of *Sphaeranthus indicus* roots (EESIR) at doses of 100 and 200 mg/kg led to a remarkable increase in levels of catalase (CAT), superoxide dismutase (SOD), and glutathione peroxidase (GPx), while decreasing thiobarbituric acid reactive substances (TBARS) compared to the diabetic control group. The 200 mg/kg dose of EESIR demonstrated significantly greater antioxidant activity ($p < 0.01$) than the 100 mg/kg dose. | Ramachandran et al. (2011) |

**Antioxidants Effects**

In an in vitro study, 1000 µg/mL of an ethanolic extract of *S. indicus* demonstrated strong scavenging activity against various free radicals, including ABTS [2,2-azinobis(3-ethylbenzothiazoline-6-sulfonate)], DPPH (1,1-diphenyl,2 picryl hydrazyl), nitric oxides, and superoxide radicals. It also shows moderate activity in chelating iron. In an in vivo study, methanolic extract of *S. indicus* significantly boosted antioxidant defenses in rats by increasing levels of catalase (CAT), superoxide dismutase (SOD), and glutathione peroxidase (GPx) by reducing malondialdehyde levels (Galani et al. 2010). The method of ethanolic extraction of phytoconstituents present in root to exert antioxidant effect is shown in Table 17.1 (Ramachandran et al. 2011).

**Larvicidal Action**

Acetone extract from both the roots and leaves of the plant led to mortality in more than 50% of Indian mosquito species that spread filarial worms. The root extract was more effective than the leaf extract. Furthermore, a refined fraction of the acetone extract proved to be effective in killing mosquito larvae. Alternately, a methanolic extract showed a repellent and feeding deterrent effect on *Tribolium castaneum*, with a significant effect at 1% concentration and full deterrent activity for feeding at 5 mL and repellent activity at 4 mL (Ramachandran 2013).

**Hepatoprotective Effects**

The methanolic extract of *Sphaeranthus indicus* has been shown to have hepatoprotective effects as verified by a remarkable reduction in liver function markers and serum bilirubin levels and also prevention of liver histopathological changes in rats induced with hepatotoxicity. Furthermore, this extract enhanced the activity of antioxidant enzymes (superoxide dismutase [SOD], catalase [CAT], and glutathione peroxidase [GPx]) and decreased levels of lipid peroxidase against acetaminophen-induced hepatotoxicity, suggesting that the reduction in oxidative stress is the key mechanism behind its hepatoprotective effect. The methanolic extract, as compared to the aqueous fraction, contains the most active hepatoprotective compound, which may include flavonols and flavonoids (Farnoosh et al. 2016).

**Anti-inflammatory Effects**

The anti-inflammatory activity is largely attributed to the inhibition of pro-inflammatory cytokines and enzymes like COX (cyclooxygenase) and LOX (lipoxygenase). Key compounds contributing to this effect include *Sphaeranthanolide*, flavonoid (quercetin and kaempferol), and phenolic acids (caffeic acid) (Basumatary and Bhimrao 2024).

#### 17.2.1.2 Extracts of Phytoconstituents

### 17.2.2 Flower

#### 17.2.2.1 Potential Medicinal Effects Exhibited by the Constituents

**Anti-inflammatory Effects**

Protein denaturation refers to the process in which the protein fails to retain its tertiary and secondary structure because of external stress or compounds. This denaturation of protein can cause inflammation. The ability of plant extract to inhibit this process of denaturation was assessed. The extract efficaciously prevented heat-induced albumin denaturation, achieving a maximum inhibition of 71% at a concentration of 500 μg/mL. In comparison, Aspirin, a conventional anti-inflammatory agent, demonstrated inhibition of 68% at a dosage of 100 μg/mL concentration (Tiwari et al. 2019). The method of standardized extraction of phytoconstituents present in flower to exert anti-inflammatory effect is shown in Table 17.2 (Chakrabarti et al. 2012).

**Immunomodulatory Activity**

The methanolic extract of *S. indicus* Linn flower heads, along with its petroleum, chloroform, and remaining methanolic fraction was observed to significantly enhance phagocytic activity, hemagglutination antibody titer, and delayed hypersensitivity in mice. In addition, it is most likely to stimulate cellular and humoral immunity (Sultana et al. 2021).

**Neuroleptic Effects**

The neuroleptic effect of petroleum ether, alcohol, and water extracts from *S. indicus* flower was tested in mice using apomorphine-induced cage climbing and catalepsy in mice models. Only the petroleum ether extract (300 mg/kg intraperitoneal [i.p]) significantly decreased the duration of time the mice engaged in climbing the cage. Both the aqueous and alcoholic extracts at (300 mg/kg i.p) caused catalepsy; however, the petroleum ether did not. Additionally, the hydroalcoholic extract (100, 200, and 500 mg/kg per os [p.o]) induced catalepsy, enhanced the cataleptic effect of haloperidol, and counteracted the stereotypic behaviors induced by apomorphine (Galani et al. 2010). The method of petroleum ether, methanolic extraction, and aqueous extraction of phytoconstituents present in flower to exert anti-inflammatory effect is shown in Table 17.2 (Ambikar and Mohanta 2013).

**Antiamnesic Effects**

Oral administration of a methanolic extract of *S. indicus* for seven continuous days improved memory in mice as compared to the control animals. In addition, pretreatment with *S. indicus* extract for 7 days protected the mice from memory deficit induced by scopolamine. Furthermore, findings indicate that oral alcoholic extract of *S. indicus* at a dose of 300 mg/kg significantly enhances memory acquisition and

**Table 17.2** Different types of phytoconstituents extracted from flower of *S. indicus* and its pharmacological effect on the animal study

| Name of activity | Extract used | Observation | Reference |
|---|---|---|---|
| Anxiolytic | Petroleum ether, alcoholic, and aqueous | In a series of tests involving mice, extracts of petroleum ether (10 mg/kg), ethanol (10 mg/kg), and water (30 mg/kg) demonstrated notable effects. Among these the petroleum ether extract showed the most significant anxiolytic activity in both elevated plus maze (EPM) and the open field test (OFT), surpassing the effects of ethanol and water extracts. However, its effectiveness was still lower compared to diazepam (1 mg/kg) | Ambavade et al. (2006) |
| Neuroprotective effect | Sphaeranthus Indicus petroleum ether(SIP), Sphaeranthus Indicus methanolic(SIM), Sphaeranthus Indicus aqueous(SIA) | SIM and SIA were found to be significantly effective neuroprotective agents capable of counteracting oxidative damage and the accelerated aging caused by D-galactose. | Ambikar and Mohanta (2013) |
| Antiamnesic effect | Petroleum ether fraction | Administering the petroleum ether fraction (10 mg/kg per os [p.o.]) led to significant improvement in the cognitive function of mice, as evidenced by the passive avoidance test ($p < 0.05$). Additionally, this treatment decreased escape latencies during training sessions and increased the time duration spent in the target quadrant during the probe trial of the water maze task ($p < 0.05$). | Patel (2012) |
| Antidiabetic | Ethanolic | The treatment with insulin and *S. indicus* extract led to a significant reduction in blood glucose levels, bringing them closer to normal values. Remarkably, the highest dose of 300 mg/kg showed effects more similar to those of insulin, whereas the lower dose of 200 mg/kg had a lesser impact. | Kannojia et al. (2007) |
| Anti-inflammatory | Standardized herbal extract (NPS31807) | Treatment with NPS31807 led to a dose-dependent decrease in pro-inflammatory cytokines produced by human macrophages and activated epidermal keratinocytes. NPS31807, derived from sphaeranthus indicus, shows potential therapeutic options for inflammatory and autoimmune conditions like psoriasis. | Chakrabarti et al. (2012) |
| Antiarthritic | Petroleum ether | The petroleum ether extract of *S. indicus* administered at a dose of 100 mg/day p.o demonstrated notable antiarthritic effect. | Badgujar et al. (2009) |

retention, demonstrating nootropic activity. This finding suggests a potential neuroprotective role of *S. indicus* (Jain and Choudhary 2019).

**Diabetic Wound Healing Properties**

The aqueous extract formulation (method given in Table 17.2) of *Sphaeranthus indicus* enhances keratinization, epithelization, and fibrosis to a level comparable to neomycin treatment in diabetic wounds. On treatment with these extracts, there was a significant increase in hydroxyproline content in 10-day-old diabetic wounds, indicating improved collagen formation, which plays an effective role in wound healing. Visual examination during the 'wound healing ability' experiment showed that the diabetic wound treated with *S. indicus* extracts was relatively clean with the minutest signs of inflammation like redness and swelling. This showed that the test formulation showed remarkable anti-inflammatory effect when applied to diabetic wound (Kumar and Nema 2017).

**Mast Cell Stabilizing Activity**

The ethanolic extract at doses of 150 and 300 mg/kg as well as the ethyl extract at doses of 100, 150, and 300 mg/kg showcased slightly superior protective action against mast cell degranulation (77–86%) and stabilizing activity (77–88%) as compared to standard drug ketotifen, which showed 75% protection and 69% stabilizing activity in sheep serum model, respectively (Singh et al. 2019).

#### 17.2.2.2 Extracts of Phytoconstituents

### 17.2.3 Leaves and Stem

#### 17.2.3.1 Potential Medicinal Effects Exhibited by the Constituents

**Antibacterial and Antifungal Effects**

Alcohol and water extract from *S. indicus* exhibits an antibacterial effect against *Alternaria solani*, *Fusarium oxysporum*, and *Penicillium pinophilum*. The fruit of *S. indicus* exhibits significant activity against both gram-positive and gram-negative bacteria. In addition, it also has antifungal properties. Petroleum ether, acetone, and 90% methanol extracts (extraction method as per Table 17.3) were evaluated for antibacterial and antifungal activities by diffusion method in bacterial and fungal test cultures. All the cultures show remarkable antibacterial and strong antifungal activities (Galani et al. 2010).

**Macrofilaricidal Effects**

The methanolic extract was effective in killing adult *Setaria digitata*, a type of cattle filarial worm, at a concentration of 4 mg/mL, as determined through a worm motility assay (Ramachandran 2013).

**Table 17.3** Phytoconstituents extracted from leaves and stem of *S. indicus*, its pharmacological properties and its effect on animal models

| Name of activity | Extract used | Observation | Reference |
|---|---|---|---|
| Bactericidal | Aqueous | In a laboratory test, a crude extract at a concentration of 0.5% demonstrated inhibition of the growth of *Ralstonia solanacearum*. Furthermore, in field circumstances, this extract substantially reduced both the incidence of the disease and the population of *Ralstonia solanacearum* in the soil. | Abhirami et al. (2024) |
| Antioxidant | Methanolic | Rats treated with 1 mg/kg carbon tetrachloride$(CCL)_4$ had a statistically substantial increase in their serum levels of alanine aminotransferase (ALT), aspartate aminotransferase (AST), and alkaline phosphatase (ALP). Notable decreases in total protein and antioxidant levels of CAT, SOD, GST were also noted in liver-injured rats. | Mathews et al. (2012) |
| Analgesic | Ethanolic | The ethanolic extract of *S. indicus*, administered at doses of (100, 200, and 400 mg/kg p.o) demonstrated a significant and dose-dependent analgesic effect in both pain models. | Meher et al. (2011a) |
| Anti-inflammatory | Ethanolic | The extract administered at various doses (100, 200, and 400 mg/kg p.o), demonstrated a significant and dose-dependent anti-inflammatory effects in both acute ($p < 0.05$) and chronic ($p < 0.05$) inflammation models. | Meher et al. (2011b) |
| Larvicidal | Dichloromethane, ethyl acetate, diethyl ether | The diethyl ether, dichloromethane (DCM), and ethyl acetate extracts of *S. indicus* showed 100% mortality of culex quinquefasciatus after 48 h exposure. The ethyl acetate extract (201.11 ppm) was found to be effective. | Tennyson et al. (2012) |

**Antioxidant Effects**

*S. indicus* holds various powerful bioactive compounds that are capable of neutralizing free radicals. The polar part of leaf methanol extract and inflorescence aqueous extract exhibits prominent antioxidant activity due to a dense accumulation of phenols and flavonoids which are well-known antioxidants (Tandon and Gupta 2020). The method of methanolic extraction of phytoconstituents present in leaves and stem to exert antioxidant effect is shown in Table 17.3 (Mathews et al. 2012).

**Anticancer Effects**

P53 is an essential cellular factor that is involved in responding to stress conditions. Its major functions under stress include interrupting the cell cycle, maintaining genomic stability, and apoptosis induction. 7-hydroxyfrullanolide an active constituent derived from *S. indicus* has demonstrated a strong antitumor effect both in vivo and in vitro. It was found that 7-HF activates both intrinsic and extrinsic apoptosis pathways in both p53-dependent and p53-independent mechanisms, thus making it significantly contributing to its antitumor activity (Pandey et al. 2019).

#### 17.2.3.2 Extracts of Phytoconstituent

### 17.2.4 Entire Plant

#### 17.2.4.1 Potential Medicinal Effects Exhibited by the Constituents

**Analgesic Effects**

The essential oil generally consists of low molecular weight compounds which assist them in their high lipophilicity and enable them to cross the blood–brain barrier and produce its action on the central nervous system. The hot plate test used here is to assess the central analgesic activity of these oils. Results showed that topical application of oil obtained from *S. indicus* at dose concentrations of 1% and 2% markedly showed their prime effect at 90 and 120 min. The analgesic activity of essential oil obtained from *S. indicus* is chiefly ascribed to the synergistic effect of terpenes and sesquiterpenes instead of the predominant constituent alone. The terpenes present in the oil are most likely to act by inhibiting endogenous substances that elicit pain in the nociceptive pathway (Padhan et al. 2017).

**Antipyretic Effects**

Various extracts of the whole plant of *Sphaeranthus indicus* were evaluated for antipyretic activity at doses of 200 and 400 mg/kg body weight. The chloroform and ethanol extract showed a significant antipyretic effect ($p < 0.05$) starting from 1 h, whereas the aqueous extract began to show activity from 2 h onwards, juxtaposed to the standard drug paracetamol. The remarkable antipyretic effect of the extracts might be attributed to the presence of flavonoids (Singh et al. 2019).

**Neuroleptic and Sedative Effects**

The hydroalcoholic extract of *S. indicus* (SIE) (method described in Table 17.4) remarkably reduced spontaneous motor activity and extended pentobarbital-induced hypnosis in mice; the lack of impact on motor coordination proposes that the effect is centrally mediated and not by neuromuscular blockade. The pharmacological action observed in SIE might be attributed to its content of steroids, saponins, tannins, flavonoids, coumarins, triterpenes, and essential oil. Thus, it can be proposed that the hydroalcoholic extract of *S. indicus* might contain a psychoactive principle with sedative and neuroleptic properties (Galani and Patel 2009).

**Broncho Dilatory Effect**

The methanolic extract (method described in Table 17.4) of the entire *S. indicus* plant along with its various fractions (petroleum ether, chloroform, ethyl acetate,

**Table 17.4** All the phytoconstituents extracted from the entire plant, its use in different pharmacological conditions and animal study of the same

| Name of activity | Extract used | Observation | Reference |
|---|---|---|---|
| Nephroprotective | Ethanolic | The extract led to a marked reduction in elevated serum creatinine and urea levels. It also helps restore renal antioxidant defense mechanism, including activities of superoxide dismutase, catalase, and glutathione peroxidase as well as the level of reduced glutathione which had been diminished by cisplatin therapy. | Mathew et al. (2012) |
| Antipyretic | Ethanol, chloroform, and aqueous | The chloroform and ethanol extracts displayed substantial antipyretic effects from 1 h after administration, while aqueous extracts showed activity starting at 2 h, relative to the standard drug, paracetamol. | Nanda et al. (2009) |
| Antiulcer | Ethanolic | The test extract exhibited noteworthy antiulcer activity. In the aspirin (ethanol) and pylorus ligation method, animals pretreated with ALSI showed a significant decrease in free acidity, total acidity, gastric volume, and ulcer index. The extract provided ulcer protection rates of 82.9% and 85.3%, compared to 92.6% for the standard drug ranitidine. | Roy et al. (2013) |
| Bronchodilatory | Methanolic | The methanolic extract and its fraction-petroleum ether, benzene, chloroform, and ethyl acetate demonstrated significant protective effects against histamine-induced bronchospasm in guinea pigs. | Sarpate et al. (2009) |
| Psychotropic/ sedative | Hydroalcoholic | The extract of *S. indicus* remarkably decreased spontaneous motor activity and extended the duration of hypnosis induced by pentobarbital, suggesting that it might act as a sedative agent. Furthermore, the hydroalcoholic extract also lowered exploratory behavior and enhanced the sedative effect. | Galani and Patel (2010) |

benzene) was assessed for bronchodilatory effect against histamine-induced bronchospasm in guinea pigs. All extracts demonstrated a remarkable protective effect against bronchospasm; the methanolic extract showed notable protective action ($p < 0.001$) compared to standard treatment, chlorpheniramine maleate 2 mg/kg (Farnoosh et al. 2016).

**Antiulcer Effects**

At a dose of 250 and 500 mg/kg body weight, the ethanol extracts of the whole plant of *S. indicus* demonstrated a notable decrease in gastric acid volume, free acidity, total acidity, and ulcer index. In contrast to the standard drug ranitidine, the antiulcer activity of the extract was 82.9% and 85.3%, which had an ulcer protection rate of 92.6%. This indicates that the extract exhibits significant antiulcer activity in a dose-dependent manner (Singh et al. 2019).

**Diuretic Effects**
The whole plant methanolic extract was found to positively increase the risk of hyperkalemia so care should be taken while using this extract alongside other diuretics as this could possibly lead to hypokalemia. The experimental results indicated that a single-dose administration of ethanolic extract of *S. indicus* Linn at 100, 200, and 300 mg/kg resulted in increased urinary output and higher concentration of sodium and potassium ions in urine. Particularly a dose of 300 mg/kg exhibited a diuretic effect equivalent to that of hydrochlorothiazide at 25 mg/kg. The diuretic effect of the methanolic extract is accreted to active phytochemicals such as saponins, flavonoids, and terpenoids. The presence of these compounds may contribute to its diuretic effect (Sundaresan et al. 2017).

#### 17.2.4.2 Extracts of Phytoconstituent

## 17.3 Phytochemicals Present in *S. indicus*

*S. indicus* contains various class of phytoconstituents. Compounds 1–10 are obtained from the essential oil obtained from the plant. Citral, eugenol, and geraniol are found in significant quantities. The hydro-distilled essential oil contains compounds 11–24. Among the various phytoconstituents present in *S. indicus*, three types of hydroxy lactones (compounds 25–27) and four types of bicyclic sesquiterpene lactones (compounds 28–31) are also identified (Figs. 17.2 and 17.3). The flower of the plant *S. indicus* contains various eudesmanolides (compounds 35–42). From the other part of the plants, eudesmanolides (compounds 48–50) are obtained. Peptide alkaloids (compound 43–46) and flavone glycoside (compound 47) are also considered as the significant phytoconstituents present in *S. indicus*. Compound 51 is the flavonoid found in all the parts of the plant (Figs. 17.4 and 17.5) (Galani et al. 2010; Ramachandran 2013).

## 17.4 Marketed Preparations

Owing to the growing awareness of *Sphaeranthus indicus*'s therapeutic potential, many *S. indicus*-based product preparations have been commercialized. These formulations of *Sphaeranthus indicus* on the market take advantage of its nutritional and medicinal qualities covered in earlier sections. A variety of marketed products have been developed to harness the potential of *Sphaeranthus indicus* for human health, given its diverse therapeutic benefits are described in the Table 17.5.

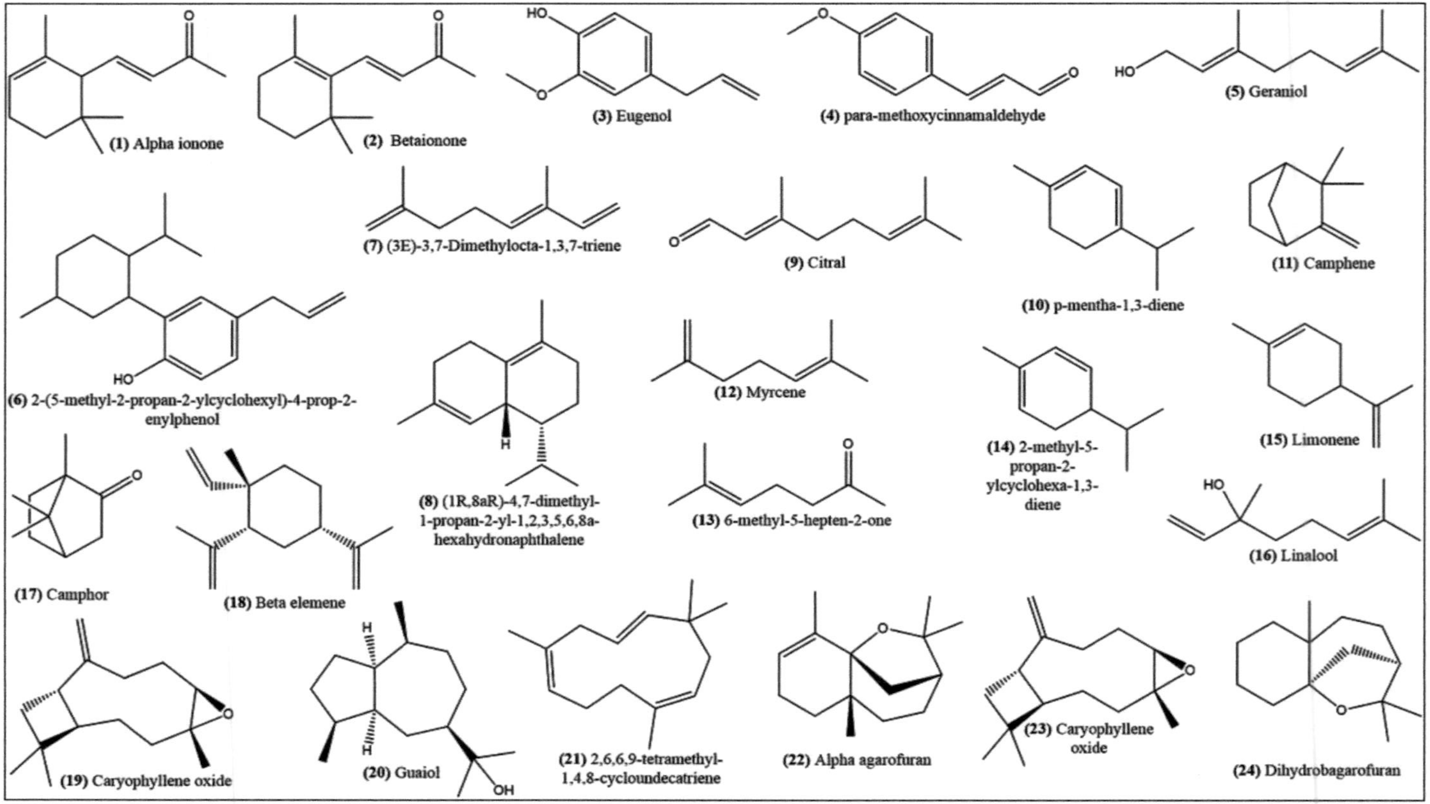


**Fig. 17.2** Phytoconstituents extracted from essential oils (compounds **1**–**10**), and hydro-distilled essential oils (compounds **11**–**24**)

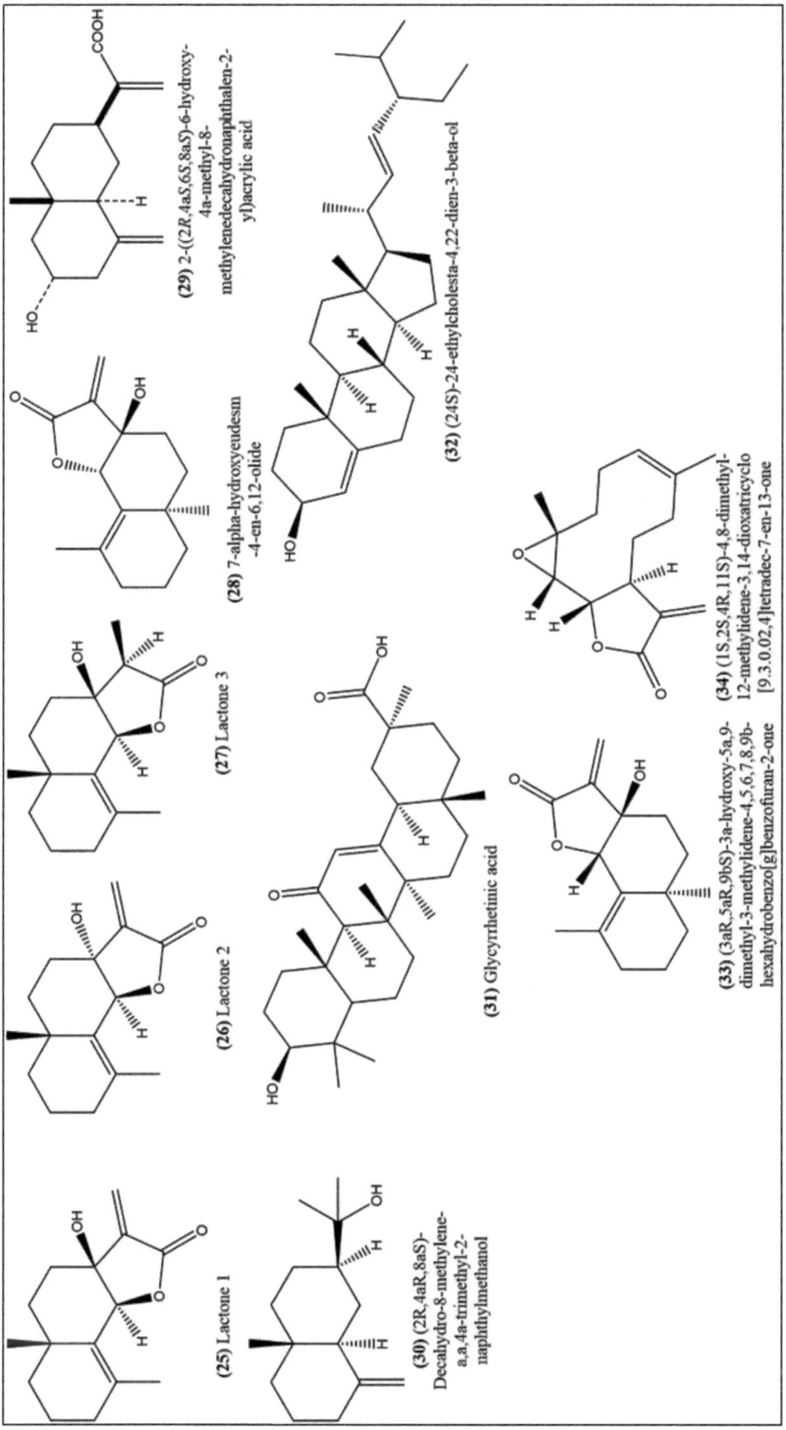


**Fig. 17.3** Phytoconstituents containing hydroxy lactones (compounds **25–27**), and bicyclic sesquiterpene lactones (compounds **28–31**)

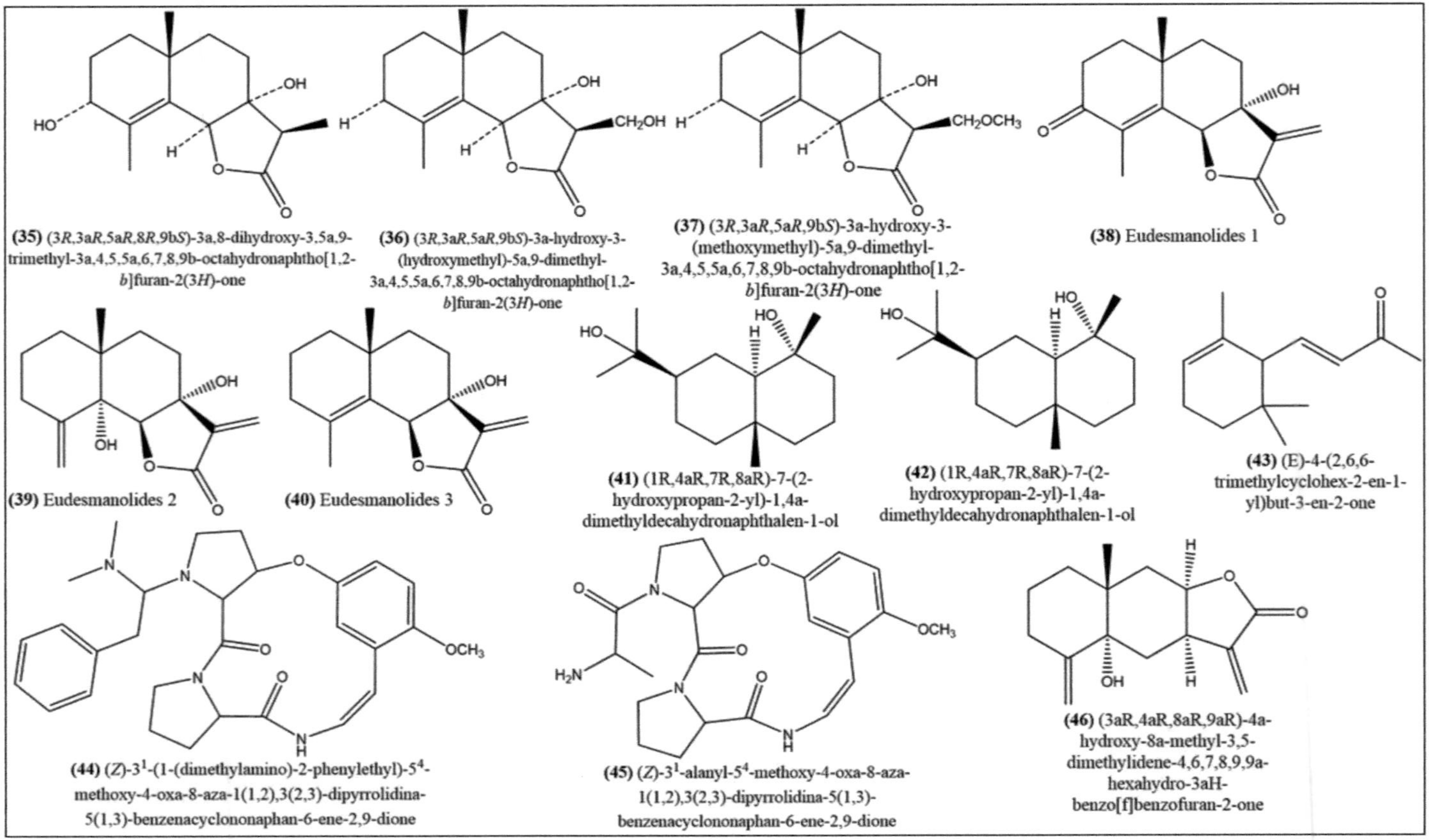


**Fig. 17.4** Phytoconstituents containing eudesmanolides (compounds **35**–**42**) and peptide alkaloids (compounds **43**–**46**)

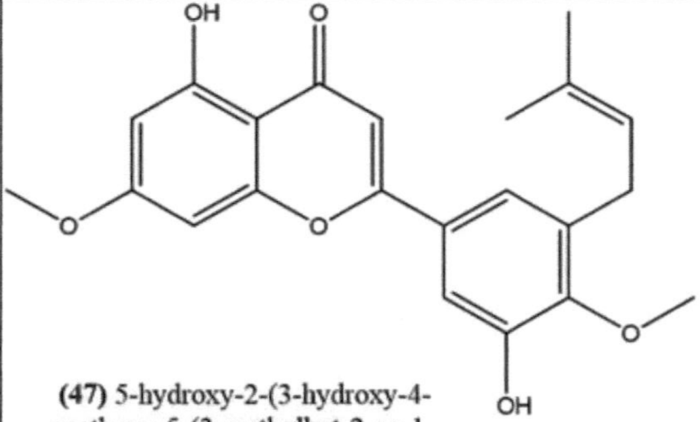

**(47)** 5-hydroxy-2-(3-hydroxy-4-methoxy-5-(3-methylbut-2-en-1-yl)phenyl)-7-methoxy-4*H*-chromen-4-one

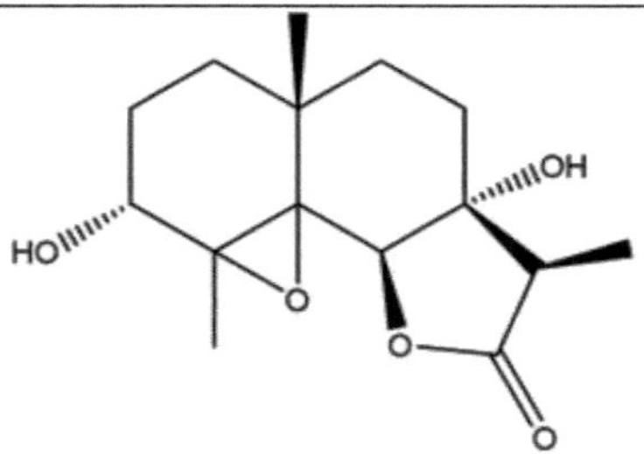

**(48)** (1*R*,3a*R*,5*R*,7a*R*,9a*R*)-5,9a-dihydroxy-1,4a,7a-trimethyloctahydro-6*H*-oxireno[2',3':8,8a]naphtho[1,2-*b*]furan-2(1*H*)-one

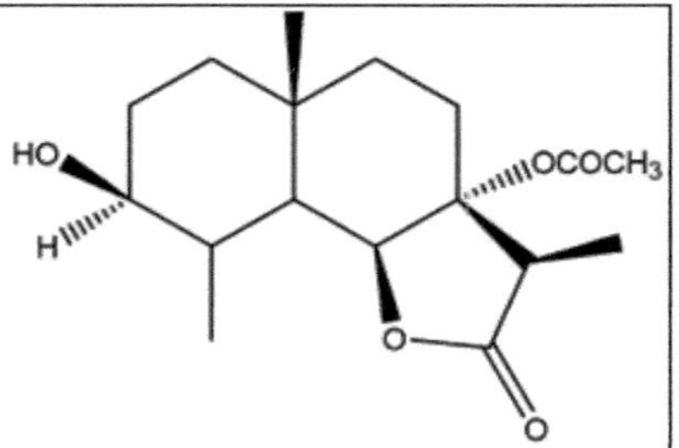

**(49)** (3*R*,3a*R*,5a*R*,8*S*,9b*S*)-8-hydroxy-3,5a,9-trimethyl-2-oxodecahydronaphtho[1,2-*b*]furan-3a(4*H*)-yl acetate

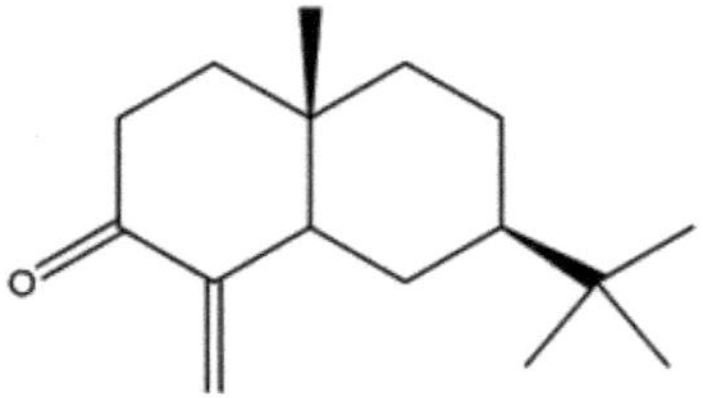

**(50)** (4a*S*,7*R*)-7-(*tert*-butyl)-4a-methyl-1-methyleneoctahydronaphthalen-2(1*H*)-one

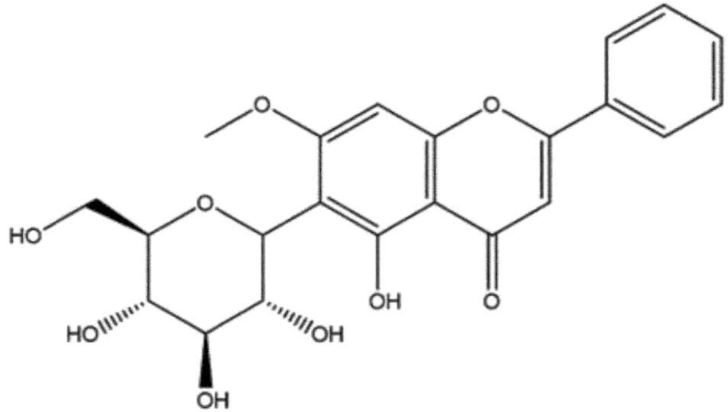

**(51)** 5-hydroxy-7-methoxy-6-c-glucosylflavone

**Fig. 17.5** Some other phytoconstituents derived from different parts of *S. indicus*

**Table 17.5** Some marketed preparations of *S. indicus* with significant therapeutic potential

| Sl. no | Name of the product | Name of manufacturer | Use |
|---|---|---|---|
| 1 | Divya Mahamanjishtharishta | Patanjali Divya Pharmacy, Haridwar, India | Blood purifier in eczema, psoriasis, and skin diseases |
| 2 | Diabecon | Himalaya Wellness Company, Makali, Bengaluru | Regenerative effect on pancreatic β-cells and insulinotropic |
| 3 | Medisun Super slim | Pavan Nutra Life Sciences | Boosts metabolism helps in burning extra fat and improves blood circulation |
| 4 | Geriforte | Himalaya Wellness Company, Makali, Bengaluru | Scavenges free radicals in the cell that are responsible for premature aging |
| 5 | Gout care pack | Planet Ayurveda | Treating high uric acid levels |
| 6 | Meratrim | RE-BODY | Effective for weight management by altering fat metabolism |
| 7 | Unjha Raktashodhak Tonic | Unjha Pharmaceuticals Pvt. Ltd., Thaleja, Ahmedabad India | Help with removing toxins, healing wounds, and dermatological disorders |
| 8 | Safi | Hamdard Laboratories (WAQF) Bangladesh | In constipation, obesity, pimple, scabies, nose bleeding, boil |
| 9 | Baidyanath Mahamanjisthadi Kadha | Shree Baidyanath Ayurved Bhawan Pvt. Ltd., Kolkata, West Bengal | Antioxidant and detoxify the entire body, help to maintain skin, hair, and eye health |
| 10 | Shiva Gutika | Basic Ayurveda, Ghaziabad | Used to improve liver health, heart health, cough, and respiratory conditions |
| 11 | Maha Vish Garbha Oil | Herbs Forever, Los Angeles, California U.S.A | Provide comfort to muscles and joints |

## 17.5 Conclusion

*Sphaeranthus indicus*, often known as the East Indian globe thistle, is a flowering plant belonging to the Sphaeranthus genera widely distributed from Northern Australia to Indomalaya. The plant's stem, leaf, flower, root, and seed are used to treat a variety of ailments and have been studied for potential antimicrobial, antiulcer, anti-inflammatory, neuroleptic, antiviral, ovicidal, and anthelmintic properties. The scientific literature reveals the pharmacological investigations of *S. indicus* that suggest its potential application in prophylaxis and treatments for numerous disorders. Despite its wide range of therapeutic and medicinal properties, further investigations on its safety and efficacy are required to prove if it can offer therapeutic benefits alone or in combination. Preclinical investigations should prioritize clinical evaluation to determine the plant's potential as a treatment for various diseases, as well as the structure–activity relationship of phytoconstituents found in *S. indicus* and other spices.

## 17.6 Summary

This chapter provided a thorough comprehension of *Sphaeranthus indicus Linn.*, a plant of significant ethnomedicinal and Ayurvedic value. Also known as Gorakhmundi, *S. indicus* is traditionally used for various therapeutic applications. This chapter consists of the plant's botanical traits, morphological attributes, and geographical distribution, summarizing all necessary information. It also highlights phytoconstituents from different chemical classes like flavonoids, alkaloids, sterols, terpenes, amino acids, and sugars which are responsible for exhibiting pharmacological effects. The pharmacological properties including anti-inflammatory, anti-oxidant, antibacterial, antihepatotoxic, neuroprotective, antidiabetic, and wound-healing properties are also discussed in detail. In a nutshell, this chapter strengthens the connection between traditional medicine and contemporary healthcare methodologies.

**Credit Authorship Contribution Statement**

Murtaza *Vandeliwala* was involved in methodology, data curation, writing-original draft, writing-review and editing. *Ruchi Pathak* and *Prinsa R. Nagar* helped in data curation, conceptualization, supervision, writing-original draft, writing-review and editing.

**Declaration of Competing Interest** The authors declare no competing financial interest.

## References

Abhirami BL, Krishna AA, Bashi MB, Abhijith BS, Varghese AW, Kumaran A (2024) Isolation of potent 5α-reductase inhibitors and antioxidants from Sphaeranthus indicus Linn. for the management of benign prostatic hyperplasia. Plant Foods Hum Nutr 79:932–938. https://doi.org/10.1007/s11130-024-01226-2

Ambavade SD, Mhetre NA, Tate VD, Bodhankar SL (2006) Pharmacological evaluation of the extracts of Sphaeranthus indicus flowers on anxiolytic activity in mice mice. Indian J Pharmacol 38(4):254–259

Ambikar DB, Mohanta GP (2013) Neuroprotective effect of petroleum ether, methanolic and aqueous extracts of flower heads of Sphaeranthus indicus on lipofuscinogenesis and fluorescence product in brain of D-galactose induced aging accelerated mice. Orient Pharm Exp Med 13:301–306. https://doi.org/10.1007/s13596-013-0110-x

Badgujar LB, Ghosh P, Gaur V, Bodhankar SL (2009) Effect of petroleum ether extract of Sphaeranthus indicus Linn. on complete freunds adjuvant induced arthritis in laboratory rats. Pharmacologyonline 2:281–291

Basumatary M, Bhimrao B (2024) Sphaeranthus indicus L.: a medicinal herb of rural and tribal India. In: Medicinal Asteraceae of India. APRF. https://doi.org/10.5281/zenodo.11465075

Chakrabarti D, Suthar A, Jayaraman G, Muthuvelan B (2012) NPS31807, a standardized extract from Sphaeranthus indicus, inhibits inflammatory, migratory and proliferative activity in keratinocytes and immune cells. Pharmacol Pharm 3:178–194

Farnoosh G, Khajeh K, Latifi AM, Aghamollaei H (2016) Engineering and introduction of de novo disulphide bridges in organophosphorus hydrolase enzyme for thermostability improvement.

J Biosci 41(4):577–588. https://link.springer.com/article/10.1007/s12038-016-9643-8 https://doi.org/10.1007/s12038-016-9643-8
Galani VJ, Patel BG (2009) Psychotropic activity of Sphaeranthus indicus Linn. in experimental animals. Pharmacog Res 1(5):307–313
Galani VJ, Patel BG (2010) Effect of hydroalcoholic extract of Sphaeranthus indicus against experimentally induced anxiety, depression and convulsions in rodents. Int J Ayurveda Res 1(2):87. https://doi.org/10.4103/0974-7788.64412
Galani VJ, Patel BG, Rana DG (2010) Sphaeranthus indicus Linn.: a phytopharmacological review. Int J Ayurveda Res 1(4):247–253. https://doi.org/10.4103/0974-7788.76790
Hassan F, Aslam B, Muhammad F, Faisal MN (2022) Hypoglycemic properties of Sphaeranthus indicus and nigella sativa in alloxan induced diabetes mellitus in rats; a new therapeutic horizon. Pak Vet J 42(2):141–146. https://doi.org/10.29261/pakvetj/2022.011
Jadhav VB, Vaghela JS (2024) Nephroprotective potential of Sphaeranthus indicus Linn extract against hyperglycemia and dyslipidemia in streptozotocin-induced diabetic nephropathy. J Health Allied Sci NU 14(2):210–218. https://doi.org/10.1055/s-0043-1769511
Jain AP, Choudhary GP (2019) Sphaeranthus indicus enhances memory and protects against amnesia in rodent models. Asian J Pharm Pharmacol 5:71–77
Kannojia P, Sharma PK, Bahlotia A, Singh VK, Jain G (2007) Antidiabetic effect of ethanolic extract of flower head of Sphaeranthus indicus in alloxan induced diabetic rats. Pharmacist 2(1):35
Kumar R, Nema RK (2017) Wound healing activity of flower head aquous extract of Sphaeranthus indicus Linn. in diabetic rats. World J Pharm Res 6(17):1144–1157. https://doi.org/10.20959/wjpr201717-10447
Mathew JE, Joseph A, Srinivasan K, Dinakaran SV, Mantri A, Movaliya V (2012) Effect of ethanol extract of Sphaeranthus indicus on cisplatin-induced nephrotoxicity in rats. Nat Prod Res 26(10):933–938. https://doi.org/10.1080/14786419.2010.534999
Mathews LA, Dhanyaraj D, Prathibhakumari PV, Prasad G (2012) Hepatoprotective and antioxidant potential of Spaeranthus indius (Linn.) on liver damage in wistar rats. Int J Pharm Pharm Sci 4(3):1–4
Meher BR, Jena J, Rath BG (2011a) Evaluation of analgesic activity of ethanolic extract of Sphaeranthus indicus. Pharm Lett 3(3):357–360
Meher BR, Rath BG, Biswal S (2011b) Evaluation of anti-inflammatory activity of ethanolic extract of Sphaeranthus indicus. J Chem Pharm Res 3(3):831–834
Nanda BK, Jena J, Rath B, Behera BR (2009) Analgesic and antipyretic activity of whole parts of Sphaeranthus indicus Linn. J Chem Pharm Res 1(1):207–212. www.jocpr.com
Padhan DK, Patel S, Mohanty A (2017) Topical analgesic activity of essential oil extracted from Sphaeranthus indicus (Asteraceae). Asian J Pharm Clin Res 10(5):275–277. https://doi.org/10.22159/ajpcr.2017.v10i5.17594
Pandey P, Singh D, Hasanain M, Ashraf R, Maheshwari M, Choyal K, Singh A, Datta D, Kumar B, Sarkar J (2019) 7-Hydroxyfrullanolide, isolated from Sphaeranthus indicus, inhibits colorectal cancer cell growth by P53-dependent and -independent mechanism. Carcinogenesis 40(6):791–804. https://doi.org/10.1093/carcin/bgy176
Patel MB (2012) Sphaeranthus indicus flower derived constituents exhibits synergistic effect against acetylcholinesterase and possess potential antiamnestic activity. J Complement Integr Med 9(1):23. https://doi.org/10.1515/1553-3840.1618
Ramachandran S (2013) Review on Sphaeranthus indicus Linn. (Koṭṭaikkarantai). Pharmacogn Rev 7(14):157–169. https://doi.org/10.4103/0973-7847.120517
Ramachandran S, Asokkumar K, Uma Maheswari M, Ravi TK, Sivashanmugam AT, Saravanan S, Rajasekaran A, Dharman J (2011) Investigation of antidiabetic, antihyperlipidemic, and in vivo antioxidant properties of Sphaeranthus indicus Linn. in type 1 diabetic rats: an identification of possible biomarkers. Evid Based Complement Alternat Med 2011:571721. https://doi.org/10.1155/2011/571721
Roy AK, Mondal S, Bera T (2013) ATP generation in Leishmania donovani amastigote form. Int J Drug Dev Res 5(2):354–361. Available online http://www.ijddr.in. Covered in Official Product of Elsevier, The Netherlands

Sarpate RV, Deore TK, Tupkari SV (2009) Bronchodilatory effect of Sphaeranthus indicus Linn against allergen induced bronchospasm in guinea pigs. Pharmacogn Mag 5(19 Suppl):74–77

Singh S, Semwal BC, Upadhaya PK (2019) Pharmacognostic study of Sphaeranthus indicus Linn.: a review. Pharm J 11(6):1376–1385

Sultana A, Fazmiya MJA, Rahman K (2021) Mundi (Sphaeranthus indicus Linn): the best blood purifier and immunomodulatory Unani herb with versatile ethnomedicinal uses and pharmacological activities. CellMed 11(2):1–10

Sundaresan PK, Prabhakaran SS, Palappallil DS, Chellappan D (2017) Diuretic activity of ethanolic extract of whole plant of Sphaeranthus indicus Linn in albino rats. Int J Basic Clin Pharmacol 6(2):265. https://doi.org/10.18203/2319-2003.ijbcp20164850

Tandon D, Gupta AK (2020) Comparative assessment of antimicrobial and antioxidant activity between whole plant and parts of Sphaeranthus indicus Linn. (Asteraceae). Clin Phytoscience 6(1):23. https://doi.org/10.1186/s40816-020-00172-1

Tennyson S, John Ravindran K, Arivoli S (2012) Screening of twenty five plant extracts for larvicidal activity against Culex quinquefasciatus say (Diptera: Culicidae). Asian Pac J Trop Biomed 2(2 Suppl):S1130–S1134. https://doi.org/10.1016/S2221-1691(12)60372-4

Tiwari S, Shukla PK, Srivastava A, Kumar M, Misra A, Srivastava S (2019) Pharmacognostic evaluation and pharmacological validation of Sphaeranthus indicus DC. Proc Natl Acad Sci, India, Sect B Biol Sci 89:1429–1438. https://doi.org/10.1007/s40011-018-01063-9

# Medicinal and Nutritional Importance of *Gloriosa superba* L. in Human Health

18

Kiran Soni, Yash Pal Sharma, Kapoor, Ankush Moran, and Pooja Sharma

**Abstract**

A medicinal plant can be defined as any plant species that contains active constituents in its parts, such as leaves, roots, bark, seeds or flowers, which can be utilized as the precursors in the synthesis of valuable drugs and for therapeutic purposes. Plants have developed the ability to produce a vast array of chemical compounds called secondary metabolites. The secondary metabolites present in medicinal plants are crucial for their importance in pharmaceuticals, drug manufacturing, cosmetics, dyes, flavours, nutrition, fragrances, dietary supplements etc. *Gloriosa superba* L. is a key medicinal plant due to the presence of colchicine, a valuable secondary metabolite with historical and contemporary significance in human health. Different chemical constituents are extracted from different parts of the plant, that is, from flowers—luterlin and its glucosides, from young leaf—cholidonic acid, N-formyl-de-Ac-colchicine and lumicolchine, from tubers and roots—colchicine, glucoside, b & g-lumiccolchicines, b-sitosterol and 2-OH-6-MeO benzoic acid, and from seeds—high level of colchicines. Colchicine has been used in different medicine systems (Ayurvedic, Unani) and is crucial for treating Familial Mediterranean Fever (FMF) and amyloidosis earlier. Colchicine inhibits neutrophil activation through β-tubulin binding, preventing microtubule assembly and mitotic spindle formation. The plant

K. Soni · Y. P. Sharma
Department of Forest Products, Dr. Yashwant Singh Parmar University of Horticulture and Forestry, Nauni, Solan, Himachal Pradesh, India

Kapoor (✉) · A. Moran
Department of Silviculture and Agroforestry, Dr. Yashwant Singh Parmar University of Horticulture and Forestry, Nauni, Solan, Himachal Pradesh, India

P. Sharma
Department of Tree Improvement and Genetic Resources, Dr. Yashwant Singh Parmar University of Horticulture and Forestry, Nauni, Solan, Himachal Pradesh, India



M. A. Ansari et al. (eds.), *Medicinal Plants and their Bioactive Compounds in Human Health: Volume 2*, https://doi.org/10.1007/978-981-95-3620-7_18

has its importance in the present scenario where its constituents play an important role in curing gout and rheumatism and have anti-cancer properties. Every part of this plant had medicinal and pharmacological importance. Its ability to address various ailments demonstrates the plants' profound impact on health and its ongoing relevance in modern medicine.

**Keywords**

Colchicine · Medicinal property · Secondary metabolites · Pharmaceutical · Modern medicine

## 18.1 Introduction

The term 'medicinal plant' includes those plants that have curing and healing properties. With the evolution of human civilization, plants have been playing a major role in healing, recovering and relieving various diseases. Medicinal plants belong to a big plant group with great interest due to their pharmaceutical, cosmetic and nutritional values (Rasool Hassan 2012). The secondary metabolites present in medicinal plants are important for pharmaceuticals, cosmetics, dyes, flavours, nutrition, fragrances, dietary supplements, etc. (Ncube and Van 2015; Guerriero et al. 2018). *Gloriosa superba* L. is a key medicinal plant due to the presence of a valuable secondary metabolite, that is, colchicine, which has been utilized in both the Ayurvedic medicine system and the Unani medicine system throughout history (Watt 1972). The importance of colchicine is understood from the fact that earlier, it was the only drug available for the treatment of FMF (Familial Mediterranean Fever) and amyloidosis (Cerquaglia et al. 2005; Rigante et al. 2006). At present, colchicine is utilized as an important drug for the treatment of rheumatoid arthritis, gout and cancer therapy (Bazylinska et al. 2016).

## 18.2 Historical and Cultural Significance

The family Colchicaceae had undergone several taxonomical changes, and based on the variations such as extrose anthers, tripartite style and septicidal capsules in the species, it was initially included in the division or suborder b of the Liliaceae family, including Asparagaceae, Liliaceae sensu stricto and Colchicaceae (Baker 1897, 1898). Further, on the basis of loculicidal capsules and corms or tubers, it was subdivided into tribes: Anguillarieae D. Don and Uvulariae A. Gray. Later on, the basis of tuberous rootstock resulted in the segregation of the new subfamily Wurmbeoideae and based on syncarpous and hemi-syncarpous nature, it was divided into two tribes (Buxbaum 1925, 1936, 1937). Lately, on the basis of the presence of colchicine as an alkaloid with tropolone ring as a unique feature, it was generated as a new separate family Colchicaceae (Dahlgren 1975; Huber 1969; Nordenstam 1982; Schulze 1975). Colchicaceae was further divided into various tribes by different researchers.

Later, Vinnersten and Manning (2007) introduced the genus concept, where *Gloriosa* and *Littonia* were the main genera of the family. However, after several morphological and phylogenetic studies, the monophyly of the genus *Gloriosa* was found in this species, including *Littonia* (Nordal and Bingham 1998; Maroyi 2012). In 1853, Hooker changed the name of the genus *Gloriosa* to *Methonica*, but currently, it was not applicable, and the genus is *Gloriosa*. Pharmacological importance includes anti-inflammatory, anti-fungal and antibacterial. Larvicidal, anthelmintic and various other ethnobotanical uses made *Gloriosa superba* a potential species of the family (Khan et al. 2007; Hemaiswarya et al. 2009; Malpani et al. 2011; Maroyi 2012).

## 18.3 Origin and Distribution

The natural habitats of *Gloriosa superba* are tropical and southern Africa, as well as tropical and temperate regions of Asia. It is also distributed throughout Myanmar, India, Burma, Bangladesh, Sri Lanka and Malaysia in Southeast Asia (Ade and Rai 2009; Padmapriya et al. 2015). Among the African countries, Senegal, Somalia, South Africa and Zambia are where the species is distributed. In Australia and some parts of USA, it has established its population widely and become naturalized and listed under the weeds category (Kavithamani et al. 2013).

It grows all over tropical India, from the Deccan peninsula and Assam to the North-West Himalayas (Kavithamani et al. 2013; Chaudhary et al. 2019). It was reported by Padmapriya et al. (2015) that in India this plant is distributed in tropical and subtropical parts including the Himalayan foothills, West Bengal, Tamil Nadu and Andhra Pradesh (Chatterjee and Ghosh 2015). Badola (2002), Alemu and Daba (2017) and Umavathi et al. (2020) reported its distribution in the Indian states of Andhra Pradesh, Karnataka, Kerala, Goa, Tamil Nadu, Bengal, Uttar Pradesh, Maharashtra, Chhattisgarh, Orissa and Himachal Pradesh. The monopoly of seed production was found in Tamil Nadu, and this state had the largest cultivation of *Gloriosa superba* spreading under seven districts, namely, Dindigul, Karur, Ariyalur, Perambalur, Tirupur, Salem and Nagapattinam. Beside Tamil Nadu its distribution was also found in other places like Palakkad, Trivandrum and Cannanore in Kerala; Savanthwadi, Nasik and Ratnagiri in Maharashtra and in Coorg, Uttara Kannada, Hassan, Mysore and Chikmangalur in Karnataka and Hazaribagh district of Jharkhand (Lal and Mishra 2011). Chauhan (2012) reported that in Himachal Pradesh, the plants were found up to 1250 m elevation from Jwala Mukhi, Bharwayin, Nauni to Kunihar division. It is the national flower of Zimbabwe and the state flower of Tamil Nadu in India (Umavathi et al. 2020).

## 18.4 Taxonomical Classification

Taxonomic classification of the species under study is detailed as below: (Chaudhary et al. 2019; Umavathi et al. 2020).

| | | |
|---|---|---|
| Kingdom | : | Plantae |
| Subkingdom | : | Tracheobionta |
| Division | : | Magnoliophyta |
| Class | : | Liliopsida |
| Order | : | Liliales |
| Family | : | Colchicaceae |
| Genus | : | *Gloriosa* |
| Species | : | *superba* |

The genus *Gloriosa* comprises of 12 species. Out of the 12 species, two, namely *G. revoilii* and *G. superba*, have been found in tropical Asia and Indochina (Maroyi 2012; Umavathi et al. 2020).

| | | | |
|---|---|---|---|
| *G. carsonii* | *G. katangensis* | *G. baudii* | *G. flavovirens* |
| *G. lindenii* | *G. littonioides* | *G. revoilli* | *G. modesta* |
| *G. simplex* | *G. rigidifolia* | *G. sessiliflora* | *G. superba* |

## 18.5 Morphological Attributes

*Gloriosa superba* L. is a medicinal plant cultivated in Tamil Nadu having high medicinal value. Along with its medicinal importance, this flower is beautiful and hence known by various names to describe its morphology thoroughly several researchers explored its morphology, as described below:

### 18.5.1 Habit

The plant is an herbaceous climber with glabrous branching and is trained to a height of 1.2 m (Farooqui and Sreeramu 2004; Lal and Mishra 2011; Lakshman and Inchal 2012; Maroyi 2012; Chaudhary et al. 2019).

### 18.5.2 Stem

The plant consists of a stem that is tall, slender and glabrous with a weak, tender and soft nature. The stem gained height from 1 up to 6.5 m. The stem of the plant is annual as it dies after the completion of the growing season and emerges again in the next growing season (Farooqui and Sreeramu 2004; Kurian and Sankar 2007; Kirtikar and Basu 2012; Vanangamudi et al. 2014; Daniels et al. 2020).

### 18.5.3 Leaves

Leaves are ovate, lanceolate, sessile and whorled in alternate and opposite fashion on the stem. The tip of the leaf is most important structure to curl around and tips are acuminated and twisted to form tendrils to grab support while climbing and the leaf base is cordate with parallel venation. Leaves are simple, sessile, lance shaped having strong blade with 7.5–15 cm leaf length and 2–4.5 cm leaf width (Kirtikar and Basu 2012; Maroyi 2012; Lakshman and Inchal 2012; Vanangamudi et al. 2014; Kumar 2019).

### 18.5.4 Flowers and Inflorescence

Flowers are actinomorphic, hermaphrodite with six free segments (yellow in the lower half and red in the upper half), superior oblong ovary (3-celled with numerous ovules), and style deflected at the base, projecting from the flower more or less horizontally. Tepals are apiculate, oblanceolate and reflexed with 7.7–15 cm long pedicels, 6 cm long style, stamens are 6.3 cm long and 8–13 mm wide, 3.8–4.5 cm long filaments, anthers are dorsifixed and 6–10 mm long. The inflorescence of the flower is lax-corymbose with six curved petals. Wavy margins of flowers change colour from green to yellow and then orange and scarlet from blossoming to till drooping (Fig. 18.1). Flowers bloom from September to October (Farooqui and Sreeramu 2004; Kurian and Sankar 2007; Lal and Mishra 2011; Kirtikar and Basu 2012; Lakshman and Inchal 2012; Maroyi 2012; Kumar 2019).

**Fig. 18.1** Life cycle of *Gloriosa superba* L.: from bloom to seed

### 18.5.5 Fruits and Seeds

Fruits are oblong, 6–12 cm long and 2–2.5 cm wide. They are equally partitioned in three lobes and change colour from green to black and open into three sections. Red to pale orange coloured, 30–150 seeds of oval shaped with spongy testa are found in loculicidal capsules. Seeds are attached to sutures (Farooqui and Sreeramu 2004; Kurian and Sankar 2007; Kirtikar and Basu 2012; Vanangamudi et al. 2014; Kumar 2019).

### 18.5.6 Tubers

Rhizomes are perennial in nature, initially white and become brown with age. They are cylindrical solid, arched, fleshy white tuberous rhizome of V-shaped limbs of two equal or unequal length with cirrhosed tips. They are bifurcated V-shaped tubes with 2.5–3.8 cm long arms, globose covered in brown tunics and yellowish internally (Farooqui and Sreeramu 2004; Kurian and Sankar 2007; Kirtikar and Basu 2012; Maroyi 2012; Kumar 2019).

## 18.6 Major Bioactive Compounds

The main chemical constituents of the plant were colchicine and gloriosine obtained from seeds and tubers. Other plant parts like leaves, flowers, roots, seeds and tubers had other constituents like isoperlolyrine, N-formylde acetylcolchicines, cornigerine, 3-demethyl colchicines, 3-demethyl-N-formyl-N-deacetyl-β-lumicolchicine, luteolin, colchicoside, gloriosine, tannins, superbine, N-Formyl-de-Me-colchicine, 3-demethyl-g-lumicolchicine, β-sitosterol, β- and γ-lumicolchicines and 3-demethylcolchicines (Srivastava and Chandra 1977; Prajapati et al. 2003; Kavina et al. 2011b; Kavithamani et al. 2013).

## 18.7 Importance of Pollination

Pollination is vital for reproduction, genetic diversity and seed production in plants. The flowers of *Gloriosa superba* L. had a peculiar structure with six tepals bent backwards, six radiating anthers and a style almost 90° at the point of attachment to the ovary, which affects its pollination. At every stage of flower development, change in perianth and stigma colour was observed. The perianth had a crimson-red colour at the top, green at the base and yellow in the middle when the stigma was receptive. However, the plant has no genetic self-incompatibility and responds well to both artificial pollination and natural pollination. The pollination method, time of pollination, stigma receptivity, pollen availability/viability and visits by pollinators affect the overall outcomes and thus studied by various researchers (Raina and Gupta 1999; Venudevan et al. 2011; Selvarasu and Kandhasamy 2012).

Raina and Gupta (1999), Venudevan et al. (2011) and Selvarasu and Kandhasamy (2012) studied the effect of stigma receptivity on pod set of *Gloriosa superba* L and maximum stigma receptivity was observed on the day of anthesis with a pod set of 80%. Raina and Gupta (1999) and Venudevan et al. (2011) studied various pollination techniques to enhance the seed yield of *Gloriosa superba*. Pollination on 1 day after anthesis leads to maximum pod set (88%). Gawali et al. (2017) studied various pollination methods, that is, self-pollination, natural open pollination and artificial cross-pollination. The maximum pod set was recorded with artificial cross-pollination (92%), followed by self-pollination (81%) and natural open pollination (70%).

## 18.8 Cultivation Practices

*Gloriosa superba* L. had medicinal significance due to the presence of phytochemicals in different parts of the plant, namely, colchicine, colchicoside and gloriosine. It is mainly used to prepare medicines for gout and rheumatoid arthritis, induce polyploidy and treat cancer. Due to its pharmacological importance, its demand increased and over-exploitation of *Gloriosa superba* from its natural habitat occurred (Gupta et al. 2013). So, to conserve and meet the demand of various industries, cultivation is done to enhance its growth, yield and phytochemical constituents. Due to the low germination ability of seeds in vitro propagation techniques for germination were performed (Table 18.1).

**Table 18.1** In vitro propagation techniques of *Gloriosa superba* L

| Sr. no. | In vitro techniques | Reference |
|---|---|---|
| 1 | Shoot cuttings and explants from the node, inter-node, leaves, flowers, pedicels and tubers | Custers and Bergervoet (1994) |
| 2 | Clonal propagation from apical bud and node segment of shoot tip | Samarajeewa et al. (1993) |
| 3 | In vitro propagation of explant of shoot and root | Somani et al. (1989) |
| 4 | Tissue culture | Finnie and Van Staden (1989, 1991) |
| 5 | Root culture | Ghosh et al. (2002) |
| 6 | Callus culture | Sivakumar and Krishnamurthy (2002) |
| 7 | Explants (corm bud, shoot, stem, leaf, pedicel) | Sivakumar and Krishnamurthy (2004) |
| 8 | Micro propagation of apical and axillary buds | Hassan Sayeed and Roy (2005) |
| 9 | In vitro tuber production | Ghosh et al. (2007) |
| 10 | Corm bud explant | Arumugam and Gopinath (2012) |
| 11 | In vitro tuberization | Kumar et al. (2015) |
| 12 | Shoot tips and tuber explants | Devi (2020) |

## 18.9 Medicinal Potential

*Gloriosa superba* L. is an important medicinal plant known for its pharmacological properties. It is widely used in indigenous health care systems, namely, Ayurvedic, Unani and Siddha, and it is used to cure skin diseases, respiratory ailments, cardiovascular issues, reproductive disorders, etc. (Balamurgan et al. 2019). Colchicine, derived from the seeds and tubers of *Gloriosa superba* L., has been used for millennia to treat conditions like gout and Familial Mediterranean Fever (FMF). It gained FDA approval in 2009 for various human disorders, primarily FMF and acute gout arthritis. Since 1972, it has been the preferred treatment to prevent FMF attacks and associated complications like amyloidosis. Colchicine inhibits neutrophil activation through β-tubulin binding, preventing microtubule assembly and mitotic spindle formation (Cerquaglia et al. 2005; Rigante et al. 2006; Ade and Rai 2009). The plant has its importance in the present scenario where its constituents play an important role in curing gout and rheumatism and have anti-cancer properties. Every part of this plant had medicinal and pharmacological importance and is described below:

### 18.9.1 Seeds

The seeds of *Gloriosa superba* L. are an economical part of the plants and a major source of colchicine and colchicoside (Jothi et al. 2019). The seeds were reported to treat chronic ulcers, acute gout, colic, infertility, cholera, scrofula, haemorrhoid, flatulence, debility, dyspepsia, helminthiasis, pruritus rheumatism and gout (Kurian and Sankar 2007; Kavithamani et al. 2013). Seeds are used as muscle relaxants and to treat rheumatic pains (Nadkarni 2002; Balamurgan et al. 2019).

### 18.9.2 Tubers

Kavina et al. (2011a) reported that the tubers were used for gonorrhoea, abortion, fever, piles, wounds, leprosy, anti-fertility and arthritis. Similarly, Ghodke and Pandhure (2020) reported its use to treat chronic ulcers, sprains, bruises, impotence, cancer and leprosy.

Jothi et al. (2019) reported that the plant is medicinally potent and the tubers were used to treat several ailments including chronic ulcers, leprosy, labour pain, cancer and haemorrhoids. It is also used to reduce pain and curing injuries of livestock. Traditionally, it showed its importance in curing parasitic skin disease by applying paste of tubers. The tubers were fed to goats and cows in small quantities as an abortifacient and used to treat other ailments like diarrhoea (Malpani et al. 2011). Kumar and Elangaimannan (2017) reported the importance of tubers in curing rheumatism, gout, fever, abdominal pain, arthritis, inflammations, leprosy, ulcers, thirst, infertility, skin disorders, piles, bruises, impotency, haemorrhoids, cancer and intestinal discomfort. Tribals use powdered rhizome and coconut oil to cure skin ailments and against scorpion and snake bites. Similarly, Balamurgan et al. (2019) reported the importance of tubers in treating fevers, leprosy, wounds, gonorrhoea, anti-fertility, piles and abortion purposes.

Tubers were used to cure blood disorders, inflammation, wounds, skin problems, poisoning and body toner. Traditionally, its use was found in the treatment of sprains, cancer, haemorrhoids, cancer, impotence, bruises, nocturnal seminal emissions and leprosy (Balamurgan et al. 2019; Ghodke and Pandhure 2020). Antifungal, mutagenic and antibacterial activities were seen with methanol, petroleum ether and aqueous extract of tubers (Kumarappan et al. 2011).

### 18.9.3 Leaves

Leaves are used for asthma, arthritis, skin and antidotes for snake bites (Kavina et al. 2011b; Kavithamani et al. 2013). All parts of this plant had a medicinal importance, and all leaves had more medicinal qualities. Leaves were used to treat asthma, pimples and other skin disorders. Leaf juice was potent against lice and uterine stimulants. It reduces arthritis pain when applied with leaf extract and sesamum oil (Kumar and Elangaimannan 2017; Ghodke and Pandhure 2020).

### 18.9.4 Roots

Roots are used in rheumatism, paralysis, labour pain, snake bite, abortifacient, purgative, stomachic, expectorant, tonic, anthelmintic, rejuvenating, dyspepsia, insect bite, thermogenic, helminthiasis, emetic and haemorrhoids (Kavina et al. 2011a; Kavithamani et al. 2013). Tuberous roots of *Gloriosa superba* were used to treat piles, parasitic affections and swollen joints and had anthelmintic, antileprotic, alternative and anti-inflammatory properties. To cure baldness tribals apply root paste on the head (Kumar and Elangaimannan 2017). Roots paste was used to cure paralysis, labour pain, rheumatism, snake bites, insect bites and expulsion of the placenta (Patil and Gavade 2012; Balamurgan et al. 2019; Kothiyal et al. 2022).

## 18.10 Pharmacological Potential

The inhibition of microtubule polymerization by colchicine restricts the binding to tubulin and affects mitosis. The inhibition of the spindle fibres interferes cell division and colchicine act as spindle poison (Ade and Rai 2009). This potential of the plant makes it a better chemotherapeutic agent that inhibits growth of cancer cells and tumours and boosts the healing process of wounds. The extract of leaves and tubers had anticoagulant, analgesic and anti-inflammatory properties (Kee et al. 2008). The crude extract of the plant showed antibacterial activity against gram-positive and gram-negative bacteria. *Gloriosa superba* showed antifungal sensitivity against *Trichophyton longifusus*, *Microsporum canis* and *Aspergillus niger* (Hemaiswarya et al. 2009). Beside that larvicidal, nematicidal and anthelmintic

**Table 18.2** Pharmacological potency of different parts of *Gloriosa superba* L

| Sr. no. | Plant part | Pharmacological potency | References |
|---|---|---|---|
| 1 | Tuber | Mutagenic property | Angunawela and Fernando (1971) |
| 2 | Leaves and leaf extract | Asthma, ulcers, piles | Jain (1991) |
| 3 | Tuber extract | Anti-inflammatory activity | Singh et al. (2007) |
| 4 | Tuber | Snake bite treatment | Samy et al. (2008) |
| 5 | Tuber paste | Rheumatoid arthritis | Veeraiah and Reddy (2012) |
| 6 | Leaf juice | Against lice and uterine stimulants | Kumar and Elangaimannan (2017) |
| 7 | Tubers | Fever, leprosy, wounds, gonorrhoea, anti-fertility, piles and abortion purposes | Balamurgan et al. (2019) |
| 8 | Seeds | Muscle relaxants | Nadkarni (2002), Balamurgan et al. (2019) |
| 9 | Tuber paste | Rheumatoid arthritis | Veeraiah and Reddy (2012) |
| 10 | Tuber | Chronic ulcers, acute gout, colic, infertility, cholera, flatulence, dyspepsia, helminthiasis and pruritus rheumatism | Kurian and Sankar (2007), Kavithamani et al. (2013) |

activities were found in the extract of *Gloriosa superba* (Zahir et al. 2009; Pawar et al. 2010). The potential pharmacological activities are detailed in Table 18.2.

## 18.11 Antioxidant Potential

Phytochemical evaluation of leaf extract had positive outcomes for the presence of carbohydrates, alkaloids, proteins, flavonoids, steroids, phenols and organic acids (Singh et al. 2012; Simon and Jayakumar 2016). Senthilkumar (2013) studied anti-oxidant potential of seeds, rhizomes, leaves and flowers and reported the presence of alkaloids (75.29–90.45%), glycosides (72.54–82.45%), steroids (50.11–56.76%), flavonoids (48.67–55.92%), starch (42.12–47.58%), terpenoids (42.10–47.58%), proteins (33.41–40.28%), tannins (30.66–36.25%), phenols (30.07–38.64%) and saponins (12.34–21.08%).

## 18.12 Toxic Effect

Toxic effects of *Gloriosa superba* L. were reported in Africa, India and Sri Lanka upon misuse of this plant. Plant consumption at higher doses was reported to be fatal for humans and livestock (Maroyi and Van der Maesen 2011). Tuber and root indigestion cause skin irritation and death, and its consumption is poisonous

(Kavithamani et al. 2013). Colchicine has a narrow therapeutic window, with 10 mg considered a toxic dose in humans, potentially leading to lethal effects. It has been documented not to reduce fertility rates or increase miscarriages and stillbirths in women; in fact, it may improve fertility outcomes (Cerquaglia et al. 2005). Side effects, particularly in older patients or those with liver or kidney issues, include vomiting, numbness, throat irritation, diarrhoea, alopecia and dermatitis. Severe toxicity can lead to multi-organ failure, including haemolytic anaemia, arrhythmias, bone marrow depression, liver and renal damage, neuromuscular disturbances, respiratory distress syndrome and death (Rigante et al. 2006; Malpani et al. 2011).

Despite its toxic effect, it has critical pharmacological potential and its different products are available in market such as goutnil, phloroglucinol (http://www.buy-pharma.md/Colchicine), colchicine tablets, USP (Drugs.com), colchicine biofarm (https://www.biofarm.ro/en/products/medications), butea superba (https://www.amazon.ca), thiocholchicoside mylan, and tiyozid (muscle relaxant drugs) (https://www.pharma-gdd.com).

## 18.13 Summary

The medicinal herb *Gloriosa superba* possesses potent therapeutic properties and can treat various ailments and diseases. Its medicinal benefits have been documented since ancient times. The seeds and tubers are significant economic components of the plant. The presence of essential polyphenols and chemical constituents underscores its medicinal importance for livelihoods. Its ability to alleviate conditions, such as rheumatoid arthritis, fever, leprosy, haemorrhoids and gonorrhoea, among other ailments, highlights this plant crucial role in promoting human health. This summary conveys the plant's significance in both historical and modern contexts, emphasizing its value in medical applications.

## References


Ade R, Rai M (2009) Review: current advances in *Gloriosa superba* L. Biodiversitas 10:210–214

Alemu ID, Daba TM (2017) *Gloriosa*, a source of colchicine: review article. Int J Biol Chem Sci 10:1888–1893

Angunawela RM, Fernando HA (1971) Acute ascending polyneuropathy and dermatitis following poisoning by tubers of Gloriosa superba. Ceylon Med J 16:233–235

Arumugam A, Gopinath K (2012) In vitro micropropagation using corm bud explants: an endangered medicinal plant of *Gloriosa superba* L. Asian J Biotechnol 4:120–128

Badola HK (2002) Endangered medicinal plant species in Himachal Pradesh. A report on the international workshop on Endangered Medicinal Plant Species in Himachal Pradesh. G.B. Plant Institute of Himalayan Environment and Development at Himachal Unit. Curr Sci 83:797–798

Baker JG (1897) Liliaceae. In: Thiselton Dyer WT (ed) Flora capensis, vol 6. Lovell Reeve & Co. Ltd, London, pp 253–567

Baker JG (1898) Liliaceae. In: Thiselton Dyer WT (ed) Flora of tropical Africa, vol 7. Lovell Reeve & Co. Ltd, London, pp 385–568

Balamurgan V, Amal TC, Karthika P, Selvakumar S, Vasanth K (2019) Somatic embryogenesis and plant regeneration in *Gloriosa superba* L.: an endangered medicinal plant. In: In vitro plant breeding towards novel agronomic traits, biotic and abiotic stress tolerance. Springer Nature, pp 27–41

Bazylinska U, Zielinski W, Kulbacka J, Samoc M, Wilk KA (2016) New diamidequat type surfactants in fabrication of long-sustained theranostic nanocapsules: colloidal stability, drug delivery and bioimaging. Colloids Surf B: Biointerfaces 137:121–132

Buxbaum F (1925) Verleichende anatomie der Melathiodeae. Feddes Report Beiheft 29:1–80

Buxbaum F (1936) Die Entwicklungslinien der Lilioideae. I. Die Wurmbaeoideae. Bot Arch 38:213–293

Buxbaum F (1937) Die Entwicklungslinien der Lilioideae. II. Die Wurmbaeoideae. Bot Arch 38:305–398

Cerquaglia C, Diaco M, Nucera G, La RM, Montalto M, Manna R (2005) Pharmacological and clinical basis of treatment of familial mediterranean fever (FMF) with colchicine or analogues: an update. Curr Drug Targets Inflamm Allergy 4:117–124

Chatterjee T, Ghosh B (2015) An efficient method of in vitro propagation of *Gloriosa superba* L. – an endangered medicinal plant. Plant Sci Res 37:18–23

Chaudhary S, El-Shorbagi AN, Shridhar B, Gupta MK, Verma HC (2019) A review on phytochemical and pharmacological profile of *Gloriosa superba* L. Int Res J Pharm 10:1–5

Chauhan NS (2012) Medicinal and aromatic plants of Himachal Pradesh. Indus Publishing Company, New Delhi, pp 212–215

Custers JBM, Bergervoet JHW (1994) Micropropagation of Gloriosa: towards a practical protocol. Sci Hortic 57:323–334

Dahlgren RMT (1975) A system of classification of the angiosperms to be used to demonstrate the distribution of characters. Bot Notiser 128:119–147

Daniels RJ, Johnson SD, Peter CI (2020) Flower orientation in *Gloriosa superba* (Colchicaceae) promotes cross-pollination via butterfly wings. Ann Bot 125:1137–1149

Devi B (2020) Assessment of genetic fidelity of in vitro raised plants of *Gloriosa superba* L. through molecular markers. Doctoral dissertation, College of Horticulture and Forestry, Dr YSP UHF, Neri, Hamirpur, pp 11–112

Farooqui AA, Sreeramu BS (2004) Cultivation of medicinal and aromatic plants. University Press, pp 131–138

Finnie JF, Van Staden J (1989) In vitro propagation of *Sandersonia* and *Gloriosa*. Plant Cell Tissue Organ Cult 19:151–158

Finnie JF, Van Staden J (1991) Isolation of colchicine from *Sudersania auranicacea* and *Gloriosa superba* variation in the alkaloid level of plant grown in vivo. J Plant Physiol 138:691–695

Gawali AS, Tripathi VD, Narkhede SS (2017) Study of sequential opening flowers and reproductive biology in *Gloriosa superba* L. a medicinal plant. Contemp Res India 4:170–174

Ghodke J, Pandhure N (2020) Qualitative and quantitative analysis of *Gloriosa Superba* L. Int J Bot Stud 5:392–395

Ghosh B, Mukherjee S, Jha TB, Jha S (2002) Enhanced colchicine production in root cultures of *Gloriosa superba* by direct and indirect precursors of the biosynthetic pathways. Biotechnol Lett 24:231–234

Ghosh S, Ghosh B, Jha S (2007) In vitro tuberisation of *Gloriosa superba* L. on basal medium. Sci Hortic 114:220–223

Guerriero G, Roberto B, Munoz-Sanchez J, Fabio A, Eslam AS, Ahmad Q, Abdulrahman A, Claudio C, Giampiero C, Francois HJ, Khawar S, Hernandez-Sotomayor S, Mohammad F (2018) Production of plant secondary metabolites: examples, tips and suggestions for biotechnologists. Genes 9:1–22

Gupta LM, Kumar S, Gupta M, Sharma V (2013) Integrated nutrient management for growth and yield in glory lily (*Gloriosa superba* L.). J Med Plant Res 7:3198–3202

Hassan Sayeed AKM, Roy SK (2005) Micropropagation of Gloriosa superba L. through high frequency shoot proliferation. Plant Tissue Cult 15:67–74

Hemaiswarya S, Raja R, Anbazhagan C, Thiagarajan V (2009) Antimicrobial and mutagenic properties of the root tubers of *Gloriosa superba* Linn. (Kalihari). Pak J Bot 41:293–299
Huber H (1969) Die Smenmerkermale und Verwandtschaftserhaltnisse der Liliiflorae. Mitt Bot Staatssamml Munchen 8:219–538
Jain SK (1991) Dictionary of Indian folk medicine and ethnobotany (a reference manual of man-plant relationship, ethnic groups and ethnobotany in India). Deep & Deep Publications, New Delhi, pp 34–45
Jothi U, Jebamalar JA, Sivakumar T (2019) Study on estimation and antioxidant activity of *Gloriosa superba* L. whole plant extract. Int J Sci Res Biol Sci 6:50–55
Kavina J, Gopi R, Panneerselvam R (2011a) Quantification of colchicine in seed and tuber samples of Gloriosa superba by high performance liquid chromatography method. J Appl Pharm Sci 1(7):116–119
Kavina J, Gopi R, Panneerselvam R (2011b) *Gloriosa superba* L. a medicinally important plant. Drug Invent Today 3:69–71
Kavithamani D, Umadevi M, Geetha S (2013) A review on *Gloriosa superba* L. as a medicinal plant. Indian J Res Pharm Biotechnol 1(4):554–557
Kee NLA, Mnonopi N, Davids H, Naude RJ, Frost CL (2008) Antithrombotic/anticoagulant and anticancer activities of selected medicinal plants from South Africa. Afr J Biotechnol 7(3):217–223
Khan H, Khan MA, Hussan I (2007) Enzyme inhibition activities of the extracts from rhizomes of *Gloriosa superba* Linn (Colchicaceae). J Enzyme Inhib Med Chem 6:722–725
Kirtikar KR, Basu BD (2012) Indian medicinal plants, vol 2. M/s Bishen Singh Mahendra Pal Singh Publisher, Dehradun, pp 2525–2528
Kothiyal S, Tiwari RC, Mittal B, Dikshit M, Sharma VB (2022) Toxicological and therapeutic applications of medicinally important plant langali (*Gloriosa superba* Linn.) – a review. World J Pharm Med Res 8:114–121
Kumar N (2019) Introduction to spices, plantation crops medicinal & aromatic plants. Scientific International, New Delhi, pp 253–256
Kumar A, Elangaimannan R (2017) Evaluation of *Gloriosa superba* L. for yield attributing characters and quantification of colchicine originated from different agro climatic zones of Tamil Nadu and Andhra Pradesh. Int J Pharm Clin Res 9:183–192
Kumar CN, Jadhav SK, Tiwari KL, Afaque Q (2015) In vitro tuberization and colchicine content analysis of Gloriosa superba L. Biotechnology 14:142
Kumarappan C, Jaswanth A, Kumarasunderi K (2011) Antihaemolytic and snake venom neutralizing effect of some Indian medicinal plants. Asian Pac J Trop Med 9:743–747
Kurian A, Sankar MA (2007) Medicinal plants. New India Publishing Agency, New Delhi, pp 105–108
Lakshman HC, Inchal RF (2012) Indigenous medicinal plants and their practical utility. New India Publishing Agency, New Delhi, pp 83–84
Lal HS, Mishra PK (2011) *Gloriosa superba* L. – an endangered plant spotted for the first time from forest of Tpchanchi, Hazaribag (Jharkhand) India. Sci Res Rep 1:61–64
Malpani AA, Aswar UM, Kushwaha SK, Zambare GN, Bodhankar SL (2011) Effect of the aqueous extract of *Gloriosa superba* L. (Langli) roots on reproductive system and cardiovascular parameters in female rats. Trop J Pharm Res 10:169–176
Maroyi A (2012) The genus *Gloriosa* (Colchicaceae)-ethnobotany, phylogeny and taxonomy. Wageningen Universirt, Wageningen, pp 13–26
Maroyi A, Van der Maesen LJG (2011) *Gloriosa superba* L. (family Colchicaceae): remedy or poison. J Med Plant Res 5:6112–6121
Nadkarni (2002) Indian materia medica. Poplar Prakashan Limited, Mumbai, pp 579–581
Ncube B, Van SJ (2015) Tilting plant metabolism for improved metabolite biosynthesis and enhanced human benefit. Molecules 20:12698–12731
Nordal I, Bingham MG (1998) Description of a new species, *Gloriosa sessiliflora* (*Colchicaceae*), with notes on the relationship between *Gloriosa* and *Littonia*. Kew Bull 53:479–482
Nordenstam B (1982) A monograp of the genus *Ornithoglossum* (Liliaceae). Opera Bot 64:1–51

Padmapriya S, Rajamani K, Sathiyamurthy VA (2015) Glory lily (*Gloriosa superba* L.) – a review. Int J Curr Pharm Rev Res 7:43–49
Patil PB, Gavade RT (2012) *Gloriosa superba* Linn – a medicinally important plant. Ann Pharm Pharm Sci 3:72–74
Pawar BM, Wavhal VP, Pawar ND, Agarwal MR, Shinde PB, Kamble HV (2010) Anthelmintic activity of *Gloriosa superba* Linn. Int J PharmTech Res 2:1483–1487
Prajapati ND, Purohit SS, Sharma AK, Kumar T (2003) A hand book of medicinal plants. Agros Bios Publishers, Jodhpur, pp 221–223
Raina R, Gupta LM (1999) Increasing seed yield in glory lily (*Gloriosa superba* L.) – experimental approaches. Acta Hortic 502:175–179
Rasool Hassan BA (2012) Medicinal plants (importance and uses). Pharm Anal Acta 3:2153–2435
Rigante D, La TI, Avallone L, Pugliese AL, Gaspari S, Stabile A (2006) The pharmacologic basis of treatment with colchicine in children with familial mediterranean fever. Eur Rev Med Pharmacol Sci 10:173–178
Samarajeewa PK, Dassanayake MD, Jayawardena SDG (1993) Clonal propagation of *Gloriosa superba*. Indian J Exp Biol 31:719–720
Samy RP, Thwin MM, Gopalakrishnakone P, Ignacimuthu S (2008) Ethnobotanical survey of folk plants for the treatment of snakebites in northern part of Tamil Nadu, India. J Ethanopharmacol 115:302–312
Schulze W (1975) Beitrage zur Taxonomie der Lilliforen. II. *Colchicaceae*. Wiss Zeitschr Friedrich Schiller-Univ Jena Math Nat R 24:417–428
Selvarasu A, Kandhasamy R (2012) Reproductive biology of *Gloriosa superba*. J Med Aromat Plants 3:5–11
Senthilkumar M (2013) Phytochemical screening of *Gloriosa superba* L. from different geographical positions. Int J Sci Res Publ 3:1–5
Simon SE, Jayakumar FA (2016) Antioxidant activity and anticancer study on phytochemicals extract from tubers of *Gloriosa superba* L. against human cancer cell (Hep-G2). J Pharmacogn Phytochem 4:1–12
Singh VK, Argal A, Aeri V, Jain G, Kannojia P (2007) Anti-inflammatory activity of *Gloriosa superba* Linn. Pharmacist 2:19–20
Singh D, Mishra M, Yadav AS (2012) Evaluation of phytocontents of extract of *Gloriosa superba* L. by thin layer chromatography and their antioxidant activity. Am J PharmTech Res 2:582–591
Sivakumar G, Krishnamurthy KV (2002) *Gloriosa superba* L. – a very useful medicinal plant. In: Singh VK, Govil JN, Hashmi S, Singh G (eds) Recent progress in medicinal plants. Ethnomedicine and pharmacognosy, vol 7. Sci Tech Pub, Houston, pp 112–132
Sivakumar S, Krishnamurthy KV (2004) In vitro organogenetic responses of *Gloriosa superba*. Russ J Plant Physiol 51:713–721
Somani VJ, John CK, Thengane RJ (1989) In vitro propagation and corm formation in Gloriosa superba. Indian J Exp Biol 27:578–579
Srivastava UC, Chandra V (1977) *Gloriosa superba* L. (Kalihari) – an important colchicine producing plant. J Res Educ Indian Med 10:92–95
Umavathi S, Gopinath K, Manjula MS, Balalakshmi C, Arumugam A (2020) *Gloriosa superba* L: a critical review of recent advances. Abasyn J Life Sci 3:48–65
Vanangamudi K, Anbukkarasi V, Prabhu M (2014) Medicinal seeds and plants. Scientific Publishers, pp 280–285
Veeraiah S, Reddy J (2012) Current strategic approaches in ethnomedicinal plants of Tinospora cordif o lia and Gloriosa superba – a review. Int J Pharma Biosci 2:1–6
Venudevan B, Sundareswaran S, Vijayakumar A (2011) Optimization of dormancy breaking treatments for germination improvement of glory lily (*Gloriosa superba* L.) seeds. Madras Agric J 97:31–32
Vinnersten A, Manning J (2007) A new classification of Colchicaceae. Taxon 56:171–178

Watt G (1972) *Gloriosa* Linn. In: A dictionary of economic products of India, vol 3. Shahdara, Delhi, pp 435–437

Zahir AA, Rahuman AA, Kamaraj C, Bagavan A, Elango G, Sangaran A, Kumar BS (2009) Laboratory determination of efficacy of indigenous plant extracts for parasites control. Parasitol Res 105:453–461

# Medicinal and Nutritional Importance of *Gmelina asiatica* in Human Health

19

Abha R. Vyas, Deep B. Sathwara, and Radhika N. Kachhadiya

**Abstract**

*Gmelina asiatica*, a plant, is revered for its many nutritional and therapeutic uses. Originating in Southeast Asia, *Gmelina asiatica* contains a wide range of bioactive components such as alkaloids, flavonoids, iridoids, phenolic acids, tannins, and terpenoids. The plant has been utilized in Ayurvedic and Unani medicine. Several cultures use *Gmelina asiatica* extensively to cure conditions including diabetes, rheumatism, and gastrointestinal issues. According to recent studies, *Gmelina asiatica* may have the potential to treat ailments like hypertension, diabetes, and chronic inflammation which highlight its potential as a functional food component as well as a medicinal agent. Clinical studies have demonstrated its anti-inflammatory, antioxidant, antibacterial, antimicrobial, and anticancer activities, and hepatoprotective effects providing a scientific basis for its traditional uses. In order to thoroughly describe its pharmacological potential that supports traditional usage and bridges the gap between traditional wisdom and modern science, this chapter addresses both in vivo and in vitro studies that support these assertions. This chapter aims to provide a comprehensive overview of the significant role of *Gmelina asiatica* in improving human health and its potential uses in modern medicine by combining the most recent scientific research with traditional knowledge. Furthermore, the nutritional value of *Gmelina asiatica* is explored in the chapter. Its rich content of vitamins, minerals, and dietary fibers emphasizes its potential as a functional food. The integration of traditional knowledge with modern scientific research offers a holistic understanding of *Gmelina asiatica's* contributions to human health, paving the way for future research and potential therapeutic applications.

A. R. Vyas · D. B. Sathwara · R. N. Kachhadiya (✉)
Department of Pharmaceutical Chemistry, LJ Institute of Pharmacy, LJ University, Ahmedabad, Gujarat, India



M. A. Ansari et al. (eds.), *Medicinal Plants and their Bioactive Compounds in Human Health: Volume 2*, https://doi.org/10.1007/978-981-95-3620-7_19

**Keywords**

Anti-inflammatory · Antioxidant · Anticancer · Diabetes · *Gmelina asiatica* · Hepatoprotective effects · Neuroprotective · Phytoconstituents

## 19.1 Introduction

Medicinal plants have long been an integral part of traditional medicine, particularly in developing countries, where they provide a rich source of secondary metabolites with significant therapeutic potential. These bioactive compounds are widely utilized in various fields, including human and veterinary medicine, agriculture, and scientific research. Plants have the remarkable ability to synthesize a diverse range of chemical substances, including proteins, amino acids, alkaloids, terpenes, flavonoids, glycosides, resins, saponins, volatile oils, gums, and tannins. These compounds are crucial in contributing to the medicinal properties of plants, making them valuable for the treatment and prevention of various diseases. As a result, there is an ongoing global search for biologically active secondary metabolites, particularly for use in dietary supplements and pharmaceuticals derived from plants.

Among the secondary metabolites, *flavonoids* and *tannins* stand out as two of the most common polyphenolic compounds in the human diet, found abundantly in plants. Polyphenolic compounds play a pivotal role in stabilizing lipid oxidation and are associated with antioxidant activity, which is essential for maintaining cellular health. Flavonoids from medicinal plants are recognized for their wide range of bioactive properties, including antioxidant, antiproliferative, antimicrobial, antiallergic, anti-inflammatory, antihepatotoxic, antiulcer, antiviral, and antispasmolytic effects. These properties make them highly beneficial in protecting the body against various diseases, including cancer. Similarly, tannins exhibit diverse biological activities such as antidysenteric, antidiarrheal, antimicrobial, antioxidant, anthelmintic, antiviral, cytotoxic, and antineoplastic effects, further highlighting their importance in traditional and modern medicine (Ksirri et al. 2023).

One such medicinal plant is *Gmelina asiatica L.* (synonym: *Gmelina parvifolia Roxb.*), a deciduous shrub that belongs to the family *Lamiaceae*, although it was previously classified under *Verbenaceae* (Zhao et al. 2021), which comprises about 35 species and 2 subspecies found across the tropical and temperate regions of Asia. Commonly referred to as the "Asiatic Bush Beech" or "Nilakumizh" in Tamil, *Gmelina asiatica* is a well-documented source of bioactive components with medicinal properties, including antimicrobial, anti-inflammatory, antioxidant, antihyperglycemic, hypoglycemic, hepatoprotective, antipyretic, nematocidal, and anticancer activities (Florence and Domettila 2016; Balijepalli et al. 2010; Binu et al. 2009). The regional names of the plant are given in Table 19.1 (Ismail et al. 1997).

The aerial parts and roots of this plant have been traditionally used to treat a variety of ailments such as jaundice, rheumatism, syphilis, gonorrhea, eye inflammation, fever, dysuria, wounds, dandruff, diabetes, hepatic diseases, and urinary tract infections. It is also known for its ability to purify the blood and reduce body heat.

**Table 19.1** Regional names of *Gmelina asiatica*

| Language and region | Pronounces of plant |
|---|---|
| Hindi | Badhara or Odhaara |
| Tamil | Nilakumil or Nilakumizh |
| Telugu | Kavvagummudu |
| Kannada | "Kirushivani" or "Kalshivani" |
| China | Ya Zhou Shi Zi |
| Indonesia | Betebet |
| Sri Lanka | Demata |
| Philippines | Kumal or Kalulut |

The significance of herbal medicine in modern civilization cannot be overstated. According to the World Health Organization (WHO), approximately 80% of the global population relies on herbal remedies as part of their primary healthcare. This underscores the importance of plants like *Gmelina asiatica*, which are rich in diverse phytoconstituents. These phytoconstituents, categorized as primary metabolites (such as lipids, proteins, and carbohydrates) and secondary metabolites (such as phenolics, flavonoids, alkaloids, resins, glycosides, tannins, and terpenoids), play a vital role in the plant's growth and survival, while also offering immense therapeutic value (Robinson and Zhang 2011).

The pharmacognostic study of medicinal plants is essential for ensuring the quality, purity, and proper identification of plant-based medicines. Due to the complex nature of their chemical constituents, quality control of crude plant drugs is a challenging but necessary task. Anatomical features, including microscopic examination of plant tissues, provide valuable diagnostic tools for identifying and assessing the quality of medicinal plants. In the case of *G. asiatica*, such studies are critical to verifying its identity and ensuring the consistency and efficacy of its medicinal uses (Florence and Domettila 2016; Florence and Regini Balasingh 2016).

### 19.1.1 Exomorphic Features

*Gmelina asiatica* is a deciduous shrub that typically grows to 1–3 m, though under optimal conditions, it can reach up to 10 m in height. The shrub is characterized by spines and minute yellowish-brown hairs on its young shoots (Mahendra et al. 2015). The petioles, which support the leaves, range from 0.5 to 4.5 cm in length. The leaves themselves are papery, ovate to obovate in shape, and typically measure 3–5 cm long and 2.2–3 cm wide, though they can grow up to 9 cm in length and 8.5 cm in width. The edges vary from entire to three to five lobed, with apex forms including acuminate, acute, rounded, or obtuse, and bases that are either cuneate or rounded. Leaves exhibit three to four pairs of veins, adorned with tiny hairs along the nerves. The adaxial surface is dark green, shiny, and devoid of hair, but the abaxial surface may be hairless or possess villi with glands. The abaxial (lower) surface of the leaves is dark brown, villous, and glandular, while the adaxial (upper) exterior is glabrescent. The leaf base is cuneate, the margin is entire or lobed, and the apex is acuminate. The veins are present in three to four pairs, giving a structured appearance to the leaf (Gilbert 1994; de Kok 2012). Terminal, pendulous, and downy inflorescences of *Gmelina asiatica* tend to have numerous blooms that droop. They can have readily detached bracteoles and leaf-like bracts, resembling a simple or compound raceme (panicle) (Kannan et al. 2012).

The main flowering stems are small and frequently have one to five flowers on them. The calyx (sepals) is usually 3–6 mm long and 2.5–4 mm wide. It has four to five pubescent lobes with a globular center and many disc-shaped glands on the outside. The yellow corolla is between 2 and 5 cm in length, exhibiting a two-lipped and four-lobed structure, with the anterior lip divided into three lobes (the center lobe being the longest) and the posterior lip consisting of a single lobe. Delicate appressed hairs adorn the external surface of the corolla and the filaments, while dense glandular hairs populate the inside. The stigma has two uneven clefts, and the ovary is globous and four-locular with a thin, slightly curved style. The fruit ripens hairless and is a smooth, ovoid to obovoid golden drupe (Adams et al. 2014; Ksirri et al. 2024).

The heartwood exhibits a light pinkish-red hue, while the roots are brown with superficial fissures (Krishnan and Gopi 2015), indicating the plant's unique morphological characteristics. *Gmelina asiatica* is unusual among plants due to its mix of floral, anatomical, and morphological traits. A transverse section of the root reveals cork tissue composed of 15–18 layers of regular radial files of tangentially elongated cells with suberized and lignified walls. The secondary cortex consists of four to five layers of parenchyma cells, interspersed with sclereids arranged in small groups. The lumen of the sclereids is often filled with tannin. Starch is present in the parenchyma cells, and only a few lipid droplets are found in some cortical cells. The secondary phloem elements, which represent more than two or three seasons of growth, are arranged in distinct bands, demarcated by phloem fibers. This structure forms 25–30 layers of narrow cells located outside the cambium. The functional phloem region consists of phloem fibers, sieve elements, and phloem parenchyma cells, with the phloem ray cells and parenchyma being filled with abundant starch grains. Occasionally, some cells may contain one or two lipid droplets. The vascular cambium of *G. asiatica* is three to four cells thick. The wood is diffuse-porous, with the pores being almost circular in shape and typically arranged in tangential rows alternating with continuous bands of fibers. The rays are two cells thick, and the axial parenchyma is paratracheal. Some of the vessels also contain tyloses. Starch grains are present in the axial parenchyma cells and in the rays, though not as abundantly as in the phloem.

This detailed anatomical description of *Gmelina asiatica* provides valuable diagnostic features for identifying the plant (Fig. 19.1) (Ksirri et al. 2024; Santhosha and Dinesh Mohan 2023), ensuring the quality and purity of its medicinal use, and offering insights into the structure of its bioactive components.

**Fig. 19.1** Morphological parts of the plant *Gmelina asiatica*

## 19.1.2 Phytochemical Features

Plants have an almost limitless ability to synthesize aromatic substances, most of which are phenols or their oxygen-substituted derivatives. Most are secondary metabolites, of which at least 12,000 have been isolated, a number estimated to be less than 10% of the total. In many cases, these substances serve as plant defense mechanisms against predation by microorganisms, insects, and herbivores. The analysis and extraction of plant material play an important role in the development, modernization, and quality control of herbal formulations. Hence, for the discovery of lead compounds for use as therapeutic drugs, the active principles in medicinal plants need to be identified.

The chloroform extract of different parts of *Gmelina asiatica* contained 22 constituents. The structures of a few phytoconstituents along with their morphological parts are given in Table 19.2.

Phytoconstituents of the plant were examined through GC-MS, while in the ethanolic extract, nine constituents were identified which are shown in Fig. 19.2.

GC-MS also identified the main constituent, mono-linoleoyl glycerol, trimethylsilyl ether (Fig. 19.3), a steroidal compound known for its antimicrobial, anti-inflammatory, antiarthritic, diuretic, and antiasthmatic properties. Based on the chemical nature and percentages of the compounds present in both the chloroform

**Table 19.2** Structures of phytoconstituents

| Name of phytoconstituent | Structure | Parts of the plant |
|---|---|---|
| Quercetagetin | 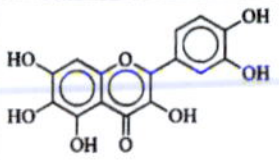 | Leaf, fruit, flower |
| Nicotiflorin | 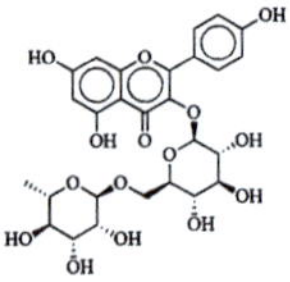 | Leaf, fruit, flower |
| Isorhoifolin | 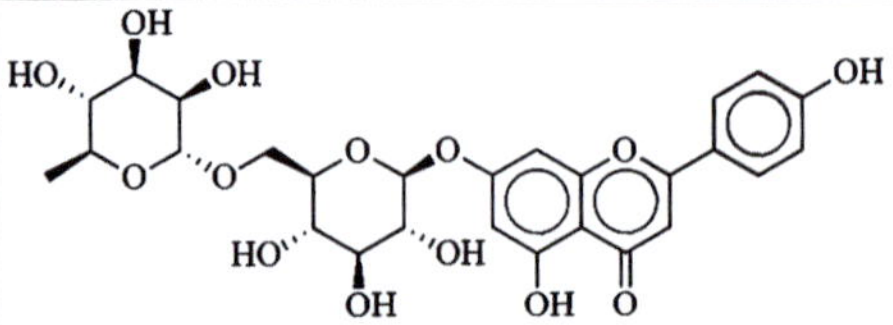 | Leaf, fruit, flower |
| Apigenin-7-glucuronide | 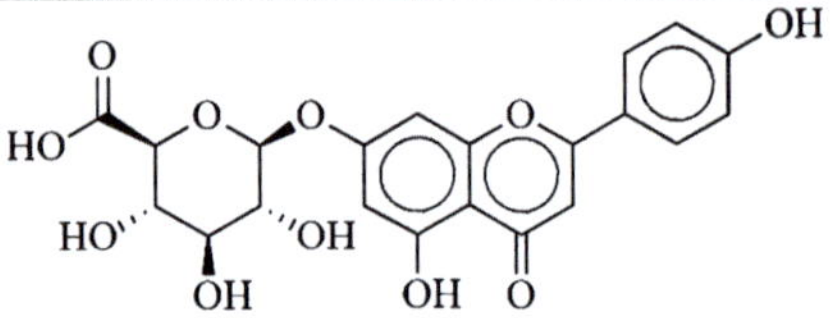 | Leaf, fruit, flower |
| Luteolin-7-glucuronide | 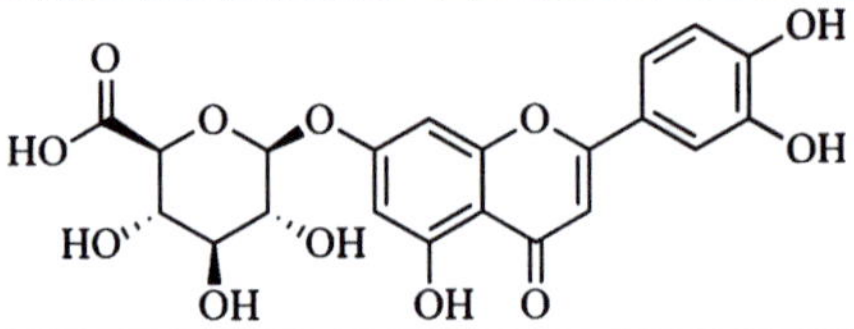 | Leaf, fruit, flower |
| Pinoresinol | 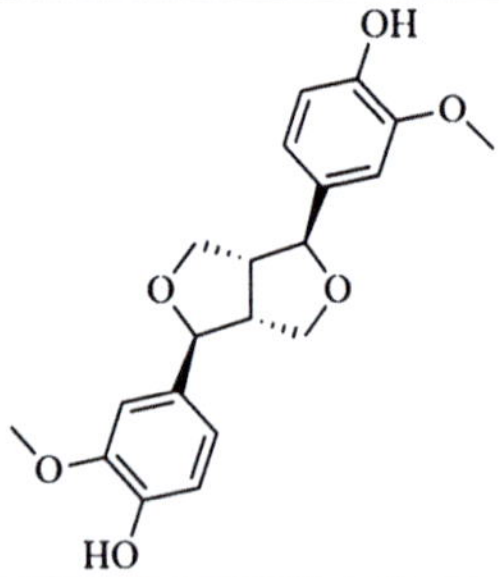 | Root |
| Sakuranetin | 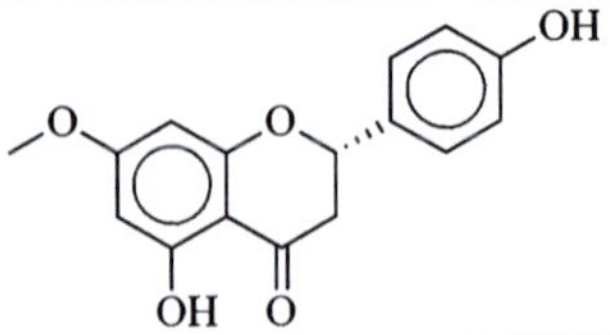 | Root |

(continued)

**Table 19.2** (continued)

| Name of phytoconstituent | Structure | Parts of the plant |
|---|---|---|
| 5-[3-(1,3-Benzodioxol-5-yl)-1,3,3a,4,6,6a-hexahydrofuro[3,4-c]furan-6-yl]-1,3-benzodioxole | 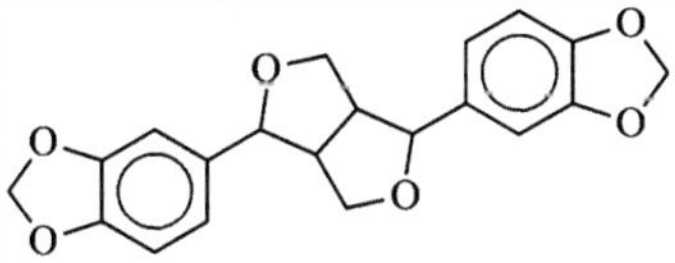 | Root |
| Gmelinol | 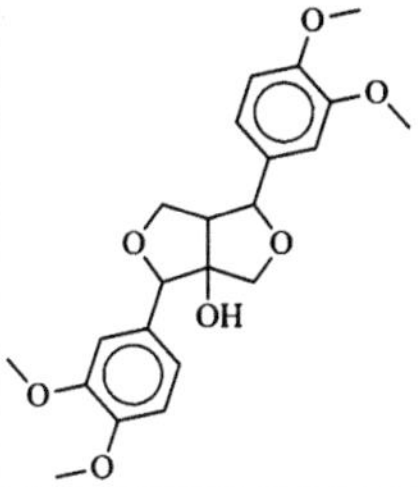 | Arial part |
| Beta-Sitosterol | 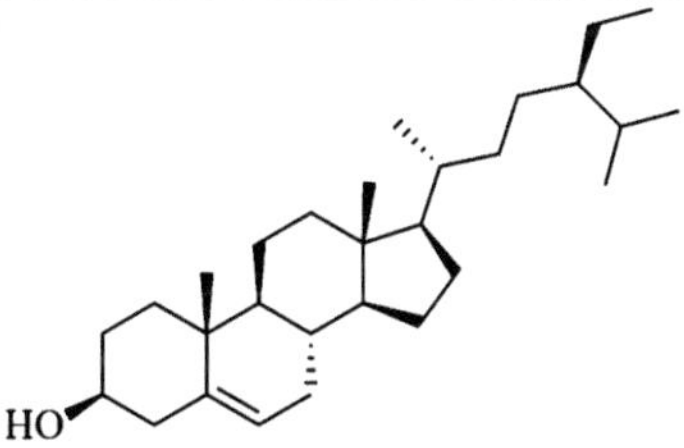 | Arial part |
| Cycloolivil | 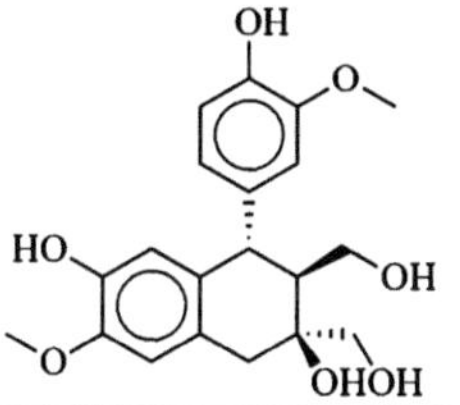 | Arial part |
| 4-Methoxycinnamic acid | 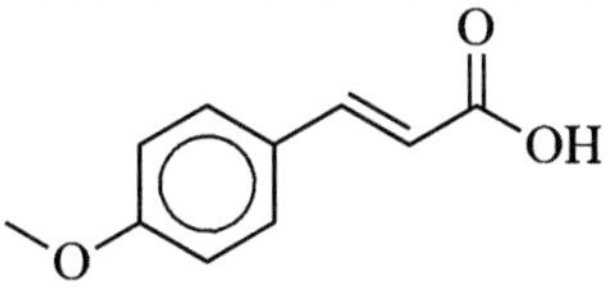 | Arial part |
| Spathulenol | 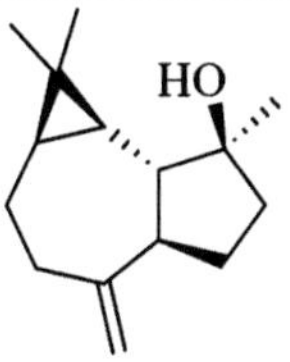 | Arial part |
| Heptacosane |  | Arial part |
| Palmitic acid | 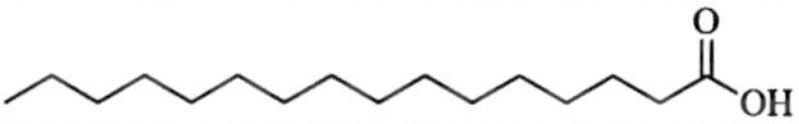 | Arial part |

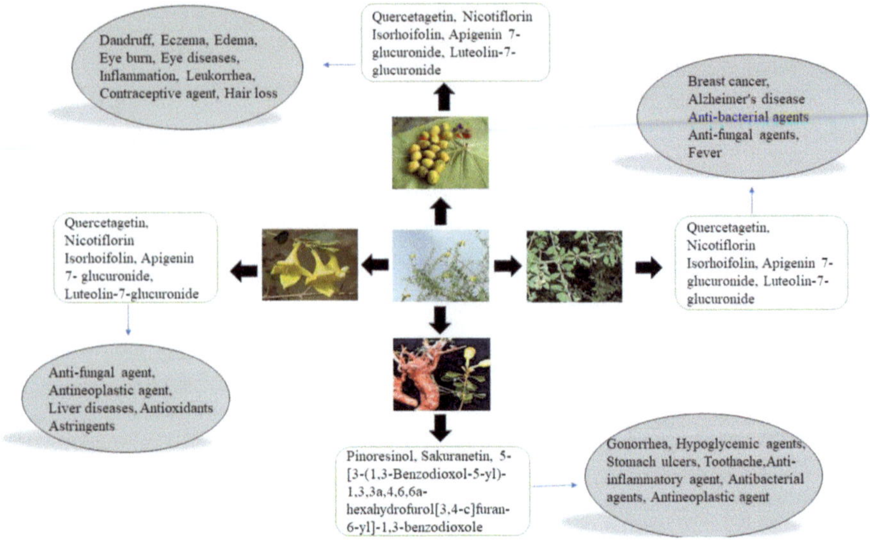


**Fig. 19.2** Phytochemical details of the plant

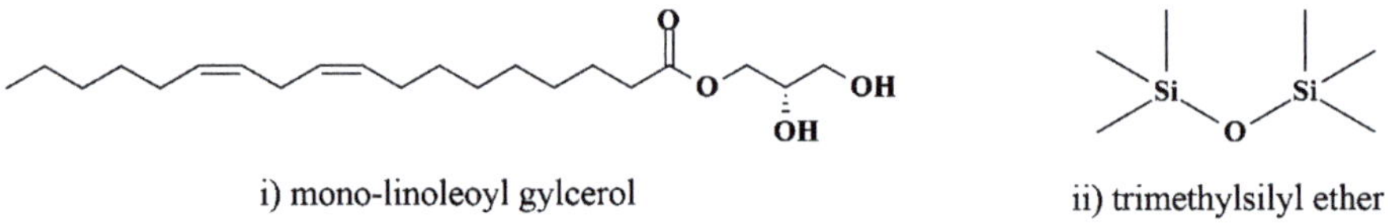


**Fig. 19.3** Structure of phytoconstituents present in *Gmelina asiatica* plant

and ethanolic extracts, these constituents may contribute to a range of pharmacological activities (Binu et al. 2009; Florence and Jeeva 2015).

## 19.2 Pharmacological Activities

The scientific evidence regarding *G. asiatica* demonstrates a range of bioactivities through both in vitro and in vivo studies. In vitro studies have shown antibacterial, antidandruff, antifungal, antioxidant, and nephroprotective activities, while in vivo studies provide evidence for antianxiety, anticancer, antidiabetic, anti-inflammatory, antipyretic, and hepatoprotective effects. Among these, the antibacterial properties have been the subject of numerous studies, with strong evidence from both in vitro and in vivo models. Researchers have used various parts of *G. asiatica*, including aerial parts, leaves, roots, and stems, with the leaf being the most frequently studied. Extracts used to identify constituents aqueous, chloroform, ethanol, ethyl acetate, methanol, and petroleum ether have been investigated, among them the most common ethanol extract (Netala 2013). Several bioactive phytoconstituents have been identified, including E-11-Hexadecanoic acid, hexadecanoic acid, linoleic acid,

(E)-9-Octadecanoic acid, heptadecanoic acid, 1,2-benzene dicarboxylic acid, benzene (1-butylhexadecyl), cholesterol trimethylsilyl ether, and nitidine. Scientific evidence now supports traditional medicinal uses of *G. asiatica* for conditions such as diabetes, edema, and rheumatism.

The most significant studies in this area provide the highest levels of scientific validation, identify the lowest effective concentrations or doses, and highlight the bioactive compounds responsible for the observed therapeutic effects.

The focus of this review is on the key biological and pharmacological activities of *Gmelina asiatica*, including its anticancer, neuroprotective, antioxidant, antidiabetic, and anti-inflammatory properties. Scientists have extensively studied these major bioactivities, providing a robust foundation of scientific evidence supporting *G. asiatica's* therapeutic potential in addressing critical health conditions.

### 19.2.1 Anticancer Activity

Cancer has long been a devastating disease affecting humanity, characterized by rapid, uncontrolled cell growth, local tissue invasion, and distant metastasis. Despite advancements in modern technology, cancer treatment remains a complex challenge. It continues to be a major public health burden worldwide, with incidence rates rising annually in both developed and developing nations. Over the past half-century, medicinal plants have been instrumental in cancer treatment, with secondary metabolites and their derivatives showing promise in therapeutic applications. Antitumor agents derived from these plants have demonstrated the ability to selectively target and destroy tumor cells while minimizing damage to surrounding healthy tissues.

Plant-derived anticancer drugs are widely used in clinical practice for cancer treatment. Vinblastine and vincristine, which come from *Catharanthus roseus* (Apocynaceae), *Taxus baccata L, Zingiber officinale, Solanum nigrum, Cassia fistula, Cinnamomum zeylanicum* to treat several cancers, such as testicular, breast, lung, and Kaposi's sarcoma. (Abebe (2016); Cragg and Newman (2005); Han et al. (2008); Ikram and Khattak (1987); Jain et al. (2016); Kooti et al. (2017); Merlin and Parthasarathy (2011); Routhu et al. (2005); Silvia and Satyanaraya (2014); van Der Heijden et al. (2004.

The anticancer potential of *Gmelina asiatica* leaf extracts was evaluated using the SiHa cell line, an HPV-16-positive human cervical cancer model linked to squamous cell carcinoma. Cytotoxic effects were assessed using MTT assays and fluorescence imaging. Extract samples (1 mg each) were dissolved in a 1:1 dimethyl sulfoxide (DMSO)-water solution and tested at concentrations of 10, 50, 100, and 200 mg/mL. The SiHa cell line, obtained from NCCS Pune, was cultured in DMEM with 10% heat-inactivated fetal bovine serum (HI-FBS) and antibiotics, maintained in 5% $CO_2$ at 37 °C. Cells at 70–80% confluence in a 96-well plate were exposed to extract concentrations for 24, 48, and 72 hours to assess dose- and time-dependent effects, with all experiments performed in triplicate (Nasir et al. 2023; Shiri Aghbash et al. 2023).

The impact of lignans and flavonoids derived from the ethyl acetate extract of *Gmelina asiatica* (Lemnaceae) roots (EGAR) was evaluated on both estrogen receptor-positive (MCF-7) and estrogen receptor-negative (MDA-MB-231) human breast cancer cell lines. The extract demonstrated a clear dose-dependent inhibition of cell proliferation, with half-maximal inhibitory concentration ($IC_{50}$) values recorded at 32.9 ± 3.8 μg/mL for MCF-7 cells and 19.9 ± 2.3 μg/mL for MDA-MB-231 cells. Notably, treatment with *G. asiatica* extracts induced apoptosis in MCF-7 cells, evidenced by observable cell morphological changes and chromatin condensation. Among the various extracts tested, the chloroform extract exhibited the highest cytotoxicity, outperforming other extracts of *G. asiatica* in inducing anticancer effects (Balijepalli et al. 2010). Prior to MTT addition, the media was replaced to wash cells, and MTT solution (100 μL, 0.5 mg/mL) was added to each well. Post 3-hour incubation in the dark, viable cells converted MTT to purple formazan crystals, which were dissolved with DMSO for OD measurement at 570 nm. $IC_{50}$ values were calculated at 48 hours. Data analysis showed a concentration-dependent reduction in cell viability over time, with significant anticancer activity ($p < 0.0001$) observed across all concentrations. GAME, a methanolic extract, demonstrated the highest cytotoxicity with an $IC_{50}$ of 74.9 mg/mL, highlighting its potential for further cancer research (Scott Christian College 2016).

Separately, cytotoxicity studies have revealed *G. asiatica's* potential against breast cancer cell lines MCF-7 and MDA-MB-231, aligning with global efforts by institutes like the National Cancer Institute, which have identified 324 plant compounds with anticancer activity from over 35,000 plant samples. Additionally, necrotic potential was explored for quercetin, quercetagetin, and patuletin in CaSki, MDA-MB-231, and SK-Lu-1 cell lines. Quercetagetin uniquely showed necrosis with a 15–30% LDH release, linked to its C6 hydroxyl group. This is the first report of quercetagetin's necrotic activity, suggesting its structural role in necrosis-inducing effects (Alvarado-Sansininea et al. 2018).

### 19.2.2 Anti-inflammatory Activity

Numerous phenolic compounds, including flavonoids, tannins, and curcumins, are thought to exert their effects by scavenging free radicals or suppressing pro-inflammatory enzymes such as cyclooxygenases (COX) and lipoxygenases (LOX) within inflammatory pathways (Jongrungraungchok et al. 2023). A biological defense mechanism, inflammation is triggered by the immune system in reaction to damage, infection, or toxic stimuli. Inflammation is chronic in diseases like arthritis, cancer, cardiovascular issues, and neurological conditions. The possible adverse effects of synthetic medications such as nonsteroidal anti-inflammatory drugs (NSAIDs) have drawn a lot of interest to the creation of innovative anti-inflammatory agents, especially from natural sources.

In vivo investigations of *G. asiatica's* anti-inflammatory effects have been performed by two different research teams utilizing the albino rat model (Ismail et al. 1997; Merlin et al. 2009). The findings from these research studies collectively suggest that *Gmelina asiatica* exhibits substantial anti-inflammatory properties, proving effective in both acute and chronic inflammation (Ksirri et al. 2023).

As a traditional medicine, the root of this plant is used to treat arthritis and inflammation. The study was made by Syed et al. in 1997 to assess the anti-inflammatory properties of *Gmelina asiatica's* root powder in male albino rats. The researchers employed two techniques to provoke inflammation in the rats: carrageenan-induced paw edema (acute inflammation) and cotton pellet granuloma (chronic inflammation). The experimental protocol indicated that *Gmelina asiatica* root powder markedly decreased edema during the different stages of acute inflammation. The treatment with *Gmelina asiatica* also diminished the weight of cotton pellet granuloma in chronic inflammation. The researchers observed that *Gmelina asiatica* diminished the lipid peroxide levels in granuloma exudate and liver, y-glutamyl transpeptidase in the granuloma, and restored the elevated blood acid and alkaline phosphatase levels to normal. The serum albumin level was likewise normalized with *Gmelina asiatica* treatment. The research indicates that the anti-inflammatory properties of *Gmelina asiatica* may be due to its strong antioxidative, antiproliferative, and membrane-stabilizing actions (Ismail et al. 1997). The findings indicate that *Gmelina asiatica* could be a beneficial supplement in managing inflammation and rheumatism.

Merlin and his research partners used both chronic and acute inflammation models in this work to assess the anti-inflammatory properties of an ethanolic extract of a plant's aerial parts in adult male and female albino rats. The extract was delivered in different doses orally, with indomethacin serving as the reference medication. Acute inflammation was assessed using carrageenan, histamine, and dextran-induced paw edema models. Also, chronic inflammation was evaluated using the cotton pellet-induced granuloma paradigm, in which both the plants extract and standard medicine indomethacin. In all models, the extract had a substantial inhibitory impact, diminishing paw edema in acute models and granuloma weight in the chronic model, thereby exhibiting dose-dependent anti-inflammatory activity. The extract's effectiveness was similar to that of indomethacin, indicating a modulation of inflammatory mediators such as histamine and prostaglandins. The results suggest that the ethanolic extract of the plant's aerial components possesses potential as a natural anti-inflammatory medicine. Additional research is advised to identify the bioactive chemicals accountable for the activity and to investigate their mechanisms of action (Merlin et al. 2009).

### 19.2.3 Antidiabetic Activity

Kasiviswanath et al. investigated the hypoglycemic effects of *G. asiatica* root in both alloxan-induced diabetic and normal rats. The experimental data demonstrated a dose-dependent decrease in blood glucose levels in rats administered the ethanolic extract, suggesting that the hypoglycemic effect of *G. asiatica* root extract intensified with greater dosages. The researchers discovered that the antidiabetic effectiveness of *G. asiatica* extract surpassed that of tolbutamide in diabetic rat models. Moreover, no indications of toxicity were detected in the rats at therapeutic levels, implying that the extract may be both efficacious and safe for diabetes management. The experimental work indicates the herb's potential as a natural remedy for diabetic management (Vivekanandarajah et al. 2022).

### 19.2.4 Neuroprotective Activity

In 2015, Kamboj evaluated *G. asiatica* for neuroprotective activity because it is an abundant source of flavonoids which is rewarded for its neuroprotective activity. The plant contains flavonoids like quercetagetin and kaempferol which are reported to have anxiolytic properties. Kamboj examined *G. asiatica* by using the Elevated Plus Maze (EPM) test in which subjected animals were administered a methanolic extract of the leaves (400 mg/kg). The results showed a significant increase in both the number of rodents entering the open arm and the duration spent in this arm, compared to the control group. This heightened activity in the open arm, as opposed to the closed arm, signifies the anxiolytic effects of *G. asiatica* leaves. Likewise, these results indicate a possible neuroprotective function of this plant (Ksirri et al. 2024; Kamboj 2015).

### 19.2.5 Anti-oxidant Activity

*Gmelina asiatica* was examined for its antioxidant properties using in vitro methods such as free radical scavenging test, reducing power assay, lipid peroxidation inhibition assay, and total antioxidant capacity method Reduced glutathione (GSH), superoxide dismutase (SOD), catalase (CAT), malondialdehyde (MDA), and hydroperoxide levels were measured in Wistar albino rats treated with $CCl_4$ in order to evaluate the antioxidant activity in vivo. The study shows a significant antioxidant effect which could be the reason for many other therapeutic activities related to this plant.

## 19.3 Conclusion

In conclusion, a wide range of pharmacological activities, including antioxidant, anti-inflammatory, antidiabetic, anticancer, and neuroprotective properties, support the prospective medicinal and nutritional value of *G. asiatica*. It emerges as a valuable medicinal plant with significant therapeutic and nutritional potential. The plant's diverse bioactive compounds like flavonoids, alkaloids, and terpenoids contribute to its demonstrated antioxidant, anti-inflammatory, anticancer, and neuroprotective effects which can contribute to preventive health and disease management. Research has validated its traditional uses for managing conditions like diabetes, hypertension, rheumatism, and chronic inflammation. The integration of in vitro and in vivo studies provides a comprehensive understanding of its pharmacological activities, underscoring the need for further exploration into its mechanisms and potential therapeutic applications. With its broad spectrum of bioactivities, *G.*

*asiatica* holds promise as a functional food and a natural remedy, paving the way for future advancements in plant-based health solutions.

This chapter concludes with a discussion of *Gmelina asiatica's* nutritional significance due to its abundant vitamin and mineral content in addition to the necessity of more research into its phytochemical characteristics and prospective therapeutic potential. These findings lay the groundwork for future research to fully leverage the potential of this remarkable plant in both preventive and therapeutic health strategies.

**Acknowledgement** The authors are thankful to L J Institute of Pharmacy, L J University, Ahmedabad, India, for providing the necessary facilities and support.

## References

Abebe W (2016) An overview of Ethiopian traditional medicinal plants used for cancer treatment. Eur J Med Plants 14(4):1–16. https://doi.org/10.9734/ejmp/2016/25670

Adams S, Shivakumar A, Krishnamurthy K, Nagarajan M, Venkatasubramanian P (2014) Comparative pharmacognostic and phytochemical study of medicinal plant species used as Gambhari in ayurveda. Inventi Rapid: Planta Activa:1–10

Alvarado-Sansininea JJ, Sánchez-Sánchez L, López-Muñoz H, Escobar ML, Flores-Guzmán F, Tavera-Hernández R, Jiménez-Estrada M (2018) Quercetagetin and patuletin: antiproliferative, necrotic and apoptotic activity in tumor cell lines. Molecules (Basel, Switzerland) 23(10). https://doi.org/10.3390/molecules23102579

Balijepalli MK, Tandra S, Pichika MR (2010) Antiproliferative activity and induction of apoptosis in estrogen receptor-positive and negative human breast carcinoma cell lines by Gmelina asiatica roots. Pharm Res 2(2):113–119. https://doi.org/10.4103/0974-8490.62949

Binu M, Parthasarathy V, Manavalan R, Kumaravel S (2009) Chemical investigation of aerial parts of Gmelina asiatica Linn by GC-MS. Pharm Res 1:152–156

Cragg GM, Newman DJ (2005) Plants as a source of anti-cancer agents. J Ethnopharmacol 100(1–2):72–79. https://doi.org/10.1016/j.jep.2005.05.011

de Kok R (2012) A revision of the genus Gmelina (Lamiaceae). Kew Bull 67:293–329. https://doi.org/10.1007/s12225-012-9382-4

Florence AR, Domettila C (2016) Anatomical studies on leaf and stem of Gmelina asiatica L.: an ethnomedicinal important plant. J Pharmacogn Phytochem 5(6):115–119

Florence AR, Jeeva S (2015) FTIR and GC-MS spectral analysis of Gmelina asiatica L. leaves. Sci Res Rep 5(2):125–136

Florence AR, Regini Balasingh GS (2016) Phytochemical analysis of Gmelina asiatica L. leaves. Int J Chem Stud 4(6):78–82

Gilbert M (1994) Flora of China. Verbenaceae 17:1–49

Han H-J, Tan N-H, Zeng G-Z, Fan J-T, Huang H-Q, Ji C-J, Jia R-R et al (2008) Natural inhibitors of DNA topoisomerase I with cytotoxicities. Chem Biodivers 5(7):1364. https://doi.org/10.1002/cbdv.200890124

Ikram M, Khattak SG, Gilani SN (1987) Antipyretic studies on some indigenous Pakistani medicinal plants: II. J Ethnopharmacol 19(2):185–192. https://doi.org/10.1016/0378-8741(87)90040-7

Ismail TS, Gopalakrishnan S, Begum VH (1997) Biochemical modes of action of Gmelina asiatica in inflammation. Indian J Pharmacol 29(5). https://journals.lww.com/iphr/fulltext/1997/29050/biochemical_modes_of_action_of_gmelina_asiatica_in.6.aspx

Jain S, Dwivedi J, Jain PK, Satpathy S, Patra A (2016) Medicinal plants for treatment of cancer: a brief review. Pharm J 8:87–102. https://doi.org/10.5530/pj.2016.2.1

Jongrungraungchok S, Madaka F, Wunnakup T, Sudsai T, Pongphaew C, Songsak T, Pradubyat N (2023) In vitro antioxidant, anti-inflammatory, and anticancer activities of mixture Thai medicinal plants. BMC Complement Altern Med 23. https://doi.org/10.1186/s12906-023-03862-8
Kamboj S (2015) Pharmacogenetic and anti-anxiety studies on leaves of Gmelina asiatica (Linn.). Med Aromat Plants 4(4):103
Kannan R, Prasant K, Babu U (2012) Botanical pharmacognosy of stem of Gmelina asiatica Linn. Anc Sci Life 31:190–193. https://doi.org/10.4103/0257-7941.107347
Kooti W, Servatyari K, Behzadifar M, Asadi-Samani M, Sadeghi F, Nouri B, Marzouni HZ (2017) Effective medicinal plant in cancer treatment, part 2: review study. J Evid Based Complementary Altern Med 22(4):982–995. https://doi.org/10.1177/2156587217696927
Krishnan V, Gopi MK (2015) Micromorphological characterisation of two simulating root drugs: Gmelina arborea Roxb. and Gmelina asiatica L. (Verbenanceae). Indian J Med Healthcare. https://doi.org/10.13140/RG.2.2.31922.22726
Ksirri R, Bhanukiran K, Maity S, Maiti P, Hemalatha S (2023) Evaluation of anticancer activity of Gmelina asiatica leaves, in-vitro and in-silico studies. J Biomol Struct Dyn:1–16. https://doi.org/10.1080/07391102.2023.2263894
Ksirri R, Khazem M, Bhanukiran K, Hemalatha S (2024) Gmelina asiatica: exploring traditional uses, pharmacological insights, and phytoconstituents—a comprehensive review (1961–2023). Phytochem Rev:1–24. https://doi.org/10.1007/s11101-024-09951-x
Mahendra C, Gowda D, Babu UV (2015) Insignificant activity of extracts of Gmelina asiatica and Ipomoea digitata against skin pathogens. J Pharm Negat Results 6. https://doi.org/10.4103/0976-9234.157383
Merlin NJ, Parthasarathy V (2011) Antioxidant and hepatoprotective activity of chloroform and ethanol extracts of Gmelina asiatica aerial parts. J Med Plants Res 5(4):533–538
Merlin NJ, Parthasarathy V, Manavalan R, Devi P, Meera R (2009) Phyto-physico chemical evaluation, anti-inflammatory and anti microbial activities of aerial parts of Gmelina asiatica. Asian J Res Chem:76–78. https://ajrconline.org/AbstractView.aspx?PID=2009-2-1-21
Nasir NM, Alsalim TA, El-Arabey AA, Abdalla M (2023) Anticancer, antioxidant activities and molecular docking study of thiazolidine-4-one and thiadiazol derivatives. J Biomol Struct Dyn 41(9):3976–3992. https://doi.org/10.1080/07391102.2022.2060306
Netala S (2013) Phytochemical and antioxidant studies on methanolic extract of Gmelina asiatica Linn stem. Int J Pharmacogn Phytochem Res 6:276–281
Robinson MM, Zhang X (2011) The world medicines situation 2011 traditional medicines: global situation, issues and challenges, 3rd edn. World Health Organization, pp 1–14
Routhu K, Alluri R, Kilari EK (2005) Hypoglycemic and antihyperglycemic effect of Gmelina asiatica Linn. in normal and in alloxan induced diabetic rats. Biol Pharm Bull 28:729–732. https://doi.org/10.1248/bpb.28.729
Santhosha D, Dinesh Mohan S (2023) Pharmacognosy, phytochemistry and pharmacological profile of Gynandropsis gynandra L.: a review. Ann Phytomed Int J 12(2):205–213. https://doi.org/10.54085/ap.2023.12.2.33
Scott Christian College (2016) Scott Christian College 7(5):2116–2121. https://doi.org/10.13040/IJPSR.0975-8232.7(5).2116-21
Shiri Aghbash P, Hemmat N, Baradaran B, Bannazadeh Baghi H (2023) SiRNA-E6 sensitizes HPV-16-related cervical cancer through Oxaliplatin: an in vitro study on anti-cancer combination therapy. Eur J Med Res 28(1):1–13. https://doi.org/10.1186/s40001-023-01014-9
Silvia N, Satyanaraya T (2014) Phytochemical and antioxidant studies on methanolic extract of Gmelina asiatica Linn stem. Int J Pharmacogn Phytochem Res 6:276–281
van Der Heijden R, Jacobs DI, Snoeijer W, Hallard D, Verpoorte R (2004) The Catharanthus alkaloids: pharmacognosy and biotechnology. Curr Med Chem 11(5):607–628. https://doi.org/10.2174/0929867043455846
Vivekanandarajah S, Vinujan S, Rajamanoharan P, Rajeev P, Sebastian PR (2022) Biological activities of extracts of Gmelina asiatica L. Doğu Fen Bilimleri Dergisi 5:1–3
Zhao F, Chen Y-P, Salmaki Y, Drew BT, Wilson TC, Scheen A-C, Celep F et al (2021) An updated tribal classification of lamiaceae based on plastome phylogenomics. BMC Biol 19(1):2. https://doi.org/10.1186/s12915-020-00931-z

# Medicinal and Nutritional Importance of *Nigella sativa* on Human Health

# 20

Ushakar Mishra, Benazir Fatima, and Abha Singh

**Abstract**

The potential health benefits of functional foods or nutraceuticals for human health are suggested by ample evidences, collected over the years. Black cumin (*Nigella sativa* L.), a highly valued nutraceutical herb with numerous health advantages, has sparked increased interest among fitness-oriented individuals, scientists, and the pharmaceutical industry. *Nigella sativa* is an annual herb with many therapeutic applications in addition to its commercial value as a plant that yields spices. It has long been a part of the folk medicine systems of Europe, the Far East Asia, and the Arabian Gulf region. Since ancient times, seeds and oils of *N. sativa* have been the main medicinal source for masses. It is one of the most widely used herbal remedies in the world and contains a variety of chemical components, including alpha-hederin, nigellicine, carvacrol, nigllimine, nigellidine, thymoquinone, dithymoquinone, and thymol. Its principal bioactive component, thymoquinone (TQ), has been shown to have pleiotropic pharmacological activities, including the ability to reduce oxidative stress and inflammation, as well as to boost immunity, cell survival, and energy metabolism, all of which contribute to a variety of health advantages such as protection against metabolic, cardiovascular, digestive, hepatic, renal, pulmonary, reproductive, and neurological illnesses, cancer, and so on. This study provides a brief explanation of the phytochemicals, pharmacological characteristics, Ayurvedic properties, and traditional uses of *Nigella sativa*.

**Keywords**

*Nigella sativa* · Functional foods · Therapeutic · Thymoquinone · Phytochemicals

U. Mishra (✉) · B. Fatima · A. Singh
Microbial Biodiversity Lab, Department of Botany, Faculty of Science, Patna University, Patna, Bihar, India



M. A. Ansari et al. (eds.), *Medicinal Plants and their Bioactive Compounds in Human Health: Volume 2*, https://doi.org/10.1007/978-981-95-3620-7_20

## 20.1 Introduction

Plant-based diets provide basic dietary requirements and support health by boosting immune to fight many kinds of disorders. The ideas of "nutraceuticals," or "functional foods," have become somewhat popular in the past few years since average life expectancy and a good diet are clearly correlated (Gupta and Prakash 2015). Also expressing interest in these concepts are doctors, dietitians, nutritionists, food scientists, representatives of food corporations, and pharmaceuticals. As the global functional food business expands, significant study is being done to pinpoint traditional foods with possible health benefits.

*Nigella sativa* L., also referred to as Black cumin or "kalonji," is an annual herb in the Ranunculaceae family (Table 20.1), characterized as a primitive dicot. It is a little prostrate plant, measuring 20–90 cm in height. These plants contain numerous isolated constituents, including proteins, peptides, volatile oils, and polyphenols, which exhibit a broad spectrum of pharmacological activities such as antibacterial, antiviral, anti-inflammatory, antioxidant, anticancer, antidiabetic, antinephrotoxic, antiparasitic, antimalarial, antinociceptive, antiulcer, and antihistaminic properties; they also possess capabilities for treating acne vulgaris, pigmentation, and wound healing, potentially aiding in the prevention of various chronic disorders. Owing to its extensive pharmacological properties, it has historically been incorporated into several traditional medical systems, including Indian Ayurveda, European practices, Far Eastern traditions, and those of the Arabian Gulf region. *N. sativa* enhances the sensory attributes, shelf-life, robustness, and freshness of several culinary products, including bread, pizza, biscuits, cookies, and cakes, establishing it as an exceptional spice (Tembhurne et al. 2014; Alu'datt et al. 2024).

## 20.2 Geographical Distribution

The plant is found throughout the world in many different places including Southern Europe, North Africa, Southeast Asia, the Middle East, Mediterranean region, India, Pakistan, Syria, Turkey, and Saudi Arabia (Tahan and Bayram 2011; Zohary et al. 2012). Like wheat, *N. sativa* is grown during the winter season. In India, it is specifically grown in Punjab, Himachal Pradesh, Bihar, Assam, the Gangetic plains, and Maharashtra (Paarakh 2010; Hajra 2011).

**Table 20.1** Taxonomic classification

| | | |
|---|---|---|
| Kingdom | : | Plantae |
| Phylum | : | Magnoliophyta |
| Class | : | Magnoliopsida |
| Order | : | Ranunculales |
| Family | : | Ranunculaceae |
| Genus | : | *Nigella* |
| Species | : | *Sativa* |

## 20.3 Botanical Description

*Nigella sativa* is a small prostrate annual herb that can reach lengths of 20–90 cm. It has bipinnately divided, alternating leaves, each up to 3 cm long. While the upper leaves are long, the lower leaves are petiolate and small. The terminal, pedunculated flowers are in whitish blue or purplish color. They have four distinct sepals: regular, deciduous, imbricate, and petaloid. There are five petals, each with small bifid limbs and long claws. There is an abundance of stamens. Carpels range in number from three to ten, are sessile, connate below, and have multiple horizontal ovules on each. Typically, there are two lengthy serrations on the suture and style. The fruit, which is a capsule, breaks open along the ventral suture of a free segment of each carpel. The primary plant components, dried fruit and seeds, are primarily used in medicine. The seeds are black in color and trigonous in shape and have a strong scent and a significant amount of oil in it (Varghese 1996; Thakur et al. 2021). The seed is roughly one-eighth of an inch long, with an oily, white kernel and a rough interior. The seeds contain two types of oil: fixed and essential. The main substance extracted from the black seed is thymoquinone (TQ), which is separated from fixed oil and essential oil (Mathur et al. 2011; Javed et al. 2012; Ahmad et al. 2013).

## 20.4 Nutritional Value

Previous researches have demonstrated the high nutritional potential of *N. sativa*, including 22% protein, 38–40% fat, and 32% carbohydrates (Rastogi 1990; Morikawa et al. 2004). Iron (105 rag), copper (18 rag), zinc (60 mg), phosphorus (527 rag), calcium (1860 rag), thiamin (15.4 rag), niacin (57 mg), pyridoxine (5.0 rag), and folic acid (160 pg) are the mineral and vitamin contents present in per kilogram of seeds (Enomoto et al. 2001; Nickavar et al. 2003). The presence of certain compounds including glucosinolates, anthraquinones, alkaloids, saponins, cardiac glycosides, triterpenes, tannins, flavonoids, volatile oils, and coumarins (Table 20.2) was also discovered in a qualitative investigation conducted on *N. sativa* (Bourgou et al. 2008).

## 20.5 Chemical Composition

Numerous active chemicals from several black seed varieties have been identified, isolated, and published thus far. Thymoquinone (30–48%), thymohydroquinone, dithymoquinone, p-cymene (7–15%), carvacrol (6–12%), 4-terpineol (2–7%), tanethol (1–4%), longifolene (1–8%), α-pinene, and thymol are the principal active chemicals (Table 20.3). Minor quantities of supplementary chemicals are also found in black seeds. Seeds include two distinct categories of alkaloids: those comprising nigellidine and nigellicine, pyrazol alkaloids or indazole ring alkaloids, and isoquinoline alkaloids, including nigellicimine and nigellicimine N-oxide. Additionally, the seeds of *N. sativa* possess saponin, a potential anticancer compound, and

**Table 20.2** Nutritional composition of *Nigella* seed (per 100 g)

| Sl. no. | Nutrients | Quantity |
|---|---|---|
| 1 | Moisture (g) | 4 |
| 2 | Protein (g) | 22 |
| 3 | Fat (g) | 41 |
| 4 | Carbohydrate (g) | 17 |
| 5 | Fiber (g) | 8 |
| 6 | Ash (g) | 4.5 |
| 7 | Potassium (g) | 0.5 |
| 8 | Calcium (g) | 1.0 |
| 9 | Phosphorous (g) | 0.5 |
| 10 | Iron (mg) | 10 |
| 11 | Thiamine (mg) | 1.5 |
| 12 | Niacin (mg) | 6 |
| 13 | Pyridoxine (mg) | 0.7 |
| 14 | Tocopherol (mg) | 34 |

Source: Huchchannanavar et al. (2019)

alpha-hederin, a water-soluble pentacyclic triterpene (Gilani et al. 2001; Benhaddou-Andaloussi et al. 2011).

Trace quantities of additional chemicals, including citronellol, limonene, and carvone, have also been identified. The predominant pharmacological properties of *N. sativa* are mostly attributed to its quinine constituents, with thymoquinone (TQ) being the most abundant. When kept, TQ yields elevated products for oligo condensation, such as dithymoquinone. *N. sativa* seeds contain protein (26.7%), fat (28.5%), carbs (24.9%), crude fiber (8.4%), and total ash (4.8%). Furthermore, the seeds are a valuable source of many vitamins and minerals, including copper, phosphorus, zinc, and iron. Carotene included in the seeds is metabolized by the liver into vitamin A. Additionally, multiple reports indicate the presence of vanillic acid in both the root and shoot (Gilani et al. 2001).

## 20.6 Phytochemicals

Research on the phytochemicals of this herb has been extremely limited. The seeds of the plant are reported to contain few chemical compounds like nigellone, nigellicine, nigellimine, nigellimine-N-oxide, avenasterol-5-ene, avenasterol-7-ene, campesterol, cholesterol, citrostadienol, cycloeucalenol, 24-ethyllophenol, obtusifoliol, sitosterol, stigmastanol, stigmasterol, stigmasterol-7-ene, beta-amyrin, butyrospermol, cycloartenol, gramisterol, lophenol, 243-methyllophenol, 24-methylene-cycloarthanol, taraxerol, tirucallol, 3-O-[β-D-xylopyranosyl (1-3)-αL-rhamnopyranosyl (1-4)-β-Dglucopyranosyl (1-6)β-D-glucopyranosyl] hederagenin, volatile oil (0.5–1.6%), fatty oil (35.6–41.6%), oleic acid, esters, esters of unsaturated fatty acids with C15 and higher terpenoids, esters of dihydroxystearic and linoleic acid, aliphatic alcohol, nigellidine, carvone, d-limonene, cymene,

**Table 20.3** Chemical composition of *N. sativa* seeds

| Sl. no. | Group | Components | References |
|---|---|---|---|
| 1 | Fixed oil (32–40%) | Arachidonic, eicosadienoic, linoleic, linolenic, oleic, palmitic, stearic, myristic acid, cholesterol, campesterol, stigmasterol, β-sitosterol, α-spinasterol, steryl esters, steryl glucosides | Lim (2012), Manoharan et al. (2021), and Rahim et al. (2022) |
| 2 | Volatile oil (0.40–0.45%) | Nigellone, TQ, thymohydroquinone, dithymoquinone, thymol, carvacrol, 2-(2-methoxypropyl)-5-methyl-1,4-benzenediol | Nickavar et al. (2003), El-Tahir and Bakeet (2006), Benkaci-Ali et al. (2010), and Manoharan et al. (2021) |
| 3 | Proteins (16.0–19.9%) | Alanine, valine, glycine, isoleucine, leucine, proline, threonine, serine, aspartic acid, methionine, phenylalanine, glutamic acid, tyrosine, lysine, arginine | Sabbah et al. (2020) and Manoharan et al. (2021) |
| 4 | Alkaloids | Nigellicine, nigellidine, nigellimine-N-oxide | Mehta et al. (2009), Adamska et al. (2019), Koshak and Koshak (2020), and Niu et al. (2020) |
| 5 | Coumarins | 6-Methoxy-coumarin, 7-hydroxy-coumarin, 7-oxy-coumarin | Manoharan et al. (2021) and Suri and Sareen (2022) |
| 6 | Saponins | Hederagenin | Kumara and Huat (2001), Mehta et al. (2009), Adamska et al. (2019), and Koshak and Koshak (2020) |
| 7 | Vitamins | DL-α-tocopherol, DL-β-tocopherol, all-trans-retinol, vitamin B1, vitamin B6, niacin, folic acid | Ansary et al. (2021) and Manoharan et al. (2021) |
| 8 | Minerals (1.8–3.7%) | Calcium, phosphorous, potassium, sodium, iron | Haron et al. (2014) and Hossain et al. (2024) |

α-β-unsaturated hydroxyl ketone, steroids, hederagenin glycoside, melanthin, melanthigenin, bitter principle, tannins, resin, protein, reducing sugar, glycosidal saponin, 3-O-β-D-glucopyranoside, cycloart-23-methyl-7,20,22-triene-3β, 25-diol, nigellidine-4-O-sulfite, nigellamines A3, A4, A5, C, and nigellidine A1, A2, B1, and B2 (Rastogi 1990; Atta-ur-Rahman et al. 1995; Sharma et al. 2005; Ali et al. 2008). The seed oil of the plant *N. sativa* contains cholesterol, campesterol, stigmasterol, β-sitosterol, α-spinasterol, (+)-citronellol, (+)-limonene, p-cymene, citronellyl acetate, carvone, nigellone, arachidic, linolenic, linoleic, myristic, oleic, palmitic, palmitoleic, and stearic acids. Fixed oil isolated from the *N. sativa* species are linoleic acid (55.6%), oleic acid (23.4%), and palmitic acid (12.5%) while volatile oils are trans-anethole (38.3%), p-cymene (14.8%), limonene (4.3%), and carvone (4.0%). It also contains 2-(2-methoxypropyl)-5-methyl-1,4-benzenediol, thymol, and carvacrol. Vanilliac acid is isolated from the root part of the plant (Rastogi 1990; Enomoto et al. 2001; Nickavar et al. 2003; Morikawa et al. 2004; Bourgou et al. 2008).

## 20.7 Ayurvedic Perspective

For many years, kalonji has been a traditional plant. Asthma, bronchitis, cough, dizziness, chest congestion, dysmenorrhea, obesity, diabetes, paralysis, hemiplegia, back pain, infection, inflammation, rheumatism, hypertension, and gastrointestinal issues like dyspepsia, flatulence, dysentery, and diarrhea are among the many diseases that *N. sativa* seeds are effective against in the Ayurvedic medical system (Forouzanfar et al. 2014). Furthermore, the use of this plant as a diuretic, emmenagogue, lactagogue, anthelmintic, and carminative has also been documented in various studies. Moreover, it is used for eczema, swollen joints, orchitis, nasal ulcers, and abscesses (Mousavi et al. 2018; Namazi et al. 2018). It is believed that the oil from kalonji seeds acts as a local anesthetic (Mousavi et al. 2018). The edible *N. sativa* seeds and seed oil are also used in traditional recipes.

## 20.8 Reported Therapeutic Studies

Numerous investigations have been carried out on this plant to determine its medicinal and pharmacological applications. To demonstrate the clinical effectiveness of this herb, particularly in the treatment of cancer, immunomodulatory disorders, and skin diseases, extensive clinical trials are still required. Some of the pharmacological properties possessed by the plant are mentioned below (Fig. 20.1):

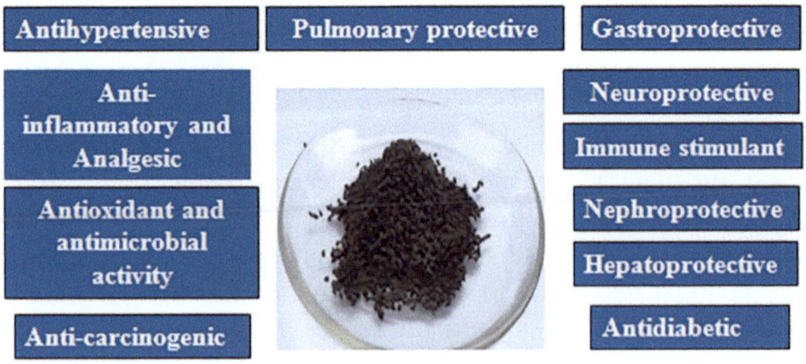


**Fig. 20.1** Pharmacological applications of *N. sativa*

**1. Antibacterial Activity** The antibacterial properties of *N. sativa* have been reported in numerous investigations. The primary chemical constituent isolated from this plant is thymoquinone. The findings indicate that most bacteria, especially Gram-positive cocci such as *Staphylococcus aureus* and *Staphylococcus epidermidis*, exhibited susceptibility to its antibacterial activity (Salem et al. 2010; Chaieb et al. 2011; Islam et al. 2013; Emeka et al. 2015).

**2. Antiviral Activity** A few recent studies have examined *N. sativa*'s antiviral properties, and significant findings regarding the protective properties of *N. sativa* oil against the hepatitis C virus (HCV) were found in a 2013 study. A clinical investigation was conducted on 30 HCV patients who were not patients for interferon (IFN)-α/ribavirin treatment. A dosage form containing 450 mg of *N. sativa* oil was administered orally, three times a day to these patients. A diabetic patient's clinical state, glycemic control, overall viral count, and oxidative stress were all improved after 3 months (Barakat et al. 2013).

**3. Antifungal Activities** It is evident from the studies that have been published that *N. sativa* has antiviral activity against the majority of pathogenic fungi (Gupta et al. 2012). Thymoquinone's antifungal activity against the majority of fungal strains was investigated (Rogozhin et al. 2011). According to a comparative study, the antifungal activity of Amphotericin B was found to be less effective than that of *N. sativa*. Using various pathogenic fungal strains, the methanolic and ethanolic extract of *N. sativa* seeds was investigated. These strains belonged to the species *Aspergillus*, *Candida*, *Cryptococcus*, and *Issatchenkia*. The decoction was made using the cold maceration method and the NCCLS (the National Committee of Clinical Laboratory Science) method was used to evaluate it. It was discovered that the plant's methanolic extract significantly inhibited fungal growth against *Candida parapsilosis* and *Issatchenkia orientalis*, while the ethanolic extract significantly inhibited the growth of *Issatchenkia orientalis* (Raval et al. 2010). In a different investigation, *Trichophyton mentagrophytes*, *Microsporum canis*, and *Microsporum gypseum* were used as dermatophyte fungal strains to test the essential oil and different *N. sativa* extracts. Using the disk diffusion method, the antifungal activity of *N. sativa* was investigated. The outcome demonstrated a strong antifungal effect against *M. canis*, *M. gypseum*, and *T. mentagrophytes* (Mahmoudvand et al. 2014). Growth of *Aspergillus fumigatus* and *Aspergillus flavus* was inhibited, according to the results of Khosravi et al. (2011). Various treatments for fungal infections using natural products were reported by some researchers. Improved antifungal effect of *N. sativa* has also been reported (Bita et al. 2012; Asdadi et al. 2014).

**4. Antiparasitic Activity** The antiparasitic activity of *N. sativa* methanolic extract was assessed through various studies. To assess *N. sativa*'s antiparasitic potential, the extract was tested on mice infected with *Plasmodium*. Excellent suppression was observed up to 94%, $p < 0.05$, in comparison to chloroquine. Owing to the *N. sativa* extract's antioxidant properties, antimalarial activity was discovered in the mice (Tariq 2008; Javed et al. 2012). According to a recent study, *N. sativa oil* reduces the total number of *Schistosoma mansoni* worms in the liver and the total number of ova found in the intestine and liver, both of which are excellent results (Sharma et al. 2005; Boskabady et al. 2007; Ahmad et al. 2018).

It was ascertained from the published researches that *N. sativa* oil had antioxidant and antischistosomal properties. Mice that were infected were given the extract. When schistosomiasis-infected mice were compared to infected, untreated mice, the results showed improvement in the hematological, biochemical, and antioxidant capacities (Tembhurne et al. 2014; Adam et al. 2016). An additional investigation was carried out to assess the efficacy of thymoquinone and *N. sativa* oil against *S. mansoni* in mice that were infected. Comparing the experimental group to the control group revealed a decrease in chromosomal abnormalities, particularly in chromosomes 2 and 6 and some in chromosomes 13 and 14 (Mahmoud et al. 2002). The antioxidative qualities of seeds inhibit the activities of some glucose metabolism-related enzymes in adult worms. These enzymes suffered damage after a few days and the parasites were later discovered to be weaker and eliminated (El Shenawy et al. 2008; Forouzanfar et al. 2014).

**5. Properties for Healing Wounds** Thymoquinone is the chemical component that shields tissues from oxidative damage and membrane lipid peroxidation. These results suggested using *Nigella sativa* topically for wound healing (Aboul-Ela 2002; Mohammed et al. 2005). The effect of the *N. sativa* plant on wound healing was investigated using wound model rats in a study. In comparison to silver sulfadiazine, the results indicated that it accelerates the healing process of wounds through an unclear mechanism (Yaman et al. 2010). According to another study, *N. sativa* seed hexanic extract exhibits good activity in promoting collagen formation and accelerating the process of epithelialization. In order to determine how well seeds oil healed second-degree burn wound models in rats, hexanic extract was applied topically in the study. For each group, there were three groups of six animal models. Silver sulfadiazine was used as a standard antiseptic medication. The outcome revealed a wound area reduction of 81.20%, which was greater than that of the control group. In the rat model, the study demonstrated total epidermis regeneration on day 12 (Sarkhail et al. 2011). In another study, the possibility that *N. sativa* seed ether extract has the ability to heal wounds was explored. The mildly dorsal skin of male BALB/c mice models was shaved and 100 μL of *Staphylococcal aureus* suspension was added, at a dose of 10 colony-forming units. Findings indicated that tissue damage, bacterial growth, and the total and absolute white blood cell counts were all reduced to improve the healing process (Abu-Al-Basal 2011). An in vitro study was carried out with a monolayer of human gingival fibroblast in which the DPPH scavenging activity of aqueous extracts was assessed and the impact on HGF proliferation was ascertained. Following the use of the oil, the result indicated an increase in the rate of proliferation and closure activity (Ab Rahman et al. 2014).

**6. Anti-inflammatory**

*Psoriasis*: The antipsoriatic potential of the ethanolic extract of *N. sativa* seeds was assessed. A mouse tail model for psoriasis was used to screen the extract. In comparison to the control group, the results indicated a rise in relative epidermal thickness (Dwarampudi et al. 2012). Ahmed Jawad et al. (2014) investigated *N. sativa*'s ethanolic extract to assess its antipsoriatic properties. There were two dosage forms for seed oil application: ointment and oral form. For this investigation, 60 patients with mild to moderate palmoplantar psoriasis and plaque were chosen. The findings demonstrated that an oral dosage of *N. sativa* has an antipsoriatic effect and helped patients to respond well.

*Acne Vulgaris*: One of the most common infections in humans is acne vulgaris. The effectiveness of *N. sativa* oil in treating acne vulgaris has been noted in numerous scientific investigations. The study conducted by Hadi and Ashor (2010) investigated the efficacy of lotion formulations that contained 20% *N. sativa* oil extract. For this investigation, 62 acne patients were chosen. The lotion used to treat mild- to middle-stage acne vulgaris, containing benzoyl peroxide, was found to be more harmful than *N. sativa*. This oil's antibiotic effect was compared to that of the common medication amoxicillin in a second study. Using carbopol 940, the oil was used to formulate the gel. Positive outcomes were achieved when this formulation was used as a therapy. Therefore, *N. sativa* is a useful herbal remedy for skin conditions like psoriasis, acne, dermatitis, rashes, and infections.

**7. Skin Pigmentation**

*Vitiligo*: A study was conducted on vitiligo disorder patients in which the researcher proposed *N. sativa* oil and fish oil as a treatment. The outcomes demonstrated a reduction in the size of the lesion and demonstrated efficacy in treating vitiligo (Ghorbanibirgani et al. 2014). *N. sativa* has the ability to disperse melanin throughout the skin, according to another study. The lyophilized seed extract was assessed in wall lizard melanophores that were separated in order to determine the cellular mechanism underlying skin darkening. The study demonstrated the mechanism of action that raised the sensitivity of cholinergic receptors within melanopsin, hence increasing the intensity of melanin. The lizard's outermost portion is called melanopsin (Ali and Meitei 2011). Thymoquinone has also been shown to be helpful in reducing vitiligo and skin pigmentation.

**8. Antioxidant Activity** *N. sativa* extracts have been used in numerous in vitro and in vivo studies to assess this activity. *N. sativa* and its derivatives were discovered to have antioxidative stress properties and the ability to scavenge radicals (Gholamnezhad et al. 2015; Abedi et al. 2017; Kadam and Lele 2017; Fahmy et al. 2018). Cherif et al. (2018) assessed oxidative activity and found that the consumption of *N. sativa* seeds affected the fatty acids in meat. For the study, 28 male Barbarine lambs were chosen. Two distinct feeding systems—higher or lower concentration—were used to administer the diet. In the end, it was found that the meat of lambs receiving NSS had fewer TBARS (thiobarbituric acid reactive substances) than the meat of other lambs.

**9. Anticancer Activity** A number of investigations have been carried out to assess *N. sativa*'s potential to fight cancer. The extract was evaluated using mouse-derived cancer cells. The findings indicated that thymoquinone had chemopreventive potential (El-Najjar et al. 2010; Aljabre et al. 2015; Thakur et al. 2021). Khalife et al. (2016) examined how thymoquinone stopped the S-phase of the cell cycle in human colon cancer cells and caused apoptosis in p53-independent p21 cells. Several cancer cell lines, such as MCF-7/Topo breast carcinoma cells, as well as decline regulators of NF-B and MMP-9 in Panc-1 cells and bcl-2 in gastric cancer cells were all shown to be positively affected by the active compound's anticancer properties. Many thymoquinone derivatives, including 6-menthoxybutyryl, 6-hencosahexanyl conjugate, 4-acylhydrazones, and 6-alkyl, have been shown by Ahmed Jawad et al. (2014) to be effective against a variety of cancer lines. According to the studies by various researchers, the *Nigella* oil nanoemulsion in MCF-7 cells caused cytoplasmic vacuolation, chromatin marginalization, nucleus fragmentation, and cell membrane blebbing (Periasamy et al. 2016; Al Asoom 2017; Czajkowska et al. 2017; Manjegowda et al. 2017).

**10. Antidiabetic Effect** In a study, it was found that *N. sativa* extract possesses strong antidiabetic properties. Mice with diabetes were injected with thymoquinone. The outcome in models demonstrated a decrease in blood glucose levels along with an increase in insulin and C-peptide (Sultan et al. 2014; Heshmati and Namazi 2015; Gray et al. 2016). An additional in vivo investigation was carried out on mice to assess the antidiabetic efficacy. The models received daily gavage administrations of the *Nigella sativa* aqueous extract. The extract was shown to have an anti-hyperglycemic, antihyperlipidemic, and antioxidant effect in addition to significantly lowering glycemia, TG, cholesterol, LDL-c, and TBARS (El Rabey et al. 2017).

**11. Impact on Nervous System** According to various studies, a rat model study showed that the methanolic extract of *N. sativa* possesses antidepressant and analgesic properties (Ahlatci et al. 2014; Bensiameur-Touati et al. 2017). Methanolic extract was shown to dramatically lower the level of hydroxyindole acetic acid and raise serotonin in the rat brain (Hobbenaghi et al. 2014).

**12. Nephroprotective Activity** A study found that ascorbic acid and *N. sativa* together have nephroprotective effects. The rabbits were given injections of the decoction. Decoction was found to have a nephroprotective effect through a reduction in blood urea nitrogen, serum creatinine, and antioxidant activity (Asif and Malik 2017; Thakur et al. 2021).

**13. Effects on Respiratory System** A study involving 63 adult male rats demonstrated the activity of nigellone and thymoquinone through in vivo investigation.

Nine groups were established based on the models. *N. sativa* has demonstrated significant efficacy in preventing epithelial damage and apoptosis induced by formaldehyde (Sapmaz et al. 2017).

**14. Impact on Reproductive System** Iranian infertile men who met the criteria of an abnormal sperm count were the subjects of a study. For 2 months, the patients were prescribed *N. sativa* oil. After 2 months, it was determined that giving infertile men a daily dosage of 5 mL of *N. sativa* oil improves abnormal semen quality without having any negative side effects (Kolahdooz et al. 2014; Mahdavi et al. 2015). An in vivo investigation using a polycystic ovary rat model proved the efficacy of *N. sativa*. Thymoquinone was discovered to be a possible treatment for PCO rats' hormone imbalance (Arif et al. 2016). Thirty male Wistar rats were used in another study to examine the adverse effects on reproduction. Over the course of 45 days, the models received gavage and an acetamiprid and *N. sativa* oil dosage. The outcome demonstrated stimulation of spermatogenesis and a significant increase in the weight of seminal vesicles and the number of spermatids (Mosbah et al. 2018).

## 20.9 Conclusion

Since ancient times, medicinal plants have been essential to preserving community health. One of the most widely used medicinal herbs in the folk, Ayurvedic, Siddha, and Yunani systems of medicine is *N. sativa*, also known as kalonji. This herb has a number of important medicinal properties that are applied to the management of various illnesses. According to the published researches, thymoquinone is regarded as an active component with a wide range of biological characteristics, including the ability to heal wounds and having anti-inflammatory, antidiabetic, antidepressant, nephroprotective, antibacterial, antiviral, and antifungal effects. Numerous scientific investigations indicate that *N. sativa* has a broad range of advantages for skin conditions.

*N. sativa* has always played a crucial part in pharmacology and food industries because their active ingredients aid in the prevention and treatment of numerous illnesses in addition to enhancing the nutritional value and shelf-life of food items. Many facets of black cumin, including their possible uses and functions, have been examined in this study. The food industries extensively utilize the antibacterial and antioxidant properties of black cumin and its essential oil. They consistently improve food products' sensory, nutritional, and storage attributes. With multiple functions like free radical scavenging, antibacterial impact, antidiabetic potency, anti-inflammatory, anticarcinogenic, and analgesic activity, among others, the phytochemicals in black cumin are highly effective when utilized in food matrix. They get all of these potencies from the bioactive ingredients, which include thymoquinone, p-cymene, terpenole, and thymol. Future drug treatments for a wide range of diseases may benefit from safer and more effective combinations resulting from chemical modifications in the molecular structures of α-Hederin, thymoquinone, and other components of *N. sativa* seeds. Therefore, it can be concluded that *N. sativa* is a useful medicinal plant that can be used to create pharmaceutical and nutraceutical products that may help to treat a wide range of illnesses.

## References

Ab Rahman MR, Abdul Razak F, Mohd Bakri M (2014) Evaluation of wound closure activity of *Nigella sativa*, *Melastoma malabathricum*, *Pluchea indica* and *Piper sarmentosum* extracts on scratched monolayer of human gingival fibroblasts. Evid Based Complement Alternat Med 2014(1):190342

Abedi AS, Rismanchi M, Shahdoostkhany M, Mohammadi A, Mortazavian AM (2017) Microwave-assisted extraction of *Nigella sativa* L. essential oil and evaluation of its antioxidant activity. J Food Sci Technol 54:3779–3790

Aboul-Ela EI (2002) Cytogenetic studies on *Nigella sativa* seeds extract and thymoquinone on mouse cells infected with schistosomiasis using karyotyping. Mutat Res/Genet Toxicol Environ Mutagen 516(1–2):11–17

Abu-Al-Basal MA (2011) Influence of *Nigella sativa* fixed oil on some blood parameters and histopathology of skin in staphylococcal-infected BALB/c mice. Pak J Biol Sci: PJBS 14(23):1038–1046

Adam GO, Rahman MM, Lee SJ, Kim GB, Kang HS, Kim JS, Kim SJ (2016) Hepatoprotective effects of *Nigella sativa* seed extract against acetaminophen-induced oxidative stress. Asian Pac J Trop Med 9(3):221–227

Adamska A, Stefanowicz-Hajduk J, Ochocka JR (2019) Alpha-hederin, the active saponin of *Nigella sativa*, as an anticancer agent inducing apoptosis in the SKOV-3 cell line. Molecules 24(16):2958

Ahlatci A, Kuzhan A, Taysi S, Demirtas OC, Alkis HE, Tarakcioglu M, Demirci A, Caglayan D, Saricicek E, Cinar K (2014) Radiation-modifying abilities of *Nigella sativa* and thymoquinone on radiation-induced nitrosative stress in the brain tissue. Phytomedicine 21(5):740–744

Ahmad A, Husain A, Mujeeb M, Khan SA, Najmi AK, Siddique NA, Damanhouri ZA, Anwar F (2013) A review on therapeutic potential of *Nigella sativa*: a miracle herb. Asian Pac J Trop Biomed 3(5):337–352

Ahmad A, Al-Abbasi FA, Sadath S, Ali SS, Abuzinadah MF, Alhadrami HA, Alghamdi AA, Aseeri AH, Khan SA, Husain A (2018) Ameliorative effect of camel's milk and *Nigella Sativa* oil against thioacetamide-induced hepatorenal damage in rats. Pharmacogn Mag 14(53):27

Ahmed Jawad H, Ibraheem Azhar Y, Al-Hamdi KI (2014) Evaluation of efficacy, safety and antioxidant effect of *Nigella sativa* in patients with psoriasis: a randomized clinical trial. J Clin Exp Investig 5(2):186–193

Al Asoom LI (2017) Coronary angiogenic effect of long-term administration of *Nigella sativa*. BMC Complement Altern Med 17:1–7

Ali SA, Meitei KV (2011) *Nigella sativa* seed extract and its bioactive compound thymoquinone: the new melanogens causing hyperpigmentation in the wall lizard melanophores. J Pharm Pharmacol 63(5):741–746

Ali Z, Ferreira D, Carvalho P, Avery MA, Khan IA (2008) Nigellidine-4-O-sulfite, the first sulfated indazole-type alkaloid from the seeds of *Nigella sativa*. J Nat Prod 71(6):1111–1112

Aljabre SH, Alakloby OM, Randhawa MA (2015) Dermatological effects of *Nigella sativa*. J Dermatol Dermatol Surg 19(2):92–98

Alu'datt MH, Rababah T, Al-u'datt DA, Gammoh S, Alkandari S, Allafi A, Alrosan M, Kubow S, Al-Rashdan HK (2024) Designing novel industrial and functional foods using the bioactive compounds from *Nigella sativa* L. (black cumin): biochemical and biological prospects toward health implications. J Food Sci 89(4):1865–1893

Ansary J, Giampieri F, Forbes-Hernandez TY, Regolo L, Quinzi D, Gracia Villar S, Garcia Villena E, Tutusaus Pifarre K, Alvarez-Suarez JM, Battino M, Cianciosi D (2021) Nutritional value and preventive role of *Nigella sativa* L. and its main component thymoquinone in cancer: an evidenced-based review of preclinical and clinical studies. Molecules 26(8):2108

Arif M, Thakur SC, Datta K (2016) Implication of thymoquinone as a remedy for polycystic ovary in rat. Pharm Biol 54(4):674–685

Asdadi A, Harhar H, Gharby S, Bouzoubaâ Z, Yadini AE, Moutaj R, Hadek ME, Chebli B, Hassani LM (2014) Chemical composition and antifungal activity of *Nigella Sativa* L. oil seed cultivated in Morocco. Int J Pharm Sci Invent 3(11):09–15

Asif S, Malik L (2017) Protective effects of *Nigella sativa* on acetylsalicylic acid-induced nephrotoxicity in albino rats. J Coll Phys Surg Pak 27(9):536

Atta-ur-Rahman SM, Hasan SS, Choudhary MI, Ni CZ, Clardy J (1995) Nigellidine—a new indazole alkaloid from the seeds of *Nigella sativa*. Tetrahedron Lett 36(12):1993–1996

Barakat EM, El Wakeel LM, Hagag RS (2013) Effects of *Nigella sativa* on outcome of hepatitis C in Egypt. World J Gastroenterol: WJG 19(16):2529

Benhaddou-Andaloussi A, Martineau L, Vuong T, Meddah B, Madiraju P, Settaf A, Haddad PS (2011) The in vivo antidiabetic activity of *Nigella sativa* is mediated through activation of the AMPK pathway and increased muscle Glut4 content. Evid Based Complement Alternat Med 2011(1):538671

Benkaci-Ali F, Baaliouamer A, Wathelet JP, Marlier M (2010) Chemical composition of volatile oils from Algerian *Nigella sativa* L. seeds. J Essent Oil Res 22(4):318–322

Bensiameur-Touati K, Kacimi G, Haffaf EM, Berdja S, Aouichat-Bouguerra S (2017) In vivo subacute toxicity and antidiabetic effect of aqueous extract of *Nigella sativa*. Evid Based Complement Alternat Med 2017(1):8427034

Bita A, Rosu AF, Calina D, Rosu L, Zlatian O, Dindere C, Simionescu A (2012) An alternative treatment for Candida infections with *Nigella sativa* extracts. Eur J Hosp Pharm 19(2):162

Boskabady MH, Javan H, Sajady M, Rakhshandeh H (2007) The possible prophylactic effect of *Nigella sativa* seed extract in asthmatic patients. Fundam Clin Pharmacol 21(5):559–566

Bourgou S, Ksouri R, Bellila A, Skandrani I, Falleh H, Marzouk B (2008) Phenolic composition and biological activities of Tunisian *Nigella sativa* L. shoots and roots. C R Biol 331(1):48–55

Chaieb K, Kouidhi B, Jrah H, Mahdouani K, Bakhrouf A (2011) Antibacterial activity of Thymoquinone, an active principle of *Nigella sativa* and its potency to prevent bacterial biofilm formation. BMC Complement Altern Med 11:1–6

Cherif M, Valenti B, Abidi S, Luciano G, Mattioli S, Pauselli M, Bouzarraa I, Priolo A, Salem HB (2018) Supplementation of *Nigella sativa* seeds to Barbarine lambs raised on low-or high-concentrate diets: effects on meat fatty acid composition and oxidative stability. Meat Sci 139:134–141

Czajkowska A, Gornowicz A, Pawłowska N, Czarnomysy R, Nazaruk J, Szymanowski W, Bielawska A, Bielawski K (2017) Anticancer effect of a novel octahydropyrazino [2, 1-a: 5, 4-a′] diisoquinoline derivative and its synergistic action with *Nigella sativa* in human gastric cancer cells. Biomed Res Int 2017(1):9153403

Dwarampudi LP, Palaniswamy D, Nithyanantham M, Raghu P (2012) Antipsoriatic activity and cytotoxicity of ethanolic extract of *Nigella sativa* seeds. Pharmacogn Mag 8(32):268

El Rabey HA, Al-Seeni MN, Bakhashwain AS (2017) The antidiabetic activity of *Nigella sativa* and propolis on streptozotocin-induced diabetes and diabetic nephropathy in male rats. Evid Based Complement Alternat Med 2017(1):5439645

El Shenawy NS, Soliman MF, Reyad SI (2008) The effect of antioxidant properties of aqueous garlic extract and *Nigella sativa* as anti-schistosomiasis agents in mice. Rev Inst Med Trop Sao Paulo 50:29–36

El-Najjar N, Chatila M, Moukadem H, Vuorela H, Ocker M, Gandesiri M, Schneider-Stock R, Gali-Muhtasib H (2010) Reactive oxygen species mediate thymoquinone-induced apoptosis and activate ERK and JNK signaling. Apoptosis 15:183–195

El-Tahir KE, Bakeet DM (2006) The black seed *Nigella sativa* Linnaeus-a mine for multi cures: a plea for urgent clinical evaluation of its volatile oil. J Taibah Univ Med Sci 1(1):1–9

Emeka LB, Emeka PM, Khan TM (2015) Antimicrobial activity of *Nigella sativa* L. seed oil against multi-drug resistant *Staphylococcus aureus* isolated from diabetic wounds. Pak J Pharm Sci 28(6):1985–1990

Enomoto S, Asano R, Iwahori Y, Narui T, Okada Y, Singab AN, Okuyama T (2001) Hematological studies on black cumin oil from the seeds of *Nigella sativa* L. Biol Pharm Bull 24(3):307–310

Fahmy HM, Saad EA, Sabra NM, El-Gohary AA, Mohamed FF, Gaber MH (2018) Treatment merits of latanoprost/thymoquinone–encapsulated liposome for glaucomatus rabbits. Int J Pharm 548(1):597–608

Forouzanfar F, Bazzaz BS, Hosseinzadeh H (2014) Black cumin (*Nigella sativa*) and its constituent (thymoquinone): a review on antimicrobial effects. Iran J Basic Med Sci 17(12):929

Gholamnezhad Z, Keyhanmanesh R, Boskabady MH (2015) Anti-inflammatory, antioxidant, and immunomodulatory aspects of Nigella sativa for its preventive and bronchodilatory effects on obstructive respiratory diseases: a review of basic and clinical evidence. J Funct Foods 17:910–927

Ghorbanibirgani A, Khalili A, Rokhafrooz D (2014) Comparing *Nigella sativa* oil and fish oil in treatment of vitiligo. Iran Red Crescent Med J 16(6):e4515

Gilani AH, Aziz N, Khurram IM, Chaudhary KS, Iqbal A (2001) Bronchodilator, spasmolytic and calcium antagonist activities of *Nigella sativa* seeds (Kalonji): a traditional herbal product with multiple medicinal uses. J Pak Med Assoc 51(3):115

Gray JP, Zayasbazan Burgos D, Yuan T, Seeram N, Rebar R, Follmer R, Heart EA (2016) Thymoquinone, a bioactive component of *Nigella sativa*, normalizes insulin secretion from pancreatic β-cells under glucose overload via regulation of malonyl-CoA. Am J Physiol Endocrinol Metab 310(6):E394–E404

Gupta C, Prakash D (2015) Nutraceuticals for geriatrics. J Tradit Complement Med 5(1):5–14

Gupta S, Satishkumar MN, Duraiswamy B, Das S, Chhajed M (2012) Potential herbs and its phytoconstituents against fungal infection: a systematic review. World J Pharm Res 1:1–20

Hadi NA, Ashor AW (2010) *Nigella sativa* oil lotion 20% vs. benzoyl peroxide lotion 5% in the treatment of mild to moderate acne vulgaris. Iraqi. Postgrad Med J 9(4):371–376

Hajra N (2011) *Nigella sativa*: the miraculous herb. Pak J Biochem Mol Biol 44(1):44–48

Haron H, Grace-Lynn C, Shahar S (2014) Comparison of physicochemical analysis and antioxidant activities of *Nigella sativa* seeds and oils from Yemen, Iran and Malaysia. Sains Malays 43(4):535–542

Heshmati J, Namazi N (2015) Effects of black seed (*Nigella sativa*) on metabolic parameters in diabetes mellitus: a systematic review. Complement Ther Med 23(2):275–282

Hobbenaghi R, Javanbakht J, Sadeghzadeh SH, Kheradmand D, Abdi FS, Jaberi MH, Mohammadiyan MR, Khadivar F, Mollaei Y (2014) Neuroprotective effects of *Nigella sativa* extract on cell death in hippocampal neurons following experimental global cerebral ischemia-reperfusion injury in rats. J Neurol Sci 337(1–2):74–79

Hossain MS, Rahman MZ, Mozumder SN, Rahman J (2024) Nutritional composition, fatty acids, bioactive compounds, and antioxidant activity of *Nigella sativa* seed grown in Bangladesh. Discov Food 4(1):52

Huchchannanavar S, Yogesh LN, Prashant SM (2019) The black seed *Nigella sativa*: a wonder seed. Int J Chem Stud 7(3):1320–1324

Islam MH, Ahmad IZ, Salman MT (2013) Antibacterial activity of *Nigella sativa* seed in various germination phases on clinical bacterial strains isolated from human patients. E3 J Biotechnol Pharm Res 4(1):8–13

Javed S, Shahid AA, Haider MS, Umeera A, Ahmad R, Mushtaq S (2012) Nutritional, phytochemical potential and pharmacological evaluation of *Nigella Sativa* (Kalonji) and *Trachyspermum Ammi* (Ajwain). J Med Plants Res 6(5):768–775

Kadam D, Lele SS (2017) Extraction, characterization and bioactive properties of *Nigella sativa* seedcake. J Food Sci Technol 54:3936–3947

Khalife R, Hodroj MH, Fakhoury R, Rizk S (2016) Thymoquinone from *Nigella sativa* seeds promotes the antitumor activity of noncytotoxic doses of topotecan in human colorectal cancer cells in vitro. Planta Med 82(04):312–321

Khosravi AR, Minooeianhaghighi MH, Shokri H, Emami SA, Alavi SM, Asili J (2011) The potential inhibitory effect of *Cuminum cyminum*, *Ziziphora clinopodioides* and *Nigella sativa* essential oils on the growth of *Aspergillus fumigatus* and *Aspergillus flavus*. Braz J Microbiol 42:216–224

Kolahdooz M, Nasri S, Modarres SZ, Kianbakht S, Huseini HF (2014) Effects of *Nigella sativa* L. seed oil on abnormal semen quality in infertile men: a randomized, double-blind, placebo controlled clinical trial. Phytomedicine 21(6):901–905

Koshak AE, Koshak EA (2020) *Nigella sativa* L as a potential phytotherapy for coronavirus disease 2019: a mini review of in silico studies. Curr Ther Res 93:100602

Kumara SS, Huat BT (2001) Extraction, isolation and characterisation of antitumor principle, α-hederin, from the seeds of *Nigella sativa*. Planta Med 67(01):29–32

Lim TK (2012) Edible medicinal and non-medicinal plants. Springer, Dordrecht

Mahdavi R, Heshmati J, Namazi N (2015) Effects of black seeds (*Nigella sativa*) on male infertility: a systematic review. J Herb Med 5(3):133–139

Mahmoud MR, El-Abhar HS, Saleh S (2002) The effect of *Nigella sativa* oil against the liver damage induced by *Schistosoma mansoni* infection in mice. J Ethnopharmacol 79(1):1–1

Mahmoudvand H, Sepahvand A, Jahanbakhsh S, Ezatpour B, Mousavi SA (2014) Evaluation of antifungal activities of the essential oil and various extracts of *Nigella sativa* and its main component, thymoquinone against pathogenic dermatophyte strains. J Mycol Med 24(4):e155–e161

Manjegowda SB, Rajagopal HM, Dharmesh SM (2017) Polysaccharide of Black cumin (*Nigella sativa*) modulates molecular signaling cascade of gastric ulcer pathogenesis. Int J Biol Macromol 101:823–836

Manoharan N, Jayamurali D, Parasuraman R, Govindarajulu SN (2021) Phytochemical composition, therapeutical and pharmacological potential of *Nigella sativa*: a review. Tradit Med Res 6(4):32

Mathur ML, Gaur J, Sharma R, Haldiya KR (2011) Antidiabetic properties of a spice plant *Nigella sativa*. J Endocrinol Metab 1(1):1–8

Mehta BK, Mehta P, Gupta M (2009) A new naturally acetylated triterpene saponin from *Nigella sativa*. Carbohydr Res 344(1):149–151

Mohammed A, Ma O, Yaro AH (2005) Analgesic and anti-inflammatory activities of ethanol seed extract of *Nigella sativa* (black cumin) in mice and rats. Editor Advis Board 18(2):277–281

Morikawa T, Xu F, Kashima Y, Matsuda H, Ninomiya K, Yoshikawa M (2004) Novel dolabellane-type diterpene alkaloids with lipid metabolism promoting activities from the seeds of *Nigella sativa*. Org Lett 6(6):869–872

Mosbah R, Djerrou Z, Mantovani A (2018) Protective effect of *Nigella sativa* oil against acetamiprid induced reproductive toxicity in male rats. Drug Chem Toxicol 41(2):206–212

Mousavi SM, Sheikhi A, Varkaneh HK, Zarezadeh M, Rahmani J, Milajerdi A (2018) Effect of *Nigella sativa* supplementation on obesity indices: a systematic review and meta-analysis of randomized controlled trials. Complement Ther Med 3(8):48–57

Namazi N, Larijani B, Ayati MH, Abdollahi M (2018) The effects of *Nigella sativa* L. on obesity: a systematic review and meta-analysis. J Ethnopharmacol 219:173–181

Nickavar B, Mojab F, Javidnia K, Amoli MA (2003) Chemical composition of the fixed and volatile oils of *Nigella sativa* L. from Iran. Zeitschrift für Naturforschung C 58(9–10):629–631

Niu Y, Zhou L, Meng L, Chen S, Ma C, Liu Z, Kang W (2020) Recent progress on chemical constituents and pharmacological effects of the genus *Nigella*. Evid Based Complement Alternat Med 2020(1):6756835

Paarakh PM (2010) *Nigella sativa* Linn.–a comprehensive review. Indian J Nat Prod Resour 1:409–429

Periasamy VS, Athinarayanan J, Alshatwi AA (2016) Anticancer activity of an ultrasonic nanoemulsion formulation of *Nigella sativa* L. essential oil on human breast cancer cells. Ultrason Sonochem 31:449–455

Rahim MA, Shoukat A, Khalid W, Ejaz A, Itrat N, Majeed I, Koraqi H, Imran M, Nisa MU, Nazir A, Alansari WS (2022) A narrative review on various oil extraction methods, encapsulation processes, fatty acid profiles, oxidative stability, and medicinal properties of black seed (*Nigella sativa*). Foods 11(18):2826

Rastogi RM (1990) Compendium of Indian medicinal plants, vol 1. Central Drug Research Institute, Lucknow, pp 388–389

Raval BP, Shah TG, Suthar MP, Ganure AL (2010) Screening of *Nigella sativa* seeds for antifungal activity. Ann Biolog Res 1:164–171

Rogozhin EA, Oshchepkova YI, Odintsova TI, Khadeeva NV, Veshkurova ON, Egorov TA, Grishin EV, Salikhov SI (2011) Novel antifungal defensins from *Nigella sativa* L. seeds. Plant Physiol Biochem 49(2):131–137

Sabbah M, Altamimi M, Di Pierro P, Schiraldi C, Cammarota M, Porta R (2020) Black edible films from protein-containing defatted cake of *Nigella sativa* seeds. Int J Mol Sci 21(3):832

Salem EM, Yar T, Bamosa AO, Al-Quorain A, Yasawy MI, Alsulaiman RM, Randhawa MA (2010) Comparative study of *Nigella sativa* and triple therapy in eradication of Helicobacter Pylori in patients with non-ulcer dyspepsia. Saudi J Gastroenterol 16(3):207–214

Sapmaz E, Sapmaz HI, Vardi N, Tas U, Sarsilmaz M, Toplu Y, Arici A, Uysal M (2017) Harmful effects of formaldehyde and possible protective effect of *Nigella sativa* on the trachea of rats. Niger J Clin Pract 20(5):523–529

Sarkhail P, Esmaily H, Baghaei A, Shafiee A, Abdollahi M, Khademi Y, Madandar M, Sarkheil P (2011) Burn healing potential of *Nigella sativa* seed oil in rats. Int J Pharm Sci Res 2(1):34–40

Sharma Y, Bashir S, Irshad M, Gupta SD, Dogra TD (2005) Effects of acute dimethoate administration on antioxidant status of liver and brain of experimental rats. Toxicology 206(1):49–57

Sultan MT, Butt MS, Karim R, Ahmad AN, Suleria HA, Saddique MS (2014) Toxicological and safety evaluation of *Nigella sativa* lipid and volatile fractions in streptozotocin induced diabetes mellitus. Asian Pac J Trop Dis 4:S693–S697

Suri K, Sareen S (2022) Bioactive compounds in *Nigella* seeds. In: Spice bioactive compounds. CRC Press, pp 187–216

Tahan M, Bayram I (2011) Effect of using black cumin (*Nigella sativa*) and parsley (*Petroselinum crispum*) in laying quail diets on egg yield, egg quality and hatchability. Arch Zootech 14(4):39

Tariq M (2008) *Nigella sativa* seeds: folklore treatment in modern day medicine. Saudi J Gastroenterol 14(3):105–106

Tembhurne SV, Feroz S, More BH, Sakarkar DM (2014) A review on therapeutic potential of *Nigella sativa* (kalonji) seeds. J Med Plants Res 8(3):167–177

Thakur S, Kaurav H, Chaudhary G (2021) *Nigella sativa* (Kalonji): a black seed of miracle. Int J Res Rev 8(4):342–357

Varghese E (1996) Applied ethnobotany: a case study among the Kharias of Central India. Deep Publications, New Delhi

Yaman I, Durmus AS, Cerıbası S, Yaman M (2010) Effects of *Nigella sativa* and silver sulfadiazine on burn wound healing in rats. Vet Med 2010(12):619–624

Zohary D, Hopf M, Weiss E (2012) Domestication of plants in the old world: the origin and spread of domesticated plants in Southwest Asia, Europe, and the Mediterranean Basin. Oxford University Press

# Medicinal and Nutritional Importance of *Enhydra fluctuans* in Human Health

# 21

Avijit Bej, Nandita Mandal, Anwesha Das, Sanjukta Das, and Arijit Nandi

**Abstract**

Helencha or harkuch with the scientific name *Enhydra fluctuans* Lour is a type of herbal plant that belongs to the tropical group and is part of the Asteraceae family. This plant, which is partly submerged in water, has an importance both in medicine and as food. This plant grows best in the marsh nearby, and the vegetable is very nutritious, containing lots of nutrients, minerals, and antioxidants, which are essential for the human body. Conventional medicine has employed plant parts, especially the leaves and stems, for a broad range of illnesses, including liver problems, skin diseases, and digestive system complications. Vitamin C, vitamin A, and iron from the plant boost the body's immunity, eyesight, and get rid of anemia problems. Furthermore, it has anti-inflammatory, antimicrobial, anti-depressant, antidiarrheal activity, and liver protective qualities attributable to the bioactive constituents of *Enhydra fluctuans* such as flavonoids, alkaloids, carotene, saponins, cholesterol, glucoside, enhydrin, and tannins. Based on the above health benefits of *Enhydra fluctuans*, it is significant to continue studying

A. Bej
Mahadeva Lal Schroff College of Pharmacy, Aurangabad, Bihar, India

N. Mandal · A. Das
Department of Pharmacy, Sanaka Educational Trust's Group of Institutions (SETGOI), Malandighi, Durgapur, West Bengal, India

S. Das
Department of Pharmacy, Birbhum Pharmacy School, Bandersole, Birbhum, West Bengal, India

A. Nandi (✉)
Institute for Molecular Bioscience, The University of Queensland, St Lucia, QLD, Australia

Department of Pharmacy, Sanaka Educational Trust's Group of Institutions (SETGOI), Malandighi, Durgapur, West Bengal, India



M. A. Ansari et al. (eds.), *Medicinal Plants and their Bioactive Compounds in Human Health: Volume 2*, https://doi.org/10.1007/978-981-95-3620-7_21

the implemented in nutrition and medicine. This chapter aims to highlight the significance of *Enhydra fluctuans*, known for its strong medicinal properties, and to review the research conducted on this plant to date.

**Keywords**

Anti-inflammatory properties · Antioxidant properties · *Enhydra fluctuans* · Helencha · Nutritional importance

**Learning Objectives**

- Learn about the plant features and environmental needs of *Enhydra fluctuans.*
- Discuss the nutritional content of *Enhydra fluctuans.*
- Unveil the medicinal properties of *Enhydra fluctuans.*
- Assess the bioactive compounds of *Enhydra fluctuans.*
- Assess the therapeutic applications supported by scientific evidence.
- Acknowledge the importance of ongoing research on *Enhydra fluctuans.*

## 21.1 Introduction

Nature has been a source of medicinal compounds for thousands of years, with recent discoveries supporting their usage in ancient medicine. We have focused on *Enhydra fluctuans* Lour, an edible, semi-aquatic, non-woody vegetable plant with distinct leaves. *Enhydra fluctuans (E. fluctuans)*, also called Indian Marshweed or Raktakarabi, is a semiaquatic herb with a rich history of use in traditional medicine, especially within Ayurveda. It is endemic to India, Bangladesh, Burma, Sri Lanka, and other Southeast Asian locations (Kuri et al. 2014). The leaves are mildly bitter and can treat inflammation, skin disorders, smallpox, and digestive issues, and are recognized for their noteworthy pharmacological properties. The leaves are also antibilious and used to treat neurological disorders and hepatic torpidity. It has nutritional value, and its methanolic extract has been shown to have analgesic and anti-diarrhea properties. The leaves of *E. fluctuans* have been shown to exhibit hypotensive action (Rathod et al. 2024; Sannigrahi et al. 2010). The various phytochemicals present in *E. fluctuans.* are responsible for its therapeutic properties. *E. fluctuans.* has been found to contain several active constituents like flavonoids, alkaloidal substances, phenolic compounds, and volatile oils (Sarma et al. 2014). These constituents contribute to its medicinal qualities, such as anti-inflammatory, analgesic, antimicrobial, antioxidant, hepatoprotective, immunomodulatory, and anticancer properties. *E. fluctuans.* is a multi-targeted enzyme inhibitor with novel therapeutic potential; this means it controls the immune function, attenuates oxidative stress-induced damage, lowers inflammatory mediators' expressions, and helps in tissue healing (Parveen et al. 2024; Kumar et al. 2023). Interestingly, a study has demonstrated that *E. fluctuans.* induces apoptosis in tumor cells as well by arresting the cell cycle and inducing expression of pathways that inhibit angiogenesis molecules phenomena in these (tumor) cells, thus limiting further metastatic processes (Banik et al. 2020). *E. fluctuans*. on the other hand, seems to be promising as a therapeutic entity, although precautions should still be taken to ensure its safety and potential side effects in this plant extract treatment (Al-Mekhlafi ct al. 2018). The standard extraction and isolation of active principles followed by their identification could be further validated with extensive clinical trials for a crude herb like *E. fluctuans*. to establish as medicine in the mainstream field (Kumar et al. 2023).

## 21.2 Botanical Description

This plant, sometimes known as Water Cress or Marsh Herb, grows in tropical settings, including ditches, watercourses, fish pond borders, and open rice fields. It can sometimes float on water, and it is found at sea level or up to 1800 m. When young, it is quite susceptible to cold. Stems are 30–60 cm long, meaty, hollow, and purplish; internodes are 7–10 cm long and root at the lowest nodes. The leaves are sessile, 2.5–7.5 cm long, opposite phyllotaxy, linear to oblong in form, with a sharp apex and sub-crenate or strongly dented margins. Heads are terminal and sessile; flowers range from white to greenish-white; the outer pair of involucral bracts are elliptical and 1–1.2 cm long, while the inner pair are smaller; fruits are achene and enclosed in hard receptacle scales. It can proliferate through fragmentation and is so plentiful that it clogs waterways (Barua et al. 2020).

## 21.3 Phytochemical Constituents

*E. fluctuans* Lour contains a variety of phytochemicals, including chlorophyll, proteins, sugar, terpenoids, flavonoids, saponins, alkaloids, phenolics, phenolic compounds, sesquiterpene lactones, phytosterols, and more. It contains a high concentration of flavonoids and a moderate number of alkaloids, tannins, and phenols. The plant also contains a lot of beta-carotene and protein (Guchhait et al. 2023). The phytochemical constituents of the plant are tabulated in Table 21.1.

## 21.4 Nutritional Value

*E. fluctuans*, commonly known as "Helencha," is a semi-aquatic herb often enjoyed as a leafy vegetable in various parts of India, prized for its nutrition and healing properties. It is packed with protein, with about 19.64% found in the stems and up to 20.58% in the leaves. The plant also provides a healthy dose of vitamins and minerals like vitamins A and C, calcium, and iron (Suraiya et al. 2023; Bhattacharya et al. 2024a, b). Along with its nutritional benefits, it's rich in antioxidants such as phenolic compounds, saponins, β-carotene, and ascorbic acid (Dua et al. 2016a, b;

**Table 21.1** Phytochemical constituents of *E. fluctuans*

| Phytochemicals | Compound name | Amount present | Extraction process | Phytochemical tests | References |
|---|---|---|---|---|---|
| Flavonoids | Baicalein 7-O-diglucoside | 1.2–2.5% | Solvent extraction using ethanol or methanol | Shinoda test | Sannigrahi et al. (2011) and Chen et al. (2003) |
| Alkaloids | – | 0.5–1.5% | Acid-base extraction with ethanol and dilute acid | Dragendorff's test Wagner's test | Aasim et al. (2019) |
| Tannins | – | 1.0–3.0% | Aqueous or ethanol extraction | Ferric chloride test ($FeCl_3$) | Chattaraj et al. (2023b) and Alachaher et al. (2018) |
| Saponins | – | 0.8–2.0% | Aqueous or methanol extraction | Foam test (froth formation) | Dua et al. (2016a, b) |
| Phenolic compounds | – | 1.5–3.5% | Solvent extraction using methanol or ethanol | Ferric chloride test ($FeCl_3$) | Alachaher et al. (2018) |
| Terpenoids | Fluctuadin, Fluctuanin, 8-desacyl enhydrin tiglate | 0.3–1.0% | Solvent extraction using petroleum ether or hexane | Salkowski test (chloroform and $H_2SO_4$) | Mukherjee and Wahile (2006) and Bohlmann et al. (1982) |
| Steroids | Stigmasta-5,22,25-trien-3beta-ol, Stigmasterol | 0.2–0.8% | Solvent extraction using ethanol or chloroform | Liebermann-Burchard test (acetic anhydride) | Alachaher et al. (2018) |
| Glycosides | 4′,5,6,7-Tetrahydroxy-8-methoxyisoflavone-7-O-beta-D galactopyranosyl-(1 → 3)-O-beta-D-xylopyranosyl-(1 → 4) O-alpha-L-rhamnopyranoside | 0.5–1.5% | Aqueous or ethanol extraction | Keller-Kiliani test (glacial acetic acid) | Alachaher et al. (2018) and Chattaraj et al. (2023b) |

**Table 21.2** Comparison with other leafy vegetables (Suraiya et al. 2023)

| Nutrient | *E. fluctuans* | Spinach | Amaranth | Fenugreek |
|---|---|---|---|---|
| Protein (%) | 19.64–20.58 | 2.9 | 4.4 | 4.2 |
| Vitamin A (IU) | High | 11,458 | 5700 | 2600 |
| Vitamin C (mg) | High | 28.1 | 40 | 22 |
| Calcium (mg) | High | 99 | 266 | 204 |
| Iron (mg) | High | 2.7 | 3.9 | 1.2 |

Rathod et al. 2024). Traditionally, Helencha has been used in folk medicine to help with inflammation, relieve pain, protect the liver from toxins, manage diabetes, fight intestinal worms, and lower blood pressure (Simlai et al. 2014). Much of its medicinal power comes from natural compounds like phenolics and saponins. When comparing it with other frequently consumed leafy vegetables, *E. fluctuans* is still a species with superior nutrient value. The protein and fat content are significantly higher than most greens, such as spinach, amaranth, and fenugreek; hence, it is a potential source of protein for vegetarians. Comparing the ratio of vitamins and minerals with spinach, *E. fluctuans* contains approximately equivalent amounts of vitamins A and C; however, they contain more iron and calcium than spinach (Suraiya et al. 2023). In addition, it contains substantial levels of antioxidants, including phenolic compounds and saponins, that would improve the body's health. The presence of proteins, essential vitamins, and antioxidants makes *E. fluctuans* one of the most nutritious varieties of green leaves. The comparison with other leafy vegetables is tabulated in Table 21.2.

## 21.5 Medicinal Properties

*Enhydra fluctuans* is an aquatic plant that has health benefits that are medicinal and nutritional for humans. The different medicinal properties of the plant are illustrated in Fig. 21.1.

### 21.5.1 Antioxidant Properties

*E. fluctuans* contains phytochemicals such as flavonoids, polyphenols, and vitamins A, C, and E, which possess antioxidant properties. These compounds are useful in scavenging these free radicals that cause added stress and protect cells from damage. The antioxidant content may also help in the prevention of chronic diseases, including cancer, cardiovascular diseases, and neurological disorders (Chattaraj et al. 2024; Banerjee et al. 2008).

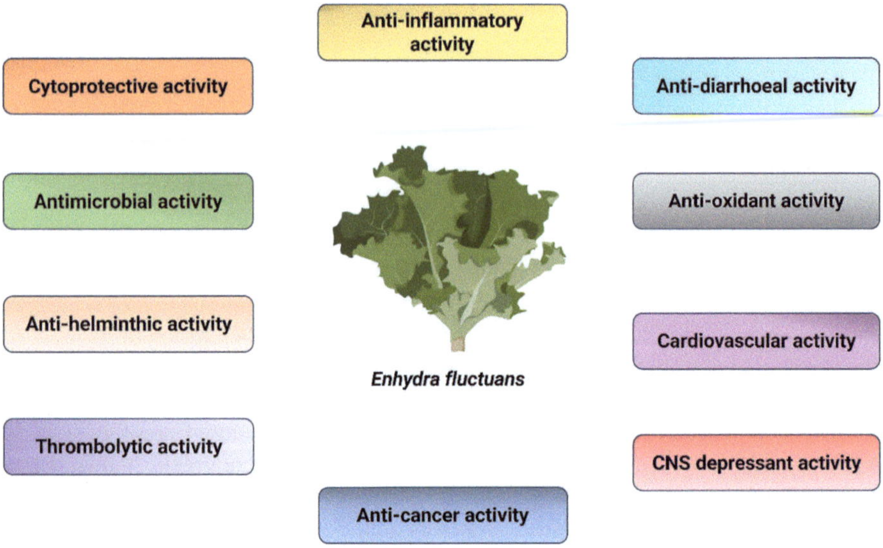


**Fig. 21.1** Different medicinal properties of *E. fluctuans*

### 21.5.2 Anti-inflammatory Effects

This herb is particularly endowed with bioactive compounds that have been found to possess strong anti-inflammatory properties. These kinds of compounds produce their impact on cytokine mediators such as interleukin 6, as well as enzymes that are associated with inflammation, including COX 2. Hence, *E. fluctuans* may significantly alleviate inflammation-based diseases such as arthritis and inflammatory bowel disease (Bhattacharya et al. 2024c; Saha and Paul 2019).

### 21.5.3 Antimicrobial Activities

The *E. fluctuans* extract has also been shown to possess antibacterial and antifungal properties. Alkaloids and terpenoids in the plant disrupt the microbial cell wall, limiting their growth, and giving a natural approach to fight against infections (Nandi et al. 2024b; Ruhul Amin et al. 2012).

### 21.5.4 Antidiabetic Effects

It has been found that *E. fluctuans* can be useful in regulating blood glucose levels by increasing insulin sensitivity and decreasing the absorption of intestinal glucose. It can also stimulate the production of pancreatic beta cells, making it an ideal natural cure for type 2 diabetes (Hasan et al. 2019).

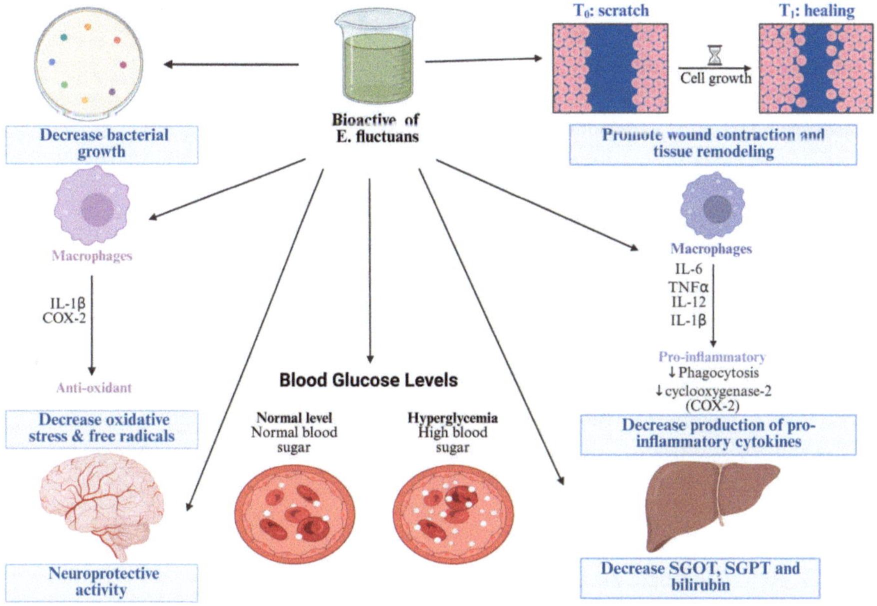


**Fig. 21.2** Different pharmacological properties of *E. fluctuans*

### 21.5.5 Cardiovascular Benefits

This plant contains nutrients that are good for the heart. It is also an antioxidant, which has the potential to lower incidences of atherosclerosis. Furthermore, *E. fluctuans* can assist with maintaining blood pressure levels and regulating cholesterol; the level of LDL cholesterol is reduced, while the levels of HDL cholesterol are increased, which decreases the potential of developing cardiovascular diseases (Panda 2021).

## 21.6 Pharmacological Activities

Various experimental studies have explored the pharmacological properties of *E. fluctuans*, highlighting its potential for therapeutic use (Fig. 21.2).

### 21.6.1 Antioxidant Activity

The scientific research also suggests that *E. fluctuans* has high antioxidant potential when analyzed both in vitro and in vivo. It contains a relatively large amount of phenol compounds that work as the most significant free radical scavengers and antioxidants. There are also types of compounds capable of mitigating the impact of reactive oxygen species (ROS), thus preventing the oxidative alteration of cells and associated diseases with the development of chronic diseases such as cancer, cardiovascular disorders, and neurodegenerative conditions. This implies that *E. fluctuans* may have potential as a therapeutic agent in diseases that cause oxidative stress, including heart and Alzheimer's diseases (Parveen et al. 2024; Haoya et al. 2018).

### 21.6.2 Anti-inflammatory Activity

Research on animals has shown us that extracts from *E. fluctuans* reduce inflammation, including arthritis and skin rashes. It achieves this by inhibiting the production of pro-inflammatory cytokines, including TNF-α and IL-6, and lowering the expression of inflammatory enzymes like cyclooxygenase-2 (COX-2). Such activity indicates that *E. fluctuans* may have potential as therapeutic agents for managing inflammatory conditions, such as rheumatoid arthritis and inflammatory bowel disease (IBD) (Das et al. 2023; Roy et al. 2021).

### 21.6.3 Antidiabetic Activity

Research on diabetic animal models has shown that *E. fluctuans* has hypoglycemic effects in reducing blood sugar. The plant enhances insulin sensitivity, facilitating better glucose uptake by cells, and stimulates key enzymes involved in glucose metabolism, such as hexokinase and glycogen synthase. These effects ratify the potential of *E. fluctuans*, which can be active in the treatment of type 2 diabetes and connected metabolic disorders (Phukan et al. 2015).

### 21.6.4 Hepatoprotective Activity

Given this, investigations have evidenced that *E. fluctuans* have the biochemical property that guards the liver in experimental animal species exposed to hepatotoxic compounds like carbon tetrachloride. Flavonoids found in the plant reduce oxidative stress in liver cells, reducing tissue damage and improving liver performance. These outcomes suggest that *E. fluctuans* are most likely to possess therapeutic use for liver disorders, which include viral hepatitis and cirrhosis (Ali et al. 2013). Studies have shown that the aerial parts of *E. fluctuans* can significantly decrease elevated levels of liver enzymes (SGOT, SGPT) and bilirubin induced by CCl4 toxicity, indicating a protective effect against hepatotoxicity (Nandi et al. 2023b; Upadhyay et al. 2024).

### 21.6.5 Neuroprotective Activity

Exploring the neuroprotective role of *E. fluctuans* through its extracts, it is useful in models of neurodegeneration in which the presence of the compound lessens the damage that occurs to brain cells. This protective effect is attributed to the ability of the plant to reduce neuroinflammation and oxidative stress in the brain tissues. These findings suggest that *E. fluctuans* might hold great promise for Alzheimer's and Parkinson's disease, since neurodegeneration is hallmarked by the dying off of brain cells (Barua et al. 2020).

### 21.6.6 Antimicrobial Activity

*E. fluctuans* demonstrates antimicrobial properties effective against various pathogens, including both bacteria and fungi. This activity results from the phytochemicals contained in plants, which inhibit the structure of bacterial cells and, therefore, restrict their overgrowth. From these insights, it can be deduced that *E. fluctuans* may have potential applications in the treatment of infections and skin diseases caused by these pathogens (Bhattacharya et al. 2024a; Ruhul Amin et al. 2012).

### 21.6.7 Wound Healing Properties

*E. fluctuans* extract on topical application has been found to enhance wound healing in animal experiments. Such advancement is due to the plant's anti-inflammatory and antioxidant activity, which fosters wound contraction and tissue remodelling. Such revelations provided evidence of the future utility of *E. fluctuans* in the treatment of wounds and skin regenerative treatment (Nandi et al. 2023a; Arya et al. 2022).

## 21.7 Traditional and Ethnomedicinal Uses

In various traditional practices, *E. fluctuans* is highly regarded for its medicinal properties. In the system of Indian Ayurveda, it is principally used in digestive disorders, skin diseases, and inflammation, due to its cold and pungent properties. The leaves are commonly used for preparing soups or included in curries, and from the fresh leaves, a paste is made and used for skin issues such as eczema and boils (Nandi et al. 2024c; Kamal et al. 2019). In rural areas of Bangladesh, it has been used as an anti-inflammatory and analgesic, commonly used in jaundice, liver diseases, or constipation, either as a green leaf vegetable or for medicinal purposes using green leaves (Chattaraj et al. 2023a; Sarma et al. 2014). Considering the properties of *E. fluctuans* used in Materia Medica as a drink called the Ice-Bamboo Wine, in Chinese folk medicine, it is also known for its function in promoting urination and its heat-clearing and swelling-reducing effects, helping in febrile diseases and edema, it is normally consumed in a soup or tea (Dua et al. 2015). In Thai traditional medicine, the plant is utilized to treat digestive system issues and malnutrition, and it can be prepared fresh in Thai dishes or salads or can also be boiled. In Nepal, it is used to treat diarrhea and dysentery and to clean the blood, and the leaves of the plant are used in preparing vegetables by boiling them (Nandi et al. 2023c; Dua et al. 2015). In ethnomedicine, it has a reputation for providing relief from inflammation and pain, this is applied directly on the skin, usually in areas that are painful, such as joint pains and arthritis, and it has the ability to protect the liver, particularly in cases of jaundice (Azubuike et al. 2018). It does this by containing a high level of antioxidants and also crushing oxidative stress, and the leaves of this plant are usually used as a vegetable or in a juice. Besides, fresh leaf paste is used in the management of skin complications like rashes, eczema, and ulcers (Nandi et al. 2024a; Phukan et al. 2015).

## 21.8 Recent Research and Developments

*E. fluctuans* has gained a lot of research focus in the current days due to its pharmacological and nutritional significance. It has been discussed in recent works due to its bioactive compounds, which provide championship in inflammation, liver diseases, and oxidative stress. Studies also suggest that *E. fluctuans* contains slight levels of flavonoids and phenolic compounds, which possess antioxidant properties and remove living organisms responsible for oxidation, leading to age-related diseases and health.

### 21.8.1 Phytochemical Analysis

Flavonoids, phenolic acids, tannins, and saponins have been established to display biochemical activity in *E. fluctuans*, among other chemical compositions. These compounds are proven to have a high impact on antioxidants and anti-inflammation, so it can be used for medicine.

### 21.8.2 Antioxidant Activity

The amounts of flavonoids and phenolic compounds were determined to be high in the plant based on select experimental measurements and showed significant free radical scavenging capacity. One of the studies revealed that the methanol extract of *E. fluctuans*, with an IC50 value of 75 μg/mL, has antioxidant activity that was analyzed by using the DPPH radical scavenging assay (Singh et al. 2024; Nasreen and Choudhury 2018).

### 21.8.3 Anti-inflammatory and Analgesic Properties

The extract of *E. fluctuans* was proven to possess powerful anti-inflammatory properties in animal experiments. For example, in an in vivo study, carrageenan-induced paw edema was used in rats, and based on this, an oral administration of the extract at a dose of 200 mg/kg significantly reduced inflammation by 45% in the control group. Here also, we have seen that there was a suppressive effect on the pain reaction in the acetic acid-induced writhing models (Sannigrahi et al. 2011).

### 21.8.4 Anti-diabetic Potential

Some studies have been conducted on how this plant can be used to control diabetes, and some of them have turned out to be positive. Another study, which was conducted on diabetes-induced rats, the result showed that there was a reduction of sugar level in the 14 days of using the *E. fluctuans* leaf extract. In the GLP-1 receptor antagonist, C33H37-treated rats, fasting blood glucose was significantly reduced by 250 mg/kg of the extract by about 28% from the baseline (Hasan et al. 2019).

### 21.8.5 Nutritional Composition

The vitamin and mineral content analysis proves that *E. fluctuans* is a plant full of vitamins and minerals. It is rich in vitamins A, C, and E, calcium, potassium, and iron, among others. For instance, 100 g of fresh leaves contains 35 mg of vitamin C, which has benefits in the support of the immune system and skin repair (Mall 2017).

### 21.8.6 Antibacterial and Antifungal Activity

The antimicrobial activity of *E. fluctuans* has been investigated in several in vitro experiments. Methanolic extracts of the plant were examined for antibacterial activities on some bacteria: *Escherichia coli* and *S. aureus*, resulting in inhibition zones of 12 and 15 mm, respectively, in the disk diffusion assay. This raises the possibility of its application in the treatment of bacterial infections (Saha and Paul 2019).

## 21.9 Potential in Modern Medicine Integration into Contemporary Practices

*E. fluctuans* represents a promising candidate for integration into contemporary therapeutic practices due to its extensive pharmacological properties and historical significance in traditional medicine. The integration of *E. fluctuans* into contemporary therapeutic practices faces both opportunities and challenges:

### 21.9.1 Standardization and Quality Control

There is a pressing need for standardized extraction techniques to ensure the consistent quality and potency of herbal products derived from *E. fluctuans* (Kumar et al. 2023; Sarma et al. 2014).

### 21.9.2 Clinical Trials

Rigorous clinical trials are necessary to substantiate its efficacy and safety in modern medicine. Current studies primarily focus on preclinical data; further research is essential to explore its therapeutic mechanisms fully (Barua et al. 2020).

### 21.9.3 Formulation Development

Innovative formulations that enhance bioavailability and therapeutic efficacy are crucial for integrating this herb into mainstream healthcare practices (Kumar et al. 2023).

## 21.10 Nutritional Applications

### 21.10.1 Nutritional Profile

The primary nutrients found in *E. fluctuans* include:

#### 21.10.1.1 Macronutrients (Hossain et al. 2016)

##### 21.10.1.1.1 Fiber

*E. fluctuans* is a good source of dietary fiber consumption, including roughly 15.37% of it.

##### 21.10.1.1.2 Carbohydrates

The carbohydrate content is found to be present, but particular levels are not consistently reported throughout studies.

##### 21.10.1.1.3 Proteins

Though exact percentages are not given, the plant does contain proteins, which add to its nutritious value.

#### 21.10.1.2 Vitamins and Minerals

Contains notable concentrations of important minerals, including magnesium, potassium, and calcium, though precise levels may fluctuate in accordance with the surrounding environment. The plant is abundant in β-carotene, an essential vitamin

A precursor that is vital for the immune system and visual health (Barua et al. 2020; Kuri et al. 2014).

#### 21.10.1.3 Phytochemicals (Hossain et al. 2016; Kuri et al. 2014)
Rich in various bioactive compounds, including:

##### 21.10.1.3.1 Flavonoids
Contributing to antioxidant properties, which help to mitigate oxidative stress in cells.

##### 21.10.1.3.2 Saponins
Popular for their advantageous health properties, which may include minimizing cholesterol.

##### 21.10.1.3.3 Tannins
Associated with astringent properties and health benefits.

##### 21.10.1.3.4 Phenolic Compounds
Providing strong antioxidant activity.

##### 21.10.1.3.5 Ascorbic Acid (Vitamin C)
Important for immune function and skin health.

#### 21.10.1.4 Antioxidant Properties
The plant contains significant antioxidant activity due to its high phenolic and flavonoid content, making it beneficial for combating oxidative stress (Dua et al. 2016a).

*E. fluctuans* is used as a food source and in traditional medicine because it is generally a highly nutritious plant that provides a variety of vital nutrients and bioactive substances.

### 21.10.2 Used in Dietary Supplements

*E. fluctuans*, also known as water celery, is utilized in dietary supplements primarily for its health-promoting properties.

It is typically available in various forms.

#### 21.10.2.1 Capsules or Tablets
Containing powdered extracts for convenient consumption.

#### 21.10.2.2 Liquid Extracts
Used in tinctures or as additives in health drinks.

#### 21.10.2.3 Powdered Form
Can be added to smoothies or other foods.

Due to its various health advantages, *Enhydra fluctuans* is included in dietary supplements designed for:

A. *Support for antioxidants*: To fight against oxidative stress (Kumar et al. 2023).
B. Reducing inflammation for individuals with chronic inflammatory conditions.
C. Managing blood sugar levels: A natural addition to diabetes treatment (Hasan et al. 2019).
D. Improving liver function: Boosting detoxification mechanisms (Dua et al. 2016a, b).
E. Cognitive Support: It is occasionally promoted as having the ability to improve memory and cognitive performance (Barua et al. 2020).

### 21.10.3 Role in Nutraceuticals

*E. fluctuans*, also known as water mimosa or swamp cabbage, is a plant that has important medicinal and nutraceutical properties. Studies have examined the bioactive compounds found in different extracts of the plant, mainly focusing on their antioxidant, anti-inflammatory, and cholinesterase inhibitory effects.

#### 21.10.3.1 Antioxidant Properties

Studies show that *E. fluctuans* has powerful antioxidant properties due to its rich concentration of polyphenolic compounds. Various solvent extracts, such as methanol, chloroform, and ethyl acetate, have shown different levels of ability in scavenging free radicals.

The fraction of *ethyl acetate* showed the greatest amount of polyphenolics (179.7 μg/mg pyrocatechol equivalent) and displayed better antioxidant properties than the other fractions (Sannigrahi et al. 2010).

Another research found that the *chloroform extract* had the strongest inhibitory effect on acetylcholinesterase and had notable antioxidant activity with an IC50 value of 113.27 μg/mL in DPPH tests (Lopa et al. 2021).

These results demonstrate how *E. fluctuans* could be a valuable natural source of antioxidants for use in functional foods and nutraceuticals to fight diseases linked to oxidative stress.

#### 21.10.3.2 Cholinesterase Inhibitory Activity

Acetylcholinesterase, the primary cholinesterase found in the brain, is responsible for breaking down acetylcholine and exhibits greater specificity towards this neurotransmitter. Due to the gradual decrease in acetylcholine levels in AD, it seems to be related to the decline in cognitive function and memory. Cholinesterase inhibitors that increase the natural amount of acetylcholine are considered the primary choice of pharmacological treatment. Ellman's colorimetric method was used to study the inhibition of brain acetylcholinesterase by varying concentrations of *E. fluctuans* extracts. In addition to the test samples, researchers used the acetylcholinesterase inhibitor donepezil as a reference compound in this study. This compound is more selective for acetylcholinesterase than butyrylcholinesterase, and its IC50 value was determined to be 5.05 µg/mL. The findings showed that all four extracts have properties that inhibit acetylcholinesterase. The chloroform extract exhibited the strongest inhibition, followed by the ethyl acetate extract and petroleum ether extract with I50 values of 83.90, 167.67, and 758.00 µg/mL, respectively. The activity of the aqueous extract was relatively low, with an IC50 greater than 800 µg/mL. The effectiveness of the extracts stood out in comparison to other medicinal plants like *Vanda roxburghii*, *Bacopa monnieri*, *Centella asiatica*, *Convolvulus pluricaulis*, and *Aegle marmelos*, which are traditionally used to improve memory and are related to Alzheimer's disease treatment. Their findings indicate that *E. fluctuans* has a significant inhibitory effect on acetylcholinesterase (Lopa et al. 2021).

#### 21.10.3.3 Cytoprotective Activity Against Toxins

In this study, the edible extract of Aralia elata ethyl acetate fraction or *E. fluctuans* aqueous extract (AEEF) would offer significant protection against lead (Pb)-acetate intoxication in both isolated (in vitro) and intact (in vivo) systems. AEEF was able to greatly reduce the generation of ROS caused by Pb and oppose the internal redox challenge. Phenolics, flavonoids, and ascorbic acid are well-known antioxidants in our diets that can play a major role in helping to protect our health from reduction during Pb-induced oxidative stress. While other functional foods like garlic and cruciferous vegetables also offer detoxifying properties, the specific protective role of *E. fluctuans* against heavy metals is particularly notable (Dua et al. 2016a, b).

#### 21.10.3.4 Anti-inflammatory and Antimicrobial Effects

The anti-inflammatory and antimicrobial effects of *E. fluctuans* have been shown to be beneficial for a range of health conditions. The variety of *E. fluctuans* uses in traditional medicine, such as treating liver diseases and hypertension, contributes to its distinctiveness, despite the common anti-inflammatory properties found in many functional foods like turmeric and ginger. The leaves of *E. fluctuans* exhibit antibacterial activity against pathogens like *Escherichia coli*, *Pseudomonas aeruginosa*, and *Staphylococcus aureus*, indicating its potential use in treating infections (Ruhul Amin et al. 2012).

#### 21.10.3.5 Anti-urolithic Activity

Recent studies have highlighted its effectiveness in inhibiting the growth of calcium phosphate crystals, which are associated with kidney stones. This suggests that *E. fluctuans* may be beneficial for individuals suffering from urinary issues (Chattaraj et al. 2023b).

The plant is also noted for its analgesic, anti-diarrheal, and neuropharmacological effects, making it versatile in treating various health conditions.

### 21.10.4 Applications in Functional Foods

Given its rich nutrient profile and health-promoting properties, *E. fluctuans* can be incorporated into various functional foods:

#### 21.10.4.1 Culinary Uses

Commonly consumed as a vegetable in salads, soups, or stir-fries, enhancing both flavor and nutritional value.

#### 21.10.4.2 Dietary Supplements

Extracts can be formulated into capsules or powders targeting specific health issues such as oxidative stress or cognitive decline.

#### 21.10.4.3 Fortified Foods

Incorporating powdered forms of *E. fluctuans* into health bars or smoothies can provide additional antioxidants and support overall health.

## 21.11 Safety and Toxicology

### 21.11.1 Toxicity Studies

Experimental in vivo investigations in *E. fluctuans* show low acute and subchronic toxicities in animals fed moderate doses of the plant. All acute and subacute toxicity studies carried out on different rodent models have revealed no poisonous effects at the dosage typically used in folk medicine. Nonetheless, increased side effects are associated with high dosages, which may cause minor side effects such as gastrointestinal discomfort, which means that dosage is crucial (Hasan et al. 2019).

### 21.11.2 Safe Consumption

Traditionally, *E. fluctuans* are used for human consumption as a vegetable without any ill health implications, as evidenced by anthropological literature on their use. The fact that the plant in its raw form is included in traditional dishes such as spices in traditional foods in Asian countries, including India, Bangladesh, and China, suggests that the plant is safe for regular use. No severe toxicity has been reported when it is consumed as a vegetable or in therapeutic Signora herbal preparations (Chattaraj et al. 2022).

### 21.11.3 Flavonoid and Phenolic Content

Several works have demonstrated that flavonoid and phenolic acid content in *E. fluctuans*, and these compounds have been shown to possess antioxidant properties.

These compounds are beneficial when consumed, but a high intake of phenolic compounds in the long run has specific toxic effects that may strain the liver or kidneys (Akhe et al. 2019).

### 21.11.4 Recommended Dosage

To be safe for use in therapies, experts suggest that patients should take no more than 200–500 milligrams of plant extract from *E. fluctuans* daily. This dosage has been effective in treating inflammation and preventing oxidative stress, all without any side effects being linked to this dosage. It is risky to use valsartan for the long term or in doses that exceed this range, according to the authors (Khushi et al. 2016).

### 21.11.5 Potential Side Effects

However, there is some risk involved; thus, using *E. fluctuans* can potentially cause side effects at a higher dosage, namely, gastrointestinal upset, such as nausea and diarrhea are possible. These effects were found only in occasional cases during trials based on high concentrations of the plant extract during toxicity tests. As for standard animal models, no distinct toxicological effects such as organ dysfunction or life-threatening manifestations were seen (Delfita and Tjong 2023).

## 21.12 Conclusion

This chapter gives a detailed overview about the literature on the pharmacological properties, phytochemicals, traditional & ethnobotanical uses, nutritional profile, and safety data of *E. fluctuans*. *E. fluctuans*, commonly known as water crest or marsh herb, belonging to the family Asteraceae, is a hydrophytic plant. It is prevalent across tropical Africa, some parts of Southeast Asia, Bangladesh, Malaysia, China, India, and other countries. This plant is used medicinally by people in North East India to cure constipation, and among the tribal people of West Bengal, North East India, and Bangladesh, it is a well-known dietary supplement. *E. fluctuans* is widely utilized in traditional remedies across various communities in North East India, particularly for its medicinal properties like gastrointestinal issues, skin conditions, dysentery, anthelmintic properties, anti-inflammatory uses, and general health tonic.

## 21.13 Chapter Summary

This chapter offers a detailed overview of the medicinal, nutritional, and ethnobotanical value of *E. fluctuans*, a tropical hydrophytic Asteraceae species, with a cosmopolitan distribution in the tropical countries of Southeast Asia, India, and Africa.

It is a bi-utility plant used both as a food supplement and a traditional medicine, mostly in North East India and nearby areas. *E. fluctuans* is also employed in the treatment of various diseases like gastrointestinal diseases, skin diseases, dysentery, and inflammation. The dense phytochemical content of the herb, with flavonoids, alkaloids, and tannins, validates its pharmacological activities in the form of anti-inflammatory, antimicrobial, hepatoprotective, and antioxidant activities. Additionally, its nutrient content, with a high percentage of vitamins and minerals, contributes to its use in treating anemia and boosting immunity. This chapter highlights the need for additional research into its bioactive components and therapeutic potential to unlock its potential in contemporary medicine.

**Acknowledgments** All the authors are thankful to the principal and head of the Department of Pharmacy, Sanaka Educational Trusts Group of Institutions, Durgapur, West Bengal, 713212, for their encouragement of this work.

**Conflict of Interest Statement** The authors declare no conflict of interest.

**Consent for Publication** Not applicable.

**Data Availability Statement** Not applicable.

## References

Aasim M, Khawar KM, Ahmed SI, Karatas M (2019) 2 Plant and human health: volume 2: phytochemistry and molecular aspects. In: Multiple uses of some important aquatic and semiaquatic medicinal plants. https://doi.org/10.1007/978-3-030-03344-6

Akhe FK, Khatun MM, Ray S, Arafat ST, Uddin SJ, Hossain H, Siddique MN (2019) Profiling of bioactive compounds in water cress (Enhydra fluctuans Lour) by high performance liquid chromatography. Pharmacol Online 1(May):65–75. https://www.researchgate.net/publication/333077253_PROFILING_OF_BIOACTIVE_COMPOUNDS_IN_WATER_CRESS_ENHYDRA_FLUCTUANS_LOUR_BY_HIGH_PERFORMANCE_LIQUID_CHROMATOGRAPHY

Alachaher FZ, Dali S, Dida N, Krouf D (2018) Comparison of phytochemical and antioxidant properties of extracts from flaxseed (Linum usitatissimum) using different solvents. Int Food Res J 25(1):75–82

Ali R, Kuri S, Das A, Islam A, Ali R (2013) Preliminary phytochemical screening and in vitro thrombolytic potential of the methanolic extract of Enhydra fluctuans Lour (leaves). Int J Pharmamedix India 1(2):270–280

Al-Mekhlafi FA, Abutaha N, Farooq M, Al-Wadaan M (2018) Insecticidal effect of Solenostemma argel extracts against Culex pipiens. J Am Mosq Control Assoc 34(3):217–223. https://doi.org/10.2987/17-6725.1

Arya AK, Durgapal M, Bachheti A, Deepti, Joshi KK, Gonfa YH, Bachheti RK, Husen A (2022) Ethnomedicinal use, phytochemistry, and other potential application of aquatic and semiaquatic medicinal plants. Evid Based Complement Alternat Med:1–19. https://doi.org/10.1155/2022/4931556

Azubuike NC, Onyemelukwe AO, Maduakor UC, Onwukwe OS, Ezeh AC (2018) Hepatoprotective effects of the leaf extracts of Cassia occidentalis against carbon tetrachloride-induced hepatotoxicity in albino rats. Pharmacology 3:68–74

Banerjee D, Chakrabarti S, Hazra AK, Banerjee S, Ray J, Mukherjee B (2008) Antioxidant activity and total phenolics of some mangroves in Sundarbans. Afr J Biotechnol 7(6):805–810

Banik B, Das S, Das MK (2020) Medicinal plants with potent anti-inflammatory and anti-arthritic properties found in eastern parts of the Himalaya: an ethnomedicinal review. Pharmacogn Rev 14(28):121–137. https://doi.org/10.5530/phrev.2020.14.16

Barua A, Alam MS, Junaid M, Akter Y, Afrose SS, Sharmin T, Akter R, Hosen SMZ (2020) Phytochemistry, traditional uses and pharmacological properties of Enhydra fluctuans Lour: a comprehensive review. Curr Pharm Biotechnol 22(8):1061–1068. https://doi.org/10.2174/1389201021666200922161529

Bhattacharya T, Nandi A, Chander S (2024a) Gallstone: a factor for acute pancreatitis. In: Gallstone formation, diagnosis, treatment and prevention, pp 117–141. https://doi.org/10.1016/B978-0-443-16098-1.00015-1

Bhattacharya T, Nandi A, Das A, El-Shazly M (2024b) Role of liver in gallstone formation. In: Gallstone formation, diagnosis, treatment and prevention, pp 51–70. https://doi.org/10.1016/B978-0-443-16098-1.00014-X

Bhattacharya T, Tiwari S, Nandi A, Lin WY, Arumugam S (2024c) Development of microRNA as prognostic markers in head and neck cancer. In: Diagnostic, prognostic, and therapeutic role of microRNAs in head and neck cancer, pp 193–205. https://doi.org/10.1016/B978-0-443-15968-8.00011-9

Bohlmann F, Ahmed M, Robinson H, King RM (1982) Melampolides from Enhydra fluctuans var. fluctuans. Phytochemistry 21(7):1675–1678. https://doi.org/10.1016/S0031-9422(82)85039-5

Chattaraj B, Nandi A, Anwesha D, Sharma A, Kumar D, Dey NY, Rajagopal M (2022) In vitro antilithiatic activity of Enhydra fluctuans Lour. on calcium oxalate crystallisation and the elucidation of the possible mechanism involved therein through in silico studies. Front Pharmacol 13:1–29

Chattaraj B, Khanal P, Nandi A, Das A, Sharma A, Mitra S, Dey YN (2023a) Network pharmacology and molecular modelling study of Enhydra fluctuans for the prediction of the molecular mechanisms involved in the amelioration of nephrolithiasis. J Biomol Struct Dyn:1–11. https://doi.org/10.1080/07391102.2023.2189476

Chattaraj B, Nandi A, Das A, Baidya A, Mahata S, Chowdhury A, Mitra S et al (2023b) Enhydra fluctuans Lour. aqueous extract inhibited the growth of calcium phosphate crystals: an in vitro study. Food Chem Adv 2(August 2022):100287. https://doi.org/10.1016/j.focha.2023.100287

Chattaraj B, Nandi A, Lin WY (2024) Role of the gallbladder in our metabolism and immune system. In: Gallstone formation, diagnosis, treatment and prevention, pp 23–38. https://doi.org/10.1016/B978-0-443-16098-1.00008-4

Chen LJ, Games DE, Jones J (2003) Isolation and identification of four flavonoid constituents from the seeds of Oroxylum indicum by high-speed counter-current chromatography. J Chromatogr A 988(1):95–105. https://doi.org/10.1016/S0021-9673(02)01954-4

Das A, Nandi A, Kumari V, Alvala M (2023) FBDD & de novo drug design. In: Applied computer-aided drug design: models and methods. Bentham Science Publishers, pp 159–201. https://doi.org/10.2174/9789815179934123010008

Delfita R, Tjong DH (2023) Ameliorative effect of Enhydra fluctuans on liver damage in diabetic rats. Iraqi J Pharm Sci 32(3):128–133. https://doi.org/10.31351/vol32iss3pp128-133

Dua TK, Dewanjee S, Khanra R, Bhattacharya N, Bhaskar B, Zia-Ul-Haq M, Feo V (2015) The effects of two common edible herbs, Ipomoea aquatica and Enhydra fluctuans, on cadmium-induced pathophysiology: a focus on oxidative defence and anti-apoptotic mechanism. J Transl Med 13(1):1–19. https://doi.org/10.1186/s12967-015-0598-6

Dua TK, Dewanjee S, Khanra R (2016a) Prophylactic role of Enhydra fluctuans against arsenic-induced hepatotoxicity via anti-apoptotic and antioxidant mechanisms. Redox Rep 21(4):147–154. https://doi.org/10.1179/1351000215Y.0000000021

Dua TK, Dewanjee S, Khanra R, Joardar S, Barma S, Das S, Zia-Ul-Haq M, De Feo V (2016b) Cytoprotective and antioxidant effects of an edible herb, Enhydra fluctuans Lour. (Asteraceae), against experimentally induced lead acetate intoxication. PLoS One 11(2):1–21. https://doi.org/10.1371/journal.pone.0148757

Guchhait KC, Chavan A, Das K, Mehra A, Walling T, Sharma BP, Kumar S (2023) Enydra fluctuans Lour. (Buffalo spinach): an unexplored leafy vegetable. In: Unexplored food plants, pp 51–58. https://doi.org/10.5281/zenodo.8307263

Haoya N, Islam M, Sultana S, Mahia S, Sukorno F, Biswas M (2018) Phytochemical Screening and evaluation of antioxidant, antibacterial and cytotoxicity activities of methanolic and ethanolic extracts of Enhydra fluctuans Lour (Helencha) (family: Asteraceae) stems and leaves. Int J Plant Soil Sci 23(6):1–14. https://doi.org/10.9734/ijpss/2018/42796

Hasan MN, Sabrin F, Rokeya B, Khan MSH, Ahmed MU, Matondo A, Billah MM, Akter S (2019) Glucose and lipid lowering effects of Enhydra fluctuans extract in cadmium treated normal and type-2 diabetic model rats. BMC Complement Altern Med 19(1):1–10. https://doi.org/10.1186/s12906-019-2667-5

Hossain E, Sarma SM, Sikder H, Kabir H (2016) Nutritive value of Helencha (Enhydra fluctuans). Online J Anim Feed Res 6(1):20–23

Kamal S, Rony SR, Sharmin S, Laboni FR, Sohrab MH (2019) Phytochemical and pharmacological potential of Enhydra fluctuans available in Bangladesh. J Pharm Res Int 29(4):1–11. https://doi.org/10.9734/jpri/2019/v29i430243

Khushi S, Akhter MJ, Hasan MM, Rahman MA, Murshid GMM, Sadhu SK (2016) Phytochemical and pharmacological screening of Enhydra fluctuans Lour. Life Sci 13(June):19–26

Kumar G, Dahiya L, Behera B (2023) Review on pharmaco-therapeutic potential of wetland herb Enhydra fluctuans Lour (Hilamochika). Int J Ayurveda Pharm Res 11(5):111–119

Kuri S, Billah MM, Rana SMM, Naim Z, Islam MM, Hasanuzzaman M, Ali MR, Banik R (2014) Phytochemical and in vitro biological investigations of methanolic extracts of Enhydra fluctuans Lour. Asian Pac J Trop Biomed 4(4):299–305. https://doi.org/10.12980/APJTB.4.2014C677

Lopa SS, Al-Amin MY, Hasan MK, Ahammed MS, Islam KM, Khurshid Alam AHM, Tanaka T, Sadik MG (2021) Phytochemical analysis and cholinesterase inhibitory and antioxidant activities of Enhydra fluctuans relevant in the management of Alzheimer's disease. Int J Food Sci 2021:1–8. https://doi.org/10.1155/2021/8862025

Mall TP (2017) Diversity of under-utilized plants for nutrition and health from Bahraich (UP), India. Int J Trop Agric 35(January 2017):69–88. https://www.researchgate.net/profile/Tej-Mall/publication/326345048_INTERNATIONAL_JOURNAL_OF_TROPICAL_AGRICULTURE_Diversity_of_Under-utilized_Plants_for_Nutrition_and_Health_from_Bahraich_UP_India/links/5b46ff0eaca272c609379591/INTERNATIONAL-JOURNAL-OF-TR

Mukherjee PK, Wahile A (2006) Integrated approaches towards drug development from Ayurveda and other Indian system of medicines. J Ethnopharmacol 103(1):25–35. https://doi.org/10.1016/j.jep.2005.09.024

Nandi A, Das A, Dey YN, Roy KK (2023a) The abundant phytocannabinoids in rheumatoid arthritis: therapeutic targets and molecular processes identified using integrated bioinformatics and network pharmacology. Life 13(3):1–19. https://doi.org/10.3390/life13030700

Nandi A, Das A, Lin W-Y, Chakraborty P (2023b) Antiviral drugs – in past and current pandemics. In: Current trends in the identification and development of antimicrobial agents. Bentham Science Publishers, pp 167–194. https://doi.org/10.2174/9789815080056123020009

Nandi A, Nigar T, Das A, Dey YN (2023c) Network pharmacology analysis of Plumbago zeylanica to identify the therapeutic targets and molecular mechanisms involved in ameliorating hemorrhoids. J Biomol Struct Dyn:1–15. https://doi.org/10.1080/07391102.2023.2280681

Nandi A, Auti PS, Jagtap UA, Paul AT (2024a) Investigating the role of indole and quinazolinone-based hybrid analogues with ketoamide fragment and alkyl extension for potential PL inhibition. J Mol Struct 1301:137337. https://doi.org/10.1016/j.molstruc.2023.137337

Nandi A, Dey YN, Maity D, Das A (2024b) Vaccine development strategies and impact. In: Diagnosis and analysis of COVID-19 using artificial intelligence and machine learning-based techniques, pp 149–180. https://doi.org/10.1016/B978-0-323-95374-0.00015-4

Nandi A, Mandal N, Das A, Dey YN (2024c) Approaches based on network pharmacology and molecular docking to predict the molecular mechanism of Plumbago zeylanica's anti-inflammatory action. J Biol Regul Homeost Agents 38(3):2055–2067

Nasreen A, Choudhury K (2018) Phytochemical analysis of Enhydra fluctuans (Fam. – Asteraceae) extracts to detect their chemical components with antioxidant property. Journal of Emerging Technologies and Innovative Research (JETIR) 5(5):31–38

Panda T (2021) Traditional herbal medicines in prevention and treatment of traditional herbal medicines in prevention and treatment of cardiovascular diseases in Odisha cardiovascular diseases in Odisha. J Phytol Res 34(2):239–251

Parveen D, Das A, Amin S, Alam MM, Akhter M, Khan MA, Ali R et al (2024) Effectiveness of estrogen and its derivatives over dexamethasone in the treatment of COVID-19. J Biomol Struct Dyn 42(4):1858–1874. https://doi.org/10.1080/07391102.2023.2205944

Phukan P, Bawari M, Sengupta M (2015) Promising neuroprotective plants from North-East India. Int J Pharm Pharm Sci 7(3):28–39

Rathod B, Joshi S, Regu S, Manikanta KVNS, Kumar H, Dubey S, Chowdhury A et al (2024) Design and synthesis of novel quinazolinone-based pyruvate kinase M2 activators as selective inhibitors of oral cancer cells. J Mol Struct 1304(January):137595–137607. https://doi.org/10.1016/j.molstruc.2024.137595

Roy KK, Mollah MKI, Reja MM, Shil D, Maji RK (2021) Evaluation of analgesic and anti-inflammatory activity of ethanolic extract of Enhydra fluctuans on male wistar rats. J Appl Pharm Res 9(1):37–42. https://doi.org/10.18231/joapr.2021.9.1.37.42

Ruhul Amin M, Ripon Mondol M, Habib R, Tofazzal Hossain M (2012) Antimicrobial and cytotoxic activity of three bitter plants – Enhydra fluctuans, Andrographis peniculata and Clerodendrum viscosum. Adv Pharm Bull 2(2):207–211. https://doi.org/10.5681/apb.2012.032

Saha S, Paul S (2019) A review on phytochemical constituents and pharmacological properties of Enhydra fluctuans Lour. J Pharmacogn Phytochem 8(2):887–893

Sannigrahi S, Mazuder UK, Pal DK, Parida S, Jain S (2010) Antioxidant potential of crude extract and different fractions of Enhydra fluctuans Lour. Iranian J Pharm Res 9(1):75–82

Sannigrahi S, Mazumder UK, Pal D, Mishra SL, Maity S (2011) Flavonoids of Enhydra fluctuans exhibits analgesic and anti-inflammatory activity in different animal models. Pak J Pharm Sci 24(3):369–375

Sarma U, Borah VV, Saikia KKR, Hazarika NK (2014) Enhydra fluctuans: a review on its pharmacological importance as a medicinal plant and prevalence and use in North–East India. Int J Pharm Pharm Sci 6:48–50

Simlai A, Chatterjee K, Roy A (2014) A comparative study on antioxidant potentials of some leafy vegetables consumed widely in India. J Food Biochem 38(3):365–373. https://doi.org/10.1111/JFBC.12062

Singh D, Patel R, Aggarwal A, Das A, Sharma S, Behera B, Panigrahy R et al (2024) Phenylthiazolidin-4-one piperazine conjugates: design, synthesis, anticancer and antimicrobial studies. Results Chem 7(October 2023):101237–101248. https://doi.org/10.1016/j.rechem.2023.101237

Suraiya S, Bristy SA, Ali MS, Biswas A, Ali MR, Haq M (2023) A green approach to valorizing abundant aquatic weeds for nutrient-rich edible paper sheets production in Bangladesh. Clean Technol 5(4):1269–1286. https://doi.org/10.3390/cleantechnol5040064

Upadhyay DB, Mokariya JA, Patel PJ, Patel SG, Das A, Nandi A, Nogales J et al (2024) Indole clubbed 2,4-thiazolidinedione linked 1,2,3-triazole as a potent antimalarial and antibacterial agent against drug-resistant strain and molecular modeling studies. Arch Pharm (December 2023):1–20. https://doi.org/10.1002/ardp.202300673

# Medicinal and Nutritional Importance of *Trachyspermum ammi* in Human Health

22

Kamilia Bireche and Asma Abid

**Abstract**

*Trachyspermum ammi* (commonly known as ajwain or carom seeds) is a traditional medicinal plant with a diverse array of bioactive compounds that offer significant health benefits. This chapter explores the therapeutic and nutritional potential of *T. ammi* in promoting human health. Rich in essential oils, phenolic compounds and flavonoids, *T. ammi* has demonstrated remarkable pharmacological activities including analgesic, antibacterial, antifungal, anti-inflammatory and antioxidant properties. Its active compounds, primarily thymol, contribute to its efficacy against various pathogens, and recent studies indicate its potential in managing conditions like hypertension, hyperlipidemia and digestive disorders. Nutritionally, *T. ammi* seeds are a valuable source of fibre, essential fatty acids and minerals, supporting gut health and aiding in cholesterol management. Emerging research also highlights its potential roles in antiviral activities and enzyme modulation, contributing to metabolic and immune support. With its broad-spectrum therapeutic effects and nutritional benefits, *T. ammi* holds promise as a complementary approach to conventional medicine and as a functional food for preventive health. This chapter delves into its bioactive mechanisms, clinical applications and future directions for research, emphasizing its value in both traditional and modern healthcare paradigms.

K. Bireche
Laboratory of Valorization and Promotion of Saharan Resources (VPRS), Faculty of Mathematics and Matter Sciences, University of Ouargla, Ouargla, Algeria

A. Abid (✉)
Laboratory of Valorization and Promotion of Saharan Resources (VPRS), Faculty of Mathematics and Matter Sciences, University of Ouargla, Ouargla, Algeria

Higher School of Saharan Agriculture, El Oued, Algeria



M. A. Ansari et al. (eds.), *Medicinal Plants and their Bioactive Compounds in Human Health: Volume 2*, https://doi.org/10.1007/978-981-95-3620-7_22

**Keywords**

*Trachyspermum ammi* · Pharmacological activities · Bioactive compounds · Nutrition · Human health

## 22.1 Introduction

Ajwain, scientifically referred to as *Trachyspermum ammi*, has a lot of historical and cultural background that has contributed to the understanding and application of this herb in contemporary medicine. Native to Egypt and widely cultivated in the Indo-Iranian region, among others, it has remained an integral traditional medicine system for many centuries in Ayurveda and Unani. Its diversified pharmacological properties were discovered and utilized within various cultures for ages, which also created a pathway for its incorporation into modern-day medicines. The next subtopic will discuss the historical background of this herb, its utilization by various cultures and its later integration into modern medicine.

*Trachyspermum ammi* has been a very important ingredient of Ayurvedic and Unani medicine, and it has been used in the stimulant, carminative and antispasmodic classes. It has been traditionally employed as a remedy for disorders of flatulence, dyspepsia, diarrhoea and bronchial complaints (Vharamble et al. 2023). In ethnomedicine, the seeds of the plant are pungent in taste and aromatic in odour and have been applied in various cultural remedies. Ajwain is one of the ingredients in various Indian culinary practices that, beyond its culinary value, acts as a medicinal component. It helps in digestion and assists in relieving respiratory diseases (Vharamble et al. 2023; Dahiya et al. 2023). During the middle age of Persia, it was known to be medicinally active and in their medical manuscripts was named Zenyan or Nankhah. Such references prove that ajwain has been used as a medication for thousands of years by different cultures (Goyal et al. 2022).

Ajwain is a reservoir of bioactive compounds such as thymol, γ-terpinene and para-cymene. These compounds are very rich in the pharmacological values they give. The compounds have great antimicrobial, antioxidant and anti-inflammatory properties (Bairwa et al. 2012). Studies have supported ajwain for potential therapeutic applications. Particularly, its antidiabetic remedy was supported because the compound shows an inhibitive role towards α-amylase and induces glucose utilization, thus proving it to be a useful assistant for the management of diabetes (Amar and Kuppusamy 2022). Ajwain has also been found to exhibit very impressive antifungal properties against Candida albicans, further solidifying its role in fighting off fungal infection. This combination of properties positions ajwain as an intriguingly multi-purpose natural remedy with great potential in health and well-being (Arasu et al. 2021).

The wide pharmacological activities of *Trachyspermum ammi* have been justified in the modern scientific literature and include bronchodilatory, antihypertensive and hepatoprotective properties. Indeed, these would be a justification for the use of this herb in current mainstream medical therapy for the treatment of asthma, hypertension and liver dysfunction (Vharamble et al. 2023; Khan and Gilani 2023). On the other hand, techniques like microwave-assisted extraction continue to reduce the effort and enhance the yield of ajwain's active ingredients, thereby making its active constituents more accessible for their pharmaceutical applications (Nabi et al. 2023). More information regarding the plant *Trachyspermum ammi* will be discussed in this chapter book to point out its importance for human health in both medical and nutritional terms.

## 22.2 Botanical Description and Classification of *Trachyspermum ammi*

The general name for *Trachyspermum ammi* is ajwain, which belongs to the Apiaceae family and is a first-class herbaceous plant. It is regarded for its well-characterized morphological features, distinguishing it from all other species in the same genus. These not only play a significant role in plant identification but are responsible for the popular use of this plant in various culinary and medicinal purposes. In the following section, important morphological features of *Trachyspermum ammi* are enumerated, distinguishing it from its congeners.

### 22.2.1 Taxonomic Classification (Goyal et al. 2022)

| | |
|---|---|
| Kingdom | Plantae, Plant |
| Subkingdom | Tracheobionta, Vascular plant |
| Phyllum | Spermatophyta, Seed plant |
| Subphyllum | Angiospermae, Flowering plant |
| Class | Magnoliopsida, Dicotyledons |
| Order | Apiales |
| Family | Apiaceae |
| Genus | Trachyspermum |
| Species | Ammi |

Synonyms: *Carum copticum* Benth. & Hook (Naquvi et al. 2022).

### 22.2.2 Distribution and Cultivation

Native Locations: *Trachyspermum ammi*, native to Egypt is now cultivated in India, Iran, Pakistan, Iraq, Afghanistan and parts of Europe (Naquvi et al. 2022).

Cultivation in India: In India, these plants are cultivated as a principal seed spice crops in the Gujarat and Rajasthan (Dahiya et al. 2023).

### 22.2.3 Botanical Description

*Trachyspermum ammi* is an annual herb, tall, with white flowers. The plant bears small, brownish seeds that are greyish-brown in colour, pungent in taste and aromatic in odour. These seeds are 2 mm in length and 1 mm in width, consisting of two mericarps, with a bifid stylopod, and five light-coloured ridges (Nabi et al. 2023; Vharamble et al. 2023).

### 22.2.4 Fruit Morphology

The fruit of *Trachyspermum ammi* is a schizocarp and divides into two mericarps when it attains maturity. Each mericarp bears bifid stylopods and five light-coloured ridges, which are its characteristic features. The fruits are ovoid, about 2 mm long and 1 mm wide. It has a greyish-brown colour with a pungent taste and aromatic odour distinctive to *T. ammi* (Nabi et al. 2023; Vharamble et al. 2023).

### 22.2.5 Seed Characteristics

The seeds of *T. ammi* are small and grey, with a bitter and peppery taste. This has great importance in distinguishing the seeds of *T. ammi* from other species since there are other species whose seeds are of different sizes, colours or tastes (Dahiya et al. 2023). Seeds contain about 5% essential oil, which is quite a high concentration and adds to the value of their medicinal and culinary uses (Nabi et al. 2023).

The botanical attributes of *Trachyspermum ammi* are well documented and form reliable identifiers; morphological variations depending on environment and cultivation practice may also be considered. Besides this, the presence of other similar species under this Apiaceae family sometimes leads to misidentification. A holistic approach involving botanical and phytochemical analyses can be suggested for proper identification and differentiation of species. This holistic approach ensures that specific attributes unique to *T. ammi* continue to be consistently recognized and utilized across traditional and contemporary applications.

## 22.3 Phytochemical and Nutritional Composition

*Trachyspermum ammi* possesses a rich bioactive principle of enormous pharmacological and medicinal value. The seeds of *T. ammi* are particularly noted for the diverse chemical constituents in them that have added to the therapeutic potentiality of this plant. This section describes some of the major bioactive compounds isolated from *T. ammi* based on several research studies.

**Rosmarinic Acid-3-O-Glucopyranoside, Kaempferol-(Coumaroyl-Glucosyl)-Rhamnoside and Quercetin-3-O-Galactoside** These phenolic compounds have been isolated from the de-oiled seeds of *T. ammi*, showing significant antioxidant and anticandidal activities (Dutta and Kundu 2022).

**3,5-Dimethylanisole and 6-Octadecenoic Acid, Methyl Ester** Identified through GC-MS analysis, these compounds are part of the volatile and semi-volatile profile of *T. ammi* seeds (Abdullah et al. 2020).

At the end of this section will find Table 22.1 which will detail more information about the bioactive and principal constituents of ajwain.

### 22.3.1 Essential Oils and Volatile Compounds

**Thymol** Thymol is one of the major ajwain essential oil constituents. Thymol constitutes 35% to 60% of the concentration of this essential oil. It exhibits antiseptic and antimicrobial properties (Ullah et al. 2024).

**Carvacrol** Carvacrol is another major constituent present in this essential oil. Very recently, this carvacrol was identified with potential inhibition of PD-1/PD-L1 axis, which plays a significant role in tumour immunity (Kumar et al. 2024).

**p-Cymene and γ-Terpinene, α- and β-Pinenes, Phytoestrogens and Catecholamines** These, too, are found in *T. ammi* essential oil and lend it its characteristic aroma and may also be responsible for some of its pharmacological properties (Quoc 2022; Chahal et al. 2017; Nabi et al. 2023).

**Essential Oil Yield** Fruits produce 2–4% of essential oil, which is accountable for the smell and taste of ajwain (Ullah et al. 2024).

### 22.3.2 Macronutrients and Fibre

**Carbohydrates** Ajwain seed contains about 38.6% carbohydrate, serving as a chief energy source (Ullah et al. 2024).

**Protein** These seeds form a moderately good source of proteins and about 15.4% of the composition of the seed (Ullah et al. 2024).

**Table 22.1** Comprehensive table of *T. ammi*'s nutritional composition (Nabi et al. 2023; Chauhan et al. 2012)

| | Category | Details |
|---|---|---|
| Nutritional composition | Fibre | 11.9% |
| | Carbohydrates | 24.6–38.6% |
| | Protein | 17.1–15.4% |
| | Fat | 18.1–21.1% |
| | Moisture | 8.9% |
| | Mineral matter | 7.1% (includes calcium, phosphorus, iron, nicotinic acid) |
| | Other compounds | Tannins, glycosides, saponins, flavones |
| Essential oil | Yield | 2–4% brownish essential oil |
| | Major constituent | Thymol (35–60%) |
| | Other components | α-thujene, α-pinene, sabinene, β-pinene, α-phyllandrene, γ-terpinene, p-cymene, β-phyllanderene, terpinene-4-ol, carvacrol, styrene, δ-3-carene |
| Identified monoterpenoids | 3,7-dimethyloct-3(10)-ene-1,2,6,7-tetrol (mixture of two stereoisomers) | Source: Water-soluble portion of the methanol extract of the fruit of *Carum copticum* |
| | A: (2S,6Z)-3,7-dimethyloct-3(10)-ene-1,2,6,7-tetrol 1-O-β-D-glucopyranoside | |
| | B: 6-hydroxythymol 6-O-β-D-glucopyranoside | |
| | C: 6-hydroxythymol 3-O-β-D-glucopyranoside | |
| | D: 7-hydroxythymol 3-O-β-D-glucopyranoside ($C_{16}H_{24}O_7$) | |
| | E: (4R,6S)-p-menth-1-ene-4,6-diol 4-O-β-D-glucopyranoside ($C_{16}H_{28}O_7$) | |
| | F: 6-hydroxythymol 3,6-di-O-β-D-glucopyranoside ($C_{22}H_{34}O_{12}$) | |
| | G: (4S)-p-menth-1-ene-4,7-diol 4-O-β-D-glucopyranoside ($C_{16}H_{28}O_7$) | |
| | H: (4R,6S)-p-menth-1-ene-4,6-diol 4-O-β-D-glucopyranoside ($C_{16}H_{28}O_7$) | |
| | I: 3β-hydroxy-p-menth-1-en-4β,5β-oxide 3-O-β-D-glucopyranoside ($C_{16}H_{26}O_7$) | |
| | p-menth-3-ene-1β,2β,5β-triol | |

(continued)

**Table 22.1** (continued)

| | Category | Details |
|---|---|---|
| Alkyl glucoside identified | 2-methyl-3-buten-2-ol-β-D-glucopyranoside | |
| Aromatic compound | Benzyl-β-D-glucopyranoside | |
| | 19-(3-hydroxy-4,5-dimethoxyphenyl)-propane-29, 39-diol | |
| | $C_{15}H_{22}O_8$ characterized as 3,4-dihydroxyphenylpropanol-3-O-β-D-glucopyranoside | |
| Nucleosides identified | Adenosine, uridine | |
| Glucides identified | (2S,3R)-2-methylbutane-1,2,3,4-tetrol and (3R)-2-hydroxymethylbutane-1,2,3,4-tetrol | |
| Glucosides identified | 1-deoxy-L-erythritol ($C_4H_{10}O_3$), 1-deoxypentitol (C5H12O4) | |
| Minerals in ajwain fruits | Aluminium, calcium, cadmium, copper, iron, lithium (nitrates and nitrites not detected) | |

**Fat** Ajwain contains about 18.1% fat, including essential fatty acids (Ullah et al. 2024).

**Fibre** Ajwain comprises 11.9% of fibre, which is good for the digestive system's health and gut (Ullah et al. 2024).

### 22.3.3 Minerals and Vitamins

The mineral content also includes calcium, phosphorus and iron, which help in nourishing bones and transporting oxygen within the system. Nicotinic acid, also known as niacin or vitamin B3, is present in ajwain. It participates in energy metabolism, besides maintaining healthy skin (Ullah et al. 2024).

Nutritional composition gives ajwain a vital place in food as well as pharmacology. It contains essential oils, minerals and bioactive compounds, which make it a candidate for various health benefits.

Ajwain, renowned for its health benefits, should be used in moderation, as excessive intake can lead to adverse effects. Further research is also warranted to investigate potential interactions between ajwain and other dietary elements or medications. While the bioactive compounds of *T. ammi* offer promising therapeutic properties, challenges remain in their extraction and practical application. Traditional extraction methods are often time-intensive and inefficient, underscoring the need for modern approaches, such as microwave-assisted extraction, to improve both yield and bioactivity. This method and its benefits will be explored further in the following section.

## 22.4 Extraction Methods

There are several techniques for extracting bioactive compounds from *Trachyspermum ammi*, with each having its ways of maximizing yield, maintaining the integrity of the compounds, and ensuring sustainability. These vary from conventional techniques developed over time to more advanced and efficient modern-day approaches. An appropriate choice of extraction method is of paramount importance because essential oils and other phytochemicals of immense therapeutic potential constitute an important group of bioactive compounds. Various discussions on the different extraction techniques of the bioactive compounds of *T. ammi* are given below.

### 22.4.1 Conventional Extraction Methods

**Solvent Extraction** The most prevalent methods of extracting bioactive compounds from *T. ammi* are through the use of solvents. The solvents most commonly utilized are methanol, ethanol, *n*-hexane and ethyl acetate. It is very important to choose the right solvent because the polarity will affect the resulting bioactivity of the extract. However, these traditional methods at large, including maceration and Soxhlet extraction, are very time-consuming and may not be feasible for a high yield and low cost (Quoc 2022; Abdullah et al. 2020).

**Hydrodistillation** The essence of extraction of essential oils from *T. ammi* seeds or water or steam is allowed to extract volatile compounds. These get condensed and eventually collected. It is a traditional technique and widely applied as it is simple in nature (Quoc 2022).

### 22.4.2 Modern Extraction Techniques

**Microwave-Assisted Extraction (MAE)** MAE is one of the novel techniques for extraction in which microwave energy is used to heat up the solvent and plant material, thereby enhancing the process of extraction. This technique is less time-consuming and more cost-effective than other conventional techniques. MAE has been reported to give a higher extractive value, and hence MAE is very effective in obtaining essential oil from *T. ammi* (Nabi et al. 2023; Dutta et al. 2021).

**Ultrasound-Assisted Extraction (UAE)** This technique depends on the ultrasonic waves for the agitation of plant material-solvent interaction, therefore enhancing the rate of the solvent's penetration into the plant matrix. Good yields in the total phenolic content and antioxidant activity have been reported using methanol as the extraction solvent (Dutta et al. 2021).

**Supercritical Fluid Extraction (SFE)** SFE uses supercritical fluids, most often carbon dioxide, to execute the extraction of bioactive compounds. This method has been characterized by effectiveness and mild treatment of the environment since less solvent and lower temperatures preserve the integrity of heat-sensitive compounds (Debnath and Sharma 2024; Quoc 2022).

**Pressurized Liquid Extraction (PLE)** In the process of PLE, high pressure and temperature are maintained to retain the solvent in a liquid state that improves extraction efficiency. Since it consumes less solvent, requires less time for preparation, and has higher reproducibility, this technique is considered favourable (Barik et al. 2024).

**High-Speed Homogenization Extraction (HHE)** HHE uses high-speed mechanical forces to disrupt plant cells, which favours the release of bioactive compounds. This technique was effective in improving the extraction yield and quality of *T. ammi* extracts (Dutta et al. 2021).

### 22.4.3 Factors Influencing Extraction

**Solvent Selection** The selection of solvent is made on the basis of the principle that the closer the polarity of the solvent to that of the compound of interest, the greater will be its extraction efficiency and bioactivity. Methanol, being highly polar, has been conventionally commonly used in the extraction of phenolic compounds (Quoc 2022; Dutta et al. 2021).

**Temperature and Pressure** Both parameters are of great importance in newer techniques of extraction such as MAE, UAE and PLE since they can amplify the solubility and diffusion of the bioactive compounds (Debnath and Sharma 2024; Barik et al. 2024).

Although modern extraction methods generally offer significant improvements related to efficiency and sustainability, they do require special equipment and technical expertise. The choice of method will follow not only from the identity of targeted bioactive compounds, desired yield and available resources but also from ecological and economic impacts to guide sustainable practices in the extraction of *T. ammi* bioactive compounds.

## 22.5 Traditional Uses of *T. ammi*

*T. ammi* is a traditional medicinal plant with a rich history of use in various cultural and medicinal practices. It is primarily cultivated in regions such as India, Iran and Egypt, and is renowned for its distinctive aroma and therapeutic properties. Ajwain seeds are particularly valued for their essential oils, which contain bioactive compounds like thymol, carvacrol and p-cymene. These compounds contribute to the plant's diverse pharmacological activities and traditional uses. Below, we explore the traditional applications of *Trachyspermum ammi* in detail.

**Digestive Health** Ajwain is traditionally used to treat a variety of digestive disorders. It is known for its carminative properties, which help alleviate flatulence, indigestion and colic. The seeds are often used in Ayurvedic and Unani medicine to stimulate appetite and treat dyspepsia and diarrhoea (Vharamble et al. 2023; Bairwa et al. 2012) The seeds are also used to address abdominal pain and piles, showcasing their broad application in gastrointestinal health (Bairwa et al. 2012).

**Respiratory Ailments** Ajwain has been traditionally employed to manage respiratory issues such as asthma and bronchitis. Its bronchodilatory effects are comparable to those of theophylline, a common asthma medication, making it a valuable natural remedy for respiratory conditions (Vharamble et al. 2023). The seeds are also used to treat coughs and colds, providing relief from symptoms associated with respiratory infections (Goyal et al. 2022).

**Pain Relief** Ajwain is used for its analgesic properties to relieve headaches, colic and rheumatism. It is also applied externally for earaches and other pains (Hanif et al. 2021; Saraswat et al. 2020).

**Antimicrobial and Antifungal Uses** The essential oils in ajwain seeds possess strong antimicrobial and antifungal properties, which have been utilized in traditional medicine to treat infections. These properties are attributed to the presence of compounds like thymol and carvacrol (Dahiya et al. 2023). Ajwain's antimicrobial activity has been validated through modern studies, which demonstrate its effectiveness against various bacterial and fungal species (Nag and Gupta 2023).

**Women's Health** In traditional medicine, ajwain is used as a galactagogue to promote lactation in nursing mothers. It is also employed to alleviate menstrual pain and regulate menstrual cycles (Naquvi et al. 2022). The seeds are believed to have abortifacient properties, although this use is approached with caution due to potential risks (Bairwa et al. 2012).

**Other Traditional Uses** Ajwain is used in traditional medicine to treat conditions such as headaches, heartburn and cholera. It is also believed to support kidney function and provide relief from arthritis and other inflammatory conditions (Goyal et al. 2022). The seeds are used as a diuretic and to manage hypertension and hyperlipidaemia, reflecting their broad therapeutic potential (Bairwa et al. 2012).

Although *T. ammi* is commonly used in traditional medicine, it's essential to recognize potential adverse effects and interactions with other medications. Its potent bioactive compounds call for cautious use, especially among vulnerable groups like pregnant women and individuals with existing health conditions. Ongoing pharmacological research is dedicated to evaluating the efficacy and safety of ajwain, with a focus on harmonizing traditional knowledge with modern medical practices (Dahiya et al. 2023; Goyal et al. 2022).

## 22.6 Pharmacological Activities

This plant, known for its broad range of traditional pharmacological uses, has attracted recent scientific attention for its potential in modern medicine. Studies have highlighted its diverse bioactive compounds and their effects. This section details the pharmacological activities of *T. ammi*, supported by recent research findings.

### 22.6.1 Antioxidant and Antimicrobial Activities

*Trachyspermum ammi* exhibits significant antioxidant properties, primarily due to its ability to scavenge free radicals. This activity is attributed to compounds such as kaempferol and sinapic acid, which were identified in nanosuspensions of the plant extract. The plant also demonstrates antibacterial activity, particularly against *Escherichia coli*, by inhibiting biofilm formation. This suggests its potential use in treating bacterial infections (Jabeen et al. 2023). Essential oils from ajwain seeds contain monoterpenes and sesquiterpenes, which contribute to their antimicrobial and antiviral properties (Vharamble et al. 2023).

### 22.6.2 Antidiabetic and Antiurolithic Effects

*T. ammi* has shown promising antidiabetic effects. Studies using methanolic extracts of the plant have demonstrated significant glucose-lowering effects in rodent models, suggesting its potential in managing diabetes (Raju et al. 2023). The plant also exhibits antiurolithic activity, which involves inhibiting calcium oxalate crystal aggregation and providing diuretic effects. This activity is beneficial in preventing kidney stone formation (Khan and Gilani 2023).

### 22.6.3 Neuroprotective and Antispasmodic Properties

Ajwain extracts have been found to offer neuroprotection by inhibiting acetylcholinesterase and reducing oxidative stress, which are crucial in managing neurodegenerative diseases like Alzheimer's disease (Sharma et al. 2023). Prevention of Amyloid-beta Fibrilization: Ajwain bioactives inhibit the oligomerization and fibrilization of amyloid-beta peptides, which are implicated in Alzheimer's pathology (Sharma et al. 2023). Promotion of Neurogenesis: Thymol in ajwain promotes neurogenesis, enhancing learning and memory in Alzheimer's disease models (Timalsina et al. 2023). The plant's antispasmodic properties are evident in its ability to relax smooth muscle contractions, which can be beneficial in treating conditions like asthma and gastrointestinal disorders (Khan and Gilani 2023; Naquvi et al. 2022).

### 22.6.4 Hepatoprotective and Cardiovascular Benefits

Ajwain has demonstrated hepatoprotective effects, providing significant protection against liver damage induced by toxic agents (Vharamble et al. 2023). The plant also exhibits antihypertensive and hypolipidemic activities, which contribute to cardiovascular health by lowering blood pressure and cholesterol levels (Naquvi et al. 2022).

### 22.6.5 Hypolipidemic Effects

Ajwain has been shown to lower cholesterol levels, including LDL cholesterol, and improve the lipid profile, which is beneficial for heart health (Ullah et al. 2024).

### 22.6.6 Antihypertensive Properties

The seeds have a blood pressure-lowering effect, which is beneficial for individuals with hypertension (Ullah et al. 2024).

### 22.6.7 Diuretic and Antilithiatic

Ajwain is known for its diuretic properties and has been used in ethnopharmacology to treat urinary stones, showing satisfactory results against calcium oxalate stones (Ullah et al. 2024).

### 22.6.8 Gastrointestinal Health

The essential oil of *T. ammi* is effective in treating gastrointestinal disorders, including peptic ulcers. Its major components, such as thymol, para-cymene and gamma-terpinene, have been shown to reduce the incidence of peptic ulcers in animal models, with thymol being particularly potent (Eftekhari et al. 2019), and additionally, *T. ammi* is traditionally used to alleviate flatulence, dyspepsia and diarrhoea (Anwar et al. 2016).

### 22.6.9 Immunomodulatory Effects

*T. ammi* extracts have been found to modulate immune responses, enhancing cell-mediated immunity. This is particularly useful in managing conditions that require immune system regulation (Siddiqui et al. 2019).

### 22.6.10 Anti-cancer of *T. ammi*

The methanolic extract and essential oil of *T. ammi* seeds have been shown to possess cytotoxic effects against different cancer cell lines, including human oral and colorectal cancer cells. These effects are primarily attributed to the plant's ability to induce apoptosis and generate reactive oxygen species (ROS), which are crucial mechanisms in cancer treatment. The following sections detail the anticancer activities of *T. ammi* as reported in the studies.

#### 22.6.10.1 Anti-cancer Activity on Oral Cancer Cells

The methanolic extract of *T. ammi* seeds exhibited cytotoxicity against the KB human oral cancer cell line with an IC50 value of 125 μg/mL.

The extract promotes ROS (generation and modulates apoptosis induction, suggesting its potential as an anticancer agent for oral cancer treatment (Aruchamy and Duraisamy 2021).

#### 22.6.10.2 Anti-cancer Activity on Colorectal Cancer Cells

*T. ammi* essential oil, when encapsulated in PLGA nanoparticles, showed significant anticancer activity against the HT-29 human colon cancer cell line, indicating its potential as a natural and efficient anticancer compound (Aruchamy and Duraisamy 2021).

In combination with red light irradiation, *T. ammi* essential oil reduced the survival of SW480 colorectal cancer cells by inducing ROS generation and causing cell cycle arrest at the G0/G1 phase (Khorsandi et al. 2022).

#### 22.6.11 Other Pharmacological Activities

*T. ammi* is known for its antimicrobial, antiviral and anti-inflammatory properties. It has been traditionally used to treat respiratory ailments, digestive disorders and as a carminative and stimulant. The plant also shows potential in *hepatoprotection, bronchodilation and antihypertensive* effects, making it a versatile medicinal herb (Vharamble et al. 2023; Naquvi et al. 2022).

*Trachyspermum ammi* is a versatile medicinal plant with a broad spectrum of pharmacological activities. Its traditional uses are supported by modern scientific research, which highlights its potential in treating various health conditions Table 22.2. However, while the plant shows promise, further studies are needed to fully understand its mechanisms and optimize its use in clinical settings. This comprehensive exploration of its pharmacological profile underscores the importance of integrating traditional knowledge with contemporary scientific approaches.

## 22.7 Practical Applications of *T. ammi*

### 22.7.1 Culinary and Flavouring Uses

Ajwain is widely used as a spice in Indian cuisine, imparting a distinct aroma and flavour to dishes. It is commonly added to curries, pickles and baked goods (Jahan et al. 2013).

### 22.7.2 Nutraceutical Applications

#### 22.7.2.1 Dietary Supplements

Ajwain is rich in bioactive compounds like thymol, carvacrol and terpinene, which contribute to its health benefits. These compounds make ajwain a valuable ingredient in dietary supplements aimed at improving digestion, reducing bloating and enhancing overall health.

#### 22.7.2.2 Functional Foods

The use of ajwain in food processing, such as in pickles and baked goods, not only adds flavour but also imparts health benefits due to its antimicrobial and digestive properties. This positions ajwain as a functional food ingredient with potential nutraceutical applications (Siddiqui et al. 2019; Vharamble et al. 2023).

**Table 22.2** Comprehensive table of *Trachyspermum ammi*'s pharmacological activities (Nabi et al. 2023; Chauhan et al. 2012)

| Activity | Used parts | Mechanism | Additional details |
|---|---|---|---|
| Analgesic and antinociceptive | Ethanolic and essential oils | Extends tail-flick latency, indicating pain relief; likely through the action of thymol. | Reduces neuropathic foot burns significantly compared to placebo; antinociceptive effects observed in formalin tests. |
| | Fruits (Ethanolic extract) | Increased tail-flick latency (TFL) in response to pain. | Measured using tail-flick analgesiometer for 2 hours post-administration. |
| Antibacterial | Essential oil, methanolic extracts | Inhibits growth of Gram-positive and Gram-negative bacteria by disrupting cell membranes or metabolic processes. | Effective against pathogens like Salmonella, Staphylococcus; methanolic extracts show significant inhibition zones. |
| | Fruits (acetone, aqueous, methanolic extracts) | Inhibits growth of bacterial strains via cell wall disruption. | Effective against *E. faecalis*, *S. aureus*, *E. coli*, *K. pneumonia*, *P. aeruginosa*, *S. typhi* and others. |
| Antifilarial | Fruits (methanolic extract) | Exhibits macrofilaricidal activity against filarial worms. | Effective against *Setaria digitata* (bovine) and *Brugia malayi* (human). |
| Antifungal | Essential oil, methanolic extracts | Exhibits antifungal activity by disrupting fungal cell walls or metabolic processes. | Active against fungi such as *Curvularia ovoidea* and *Aspergillus niger*; higher activity in methanolic extracts than aqueous ones. |
| Insecticidal | Essential oil, seeds | Causes mortality in target insects; may disrupt nervous system function or feeding mechanisms. | Significant mortality observed in *P. interpunctella*; increased concentrations of essential oil lead to higher mortality rates. |
| | Essential oil | Inhibits oviposition, egg hatching and development of insects. | Causes chronic toxicity in *Callosobruchus chinensis*, protecting stored grains. |

(continued)

**Table 22.2** (continued)

| Activity | Used parts | Mechanism | Additional details |
|---|---|---|---|
| Anthelmintic | Methanolic extract | Interferes with energy metabolism of parasites; affects ATPase activity, leading to impaired movement and expulsion of parasites. | Effective against *Setaria digitata* and *Brugia malayi*; contains thymol, which disrupts membrane functions and energy metabolism. |
| | Essential oil | Interferes with ATPase activity, depleting energy reserves in parasites. | Effective against *Haemonchus contortus* in sheep and Ascaris lumbricoides in humans. |
| Antiplatelet activity | Dried ethereal extract | Prevents platelet aggregation induced by arachidonic acid, collagen and epinephrine. | In vitro study using human blood samples demonstrated effective inhibition of platelet aggregation. |
| Anti-inflammatory effects | *n*-Hexane, chloroform, methanol, aqueous extracts | Inhibits production of inflammatory mediators (e.g., leukotrienes, prostaglandins) and enhances GABA neurotransmission; reduces carrageenan-induced oedema. | Effective at doses of 500 and 1000 mg/kg; *n*-hexane showed superior anti-inflammatory effects compared to other extracts. |
| Antitussive and bronchodilatory effects | Macerated and aqueous extracts, essential oils | Reduces cough frequency through anti-inflammatory properties; inhibits H1 histamine receptors. | Significant cough reduction in clinical tests; essential oils, particularly carvacrol, exhibit bronchodilatory effects comparable to theophylline. |
| Diuretic and anti-lithiasis activity | Ajwain seeds | Enhances urine production and excretion of salts and potassium; prevents calcium oxalate stone formation by reducing urinary calcium, oxalate and phosphate levels. | Ethnopharmacological claims supported by human and animal studies; effective in managing urinary incontinence and preventing kidney stones. |
| Antihyperlipidemic properties | Ajwain seeds (powder, extracts) | Reduces levels of LDL cholesterol, total lipids and triglycerides; increases HDL cholesterol levels. | Petroleum ether extract showed more significant results than methanolic extract; effective in rat models of hyperlipidaemia. |
| | Powder | Lowers cholesterol, triglycerides, LDL and total lipid levels. | Prevented CCl4-induced hepatotoxicity and normalized liver enzyme levels in rats. |

(continued)

**Table 22.2** (continued)

| Activity | Used parts | Mechanism | Additional details |
|---|---|---|---|
| Detoxification activity | Ajwain seeds | Reduces aflatoxin G1 levels; thymol enhances liver function, promotes toxin elimination and aids digestion. | Significant reduction of aflatoxin observed in a randomized controlled trial; cleansing effect attributed to thymol. |
| | Seeds | Degrades aflatoxins through enzymatic breakdown. | Effective against aflatoxins AFG1, AFB1, AFB2 and AFG2. |
| Antioxidant properties | Essential oil, extracts | Neutralizes free radicals, preventing cellular damage; polyphenols and thymol exert antioxidant and anti-inflammatory effects. | High antioxidant activity demonstrated through methods like FRAP, DPPH and ABTS; Ajwain extract reduces oxidative stress from toxins. |
| Antiulcer activity | Ethanolic extract | Reduces ulcerative lesions and wound index in various ulcer models; protective effect on gastric mucosa. | Significant reduction in ulcer index with doses of 100 mg/kg and 200 mg/kg in rat models; shows potent anti-ulcer properties. |
| Antihypertensive activity | Aqueous-methanolic extract | Lowers arterial blood pressure; inhibits K-shrinkage in isolated rabbit aorta and jejunum preparations. | Supports antihypertensive and antispasmodic properties; demonstrated in sedated animal models. |
| | Whole plant | Acts as a calcium channel blocker, relaxing smooth muscles. | Reduces blood pressure, relieves gut spasms and dilates bronchi. |
| Digestive stimulant activity | Ajwain seeds | Increases bile and stomach acid production, enhancing digestive enzyme activity; promotes overall digestion. | Stimulates pancreatic lipase and amylase; considered a traditional remedy for digestive issues. |
| Gastroprotective | Fruits (Ethanolic extract) | Reduces gastric transit time and inhibits *H. pylori*. | Gastroprotective effects prevent peptic ulcer and gastric cancer. |
| Viability of spermatogonia stem cells | Essential oil (thymol) | Promotes spermatogonia stem cell growth and gene expression; decreases cell apoptosis. | Increased colony size and number with oil treatments; thymol exhibited antioxidant properties, affecting gene expressions related to cell survival. |

(continued)

**Table 22.2** (continued)

| Activity | Used parts | Mechanism | Additional details |
|---|---|---|---|
| Spermicide activity | Essential oil (*T. ammi*) | Damages human spermatozoa's membrane and DNA; inhibits motility; leads to decreased viability and altered cell morphology. | Potent spermicidal action; potential natural contraceptive. |
| Toxicity and teratogenicity | Seeds | Reported teratogenic effects in rat foetuses; associated with risks of congenital abnormalities in humans. | Use of *T. ammi* seeds linked to abortion attempts; potential for human fetotoxicity. |
| Oestrogenicity | Dry ajwain seeds | Contains phytoestrogens; historically used as a galactagogue. | Total phytoestrogen concentration of 473 ppm; second highest among tested plants. |
| Antidiabetic activity | Thymol | Reduces blood glucose levels by stimulating insulin secretion; protective effects against diabetic neuropathy. | Significant blood glucose reduction at doses of 10 and 20 mg/kg; progressive healing activity observed in treated groups. |
| Antimicrobial activity | Essential oil, extracts | Effective against gram-negative bacteria (e.g., *E. coli*, *Klebsiella pneumonia*) and certain fungi; antiviral effects against HCV. | Minimum inhibitory concentrations (MIC) of 100 ppm for E. coli and 250 ppm for *Klebsiella*; antifungal activity against *Candida* and *Aspergillus* species at low concentrations. |
| Anti-amnesic | Seed powder (*T. ammi*) | Improved cognitive function by increasing brain GSH levels, reducing MDA and enhancing step-down latency in passive avoidance. | Conducted in mouse models; showed significant effects against scopolamine, electroshock and alprazolam-induced amnesia. |
| Anti-Candida activity | Methanol extract, aromatic water | Antifungal activity against *Candida* species by altering membrane permeability and inhibiting ergosterol synthesis. | Maximum inhibition zones against *C. albicans*: 38.3 mm; aids in healing necrotic tissue; possesses antibacterial properties. |

(continued)

**Table 22.2** (continued)

| Activity | Used parts | Mechanism | Additional details |
|---|---|---|---|
| Anti-rheumatic potential | Selenium nanoparticles (SeNPs) | Reduces inflammation and improves redox state; inhibits MMP9, IL-1, IL-6, TNF- and CRP. | SeNPs showed no toxicity in mice; decreased paw oedema and lymphocyte infiltration; beneficial for rheumatoid arthritis. |
| Antispasmodic activity | *Trachyspermum ammi* essence | Induces relaxation of ileal smooth muscle through cholinergic and histaminergic pathways. | Exhibited a stronger effect compared to aqueous extracts; effective against Ach-induced contractions. |
| Antiviral activity | Essential oil (*T. ammi*) | Demonstrated antiviral effects against Japanese encephalitis virus (JEV) in vitro. | Tested using plaque assays and neutralization tests; shows potential as an antiviral agent. |
| | Seeds | Inhibits HCV protease activity, preventing viral replication. | Significant inhibition (≥60%) at 100 µg/mL against HCV protease. |
| Antihistaminic | Aqueous, ethanolic extracts | Acts as H1 receptor antagonist, competing with histamine. | Observed in histamine response curves on guinea pig tracheal chains. |
| Enzyme modulation | Seeds (thymol) | Alters acetylcholinesterase, lactic dehydrogenase, succinic dehydrogenase and cytochrome oxidase activities. | Proven molluscicidal against *Lymnaea acuminata*. |

### 22.7.3 Food Preservation and Safety

#### 22.7.3.1 Antifungal Properties

The essential oil of *Trachyspermum ammi* has demonstrated significant antifungal activity against *Aspergillus flavus*, a common food contaminant. It effectively inhibits aflatoxin B1 production, thereby protecting food products like Sorghum bicolour from spoilage and nutritional degradation (Singh et al. 2024).

#### 22.7.3.2 Antimicrobial Activity

The seed extract of *Trachyspermum ammi* has been identified as a potential natural preservative due to its ability to inhibit pathogenic microorganisms such as *Pseudomonas aeruginosa* and *Escherichia coli*, which are common in food spoilage (Quoc 2022).

#### 22.7.3.3 Preservative

Due to its antimicrobial properties, *T. ammi* is used as a natural preservative in food products, helping to extend shelf life and prevent spoilage (Jahan et al. 2013).

### 22.7.4 Industrial and Other Applications

#### 22.7.4.1 Insecticidal and Wormicidal Activities

*T. ammi* exhibits insecticidal properties, making it useful in protecting stored grains from pests. It also has germicidal effects, which can be applied in agricultural settings (Jahan et al. 2013).

#### 22.7.4.2 Pharmaceutical Formulations

The plant's essential oils and extracts are used in various pharmaceutical formulations for their therapeutic properties. Recent patents have been filed for novel formulations utilizing *T. ammi*'s bioactive compounds (Nabi et al. 2023).

### 22.7.5 Novel Drug Delivery Systems

#### 22.7.5.1 Nanoparticles and Microemulsions

To enhance the bioavailability and therapeutic efficacy of ajwain's essential oil, novel drug delivery systems such as nanoparticles, microemulsions, niosomes and liposomes have been explored. These systems help overcome the limitations of conventional formulations by improving the stability and controlled release of active compounds (Siddiquie et al. 2024).

#### 22.7.5.2 Nanoparticle Synthesis

*Trachyspermum ammi* extracts serve as capping and reducing agents in the synthesis of non-toxic nanoparticles, which have applications in pharmaceuticals and biological systems. This green synthesis approach is environmentally friendly and cost-effective.

### 22.7.6 Topical Formulations

Ajwain essential oil has been formulated into topical preparations for its antimicrobial properties. These formulations, including hydrophilic ointments, have shown effectiveness in managing skin infections, with minimal skin sensitization potential.

### 22.7.7 Clinical Application

#### 22.7.7.1 Ventilator-Associated Pneumonia (VAP)

A clinical trial investigated the use of *Trachyspermum ammi* syrup in preventing VAP. Although the results showed a reduction in the incidence of VAP in the treatment group compared to the placebo, the difference was not statistically significant. This suggests a potential benefit, warranting further research with varied dosages (Ahmadipour et al. 2024).

#### 22.7.7.2 Vulvovaginal Candidiasis (VVC)

The essential oil of *Trachyspermum ammi* has been evaluated for its antifungal properties, showing effectiveness against fungal strains responsible for VVC. This highlights its potential as a natural treatment option for fungal infections (Usman et al. 2022).

### 22.7.8 Emerging Applications and Research

Recent advancements in drug delivery systems, such as nanoparticles and micro-emulsions, are being explored to enhance the bioavailability and efficacy of *T. ammi*'s active compounds (Siddiquie et al. 2024).

### 22.7.9 Potential for Drug Development

**Novel Drug Delivery Systems** To enhance the therapeutic efficacy of *Trachyspermum ammi*, novel drug delivery systems such as nanoparticles, micro-emulsions and liposomes are being explored. These systems aim to overcome the limitations of conventional formulations and improve the bioavailability of active compounds (Siddiquie et al. 2024).

**Clinical Trials and Formulations** Clinical trials have been conducted to assess the efficacy of ajwain in various conditions, such as ventilator-associated pneumonia, where it showed a reduction in infection rates, although not statistically significant. This indicates potential for further exploration in clinical settings (Ahmadipour et al. 2024).

**Patents and Innovations** Recent reviews have highlighted the filing of patents for novel formulations of *Trachyspermum ammi*, indicating ongoing research and interest in developing new therapeutic agents from this plant (Nabi et al. 2023).

## 22.8 Safety and Toxicity

*Trachyspermum ammi* is generally regarded as safe, with studies indicating low toxicity at therapeutic doses. Its use in traditional medicine and culinary practices further supports its safety profile(Dahiya et al. 2023).

The acute toxicity study of its methanolic leaf extract in rodents confirmed its safety at high doses, reinforcing its potential for therapeutic applications (Raju et al. 2023).

## 22.9 Research Gaps

**Limited Clinical Trials** While there are numerous studies on the pharmacological properties of *Trachyspermum ammi*, clinical trials are limited. For instance, a study on its effect on ventilator-associated pneumonia (VAP) showed potential benefits, but the results were not statistically significant, indicating a need for more comprehensive clinical trials with varied dosages and larger sample sizes (Ahmadipour et al. 2024).

**Extraction Methods** The extraction processes for *Trachyspermum ammi*'s bioactive compounds are often time-consuming and inefficient. Traditional methods like maceration and Soxhlet extraction are still prevalent, which are not optimal for large-scale production. Modern extraction techniques such as microwave-assisted or ultrasound-assisted extraction could enhance yield and efficiency, yet these methods are underexplored (Quoc 2022).

**Mechanistic Studies** Although the plant exhibits various pharmacological activities, the underlying mechanisms are not fully understood. For example, its antidiabetic properties have been attributed to enzyme inhibition, but further studies are needed to confirm the specific bioactive compounds responsible and their mechanisms of action (Raju et al. 2023).

The exploration of *Trachyspermum ammi* reveals its remarkable medicinal potential alongside key limitations and avenues for future research. Rich in essential oils and bioactive compounds, this plant is widely utilized in traditional medicine and has been extensively studied for its pharmacological properties. However, challenges remain in fully optimizing its therapeutic applications and unlocking its complete potential. The following sections will examine these limitations and suggest directions for advancing research on *T. ammi*.

## 22.10 Areas for Future Research

Research on *Trachyspermum ammi* (ajwain) highlights potential advancements in modern healthcare through innovative drug delivery systems, detailed phytochemical analyses and optimized extraction techniques to enhance bioavailability, efficacy and sustainability. Studies on environmental impacts and the integration of traditional medicine could support its therapeutic acceptance. In-depth clinical trials and mechanistic studies are needed to validate its effects, particularly for neurodegenerative and metabolic conditions. Exploring new therapeutic areas, such as antiurolithic potential, may further expand *T. ammi*'s applications in integrative medicine.

## 22.11 Conclusion

Thus, the historical and cultural origins of *T. ammi* or ajwain not only lie in its current conventional medicinal application but also are an integral part of the global awakening toward the importance of ethnomedicine and traditional knowledge systems. It is this trend that perhaps opens up the chance for medicinal plants such as ajwain to contribute value to healthcare, even though the actual integration of these plants into modern practice requires rigorous scientific validation in terms of safety and efficacy. The complete therapeutic potential of *T. ammi* has yet to be tapped since ongoing research has definite limitations. Overcoming these challenges with advanced extraction techniques, clinical trials and innovative drug delivery systems may increase its applications; further, knowledge of its environmental impacts and traditional uses may supplement its status in modern medicine. These steps might mean that *T. ammi* becomes one of the cornerstones in the field of natural medicine and integrative health solutions.

## References


Abdullah BM, Mehdi MAH, Khan AR, Pathan JM (2020) Gas chromatography-mass spectrometry (GC-MS) analysis of ajwain (Trachyspermum ammi) seed extract. Int J Pharm Qual Assur 11:228–231

Ahmadipour M, Razban F, Rashidipour N, Ahmadi F (2024) Effect of Trachyspermum ammi syrup on the prevention of ventilator-associated pneumonia. J Herb Med 43:100818

Amar G, Kuppusamy A (2022) In-vitro study of Trachyspermum ammi extract against diabetes mellitus. Trends Sci 19:5698–5698

Anwar S, Ahmed N, Habibatni S, Abusamra Y (2016) Ajwain (Trachyspermum ammi L.) oils. In: Essential oils in food preservation, flavor and safety. Elsevier, pp 181–192

Arasu A, Pingley V, Prabha N, Ravikumar O, Annathurai K, Kasirajan S, Govindasamy A, Alwahibi MS, Elshikh MS, Gawwad MRA, Arockiaraj J (2021) Impact and fungitoxic spectrum of Trachyspermum ammi against Candida albicans, an opportunistic pathogenic fungus commonly found in human gut that causes Candidiasis infection. J Infect Public Health 14:1854–1863

Aruchamy M, Duraisamy P (2021) In vitro antioxidant and anticancer activity of methanolic extract of Trachyspermumammiseed on human oral cancer (KB) cell line. J Univ Shanghai Sci Technol 23:426–436

Bairwa R, Sodha R, Rajawat B (2012) Trachyspermum ammi. Pharmacogn Rev 6:56

Barik R, Sugunan S, Shafri MABM (2024) Pressurized liquid extraction for the isolation of bioactive compounds. In: Bioactive extraction and application in food and nutraceutical industries. Springer, pp 275–298

Chahal K, Dhaiwal K, Kumar A, Kataria D, Singla N (2017) Chemical composition of Trachyspermum ammi L. and its biological properties: a review. J Pharmacogn Phytochem 6:131–140

Chauhan B, Kumar G, Ali M (2012) A review on phytochemical constituents and activities of Trachyspermum ammi (L.) Sprague fruits. AJPTR 2:329–340

Dahiya H, Sonia K, Boora S, Yadav S, Kaushik S, Singh S, Yadav JP, Kaushik S (2023) Trachyspermum ammi (Ajwain): a sacred plant with high medicinal and therapeutic potential. Res J Pharm Technol 16:3285–3288

Debnath S, Sharma A (2024) Exploring Trachyspermum ammi and Foeniculum vulgare in hydroponic system and compare its chemical constituents with soil-based method: a prospective in agriculture. Recent Pat Biotechnol 18:257–266

Dutta S, Kundu A (2022) Macroporous resin-assisted enrichment, characterizations, antioxidant and anticandidal potential of phytochemicals from Trachyspermum ammi. J Food Biochem 46:e13847

Dutta S, Kundu A, Dutta A, Saha S, Banerjee K (2021) A comprehensive chemical profiling of phytochemicals from Trachyspermum ammi and encapsulation for sustained release. LWT 147:111577

Eftekhari M, Hoseinsalari A, Mansourian M, Farjadmand F, Shams Ardekani MR, Sharifzadeh M, Hassanzadeh G, Khanavi M, Gholami M (2019) Trachyspermum ammi (L.) Sprague, superb essential oil and its major components on peptic ulcers: in vivo combined in silico studies. DARU J Pharm Sci 27:317–327

Goyal S, Chaturvedi V, Dhingra G, Tanwar S, Sharma K, Singh S (2022) Trachyspermum ammi: a review on traditional and modern pharmacological aspects. Biol Sci 2:324–337

Hanif MA, Hassan SM, Mughal SS, Rehman A, Hassan SK, Ibrahim A, Hassan H (2021) An overview on ajwain (Trachyspermum Ammi) pharmacological effects: current and conventional. Technology 5:1–6

Jabeen R, Ali T, Naeem M, Hussain F, Danish S, Alahmadi TA, Samdani MS, Ansari MJ (2023) An insight into biochemical characterization and explorations of antioxidant, antibacterial, cytotoxic, and antidiabetic activities by trachyspermum ammi nanosuspensions. Front Biosci Landmark 28:340

Jahan N, Ahmad M, Mehjabeen. (2013) Toxicity studies of Trachyspermum ammi (L.) Sprague ex Turrill and its smooth muscles effects. In: Chemistry: the key to our sustainable future. Springer, pp 311–320

Khan A, Gilani AH (2023) An insight investigation to the antiurolithic activity of Trachyspermum ammi using the in vitro and in vivo experiments. Urolithiasis 51:43

Khorsandi K, Kianmehr Z, Ghelichkhani E (2022) Combination effect of red light irradiation and Traychspermum ammi essential oil on colorectal cancer cells (SW480). Lasers Med Sci 37:1031–1040

Kumar V, Bahuguna A, Kim MD (2024) Molecular insights into binding of bioactive compounds from essential oil of Trachyspermum ammi with human programmed cell death protein 1. J Biomol Struct 42:6871–6881

Nabi K, Imanshu SS, Bhatia D, Bhatti M, Singh L (2023) Updated detailed review of Trachyspermum ammi: composition, applications and pharmacological profile. Int J Life Sci Pharm Res 13:P221–P238

Nag K, Gupta C (2023) Antimicrobial and antioxidant properties of Trachyspermum ammi leaf extracts against some plant pathogens. J Nutr Health Food Eng 13:1–4

Naquvi KJ, Ansari S, Salma A, Ahamad J, Najib S (2022) A review on phytochemical investigations and biological activities of Trachyspermum Ammi (L.) Sprague. Res J Pharm Technol 15:2364–2370

Quoc L (2022) Restrictions of the extraction process and bioactivities of Trachyspermum ammi seed extract. Malays J Med Sci: MJMS 29:173–175

Raju MG, Priya GS, Edunoori K (2023) Antidiabetic activity of methanolic leaf extract of Trachyspermum ammi in rodents. Int J Biochem Res 32:27–37

Saraswat N, Sachan N, Chandra P (2020) A review on ethnobotanical, phytochemical, pharmacological and traditional aspects of indigenous Indian herb Trachyspermum ammi (L). Curr Tradit Med 6:172–187

Sharma H, Yang H, Sharma N, An SSA (2023) Trachyspermum ammi bioactives promote neuroprotection by inhibiting acetylcholinesterase, Aβ-oligomerization/fibrilization, and mitigating oxidative stress in vitro. Antioxidants 13:9

Siddiqui MJ, Aslam A, Khan T (2019) Comparison and evaluation of different seed extracts of Trachyspermum ammi for immunomodulatory effect on cell-mediated immunity through delayed-type hypersensitivity assay skin thickness method. J Pharm Bioallied Sci 11:43–48

Siddiquie F, Ahsan F, Mahmood T, Ahmad MA, Singh A, Bano S (2024) Unlocking the food treasures: Trachyspermum ammi–a comprehensive exploration from field to pharmacology. Food Saf Health 2:322–343
Singh PP, Jaiswal AK, Singh R, Kumar A, Gupta V, Raghuvanshi TS, Sharma A, Prakash B (2024) Assessment of Trachyspermum ammi essential oil against Aspergillus flavus, aflatoxin B1 contamination, and post-harvest quality of Sorghum bicolor. Food Chem 443:138502
Timalsina B, Haque MN, Choi HJ, Dash R, Moon IS (2023) Thymol in Trachyspermum ammi seed extract exhibits neuroprotection, learning, and memory enhancement in scopolamine-induced Alzheimer's disease mouse model. Phytother Res 37:2811–2826
Ullah MA, Hassan A, Hamza A (2024) Medical constituents of ajwain (Trachyspermum ammi) for human benefits. Middle East Res J Biol Sci 4:1–8
Usman MRM, Vadnere G, Pawar S (2022) Pharmacognostic evaluation of Trachyspermum ammi (ajwain seeds) seed extract for vulvovaginal candidiasis (VVC). Int J Med Pharm Sci 12:6
Vharamble NR, Patil SB, Naik OS, Shah R (2023) Overview of Trachyspermum ammi and its medicinal applications. Res J Pharmacogn Phytochem 15:319–323

# Medicinal and Nutritional Importance of *Nardostachys jatamansi* in Human Health

# 23

Sharmistha Ganguly, Dwaipayan Sinha, Ratul Bhattacharya, Dhavalkumar Patel, Abdullah Salik, Srijan Mukhopadhyay, Prajwal Kiran Saldanha, Somnath Kar, Shalini Kumari, and Mohammad Azam Ansari

**Abstract**

*Nardostachys jatamansi* is a critically endangered medicinal plant which is immensely used in traditional medicine. Belonging to the Valerianaceae family, it's a miniature, perennial herbaceous plant with hairy rhizomes and found in alpine and sub-alpines regions of the Himalayan range. It has been used in traditional Indian medicine system since ages and has prominent evidences in disease curing ability. The Ayurvedic, homeopathic, and ethnomedicinal practices also pave pathway to modern pharma industry for its enlarged antiquity usage. It has been reported by several researchers as antispasmodic, antiepileptic, hepatopro-

S. Ganguly (✉) · S. Kumari
University Department of Botany, Ranchi University, Ranchi, Jharkhand, India

D. Sinha
Department of Botany, Government General Degree College, Mohanpur, Paschim Medinipur, West Bengal, India

R. Bhattacharya
Department of Botany, Raja Ram Mohan Roy Mahavidyalaya, Hooghly, West Bengal, India

D. Patel
Department of Microbiology & Biotechnology, University School of Sciences, Gujarat University, Ahmedabad, Gujarat, India

Department of Research & Development, Latambarcem Brewers Private Limited, Dicholi, Goa, India

A. Salik
Department of Pharmaceutical Sciences, Birla Institute of Technology, Ranchi, Jharkhand, India



M. A. Ansari et al. (eds.), *Medicinal Plants and their Bioactive Compounds in Human Health: Volume 2*, https://doi.org/10.1007/978-981-95-3620-7_23

tective, anticonvulsant, neuroprotective, anti-Alzheimer's, anti-Parkinson's and antidepressant. The major secondary metabolites present in the plant are mainly nardostachone, jatamansone and actinidine which are solely responsible for its so far known efficacies. They are excellent vasodilators, bronchodilators and inhibit aggregation of platelets. Among these, jatamansone showed antiarrhythmic and antihypertensive activities. This chapter discusses on the holistic information of the plant *Nardostachys jatamansi* where its geographical distribution, habitat loss, morphological description, microscopical description, nutrient analysis, phytochemical attributes, traditional and enthomedicinal significance and different pharmacological efficacies are discussed elaborately. Furthermore, future prospects, significance and judicious utilization are also demonstrated which can help the researchers in exploring the plant more for its utilization in human health besides, thinking upon different strategies for its conservation.

**Keywords**

*Nardostachys jatamansi* · Medicinal plant · Phytochemistry · Traditional medicine · Bioactivity

## Abbreviations

| | |
|---|---|
| ALT | Alanine aminotransferase |
| AST | Aspartate aminotransferase |
| CNS | Central nervous system |
| DOX | Doxorubicin |
| ECD | Electron capture detector |
| ERK | Extracellular signal regulated kinase |
| FT-IR | Fourier transform infrared spectroscopy |
| GC-MS | Gas chromatography-mass spectroscopy |
| HaCaT | Human immortalized keratinocyte cell line |
| HCC | Hepatocellular carcinoma |
| HPLC | High-performance liquid chromatography |

S. Mukhopadhyay
Taxonomy and Biosystematic Laboratory, PG Department of Botany, Ramakrishna Mission Vivekananda Centenary College, Rahara, Kolkata, West Bengal, India

P. K. Saldanha
Department of Botany, Bangalore University, Jnana Bharathi Campus, Bangalore, Karnataka, India

S. Kar
Department of Botany, Holy Cross College, Tripura University, Agartala, Tripura, India

M. A. Ansari
Department of Epidemic Disease Research, Institutes for Research and Medical Consultation (IRMC), Imam Abdulrahman Bin Faisal University, Dammam, Saudi Arabia
e-mail: maansari@iau.edu.sa

| | |
|---|---|
| $IC_{50}$ | Inhibitory concentration at 50% |
| LC–MS/MS | Liquid chromatography-tandem mass spectroscopy |
| MTT | 3-(4,5-dimethylthiazol-2-yl)-2,5- diphenyltetrazolium bromide assay |
| NMR | Nuclear magnetic resonance |
| PCNA | Proliferation cell nuclear antigen |
| RANTES | Regulated on activation normal T cell expressed and secreted |
| RP-HPLC | Reverse phase high-performance liquid chromatography |
| STAT3 | Signal transducer and activator of transcription 3 |
| TLC | Thin layer chromatography |
| UPLC-PDA | Ultra performance liquid chromatography with a photodiode array |
| UPLC-QTOF MS | Ultra performance liquid chromatography with quadrupole time-of-flight mass spectrometry |

**Learning Objectives**

- *Nardostachys jatamansi* is a critically endangered herbaceous medicinal plant which is immensely used in traditional medicine belonging to the Valerianaceae family.
- It has been reported to have several bioactivities as antispasmodic, antiepileptic, hepatoprotective, anticonvulsant, neuroprotective, anti-Alzheimer's, anti-Parkinson's and antidepressant.
- It acts as excellent vasodilators, bronchodilators and inhibits aggregation of platelets.
- It is a hub of secondary metabolites like nardostachone, jatamansone and actinidine.
- It is threatened with extinction due to restricted and isolated habitats with low-density populations and hence requires effective conservation.

## 23.1 Introduction

*Nardostachys jatamansi* DC, recognized as "Spikenard," is a perennial herb of the family Caprifoliaceae (previously Valerianaceae) (Dhiman and Bhattacharjee 2020). This species has become an essential subject of scientific and ethnopharmacological research due to its pronounced medicinal and nutritional benefits (Wang et al. 2021). The plant has extensive use in old medicine for its extensive series of biochemical activities, essentially neuroprotection, antianxiety, and anti-inflammatory (Sahu et al. 2016). In recent research, the bioactive composites from *N. jatamansi* are studied for their pharmacological prospects against prolonged diseases such as neurodegenerative disorders, cancer, and cardiovascular diseases (Pathak and Godela 2023).

*Nardostachys jatamansi* grows mainly in the ranges of Himalayas across India, Bhutan, Nepal, Tibet, and southwest China (Kaur et al. 2020; Ved et al. 2015). It propagates 3000 to 5000 m exceeding sea level in cold, moist places with well-drained organic soils. It prefers to grow on shaded slopes of forests so that atmospheric humidity and temperature variances are optimal for its growth cycle (Wang et al. 2021). These specific ecological conditions thus play a vital role in the improvement of secondary metabolites in the plant; the development of secondary metabolites in the plant appears to be influenced by altitude and other environmental stressors (Wang et al. 2021).

For centuries, *N. jatamansi* has been the cornerstone of traditional medicinal systems; specifically, it has found a place within the context of Ayurvedic, Unani, and Siddha practices (Thakur et al. 2021). In Ayurveda, *N. jatamansi* is classified as one of the Rasayana drugs, where such medicines have been documented to induce longevity, a healthy mental state, and vigor (Raina et al. 2023). In this ancient medicinal system, the plant was utilized to cure a diversity of neurological and psychiatric disorders—anxiety (Pant et al. 2022), depression (Sahu et al. 2016), and insomnia (Pandey et al. 2013).

Its roots were also used for treating digestion disorders (Pant et al. 2022), liver dysfunctions (Ali et al. 2000), and inflammatory diseases (Yoon et al. 2024). Documentation of Unani medicine suggests *N. jatamansi* serves as a crucial pharmacological agent in managing diseases pertaining to the central nervous system. Even in Siddha medicine, its sedative and stimulating action on cognition can be noted (Pathak and Godela 2023).

In ethnobotanical terms, indigenous populations of this Himalayan region have been using *N. jatamansi* for years to treat wounds (Pant et al. 2022), fever (Pathak and Godela 2023), and skin infections (Kaur et al. 2020), besides its religious and ceremonial significance as aromatic incense in worshipping rituals (Kaur et al. 2020; Chandra 2022). Traditional uses of *N. jatamansi* provide tremendous value towards understanding the pharmacological potential, forming the foundation for modern scientific investigation into its healing efficacy (Shah 2023).

The medicinal efficacy of *N. jatamansi* is primarily attributed to its rich phytochemical profile. The plant contains various bioactive compounds, including sesquiterpenes, flavonoids, alkaloids, and essential oils, contributing to its varied medicinal properties (Shukla et al. 2021). Among these compounds, jatamansone, jatamansin, and nardosinone (Chatterjee et al. 2005; Bose et al. 2019) have been recognized as the primary components which are active and accountable for the plant's pharmacological effects (Pathak and Godela 2023). Recent studies have provided substantial evidence supporting the neuroprotective effects of *N. jatamansi*, particularly in neurodegenerative diseases (Li et al. 2023). Major constituents include nardostachysin, jatamansin, jatamol A and B, and nardosinone, where the former is especially important in terms of a terpenoid ester (Chatterjee et al. 2000). Recent studies have identified novel sesquiterpenoid dimers like dinardokanshones C–E and isonardoeudesmols that are particular serotonin transporter (SERT) inhibitors and have been considered in psychiatry, neurology, and gastric disorders (Wu et al. 2018).

This comprehensive overview underscores the nutritional and medicinal significance of *N. jatamansi*, highlighting its bioactive compounds, pharmacological activities, and potential applications in human health.

## 23.2 Global Distribution

*Nardostachys jatamansi* has relatively limited distribution and mainly thrives in the Himalayan belt, cutting across several countries. It occurs explicitly in Nepal, India, Bhutan, Myanmar, and the southern China regions, indicating its existence in those high-altitude ecosystems (Kaur et al. 2020; Ved et al. 2015). The species grows in the deep environments of alpine and sub-alpine sites. It is best distributed to altitude ranging from 3000 to 5000 m, in which the environment is more favorable to its growth (Wang et al. 2021). The competence of the species to acclimatize to a variety of high-altitude environments is demonstrated by its presence in the Yunnan region in southern China, which serves as the southernmost limit of its range (Li et al. 2019). Significant populations of *N. jatamansi* are present in Nepal, especially in regions characterized by rich biodiversity at high altitudes, suggesting a predilection of the species for mountainous environments (Kunwar et al. 2023).

In India, several Himalayan states in India, comprising Uttarakhand, Sikkim, Himachal Pradesh, Arunachal Pradesh, and Jammu & Kashmir, are native to *N. jatamansi* (https://indiabiodiversity.org). These areas offer a variety of ecosystems with distinct elevation gradients and climatic conditions (Shrestha et al. 2022; Ved et al. 2015). The species' broad and specialized distribution pattern suggests that the species is very resilient to changes in the environment, allowing it to survive in various microclimatic niches across the Himalayan region (Kunwar et al. 2023).

## 23.3 Habitat Loss

Habitat loss has been the primary determinant of the loss of many species, especially those like *Nardostachys jatamansi* and *Nardostachys grandiflora*, species that are endemic to the Himalayas and have, therefore, gained significant medicinal values. The unchecked harvesting of these plants, together with some rapid environmental changes, has come at such a great cost that the populations of these plants have come under great strain and have subsequently become endangered (Shin et al. 2022) According to IUCN, *Nardostachys jatamansi* falls under critically endangered category (Chauhan 2021).

*Nardostachys jatamansi* is used in traditional medicine but is threatened by over-collection and habitat destruction. Mainly alpine-distributed, this species is strongly demanded for its rhizomes, which are used to treat many disorders, including insomnia, neurosis, and epilepsy (Chauhan et al. 2021). This combined with accelerating climate change has greatly reduced its population in natural habitats. Ecological studies have revealed unsustainable harvesting practices and slow growth coupled with low population density as threats to the long-term survival of this species. Emergency conservation measures through ex situ propagation and sustainable utilization are needed presently (Reid et al. 2022).

Similarly, *Nardostachys grandiflora* is threatened with extinction due to restricted and isolated habitats with low-density populations along with climatic changes. The other species used for medicinal and aromatic purposes have been found to decline in the alpine regions of Sikkim (Sherpa et al. 2023). The population dynamics of this species are also not well studied; hence, call for immediate attention towards its conservation is required. It has been found that the whole world market demands these species commercially in large quantities, and thus there are threats regarding their sustainability. Thus, it has been put forward that change in climate may decrease the future suitability of habitat for these species. Consequently, the threat to survival will be intensified (Rana et al. 2020). Novel conservation tactics include the formulation of biotechnological methods and mixed culture for dealing with such threats. Such vital plants can hardly survive without evidence-based and integrated conservation strategies (Wen et al. 2022; Kaur et al. 2020). Therefore, their holistic reproductive strategies, habitat requirements, and population dynamics need to be understood for effective conservation.

## 23.4 Botanical Description

*Nardostachys jatamansi* is a perennial herbaceous species, 16–50 cm in height, rhizomatous, rhizome stout, spindle-shaped, long and aromatic, and covered with the remnants of old leaf fibers on the outer surface. The stem is erect, slender, 5–50 cm long, pubescent towards the upper portion while glabrous at the base. Leaves are dimorphic in nature likey radical and cauline which preliminarily forms a rosette at the base, sessile and elliptic to obovate in shape. Cauline leaves are sessile, narrowly spatulate to linear-oblanceolate in sha (3-25 cm x 0.5-2.5 cm), glabrous to sparsely pubescent, displaying three conspicuous veins, having an attenuatee base, margins are entire and apex s obtuse. The lower ones adopt a form, while the uppermost ones tend towards lanceolate to oblanceolate, occasionally presenting sparse serrations. These cauline leaves are smaller and sessile and contribute to the plant's distinctive morphology (Fig. 23.1).

The inflorescence is compound heads or capitula, with bracts configured in 3–5 whorls, linear-lanceolate in shape, pubescent. Flowers are typically arranged in groups, 1–5 per bract, actinomorphic, bisexual. Sepals are 5, semi-orbicular to triangular-lanceolate, lobed, persistent, and expanding during the fruiting stage, ciliate, pubescent, and veined. Petals 5, gamopetalous, campanulate, limb 5 lobed, corolla tube 0.5–0.8 cm (length), 0.2–0.4 cm (diameter), purplish-red to pink in color. Stamens are 4, polyandrous, filament villous. Style is subequal to stamens, capitate stigma, and locule 3. The fruit is achene, obovate in shape, 0.3–0.4 cm in size, pubescent, white in color, compressed. Seeds are one per fruit (https://indiabiodiversity.org; Mabberley and Noltie 2014; Shah 2023).

**Fig. 23.1** Photograph showing *Nardostachys jatamansi*

## 23.5 Chemical Attributes of *Nardostachys jatamansi*

*Nardostachys jatamansi* is renowned for its rich chemical composition and therapeutic properties. Traditionally used in Ayurvedic medicine, it has garnered significant interest due to its varied series of bioactive compounds, primarily essential oils, flavonoids as well as alkaloids (Singh et al. 2013; Bhatt and Kothiyal 2015). The essential oil contains numerous bioactive constituents which are sesquiterpenes, such as β-nerolidol, patchoulol, and calarene and these compounds contribute to the oil's aromatic profile and are believed to hold various health benefits which include anti-inflammatory, antimicrobial, as well as sedative properties. The unique combinations of these sesquiterpenes were traditional use as a calming agent and for promoting relaxation (Disket et al. 2012; Li et al. 2021).

Later on, with growth and development in science, the curiosity in the researcher arose to study the properties of such bioactive compounds. Arora et al. (1958), first studied the chemical attributes of *N. jatamansi*; later on with development many researcher bridged the gap. Rahman et al. (2011) outlined the biological active compounds like sesquiterpenes and coumarins. It was also stated that the roots contain valeranal, valeranone, calarenol, nardone, calarene, nardostachone, 8-n-hexaconsanol, and many more components. Multiple studies were conducted and the chemical attitude were studied and found that root oil contains terpenic coumarins, jatamansin, oroselol, hydrocarbons, ß-eudesmol, ß-sitosterol, elemol, angelicin, and jatamansino (Singh 2020; Mohammad and Kushwaha 2021; Gupta et al. 2024). Flavonoids present in *Nardostachys jatamansi*, such as kaempferol and quercetin, are renowned for their antioxidant efficacies, which help in combating

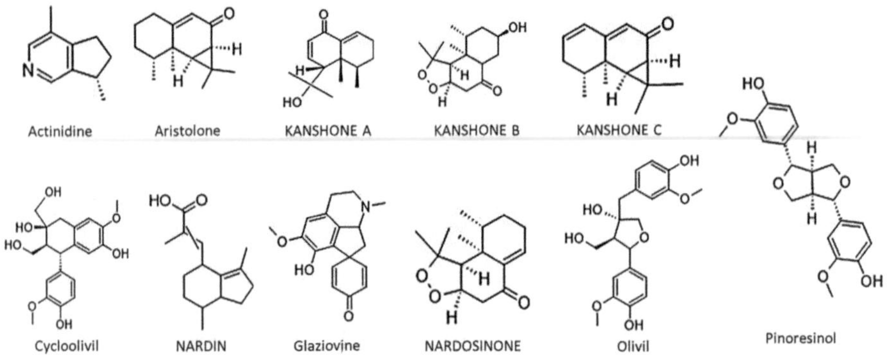


**Fig. 23.2** The phytochemical structure of bioactive chemicals

reactive oxygen species in the body (Feng et al. 2024). The structure of chemical attributes is displayed in Fig. 23.2 and summarized in Table 23.1.

The chemical profile of *Nardostachys jatamansi* highlights its potential in modern pharmacology. The synergy between its various constituents may lead to enhanced therapeutic effects. In addition to essential oils, *Nardostachys jatamansi* also comprises a range of alkaloids like jatamansone and nardostachysin (Ma et al. 2022). They have been studied for their potential pharmacological activities, including neuroprotective (Krishnan et al. 2023) and anxiolytic effects (Priyadarshini et al. 2020). These alkaloids are thought to interact with various neurotransmitter systems, further validating the plant's use in traditional medicine for treating anxiety and insomnia (Puri et al. 2022).

The chemical attributes of *Nardostachys jatamansi* underline its significance in both traditional and in modern medicinal practices. The various groups of bioactive composites ranging from essential oils to flavonoids and alkaloids offer a promising avenue for research and possible healing applications. As scientists endure to explore these compounds, the plant's legacy as a treasured natural resource for healthiness and wellness is likely to endure and expand.

## 23.6 Macroscopical Characters

Perennial herb; tap-root and adventitious roots both occur; rhizome stout, cylindrical, branched or unbranched, covered with reddish-brown fibers, aromatic. Rosulate leaves spathulate-lanceolate, shortly petiolate; ramal leaves elliptic-ovate or oblanceolate to lanceolate, petiolate to sessile; apex obtuse; margin entire, rarely sparsely serrate; base of radical leaves attenuate-connate, persistent as reddish-brown fibers, attenuate-decurrent in ramal leaves; surface sparsely pubescent or glabrous; veins parallelodromous; phyllotaxy opposite-decussate or opposite-distichous, rarely alternate, 3–25 × 0.5–2.5 cm. *Inflorescence* axillary, contracted, compound or simple truncated thyrses; bracts 4–6, lanceolate; floral axes erect, 5–50 cm long. *Flower* bisexual, zygomorphic, epigynous, sessile; bracts 2, linear, 6 mm long; *Calyx* united, 5-partite, foliaceous, ciliate, persistent; lobes ovate-lanceolate; apex acute;

**Table 23.1** Phytochemical constituents of *Nardostachys jatamansi*

| Sl. no. | Plant Part | Type of compound | Name of the compound detected/ isolated | Instrumentation involved | References |
|---|---|---|---|---|---|
| 1. | Rhizome | Polyphenols | 8α–hydroxypinoresinol | HPLC | Kim et al. (2021b) |
| | | Sesquiterpenoid | Desoxo-narchinol | | |
| | | Polyphenols | Pinoresinol | | |
| | | Nardosinone-type sesquiterpenes | Nardosinonediol | | |
| | | Sesquiterpenes | Kanshone A | | |
| | | Nardosinone-type sesquiterpenes | Isonardosinone | | |
| | | Sesquiterpenoid | 1α-hydroxy-(-)-aristolone | | |
| | | Sesquiterpenoid | Nardoaristolone B | | |
| 2. | Rhizome and roots | Sesquiterpenoids | Kanshone L | HPLC & NMR | Yoon et al. (2018) |
| | | Sesquiterpenoids | Kanshone M | | |
| | | Nardosinone-type sesquiterpenes | 7-methoxydesoxo-narchinol | | |
| | | Sesquiterpenoids | Kanshone N | | |
| | | Sesquiterpenoids | Nardosdaucanol | | |
| | | Sesquiterpenoids | Kanshone D | | |
| | | Sesquiterpene | Nardosinone G | | |
| | | Sesquiterpenoids | Narchinol A | | |
| | | Sesquiterpene | Nardoaristolone B | | |
| 3. | Whole plant | Nardosinone-type sesquiterpenes | Nardosinone | Column chromatography & NMR | Ko et al. (2018) |
| | | Sesquiterpene | Isonardosinone | | |
| | | Sesquiterpenoids | Kanshone E | | |
| | | Sesquiterpenoids | Kanshone B | | |

(continued)

**Table 23.1** (continued)

| Sl. no. | Plant Part | Type of compound | Name of the compound detected/ isolated | Instrumentation involved | References |
|---|---|---|---|---|---|
| 4. | Whole plant | Polyphenols | Cycloolivil | Column chromatography & NMR | Yoon et al. (2024) |
| | | Polyphenols | Olivil | | |
| | | Alpha-hydroxy acid (AHA) family | 2-(3′-hydroxy-5′-ethoxyphenyl)-3-hydroxylmethyl-7-methoxy-2,3-dihydrobenzofuran-5-carboxylic acid | | |
| | | Sesquiterpene | Teuclatriol | | |
| | | Sesquiterpene | Bullatantriol | | |
| | | Eudesmane sesquiterpenoid | 1β,4β,7α-trihydroxy eudesmane | | |
| | | Sesquiterpenoid | Nardoeudesmol A | | |
| | | Diterpene compounds | Debilon | | |
| | | Sesquiterpene | Nardonoxide | | |
| | | Sesquiterpene | Nardosinone | | |
| | | Amino methyl group decane derivatives | 4β-hydroxy-8β-methoxy-10-methylene-2,9-dioxatricyclo[4.3.1.03,7]decane | | |
| | | Sesquiterpenes | Jatamanin A | | |
| | | Octahydro-naphthalene derivatives | 4α,5-dimethyl-1,3-dioxo1,2,3,4,4α,5,6,7-octahydronaphthalene | | |
| | | Sesquiterpenoids | Kanshone M | | |
| 5. | Rhizomes | Terpenoid | Nardal | UV, IR,1H and13C NMR, 2D NMR, Mass | Gottumukkala et al. (2011) |
| | | | Jatamansic acid | | |
| | | | Nardin | | |

| | | | | | |
|---|---|---|---|---|---|
| 6. | Rhizomes | Terpenoid | Nardal | IR, MS, 1D, and 2D NMR | Rekha et al. (2013) |
| | | Terpenoid | Nardin | | |
| | | Sesquiterpenoids | Taraxerone | | |
| | | Triterpenoid | Ursolic acid | | |
| | | Triterpenoid | 3-O-arabinosyl ursolic acid | | |
| 7. | Rhizome | 8,9-dicarboxylic acid derivative of aristolane-type sesquiterpenoid | (4R,5S,6S,7R)-1(10)-aristolane-8,9-diacid | Spectroscopic data and electronic circular dichroism (ECD) spectra analyses, single-crystal X-ray diffraction | Ma et al. (2022) |
| | | Analogue of nardoguaianone K | (4R,6S,7R,10S)-10-Hydroxyguaia-1(5)-6,7-epoxy-2-one | | |
| | | Isonardosinane-type sesquiterpene | (4R,5R,6R,8R)-1(10)-Isonardosinone-8-ol-9-one-7,11-lactone | | |
| | | Sesquiterpene dimer | Dinardokanshone F | | |
| 8. | Root | Alkaloid | Actinidine | LC–MS/MS fragmentation pattern and purified by semi-preparative HPLC | Krishnan et al. (2023) |
| | | Alkaloid | Glaziovine | | |
| 9. | Rhizomes | Polyphenols | Jatamansone | TLC | Amit Parekh and Jadhav (2009) |

(continued)

**Table 23.1** (continued)

| Sl. no. | Plant Part | Type of compound | Name of the compound detected/ isolated | Instrumentation involved | References |
|---|---|---|---|---|---|
| 10. | Root | Polyphenols | Gallic acid | RP-HPLC | Razack et al. (2015) |
| | | Flavonoids | Catechin | | |
| | | Polyphenols | Chlorogenic acid | | |
| | | Polyphenols | Homovanillin | | |
| | | Flavonoids | Epicatechin | | |
| | | Flavonoids | Rutin hydrate | | |
| | | Flavonoids | Quercetin-3-Rhamnoside | | |
| | | Hydrocarbon | Dodecane | GC-MS | |
| | | Terpenoids | Linalool | | |
| | | Terpenoids | Alpha-Muurolene | | |
| | | Terpenoids | L-calamenene | | |
| | | Terpenoids | A-ionone | | |
| | | Aliphatic hydrocarbon | (−)-à-Panasinsen | | |
| | | Sesquiterpene alcohol | *trans*-Nerolidol | | |
| | | Piperazine derivative | Para-methoxyphenyl piperazine | | |
| | | Fatty aldehyde | 2-tetradecenal | | |
| | | Fatty acid | Palmitic acid | | |
| | | Fatty acid | Oleic acid | | |
| | | Straight-chain alkane with 27 C | Heptacosane | | |
| 11. | Rhizome | Iridoid glycoside | Patrinalloside | NMR | Ngo et al. (2022) |
| | | Iridoid glycoside | 6″-acetyl patrinalloside | | |
| | | Iridoid glycoside | 2′-O-acetylpatrinoside | | |

| | | | | | |
|---|---|---|---|---|---|
| 12. | Root | Pyranocoumarin | Jatamansinol | FT-IR and NMR | Anupama et al. (2022) |
| 13. | Rhizomes (essential oil from rhizome) | Alkyl aldehyde | Heptanal | TLC and GC-MS | Bhagwan et al. (2024) |
| | | Fatty aldehyde | cis-2-Nonenal | | |
| | | Fatty aldehyde | Nonanal | | |
| | | Secondary alcohol | 3-Decyn-2-ol | | |
| | | Secondary alcohol | (9-Oxabicyclo[3.3.1]non-6-en-3-yl)methanol | | |
| | | Fatty aldehyde | 2,4-Decadienal, (E,Z)- | | |
| | | Organooxygen compound. | 3-Cyclohexene-1-carboxaldehyde,1,3,4-trimethyl- | | |
| | | Furanone derivative | 5-pentyl-2(5H)-furanone | | |
| | | Ester derivative | 4-Hydroxy-4-methylhex-5-enoic acid, tert.-butyl ester | | |
| | | Fatty aldehyde | cis-4,5-Epoxy-(E)-2-decenal | | |
| | | Phenolic aldehyde | Isovanillin | | |
| | | Phenylpropanoid | Methyleugenol | | |
| | | Fatty aldehyde | Cedren-13-ol, 8- | | |
| | | Dimer of nuciferol B | trans-Nuciferol | | |
| | | Sesquiterpene alcohol | Carotol | | |
| | | Cycloaromadendrane sesquiterpenoid | Isospathulenol | | |

(continued)

**Table 23.1** (continued)

| Sl. no. | Plant Part | Type of compound | Name of the compound detected/ isolated | Instrumentation involved | References |
|---|---|---|---|---|---|
| 14. | Underground parts | Sesquiterpenoid | Desoxo-narchinol A | UPLC-PDA | Zhang et al. (2018) |
| | | Flavonoid | Buddleoside | | |
| | | Sesquiterpene | Isonardosinone | | |
| | | Sesquiterpene | Nardosinone | | |
| | | Sesquiterpene | Kanshone H | | |
| | | Sesquiterpene | (-)-aristolone | | |
| 15. | Rhizome | Terpenoid | Nardin | UPLC-QTOF MS | Mallavadhani et al. (2011) |
| 16. | Rhizome (volatile oil) | Sesquiterpenoid | Calarene | GC-MS | Singh et al. (2018) |
| | | Sesquiterpene | Varidiflorene | | |
| | | Sesquiterpenoid | Alpha-Panasinsen | | |
| | | Sesquiterpene | Alpha-Santalene | | |
| | | Sesquiterpenoid | Gamma-Himachalene | | |
| | | Sesquiterpenoid | Jatamansone | | |
| | | Sesquiterpenoid | Epiglobulol | | |
| | | Cyclic carboxylic esters | Resibufogenin | | |
| | | Polyphenols | Ionol 4 | | |
| | | Terpenoids | 2,2,7,7-Tetramethyltricyclo [6,2,1,0(1,6)]undec-4-en 3-one | | |

| | | | | | |
|---|---|---|---|---|---|
| 17. | Roots (essential oils) | Monoterpene | Alpha-thujene | GC-MS | Sharma et al. (2016) |
| | | Terpene | Alpha-pinene | | |
| | | Monoterpenoid | 1,8-Cineole | | |
| | | Terpinene | Gamma-terpinene | | |
| | | Monoterpenoid | Isopinocarveol | | |
| | | Terpenoid | Pinocarvone | | |
| | | Beta-pinene | Myrtenol | | |
| | | Terpinolene | Alpha-terpinolene | | |
| | | Monoterpenoid | Beta-citronellol | | |
| | | Monoterpene | Delta-3-carene | | |
| | | Sesquiterpenes | Gamma-cadinene | | |
| | | Polycyclic hydrocarbons | Beta-patchoulene | | |
| | | Sesquiterpenoids | (-)-Aristolene | | |
| | | Sesquiterpene | Alpha-gurjunene | | |

Abbreviations: *HPLC* high-performance liquid chromatography, *NMR* nuclear magnetic resonance, *ECD* electron capture detector, *LC–MS/MS* liquid chromatography-tandem mass spectroscopy, *TLC* thin layer chromatography, *RP-HPLC* reverse phase high-performance liquid chromatography, *GC-MS* gas chromatography-mass spectroscopy, *FT-IR* Fourier transform infrared spectroscopy, *UPLC-PDA* ultra performance liquid chromatography with a photodiode array, *UPLC-QTOF MS* ultra performance liquid chromatography with quadrupole time-of-flight mass spectrometry margin dentate, 1.5–2 mm long; aestivation quincuncial. *Corolla* united, campanulate-tubular, saccate, villous, rarely glabrous, pale pinkish-purple in color, 4.5–9.0 mm long, 5-lobed; lobes obtuse; aestivation imbricate. *Stamens* 4, epipetalous, antisepalous; filament villous, pinkish white, 9 mm long; anther dorsifixed, bithecous, exserted, dehiscence longitudinal, violet in color, 2 mm long; *Carpels* 3, united; ovary syncarpous, incompletely 3-locular, inferior; placentation apical; ovules 3, 2 sterile and 1 fertile; style pink, 9 mm long; stigma papillate, capitate or trilobed, purple. *Achenes* obovoid, hispid, pale brown, 3–4 mm long, bearing single seed. (de Candolle 1824; Weberling 1975; Deyuan et al. 2011; Gautam and Raina 2016)

*Flowering time*: June–August
*Fruiting time*: August–October (Gautam and Raina 2016)

## 23.7 Microscopical Characters

The rhizome's transverse section reveals many layers of polyhedral cells which are flattened. It also has the prismatic crystals. Brown-colored oil cells are noticeable, along with vascular bundles, a sizable pith with sporadic clusters of sclerenchymatous cells, numerous spiral vessels around the periphery, endosperm, parenchymatous cells, idioblasts, starch grains, and an endodermis with volatile oil globules are present (Singh et al. 2011).

## 23.8 Nutrient Analysis

*Nardostachys jatamansi*, generally recognized as spikenard or jatamansi, is an herb traditionally used in medicines of Ayurveda. It has been esteemed for its potential health benefits and nutritional importance. The nutritive content of leaves *of N. jatamansi* is represented in Table 23.2.

**Table 23.2** Nutritive content of leaves *of N. jatamansi* (Lokhande et al. 2009)

| *Macronutrient content of leaves of Nardostachys jatamansi (mg/g)* | |
|---|---|
| Sodium | 1.80 ± 0.07 |
| Potassium | 12.1 ± 1.0 |
| Calcium | 10.1 ± 0.9 |
| Chlorine | 4.7 ± 0.4 |
| *Macronutrient content of leaves of Nardostachys jatamansi (μg/g)* | |
| Iron | 770 ± 155 |
| Zinc | 16.3 ± 1.3 |
| Chromium | 17.44 |
| Manganese | 36 ± 3 |
| Copper | 6 ± 0.4 |
| Cobalt | 5.5 ± 0.4 |
| Nickel | 22.3 ± 1.9 |
| Bromine | 6.1 ± 0.4 |

## 23.9 Physicochemical Analysis of Rhizome Powder

The physicochemical analysis involves the examination of its physical and chemical properties, which is essential to regulate the purity, quality, and uniformity of the plant material utilized in pharmaceutical formulations. The percentage of ash value signifies the total amount of inorganic salts (minerals) exists in the plant, including impurities and extractive values that determine the solubility of phytochemicals in non-polar and polar compounds like water, alcohol or ether, which is crucial for the extraction and formulation processes (Latif and Rehman 2014). Moisture content helps assess the shelf life and stability of the herbal product (Mohammad et al. 2014). Earlier studies data is revealed in Table 23.3.

## 23.10 Phytochemical Screening of Rhizome Powder

Phytochemical screening is necessary to identify the bioactive constituents responsible for its healing properties and drug discovery. *N. jatamansi* possesses steroids, flavonoids, volatile oils, essential oils, sterols, sugar, starch, gum, resin, glycosides, bitter extract, ketone, tannins, mucilage, carbohydrates, gums, terpenes, and alkaloids (Chawla et al. 2023; Shukla et al. 2024; Shriram et al. 2023). The powder is exposed to solvent extraction by usage of different solvents like ethanol, methanol, chloroform, or aqueous solutions using Soxhlet extraction or maceration method. The quality of the crude medication is also indicated by the presence or lack of certain phytoconstituents (Sumitra 2014). Table 23.4 indicates the presence (+) and absence (−) of some phytochemicals in rhizome *of N. jatamansi,* while quantity of each representative of phytochemicals is represented in Table 23.5.

**Table 23.3** Physicochemical analysis of rhizome powder of the plant *Nardostachys jatamansi* (% w/w)

| Rhizome powder of *Nardostachys jatamansi* (% w/w) | | | |
|---|---|---|---|
| | Kumar and Pal (2011) | Jha et al. (2012) | Rashid et al. (2018) |
| Total ash | 8.5 | 20.53 | 7.08 ± 0.03 |
| Acid insoluble ash | 3.8 | 0.76 | 3.71 ± 0.18 |
| Water-soluble ash | 0.5 | 1.005 | 0.58 ± 0.07 |
| Sulphated ash | 8.7 | – | – |
| Water-soluble extractive | 7.3 | 17.1 | 2.5 ± 0.06 |
| Alcohol-soluble extractive | 5.2 | 2.88 | 3.54 ± 0.72 |
| Moisture content | – | – | 8.6 ± 0.25 |

**Table 23.4** Phytochemical screening of rhizome powder of *Nardostachys jatamansi* (Kumar and Pal 2011)

| Sl. no. | Phytochemicals | Petroleum ether extract | Chloroform extract | Ethanol extract | Water extract |
|---|---|---|---|---|---|
| 1. | Alkaloid | – | – | + | + |
| 2. | Flavonoids | – | – | + | + |
| 3. | Glycosides | – | – | – | – |
| 4. | Carbohydrate | – | – | + | + |
| 5. | Saponins | – | – | + | + |
| 6. | Tannins and phenols | – | + | + | + |
| 7. | Fixed oils | + | – | – | – |
| 8. | Steroids | + | + | – | – |
| 9. | Protein and amino acid | – | – | + | + |

"+" = Present; "–" = Absent

**Table 23.5** Phytochemical analysis of rhizome powder of *Nardostachys jatamansi* (Dubey et al. 2023)

| Quantification of phytochemicals | Amount (mg/g) |
|---|---|
| Total phenolic content (mg GAE/g) | 8.631 |
| Total flavonoid content (mg rutin/g) | 30.267 |
| Total flavonol content(mg rutin/g) | 72.971 |
| Total tannin content (mg catechin/g) | 8.875 |
| Total alkaloid content (mg/g) | 9.19 |
| Total saponin content (mg/g) | 0.927 |

## 23.11 Traditional and Ethnomedicinal Significance

### 23.11.1 Ethnobotanical Usage

The enormous therapeutic value of *N. jatamansi* has long been recognized by historic usage in traditional medicines like Unani in ancient Arabia and Greece, Ayurveda in India and the ancient Egyptian and Roman medical systems. In traditional medical system, this herb is found to be used as tranquilizing agent, hypotensive (Toolika et al. 2015), anticancerous (Ma et al. 2022), antidiabetic, anti-inflammatory (Bose et al. 2019), antidepressor (Garg et al. 2023), for hematopoietic activity (Gu et al. 2022) and cardioprotective agent (Subashini et al. 2007), in traditional medical systems. In addition to its usage as a flavoring component in medicinal oil formulation, jatamansi oil has antiarrhythmic properties (Bhatt and Kothiyal 2015). It treats heart conditions, blood pressure (Bhat and Malik 2020), sleep disorders, and mental illnesses (Castillo et al. 2018), seminal debility, palpitations and seizures (Shah and Gilani 2011; Prabhu and Panneerselvam 2013) hysteria and acts as an antispasmodic (Gowda et al. 2013). Table 23.6 illustrates the various uses of *N. jatamansi* in ethnomedicine.

**Table 23.6** Ethnomedicinal use of *N. jatamansi* as claimed by locals and tribes from various regions

| Sl. no. | Country | Region | Tribes | Local name | Plant part used | Type of preparation | Type of application | Ethnomedicinal property | Reference |
|---|---|---|---|---|---|---|---|---|---|
| 1. | India | Uttarakhand, Western Himalaya | Jaad Bhotiya tribe | Jatamansi, Bhootkeshi, Maansi | Roots, leaves | Infusion | Juice | It is used in internal pain | Balodi et al. (2018) |
| 2. | India | Arunachal Pradesh, Eastern Himalaya | Monpa tribe | Pang- po | Root stock, rhizome | Extract | Paste, juice | It is general tonic; treats dysentery, skin diseases, cough, fever, and cold. It acts as stimulant and diuretic | Tiwari et al. (2009) and Bhuyan and Pangu (2018) |
| 3. | India | Garhwal, Himalaya, Uttaranchal | Jaunasri tribe | Jatamansi, Dhoop | Rhizome | Extract | Juice | It treats epilepsy, hysteria | Bhatt and Negi (2006) |
| 4. | Nepal | Humla District, Western Nepal | Chetri, Thakuri tribe | Jatamansi, Bhulte, Pangpoe | Rhizome | Powdered infusions | Juice | It treats headaches, fever, cholera, leprosy, food poisoning, intestinal worms, cough, high altitude sickness | Rokaya et al. (2010) and Kunwar et al. (2006) |
| 5. | India | Western Himalaya, Uttarakhand | Bhotias, Tolchhas and Garhwalis | Jatamansi, Bhootkeshi, Maansi | Whole plant | Powder | Juice, Paste | It treats cough, cold, arthritis, heart disease, urinary problems | Negi et al. (2019) |
| 6. | India | Niti Valley, Central Himalaya | Bhotias | Jatamansi, Bhootkeshi, | Roots | Powder | paste | It treats rheumatism and joint pain | Phondani et al. (2010) |

(continued)

**Table 23.6** (continued)

| Sl. no. | Country | Region | Tribes | Local name | Plant part used | Type of preparation | Type of application | Ethnomedicinal property | Reference |
|---|---|---|---|---|---|---|---|---|---|
| 7. | Nepal | Far-West Nepal | Rana tharu tribe | Tapre, Jatamansi | Leaves, rhizomes | Extract | Paste | It treats altitude sickness, fever healing wounds, cough, high blood pressure, stomach diseases and swelling | Kunwar et al. (2009) |
| 8. | Nepal | Rasuwa District, Central Nepal | Tamang people | Jatamansi, Bhulti | Whole plant | Infusion | Juice | It treats fever, headache, weakness, dizziness | Uprety et al. (2010) |
| 9. | India | North Sikkim (Dzongu area) | Lepchas, Nepalese, Bhutias | Pangbu, Pung-zong | Rhizome | Powder | Juice | It treats nervous and gastric disorder | Maity et al. (2004) |
| 10. | China | Yadong River Valley, Tibet, China | Burang Tibetian people | Gansong,Songban, Gansongxiang | Rhizome | Powder | Fumes | It treats headaches, relieves stress and anxiety; also used as mosquito repellant | Ding et al. (2022) and Rehan et al. (2015) |

### 23.11.2 Ayurvedic Usage

The foundation of Ayurveda medicine is comprised of priceless medicinal plants. Since ancient Vedic times (1000–800 B.C.), *Nardostachys jatamansi* DC. has been treating several ailments like epilepsy, hysteria, and insomnia and has also been utilized as an antispasmodic agent (Das et al. 2015; Gupta et al. 2024). The plant is stated in various classical literatures like Charaka Samhita, Sushruta Samhita, and Sarangdhar Samhita (Sama et al. 2024). The Charaka Samhita includes *Nardostachys jatamansi* among the categories of herbs that are used to treat urinary calculi (ashmari), medicate smoking (dhumapana), regain consciousness (sangya sthapana), treat dermatitis (kandughna), and make bitter tonics (tikta skandha) (Tripathi 2004). Based on Sushrut Samhita *Nardostachys jatamansi* is used for fumigating infected lesions (Shastri 2005). Several dose forms of have been described in classical literature. These include Bala tail beneficial in arthritis and chronic pains, Amritadi tail for skin care, Brahami vati for brain booster (Sama et al. 2024), Mansi Tail for skin illnesses (Bhagwan et al. 2024), Mansyadi Ghrut for orchiditis (Shastri and Shastri 1999), Dashanga Lepa for inflammatory diseases, Mansyadi Lepa for skin problems (Shastri 2006), and Mansyadi Kwath for nervous disorders (Bhagwan et al. 2024).

Throughout Indian history, the rhizomes of this herb have been engaged in medicines used in Ayurveda, as an antispasmodic, diuretic, bitter tonic, stimulant, and therapy for convulsions, hysteria, and epilepsy (Gupta et al. 2024). The root extract has been utilized in Ayurvedic medicine for treating nervous convulsive disorders, jaundice, kidney stones, heart palpitations, flatulence, headaches, digestive and respiratory disorders, gastric diseases, skin diseases, impurities of blood and debility in semens (Pandey et al. 2013; Gowda et al. 2013). The extracts of the plant have been reported to have deobstruent, neuroprotective, cardioprotective, hepatoprotective, exhilarating, antioxidant, anticonvulsant, anti-parkinsonian, antihypertensive, sedative, antidiabetic, and antidepressant properties in the Unani medicinal system (Bhat and Malik 2020).

In the market, there are several herbal formulations which have *Nardostachys jatamansi* as its key ingredient, such as, *Anzocare* (for anxiety), *Aruna* (postmenopausal pain), *B-pex* (cardiac neurosis), *Bai Baidyanath Kesh Oil* (for hair growth), *Braintone syrup* (stress and anxiety), *Chaitanya* (immunity booster), *Dimag paushtik rasayan* (for memory loss), *Eladhi Choorana* (treats pittaja disorder), *Hartone* (cardiac problems and hypertension), *Histanin* (allergies and sneezing), *Iqwin* (antistress), *Liverson* (jaundice and hepatitis), *Myocare Liniment* (pain relief), *Nstim* (neurological disorder), *Multani Mulcurd* (cardiac disorder), *Sadamanjil Chooranam* (treats insomnia), *Stresnil* (for depression), *Vire Virechana Gulika* (for constipation), and *Tengin pushpadi Tailam* (for headache and eye disease) (Kaur et al. 2020).

## 23.12 Different Medicinal Properties

*Nardostachys jatamansi* has attracted focus for its immense medicinal activities. Several researchers have explained its significant anticonvulsant and antianxiety in animal models. It also exhibits anti-Alzheimer's, anti-Parkinson's diseases and also works against neurodegenerative disorders. Several studies also reported its anti-inflammatory and anticancerous efficacies. Recent scientific investigations have aimed to substantiate these traditional uses through rigorous pharmacological studies. The properties enlisted below focus on current research findings which provide a complete knowledge on the bioactive compounds, mechanisms of action, and clinical efficacy of this *Nardostachys jatamansi*. Figure 23.3 depicts different medicinal properties of *Nardostachys jatamansi*.

### 23.12.1 Anticancerous Activity of *Nardostachys jatamansi*

The traditional Unani and Ayurvedic system of medicine claims the anticancer potential of *Nardostachys jatamansi* which is being validated by several studies made by researchers throughout the globe especially focusing on the plants of the Himalayas (Rasheed et al. 2010; Pandey et al. 2013; Dhiman and Bhattacharya 2020; Pathak and Godela 2023). A study from India endorsed the antiproliferative efficiency of the root extract of this plant against breast carcinoma cell lines, namely, M. D. Anderson - Metastatic Breast - 231 (MDA-MB-231 and MCF-7) (Chaudhary et al. 2015). The methanolic root extract of the plant exhibited highest antiproliferative potential with Inhibitory Concentration at 50% ($IC_{50}$) value 58.01 ± 6.13μg/mL in Michigan Cancer Foundation-7 (MCF-7) cell line and 23.83 ± 0.69μg/mL in

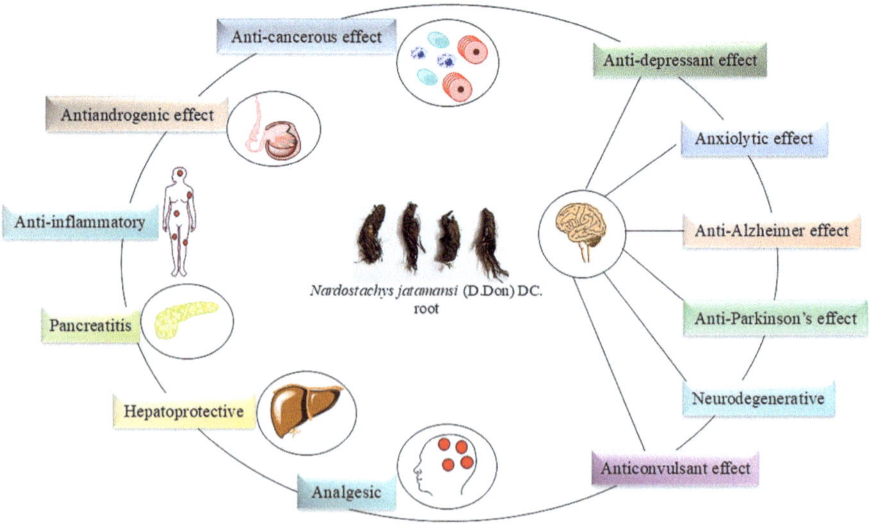


**Fig. 23.3** Schematic diagram showing different types of activity of *Nardostachys jatamansi*

M. D. Anderson - Metastatic Breast - 231 (MDA-MB-231 and MCF-7) cell line, respectively (Chaudhary et al. 2015). Again the fractions of petroleum ether root extract and diethyl ether root extract showed potency against Michigan Cancer Foundation-7 (MCF-7) cell line with $IC_{50}$ 60.59 ± 4.78μg/mL and M. D. Anderson - Metastatic Breast - 231 (MDA-MB-231 and MCF-7) cell line with $IC_{50}$ 25.04 ± 0.90μg/mL, respectively, which was analyzed by 3-(4,5-dimethylthiazol-2-yl)-2,5- diphenyltetrazolium bromide assay (MTT) assay (Chaudhary et al. 2015). Cell arrest was observed at G2/M phase in methanolic, petroleum ether and ethyl acetate root extracts whereas $G_0/G_1$ phase arrest was observed in diethyl ether root extract in M. D. Anderson - Metastatic Breast - 231 (MDA-MB-231 and MCF-7) cell line where the compounds like lupeol and β-sitosterol were found to be responsible for this efficacy which was determined by HPTLC method (Chaudhary et al. 2015). Another study from a China vouchsafes that a compound named nardostachin from the plant *Nardostachys jatamansi* showed strong cytotoxicity counter to pancreatic cancer cell lines (Ma et al. 2022). Human pancreatic cancer cell lines like Cystic Fibrosis Pancreatic Adenocarcinoma-1 (CFPAC-1), Human Pancreatic Adenocarcinoma Cell Line - 2 (CAPAN-2), Human pancreatic adenocarcinoma cell line-SW1990 Cell Line (SW1990), and PANCreas 1 (PANC-1) were used for cytotoxicity in which compounds like epoxy nardosinone unveiled cytotoxicity in contradiction of Human Pancreatic Adenocarcinoma Cell Line - 2 (CAPAN-2) cell lines with $IC_{50}$ value of 2.60 ± 1.85 μM (Ma et al. 2022). In comparison to Taxol with $IC_{50}$ 0.32 ± 0.13 μM, 1-hydroxylaristolone showed more cellular toxicity on Cystic Fibrosis Pancreatic Adenocarcinoma-1 (CFPAC-1) cell lines with $IC_{50}$ 1.12 ± 1.19 μM (Ma et al. 2022). Other sesquiterpenes like 1(10)-aristolane-9*β*-ol, 1-hydroxylaristolone, alpinenone, 1(10)-aristolen-2-one, nardostachin, and valtrate isovaleroyloxyhydrine display robust cytotoxic effect in contradiction of PANCreas 1 (PANC-1) cell lines where $IC_{50}$ value ranges from 0.01 ± 0.01 to 6.50 ± 1.10 μM along with nardonoxide, 10-hydroxyguaia-1(5)-6,7-epoxy-2-one, nardoguaianone K, 1(10)-isonardosinone-8-ol-9-one-7,11-lactone, and epoxy nardosinone in contradiction of Human pancreatic adenocarcinoma cell line-SW1990 Cell Line (SW1990) cell lines where $IC_{50}$ value ranges from 0.07 ± 0.05 to 4.82 ± 6.96 μM (Ma et al. 2022). Nardostachin also triggered cellular apoptosis through the mitochondria-dependent pathway and arrested the cells of SW1900 cell line at $G_2$/M phase of the cell cycle (Ma et al. 2022). The hydroalcoholic root extract of the plant exhibited antiproliferative efficacies against mouse hepatocellular carcinoma (HCC) cells Hepa1–6 cell lines by inhibiting the tumor formation and reducing STAT3 activation by preventing the ERK expression and diminished the HCC cell proliferation at the series of 0–100μg/mL by downregulating the PCNA expression (Lee et al. 2021). A study in China revealed that nardoguaianone L (G-6) which was isolated from the plant displayed effective cytotoxicity on pancreatic cancer cell line (SW1990 cells) with the $IC_{50}$ value was 2.1 ± 0.3μM confirmed by MTT assay (Sang et al. 2022). It inhibited in the formation of colony and migration of the SW1990 cells, besides inducing cell apoptosis which was endorsed by flow cytometric Annexin V/PI assay, thereby regulating the MET/PTEN/TGF-β pathway (Sang et al. 2022). Nardoguaianone L (G-6), the plant isolated compound, elevated the expression of Bax, simultaneously downregulating the Bcl-2 in the mitochondrial dependent apoptotic signaling pathway in SW1990 cell line (Sang et al. 2022). Matrix metalloproteinases 2 and 9 gets impaired by 20 and 40μM concentration of nardoguaianone L (G-6) and causes inhibition of Human pancreatic adenocarcinoma cell line-SW1990 Cell Line (SW1990) cell migration which was confirmed by immunofluorescent staining (Sang et al. 2022). In another study, the ethanolic root extract of *Nardostachys jatamansi* exhibited decrease in viability of human neuroblastoma cell lines 40μg/mL onwards and is unaffected on non-cancerous HEK-293 cells (Suryavanshi et al. 2017). It also delimited the growing kinetics of human neuroblastoma cancer cell lines, namely, Human cell line derived from a metastatic site of an Askin's tumor (SKNMC) and Institute for Medical Research-32 (IMR-32) with $IC_{50}$ values of 61.43 and 63.42μg/mL, respectively, further diminishing the expression of some proteins like MYCN (v-myc myelocytomatosis viral-related oncogene, neuroblastoma derived) and MDM2 (murine double minute 2) by inducing $G_2$/M phase arrest in cell cycle and apoptosis in SKNMC and Institute for Medical Research-32 (IMR-32) cell lines, respectively. Besides, it also enhanced the tumor suppressor protein p53 (Suryavanshi et al. 2017). In 2013 another study was conducted where the antiproliferative efficacy was reported in the n-hexane and alcoholic fraction of the root extract of the plant against SK-N-SH and Institute for Medical Research-32 (IMR-32) neuroblastoma cell lines at 100μg/mL dose (Monga and Kumar 2013). In alcoholic fraction the reduction was seen by 71% for Institute for Medical Research-32 (IMR-32) and 85% for SK-N-SH, whereas, in hexane fraction the reduction was seen by 91% for Institute for Medical Research-32 (IMR-32) and 82% for SK-N-SH cell lines (Monga and Kumar 2013).

### 23.12.2 Neuromodulation, Antidepressant, and Neuroprotective Activity of *Nardostachys jatamansi*

*Nardostachys jatamansi*, generally recognized as spikenard or *jatamansi*, is an herb which is perennial and traditionally utilized in Indian Ayurveda as well as Tibetan medicine for its neuroprotective and antidepressant properties. Ethnomedicinal practices suggest that its calming and mood-enhancing effects make it a powerful natural remedy for mental health disorders, particularly depression (Rahman and Muralidharan 2010).

(a) *Antidepressant and Neuroprotective Effects of N. jatamansi*

Preclinical studies offer robust proof for the antidepressant potential of *Nardostachys jatamansi* (Li et al. 2020, 2021; Panchal and Goswami 2020). Animal models such as the Forced Swim Test (FST) and the Tail Suspension Test (TST), conducted in mice, have demonstrated significant antidepressant-like effects (Panchal and Goswami 2020; Rao et al. 2013). A study revealed a dose-dependent reduction in immobility time in both tests when methanolic extract of the plant (125, 250, 500 mg/kg) was administered to Bagg Albino (BALB/c) mice. Notably, the results were more pronounced when *N. jatamansi* was combined with fluoxetine, indicating a synergistic effect (Saleem et al. 2021). Another study involving electron-beam-irradiated mice showed that *N. jatamansi* reduced immobility time and improved mood, confirming its antidepressant efficacy (Deepa et al. 2013).

Several preclinical experiments have highlighted the antidepressant potential of *N. jatamansi* (Saleem et al. 2021; Li et al. 2020, 2021; Panchal and Goswami 2020). A study evaluated the antidepressant effects of methanolic extract of the plant (125, 250, 500 mg/kg) on Bagg Albino (BALB/c) mice using the TST and FST. The outcomes revealed a dose-dependent reduction in immobility time in both tests, indicative of antidepressant-like activity. Notably, the 500 mg/kg dose outperformed fluoxetine, a commonly prescribed selective serotonin reuptake inhibitor (SSRI), in reducing the time of immobility in the FST model (Saleem et al. 2021).

In another study involving electron beam-irradiated mice, ethanolic root extract significantly reduced immobility time, indicating that the herb could alleviate depression-like symptoms even under extreme stress conditions (Deepa et al. 2013).

(b) *Mechanism of Action*

The antidepressant effects of jatamansi are attributed to several bioactive compounds, including sesquiterpenes like jatamansone (Kumar and Behera 2023). These compounds have shown significant activity in modulating neurotransmitters like serotonin and dopamine pivotal for mood regulation (Dhingra and Goyal 2008). Nardosinone, another compound, has been linked to neuroprotective and serotonin-modulating effects (Li et al. 2020, 2021). Valeranone and jatamansinol are linked to mood stabilization and exhibit mild opiate properties, making them effective in managing anxiety and depression (Saleem et al. 2021).

The antidepressant activity of *N. jatamansi* primarily involves its action on neurotransmitter pathways, mainly through the reticence of monoamine oxidase (MAO), an enzyme which is responsible for the neurotransmitter breakdown like serotonin, norepinephrine and dopamine (Dhingra and Goyal 2008). Besides inhibiting MAO, the accessibility of these neurotransmitters elevates, improving mood and reducing depressive symptoms (Dhingra and Goyal 2008). Additionally, the antioxidizing property protects neural tissues from oxidative stress, often linked to depression (Saleem et al. 2021). Jatamansone has also been found to enhance GABA (gamma-aminobutyric acid) levels, contributing to its calming and antidepressant effects (Deepa et al. 2013).

*N. jatamansi* the serotonin transporter (SERT) plays a crucial role in serotonin reuptake (Li et al. 2020). Enhancing SERT activity is a recognized mechanism for selective serotonin reuptake inhibitors (SSRIs), a typical class of antidepressants (Li et al. 2021). Interestingly, *N. jatamansi* functions as both an SSRI and a selective serotonin reuptake enhancer (SSRE), depending on the pathophysiological status of depression (Li et al. 2020). This dual modulation makes it adaptable to varying neural environments, regulating serotonin levels more effectively than conventional SSRIs, focusing solely on inhibiting serotonin reuptake (Li et al. 2020).

(c) *Future Prospects*

The clinical potential of *N. jatamansi* is substantial. Given its ability to modulate key neurotransmitters and work synergistically with conventional antidepressants, it could play a crucial role in treating treatment-resistant depression. More clinical trials in humans are needed to establish its efficacy, safety, and dosage for treating depression. Additionally, exploring its use in combination therapies could lead to more effective treatment protocols, particularly for patients who experience side effects from standard antidepressants (Saleem et al. 2021).

Ephemeral use of *N. jatamansi* is commonly called safe, with few reported side effects. However, in higher doses, it may cause mild sedation, dizziness, or gastrointestinal discomfort. Long-term use, particularly at higher doses, could potentially lead to dependency due to its sedative properties, although this has not been extensively studied (Deepa et al. 2013). Added research is required to completely know the enduring effects of *N. jatamansi*, especially in clinical populations.

*N. jatamansi* presents a promising natural alternative for treating depression, with preclinical evidence supporting its antidepressant activity through multiple mechanisms, including serotonin modulation, MAO inhibition, and neuroprotection. Its synergistic potential with conventional antidepressants makes it an appealing candidate for combination therapies. While more research is needed to confirm its clinical efficacy, *N. jatamansi* holds significant promise as a safe and effective treatment for depression, especially for those seeking natural alternatives or adjunct therapies.

### 23.12.3 Anti-inflammatory Activity

Inflammation is a multidimensional condition which is pathological and characterized by the proliferation of many immune cells, including macrophages (Shin et al. 2015). In the course of inflammatory reactions, pro-inflammatory cells, primarily macrophages which are active, are accountable for the majority of the molecular and cellular pathophysiology of sepsis due to their production of cytokines along with other pro-inflammatory substances, inclusive of interleukin IL-1β, IL-6, tumor necrosis factor alpha (TNF-α), nitric oxide (NO), and prostaglandin E2 (PGE2) (Kang et al. 2006; Noguchi et al. 2003; Szekanecz and Koch 2007; Vane et al. 1994). Inflammation is just one of the many ailments addressed by *Nardostachys jatamansi*, a plant native to the Himalayas that is popular in the Asian subcontinent (Kwon et al. 2013).

A research stated that the methanolic extract of the plant reduced inflammatory mediators and intervened with Doxorubicin (DOX)-induced inflammatory sequences in heart tissue. In the midst of inflammation, T cells and macrophages release IL-6, an interleukin with complicated effects on inflammatory signaling, and TNF-α, a mediator of early inflammatory processes, is found to be involved in Doxorubicin (DOX)-induced inflammation. The treatment with methanolic extract of *N. jatamansi* at doses of 250 mg/kg and 500 mg/kg for 3 weeks duration concluded in decreasing levels of IL-6. Analogous to IL-6, the TNF-α concentration in the same extract and similar concentration exhibited a considerable decrease in the inflammatory response (Singh et al. 2020). Desoxo-narchinol-A (DN), yet another compound derived from *N. jatamansi*, demonstrates anti-inflammatory capabilities towards lipopolysaccharide (LPS). The implications of DN in the regulation of NO, iNOS, PGE2, and cyclooxygenase-2 (COX-2) in LPS-stimulated macrophages were observed at concentrations ranging from 50 to 500 nM. The preliminary treatment with DN impeded LPS-induced inducible nitric oxide synthase (iNOS), NO, and the upregulation of COX-2 levels in a dose-responsive manner. It inhibits other mediators of inflammation including such as TNF-α, IL-6, and IL-1β (Shin et al. 2015).

Another research was conducted with different fractions of the methanolic extract (20, 40, 60, 80, and 100%) of *N. jatamansi* which not only restricted LPS-induced manufacture of nitric oxide but also prohibited the manufacturing of cytokines, like IL-6, IL-1*β*, along with TNF-α demonstrating the beneficial properties of the plant in LPS-Induced inflammatory response (Bae et al. 2014). Cycloolivil, a phytoconstituent extracted from *N. jatamansi*, was examined for its anti-inflammatory activities on HaCaT keratinocytes and for its ability to mitigate allergic reactions to the skin. It was found that cycloolivil at a concentration ranging from 10 to 40μM significantly reduced IL-6 levels with an $IC_{50}$ value of 31.05 ± 0.93 μMn and restricted the manufacturing of both RANTES and IL-8 at all verified concentrations (Yoon et al. 2024). Studies have found that narchinol B and desoxo-narchinol A derived from *N. jatamansi* impeded the generation of PGE2 and NO, as well as the levels of induced COX-2 protein and NO synthase in LPS-stimulated BV2 and microglial cells in a dose-responsive manner (Yoon et al. 2018).

Additionally, the possible function of *N. jatamansi* in neuroinflammatory diseases was assessed in which the methanolic roots extract ($IC_{50}$ 14.58 ± 0.9μg/mL) and leaves ($IC_{50}$ 23.54 ± 1.6μg/mL) were evaluated. The methanolic extract of *N. jatamansi* exhibited substantial anti-5-LOX activity relative to nordihydroguaiaretic acid (NDGA) at concentrations extending from 2 to 50μg/mL, with NDGA being a recognized inhibitor of 5-LOX ($IC_{50}$ 2.56 ± 0.4μg/mL) (Bose et al. 2019). The ethanolic extracts of the plant were also examined for their potentiality against inflammation on different animal models. The plant extract in ethanol at concentrations of 150 and 300 mg/kg demonstrated substantial anti-inflammatory effects in a dose-dependent method, with the percentage of prohibition of paw edema recorded at 29.06 and 55.81%, respectively, in comparison to the group-control (Singh et al. 2014). The methanolic extract of the plant was investigated for anti-inflammatory activities employing the LPS-stimulated RAW2647 inflammatory cellular model. The anti-inflammatory result of the plant extract was evaluated through their inhibitory impact on the production of inducible nitric oxide synthase (iNOS), nitric oxide (NO), and cyclooxygenase-2 (COX-2). The study observed a reduction in NO levels from 100% in the LPS-incubated category lacking *Nardostachys jatamansi* to 55.1, 24.0, and 5.8% following evolution with *N. jatamansi* extracts at concentrations of 25, 50, and 100μg/mL, respectively (Li and Wang 2014).

### 23.12.4 Anti-convulsant Activity

Epilepsy and seizures are recurrent neurological disorders resulting from aberrant electrical responses in the CNS (Sumadewi et al. 2023). It is a persistent, paroxysmal, and repetitive disorder characterized by neurological, cognitive, social and psychological elements that cumulatively lead to repeated seizures, as well as persistent warning signs for seizures of any kind (Fisher et al. 2014; Stafstrom and Carmant 2015). Investigating a correlation between oxidative stress and neurological illnesses has been relatively uncomplicated, as seen by elevated levels of oxidative stress markers in tissues throughout disease course (Simonian and Coyle 1996).

The ethanolic concentrate of *N. jatamansi*, derived by refluxing 95% ethanol, was examined for its anticonvulsant properties. The effectiveness of the extract in seizures management was assessed using the maximal electroshock seizure (MES) model and pentylenetetrazole (PTZ)-induced seizures. In order to evaluate the anticonvulsant efficacy of *N. jatamansi* extract utilizing the PTZ model, various experimental groups of rats, consisting of 5–8 individuals, were administered doses of 50, 100, and 400 mg/kg body weight (b.w.), intraperitoneally (i.p.) and subsequently subjected to pentylenetetrazole (70 mg/kg, i.p.) treatment (Goodman et al. 1953). To further assess the anticonvulsant efficacy via the MES model, the subjects were categorized into two groups (EA 1972). A group of animals, consisting of 6–12 individuals, received oral administration of phenytoin from dosages lower to higher. A separate group, consisting of 5–12 individuals, received the aforementioned eight doses of phenytoin in conjunction with 50 mg of the plant extract. The results indicated a considerable disparity between the groups administered *N. jatamansi* extract from roots and the untreated group receiving only the vehicle. Post hoc evaluation (Scheffe approach) indicated that intraperitoneal administration of lower dosages (50 and 100 mg/kg) of the extract failed to safeguard animals from MES seizures or diminish the extension/flexion ratio. The elevated doses administered (200 and 400 mg/kg) demonstrated a substantial decrease in the extension/flexion ratio relative to the group-control. Moreover, the extract delivered intraperitoneally exhibited no impact on PTZ seizures (Rao et al. 2005).

Another phytoconstituent produced from *N. jatamansi*, jatamansone, was the subject of an investigation into its potential anticonvulsant properties. The activity of jatamansone was determined by administering it orally to six rats at a range of doses, and the MES test, the cPTZ test, and the minimum clonic seizure test were used to evaluate its effects. It demonstrated significant efficacy in the MES test at a dosage of 50 mg/kg, as seen by complete protection at both one and four hours point in time. Complete protection was noted at 125 mg/kg; however, there was a noteworthy decline in defense at 150 and 200 mg/kg. The efficacy of jatamansone was determined to be superior to that of conventional phenytoin at a dosage of 50 mg/kg. In the PTZ screen, jatamansone demonstrated ineffectiveness, as no protective effects were noted even at a dosage of 75 mg/kg. Likewise, jatamansone was determined to be ineffective in the minimum clonic seizure test at a dosage of 50 mg/kg (Pallavi and Kumar 2012).

### 23.12.5 Antiandrogenic Property

Antiandrogenic properties refer to the capability of certain substances to constrain the effects of androgens, which are male hormones like testosterone. This can be beneficial in treating conditions such as prostate cancer (Prutianu ct al. 2021), polycystic ovary syndrome (PCOS) (Sadeghi et al. 2022), and androgenic alopecia (Nestor et al. 2021), as it helps to regulate overall hormone levels and mitigate their effects on the body.

Studies have also shown that *Nardostachys jatamansi* may inhibit the action of androgens by involving several mechanisms, like blocking androgen receptors or, on the other hand, reducing the synthesis of testosterone (Chavan 2020). Research has indicated that *Nardostachys jatamansi* influences hormonal pathways (reducing testosterone levels) and this effect helps lighten symptoms associated with hyperandrogenism (Onder et al. 2024). The bioactive compounds of the plant, like sesquiterpenes and flavonoids, gradually decreases its efficacy as they interfere by blocking the receptor site of androgens (Bayala et al. 2020). Furthermore, the plant's antioxidant properties help mitigate inflammation linked to androgen activity (Valmiro et al. 2021). Another study also stated that *Nardostachys jatamansi* contains jatamansone (a sesquiterpene) along with other flavonoids that interact with androgen receptors, and these bioactive components potentially block the androgen activity (Dhiman and Bhattacharya 2020).

RIKILT yeast Androgen bioAssay (RAA) was used to study the antiandrogenic efficacy of the plant. The plants were assessed for their potential in combating PCOS prompted by estradiol valerate (EV) in rat models. The result showed that after the treatment with extract, the steroidal hormonal levels and estrous cyclicity were normalized and the growth of ovarian follicular was regularized that exhibited androgenic properties and mild androgenic activity (Sandeep et al. 2015). A methanolic extract of *Nardostachys jatamansi* was prepared to study the androgenic properties in another study. These methanolic extracts restored the distressed levels of hormone and ovarian follicular growth in adult Sprague-Dawley female rats (Dhiman and Bhattacharya 2020).

Some studies were made to reveal the potential of *Nardostachys jatamansi* on polycystic ovarian syndrome. *Nardostachys jatamansi* DC and a few plants were used to study the impact on PCOS-induced rat models. This plant regularized reduced steroid hormone levels, estrous cyclicity dose-dependently and improved ovarian dynamics (Khan et al. 2024) *Jatamansi* extract exhibited to normalize estrus cycle and steroid hormone levels in estradiol valerate-induced PCOS rats and reduce cyclicity in all rats. In addition, the extract has shown a positive effect in regulating ovarian follicular growth, producing an amplified ovulation rate in the animal model (Ahmad and Safuan 2019).

A research was conducted to evaluate "DXB-2030," a polyherbal blend which has *Nardostachys jatamansi* (20%) extract in an investigational model of testosterone propionate (TP), responsible for inducing PCOS in female rats. The findings showed that "DXB-2030" could reverse the TP-induced changes. This was an outcome of an increased GLUT4 expression, which further led to decreased body weight, T levels, and the cystic follicles of the ovaries. The conclusion was that "DXB-2030" affects the reproductive system (women with PCOS) it helps in the reduction in size and volume of the cyst and ovary, respectively. It has an antiandrogenic effect, reversion of estrus cyclicity, and repair of the histology of all ovarian tissue (Zhang et al. 2021; Azeemuddin et al. 2019). All the research studies reveal that *Nardostachys jatamansi* has antiandrogenic properties and can be utilized for diverse applications.

## 23.12.6 Analgesic Property

Analgesic is one type of medication that relieves the pain without affecting the consciousness or altering any sensory perception. Their mode of mechanism is reducing inflammation or by altering how the brain processes pain (Draguhn et al. 2021). Donguibogam, a traditional medical manual (Korean) published around 400 years ago, mentioned *Nardostachys jatamansi* (Kim et al. 2021a, b). The roots of the plant have analgesic (Kim et al. 2021a, b) and soothing (Pant et al. 2022) potential, which treats pain and neurological symptoms. *Nardostachys jatamansi* contains terpenoids and lignans, which are secondary metabolites, and are known to reduce pain (Liu et al. 2018).

Many researches have stated the well-known effect of roots of *Nardostachys jatamansi* having neuroprotective properties, sedative properties, and analgesic efficacies (Kim et al. 2021a, b; Pant et al. 2022; Anupama et al. 2023). This is because of the bioactive ingredients which are present in it, such as essential oils, sesquiterpenoids, triterpenoids, iridoids, coumarin, flavonoid, and lignin are well-known constituents (Ahmad et al. 2006; Dhiman and Bhattacharya 2020). The anti-inflammatory activities of terpenic coumarin, sesquiterpenoids, and phytosterols like narchinol and nardosinone subsidize to the analgesic properties delivered by *N. jatamansi* (Gómez-Betancur et al. 2015). Few studies have shown positive effects like prohibition of the manufacture of inflammatory cytokines (PEG2, IL-6, IFN-a/β, and TNF-α) and phosphorylation of MAPK (mitogen-activated protein kinases) signaling is associated in the analgesic property (Yoon et al. 2018; Kim et al. 2021a, b). Most flavonoids that are present in *Nardostachys jatamansi* are analgesic, antioxidant, and anti-inflammatory, besides extensively used for reliable preclinical and clinical profiles (Ferraz et al. 2020). For instance, quercetin and troxerutin act as an anti-inflammatory agent which helps in reducing pain in clinics (Xue et al. 2019).

In another study, the efficacy of analgesic property of naringenin 4′,7-dimethyl ether (NRG-DM), a flavonoid present in *Nardostachys jatamansi,* was studied along with the mechanism associated. The analgesic property of NRG-DM was evaluated in a formalin-persuaded mouse inflammatory pain model along with using mustard oil-induced mouse as a colorectal pain model. In the study, the mice were intraperitoneally regulated by vehicle or NRG-DM. The result showed that the neuronal delayed rectifier potassium currents ($I_K$) and voltage-gated sodium currents ($I_{Na}$) were repressed considerably, which helped alter the neuron pathway for signaling. In conclusion to the above study, the author claimed that NRG-DM is a potent analgesic contestant which inhibits multiple neuronal channels (mediating $I_K$ and $I_{Na}$) (Gu et al. 2022).

In conclusion, the extensive research on *Nardostachys jatamansi* highlights its potential as a significant analgesic agent, driven by its diverse bioactive components like terpenoids and flavonoids, which works through numerous mechanisms, and which includes the inhibition of inflammatory cytokines and modulation of neuronal signaling pathways. This underscores the ultimate promise of the potential use of *Nardostachys jatamansi* in analgesic properties.

### 23.12.7 Anxiolytic Property

Anxiolytic properties refer to the ability of a constituent to decrease anxiety and promote calmness (Kenda et al. 2022). Medications or compounds with anxiolytic effects assist in alleviating symptoms of anxiety disorders. This is done by modulating neurotransmitters like Gamma-aminobutyric acid (GABA) or serotonin in the brain (Panara et al. 2020). This can lead to a sense of relaxation and emotional stability. *Nardostachys jatamansi* is known for its depressant activity, slowing down brain activity, which causes the muscles to relax, calm, and soothe a person (Jha and Thakur 2020). The *Nardostachys jatamansi* extract, combined with other herbs, has shown depressant activity (Saleem et al. 2021). *Nardostachys jatamansi* has been extensively used in numerous Asian countries as a bitter tonic, an antispasmodic, and a stimulant plus to treat hysteria, epilepsy, palpitation, coria, palpitation, and convulsion (Mude et al. 2020).

*Nardostachys jatamansi* has reflective claims hostile to pharmacological mediations and is characterized as a hypnosedative drug as per Ayurveda (Das et al. 2021). The anxiolytic action of the plant was deliberated by using behavioral anxiolytic tests. *Nardostachys jatamansi* rhizome has been suggested for insomnia in the Indian medicine. Animals (mice model) were divided into three groups in which they were treated with water (control), diazepam (standard), and powder of *Nardostachys jatamansi* for 7 days. Mice were treated with the plant extract which showed elevated level of GABA and monoamine neurotransmitters in brain and tissue, respectively, and parameters of antioxidant were increased (Panara et al. 2020).

A study showed that 70% ethanolic plant extract has the presence of metabolites which exhibits anxiolytic properties by using liquid chromatography-electrospray ionization-tandem mass spectrometry (LC-ESI-MS/MS) method. Oxidative stress was triggered using buthionine sulfoximine (BSO) at a concentration of 300 mg/kg for 7 days after pretreatment with the plant extract in a concentration of 250 mg/kg. Anxiety levels were assessed in mice utilizing behavioural indices and biochemical parameters along with western blot (glutathione reductase and glyoxalase 1; 2 important biomarkers which link oxidative stress with anxiety). The result showed that the extract reduced oxidative stress-mediated markers in serum viz. lipid peroxides, cortisol, and protein carbonyls in brain which demonstrated amended brain antioxidant status. The extract also alleviated the levels of glutamate and choline in the brain. Apart from this, the extract increased the level of GABA in the brain and monoamine levels, thus efficiently seizing the anxiety-like effects of BSO. This study demonstrated that *Nardostachys jatamansi* is effective in remedying oxidative stress-mediated anxiety as well as neuropsychiatric disorders (Razack et al. 2021).

Based on research in past and ongoing, *Nardostachys jatamansi* has exhibited many significant anxiolytic properties. It can modulate neurotransmitters like GABA and enhance antioxidant status in the brain, as confirmed by various studies. Its traditional use in treating anxiety-related conditions, combined with progressive research validating its effects, emphasizes its potential as an effective natural remedy for managing anxiety and related neuropsychiatric disorders.

### 23.12.8 Anti-Alzheimer's Effect

The history and record of working with various active constituents of *Nardostachys jatamansi* (D. Don) in Ayurveda date back a long time. A rich combination of antioxidants and anti-inflammatory chemicals in *Nardostachys jatamansi* confirms it as a capable contender for ancient medical systems and modern therapeutic processes (Tripathi et al. 1996). In traditional medicine of Ayurveda, it is documented that *N. jatamansi* treats brain ailments (as Medhya), sleep disorders (as Nidhrajnana), mental disorders (as Manasrogaghna), digestive disorders (as Pachana), etc. (Goyal and Rath 2021). Several potent bioactive compounds and phytochemicals such as coumarins like jatamansin and jatamansinol, alkaloids like actinidine, sesquiterpenes like valeranone, jatamansinone, nardosinone, etc., and their derivatives are present which are responsible for its extensive use and are also exploited commercially in therapeutics and pharmaceutical industries (Chatterjee et al. 2005; Saroya 2011; Purnima et al. 2015). Multi-target potency and efficacy studies of the bioactive phytochemicals and various molecular studies like molecular docking, molecular mechanics/generalized born surface area (MM-GBSA) analysis, etc., justify the prospective of *N. jatamansi* in the treatment of Alzheimer's disease (AD) (Kizhakke et al. 2023; Anupama et al. 2021).

Alzheimer's disease (AD) is an age-related irreversible neurodegenerative and progressive disorder that results in impaired intellectual skills and memory loss (Jiang and MacNeil 2023). Sleep disorders, memory impairment, various cognitive imbalances, dementia, etc., which are associated with aging increase the chance of developing Alzheimer's disease even in mid-age group persons (Nasrouei et al. 2020). A sharp increase in the cases of AD is expected by 2050 and the most concerning part is that AD is becoming a severe medical issue as the scope of actual actions needed to preclude and cure impairments due to AD is still restricted (Varinthra et al. 2023). Many published studies have shown the protective action of *Nardostachys jatamansi* against neurotoxicity, sleep deprivation, and cognitive disturbances (Anupama et al. 2022; Krishnan et al. 2023; Razack et al. 2021). Sesquiterpenes like valeranone found in *Nardostachys jatamansi* have been established by various studies to reverse diazepam and scopolamine-induced amnesia and exhibit the potential of memory retention in diazepam-induced amnestic young and aged mice (Karkada et al. 2012; Chennaiah et al. 2016; Malabadi et al. 2021).

Pretreatment with methanolic extract of the plant continuously for 15 days has been exhibited to induce a noteworthy development in motor coordination which can be attributed to its anticerebral ischemic action because of the occurrence of a bioactive compound called jatanolides in *Nardostachys jatamansi* (Joshi and Parle 2006). The pathological features of AD such as non-cognitive motor impairment and age-related macular degeneration (AMD) are attributed to $A\beta_{42}$ protein-induced oxidative stress (Buchman et al. 2024). The role of neurotransmitter acetylcholine (ACh) in AD pathogenesis can be a result of $A\beta_{42}$ protein deposition in the brain which disturbs neurotransmission, decreasing ACh and causing reciprocal accumulation of acetylcholinesterase/butyrylcholinesterase (AchE/BuChE) in the synaptic cleft (Ferreira-Vieira et al. 2016). Jatamansinol eases locomotor activity, accelerates learning and memory, and decreases $A\beta_{42}$ protein levels in AD flies in experimental sets by boosting the antioxidant properties, restricting $A\beta_{42}$ protein-induced oxidative stress, and stopping cholinesterase properties in the AD *Drosophila* model indicating its possible clinical and therapeutic efficacy against AD (Kizhakke et al. 2022).

### 23.12.9 Hepatoprotective Activity

*Nardostachys jatamansi* is a well-known medicinally potent herb of Indian origin frequently utilized for eras in Unani (Greeko-Arabic system) and Ayurveda schools of the Indian medicine system. The decoction of the roots and alcoholic root extract of *N. jatamansi* were used as hepatotonic (Gupta et al. 2024), diuretic (Bhat and Malik 2020), cardiotonic (Subashini et al. 2006), antispasmodic (Pathak and Godela 2023), analgesic, etc. (Mehra and Garg 1962). Animals when exposed to hepatotoxin called thioacetamide face pronounced liver damage evidenced by enzymatic alterations in the levels of serum aspartate aminotransferase (AST), alanine aminotransferase (ALT), and Acylation stimulating protein (ASP) (Rawat et al. 1997). An established experimental study proved that 50% ethanolic extract of *N. jatamansi* exhibited a promising hepatoprotection against thioacetamide-induced liver damage in rats which was confirmed by the lowered levels of serum AST, ALT, and Acylation stimulating protein (ASP) (Shakir et al. 2000). The efficiency of the herb in hepatoprotection can be attributed to the availability of an extensive series of bioactive compounds in *N. jatamansi*, like sesquiterpenes, neolignans, and lignans (Shakir et al. 2000). Alcoholic extracts of *N. jatamansi* have also been shown by a study to display a prominent reduction in the concentration of certain hepatotoxic indicator enzyme like serum alkaline phosphatase (SALP), serum glutamic pyruvic transaminase (SGPT), serum glutamic oxaloacetic transaminase (SGOT), bilirubin & other hepatic enzymes, representing elevated quantity of hepatocytes with extra rejuvenating and regenerative activity, which indicates improved hepatoprotection as evaluated through the experimental studies (Sharma et al. 2021). When assessed by carbon tetra chloride ($CCl_4$)-induced hepatotoxicity on Wister rats of both sexes and the hepatohistopathology was studied, it was found that the quantity of biochemical markers such as glutamate oxaloacetate transaminase (SGOT), alkaline phosphates (SALP), and glutamate pyruvate transaminase (SGPT) was significantly altered which was again confirmed by the positive stability study of some physiochemical markers like particle size, sedimentation volume, pH, viscosity, Zeta potential, etc., at variable temperatures (Dandagi et al. 2008). When Nardosinone from *N. jatamansi* was tested for the development of drugs for liver-related metabolic diseases, a novel direction was explored when it showed a positive result in hepatic lipid metabolism (Ma et al. 2024).

### 23.12.10 Anti-Parkinson's Disease Effect

Parkinson's disease (PD) is the most relevant disease in the mid-age group and aged population and is the second most significant neurodegenerative disorder after Alzheimer's disease (AD) (Wakabayashi et al. 2013). Its histopathology includes extensive impairment and removal of dopamine (DA) neurons in the dense region of substantia nigra and the expression of $\alpha$-synuclein (a major constituent of Lewy bodies), which leads to excessive damage of dopaminergic neurons thereby decreasing the expression levels of dopamine in the striatum region, which indicates to a sequence of signs such as severe tremor, muscular dyskinesia, abnormal posture, constipation, depression, and hyposmia (Kakkar et al. 2018; Shenoy and Karkada 2024; Deka et al. 2023). Currently, the main medicine used for treating Parkinson's disease is levodopa, which increases the dopamine level in the brain and imitates the effect of dopamine (Li et al. 2023). However, some side effects are noticed by the chronic use of the medicine, such as some cognitive disorders and mental abnormalities which restrain its application for PD treatment (Olanow et al. 2001). Earlier studies have suggested that Nardostachone which is mined from the rhizomes and root of *Nardostachys jatamansi* exhibits some pharmacological actions on the nervous system like anticonvulsant, antidepressant, and antianxiety effects and could lessen the PD symptoms in rats (Ahmad et al. 2006; Razack et al. 2015). In the case of diseases like PD, the drug-designing is manipulated in such a way that the blood–brain barrier can be penetrated by the drug to exert its effect which is substantially a difficult pharmacological process (Swati and More 2008). Nardosinone, a sesquiterpene peroxide, is a prime component of *Nardostachys jatamansi* which when experimented on mouse embryonal cells showed proliferation of embryonic neuronal stem cells and exhibited the least toxicity and thus considered a potent drug with minimal side effects for treating neurodegenerative disorders like PD (Qin 2011). On performing pathological and behavioral tests with rat models, it has been seen that the treatment of rats with PD successfully minimized the degree of damage to brain function when administered with a combination of levodopa-dopamine and Nardosinone, established by better performance in behavioral characteristics than the other experimental formulas (Liu et al. 2024). Hydro-alcoholic extract of *N. jatamansi* reverses the symptoms of haloperidol-induced Parkinson's disease in rats and according to some studies, the process by which the enhancement occurs can be accredited to pharmacological or biochemical mechanisms that may may accelerate the circulation process of dopamine by upregulation of dopamine signaling pathway (Ma et al. 2024).

## 23.13 Neuroprotective Property

Neurodegeneration refers to the loss of neurons characterized by gradual cell death along with motor and cognitive dysfunctions specific to the area of the brain which experiences neuronal damage (Peña-Bautista et al. 2020). The most prevalent diseases that have well-characterized neurodegeneration are Alzheimer's disease and Parkinson's disease, with 55 million people having dementia worldwide, out of which 60–70% of the cases can be recognized as Alzheimer's disease alone (WHO 2023).

Human cerebral ischemia is one of the principal reason of many neurological deficits and death (Wei and Wei 2017). The extract of *Nardostachys jatamansi* (D.Don) DC. enhances $\gamma$-amino butyric acid (GABA), which is known to be therapeutic in treating ischemia (Prabhu et al. 1994). The alcoholic extract of *N. jatamansi*, suspended in a mixture of ethanol: Tween 80: distilled water (1:2:5), was administered to mice at 250 mg/10 mL (vehicle)/kg b.w. for 15 days before occluding the mid cerebral artery (thereby inducing ischemia) for 2 h and allowing 22 h of reperfusion. This improved glutathione levels and significantly increased catalase and $Na^+$-$K^+$ ATPase activities, which enhanced spontaneous motor activity and coordination (Salim et al. 2003). *Majun Khadar*, an Unani polyherbal formulation containing *N. jatamansi*, also shows the same therapeutic effects (Yousuf et al. 2011). Ethanolic extract of the plant also bolstered memory and learning in young mice and was effective in reversal of amnesia, whether chemically induced (by diazepam or scopolamine) or age-related (Joshi and Parle 2006).

The ability of a compound to cross the blood–brain barrier is a basic requirement for successful drug delivery in case of neurodegenerative disorders. Evaluation of pyranocoumarins in this regard reveals their permeability and activity in the nervous system. Molecular docking studies have further elucidated the neuroprotective activity of pyranocoumarins from *N. jatamansi* that operate via inhibition of enzymes which degrade neurotransmitters or promote neurodegeneration through their activity, for example, dihydro jatamansi inhibits acetylcholinesterase (AChE), jatamansinol inhibits butyrylcholinesterase (BuChE), glycogen synthase kinase 3b (GSK3b), and kelch-like ECH-associating protein 1 (Keap1) which are all targets of pre-existing drugs (like galantamine and rivastigmine) used in treating Alzheimer's disease (Anupama et al. 2021).

Selenoproteins are known to perform antioxidant functions and aid in maintaining redox homeostasis (Bräuer and Savaskan 2004). Crocetin protects neurons from the deleterious effects of 6-hydroxydopamine, a neurotoxin (Ahmad et al. 2005). In one study, *N. jatamansi* extract (200 mg/kg body weight), in conjunction with Sodium selenite (0.05 mg/kg b.w.) and crocetin (25µg/kg b.w.), was administered via gavage for 15 days to male albino Wistar rats. The behavior of the rats, which were cognitively impaired by streptozotocin induction, showed significant improvement (Khan et al. 2012).

Microglial cells are specialized macrophages that clean nerve tissue (Colonna and Butovsky 2017). They release pro-inflammatory and neurotoxic mediators, which are activated by a signaling cascade initiated by toll-like receptor (TLR) family receptors, which in turn rely on the binding of lipopolysaccharides (LPSs) (Medzhitov 2001). One pathway that the cascade takes is the TLR4-mediated myeloid differentiation factor 88 (MyD88)-dependent pathways. MyD88 is responsible for regulating the signaling pathway for most TLRs, which activates nuclear factor κB (NF-κB) and mitogen-activated protein kinases (MAKs) (Kaltschmidt and Kaltschmidt 2009). Nardostachin, an active principle of *N. jatamansi* extract, hinders TLR4 and MyD88 protein expression and prevents phosphorylation and translocation of NF-κB, in LPS-stimulated BV2 cells and rat primary microglial cells. Additionally, it also reduces LPS-induced overproduction of nitric oxide, prostaglandin E2, as well as pro-inflammatory cytokines, like interleukin (IL)-1β, IL-6, IL-12 and tumor necrosis factor-α in the same (Kim et al. 2020).

Acquisition and retention of spatial memory tasks are impaired under the effects of chronic restraint stress (Sunanda and Raju 2000). Rats subjected to such stresses were evaluated in their ability to comprehend the eight-arm radial maze task. Rats treated with *N. jatamansi* extract (200 mg/kg body weight) had learning curves comparable to unstressed individuals and also gradually improved on their choices and minimized reference memory errors by the eighth day of training (Karanth et al. 2012).

## 23.14 Pancreatitis

Pancreatitis is simply the pancreas inflammation which can be acute, where the pancreas shows inflammation ephemerally (NHS 2024) or chronic, where the inflammation is prolonged and has caused permanent damage to the organ (UCM 2024).

There exists no specific treatment or cure for acute pancreatitis as the pathophysiology of this disease is not yet fully understood (Banks 1993). Local necrosis of pancreatic cells, usually triggered by digestive enzymes, causes the release of pro-inflammatory cytokines like tumor necrosis factor (TNF)-α, interleukin (IL)-1β, and IL-6 which further aggravates the disease (Büchler et al. 2000; Zhang et al. 2014).

Cerulein, a cholecystokinin analogue, is well known to trigger acute edematous pancreatitis in rats, mice, and healthy humans (Willemer et al. 1992). Excessive stimulation of leucocytes is a downstream result of cerulein activity which leads to sterile inflammation and finally pancreatitis (Lesina et al. 2014). NF-κB (nuclear factor kappa B) and MAPK (mitogen-activated protein kinase) are crucial transcription factors in the pancreas which trigger the production of inflammatory mediators (Yubero et al. 2009). Iκ-Bα (nuclear factor of kappa light polypeptide gene enhancer in B cells inhibitor alpha; NFKBIA) is an inhibitor of the transcriptional regulator NF-κB (Wang et al. 2020). Earlier studies showed that pretreatment with the plant extract failed to restricts degradation of Iκ-Bα in pancreatic acinar cells which is cerulein-induced (Bae et al. 2010). However, pretreatment with 8α-hydroxy pinoresinol (HP), isolated from *N. jatamansi,* stopped Iκ-Bα degradation and nuclear localization of NF-κB p65 into the nucleus of cerulein-treated mice. Even so HP in particular could not inhibit MAPK in the same mice (Choi et al. 2019). When acute pancreatitis afflicted cells were treated with the inhibitor of NF-κB, NAC (N-acetyl-L-cysteine), the cells showed a decline in morphological damage. Moreover, the pancreatic mRNA and serum levels of IL-1β, TNF-α, and IL-6 peaked during acute pancreatitis but showed promising decline upon administration of HP (Choi et al. 2019). The translocation of NF-κB into the nucleus can also be prevented by the activation of the enzyme heme oxygenase-1 (HO-1) (Bellezza et al. 2012). Free heme causes oxidative and inflammatory damage which can be prevented by the catalytic activity of HO-1 in the first and rate-limiting step of the oxidative degradation of free heme (Pae and Chung 2009). The intraperitoneal administration of the biologically active fourth fraction of *N. jatamansi* in cerulein-treated mice significantly increases the production of HO-1and by extension reduces oxidative and inflammatory damage (Bae et al. 2012a).

In mice, severe acute pancreatitis (SAP) was triggered by a choline-deficient diet supplemented with ethionine (CDE), and *N. jatamansi* extract was administered orally for 3 days. Histological evidence showed that treatment with *N. jatamansi* extract alleviated SAP (Bae et al. 2012c). Furthermore, neutrophil entry into the pancreas was prevented by introducing *N. jatamansi* extract. Since *N. jatamansi* extract was also found to inhibit the discharge of digestive enzymes and cytokine manufacture, it further lowered the risk of inflammation of the pancreas. MAPKs were also found to be deactivated by the effects of *N. jatamansi* extract, which further protected the pancreas (Bae et al. 2012c). Pretreatment with *N. jatamansi* extract also proved to be effective in reducing acute pancreatitis-associated lung damage (Bae et al. 2013).

Certain hallmarks in the histology of chronic pancreatitis-afflicted cells are prominent fibrosis, ductal changes, glandular atrophy, and inflammation (Apte et al. 2011). Alcohol abuse certainly doesn't help with the symptoms (Lankisch et al. 1993). With inhibition of the destruction of pancreatic acinar cell, deposition of collagen, and activation of pancreatic stellate cell (PSC) as evidence, it is clear that *N. jatamansi* extracts successfully alleviate the fibrosis, previously deemed an unresolvable issue (Bae et al. 2012b).

## 23.15 Conclusion

For many years, *Nardostachys jatamansi* has been utilized for treating an array of pathological disorders. It is made up of different kinds of compounds, including proteins and amino acids, carbohydrates, terpenoids, glycosides, fixed oils, alkaloids, flavonoids, tannins, and saponin. Exploring these substances further may result in the creation of novel functional foods and dietary supplements. *Nardostachys jatamansi* may find a role in dietary supplements and functional foods that are intended to improve health and prevent disease as demand for natural and holistic health products rises. The plant's compounds have demonstrated anti-inflammatory and antioxidant activities. This could make it valuable for managing chronic inflammatory conditions and preventing oxidative damage. Research has shown that *Nardostachys jatamansi* has potential neuroprotective properties. It may benefit in managing neurodegenerative disorders such as Alzheimer's and Parkinson's by reduction in inflammation and oxidative stress. Continued research could lead to new treatments for these conditions. Some studies suggest that spikenard might support cardiovascular health by improving circulation and reducing blood pressure. Further research could solidify its role in cardiovascular disease prevention. *Nardostachys jatamansi* has shown promise in combating various pathogens. This could lead to the improvement in new antimicrobial agents, in an antibiotic resistance environment.

Its prospects are promising due to ongoing research and the increasing global interest in natural products as many of them have therapeutic significance. Studies have demonstrated that the plant owns an extensive series of medicinal efficacies, including antibacterial and antifungal, anticonvulsant, antioxidant and anticancer, antidepressant and anxiety-reducing, anti-Parkinson, antidiabetic, antihypertensive, antiulcer, hepatoprotective, hair growth, and neuroprotective properties. Further investigation and advancement may reveal novel health advantages and uses, potentially establishing it as an asset in both conventional and contemporary medicine. In order to do so, more clinical trials are required to authenticate the medicinal claims and understand the security and effectiveness of *Nardostachys jatamansi* in various therapeutic applications. Acceptance of jatamansi in conventional medicine and nutrition will depend on the standardization of active ingredients and quality assurance in products derived from it. Integrating spikenard into modern medical procedures may be possible through the development of novel formulations and therapeutic approaches that combine traditional knowledge with cutting-edge scientific techniques. As demand for natural products increases, there will be a need for sustainable cultivation practices to prevent overharvesting and ensure the long-term availability of *Nardostachys jatamansi*. Implementing organic farming can help maintain biodiversity and ecosystem balance. Sustainability can be increased by incorporating jatamansi farming into agroforestry systems. Sponge planting with other crops and trees improves soil fertility, reduces erosion, and builds ecosystem resilience. Planting material can be reliably supplied by creating seed banks and creating propagation methods. Additionally, this can benefit conservation efforts and help restore wild populations. Productivity can be increased and the need for wild harvesting can be decreased by using selective breeding to create cultivars with desired features such as disease resistance and higher yield.

## 23.16 Summary

*Nardostachys jatamansi* is a critically endangered medicinal plant which has been used in traditional medicines for ages and has been utilized for treating an array of pathological disorders in modern-day medicines of herbal origin. It constitutes proteins and amino acids, carbohydrates, terpenoids, glycosides, fixed oils, alkaloids, flavonoids, tannins, and saponin. *Nardostachys jatamansi* may find a role in dietary supplements and functional foods that are intended to improve health and prevent disease as demand for natural and holistic health products rises. The plant's compounds have demonstrated anti-inflammatory and antioxidant activities. This could make it valuable for managing chronic inflammatory conditions and preventing oxidative damage. Research has shown that *Nardostachys jatamansi* has potential neuroprotective properties. It may benefit in managing neurodegenerative disorders such as Alzheimer's and Parkinson's by reduction of inflammation and oxidative stress. Studies have demonstrated that the plant owns an extensive series of medicinal efficacies, including antibacterial and antifungal, anticonvulsant, antioxidant and anticancer, antidepressant and anxiety-reducing, anti-Parkinson, antidiabetic, antihypertensive, antiulcer, hepatoprotective, hair growth, and neuro-protective properties. Further investigation and advancement may reveal novel health advantages and uses, potentially establishing it as an asset in both conventional and contemporary medicine. In order to do so, more clinical trials are required to authenticate the medicinal claims and understand the security and effectiveness of *Nardostachys jatamansi* in various therapeutic applications.

**Authors' Contribution** SG was involved in conceptualization, writing, data compilation, editing, correction of manuscript; DS helped in conceptualization, pictorial contribution (schematic), expert suggestion and correction of manuscript; RB, DP, AS, SM, SKu, and PS wrote the manuscript; SK was involved in writing of manuscript and pictorial contribution (original); MA helped in expert suggestion and editing of manuscript.

## References


Ahmad F, Safuan S (2019) Assessing the effectiveness of plant extracts in polycystic ovarian syndrome: a systematic review. Malays J Med Health Sci 15(2):120–129

Ahmad AS, Ansari MA, Ahmad M, Saleem S, Yousuf S, Hoda MN, Islam F (2005) Neuroprotection by crocetin in a hemi-parkinsonian rat model. Pharmacol Biochem Behav 81(4):805–813. https://doi.org/10.1016/j.pbb.2005.06.007

Ahmad M, Yousuf S, Khan MB, Hoda MN, Ahmad AS, Ansari MA, Ishrat T, Agrawal AK, Islam F (2006) Attenuation by Nardostachys jatamansi of 6-hydroxydopamine-induced parkinsonism in rats: behavioral, neurochemical, and immunohistochemical studies. Pharmacol Biochem Behav 83(1):150–160. https://doi.org/10.1016/j.pbb.2006.01.005

Ali S, Ansari KA, Jafry MA, Kabeer H, Diwakar G (2000) Nardostachys jatamansi protects against liver damage induced by thioacetamide in rats. J Ethnopharmacol 71(3):359–363. https://doi.org/10.1016/s0378-8741(99)00153-1

Amit Parekh AP, Jadhav VM (2009) Development of validated HPTLC method for quantification of jatamansone in jatamansi oil. J Pharm Res 2(5):975–977

Anupama KP, Shilpa O, Antony A, Raghu SV, Gurushankara HP (2022) Jatamansinol from Nardostachys jatamansi (D. Don) DC. protects Aβ42-induced neurotoxicity in alzheimer's disease Drosophila model. Neurotoxicology 90:62–78. https://doi.org/10.1016/j.neuro.2023.04.002

Anupama KP, Shilpa O, Antony A, Gurushankara HP (2021) Jatamansinol from Nardostachys jatamansi: a multi-targeted neuroprotective agent for Alzheimer's disease. J Biomol Struct Dyn, 41(1), 200–220

Anupama KP, Shilpa O, Antony A, Gurushankara HP (2023) Jatamansinol from Nardostachys jatamansi: a multi-targeted neuroprotective agent for Alzheimer's disease. J Biomol Struct Dyn 41(1):200–220. https://doi.org/10.1080/07391102.2021.2005681

Apte M, Pirola R, Wilson J (2011) The fibrosis of chronic pancreatitis: new insights into the role of pancreatic stellate cells. Antioxid Redox Signal 15(10):2711–2722. https://doi.org/10.1089/ars.2011.4079

Arora RB, Singh KP, Das PK, and Mistry PN (1958) Prolonged hypotensive effect of the essential oil of Nardostachys jatamansi. Arch Int Pharmacodyn Ther, 113:3–4, pp 367–376

Azeemuddin M, Anturlikar SD, Onkaramurthy M, Baig MR, Ashok BK, Rao RP, Rafiq M, Rangesh P (2019) Effect of "DXB-2030," a polyherbal formulation, on experimental polycystic ovary syndrome associated with hyperandrogenism. Adv Pharmacol Pharm Sci 2019(1):8272850. https://doi.org/10.1155/2019/8272850

Bae GS, Park HJ, Kim DY, Song JM, Kim TH, Oh HJ, Yun KJ, Park RK, Lee JH, Shin BC, Sim HJ (2010) Nardostachys jatamansi protects against cerulein-induced acute pancreatitis. Pancreas 39(4):520–529. https://doi.org/10.1097/MPA.0b013e3181bd93ce

Bae GS, Kim MS, Park KC, Koo BS, Jo IJ, Choi SB, Lee DS, Kim YC, Kim TH, Seo SW, Shin YK (2012a) Effect of biologically active fraction of Nardostachys jatamansi on cerulein-induced acute pancreatitis. World J Gastroenterol: WJG 18(25):3223. https://doi.org/10.3748/wjg.v18.i25.3223

Bae GS, Park KC, Koo BS, Choi SB, Jo IJ, Choi CM, Song HJ, Park SJ (2012b) The inhibitory effects of Nardostachys jatamansi on alcoholic chronic pancreatitis. BMB Rep 45(7):402–407. https://doi.org/10.5483/BMBRep.2012.45.7.076

Bae GS, Park KC, Koo BS, Jo IJ, Choi SB, Song HJ, Park SJ (2012c) Nardostachys jatamansi inhibits severe acute pancreatitis via mitogen-activated protein kinases. Exp Ther Med 4(3):533–537. https://doi.org/10.3892/etm.2012.612

Bae GS, Park KC, Koo BS, Jo IJ, Choi SB, Lee DS, Kim YC, Kim JJ, Shin YK, Hong SH, Kim TH (2013) The beneficial effects of Nardostachys jatamansi extract on diet-induced severe acute pancreatitis. Pancreas 42(2):362–363. https://doi.org/10.1097/MPA.0b013e3182592cac

Bae GS, Heo KH, Choi SB, Jo IJ, Kim DG, Shin JY, Seo SH, Park KC, Lee DS, Oh H, Kim YC (2014) Beneficial effects of fractions of Nardostachys jatamansi on lipopolysaccharide-induced inflammatory response. Evid Based Complement Alternat Med 2014(1):837835

Balodi KN, Purohit MV, Shridhar V, Arunachalam K (2018) Ethno-medicinal uses of various plants species among the Jaad Bhotiya community of Uttarakhand, western Himalaya. Stud Ethno-Med 12(3). https://doi.org/10.31901/24566772.2018/12.03.558

Banks PA (1993) Medical management of acute pancreatitis and complications. In: The pancreas: biology, pathobiology, and disease. Raven, New York, pp 593–611

Bayala B, Zoure AA, Baron S, de Joussineau C, Simpore J, Lobaccaro JM (2020) Pharmacological modulation of steroid activity in hormone-dependent breast and prostate cancers: effect of some plant extract derivatives. Int J Mol Sci 21(10):3690. https://doi.org/10.3390/ijms21103690

Bellezza I, Tucci A, Galli F, Grottelli S, Mierla AL, Pilolli F, Minelli A (2012) Inhibition of NF-κB nuclear translocation via HO-1 activation underlies α-tocopheryl succinate toxicity. J Nutr Biochem 23(12):1583–1591. https://doi.org/10.1016/j.jnutbio.2011.10.012

Bhagwan SC, Gupta VS, Deshmukh VV, Sardeshmukh SS, Sardeshmukh SP (2024) Standardization and preliminary characterization of an ayurvedic stress-relieving head massage oil of Nardostachys jatamansi DC. J Ayurveda Integr Med 15(2):100900

Bhat MD, Malik SA (2020) Efficacy of Nardostachys jatamansi (D. Don) DC in essential hypertension: a randomized controlled study. Complement Ther Med 53:102532. https://doi.org/10.1016/j.ctim.2020.102532

Bhatt M, Kothiyal P (2015) A review article on phytochemistry and pharmacological profiles of Nardostachys jatamansi DC-medicinal herb. J Pharmacognosy Phytochem 3(5):102–106

Bhatt VP, Negi GC (2006) Ethnomedicinal plant resources of Jaunsari tribe of Garhwal Himalaya, Uttaranchal. Indian J Tradit Knowl 5(3):331–335

Bhuyan LR, Pangu Y (2018) High altitude medicinal plants of Tawang district, Arunachal Pradesh. Bull Arunachal Forest Res 33(1):35–43

Bose B, Tripathy D, Chatterjee A, Tandon P, Kumaria S (2019) Secondary metabolite profiling, cytotoxicity, anti-inflammatory potential and in vitro inhibitory activities of Nardostachys jatamansi on key enzymes linked to hyperglycemia, hypertension and cognitive disorders. Phytomedicine 55:58–69. https://doi.org/10.1016/j.phymed.2018.08.010

Bräuer AU, Savaskan NE (2004) Molecular actions of selenium in the brain: neuroprotective mechanisms of an essential trace element. Rev Neurosci 15(1):19–32. https://doi.org/10.1515/REVNEURO.2004.15.1.19

Büchler MW, Gloor B, Müller CA, Friess H, Seiler CA, Uhl W (2000) Acute necrotizing pancreatitis: treatment strategy according to the status of infection. Ann Surg 232(5):619–626

Buchman AS, Yu L, Oveisgharan S, Zammit AR, Wang T, Shulman JM, VanderHorst V, Nag S, Bennett DA (2024(Preprint)) Parkinson's disease and other Alzheimer's disease and related dementia pathologies and the progression of parkinsonism in older adults. J Alzheimers Dis 100:S197–S209

Castillo MA, Carrero Y, Urdaneta KE, Renouf M, Lubin C, Nola M, Semprún-Hernández N (2018) Essential oils as modifiers of human behavior. Trop Subtrop Agroecosystems 21(1):69–89

Chandra S (2022) Documentation of the plants used in different Hindu rituals in Uttarakhand, India. Asian J Ethnobiol 5(2). https://doi.org/10.13057/asianjethnobiol/y050203

Chatterjee A, Basak B, Saha M, Dutta U, Mukhopadhyay C, Banerji J, Konda Y, Harigaya Y (2000) Structure and stereochemistry of Nardostachysin, a New Terpenoid Ester constituent of the rhizomes of Nardostachys jatamansi. J Nat Prod 63(11):1531–1533. https://doi.org/10.1021/np990503m

Chatterjee A, Basak B, Datta U, Banerji J, Neuman A, Prange T (2005) Studies on the chemical constituents of Nardostachys jatamansi DC (Valerianaceae). Indian J Chem 36:430–433. https://doi.org/10.1002/chin.200525180

Chaudhary S, Chandrashekar KS, Pai KS, Setty MM, Devkar RA, Reddy ND, Shoja MH (2015) Evaluation of antioxidant and anticancer activity of extract and fractions of Nardostachys jatamansi DC in breast carcinoma. BMC Complement Altern Med 15:1–3. https://doi.org/10.1186/s12906-015-0563-1

Chauhan HK (2021) Nardostachys Jatamansi. The IUCN Red List of Threatened Species 2021. e. T50126627A88304158 [Internet]. https://doi.org/10.2305/iucn.uk.2021-3.rlts.t50126627a88304158.en

Chauhan P, Puri S, Thakur M, Rathour S, Sharma AK, Pundir A (2021) A study of wild medicinal plants used in Nargu Wildlife Sanctuary of district Mandi in Himachal Pradesh, India. J Appl Pharm Sci 11(4):135–144

Chavan PM (2020) An experimental evaluation of shukrajanan activity of Jatamansi Phanta (Nordostachys Jatamansi DC) WSR to spermatogenic activity. Doctoral dissertation, Rajiv Gandhi University of Health Sciences (India)

Chawla A, Devhare LD, Dharmamoorthy G, Ritika TS (2023) Synthesis and in-vivo anticancer evaluation of N-(4-oxo-2-(4-((5-aryl-1, 3, 4 thiadiazole-2yl) amino) Phenyl thiazolidine-3-yl)) Benzamide derivative. Int J Pharm Qual Assur 14(3):470–474. https://doi.org/10.1016/S0162-0134(98)00003-8

Chennaiah K, Jayachandrudu M, Kamakshamma J (2016) Modulations in oxidative enzymes by Nardostachys jatamansi root extract against nicotine induced oxidative stress in the skeletal muscle tissue of male albino rat. https://doi.org/10.17812/ajsmr125

Choi JW, Shin JY, Jo IJ, Kim DG, Song HJ, Yoon CS, Oh H, Kim YC, Bae GS, Park SJ (2019) 8α-Hydroxypinoresinol isolated from Nardostachys jatamansi ameliorates cerulein-induced acute pancreatitis through inhibition of NF-κB activation. Mol Immunol 114:620–628. https://doi.org/10.1016/J.MOLIMM.2019.09.002

Colonna M, Butovsky O (2017) Microglia function in the central nervous system during health and neurodegeneration. Annu Rev Immunol 35(1):441–468. https://doi.org/10.1146/annurev-immunol-051116-052358

Dandagi PM, Patil MB, Mastiholimath VS, Gadad AP, Dhumansure RH (2008) Development and evaluation of hepatoprotective polyherbal formulation containing some indigenous medicinal plants. Indian J Pharm Sci 70(2):265. https://doi.org/10.4103/0250-474X.41474

Das SK, Bishnupriya Mohanty BM, Panda PK, Joshi AB (2015) A study of various market samples of Spikenard (Nardostachys jatamansi DC.) with special reference to its pharmacognostic & phytochemical aspects. Glob J Res Med Plants Indig Med 4(3):46

Das G, Shin HS, Tundis R, Gonçalves S, Tantengco OA, Campos MG, Acquaviva R, Malfa GA, Romano A, Robles JA, Clores MQ (2021) Plant species of sub-family Valerianaceae—a review on its effect on the central nervous system. Plants 10(5):846. https://doi.org/10.3390/plants10050846

de Candolle AP (1824) Prodromus systematis naturalis regni vegetabilis, sive, Enumeratio contracta ordinum generum specierumque plantarum huc usque cognitarium, juxta methodi naturalis, normas digesta, vol IV. Sumptibus Sociorum Treuttel et Würtz, Paris, p 624

Deepa B, Suchetha K, Satheesh Rao SR (2013) Antidepressant activity of Nardostachys jatamansi in electron beam irradiated mice. Int J Res Ayurveda Pharma 4(1):101–103

Deka B, Bhattacharjee B, Shivavedi N, Singh GK, Bhat HR, Ghosh SK, Shakya A (2023) Parkinson's disease: a phytotherapeutic prospective. In: Traditional medicine for neuronal health. Bentham Science Publishers, pp 1–23. https://doi.org/10.2174/97898150401971230101

Deyuan H, Barrie FR, Bell CD (2011) *Nardostachys jatamansi* (D.Don) DC. In: Wu ZY, Raven PH, Hong DY (eds) Flora of China volume 19 (Cucurbitaceae through Valerianaceae, with Annonaceae and Berberidaceae). Science Press, Beijing, and Missouri Botanical Garden Press, St. Louis, p 661

Dhiman N, Bhattacharya A (2020) Nardostachys jatamansi (D. Don) DC.-challenges and opportunities of harnessing the untapped medicinal plant from the Himalayas. J Ethnopharmacol 246:112211. https://doi.org/10.1016/j.jep.2019.112211

Dhingra D, Goyal PK (2008) Inhibition of MAO and GABA: probable mechanisms for antidepressant-like activity of Nardostachys jatamansi DC in mice. Indian J Exp Biol 46(4):212

Ding X, Guo C, Zhang X, Li J, Jiao Y, Feng H, Wang Y (2022) Wild plants used by tibetans in Burang Town, characterized by alpine desert meadow, in Southwestern Tibet, China. Agronomy 12(3):704. https://doi.org/10.3390/agronomy12030704

Disket J, Mann S, Gupta RK (2012) A review on spikenard (Nardostachys jatamansi DC.)-an 'endangered'essential herb of India. Int J Pharm Chem 2(3):52–60. https://doi.org/10.31901/24566772.2018/12.03.558

Draguhn A, Mallatt JM, Robinson DG (2021) Anesthetics and plants: no pain, no brain, and therefore no consciousness. Protoplasma 258(2):239–248. https://doi.org/10.1007/s00709-020-01550-9

Dubey A, Bhushan D, Kawatra N (2023) Phytochemical analysis of Nardostachys Jatamansi DC, Chlorophytum Borivilianum (Soil based) and Chlorophytum Borivilianum (Commercial). Int J Darshan Inst Eng Res Emerg Technol 11(1):42–46. https://doi.org/10.32692/IJDI-ERET/11.1.2022.2207

EA S. Electrically induced convulsions. Experimental models of epilepsy-a manual for the laboratory worker 1972:433–458

Feng M, Liu Y, He B, Zhong H, Qu-Bie A, Li M, Luo M, Bao X, Li Y, Yan X, Sheng H (2024) An efficient flavonoid glycosyltransferase NjUGT73B1 from Nardostachys jatamansi of alpine Himalayas discovered by structure-based protein clustering. Phytochemistry 227:114228. https://doi.org/10.1016/j.phytochem.2024.114228

Ferraz CR, Carvalho TT, Manchope MF, Artero NA, Rasquel-Oliveira FS, Fattori V, Casagrande R, Verri WA Jr (2020) Therapeutic potential of flavonoids in pain and inflammation: mechanisms of action, pre-clinical and clinical data, and pharmaceutical development. Molecules 25(3):762. https://doi.org/10.3390/molecules25030762

Ferreira-Vieira TH, Guimaraes IM, Silva FR, Ribeiro FM (2016) Alzheimer's disease: targeting the cholinergic system. Curr Neuropharmacol 14(1):101–115. https://indiabiodiversity.org

Fisher RS, Acevedo C, Arzimanoglou A, Bogacz A, Cross JH, Elger CE, Engel J Jr, Forsgren L, French JA, Glynn M (2014) ILAE official report: a practical clinical definition of epilepsy. Epilepsia 55(4):475–482

Garg P, Alambayan J, Garg V (2023) Herbal approaches in the management of mental depression. CNS Neurol Disord Drug Targets 22(1):98–124

Gautam K, Raina R (2016) New insights into the phenology, genetics and breeding system of critically endangered Nardostachys grandiflora DC. Caryologia 69(2):91–101. https://doi.org/10.1080/00087114.2015.1109939

Gómez-Betancur I, Cortés N, Benjumea D, Osorio E, León F, Cutler SJ (2015) Antinociceptive activity of extracts and secondary metabolites from wild growing and micropropagated plants of Renealmia alpinia. J Ethnopharmacol 165:191–197. https://doi.org/10.1016/j.jep.2015.02.012

Goodman LS, Brown WC, Swinyard EA, Grewal MS (1953) Comparison of maximal seizures evoked by pentylene-tetrazol (Metrazol) and electroshock in mice, and their modification by anticonvulsants. J Pharmacol Exp Ther 108(2):168–176

Gottumukkala VR, Annamalai T, Mukhopadhyay T (2011) Phytochemical investigation and hair growth studies on the rhizomes of Nardostachys jatamansi DC. Pharmacogn Mag 7(26):146

Gowda DK, Shetty L, Krishna AP, Kumari SN, Sanjeev G, Naveen P (2013) The efficacy of nardostachys jatamansi against the radiation induced haematological damage in rats. J Clin Diagn Res 7(6):982

Goyal S, Rath S (2021) A comprehensive review of Jatamansi (Nardostachys Jatamansi). https://doi.org/10.47070/ayushdhara.v8i4.787

Gu RR, Meng XH, Zhang Y, Xu HY, Zhan L, Gao ZB, Yang JL, Zheng YM (2022) (−)-Naringenin 4′, 7-dimethyl ether isolated from Nardostachys jatamansi relieves pain through inhibition of multiple channels. Molecules 27(5):1735. https://doi.org/10.3390/molecules27051735

Gupta P, Naudiyal N, Raypa P, Kumar VA (2024) Pharmacognostic profile of a potent medicinal herb: Nardostachys Jatamansi. In: The chemistry inside spices & herbs: research and development, vol 4. Bentham Science Publishers, pp 117–146. https://doi.org/10.2174/9789815196832124040l

Jha S, Thakur ND (2020) Efficacy and toxicity study of Nardostachys Jatamansi DC and Valeriana Wallichii DC in sleep disorders. Int J Res Eng Sci Manag 3(10):100–106. https://doi.org/10.47607/ijresm.2020.343

Jha SV, Bhagwat AM, Pandita NS (2012) Pharmacognostic and phytochemical studies on the rhizome of Nardostachys jatamansi DC. using different extracts. Pharm J 4(33):16–22. https://doi.org/10.5530/pj.2012.33.3

Jiang Y, MacNeil LT (2023) Simple model systems reveal conserved mechanisms of Alzheimer's disease and related tauopathies. Mol Neurodegener 18(1):82. https://doi.org/10.1186/s13024-023-00664-x

Joshi H, Parle M (2006) Nardostachys jatamansi improves learning and memory in mice. J Med Food 9(1):113–118. https://doi.org/10.1089/jmf.2006.9.113

Kakkar AK, Singh H, Medhi B (2018) Old wines in new bottles: repurposing opportunities for Parkinson's disease. Eur J Pharmacol 830:115–127. https://doi.org/10.1016/j.ejphar.2018.04.023

Kaltschmidt B, Kaltschmidt C (2009) NF-κB in the nervous system. Cold Spring Harb Perspect Biol 1(3):a001271. https://doi.org/10.1101/cshperspect.a001271

Kang YJ, Wingerd BA, Arakawa T, Smith WL (2006) Cyclooxygenase-2 gene transcription in a macrophage model of inflammation. J Immunol 177(11):8111–8122

Karanth K, Karkada G, Shenoy K, Halahalli H (2012) Nardostachys jatamansi extract prevents chronic restraint stress-induced learning and memory deficits in a radial arm maze task. J Nat Sci Biol Med 3:125. https://doi.org/10.4103/0976-9668.101879

Karkada G, Shenoy KB, Halahalli H, Karanth K (2012) Nardostachys jatamansi extract prevents chronic restraint stress-induced learning and memory deficits in a radial arm maze task. J Nat Sci Biol Med 3(2):125. https://doi.org/10.4103/0976-9668.101879

Kaur H, Lekhak MM, Chahal S, Goutam U, Jha P, Naidoo D, Ochatt SJ, Kumar V (2020) Nardostachys jatamansi (D. Don) DC.: an invaluable and constantly dwindling resource of the Himalayas. S Afr J Bot 135:252–267. https://doi.org/10.1016/j.sajb.2020.08.010

Kenda M, Kočevar Glavač N, Nagy M, Sollner Dolenc M (2022) Medicinal plants used for anxiety, depression, or stress treatment: an update. Molecules 27(18):6021. https://doi.org/10.3390/molecules27186021

Khan MB, Hoda MN, Ishrat T, Ahmad S, Moshahid Khan M, Ahmad A, Yusuf S, Islam F (2012) Neuroprotective efficacy of Nardostachys jatamansi and crocetin in conjunction with selenium in cognitive impairment. Neurol Sci 33:1011–1020. https://doi.org/10.1007/s10072-011-0880-1

Khan MA, Ahmed MS, Farooq SF, Jan U (2024) Review on pharmacological action and therapeutic uses of Balchar (Nardostachys jatamansi DC): Unani perspective. J Complement Altern Med Res 25(8):47–56. https://doi.org/10.9734/jocamr/2024/v25i8559

Kim DC, Park JS, Yoon CS, Kim YC, Oh H (2020) Nardostachin from Nardostachys jatamansi exerts anti-neuroinflammatory effects through TLR4/MyD88 related suppression of the NFκB and JNK

Kim DC, Park JS, Yoon CS, Kim YC, Oh H (2021a) Nardostachin from Nardostachys jatamansi exerts anti-neuroinflammatory effects through TLR4/MyD88-related suppression of the NF-κB and JNK MAPK signaling pathways in lipopolysaccharide-induced BV2 and primary microglial cells. Mol Med Rep 23(1):1. https://doi.org/10.3892/mmr.2020.11720

Kim KW, Yoon CS, Park SJ, Bae GS, Kim DG, Kim YC, Oh H (2021b) Chemical analysis of the ingredients of 20% aqueous ethanol extract of Nardostachys jatamansi through phytochemical study and evaluation of anti-neuroinflammatory component. Evid Based Complement Alternat Med 2021(1):5901653. https://doi.org/10.1155/2021/5901653

Kizhakke P, Olakkaran A, Antony S, Raghu AV, Shamprasad, Gurushankara HP (2022) Jatamansinol from Nardostachys jatamansi (D.Don) DC. Protects Aβ42-Induced Neurotoxicity in Alzheimers Disease Drosophila Model. Neurotoxicology. 90. https://doi.org/10.1016/j.neuro.2022.02.011

Kizhakke P, Anupama Olakkaran, Shilpa Anet, Antony Hunasanahally P, Gurushankara (2023) Jatamansinol from Nardostachys jatamansi : a multi-targeted neuroprotective agent for Alzheimer's disease. J Biomol Struct Dyn 41(1) 200–220. https://doi.org/10.1080/07391102.2021.2005681

Ko W, Park JS, Kim KW et al (2018) Nardosinone-type Sesquiterpenes from the hexane fraction of Nardostachys jatamansi attenuate NF-κB and MAPK signaling pathways in lipopolysaccharide-stimulated BV2 microglial cells. Inflammation 41:1215–1228. https://doi.org/10.1007/s10753-018-0768-9

Krishnan N, Mariappanadar V, Dhanabalan AK, Devadasan V, Gopinath SC, Raman P (2023) Purification, identification and in silico models of alkaloids from Nardostachys jatamansi—bioactive compounds for neurodegenerative diseases. Biomass Convers Biorefin 13(16):14889–14900. https://doi.org/10.1007/s13399-022-03237-y

Kumar G, Behera B (2023) Exploring the benefits of Jatamansi (Nardostachys jatamansi): a comprehensive review. RRJ Botanical Sci 12:005. https://doi.org/10.4172/2320-0189.12.1.005

Kumar SS, Pal SA (2011) Standardization and preliminary phytochemical screening of Nardostachys jatamansi DC. rhizome. IJRAP 2(3):978–982

Kunwar RM, Nepal BK, Kshhetri HB, Rai SK, Bussmann RW (2006) Ethnomedicine in Himalaya: a case study from Dolpa, Humla, Jumla and mustang districts of Nepal. J Ethnobiol Ethnomed 2:1–6

Kunwar RM, Uprety Y, Burlakoti C, Chowdhary CL, Bussmann RW (2009) Indigenous use and ethnopharmacology of medicinal plants in far-west Nepal. Ethnobot Res Appl 7:005–028

Kunwar RM, Thapa-Magar KB, Subedi SC, Kutal DH, Baral B, Joshi NR, Adhikari B, Upadhyaya KS, Thapa-Magar S, Ansari AS, Thapa GJ (2023) Distribution of important medicinal plant species in Nepal under past, present, and future climatic conditions. Ecol Indic 146:109879. https://doi.org/10.1016/j.ecolind.2023.109879

Kwon DJ, Ju SM, Youn GS, Choi SY, Park J (2013) Suppression of iNOS and COX-2 expression by flavokawain A via blockade of NF-κB and AP-1 activation in RAW 264.7 macrophages. Food Chem Toxicol 58:479–486

Lankisch PG, Löhr-Happe A, Otto J, Creutzfeldt W (1993) Natural course in chronic pancreatitis: pain, exocrine and endocrine pancreatic insufficiency and prognosis of the disease. Digestion 54(3):148–155. https://doi.org/10.1159/000201029

Latif A, Rehman S (2014) Standardization of a herbal medicine-Swertia Chirayita Linn. Pharmacophore 5(1–2014):98–108

Lee GW, Hur W, Kim JH, Park DJ, Kim SM, Kang BY, Sung PS, Yoon SK (2021) Nardostachys jatamansi root extract attenuates tumor progression in hepatocellular carcinoma via inhibition of ERK/STAT3 pathways. Anticancer Res 41(4):1883–1893. https://doi.org/10.21873/anticanres.14954

Lesina M, Wörmann SM, Neuhöfer P, Song L, Algül H (2014) Interleukin-6 in inflammatory and malignant diseases of the pancreas. Semin Immunol 26(1):80–87. Academic Press. https://doi.org/10.1016/j.smim.2014.01.002

Li C, Wang MH (2014) Nardostachys jatamansi (D. Don) DC prevents LPS-induced inflammation in RAW 264.7 macrophages by preventing ROS production and down-regulating inflammatory gene expression. Food Sci Biotechnol 23:903–909

Li J, Wu J, Peng K, Fan G, Yu H, Wang W, He Y (2019) Simulating the effects of climate change across the geographical distribution of two medicinal plants in the genus Nardostachys. PeerJ 7:e6730. https://doi.org/10.7717/peerj.6730

Li X, Zhang Y, Wang Q (2020) Antidepressant effects of Nardostachys jatamansi on serotonin regulation. J Med Herbs, 15(3):200–207

Li R, Wang ZM, Wang Y, Dong X, Zhang LH, Wang T, Zhu Y, Gao XM, Wu HH, Xu YT (2021) Antidepressant activities and regulative effects on serotonin transporter of Nardostachys jatamansi DC. J Ethnopharmacol 268:113601. https://doi.org/10.1016/j.jep.2020.113601

Li J, Yu J, Guo J, Liu J, Wan G, Wei X, Yang X, Shi J (2023) Nardostachys jatamansi and levodopa combination alleviates Parkinson's disease symptoms in rats through activation of Nrf2 and inhibition of NLRP3 signaling pathways. Pharm Biol 61(1):1175–1185. https://doi.org/10.1080/13880209.2023.2244176

Liu QF, Jeon Y, Sung YW, Lee JH, Jeong H, Kim YM, Yun HS, Chin YW, Jeon S, Cho KS, Koo BS (2018) Nardostachys jatamansi ethanol extract ameliorates Aβ42 cytotoxicity. Biol Pharm Bull 41(4):470–477. https://doi.org/10.1248/bpb.b17-00750

Liu M, Xue J, Cao Y, Hao Z, Wang Y, Li J, Jiang T, Shi J (2024) The effects of Nardosinone on levodopa intervention in the treatment of Parkinson's disease. Biomed Pharmacother 174:116448. https://doi.org/10.1016/j.biopha.2024.116448

Lokhande RS, Singare PU, Andhele ML, Acharya R, Nair AG, Reddy AV (2009) Analysis of mineral content of some medicinal plants by NAA and AAS techniques. Radiochemistry 51:321–325

Ma LM, Wang K, Meng XH, Zheng YD, Wang CB, Chai T, Naghavi MR, Sang CY, Yang JL (2022) Terpenoids from Nardostachys jatamansi and their cytotoxic activity against human pancreatic cancer cell lines. Phytochemistry 200:113228. https://doi.org/10.1016/j.phytochem.2022.113228

Ma Z, Liang H, Wang S, Miao W, Yu L, Liu S, Luo Z, Su S, Wang J, Liu S, Li Y (2024) Nardosinone relieves metabolic-associated fatty liver disease and promotes energy metabolism through targeting CYP2D6. Phytomedicine 130:155748. https://doi.org/10.1016/j.phymed.2024.155748

Mabberley DJ, Noltie HJ (2014) A note on Valeriana jatamansi Jones (Caprifoliaceae sl). In: Blumea-biodiversity, evolution and biogeography of plants, vol 59. Naturalis Biodiversity Center, pp 37–41. https://doi.org/10.3767/000651914x683476

Maity D, Pradhan N, Chauhan AS (2004) Folk uses of some medicinal plants from North Sikkim. Indian J Tradit Knowl 3(1):66–71

Malabadi RB, Kolkar KP, Meti NT, Chalannavar RK (2021) Biosciences and plant biology. Int J Curr Res Biosci Plant Biol 8(1):14–32. https://doi.org/10.20546/ijcrbp.2021.801.002

Mallavadhani UV, Panigrahi R, Pattnaik B (2011) A rapid and highly sensitive UPLC–QTOF MS method for quantitative evaluation of Nardostachys jatamansi using Nardin as the marker. Biomed Chromatogr 25(8):902–907. https://doi.org/10.1002/bmc.1540

Medzhitov R (2001) Toll-like receptors and innate immunity. Nat Rev Immunol 1(2):135–145. https://doi.org/10.1038/35100529

Mehra PN, Garg LC (1962) Pharmacognostic study of Nardostachys Jatamansi Dc. Planta Med 10(04):433–441

Mohammad E, Kushwaha AK (2021) Jatamansi (nardostachys jatamansi): insight of its morphological, ethno-pharmacological and therapeutical aspect. World J Curr Med Pharm Res 16:127–130. https://doi.org/10.37022/wjcmpr.v3i5.194

Mohammad R, Shariq S, Roohi Z (2014) Preliminary physico-chemical evaluation of sunun poste mughilan. Hippocratic J Unani Med 2:9–19

Monga AK, Kumar S (2013) A phytopharmacological review on Jatamansi. Pharm Res 9:21–32

Mude G, Pise S, Makade K, Fating R, Wakodkar S (2020) Potentiating effect of N. Jatamansi root extract by evaluating anti-depression and anxiolytic activity in rats. J Pharmacognosy Phytochem 9(3):1734–1738

Nasrouei S, Rattel JA, Liedlgruber M, Marksteiner J, Wilhelm FH (2020) Fear acquisition and extinction deficits in amnestic mild cognitive impairment and early Alzheimer's disease. Neurobiol Aging 87:26–34. https://doi.org/10.1016/j.neurobiolaging.2019.11.003

Negi VS, Maikhuri RK, Maletha A, Phondani PC (2019) Ethnobotanical knowledge and population density of threatened medicinal plants of Nanda Devi Biosphere Reserve, Western Himalaya, India. Iran J Sci Technol Trans A: Sci 43:63–73. https://doi.org/10.1007/s40995-018-0545-5

Nestor MS, Ablon G, Gade A, Han H, Fischer DL (2021) Treatment options for androgenetic alopecia: efficacy, side effects, compliance, financial considerations, and ethics. J Cosmet Dermatol 20(12):3759–3781. https://doi.org/10.1111/jocd.14537

Ngo TH, Uprety A, Ojha M, Kil YS, Choi H, Kim SY, Nam JW (2022) Stability of valeriana-type iridoid glycosides from rhizomes of Nardostachys jatamansi and their protection against H2O2-induced oxidative stress in SH-SY5Y cells. Phytochemistry 203:113375. https://doi.org/10.1016/j.phytochem.2022.113375

NHS (2024) Acute pancreatitis. https://www.nhs.uk/conditions/acute-pancreatitis/. Accessed 8 Oct 2024

Noguchi S, Nakatsuka M, Konishi H, Kamada Y, Chekir C, Kudo T (2003) Nafamostat mesilate suppresses NF-κB activation and NO overproduction in LPS-treated macrophages. Int Immunopharmacol 3(9):1335–1344

Olanow CW, Watts RL, Koller WC (2001) An algorithm (decision tree) for the management of Parkinson's disease (2001): treatment guidelines. Neurology 56(suppl_5):S1–S88. https://doi.org/10.1212/wnl.50.3_suppl_3.s1

Onder A, Yilmaz O, Koc AS, Kizilay H (2024) Medicinal plants including spices for the treatment of polycystic ovary syndrome (PCOS) with a preclinical-clinical perspective and phytotherapeutic approaches. In: The chemistry inside spices & herbs: research and development, vol 3, pp 1–45. https://doi.org/10.2174/9789815196801124030003

Pae HO, Chung HT (2009) Heme oxygenase-1: its therapeutic roles in inflammatory diseases. Immune Netw 9(1):12–19. https://doi.org/10.4110/in.2009.9.1.12

Pallavi JJ, Kumar P (2012) Targeting oxidative stress in epilepsy: study of effect of antioxidants on anticonvulsant profile of Jatamansone. J Biol Active Prod Nature 2(5):292–301

Panara K, Nariya M, Karra N (2020) Central nervous system depressant activity of Jatamansi (Nardostachys jatamansi DC.) rhizome. AYU 41(4):250–254. https://doi.org/10.4103/ayu.AYU_251_20

Panchal R, Goswami NN (2020) Evaluation of antidepressant activity of hydro-alcoholic extract of rhizomes of Nardostachys jatamansi DC per se and in combination with fluoxetine in wistar albino rats and swiss albino mice. Int J Basic Clin Pharmacol 9(1):32–36

Pandey MM, Katara A, Pandey G, Rastogi S, Rawat AK (2013) An important Indian traditional drug of ayurveda jatamansi and its substitute bhootkeshi: chemical profiling and antioxidant activity. Evid Based Complement Alternat Med 2013(1):142517. https://doi.org/10.1155/2013/142517

Pant HC, Jalal S, Rautela I, Ali Y, Thapa A, Verma P, Pant HV, Gaurav N (2022) A review on endangered medicinal plant Nardostachys jatamansi: an important Himalayan herb. Sci Temper 13(01):82–88. https://doi.org/10.58414/scientifictemper.2022.13.1.12

Pathak S, Godela R (2023) Nardostachys jatamansi: phytochemistry, ethnomedicinal uses, and pharmacological activities: a comprehensive review. Fitoterapia 172:105764. https://doi.org/10.1016/j.fitote.2023.105764

Peña-Bautista C, Casas-Fernández E, Vento M, Baquero M, Cháfer-Pericás C (2020) Stress and neurodegeneration. Clin Chim Acta 503:163–168

Phondani PC, Maikhuri RK, Rawat LS, Farooquee NA, Kala CP, Vishvakarma SR, Rao KS, Saxena KG (2010) Ethnobotanical uses of plants among the Bhotiya tribal communities of Niti Valley in Central Himalaya, India. Ethnobot Res Appl 8:233–244

Prabhu N, Panneerselvam K (2013) Cardioprotective and antioxidant activity of *Nardostachys jatamansi* in hypoxic rats. Chin J Nat Med 11(1):68–73. https://doi.org/10.1016/S1875-5364(13)60007-2

Prabhu V, Karanth KS, Rao A (1994) Effects of Nardostachys jatamansi on biogenic amines and inhibitory amino acids in the rat brain. Planta Med 60(02):114–117. https://doi.org/10.1055/S-2006-959429/BIB

Priyadarshini D, Tripathy B, Nanda GC (2020) Perspective approach to psycho-somatic disorders in elderly patients with certain medicinal plants in Covid pandemic. Int J Ayurveda Pharma Res 25:77–83

Prutianu I, Căruntu ID, Manole MB, Giuşcă SE, Profire B, Timofte AD, Profire L, Gafton B (2021) Developments in prostate cancer therapy: from the androgen receptor to antiandrogens and beyond. Farmacia 69(3). https://doi.org/10.31925/farmacia.2021.3.1

Puri V, Kanojia N, Sharma A, Huanbutta K, Dheer D, Sangnim T (2022) Natural product-based pharmacological studies for neurological disorders. Front Pharmacol 13:1011740. https://doi.org/10.3389/fphar.2022.1011740

Purnima, Bhatt M, Kothiyal P (2015) A review article on phytochemistry and pharmacological profiles of Nardostachys jatamansi DC-medicinal herb. J Pharmacogn Phytochem 3, 102–106

Qin L (2011) Antidepressant effect and mechanism of nardosinone. Doctoral dissertation, MA thesis, Beijing University of Chinese Medicine, Beijing, in Chinese

Rahman H, Muralidharan P (2010) Comparative study of antidepressant activity of methanolic extract of Nardostachys jatamansi DC rhizome on normal and sleep deprived mice. Der Pharmacia Letter 2:441–449

Rahman H, Shaik HA, Madhavi P, Eswaraiah MC (2011) A review: pharmacognostics and pharmacological profiles of Nardastachys jatamansi DC. Elixir Pharm 23(39):5017–5020

Raina K, Kumari R, Thakur P, Sharma R, Singh R, Thakur A, Anand V, Chaudhary A (2023) Mechanistic role and potential of Ayurvedic herbs as anti-aging therapies. In: Drug metabolism and personalized therapy. https://doi.org/10.1515/dmpt-2023-0024

Rana SK, Rana HK, Ranjitkar S, Ghimire SK, Gurmachhan CM, O'Neill AR, Sun H (2020) Climate-change threats to distribution, habitats, sustainability and conservation of highly traded medicinal and aromatic plants in Nepal. Ecol Indic 115:106435. https://doi.org/10.1016/j.ecolind.2020.106435

Rao VS, Rao A, Karanth KS (2005) Anticonvulsant and neurotoxicity profile of Nardostachys jatamansi in rats. J Ethnopharmacol 102(3):351–356

Rao S, Deepa B, Suchetha K (2013) Antidepressant activity of Nardostachys jatamansi in electron beam irradiated mice. Int J Res Ayurveda Pharm 4(1):101–103

Rasheed AS, Venkataraman S, Jayaveera KN, Fazil AM, Yasodha KJ, Aleem MA, Mohammed M, Khaja Z, Ushasri B, Pradeep HA, Ibrahim M (2010) Evaluation of toxicological and antioxidant potential of Nardostachys jatamansi in reversing haloperidol-induced catalepsy in rats. Int J Gen Med:127–136. https://doi.org/10.2147/ijgm.s9156

Rashid M, ur Rahman A, Ahmad QZ, Mian SS (2018) Physicochemical analysis of Sumbul-al–Teeb (Nardostachys jatamansi DC) rhizome along with its HPLC profile. Pharm J 10(2):278–284

Rawat AK, Mehrotra S, Tripathi SC, Shome U (1997) Hepatoprotective activity of Boerhaavia diffusa L. roots—a popular Indian ethnomedicine. J Ethnopharmacol 56(1):61–66. https://doi.org/10.1016/S0378-8741(96)01507-3

Razack S, Hemanth Kumar K, Nallamuthu I, Naika M, Khanum F (2015) Antioxidant, biomolecule oxidation protective activities of Nardostachys jatamansi DC and its phytochemical analysis by RP-HPLC and GC-MS. Antioxidants 4(1):185–203. https://doi.org/10.3390/antiox4010185

Razack S, Kandikattu HK, Amruta N, Khanum F (2021) Neuroprotective effects of Nardostachys jatamansi against BSO induced anxiety via its antioxidant machinery and by elevating Catecholamines and GABA levels in mice. Curr Tradit Med 7(2):222–245. https://doi.org/10.2174/2215083806999201202153612

Rehan HS, Chopra D, Yadav M, Wardhan N, Manak S, Siddiqui KM, Aslam M (2015) Safety and efficacy of Qurse-e-istisqua in chronic hepatitis C infection: an exploratory study. Indian J Pharmacol 47(1):72–79

Reid JM, Bignal E, Bignal S, McCracken DI, Fenn SR, Trask AE, Monaghan P (2022) Integrating advances in population and evolutionary ecology with conservation strategy through long-term studies of red-billed choughs. J Anim Ecol 91(1):20–34. https://doi.org/10.1111/1365-2656.13615

Rekha K, Rao RR, Pandey R, Prasad KR, Babu KS, Vangala JR, Kalivendi SV, Rao JM (2013) Two new sesquiterpenoids from the rhizomes of Nardostachys jatamansi. J Asian Nat Prod Res 15(2):111–116. https://doi.org/10.1080/10286020.2012.738673

Rokaya MB, Münzbergová Z, Timsina B (2010) Ethnobotanical study of medicinal plants from the Humla district of western Nepal. J Ethnopharmacol 130(3):485–504. https://doi.org/10.1016/j.jep.2010.05.036

Sadeghi HM, Adeli I, Calina D, Docea AO, Mousavi T, Daniali M, Nikfar S, Tsatsakis A, Abdollahi M (2022) Polycystic ovary syndrome: a comprehensive review of pathogenesis, management, and drug repurposing. Int J Mol Sci 23(2):583. https://doi.org/10.3390/ijms23020583

Sahu R, Dhongade HJ, Pandey A, Sahu P, Sahu V, Patel D, Kashyap P (2016) Medicinal properties of Nardostachys jatamansi (a review). Orient J Chem 32(2):859–866. https://doi.org/10.13005/ojc/320211

Saleem F, Ismail MO, Memon Z, Arif F (2021) Antidepressant activity of Nardostachys jatamansi extract in animal models of depression. J Pharm Res Int 32(40):102–111. https://doi.org/10.9734/jpri/2020/v32i4031037

Salim S, Ahmad M, Zafar KS, Ahmad AS, Islam F (2003) Protective effect of Nardostachys jatamansi in rat cerebral ischemia. Pharmacol Biochem Behav 74(2):481–486. https://doi.org/10.1016/S0091-3057(02)01030-4

Sama G, Ankita, Minakshi K, Chahal RK, Thakur PS (2024) Jatamansi (nardostachys jatamansi dc.): insight of its morphology and its medicinal use in ayurveda. Int Ayurvedic Med J. https://doi.org/10.46607/iamj1012012024

Sandeep PM, Bovee TF, Sreejith K (2015) Anti-androgenic activity of Nardostachys jatamansi DC and Tribulus terrestris L. and their beneficial effects on polycystic ovary syndrome–induced rat models. Metab Syndr Relat Disord 13(6):248–254. https://doi.org/10.1089/met.2014.0136

Sang CY, Zheng YD, Ma LM, Wang K, Wang CB, Chai T, Eshbakova KA, Yang JL (2022) Potential Anti-Tumor Activity of Nardoguaianone L Isolated from Nardostachys jatamansi DC. in SW1990 Cells. Molecules. 27(21):7490. https://doi.org/10.3390/molecules27217490

Saroya AS (2011) Herbalism. In: Phytochemistry and ethnopharmacology. Science Publishers, Enfield

Sastry JLN (2018) Illustrated Dravyaguna Vijnana: Knowledge on Vedic Herbs, Controversial and IgnoredMedicinal Plants, Vol V. Chaukhambha Orientalia Publications, Varanasi

Shah NC (2023) Historical profile of Nardostachys jatamansi: an ancient incense & aromatic medicinal herb from Kumaon, Uttarakhand. Indian J Hist Sci 58(2):111–118. https://doi.org/10.1007/s43539-023-00081-5

Shah AJ, Gilani AH (2011) Bronchodilatory and spasmolytic effects of *Nardostachys jatamansi*. Evid Based Complement Alternat Med:560168. https://doi.org/10.1093/ecam/nep065

Shakir Ali, Khursheed A, Ansari MA, Jafry H, Kabeer G, Diwakar (2000) Nardostachys jatamansi protects against liver damage induced by thioacetamide in rats. 71:(3), 0–363. https://doi.org/10.1016/s0378-8741(99)00153-1

Sharma N, Sharma AR, Patel BD, Shrestha K (2016) Investigation on phytochemical, antimicrobial activity and essential oil constituents of Nardostachys jatamansi DC. in different regions of Nepal. J Coast Life Med 4(1):56–60

Sharma GK, Dhanawat M, Lodha R, Goyal KK, Pareta JK (2021) Assessment & comparison of hepatoprotective potential of indigenous therapeutic flora by using tetrachloromethane encouraged liver toxicity in Albino (Wister strain) rodent model. Int J Pharm Res (09752366) 13(1). https://doi.org/10.31838/ijpr/2021.13.01.389

Shastri AR (ed) (2005) Sushrut Samhita of Sushrutacharya, Vranitopasaniya Adhyaya: chapter 19, verse 29. Chaukhamba Publications, New Delhi, p 81

Shastri P (2006) Sharangadhara samhita of sharangadhara. In: Madhyam Khanda, vol 126 Ch. 1, Ver. 1, 7th edition, Chaukhamba Orientalia Prakashan, Varanasi

Shastri L, Shastri B (eds) (1999) Yogratnakar of anonymous, Uttarardha; Mushka Antravruddhi Vardhma Nidan Chikitsa Adhyaya, 7th edn. Chaukhamba Publications, New Delhi, p 139

Shenoy BK, Karkada G (2024) Reversal of restraint stress caused dendritic atropy in rats by Nardostachys jatamansi. Indian J Tradit Knowl 23(4):353–361. https://doi.org/10.56042/ijtk.v23i4.47449

Sherpa P, Bhattarai B, Rana M. Environment and ecology. Environment and ecology [Internet]. 2023. Available from: https://doi.org/10.60151/envec

Shin JY, Bae GS, Choi SB, Jo IJ, Kim DG, Lee DS, An RB, Oh H, Kim YC, Shin YK, Jeong HW (2015) Anti-inflammatory effect of desoxo-narchinol-A isolated from Nardostachys jatamansi against lipopolysaccharide. Int Immunopharmacol 29(2):730–738

Shin YJ, Midgley GF, Archer ER, Arneth A, Barnes DK, Chan L, Hashimoto S, Hoegh-Guldberg O, Insarov G, Leadley P, Levin LA (2022) Actions to halt biodiversity loss generally benefit the climate. Glob Chang Biol 28(9):2846–2874. https://doi.org/10.1111/gcb.16109

Shrestha UB, Lamsal P, Ghimire SK, Shrestha BB, Dhakal S, Shrestha S, Atreya K (2022) Climate change-induced distributional change of medicinal and aromatic plants in the Nepal Himalaya. Ecol Evol 12(8):e9204. https://doi.org/10.1002/ece3.9204

Shriram BK, Devhare LD, Mehrotra A, Deokar SS, Singh SP (2023) Formulation and evaluation of mosquito repellent stick. Int J Drug Deliv Technol 13(4):1283–1286

Shukla V, Singh P, Konwar R, Singh B, Kumar B (2021) Phytochemical analysis of high value medicinal plant Valeriana jatamansi using LC-MS and it's in-vitro anti-proliferative screening. Phytomedicine Plus 1(2):100025. https://doi.org/10.1016/j.phyplu.2021.100025

Shukla JK, Dhakal P, Thapa S, Kharlyngdoh E, Thorat SS, Bhardwaj PK, Chaudhary SK, Sathyanarayan N, Gour VS, Mukherjee PK (2024) Growth, yield and metabolite substantiated, eco-physiological evaluation of overcoming accessional and agro-climatic differences in ex-situ raised Valeriana jatamansi Jones. at lower altitude of Sikkim Himalayan region. Ind Crop Prod 207:117730

Simonian N, Coyle JT (1996) Oxidative stress in neurodegenerative diseases. Annu Rev Pharmacol Toxicol 36:83–106

Singh TG (2020) Phytochemical activities and pharmacology of herbal drug: Nardostachys jatamansi. Plant Arch 20(1):3842–3848

Singh V, Dubey P, Srivastava S, Rawat AKS (2011) Botanical standardization of the jatamansi, their substitute and adulterant species. Indian J Tradit Knowl 10(4):599–603

Singh UM, Vijayta Gupta VG, Rao VP, Sengar RS, Yadav MK (2013) A review on biological activities and conservation of endangered medicinal herb Nardostachys jatamansi. Int J Med Aromat Plants 3:113–124

Singh RK, Vaishali PS, Murthy PN, Panigrahi G, Sharma PK (2014) Evaluation of anti-inflammatory potential of Nardostachys jatamansi rhizome in experimental rodents. J Coast Life Med 2(1):38–43

Singh V, Rana N, Ali M (2018) GC-MS analysis and anti-microbial activities of volatile oil of nardostachys jatamansi DC rhizomes obtained from Haridwar region, uttrakhand. Pharm J 10(2):230–234

Singh M, Khan MA, Kamal YT, Ahmad J, Fahmy UA, Kotta S, Alhakamy NA, Ahmad S (2020) Effect of Nardostachys jatamansi DC. on apoptosis, inflammation and oxidative stress induced by doxorubicin in wistar rats. Plants 9(11):1579

Stafstrom CE, Carmant L (2015) Seizures and epilepsy: an overview for neuroscientists. Cold Spring Harb Perspect Med 5(6):a022426

Subashini R, Yogeeta S, Gnanapragasam A, Devaki T (2006) Protective effect of Nardostachys jatamansi on oxidative injury and cellular abnormalities during doxorubicin-induced cardiac damage in rats. J Pharm Pharmacol 58(2):257–262. https://doi.org/10.1211/jpp.58.2.0014

Subashini R, Gnanapragasam A, Senthilkumar S, Yogeeta SK, Devaki T (2007) Protective efficacy of Nardostachys jatamansi (rhizomes) on mitochondrial respiration and lysosomal hydrolases during doxorubicin induced myocardial injury in rats. J Health Sci 53(1):67–76

Sumadewi KT, Harkitasari S, Tjandra DC (2023) Biomolecular mechanisms of epileptic seizures and epilepsy: a review. Acta Epileptol 5(1):28

Sumitra C (2014) Importance of pharmacognostic study of medicinal plants: An overview. J Pharmacogn Phytochem. 2(5):69–73

Sunanda RBS, Raju TR (2000) Chronic restraint stress impairs acquisition and retention of spatial memory task in rats. Curr Sci 10:1581–1584

Suryavanshi S, Raina P, Deshpande R, Kaul-Ghanekar R (2017) Nardostachys jatamansi root extract modulates the growth of IMR-32 and SK-N-MC neuroblastoma cell lines through MYCN mediated regulation of MDM2 and p53. Pharmacogn Mag 13(49):21. https://doi.org/10.4103/0973-1296.197645

Swati S, More R (2008) Design, synthesis and biological evaluation of glutathione peptidomimetics as components of anti-Parkinson prodrugs, J Med Chem. 51:15:4581–4588

Szekanecz Z, Koch AE (2007) Macrophages and their products in rheumatoid arthritis. Curr Opin Rheumatol 19(3):289–295

Thakur S, Kaurav H, Chaudhary G (2021) Nardostachys jatamansi: importance of highly significant and endangered plant in Ayurveda. Int J Res Ayurveda Pharm 12(3):124–130. https://doi.org/10.7897/2277-4343.120387

Tiwari UL, Kotia A, Rawat GS (2009) Medico-ethnobotany of the Monpas in Tawang and West Kameng districts of Arunachal Pradesh, India. Pleione 3(1):1–8

Toolika E, Ravi M, Bhat NP, Shetty SK, Ravishankar B, Savitha HP (2015) A comparative pharmacological study on the effect of Tagara (Valeriana wallichii) AND Jatamansi (Nardostachys jatamansi) in the management of Anidra WSR to primary insomnia. J Phytopharmacol 4:147–151

Tripathi B. In: Reprint, editor. Charaka Samhita of Agnivesha. Delhi: Chaukhamba Surbharati Prakashan; 2004. p. 82, 96, 118, 782, 634

Tripathi YB, Tripathi E, Upadhyay A (1996) Antilipid peroxidative property of Nardostachys jatamanasi. Indian J Exp Biol 34(11):1150–1151

UCM (2024) University of Chicago, Medicine. Chronic Pancreatitis. https://web.archive.org/web/20170509183753/http://www.uchospitals.edu/specialties/gi/pancreas/pancreatitis/chronic.html Accessed 8 Oct 2024

Uprety Y, Asselin H, Boon EK, Yadav S, Shrestha KK (2010) Indigenous use and bio-efficacy of medicinal plants in the Rasuwa District, Central Nepal. J Ethnobiol Ethnomed 6:1. https://doi.org/10.1186/1746-4269-6-3

Valmiro DJ, Pinto LM, Barbosa LN, de Araújo LB, Gonzaga RV (2021) Drug interactions between herbal medicines and oral contraceptives. Res Soc Dev 10(13):e372101321331. https://doi.org/10.33448/rsd-v10i13.21331

Vane JR, Mitchell JA, Appleton I, Tomlinson A, Bishop-Bailey D, Croxtall J, Willoughby DA (1994) Inducible isoforms of cyclooxygenase and nitric-oxide synthase in inflammation. Proc Natl Acad Sci 91(6):2046–2050

Varinthra P, Suresh P, Chokkalingam P, Ibiayo AG, Liu IY (2023) Anti-inflammatory and memory-enhancing properties of Chinese herbal extracts: the possible application in Alzheimer's disease. Tzu Chi Med J 35(2):103–110. https://doi.org/10.4103/tcmj.tcmj_139_22

Ved D, Saha D, Ravikumar K, Haridasan K (2015) *Nardostachys jatamansi*. The IUCN red list of threatened species 2015: e.T50126627A50131395. https://doi.org/10.2305/IUCN.UK.2015-2.RLTS.T50126627A50131395.en

Wakabayashi K, Tanji K, Odagiri S, Miki Y, Mori F, Takahashi H (2013) The Lewy body in Parkinson's disease and related neurodegenerative disorders. Mol Neurobiol 47:495–508. https://doi.org/10.1007/s12035-012-8280-y

Wang X, Peng H, Huang Y, Kong W, Cui Q, Du J, Jin H (2020) Post-translational modifications of IκBα: the state of the art. Front Cell Dev Biol 8:574706. https://doi.org/10.3389/fcell.2020.574706

Wang M, Yang TT, Rao Y, Wang ZM, Dong X, Zhang LH, Han L, Zhang Y, Wang T, Zhu Y, Gao XM (2021) A review on traditional uses, phytochemistry, pharmacology, toxicology and the analytical methods of the genus Nardostachys. J Ethnopharmacol 280:114446. https://doi.org/10.1016/j.jep.2021.114446

Wei ZZ, Wei L (2017) Mitochondrial mechanisms during ischemia and reperfusion. In: Primer on cerebrovascular diseases. Academic Press, pp 230–234. https://doi.org/10.1016/B978-0-12-803058-5.00047-3

Wen J, Zhou L, Liu L, He Y (2022) Analysis of the impact of climate change on the distribution and active compound content of the plateau medicinal plant Nardostachys jatamansi (D. Don) DC. Ind Crop Prod 187:115438. https://doi.org/10.1016/j.indcrop.2022.115438

Weberling F (1975) On the systematics of Nardostachys (Valerianaceae). Taxon. 24(4):443–452

WHO (2023) World Health Organisation, Newsroom, Dementia. https://www.who.int/news-room/fact-sheets/detail/dementia. Accessed 8 Oct 2024

Willemer S, Elsasser H-P, Adler G (1992) Hormone-induced pancreatitis. Eur Surg Res 24:29–39. https://doi.org/10.1159/000129237

Wu HH, Deng X, Zhang H, Chen YP, Ying SS, Wu YJ, Liu YT, Zhu Y, Gao XM, Xu YT, Li L (2018) Dinardokanshones C–E, isonardoeudesmols A–D and nardoeudesmol D from Nardostachys jatamansi DC. Phytochemistry 150:50–59. https://doi.org/10.1016/j.phytochem.2018.03.002

Xue X, Chen Y, Wang Y, Zhan J, Chen B, Wang X, Pan X (2019) Troxerutin suppresses the inflammatory response in advanced glycation end-product-administered chondrocytes and attenuates mouse osteoarthritis development. Food Funct 10(8):5059–5069. https://doi.org/10.1039/C9FO01089K

Yoon CS, Kim DC, Park JS, Kim KW, Kim YC, Oh H (2018) Isolation of novel sesquiterpeniods and anti-neuroinflammatory metabolites from Nardostachys jatamansi. Molecules 23(9):2367. https://doi.org/10.3390/molecules23092367

Yoon CS, Lee H, Liu Z, Dong L, Lee G, Kim N, Oh H, Lee DS (2024) Cycloolivil isolated from Nardostachys jatamansi inhibits TNF-α/IFN-γ-induced chemokine production by blocking NF-κB and JAK/STAT activation in HaCaT keratinocytes. Int J Mol Sci 25(6):3342. https://doi.org/10.3390/ijms25063342

Yousuf S, Atif F, Ahmad M, Ishrat T, Khan B, Islam F (2011) Neuroprotection offered by Majun Khadar, a traditional unani medicine, during cerebral ischemic damage in rats. Evid Based Complement Alternat Med 2011(1):754025. https://doi.org/10.1093/ecam/nep224

Yubero S, Ramudo L, Manso MA, De Dios I (2009) Mechanisms of dexamethasone-mediated chemokine down-regulation in mild and severe acute pancreatitis. Biochim Biophys Acta 1792(12):1205–1211. https://doi.org/10.1016/J.BBADIS.2009.10.001

Zhang J, Niu J, Yang J (2014) Interleukin-6, interleukin-8 and interleukin-10 in estimating the severity of acute pancreatitis: an updated meta-analysis. Hepato-Gastroenterology 61(129):215–220

Zhang W, Nan G, Wu HH, Jiang M, Li TX, Wang M, Gao XM, Zhu Y, Song YS, Wang J, Xu YT (2018) A simple and rapid UPLC-PDA method for quality control of Nardostachys jatamansi. Planta Med 50(08):536–543. https://doi.org/10.1055/s-0043-123655

Zhang Y, Guo X, Ma S, Ma H, Li H, Wang Y, Qin Z, Wu X, Han Y, Han Y (2021) The treatment with complementary and alternative traditional Chinese medicine for menstrual disorders with polycystic ovary syndrome. Evid Based Complement Alternat Med 2021(1):6678398. https://doi.org/10.1155/2021/6678398

# Medicinal and Nutritional Importance of *Andrographis paniculata* in Human Health

24

Benazir Fatima, Ushakar Mishra, and Abha Singh

**Abstract**

*Andrographis paniculata* (Burm.f.) Wall. ex Nees is a plant of significant interest in various studies due to its potential therapeutic properties and applications in traditional medicine. This is a medicinal herb found in Asian countries, utilized in various traditional medicinal systems for addressing conditions such as diarrhea, flu, leprosy, leptospirosis, malaria, rabies, upper respiratory infections, sinusitis, syphilis, tuberculosis, and HIV/AIDS, among others. This chapter seeks to deliver thorough, precise, and reliable information regarding the traditional applications, phytochemical composition, and pharmacological characteristics of different extracts and fractions, along with the phytoconstituents of *A. paniculata*. Terpenoid lactones and flavonoids represent the primary bioactive classes of compounds found in *A. paniculata*, contributing to its pharmacological properties, including anticancer and antioxidant activities. Various activities such as analgesic, anticancer, antidiabetic, antifertility, anti-inflammatory, antimalarial, antimicrobial, antioxidant, antipyretic, antiviral, antiretroviral, antivenom, cardioprotective, hepatoprotective, immunomodulatory, and neuroprotective have also been documented. Andrographolide serves as a significant active compound and is accountable for the majority of the pharmacological effects observed. Compelling evidence demonstrated that the pharmacological activities observed in crude extracts and chemical markers reinforce and instill confidence in the traditional application of *A. paniculata* as a herbal remedy.

**Keywords**

*Andrographis paniculata* · Secondary metabolites · Andrographolide · Pharmacological applications

B. Fatima (✉) · U. Mishra · A. Singh
Microbial Biodiversity Laboratory, Department of Botany, Patna University, Patna, Bihar, India



M. A. Ansari et al. (eds.), *Medicinal Plants and their Bioactive Compounds in Human Health: Volume 2*, https://doi.org/10.1007/978-981-95-3620-7_24

## 24.1 Introduction

Plants and their derivatives have been used as a supplemental medicine for millennia, successfully treating a variety of medical conditions. Approximately 80,000 plant species have been identified and used for their medicinal properties worldwide. Since it is widely believed that plants have great potential for healing a wide range of illnesses, they continue to play a crucial role in modern society's healthcare system. They are commonly thought to be dependable, effective, and simple to use, with few or no negative side effects. *Andrographis paniculata* (Fig. 24.1) is ranked 17th on the Indian National Medicinal Plants Board's prioritized list of 32 medicinal plants, which is based on the plants' numerous therapeutic applications. Native to South Asian nations like India and Sri Lanka, this medicinal plant grows well in tropical and subtropical regions during the Kharif season. This herbaceous plant, which is a member of the Acanthaceae family (Table 24.1), is distinguished by its small white flowers, lance-shaped leaves, and slender stems. Some more names for *A. paniculata* include "Kalmegh" in Hindi and "Hempedu bumi" in Malay. However, the plant's bioactive compounds—rather than its outward look—have attracted research from all around the world (Hossain et al. 2014; Dai et al. 2019). Morphological features of *A. paniculata* has been shown in Table 24.2.

Plants contain a variety of secondary metabolites that give them inherent antioxidant properties. These secondary metabolites are notably prevalent in *A. paniculata*. Because of its extensive application in the food processing and pharmaceutical industries, the aforementioned chemical has attracted a lot of interest. In recent

**Fig. 24.1** *Andrographis paniculata*. (Fatima et al. 2023)

**Table 24.1** Taxonomic hierarchy (Hossain et al. 2014)

| | |
|---|---|
| *Domain* | Eukaryota |
| *Kingdom* | Plantae |
| *Division* | Angiosperma |
| *Class* | Dicotyledonae |
| *Series* | Bicarpellatae |
| *Order* | Personales |
| *Family* | Acanthaceae |
| *Genus* | *Andrographis* |
| *Species* | *Paniculata* (Burm. f.) Wall. ex Nees |

**Table 24.2** Morphological features of *A. paniculata* (Hossain et al. 2014)

| Traits | Features |
|---|---|
| Plant height | 30–110 cm |
| Leaves | Glabrous; length: 2–12 cm; width: 1–3 cm; arrangement: lanceolate; shape: pinnate, acute apex, entire margin |
| Stems | Color: dark green; length: 30–100 cm; diameter: 2–6 mm; shape: quadrangular with longitudinal furrows and wings on the angles of the young parts; slightly enlarged at the nodes |
| Flowers | Color: white with rose-purple spots on the petals; size: small, in lax spreading axillary and terminal racemes or panicles |
| Seeds | Capsules linear-oblong, acute at both ends; size: 1.9 cm × 0.3 cm; color: yellowish brown; shape: subquadrate, numerous |
| Flowering and fruiting | December to April |

years, natural goods and therapeutic plants have attracted a lot of attention from pharmacy researchers. This is mostly due to its affordability and the fact that these plants have bioactive compounds that can be used to cure a variety of contemporary illnesses (Elshafie et al. 2023). Furthermore, compared to synthetic alternatives, these natural therapies have the advantage of having less negative side effects. Over the course of its long history, this plant has been used extensively in Indian and Ayurvedic medicine as well as traditional Chinese medicine (Sivananthan and Elamaran 2013; Pawar et al. 2016). In order to treat a variety of illnesses, including sore throat, fever, liver problems, jaundice, antispasmodic conditions, malaria, eczema, intestinal worm infestation, and gonorrhea, Ayurvedic practitioners frequently use *A. paniculata* (Thokchom et al. 2018). According to Chinese medicine, this specific plant can be used to treat body heat, particularly in cases of pyretic and toxicological disorders. Furthermore, it is used to treat gastrointestinal disorders such diarrhea, inflammation of the larynx, gastrointestinal tract infections, and inflammatory processes (Hu et al. 2017; Yang et al. 2020). *A. paniculata* has been found to have immunostimulatory activities and therapeutic promise in the realm of modern pharmacology (Ahmad et al. 2022). Additionally, this specific plant species has shown encouraging qualities, including anti-inflammatory, antibacterial, anticancer, and anti-HIV activities (Hidalgo et al. 2013). According to recent studies by Samtiya et al. (2021), *A. paniculata* may have therapeutic benefits for the management of Severe Acute Respiratory Syndrome Coronavirus 2 (SARS-CoV-2).

## 24.2 Origin and Distribution

*A. paniculata* is an indigenous plant species found in the regions of Taiwan, China, and India. The geographical distribution of this species encompasses tropical and subtropical regions in Asia, particularly Southeast Asia, as well as several other countries including Cambodia, various Caribbean islands, Indonesia, Laos, Malaysia, Myanmar, Sri Lanka, Thailand, and Vietnam (Niranjan et al. 2010; Kabir et al. 2014). The presence of this particular plant has been observed in diverse phytogeographical and edaphic zones, across China, America, West Indies, and Christmas Island, as documented by Benoy et al. (2012).

## 24.3 Nutritional Composition

As per the research conducted by Tripathi et al. (2012), the proximate analysis of *A. paniculata* indicated moisture content of 73.02%, carbohydrate levels of 12.16%, and vitamin C values of 45.01 mg/100 g. The mineral analysis of *A. paniculata* leaves revealed phosphorus (P) at 250.13 mg/100 g, sodium (Na) at 152.50 mg/100 g, calcium (Ca) at 318.62 mg/100 g, and magnesium (Mg) at 7.68 mg/100 g. The findings indicate that the leaves of this plant are abundant in vitamin C and their macro- and micronutrient composition render them very appropriate for inclusion in human nutrition.

## 24.4 Phytochemistry of *A. paniculata*

According to the literature, *A. paniculata* primarily contains diterpenoid lactones (34.95%) and flavonoids (46.23%), especially from its aerial parts (61.93%). Other chemical classes identified include terpenoids (10.22%), phenolic acids (4.30%), chalconoids (2.15%), xanthones (2.15%), and volatile compounds. The biosynthesis of these compounds occurs through the mevalonic acid (MVA) and methyl erythritol phosphate (MEP)/deoxyxylulose 1-deoxy-D-xylulose 5-phosphate (DXP) pathways, with the DXP pathway being more dominant for diterpenoid lactones (Wang et al. 2009; Srivastava and Akhila 2010; Tran et al. 2021).

- *Diterpene Lactones*

Andrographolide (Fig. 24.2) is one of the most frequent chemicals, present in both the aerial and the whole plant. According to various studies (Reddy et al. 2007; Subramanian et al. 2012; Hapuarachchi et al. 2013; Hanh et al. 2020), andrographolide has a variety of biological activities, including antiangiogenic, anticancer, antidiabetic, anti-inflammatory, anti-influenza, antileishmaniasis, antipyretic, antiretroviral, antithrombotic, antiulcerogenic, proapoptotic, antiurothelial, cardioprotective, cholestatic, hepatoprotective, immunomodulatory, and radiosensitive effects. The aerial parts include a significant component, 3-O-β-D-glucopyranosyl andrographolide, which has antibacterial and antifungal properties (Shen et al. 2006; Xu et al. 2012).

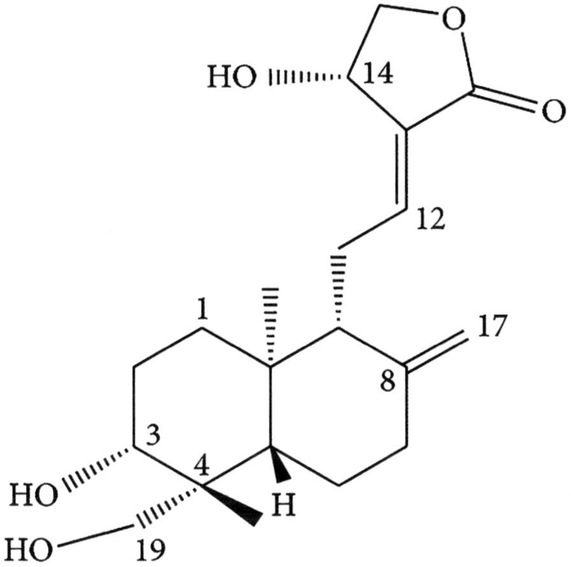


**Fig. 24.2** Chemical structure of bioactive metabolite andrographolide. (Reyes et al. 1923; Jayakumar et al. 2013)

According to Arpini et al. (2008) and Wang et al. (2014), andropanolide, which was identified in aerial parts, possesses antiviral activity, cytotoxicity, and efficacy against respiratory syncytial virus. Furthermore, among one of the compounds reported, andropanolide B has cytopathic effects (Xu et al. 2012; Wang et al. 2014). Andrographiside, a substance derived from aerial parts, possesses hepatoprotective properties (Subramanian et al. 2012; Hapuarachchi et al. 2013). Furthermore, Chen et al. (2006) and Xu et al. (2012) looked into the therapeutic potential of 3,19-dihydroxy-ent-labda-8(17),12-dien-16,15-olide. Other important chemicals include 19-nor andrographolide A, which is found in the aerial parts (Zhang et al. 2006; Subramanian et al. 2012), and 14-deoxy-11,12-didehydro andrographolide, which is found in both the aerial parts and the entire plant and has anticancer properties. Researchers (Reddy et al. 2007; Subramanian et al. 2012; Xu et al. 2012) have found that the presence of the isomers 15-methoxy-3,19-dihydroxy-8(17)11,13-entlabda-trien-16,15-olide and 3,19-dihydroxy-15-methoxy-entlabda-8(17),11,13-trien-16,15-olide in the whole plant had antiproliferative effects. Zhou et al. (2008) and Xu et al. (2012) discovered 3-O-β-D-glucosyl-14-deoxy-11,12-didehydro andrographiside to be a strong anti-inflammatory compound in the aerial parts.

The aerial portions also include antiproliferative and cytotoxic chemicals such as 3,18,19-trihydroxy-ent-labda-8(17),13-dien-16,15-olide, 14-deoxy-11,12-didehydro andrographiside, and 3-oxo-14-deoxy-11,12 didehydro andrographolide (Chen et al. 2008; Hapuarachchi et al. 2013). Neoandrographolide extracted from aerial parts contains a number of bioactivities, including anti-herpes simplex virus, anti-inflammatory, antioxidant, antiparasitic, chemosensitizer, and hepatoprotective effects (Arpini et al. 2008).

Furthermore, 14-deoxy andrographolide has antibacterial, antifungal, antihypertensive, hepatoprotective, immunomodulatory, platelet-activating factor antagonist, uterine smooth muscle relaxant, and vasodilator properties (Arpini et al. 2008; Subramanian et al. 2012). Subramanian et al. (2012) reported 3-O-β-D-glucosyl-14-deoxy andrographolide in the aerial part, an antibacterial compound, and another compound andrograpanin that possesses anti-inflammatory and cytotoxic properties found throughout the plant. Bis andrographolides A-F, have been identified in aerial parts and exhibit anticancer, anti-HIV, anti-inflammatory, cardioprotective, and venom detoxifying properties. Bis andrographolide E and G are thought to offer cardiomyocyte protection (Lim et al. 2012; Gao et al. 2019; Sun et al. 2020).

- *Chalconoides*

*A. paniculata* includes chalconoides, an important class of acetophenone chemicals in which one of the methyl hydrogens has been replaced by a benzylidene group, which have been found throughout the plant. Specifically, 2′,4′,6′,2,3,4-hexa methoxychalcone, 2′-hydroxy-2,4′,6′-trimethoxychalcone, and 2′-hydroxy-2,3,4′-trimethoxychalcone were recovered from the plant's aerial sections. These compounds are mostly acknowledged for their antibacterial properties, as demonstrated by research conducted by Rao et al. (2004) and Niranjan Reddy et al. (2005), which confirmed their presence in both the whole plant and its aerial parts. Hanh et al. (2020) identified a novel chalconoid, pashanone glucoside, in the aerial components of *A. paniculata*, hence enhancing the plant's therapeutic properties.

- *Flavanones*

A significant category of secondary metabolites is flavanones, characterized by a flavan with an oxo substituent at position 4, present in *A. paniculata* having therapeutic applications. Rao et al. (2004) identified (2R)-hydroxy-7, 2′-dimethoxyflavone in the aerial parts such as leaves and stems. The investigation by Hapuarachchi et al. (2013) revealed that 5-hydroxy-7,8-dimethoxyflavanone (2R)-5-O-β-D-glucopyranoside and 5-hydroxy-7,8,2′,5′-tetramethoxyflavanone (2R)-5-O-β-D-glucopyranoside are present in the aerial parts, which have anticancer properties and (2R)-5-hydroxy-7,2′,3′-trimethoxyflavanone are found in the whole plants. Additionally, Gupta et al. (1983) and Reddy et al. (2003) reported isolation of (2S)-5,7,2′,3′,4′-pentamethoxyflavanone and (2S)-5,7,8-trimethoxydihydroflavanone from the whole plant and aerial parts. Dihydroskullcapflavone I, another notable flavanone, has been found in the plant and has antidiabetic properties, as reported by Chen et al. (2014). Additionally andrographidin A, a compound present in the aerial parts and entire plant, exhibits antibacterial, antifungal, and cytotoxic effects, as reported by Geethangili et al. (2008) and Hapuarachchi et al. (2013).

- *Flavones*

This herb is a source of flavones, a class of compounds that exhibit significant pharmacological effects, including antibacterial, antifungal, and cytotoxic properties. Xu and Wang (2011) published 5-hydroxy-7,2′,5′-trimethoxyflavone, which was extracted from the roots. Shen et al. (2006) described 7,2′,5′-trimethoxyflavone and 5,7,2′,3′-tetramethoxyflavone, which are both found in the entire plant and have anti-inflammatory properties. Cytotoxic 5-hydroxy-7,8,2′-trimethoxyflavone has been detected in the aerial parts, leaves, and roots of various plants (Gupta et al. 1983; Geethangili et al. 2008; Xu and Wang 2011; Chen et al. 2014).

Another flavone that is cytotoxic and is present throughout the plant is skullcapflavone-2′-methoxylether. However, 5-hydroxy-7,8-dimethoxyflavone, which is also present throughout the plant, has antibacterial properties (Chen et al. 2014). Andrographidine A, a flavone glucoside with antiplatelet and vasorelaxing properties, is present in the apical portions (Wu et al. 2008). Additionally, skullcapflavone I has been shown to possess antivenometic properties rendering it a potent bioactive molecule (Xu and Wang 2011). Lutein, which possesses antiallergic, anticancer, anti-inflammatory, and neuroprotective properties, is another notable flavone (Chen et al. 2014; Chandra et al. 2016). Additionally, apigenin is recognized for its antioxidant, anticancer, and anti-inflammatory properties. These diverse flavones significantly enhance the therapeutic potential of *A. paniculata*, thereby enhancing its status as a medicinal plant with a plethora of health benefits (Chandra et al. 2016).

- *Flavanols*

Hapuarachchi et al. (2013) identified a substantial flavonol in the aerial portions of this herb named as 5-hydroxy-3,7,8,2′-tetramethoxyflavonol. Flavonols such as this are frequently associated with antioxidant and anti-inflammatory properties, which contribute to the plant's therapeutic profile, despite the fact that their specific bioactivities have not been thoroughly investigated. The aerial portions contain another well-known flavonol, quercetin, which is recognized for its numerous pharmacological properties. It is a critical chemical in plant defense systems and therapeutic applications due to its antiallergic, anticancer, antidiabetic, anti-inflammatory, antioxidant, and antiviral properties (Wu et al. 2008). The value of quercetin in the management of chronic illnesses and the promotion of overall health is underscored by its ability to modify cellular pathways.

- *Phenolics*

Cinnamic acid, which is present in the aerial parts, foliage, and roots of this herb, possesses antioxidant, anticancer, and antifungal properties (Wu et al. 2008; Xu and Wang 2011). Ferulic acid, which is abundant in the aerial parts, leaves, and roots, also possesses a diverse array of pharmacological properties, such as antibacterial, anticancer, anti-inflammatory, antimicrobial, antioxidant, and cholesterol-lowering properties. In the same vein, protocatechuic acid, which is present in the aerial portions, leaves, and roots, has been demonstrated to possess analgesic, antiaging, antiatherosclerotic, antibacterial, anticancer, antifibrotic, anti-inflammatory, antioxidant, antiulcer, antiviral, and neurological properties (Chandra et al. 2016). The medicinal potential of this herb is also underscored by the antianalgesic, antiangiogenic, anticolitis, antimutagenic, antioxidant, hepatoprotective, and antisickling properties due to the presence of another bioactive component, vanillic acid (Rao et al. 2004; Chandra et al. 2016). In addition, 3,4-dicaffeoylquinic acid, which is present in the aerial portions, leaves, and roots, has been shown to possess hepatoprotective, antiviral, antihyperlipidemic, and antidiabetic properties (Chao and Lin 2010). Consequently, it is a promising molecule for the treatment of metabolic and liver issues. Another significant component is chlorogenic acid, which exhibits analgesic, antibacterial, antidiabetic, antifungal, antihyperlipidemic, antihyperalgesic, anti-inflammatory, antigout, antipyretic, antiulcerogenic, antiviral, and hepatoprotective properties. Additionally, methyl-3,4-dicaffeoylquinate, which was identified in the aerial parts, possesses antihyperlipidemic, antidiabetic, and antihypertensive properties (Chao and Lin 2010; Chandra et al. 2016). Consequently, it is an effective treatment for metabolic syndrome. The medicinal profile of *A. paniculata* is substantially influenced by its phenolic acids, which offer a diverse array of bioactivities.

- *Xanthones*

Xanthones such as 1,8-dihydroxy-3,7-dimethoyxanthone, 4,8-dihydroxy-2,7-di methoxyxanthone, 1,2-dihydroxy-6,8-dimethoxyxanthone, and 3,7,8-trimethoxy-8-hydroxyxanthone are largely found in the root sections of this particular plant and have antimalarial action, according to Dua et al. (2004).

## 24.5 Pharmacological Implications

- *Analgesic Activity*

According to the findings of Lin et al. (2009), aqueous extract of the entire plant demonstrated superior analgesic effects compared to the ethanolic extract, as evidenced by a decrease in the acetic acid-induced writhing response in mice. In a similar manner, the methanolic extract demonstrated notable analgesic effects across various experimental animal models, employing both acetic acid-induced writhing and the tail clip method (Mandal et al. 2001).

- *Antidiabetic Activity*

The ethanolic extract significantly decreased fasting serum glucose levels in streptozotocin-induced diabetic rats, with no observable effect in normal rats (Zhang and Tan 2000). Hossain et al. (2007) demonstrate that aqueous and ethanolic extracts of this plant show considerable hypoglycemic activity, surpassing that of chloroform and petroleum ether extracts. Furthermore, a decoction of the plant's aerial portions showed considerable antidiabetic activity in alloxan-induced diabetic mice (Reyes et al. 2006). The presence of diterpenoid lactones in this plant may have an antidiabetic effect, or flavonoids may have a synergistic impact (Zhang et al. 2009).

- *Anticancer Activity*

Leaf extracts (aqueous, ethanolic, and acetone) were tested in vitro for anticancer activity on neuroblastoma (IMR-32) and human colon (HT-29) cell lines. The ethanolic extract had an $IC_{50}$ value of 200 μg/mL for IMR-32 and HT-29 cell lines, while other extracts had a 250 μg/mL value for HT-29 cells. On several cancer cell lines—including SiHa, Michigan Cancer Foundation-7 (MCF), Ovcar-5, HT-29, and HepG2—the methanolic extract of *A. paniculata* has been shown to exhibit notable antiproliferative properties (Kumar et al. 2004; Agrawal et al. 2017). Studying the ethanolic extract and diterpenoid lactones against human acute myeloid leukemic HL-60 cells, shows the highest degree of cytotoxicity (Cheung et al. 2005; Singh et al. 2013). According to the study by Tan et al. (2005), the diterpenoid lactones isolated from the leaf such as andrographolide, neoandrographolide, andrographiside, deoxyandrographiside, and 14-deoxy-12-methoxyandrographolide appeared to be noncytotoxic against Caov-3 (human ovarian carcinoma) and T-47D (human breast carcinoma). Andrographolide has been investigated for stimulating action on cytotoxic lymphocyte (CTL) generation in Bagg albino mouse lineage (BALB/c) mice (Sheeja et al. 2006). Human leukemia HL-60 cells have been used for screening four new stereoisomers, 7R-hydroxy-14-deoxy andrographolide, 7S-hydroxy-14-deoxy andrographolide, 12S,13S-hydroxy, and 12R,13R-hydroxy andrographolide (Chen et al. 2008).

- *Antifertility*

Zoha et al. (1989) demonstrated that leaf powder markedly influences antispermatogenic and antiandrogenic effects in male albino rats by elevating lactate dehydrogenase, alkaline phosphatase, glycogen, and cholesterol concentrations in the testis and ventral prostate at a dosage of 20 mg/kg. A study by Shinde et al. (2015) investigated the antifertility effects of curcumin and andrographolide in female rats, focusing on implantation, estrous cycle, and hormonal modulation. The results showed that this combination significantly altered the duration of estrous phases, and reduced litter sizes, ovarian follicle counts, and implantation relative to the control animals.

- *Antimicrobial Activity*

*Andrographis paniculata* demonstrates significant antibacterial activity in many extracts and plant components. Methanolic leaf extracts at doses of 30–120 mg/mL showed notable effectiveness against *Escherichia coli* and *Salmonella typhi*, while exhibiting diminished efficiency against *Staphylococcus aureus* (Singh et al. 2017). Ethanolic leaf extracts at 300–500 μg/mL showed significant antibacterial action

(Molla et al. 2010), whereas ethanol extracts of aerial portions at 10–200 μg/mL similarly inhibited both Gram-positive and Gram-negative bacteria (Mishra et al. 2009). A 75% methanolic extract of leaves showed activity against methicillin-resistant *S. aureus*, *Enterococcus faecalis*, and *Mycobacterium tuberculosis* (Mishra et al. 2013). Furthermore, ethanol extracts from leaves and the active component andrographolide showed considerable antibacterial efficacy (Abubacker and Vasantha 2010). Whole-plant extracts utilizing dichloromethane, methanol, and aqueous solvents at concentrations of 250–1000 μg/disk showed significant antibacterial activity against both Gram-positive and Gram-negative microorganisms (Sule et al. 2010). As reported by Fatima et al. (2023), methanolic leaf extracts of Plant growth promoting bacteria (PGPB) eliciting *A. paniculata* demonstrated a significant bacterial log reduction, with *S. aureus* showing a reduction of 21.25% (0.187 log units) and *E. coli* showing 18.75% (0.09 log units), both notably different from the control. The aqueous whole-plant extract had notable effectiveness at 10 mg/disk (Singha et al. 2003), but hexane, chloroform, and methanol extracts at concentrations of 0.1–2.5 mg/mL resulted in full reduction of parasite proliferation (Rahman et al. 1999). Ethanolic leaf extracts at 50–150 μL showed significant antibacterial effectiveness (Rajalakshmi and Cathrine 2016), whereas methanol extracts from the entire plant, at 100–250 mg/mL, prevented the development of 95% of the tested organisms (Bobbarala et al. 2009). The leaf, stems, and roots of this herb exhibit potent antibacterial and antifungal activities. Andrographolide, at concentrations of 0.1–0.75 mg/mL, showed significant antibacterial activity, especially against Gram-positive bacteria (Shao et al. 2012; Banerjee et al. 2017). Chloroform extracts of the stem and root, evaluated at concentrations of 50 and 100 μg/mL, had significant antibacterial and antifungal properties (Radhika and Lakshmi 2010). Whole-plant extracts utilizing dichloromethane and methanol at a concentration of 50 μg/mL are employed as therapies for fungal dermal infections (Sule et al. 2012). Methanolic leaf extracts at concentrations between 1000 and 7000 ppm showed notable antifungal activity (Singh et al. 2000), while aqueous leaf extracts demonstrated efficacy against *Microsporum nanum*, *Curvularia* species, and *Fusarium dimerium* (Noorkorina and Noorfatiehah 2012). Young leaves at 1–20 μL exhibited antifungal properties against *Aspergillus flavus* and *Macrophomina phaseolina* (Ghosh et al. 2004).

- *Anti-Edema Activity*

The whole-plant extracts of this particular herb have shown anti-edema effects. At dosages between 20 and 200 mg/kg, both ethanol and aqueous extracts successfully decreased edema, with the aqueous extract having the most notable benefits. Furthermore, andrographolide showed significant anti-edema action when given at doses of 25, 50, and 100 mg/kg, confirming the plant's promise for treating inflammation (Lin et al. 2009).

- *Antioxidant Activity*

*Andrographis paniculata* exhibits strong antioxidant properties across various extracts, with leaves and whole-plant preparations frequently studied. Methanolic leaf extracts at 1 g/kg body weight demonstrated a significant reduction in malondialdehyde (MDA) levels and an increase in total antioxidant status in rat urine, indicating its effectiveness against oxidative stress (Akowuah et al. 2008). Andrographolide, at 10–20 μg/mL, was shown to inhibit the proliferation of 3 T3-L1 preadipocytes by suppressing GPX1 and reducing glutathione (GSH) depletion (Chen et al. 2016). In comparison, methanolic extracts were found to be more potent in antioxidant activity than aqueous and ethanol extracts (Sharma and Joshi 2011). Additional studies used CUPric Reducing Antioxidant Capacity (CUPRAC), Ferric Reducing Antioxidant Pow (FRAP), and 2,2-Diphenyl-1-picrylhydrazyl (DPPH) assays to confirm the antioxidant activity of methanolic extracts at 0.02–0.10 μmol/mL (Kurzawa et al. 2015). Ethanol and methanol extracts, as well as isolated andrographolide and dehydroandrographolide, showed substantial DPPH radical-scavenging and ferric reducing capacities (Zhao et al. 2014). Aqueous extracts of fresh leaves, tested at 200–500 μg/mL, displayed significant antioxidant effects (Sangeetha et al. 2014), while hydroalcoholic extracts of leaves and stems exhibited antioxidant activities at 50–100 mg/kg (Singh et al. 2001). The efficiency of chloroform in extracting andrographolide from aerial parts was enhanced at concentrations of 0.025–1000 mg/mL (Vasu et al. 2010), and methanolic extracts of the whole plant showed substantial antioxidant properties as well (Tewari et al. 2010).

- *Immunomodulatory Activity*

Andrographolide, derived from the aerial components of *Andrographis paniculata*, has exhibited immunomodulatory properties. Andrographolide at dosages of 10 and 100 μg/mL was observed to affect both innate and adaptive immune responses, chiefly by modulating macrophage phenotypic polarization. This indicates its possible function in regulating immunological responses to enhance overall immune health (Wang et al. 2010).

- *Hepatoprotective Activity*

*Andrographis paniculata* demonstrates hepatoprotective qualities, particularly evident in leaf and whole-plant extracts. The methanolic leaf extract, given at doses of 10 and 100 mg/kg, exhibited hepatoprotective properties (Sutha 2010). The ethanolic extract from leaves at 500 mg/kg showed notable antihepatotoxic efficacy, especially against paracetamol-induced hepatic injury in rats (Rajalakshmi et al. 2012). Additionally, andrographolide derived from the entire plant, at a concentration of 500 ppm/kg, exhibited antioxidant activity that activates enzymes capable of neutralizing oxidants, which is advantageous in instances of significant liver damage (Trivedi et al. 2007). Research indicates that the andrographolide concentration differs markedly among whole-plant extracts (Raina et al. 2013). Aqueous extracts from whole plants demonstrate hepatoprotective effects and may contribute to the retardation of liver tumor advancement (TRIVEDI and Trivedi and Rawal 1998). Andrographolide, delivered at doses of 25–50 mg/kg, demonstrates significant hepatoprotective effects (Maiti et al. 2010), whereas 70% ethanol extracts from aerial parts, given at 100–200 mg/kg, provide protection against paracetamol-induced liver toxicity (Nagalekshmi et al. 2011). The powdered leaves have safeguarded hepatic cells from paracetamol-induced damage (Reeta Johri and Johri 2010), and methanolic extracts at 100 mg/kg, which include andrographolide, demonstrated efficacy against liver toxicity (Handa and Sharma 1990). Whole-plant ethanol extracts at 200 mg/kg facilitated recovery from $CCl_4$-induced hepatic damage owing to their antioxidant and hepatoprotective attributes (Subramaniam et al. 2015).

- *Antiviral Activity*

*Andrographis paniculata* exhibits notable antiviral and antiretroviral characteristics. The ethanol extracts derived from the aerial parts at a concentration of 25 mg/mL exhibited promising properties as a treatment for Epstein-Barr virus, showing efficacy at concentrations below 5 mg/mL (Lin et al. 2008). Furthermore, methanol extracts from the aerial parts, at concentrations of 1–3 mg/mL, demonstrated an inhibitory effect on the activity of dengue virus serotype 1 (DENV-1) in vitro. Particular diterpenoid lactones, such as andrographolide, neoandrographolide, and 14-deoxy-11,12-didehydroandrographolide derived from dried leaves, exhibited viricidal properties against herpes simplex virus 1 (HSV-1) while showing minimal cytotoxicity at effective concentrations, underscoring their promise as safe antiviral agents (Tang et al. 2012).

## 24.6 Summary

Finally, *Andrographis paniculata* turns out to be a strong medicinal plant with a wide spectrum of bioactive elements offering therapeutic advantages in many spheres of medicine. Especially andrographolide and related diterpenoids, its extensive phytochemistry confers amazing anti-inflammatory, hepatoprotective, antioxidant, anticancer, antibacterial, and immunomodulatory properties. Apart from its therapeutic qualities, *A. paniculata* provides nutrients since various elements assist general cellular health and immune system. The plant's components not only enhance immune resilience and good liver function but also help avoid oxidative stress and microbial infections, therefore highlighting its possible uses both preventively and therapeutically. Complete realization of the health benefits of *A. paniculata* and its position in integrative healthcare practices worldwide depends on future studies concentrating on the standardizing of active chemicals and clinical trials.

## References


Abubacker M, Vasantha S (2010) Antibacterial activity of ethanolic leaf extract of *Andrographis paniculata* Nees (Acanthaceae) and its bioactive compound Andrographolide. Drug Invent Today 2:16.

Agrawal R, Agrawal N, Mishra D (2017) Chemopreventive and antiproliferative effect of *Andrographis Paniculata* extract. Med One 2:e170011

Ahmad HM, Fiaz S, Hafeez S et al (2022) Plant growth-promoting rhizobacteria eliminate the effect of drought stress in plants: a review. Front Plant Sci 13:875774

Akowuah GA, Zhari I, Mariam A (2008) Analysis of urinary andrographolides and antioxidant status after oral administration of *Andrographis paniculata* leaf extract in rats. Food Chem Toxicol 46:3616–3620. https://doi.org/10.1016/j.fct.2008.09.008

Arpini S, Fuzzati N, Giori A et al (2008) HPLC-DAD-MS fingerprint of *Andrographis paniculata* (Burn. f.) Nees (Acanthaceae). Nat Prod Commun 3:1934578X0800301208

Banerjee M, Parai D, Chattopadhyay S, Mukherjee SK (2017) Andrographolide: antibacterial activity against common bacteria of human health concern and possible mechanism of action. Folia Microbiol (Praha) 62:237–244

Benoy GK, Animesh DK, Mandal Aninda MA, Priyanka DK Halder Sandip HS (2012) An overviewon Andrographis paniculata (Burm. f.) Nees. Sci World J, 2014, 274905

Bobbarala V, Rao PK, Rao GS, Aryamithra D (2009) Bioactivity of *Andrographis paniculata* against selected phytopathogens. J Pharm Res 2:480–482

Chandra P, Kannujia R, Pandey R et al (2016) Rapid quantitative analysis of multi-components in *Andrographis paniculata* using UPLC-QqQLIT-MS/MS: application to soil sodicity and organic farming. Ind Crops Prod 83:423–430

Chao W-W, Lin B-F (2010) Isolation and identification of bioactive compounds in *Andrographis paniculata (*Chuanxinlian). Chin Med 5:1–15

Chen L, Qiu F, Wei H et al (2006) Nine new ent-Labdane Diterpenoids from the aerial parts of *Andrographis paniculata*. Helv Chim Acta 89:2654–2664

Chen L, Zhu H, Wang R et al (2008) ent-Labdane diterpenoid lactone stereoisomers from *Andrographis paniculata*. J Nat Prod 71:852–855

Chen L-X, He H, Xia G-Y et al (2014) A new flavonoid from the aerial parts of *Andrographis paniculata*. Nat Prod Res 28:138–143

Chen W, Su H, Feng L, Zheng X (2016) Andrographolide suppresses preadipocytes proliferation through glutathione antioxidant systems abrogation. Life Sci 156:21–29

Cheung H-Y, Cheung S-H, Li J et al (2005) Andrographolide isolated from *Andrographis paniculata* induces cell cycle arrest and mitochondrial-mediated apoptosis in human leukemic HL-60 cells. Planta Med 71:1106–1111

Dai Y, Chen S-R, Chai L et al (2019) Overview of pharmacological activities of *Andrographis paniculata* and its major compound andrographolide. Crit Rev Food Sci Nutr 59:S17–S29

Dua V, Ojha V, Roy R et al (2004) Anti-malarial activity of some xanthones isolated from the roots of *Andrographis paniculata*. J Ethnopharmacol 95:247–251

Elshafie HS, Camele I, Mohamed AA (2023) A comprehensive review on the biological, agricultural and pharmaceutical properties of secondary metabolites based-plant origin. Int J Mol Sci 24:3266

Fatima B, Nayak S, Singh A, Das P (2023) Antibacterial photodynamic inactivation with elicited *Andrographis paniculata* (Burm. f.) Wall. ex Nees plant extract derived carbonaceous nanoparticle. Phytomedicine Plus 4:100512. https://doi.org/10.1016/j.phyplu.2023.100512

Gao S, Wang D, Chai H et al (2019) Unusual ent-labdane diterpenoid dimers and their selective activation of TRPV channels. J Org Chem 84:13595–13603

Geethangili M, Rao YK, Fang S, Tzeng Y (2008) Cytotoxic constituents from *Andrographis paniculata* induce cell cycle arrest in jurkat cells. Phytother Res Int J Devoted Pharmacol Toxicol Eval Nat Prod Deriv 22:1336–1341

Ghosh M, Thangamani D, Thapliyal M et al (2004) Isolation of *Andrographis paniculata* leaf protein with antifungal property. Acta Phytopathol Entomol Hung 39:377–381

Gupta K, Taneja S, Dhar K, Atal C (1983) Flavonoids of *Andrographis paniculata*. Phytochemistry 22:314–315

Handa S, Sharma A (1990) Hepatoprotective activity of andrographolide from *Andrographis paniculata* against carbontetrachloride. Indian J Med Res 92:276–283

Hanh TTH, My NTT, Cham PT et al (2020) Diterpenoids and flavonoids from *Andrographis paniculata*. Chem Pharm Bull (Tokyo) 68:96–99

Hapuarachchi SD, Ali Z, Abe N et al (2013) Andrographidine G, a new flavone glucoside from *Andrographis paniculata*. Nat Prod Commun 8:1934578X1300800314

Hidalgo MA, Hancke JL, Bertoglio JC, Burgos RA (2013) Andrographolide a new potential drug for the long term treatment of rheumatoid arthritis disease. Innov Rheumatol 1:247–270

Hossain MA, Roy B, Ahmed K et al (2007) Antidiabetic activity of *Andrographis paniculata*. Dhaka Univ J Pharm Sci 6:15–20

Hossain MS, Urbi Z, Sule A, Rahman KMH (2014) *Andrographis paniculata* (Burm. f.) Wall. ex Nees: a review of ethnobotany, phytochemistry, and pharmacology. Sci World J 2014:274905. https://doi.org/10.1155/2014/274905

Hu X-Y, Wu R-H, Logue M et al (2017) *Andrographis paniculata* (Chuān Xīn Lián) for symptomatic relief of acute respiratory tract infections in adults and children: a systematic review and meta-analysis. PLoS One 12:e0181780

Jayakumar T, Hsieh CY, Lee JJ, Sheu JR (2013) Experimental and clinical pharmacology ofAndrographis paniculata and its major bioactive phytoconstituent andrographolide. Evidence-BasedComplementary and Alternative Medicine: 1, 846740

Kabir MH, Hasan N, Rahman MM et al (2014) A survey of medicinal plants used by the Deb barma clan of the Tripura tribe of Moulvibazar district, Bangladesh. J Ethnobiol Ethnomed 10:1–28

Kumar RA, Sridevi K, Kumar NV et al (2004) Anticancer and immunostimulatory compounds from *Andrographis paniculata*. J Ethnopharmacol 92:291–295

Kurzawa M, Filipiak-Szok A, Kłodzińska E, Szłyk E (2015) Determination of phytochemicals, antioxidant activity and total phenolic content in *Andrographis paniculata* using chromatographic methods. J Chromatogr B 995:101–106

Lim JCW, Chan TK, Ng DS et al (2012) Andrographolide and its analogues: versatile bioactive molecules for combating inflammation and cancer. Clin Exp Pharmacol Physiol 39:300–310

Lin T-P, Chen S-Y, Duh P-D et al (2008) Inhibition of the Epstein–Barr virus lytic cycle by andrographolide. Biol Pharm Bull 31:2018–2023

Lin F, Wu S, Lee S, Ng L (2009) Antioxidant, antioedema and analgesic activities of *Andrographis paniculata* extracts and their active constituent andrographolide. Phytother Res Int J Devoted Pharmacol Toxicol Eval Nat Prod Deriv 23:958–964

Maiti K, Mukherjee K, Murugan V et al (2010) Enhancing bioavailability and hepatoprotective activity of andrographolide from *Andrographis paniculata*, a well-known medicinal food, through its herbosome. J Sci Food Agric 90:43–51

Mandal SC, Dhara A, Maiti BC (2001) Studies on psychopharmacological activity of *Andrographis paniculata* extract. Phytother Res Int J Devoted Pharmacol Toxicol Eval Nat Prod Deriv 15:253–256

Mishra U, Mishra A, Kumari R et al (2009) Antibacterial activity of ethanol extract of *Andrographis paniculata*. Indian J Pharm Sci 71:436

Mishra PK, Singh RK, Gupta A et al (2013) Antibacterial activity of *Andrographis paniculata* (Burm. f.) wall ex Nees leaves against clinical pathogens. J Pharm Res 7:459–462. https://doi.org/10.1016/j.jopr.2013.05.009

Molla M, Ahsan M, Alam M, Haque M (2010) Antibacterial activity in the leaves of seven bitter medicinal plants of Bangladesh. J Biosci 18:128–133

Nagalekshmi R, Menon A, Chandrasekharan DK, Nair CKK (2011) Hepatoprotective activity of *Andrographis paniculata* and *Swertia chirayita*. Food Chem Toxicol 49:3367–3373

Niranjan Reddy V, Malla Reddy S, Ravikanth V et al (2005) A new bis-andrographolide ether from *Andrographis paniculata* nees and evaluation of anti-HIV activity. Nat Prod Res 19:223–230

Niranjan A, Tewari S, Lehri A (2010) Biological activities of Kalmegh (*Andrographis paniculata* Nees) and its active principles-a review. Indian J Nat Prod Resour 1:125–135

Noorkorina TT, Noorfatiehah TT (2012) Antifungal activity of *Andrographis paniculata*: an in vitro study. Int Med J 19:382–385

Pawar A, Rajalakshmi S, Mehta P et al (2016) Strategies for formulation development of andrographolide. RSC Adv 6:69282–69300

Radhıka P, Lakshmi KR (2010) Antimicrobial activity of the chloroform extracts of the root and the stem of *Andrographis paniculata* nees. Int Res J Microbiol 1:037–039

Rahman NNNA, Furuta T, Takane K, Mohd MA (1999) Antimalarial activity of extracts of Malaysian medicinal plants. J Ethnopharmacol 64:249–254

Raina AP, Gupta V, Sivaraj N, Dutta M (2013) *Andrographis paniculata* (Burm. f.) wall. ex Nees (kalmegh), a traditional hepatoprotective drug from India. Genet Resour Crop Evol 60:1181–1189

Rajalakshmi V, Cathrine L (2016) Phytochemical screening and antimicrobial activity of ethanolic extract of *Andrographis paniculata*. J Pharmacogn Phytochem 5:175–177

Rajalakshmi G, Jothi KA, Venkatesan R, Jegatheesan K (2012) Hepatoprotective activity of *Andrographis paniculata* on paracetamol induced liver damage in rats. J Pharm Res 5:2983–2986

Rao YK, Vimalamma G, Rao CV, Tzeng Y-M (2004) Flavonoids and andrographolides from *Andrographis paniculata*. Phytochemistry 65:2317–2321

Reddy MK, Reddy MV, Gunasekar D et al (2003) A flavone and an unusual 23-carbon terpenoid from *Andrographis paniculata*. Phytochemistry 62:1271–1275

Reddy NP, Reddy BA, Gunasekar D et al (2007) A novel acylated flavone glycoside from *Andrographis nallamalayana*. Nat Prod Commun 2:1934578X0700200512

Reeta Johri RJ, Johri P (2010) Hepatoprotective activity of *Andrographis paniculata* (Nees) leaves on paracetamol induced hepatic damage in rats. J Exp Zool 13:51

Reyes V, Martínez O, Hernández G (1923) National center for biotechnology information. Plant Breed Univ Autónoma Agrar Antonio Narro Calzada Antonio Narro

Reyes B, Bautista N, Tanquilut N et al (2006) Anti-diabetic potentials of *Momordica charantia* and *Andrographis paniculata* and their effects on estrous cyclicity of alloxan-induced diabetic rats. J Ethnopharmacol 105:196–200

Samtiya M, Aluko RE, Dhewa T, Moreno-Rojas JM (2021) Potential health benefits of plant food-derived bioactive components: an overview. Foods 10:839

Sangeetha S, Archit R, SathiaVelu A (2014) Phytochemical testing, antioxidant activity, HPTLC and FTIR analysis of antidiabetic plants *Nigella sativa*, *Eugenia jambolana*, *Andrographis paniculata* and *Gymnema sylvestre*. J Biotechnol 9:1–9

Shao Z-J, Zheng X-W, Feng T et al (2012) Andrographolide exerted its antimicrobial effects by upregulation of human β-defensin-2 induced through p38 MAPK and NF-κB pathway in human lung epithelial cells. Can J Physiol Pharmacol 90:647–653

Sharma M, Joshi S (2011) Comparison of anti-oxidant activity of *Andrographis paniculata* and *Tinospora cordifolia* leaves. J Curr Chem Pharm Sc 1:1–8

Sheeja K, Shihab P, Kuttan G (2006) Antioxidant and anti-inflammatory activities of the plant *Andrographis paniculata* Nees. Immunopharmacol Immunotoxicol 28:129–140

Shen Y-H, Li R-T, Xiao W-L et al (2006) ent-Labdane diterpenoids from *Andrographis paniculata*. J Nat Prod 69:319–322

Shinde N, Chauhan AS, Gupta SK et al (2015) Antifertility studies of curcumin and andrographolide combination in female rats. Asian Pac J Reprod 4:188–194

Singh P, Singh U, Singh J (2000) Antifungal activity of methanolic extracts of Centella asiatica and *Andrographis paniculata*. Mycobiology 28:185–189

Singh RP, Banerjee S, Rao AR (2001) Modulatory influence of *Andrographis paniculata* on mouse hepatic and extrahepatic carcinogen metabolizing enzymes and antioxidant status. Phytother Res 15:382–390

Singh S, Mehta A, Baweja S et al (2013) Anticancer activity of *Andrographis paniculata* and Silybum marianum on five human cancer cell lines. J Pharmacol Toxicol 8(1):42–48

Singh A, Khan MMA, Sahu D et al (2017) Pharmacological and anti-bacterial activities of the leaves of *Andrographis paniculata* Nees. J Pharmacogn Phytochem 6:418–420
Singha PK, Roy S, Dey S (2003) Antimicrobial activity of *Andrographis paniculata*. Fitoterapia 74:692–694
Sivananthan M, Elamaran M (2013) Medicinal and pharmacological properties of *Andrographis paniculata*. Int J Biomol Biomed 3:1–12
Srivastava N, Akhila A (2010) Biosynthesis of andrographolide in *Andrographis paniculata*. Phytochemistry 71:1298–1304. https://doi.org/10.1016/j.phytochem.2010.05.022
Subramaniam S, Hedayathullah Khan HB, Elumalai N, Sudha Lakshmi SY (2015) Hepatoprotective effect of ethanolic extract of whole plant of *Andrographis paniculata* against $CCl_4$-induced hepatotoxicity in rats. Comp Clin Pathol 24:1245–1251
Subramanian R, Zaini Asmawi M, Sadikun A (2012) A bitter plant with a sweet future? A comprehensive review of an oriental medicinal plant: *Andrographis paniculata*. Phytochem Rev 11:39–75
Sule A, Ahmed Q, Samah O, Omar M (2010) Screening for antibacterial activity of *Andrographis paniculata* used in Malaysian folkloric medicine: a possible alternative for the treatment of skin infections. Ethnobot Leafl 2010:8
Sule A, Ahmed QU, Latip J et al (2012) Antifungal activity of *Andrographis paniculata* extracts and active principles against skin pathogenic fungal strains in vitro. Pharm Biol 50:850–856
Sun C-P, Yang Z-J, Zhao W-Y et al (2020) A novel 15-spiro diterpenoid dimer from *Andrographis paniculata* with inhibitory potential against human carboxylesterase 2. Bioorg Chem 97:103680
Sutha D (2010) A study on the hepatoprotective effect of *Andrographis paniculata* (Burm. F) Nees on mice. J Phytol 2. https://updatepublishing.com/journal/index.php/jp/article/view/2205
Tan M, Kuroyanagi M, Sulaiman S et al (2005) Cytotoxic activities of major diterpenoid constituents of *Andrographis paniculata*. In a panel of human tumor cell lines. Pharm Biol 43:501–508
Tang LI, Ling AP, Koh RY et al (2012) Screening of anti-dengue activity in methanolic extracts of medicinal plants. BMC Complement Altern Med 12:1–10
Tewari S, Niranjan A, Lehri A (2010) Variations in yield, quality, and antioxidant potential of kalmegh (*Andrographis paniculata* Nees) with soil alkalinity and season. Int J Geogr Inf Syst 16:41–50
Thokchom A, Talukdar A, Chingakham B et al (2018) Traditional usage of ethnomedicinal plants in treating liver disorders at Manipur, North East India. Eur J Med Plants 23:1–10
Tran QT, Tan WD, Wong WF, Chai CL (2021) Polypharmacology of andrographolide: beyond one molecule one target. Nat Prod Rep 38:682–692
Tripathi VPDK, Kaur D, Chauhan DK (2012) Nutritional evaluation of leaves of *Boerhaavia diffusa* L. and *Andrographis paniculata* (Burm. F.) wall. ex Nees: implications for nutraceutical applications. Int J Pharma BioSci 3(4):315–321
Trivedi N, Rawal U (1998) Effect of aqueous extract of *Andrographis paniculata* on liver tumor. Indian J Pharmacol 30:318–322
Trivedi NP, Rawal UM, Patel BP (2007) Hepatoprotective effect of andrographolide against hexachlorocyclohexane-induced oxidative injury. Integr Cancer Ther 6:271–280
Vasu S, Palaniyappan V, Badami S (2010) A novel microwave-assisted extraction for the isolation of andrographolide from *Andrographis paniculata* and its *in vitro* antioxidant activity. Nat Prod Res 24:1560–1567
Wang G-C, Wang Y, Williams ID et al (2009) Andrographolactone, a unique diterpene from *Andrographis paniculata*. Tetrahedron Lett 50:4824–4826. https://doi.org/10.1016/j.tetlet.2009.05.097
Wang W, Wang J, Dong S et al (2010) Immunomodulatory activity of andrographolide on macrophage activation and specific antibody response. Acta Pharmacol Sin 31:191–201
Wang C-H, Li W, Qiu R-X et al (2014) A new diterpenoid from the aerial parts of *Andrographis paniculata*. Nat Prod Commun 9:1934578X1400900105
Wu T-S, Chern H-J, Damu AG et al (2008) Flavonoids and ent-labdane diterpenoids from *Andrographis paniculata* and their antiplatelet aggregatory and vasorelaxing effects. J Asian Nat Prod Res 10:17–24

Xu C, Wang Z-T (2011) Chemical constituents from roots of *Andrographis paniculata*. Yao Xue Xue Bao 46:317–321

Xu C, Chou G-X, Wang C-H, Wang Z-T (2012) Rare noriridoids from the roots of *Andrographis paniculata*. Phytochemistry 77:275–279

Yang Y, Cao T, Guo Q (2020) Characterization of a traditional Chinese medicine plant the chloroplast genome of *Andrographis paniculata*. Mitochondr DNA Part B 5:1949 1951

Zhang X-F, Tan B-H (2000) Anti-diabetic property of ethanolic extract of *Andrographis paniculata* in streptozotocin-diabetic rats. Acta Pharmacol Sin 21:1157–1164

Zhang X, Wang G, Ye W et al (2006) New diterpenoids from *Andrographis paniculata* (Burm. f.) Nees. J Integr Plant Biol 48:1122–1125

Zhang Z, Jiang J, Yu P et al (2009) Hypoglycemic and beta cell protective effects of andrographolide analogue for diabetes treatment. J Transl Med 7:1–13

Zhao Y, Kao C-P, Wu K-C et al (2014) Chemical compositions, chromatographic fingerprints and antioxidant activities of *Andrographis* Herba. Molecules 19:18332–18350

Zhou K-L, Chen L-X, Zhuang Y-L et al (2008) Two new ent-labdane diterpenoid glycosides from the aerial parts of *Andrographis paniculata*. J Asian Nat Prod Res 10:939–943

Zoha M, Hussain A, Choudhury S (1989) Antifertility effect of *Andrographis paniculata* in mice. Bangladesh Med Res Counc Bull 15:34–37

# Medicinal and Nutritional Importance of *Mentha piperita* in Human Health

# 25

Ayman M. Al-Qaaneh, Mohammad A. Aburumman, Mounir M. Bekhit, Heba A. El-Nashar, Moulay H. Youssoufi, Suriya Rehman, and Shafiq Ur Rahman

**Abstract**

*Mentha piperita*, commonly known as peppermint, is a hybrid plant renowned for its therapeutic properties and rich phytochemical profile. This chapter explores the medicinal and nutritional applications of peppermint, supported by an extensive body of scientific literature. Peppermint's primary bioactive constituents—menthol, menthone, and rosmarinic acid—contribute to its diverse health benefits, which include antioxidant, anti-inflammatory, antimicrobial, and

A. M. Al-Qaaneh (✉)
Faculty of Allied Medical Sciences, Al-Balqa Applied University (BAU), Al-Salt, Jordan

Department of Pharmaceutical Technology, Faculty of Pharmacy, Jordan University of Science and Technology (JUST), Irbid, Jordan
e-mail: ayman.alqennh@bau.edu.jo

M. A. Aburumman
Department of Nursing, Faculty of Nursing, Al-Balqa Applied University (BAU), Al-Salt, Jordan

M. M. Bekhit
Department of Pharmaceutics, College of Pharmacy, King Saud University, Riyadh, Saudi Arabia

H. A. El-Nashar
Department of Pharmacognosy, Faculty of Pharmacy, Ain Shams University, Cairo, Egypt

M. H. Youssoufi
Laboratory of Applied Chemistry & Environment, Faculty of Sciences, Mohammed First University, Oujda, Morocco

S. Rehman
Department of Epidemic Diseases Research, Institute for Research and Medical Consultation, Imam Abdulrahman Bin Faisal University, Dammam, Saudi Arabia

S. U. Rahman
Department of Pharmacy, Shaheed Benazir Bhutto University, Sheringal, Khyber Pakhtunkhwa, Pakistan



M. A. Ansari et al. (eds.), *Medicinal Plants and their Bioactive Compounds in Human Health: Volume 2*, https://doi.org/10.1007/978-981-95-3620-7_25

analgesic effects. These compounds interact with various biological pathways, providing relief from gastrointestinal, respiratory, and dermatological conditions.

The nutritional profile of peppermint reveals its contributions to essential vitamin and mineral intake, offering vitamins A and C, calcium, magnesium, potassium, and dietary fiber. These nutrients support immune function, bone health, and digestive wellness, highlighting peppermint's role as a functional food in health promotion and disease prevention. Clinical trials have demonstrated peppermint's efficacy in managing irritable bowel syndrome (IBS), respiratory conditions, cognitive performance, and skin disorders, affirming its place in both traditional and modern medicine.

In modern therapeutic applications, peppermint is gaining attention for its potential cardiovascular benefits, cognitive enhancement, and immune-modulatory effects. With a favorable safety profile and well-defined dosing guidelines, peppermint serves as a safe and effective complementary therapy. Future research directions include exploring its pharmacokinetics, optimizing delivery systems, and investigating emerging applications such as antiviral and mental health support. This chapter underscores peppermint's significance as a versatile herb, bridging traditional practices and evidence-based healthcare, with promising implications for integrative medicine and preventative health strategies.

**Keywords**

*Mentha piperita* · Phytochemicals · Integrative medicine · Essential oils · Nutritional value

## 25.1 Introduction

*Mentha piperita*, commonly known as peppermint, is a hybrid plant derived from a cross between *Mentha spicata* (spearmint) and *Mentha aquatica* (water mint), with a long history of use in both traditional and modern medicine (Gholamipourfard et al. 2021). This versatile plant, recognizable by its refreshing aroma and cooling sensation, has been utilized globally for centuries to treat ailments ranging from digestive discomfort to respiratory infections, skin conditions, and headaches (Ben-Arye et al. 2011). Its use has transcended cultures and centuries, solidifying peppermint's role as one of the most widely recognized and utilized medicinal plants (Mahendran and Rahman 2020; Sargin 2021).

In contemporary pharmacognosy, the study of drugs derived from natural sources, peppermint's bioactive compounds—particularly menthol and menthone—is known to interact with multiple biological system. These interactions have diverse physiological effects, from analgesic (pain-relieving) to antimicrobial and anti-inflammatory properties (Pergolizzi Jr. et al. 2018; Zhao et al. 2022). Menthol, the most studied component, is responsible for activating cold-sensitive TRPM8 receptors, which send cooling signals to the brain, making it a popular ingredient in topical analgesics and cough drops (Yin and Lee 2020). Additionally, the plant's polyphenolic profile, rich in compounds like rosmarinic acid, luteolin, and apigenin, provides robust antioxidant effects that counteract oxidative stress, a known contributor to various chronic diseases (Aldoghachi et al. 2021).

The scientific interest in peppermint's therapeutic properties has only intensified over the past decade. An increasing body of clinical evidence has examined peppermint's efficacy in addressing digestive health, respiratory conditions, mental alertness, and skin disorders (Kazemi et al. 2024; Lv et al. 2022; Mahendran and Rahman 2020). Recent advancements in analytical chemistry have facilitated a deeper understanding of peppermint's phytochemical complexity, enabling researchers to identify more precise therapeutic applications (Mahendran and Rahman 2020). Given the demand for natural and integrative health solutions, peppermint is well-suited to meet the growing need for plant-based medicine in modern healthcare (Kazemi et al. 2025).

In this chapter, we examine *Mentha piperita* from multiple perspectives, exploring its phytochemical composition, nutritional value, and diverse medicinal applications. By delving into clinical studies, biochemical mechanisms, and ethnobotanical history, this chapter aims to provide a comprehensive overview of peppermint as a valuable agent in health promotion, disease prevention, and symptom management. This chapter also explores future research directions that hold the potential to further enhance peppermint's utility in integrative healthcare.

## 25.2 Phytochemical Profile of *Mentha piperita*

Peppermint is composed of a wide array of bioactive compounds, each of which contributes uniquely to its therapeutic properties (Alexa et al. 2018). These compounds include essential oils, polyphenols, flavonoids, and other secondary metabolites that interact synergistically to produce a range of physiological effects (Marwa et al. 2017). The primary components in peppermint oil—menthol and menthone—are largely responsible for its analgesic and antimicrobial activities (Iftinca et al. 2020). However, other compounds such as rosmarinic acid, menthyl acetate, and pulegone contribute additional antioxidant and anti-inflammatory properties, enriching peppermint's medicinal potential (Mucciarelli et al. 2001).

### 25.2.1 Essential Oils and Major Constituents

The essential oil of peppermint contains a concentrated blend of terpenoids, notably menthol (30–50%) and menthone (10–30%) (Golestannejad et al. 2018). Menthol is peppermint's most prominent and bioactive compound, responsible for its cooling sensation and a significant portion of its therapeutic effects. When applied topically, menthol interacts with the TRPM8 receptor, a cold-sensitive receptor found on sensory neurons, which triggers a cooling sensation that alleviates pain (Trif et al. 2024). Studies indicate that menthol's ability to activate these receptors provides both analgesic and antispasmodic effects, making it highly effective in treating headaches, muscle soreness, and digestive cramps (Tayarani-Najaran et al. 2013).

Menthone, the second major constituent, adds to peppermint's efficacy by enhancing the cooling effect of menthol through synergistic action (Leffingwell 2009). It also possesses antimicrobial properties, particularly against pathogens like *Escherichia coli* and *Staphylococcus aureus*, which contribute to its value in treating gastrointestinal infections (Al-Khattaf et al. 2021).

Other essential oils present in smaller quantities include menthyl acetate (3–10%), which contributes to peppermint's aroma, and pulegone, which is known for its antimicrobial properties but must be carefully regulated due to potential toxicity (Scavroni et al. 2005). The essential oil profile of peppermint varies based on factors such as soil quality, climate, and harvesting methods, which can influence both the concentration and composition of these bioactive compounds (Hudz et al. 2023) (Table 25.1).

## 25.2.2 Polyphenols, Flavonoids, and Other Bioactive Compounds

Polyphenolic compounds in peppermint, such as rosmarinic acid, luteolin, and apigenin, are integral to its antioxidant and anti-inflammatory effects (Riachi and De Maria 2015). These compounds play a critical role in neutralizing free radicals—unstable molecules that contribute to cellular damage, aging, and disease progression (McKay and Blumberg 2006). Rosmarinic acid, for instance, has been shown to inhibit the production of reactive oxygen species (ROS) and enhance the activity of antioxidant enzymes such as superoxide dismutase (SOD) and catalase (Adomako-Bonsu et al. 2017). This compound has demonstrated protective effects in models of oxidative stress-induced diseases, including cardiovascular and neurodegenerative disorders (Shamsi et al. 2020).

**Table 25.1** Major phytochemical constituents of *Mentha piperita* essential oil

| Compound | Functional groups | Concentration (%) | Biological activity | References |
|---|---|---|---|---|
| Menthol | Hydroxyl (–OH) | 30–50 | Analgesic, antispasmodic, cooling effect | de Sousa et al. (2010) and Hilfiger et al. (2021) |
| Menthone | Carbonyl (C=O) | 10–30 | Antimicrobial, mild analgesic | Al-Khattaf et al. (2021) |
| Menthyl acetate | Ester (–COO–) | 3–10 | Aromatic, soothing | Afkar and Karimzadeh (2024) and Scavroni et al. (2005) |
| Pulegone | Ketone (C=O) | <5 | Aromatic, antimicrobial; toxic in high doses | Hilfiger et al. (2021) |
| Carvone | Ketone (C=O) | <1 | Antimicrobial, flavoring agent | Piasecki et al. (2023) |

Flavonoids like luteolin and apigenin further add to peppermint's therapeutic profile. Luteolin has garnered attention for its neuroprotective effects, as it is capable of crossing the blood–brain barrier and reducing inflammation in the central nervous system. This property is particularly relevant for the management of neurodegenerative diseases such as Alzheimer's and Parkinson's disease. In vitro studies have shown that luteolin can inhibit key enzymes in neuroinflammatory pathways, potentially slowing down neuronal damage associated with these conditions (Azhar et al. 2023; He et al. 2023; Hudz et al. 2023).

The broad spectrum of bioactive compounds in peppermint reflects its complexity as a therapeutic agent. Each constituent contributes uniquely, yet synergistically, to peppermint's effects, supporting its use in diverse health applications (Rosato et al. 2018). This phytochemical complexity also suggests that peppermint may exhibit enhanced efficacy when used as a whole extract rather than isolated compounds, as the constituents may work together to amplify therapeutic effects (Jeyakumar et al. 2011; Singh et al. 2015).

### 25.2.3 Terpenoids and Minor Compounds in Peppermint's Pharmacology

In addition to major constituents, peppermint contains a range of terpenoids and minor compounds that contribute to its pharmacological profile (Padalia et al. 2011). Terpenoids, a class of naturally occurring organic chemicals derived from terpenes, are known for their antimicrobial and anti-inflammatory properties (Ge et al. 2022). In peppermint, terpenoids like limonene and pinene have demonstrated efficacy against various pathogens and may support respiratory health by reducing inflammation in the airways (Kennedy et al. 2018).

**Limonene:** Found in trace amounts in peppermint, limonene possesses anti-inflammatory and antioxidant properties (Camele et al. 2021). It is also thought to have mood-enhancing effects, as it modulates serotonin and dopamine levels in the brain, which may help alleviate symptoms of mild depression and anxiety (de Almeida et al. 2017).

**β-Pinene:** This terpenoid has shown antibacterial activity against Gram-positive bacteria and contributes to peppermint's overall antimicrobial action. Pinene also has bronchodilatory effects, making it useful in conditions involving respiratory congestion and bronchoconstriction, such as asthma (Buleandra et al. 2016).

### 25.2.4 Factors Affecting Phytochemical Composition and Potency

The phytochemical composition and potency of peppermint are influenced by several factors, including geographic origin, growth conditions, and processing methods (Al-Mijalli et al. 2022). Studies have shown that peppermint grown in regions with higher altitudes tends to have a higher concentration of menthol, enhancing its analgesic and cooling effects (Del Rosario Cappellari et al. 2020). Soil composition, water availability, and sunlight exposure also play crucial roles in the biosynthesis of peppermint's essential oils, polyphenols, and other bioactive compounds (Walker and Melin 1996).

Harvesting time is another key factor; peppermint harvested during the flowering stage typically has a higher yield of essential oil (Rohloff et al. 2005). Drying methods, storage conditions, and extraction techniques further impact the final chemical profile of peppermint products (Zheljazkov et al. 2010). Steam distillation is the preferred method for extracting peppermint oil, as it preserves the delicate volatile compounds better than other methods such as cold pressing or solvent extraction (Radivojac et al. 2021).

Understanding these factors allows manufacturers to optimize growing and processing techniques, resulting in peppermint products with consistent quality and potency, which is particularly important in therapeutic applications.

## 25.3 Nutritional Value of *Mentha piperita*

In addition to its essential oils and polyphenolic compounds, peppermint leaves offer valuable nutrients that support general health and enhance its therapeutic effects (Saqib et al. 2022). Consumed as a tea, dietary supplement, or culinary ingredient, peppermint provides essential vitamins, minerals, and dietary fiber that contribute to immune function, bone health, and digestive wellness (Adel et al. 2015, 2024; Tufarelli et al. 2023; Uribe et al. 2016) (Table 25.2).

**Table 25.2** Nutritional composition of *Mentha piperita* per 100 g I

| Nutrient | Amount per 100 g | Health benefits | References |
|---|---|---|---|
| Vitamin A | 2124 IU | Vision, immune function | de Almeida-Muradian et al. (1998) |
| Vitamin C | 31.8 mg | Antioxidant, skin health | Riachi and De Maria (2015) |
| Calcium | 243 mg | Bone health, muscle function | Raczuk et al. (2008) |
| Potassium | 569 mg | Blood pressure regulation, heart health | Anyaoku (2023) |
| Magnesium | 80 mg | Muscle function, energy metabolism | Lemjallad et al. (2019) |
| Fiber | 8 g | Digestive health, blood sugar regulation | Abdel-Wareth et al. (2019) |

### 25.3.1 Vitamins and Their Health Benefits

Peppermint leaves are a natural source of vitamins A, C, and several B-complex vitamins, all of which play crucial roles in human health (Anyaoku 2023). Vitamin A is essential for maintaining vision, immune function, and skin health. It contributes to the production of retinal pigments necessary for low-light vision, enabling the eyes to adapt to dim lighting conditions (Sajovic et al. 2022). Additionally, vitamin A strengthens epithelial barriers, which are part of the immune system's first line of defense, protecting against pathogens and reducing the risk of infections (de Almeida-Muradian et al. 1998).

Vitamin C, a potent antioxidant, is crucial for immune support and skin integrity (Carr and Maggini 2017). It is involved in collagen synthesis, a process essential for maintaining the skin's structural integrity, elasticity, and wound healing capabilities. Collagen is also a significant component of blood vessels, bones, and connective tissues (Ghahremani-Nasab et al. 2023). Vitamin C's role in immune function is well-documented, as it enhances the activity of white blood cells, promoting an effective defense against viral and bacterial infections (Carr and Maggini 2017). Studies suggest that a sufficient intake of vitamin C can reduce the duration and severity of common colds and improve recovery rates in patients with respiratory infections (Hemilä and Chalker 2023, 2025; Riachi and De Maria 2015).

Peppermint also contains B-complex vitamins such as folate, riboflavin, and niacin (Anyaoku 2023). Folate is particularly important for DNA synthesis and cell division, making it crucial for fetal development during pregnancy (Seyoum Tola 2024). Riboflavin and niacin support cellular energy production by participating in metabolic processes that convert food into usable energy (Gonçalves and Portari 2021). Additionally, riboflavin contributes to skin health, while niacin supports DNA repair and cell signaling, making these vitamins integral to overall wellness (Bellassoued et al. 2018; Mahanom Jr. et al. 1999).

### 25.3.2 Mineral Content and Health Implications

The mineral content in peppermint, including calcium, magnesium, potassium, and iron, contributes to various physiological functions (Arzani et al. 2007). Calcium is essential for bone density and strength, muscle contraction, and blood clotting. Calcium intake is particularly important during childhood, adolescence, and older adulthood to prevent osteoporosis and ensure healthy bone growth. Peppermint provides a natural source of calcium, complementing dietary sources of this essential mineral (Raczuk et al. 2008).

Magnesium, another mineral found in peppermint, is involved in over 300 enzymatic reactions in the body. It supports muscle relaxation, nerve function, and the synthesis of proteins and DNA (Souza et al. 2023). Magnesium is also essential for energy production, as it plays a role in the formation of ATP, the molecule that provides energy to cells. Deficiency in magnesium is associated with muscle cramps, fatigue, and even mood disorders (Lemjallad et al. 2019).

Potassium, a key electrolyte, helps regulate fluid balance, nerve transmission, and muscle contractions. It works with sodium to maintain proper blood pressure levels, contributing to cardiovascular health (Anyaoku 2023). Iron, which is also present in peppermint, is critical for oxygen transport in the blood. Iron deficiency can lead to anemia, a condition characterized by fatigue, weakness, and cognitive impairment. Peppermint's iron content, while modest, can support dietary iron intake and help prevent iron-deficiency anemia in vulnerable populations, such as children, pregnant women, and older adults (Anyaoku 2023).

### 25.3.3 Fiber and Digestive Health

Peppermint leaves are a rich source of dietary fiber, which provides significant digestive health benefits (Abdel-Wareth et al. 2019). Fiber promotes regular bowel movements, prevents constipation, and supports the smooth passage of waste through the digestive tract (Gill et al. 2021). Dietary fiber also acts as a prebiotic, nourishing the beneficial bacteria in the gut, which play essential roles in nutrient absorption, immune function, and even mental health through the gut–brain axis (Martin et al. 2023; Valdes et al. 2018). Studies indicate that a high-fiber diet is associated with a lower incidence of colorectal cancer and other digestive diseases, reinforcing peppermint's value as a functional food (Anyaoku 2023; Halmos 2017; Spadaccino et al. 2021).

Fiber's ability to stabilize blood sugar levels and reduce cholesterol also contributes to metabolic health. Soluble fiber, in particular, binds with cholesterol in the intestines and reduces its absorption, leading to improved cardiovascular health (Aller et al. 2004). Furthermore, fiber supports weight management by promoting satiety and reducing overall calorie intake, making peppermint a beneficial addition to weight-conscious diets (Di Ciaula et al. 2019).

## 25.4 Medicinal Benefits of *Mentha piperita*

The medicinal benefits of peppermint are rooted in its complex phytochemical composition, which includes essential oils, polyphenols, flavonoids, and other bioactive compounds. Each of these constituents interacts with specific biological pathways to produce a range of therapeutic effects, from antioxidant and anti-inflammatory actions to antimicrobial and analgesic effects. This section explores the key medicinal benefits of peppermint, supported by clinical evidence and biochemical mechanisms.

### 25.4.1 Antioxidant Properties

Peppermint's antioxidant effects are attributed primarily to its polyphenolic compounds, such as rosmarinic acid, luteolin, and apigenin. Antioxidants play a vital role in neutralizing free radicals—unstable molecules that damage cells and contribute to chronic diseases. Oxidative stress, caused by an imbalance between free radicals and antioxidants, is linked to aging and the development of diseases like cardiovascular disease, cancer, and neurodegenerative disorders (Mahendran and Rahman 2020; McKay and Blumberg 2006).

Rosmarinic acid, a key antioxidant in peppermint, scavenges free radicals and prevents lipid peroxidation, a process that can damage cell membranes. Studies have shown that rosmarinic acid enhances the activity of endogenous antioxidant enzymes, such as superoxide dismutase (SOD) and catalase, which protect cells from oxidative damage. In animal studies, rosmarinic acid has demonstrated protective effects against oxidative stress-induced tissue damage in the liver, kidneys, and brain, highlighting its potential as a preventive agent in chronic diseases (Aldoghachi et al. 2021; Diniz do Nascimento et al. 2020; Silva et al. 2022).

### 25.4.2 Anti-inflammatory Effects

The anti-inflammatory effects of peppermint are mediated primarily by rosmarinic acid, which inhibits key inflammatory pathways. Specifically, rosmarinic acid downregulates the activity of cyclooxygenase (COX) and lipoxygenase (LOX) enzymes, reducing the production of pro-inflammatory cytokines such as tumor necrosis factor-alpha (TNF-α) and interleukin-6 (IL-6). By suppressing these inflammatory mediators, peppermint may help alleviate symptoms in conditions characterized by chronic inflammation, such as arthritis, inflammatory bowel disease, and autoimmune disorders (Chumpitazi et al. 2018; Li et al. 2017; Mahendran and Rahman 2020).

In an animal study investigating the anti-inflammatory effects of peppermint, topical application of peppermint oil was found to reduce inflammation in models of arthritis and colitis (Zhao et al. 2023). These findings suggest that peppermint could serve as a natural anti-inflammatory agent for managing pain and inflammation in both musculoskeletal and gastrointestinal disorders. Given the side effects associated with long-term use of conventional anti-inflammatory drugs, peppermint offers a promising alternative with a favorable safety profile (Xia et al. 2021; Zaia et al. 2016).

### 25.4.3 Gastrointestinal Health

Peppermint oil has been extensively studied for its effectiveness in treating gastrointestinal conditions, particularly irritable bowel syndrome (IBS) (Cappello et al. 2007). Its ability to relax smooth muscles in the gastrointestinal tract and reduce spasms makes it effective for alleviating symptoms like abdominal pain, bloating, and discomfort (Hills and Aaronson 1991). The enteric-coated form of peppermint oil is designed to bypass the stomach and release directly into the intestines, maximizing its therapeutic effect for gastrointestinal issues (Weerts et al. 2020). A meta-analysis reviewing 12 randomized controlled trials (RCTs) involving over 800 IBS patients found that enteric-coated peppermint oil capsules significantly reduced abdominal pain, bloating, and flatulence. Patients who received peppermint oil reported up to a 50% reduction in symptom severity compared to those who took a placebo. The effects are attributed to peppermint's ability to relax the smooth muscle in the digestive tract, reducing spasticity and promoting digestive comfort (Alammar et al. 2019).

### 25.4.4 Respiratory Health

Peppermint has long been used to alleviate respiratory symptoms, due to menthol's cooling and decongestant properties (Eccles 1994). Menthol interacts with cold-sensitive TRPM8 receptors in the nasal passages and airways, providing a sensation of improved airflow and reduced congestion (Stinson et al. 2023). This mechanism makes peppermint a popular choice for respiratory relief, particularly in conditions involving nasal congestion and bronchial irritation. A randomized controlled trial found that oral peppermint oil significantly improved respiratory function ($FVC_1$, PEF, PIF), grip strength, and jump performance within minutes of ingestion. Neuromotor responses and physiological parameters also improved. These effects are likely due to peppermint's bronchodilatory and possible neuromuscular-modulating properties (Meamarbashi 2014). Another randomized, double-blind, placebo-controlled clinical trial investigated the effects of an essential oil spray containing peppermint (Mentha × piperita), eucalyptus (Eucalyptus globulus), lemon (Citrus limon), lavender (Lavandula angustifolia), and tea tree (Melaleuca alternifolia) oils on symptoms of upper respiratory tract infections. Participants experiencing symptoms such as sore throat, hoarseness, or cough applied the spray five times daily for 3 days. The study found that the essential oil spray significantly alleviated sore throat and fatigue compared to placebo, suggesting its efficacy in relieving symptoms associated with upper respiratory tract infections (Mammari et al. 2023).

### 25.4.5 Cognitive Benefits

Recent research has explored peppermint's cognitive-enhancing effects, particularly its ability to improve alertness, memory, and overall mental performance (Moss et al. 2008a). This property is largely attributed to menthol, which acts as a mild central nervous system stimulant (Umezu et al. 2021). Studies suggest that peppermint's stimulating effects on the brain may improve cognitive functions, making it a potential natural aid for tasks requiring concentration and mental clarity (Moss et al. 2008b).

A double-blind, placebo-controlled crossover study evaluated the cognitive effects of peppermint essential oil (rich in menthol/menthone) in 24 healthy adults. Following a single oral dose (50 μL or 100 μL), the 100 μL dose significantly improved performance on the Rapid Visual Information Processing task at 1 and 3 h post-dose. Both doses reduced mental fatigue and enhanced performance on serial subtraction tasks. These cognitive benefits are likely linked to the oil's cholinergic inhibition, calcium regulation, and GABA<sub>A</sub>/nicotinic receptor-binding properties (Kennedy et al. 2018). In another A randomized, placebo-controlled study in 25 adults investigated the cognitive and cerebrovascular effects of 200 mL peppermint tea. The peppermint group showed greater improvements in cognitive performance and cerebral blood flow (as measured by NIRS) compared to placebo. However, increased cerebral blood flow did not mediate the cognitive benefits, suggesting other mechanisms may underlie peppermint's nootropic effects (Netzler and Lovell 2025).

### 25.4.6 Skin Health and Dermatological Applications

Peppermint's antimicrobial and anti-inflammatory properties make it a popular remedy for skin conditions, particularly those involving itching, inflammation, and minor infections (Unalan et al. 2019). The cooling effect of menthol, combined with peppermint's ability to inhibit bacterial growth, makes it beneficial for managing conditions such as acne, pruritus (itching), and mild inflammatory skin disorders (Oliveira et al. 2023). A randomized controlled study evaluated the efficacy of topical peppermint oil in 50 patients with chronic pruritus of hepatic, renal, or diabetic origin. Patients applying peppermint oil twice daily for 2 weeks showed significant improvement in all domains of the 5-D itch scale compared to the placebo group using petrolatum ($p < 0.05$). The study concluded that peppermint oil is an effective, safe, and well-tolerated topical treatment for chronic pruritus (Elsaie et al. 2016).

These dermatological studies highlight peppermint's potential in natural skincare, particularly for conditions where inflammation, bacterial overgrowth, or discomfort is present. Peppermint oil's compatibility with other skincare ingredients also makes it a valuable addition to a variety of skincare formulations, from creams and lotions to cleansers and toners (Kashyap et al. 2022; Skalicka-Woźniak and Walasek 2014).

### 25.4.7 Cardiovascular Health and Blood Pressure Regulation

Recent research has explored peppermint's potential benefits in cardiovascular health, especially regarding its effects on blood pressure regulation and stress-related hypertension (Cam et al. 2020; Meamarbashi 2014; Nematolahi et al. 2024). Cardiovascular diseases are a leading cause of morbidity and mortality worldwide, and oxidative stress is a contributing factor in the development of hypertension, atherosclerosis, and other heart-related conditions. Peppermint's antioxidant properties, attributed to its polyphenols like rosmarinic acid and luteolin, may help reduce oxidative damage within the vascular system (Al-Qaaneh et al. 2022b; Cam et al. 2020; Saqib et al. 2022). In a 21-day randomized study on 50 male Wistar rats, various doses of peppermint alcoholic extract (75–600 mg/kg) were evaluated for effects on body weight and blood biochemistry. Rats receiving 150 mg/kg or more showed significant weight gain and reductions in serum glucose, triglycerides, total cholesterol, and low-density lipoprotein (LDL) compared to controls ($p < 0.05$). The study concludes that peppermint extract positively influences metabolic parameters and may have therapeutic potential for managing dyslipidemia and hyperglycemia (Mesbahzadeh et al. 2015). Furthermore, A recent systematic review and meta-analysis of seven randomized controlled trials evaluated the effects of Mentha consumption on blood pressure. While overall changes in systolic and diastolic blood pressure were not statistically significant, subgroup analyses revealed clinically meaningful reductions among individuals over 30 years of age and those with elevated baseline blood pressure. These findings suggest that Mentha may have potential as an adjunct intervention for pre-hypertensive or hypertensive populations, warranting further investigation through well-designed clinical trials (Nematolahi et al. 2024).

### 25.4.8 Antimicrobial Applications in Dentistry

Peppermint's antimicrobial properties make it a valuable agent in dental care (Dagli et al. 2015). Oral bacteria such as *Streptococcus mutans*, which contributes to plaque formation and dental caries, are effectively reduced by peppermint oil (Fayed 2019). In a clinical trial, peppermint mouthwash reduced *S. mutans* levels by up to 60%, suggesting that it could serve as an alternative to chemical mouthwashes (Doddawad et al. 2023). The refreshing scent and natural antibacterial properties of peppermint also make it an effective remedy for halitosis (bad breath), as it helps neutralize odor-causing bacteria (Golestannejad et al. 2018; Quintana Soares Lopes et al. 2024). Peppermint has thus become a popular ingredient in various oral hygiene products, including toothpastes, mouthwashes, and dental floss. Its benefits in dental care extend beyond freshening breath to potentially preventing periodontal diseases by controlling microbial populations in the oral cavity (Thosar et al. 2013).

### 25.4.9 Pain Management and Analgesic Applications

Menthol's analgesic properties are widely used in pain management (Behm et al. 2022). Menthol activates the cold-sensitive TRPM8 receptors in the skin, providing a cooling sensation that helps alleviate pain. This cooling effect is beneficial for treating musculoskeletal pain, tension headaches, and joint discomfort (DeGeorge et al. 2019). Clinical studies have shown that menthol-based topical preparations provide pain relief comparable to that of certain over-the-counter analgesics, making peppermint a valuable option for individuals seeking nonpharmaceutical pain relief (Toumia et al. 2025).

## 25.5 Safety, Toxicology, and Dosage

Peppermint oil and peppermint-derived products are generally recognized as safe for most people at recommended dosages (Fuchs et al. 1999). However, some safety concerns arise when peppermint is used in concentrated forms or at high doses, particularly due to compounds like pulegone and menthofuran, which have shown potential hepatotoxicity in animal studies (Hilfiger et al. 2021; Mucciarelli et al. 2001).

### 25.5.1 Recommended Dosage for Therapeutic Use

For adults, the typical dosage of peppermint oil ranges from 0.2 to 0.4 mL per day, especially when used in enteric-coated capsules designed for gastrointestinal relief (Liu et al. 1997). Enteric-coated forms prevent the oil from being released in the stomach, reducing the risk of irritation or reflux (Liu et al. 1997). For respiratory and cognitive benefits, lower doses via inhalation or aromatherapy are generally effective and well-tolerated (Nishi et al. 2025). Lower doses are recommended for children, and peppermint oil should be used with caution in infants and young children due to the risk of respiratory distress from menthol exposure (Valussi et al. 2021).

### 25.5.2 Safety Concerns with High Dosages

High doses of peppermint oil can cause adverse effects, including gastrointestinal symptoms such as nausea, heartburn, and stomach upset (Zhou et al. 2004). The compound pulegone, present in small amounts in peppermint, has shown hepatotoxic potential in high concentrations (Ali et al. 2025). As a result, peppermint oil products are regulated to limit pulegone content, ensuring that doses remain within safe limits (Ali et al. 2025). Additionally, excessive consumption of menthol has been associated with side effects such as dizziness, nausea, and even ataxia in extreme cases (Chumpitazi et al. 2018).

### 25.5.3 Contraindications and Cautions

Individuals with gastroesophageal reflux disease (GERD) should use peppermint oil cautiously, as it can relax the lower esophageal sphincter, potentially worsening reflux symptoms (Jarosz and Taraszewska 2014). Peppermint oil should also be avoided in individuals with significant liver or kidney disease, as their bodies may have difficulty metabolizing and excreting the compounds effectively (Marjani et al. 2012). Pregnant and breastfeeding women are advised to consult a healthcare provider before using peppermint products due to limited research on its effects during pregnancy and lactation (Terzioglu Bebitoglu 2020).

## 25.6 Future Research Directions

Despite the significant body of evidence supporting peppermint's health benefits, further research is necessary to fully understand its applications and maximize its therapeutic potential. Emerging research areas are focused on investigating peppermint's molecular mechanisms, optimizing dosage forms, and exploring its benefits in under-researched areas such as chronic pain, mental health, and immune modulation.

### 25.6.1 Immunomodulatory and Anti-inflammatory Mechanisms

Recent studies suggest that peppermint's polyphenols, particularly rosmarinic acid, may have immunomodulatory effects (Li et al. 2023). Understanding how peppermint interacts with immune cells and signaling pathways could reveal additional therapeutic applications for inflammatory and immune-related disorders. Research on peppermint's effects on cytokine production and inflammatory mediators may lead to new uses for autoimmune conditions, where immune regulation is critical for symptom management (Pelvan et al. 2022).

### 25.6.2 Pharmacokinetics, Bioavailability, and Controlled-Release Formulations

Understanding the pharmacokinetics and bioavailability of peppermint's active compounds, such as menthol, is essential for optimizing its therapeutic efficacy. Developing controlled-release formulations could improve the delivery and duration of peppermint's effects, particularly for chronic conditions that require sustained relief, like IBS (Grigoleit and Grigoleit 2005). Advanced delivery systems, including transdermal patches and slow-release oral capsules, could enhance patient compliance and make peppermint-based therapies more effective for long-term management (Zhao et al. 2022).

### 25.6.3 Peppermint's Potential in Mental Health and Cognitive Support

Peppermint's potential role in mental health and cognitive support is a promising area for future research. Preliminary studies have shown that peppermint aroma can improve focus, attention, and memory (Kennedy et al. 2018; Moss et al. 2008a; Netzler and Lovell 2025). Further research could investigate peppermint's effects on neurotransmitter pathways and its potential as a complementary therapy for cognitive disorders such as attention-deficit/hyperactivity disorder (ADHD), age-related cognitive decline, and mood disorders. Additionally, clinical trials involving populations with diagnosed cognitive impairments could provide deeper insights into peppermint's efficacy in mental health support (Deivanayagame et al. 2020).

### 25.6.4 Potential for Antiviral Applications

Peppermint's antiviral properties, particularly against respiratory viruses such as influenza and the herpes simplex virus, present an exciting avenue for future research (Schuhmacher et al. 2003). Laboratory studies suggest that peppermint oil may inhibit viral replication and provide complementary support during viral infections (Swarup et al. 2007). Given the ongoing global interest in natural antiviral agents, further studies could clarify peppermint's role in preventing or mitigating viral infections. If proven effective, peppermint could be incorporated into preventive strategies and integrated into antiviral treatments (Al-Qaaneh and Al-Ghamdi 2021; Al-Qaaneh et al. 2022a).

### 25.6.5 Advances in Combination Therapies Involving Peppermint

Combination therapies involving peppermint and other natural or pharmaceutical agents could enhance therapeutic efficacy and broaden its clinical applications. For instance, peppermint's anti-inflammatory and analgesic effects may complement the effects of nonsteroidal anti-inflammatory drugs (NSAIDs) in pain management, potentially reducing the required dosage of these medications and minimizing their side effects (Goudarzi et al. 2024). Research into synergistic interactions between peppermint and other botanicals could lead to more effective treatments for complex conditions such as arthritis, metabolic syndrome, and chronic inflammatory disorders (Goudarzi et al. 2024).

Future research pathways highlight peppermint's promise as a versatile therapeutic option. Gaining deeper pharmacological understanding, assessing long-term safety, and exploring broader clinical uses will support its inclusion in evidence-based medical practices and reinforce its role in contemporary healthcare.

## 25.7 Conclusion

*Mentha piperita*, or peppermint, is a multifaceted herb that serves as a bridge between traditional medicine and modern pharmacotherapy. Its rich phytochemical profile, primarily comprising menthol, menthone, and rosmarinic acid, offers a wide range of biological activities, including analgesic, antimicrobial, antioxidant, anti-inflammatory, and cognitive-enhancing effects. Peppermint's therapeutic benefits are well-supported by clinical trials, particularly in managing irritable bowel syndrome (IBS), respiratory congestion, cognitive enhancement, and skin health.

The scope of peppermint's applications continues to grow as emerging research reveals its potential in additional health areas, such as cardiovascular support, metabolic health, immune modulation, and mental health. Safety considerations, particularly with respect to dosage and contraindications, ensure that peppermint can be used responsibly within established guidelines.

With ongoing research into its pharmacodynamics, optimal dosing, and combination therapies, peppermint is poised to expand its role in preventive and therapeutic medicine. Future advances in delivery systems and controlled-release formulations will further enhance its efficacy and accessibility, making peppermint an invaluable asset in holistic and integrative healthcare.

In conclusion, *Mentha piperita* represents a natural, accessible, and scientifically validated solution for a variety of health challenges. Its unique combination of therapeutic properties aligns well with modern healthcare's increasing focus on plant-based remedies, preventive strategies, and patient-centered approaches. Continued exploration into peppermint's benefits and mechanisms will no doubt strengthen its place in evidence-based medicine and promote its use in both clinical and everyday health settings.

## References

Abdel-Wareth AAA, Kehraus S, Südekum K-H (2019) Peppermint and its respective active component in diets of broiler chickens: growth performance, viability, economics, meat physicochemical properties, and carcass characteristics. Poult Sci 98(9):3850–3859. https://doi.org/10.3382/ps/pez099

Adel M, Safari R, Pourgholam R, Zorriehzahra J, Esteban M (2015) Dietary peppermint (Mentha piperita) extracts promote growth performance and increase the main humoral immune parameters (both at mucosal and systemic level) of Caspian brown trout (Salmo trutta caspius Kessler, 1877). Fish Shellfish Immunol 47(1):623–629. https://doi.org/10.1016/j.fsi.2015.10.005

Adel M, Sakhaie F, Hosseini Shekarabi SP, Gholamhosseini A, Impellitteri F, Faggio C (2024) Dietary Mentha piperita essential oil loaded in chitosan nanoparticles mediated the growth performance and humoral immune responses in Siberian sturgeon (Acipenserbaerii). Fish Shellfish Immunol 145:109321. https://doi.org/10.1016/j.fsi.2023.109321

Adomako-Bonsu AG, Chan SL, Pratten M, Fry JR (2017) Antioxidant activity of rosmarinic acid and its principal metabolites in chemical and cellular systems: importance of physico-chemical characteristics. Toxicol In Vitro 40:248–255. https://doi.org/10.1016/j.tiv.2017.01.016

Afkar S, Karimzadeh G (2024) Changes in physiological traits, gene expression and phytochemical profile of Mentha piperita in response to elicitor. Biochem Genet. https://doi.org/10.1007/s10528-024-10805-6

Alammar N, Wang L, Saberi B, Nanavati J, Holtmann G, Shinohara RT, Mullin GE (2019) The impact of peppermint oil on the irritable bowel syndrome: a meta-analysis of the pooled clinical data. BMC Complement Altern Med 19(1):21. https://doi.org/10.1186/s12906-018-2409-0

Aldoghachi FEH, Noor Al-Mousawi UM, Shari FH (2021) Antioxidant activity of Rosmarinic acid extracted and purified from Mentha piperita. Arch Razi Inst 76(5):1279–1287. https://doi.org/10.22092/ari.2021.356072.1770

Alexa E, Danciu C, Radulov I, Obistioiu D, Sumalan RM, Morar A, Dehelean CA (2018) Phytochemical screening and biological activity of Mentha × piperita L. and Lavandula angustifolia mill. extracts. Anal Cell Pathol (Amst) 2018:2678924. https://doi.org/10.1155/2018/2678924

Ali K, Al Snih GM, Beyder A, Abdelmalek MF (2025) Peppermint oil associated hepatotoxicity. J Hepatol. https://doi.org/10.1016/j.jhep.2025.04.004

Al-Khattaf FS, Mani A, Atef Hatamleh A, Akbar I (2021) Antimicrobial and cytotoxic activities of isoniazid connected menthone derivatives and their investigation of clinical pathogens causing infectious disease. J Infect Public Health 14(4):533–542. https://doi.org/10.1016/j.jiph.2020.12.033

Aller R, de Luis DA, Izaola O, La Calle F, del Olmo L, Fernandez L, Arranz T, Hernandez JMG (2004) Effect of soluble fiber intake in lipid and glucose levels in healthy subjects: a randomized clinical trial. Diabetes Res Clin Pract 65(1):7–11. https://doi.org/10.1016/j.diabres.2003.11.005

Al-Mijalli SH, Mrabti NN, Ouassou H, Sheikh RA, Abdallah EM, Assaggaf H, Bakrim S, Alshahrani MM, Awadh AAA, Qasem A, Attar A, Lee LH, Bouyahya A, Goh KW, Ming LC, Mrabti HN (2022) Phytochemical variability, in vitro and in vivo biological investigations, and In Silico antibacterial mechanisms of Mentha piperita essential oils collected from two different regions in Morocco. Foods 11(21). https://doi.org/10.3390/foods11213466

Al-Qaaneh AM, Al-Ghamdi FH (2021) Tocilizumab prescribing criteria for COVID-19 patients. Hum Vaccin Immunother 17(4):1128. https://doi.org/10.1080/21645515.2020.1822137

Al-Qaaneh AM, Al-Ghamdi FH, AbdulAzeez S, Borgio JF (2022a) Safety of Tocilizumab in COVID-19 patients and benefit of single-dose: the largest retrospective observational study. Pharmaceutics 14(3). https://doi.org/10.3390/pharmaceutics14030624

Al-Qaaneh AM, Obaid WT, Al-Mohammadi OS, Al-Qaaneh AM, Rabaan AA, Mustafa SM (2022b) Dose-dependent atorvastatin associated with angioedema. Int J Clin Pharmacol Ther 60(2):106–110. https://doi.org/10.5414/cp204025

Anyaoku I (2023) Phytomedicinal and nutritional values of Mentha Piperita. Int J Res Publ Rev 4:3029–3037

Arzani A, Zeinali H, Razmjo K (2007) Iron and magnesium concentrations of mint accessions (Mentha spp.). Plant Physiol Biochem 45(5):323–329. https://doi.org/10.1016/j.plaphy.2007.03.023

Azhar MK, Anwar S, Hasan GM, Shamsi A, Islam A, Parvez S, Hassan MI (2023) Comprehensive insights into biological roles of Rosmarinic acid: implications in diabetes, cancer and neurodegenerative diseases. Nutrients 15(19):4297. https://www.mdpi.com/2072-6643/15/19/4297, https://pmc.ncbi.nlm.nih.gov/articles/PMC10574478/pdf/nutrients-15-04297.pdf

Behm DG, Herat N, Power GMJ, Brosky JA, Page P, Alizadeh S (2022) Menthol-based topical analgesic induces similar upper and lower body pain pressure threshold values: a randomized trial. J Sport Rehabil 31(1):24–30. https://doi.org/10.1123/jsr.2021-0144

Bellassoued K, Ben Hsouna A, Athmouni K, van Pelt J, Makni Ayadi F, Rebai T, Elfeki A (2018) Protective effects of Mentha piperita L. leaf essential oil against CCl(4) induced hepatic oxidative damage and renal failure in rats. Lipids Health Dis 17(1):9. https://doi.org/10.1186/s12944-017-0645-9

Ben-Arye E, Dudai N, Eini A, Torem M, Schiff E, Rakover Y (2011) Treatment of upper respiratory tract infections in primary care: a randomized study using aromatic herbs. Evid Based Complement Alternat Med 2011:690346. https://doi.org/10.1155/2011/690346

Buleandra M, Oprea E, Popa DE, David IG, Moldovan Z, Mihai I, Badea IA (2016) Comparative chemical analysis of Mentha piperita and M. spicata and a fast assessment of commercial peppermint teas. Nat Prod Commun 11(4):551–555

Cam M, Basyigit B, Alasalvar H, Yilmaztekin M, Ahhmed A, Sagdic O, Konca Y, Telci I (2020) Bioactive properties of powdered peppermint and spearmint extracts: inhibition of key enzymes linked to hypertension and type 2 diabetes. Food Biosci 35:100577. https://doi.org/10.1016/j.fbio.2020.100577

Camele I, Gruľová D, Elshafie HS (2021) Chemical composition and antimicrobial properties of Mentha × piperita cv. 'Kristinka' essential oil. Plants (Basel) 10(8). https://doi.org/10.3390/plants10081567

Cappello G, Spezzaferro M, Grossi L, Manzoli L, Marzio L (2007) Peppermint oil (Mintoil®) in the treatment of irritable bowel syndrome: a prospective double blind placebo-controlled randomized trial. Dig Liver Dis 39(6):530–536. https://doi.org/10.1016/j.dld.2007.02.006

Carr AC, Maggini S (2017) Vitamin C and immune function. Nutrients 9(11):1211. https://doi.org/10.3390/nu9111211. https://www.mdpi.com/2072-6643/9/11/1211, https://pmc.ncbi.nlm.nih.gov/articles/PMC5707683/pdf/nutrients-09-01211.pdf

Chumpitazi B, Kearns G, Shulman R (2018) Review article: the physiological effects and safety of peppermint oil and its efficacy in irritable bowel syndrome and other functional disorders. Aliment Pharmacol Ther 47. https://doi.org/10.1111/apt.14519

Dagli N, Dagli R, Mahmoud RS, Baroudi K (2015) Essential oils, their therapeutic properties, and implication in dentistry: a review. J Int Soc Prev Community Dent 5(5):335–340. https://doi.org/10.4103/2231-0762.165933

de Almeida AA, Silva RO, Nicolau LA, de Brito TV, de Sousa DP, Barbosa AL, de Freitas RM, Lopes LD, Medeiros JR, Ferreira PM (2017) Physio-pharmacological investigations about the anti-inflammatory and antinociceptive efficacy of (+)-limonene epoxide. Inflammation 40(2):511–522. https://doi.org/10.1007/s10753-016-0496-y

de Almeida-Muradian LB, Rios MD, Sasaki R (1998) Determination of provitamin A of green leafy vegetables by high performance liquid chromatography and open column chromatography. Boll Chim Farm 137(7):290–294

de Sousa AA, Soares PM, de Almeida AN, Maia AR, de Souza EP, Assreuy AM (2010) Antispasmodic effect of Mentha piperita essential oil on tracheal smooth muscle of rats. J Ethnopharmacol 130(2):433–436. https://doi.org/10.1016/j.jep.2010.05.012

DeGeorge KC, Ring DJ, Dalrymple SN (2019) Treatment of the common cold. Am Fam Physician 100(5):281–289. https://www.aafp.org/pubs/afp/issues/2019/0901/p281.pdf

Deivanayagame B, Siva Kumar AV, Maruthy K, Kareem S (2020) Effect of peppermint aroma on short term memory and cognition in healthy volunteers. Int J Physiol 8:16–20. https://doi.org/10.5958/2320-608X.2020.00004.9

Del Rosario Cappellari L, Chiappero J, Palermo TB, Giordano W, Banchio E (2020) Impact of soil Rhizobacteria inoculation and leaf-chewing insect herbivory on Mentha piperita leaf secondary metabolites. J Chem Ecol 46(7):619–630. https://doi.org/10.1007/s10886-020-01193-3

Di Ciaula A, Garruti G, Frühbeck G, De Angelis M, de Bari O, Wang DQ, Lammert F, Portincasa P (2019) The role of diet in the pathogenesis of cholesterol gallstones. Curr Med Chem 26(19):3620–3638. https://doi.org/10.2174/0929867324666170530080636

Diniz do Nascimento L, Moraes AAB, Costa KSD, Pereira Galúcio JM, Taube PS, Costa CML, Neves Cruz J, de Aguiar Andrade EH, Faria LJG (2020) Bioactive natural compounds and antioxidant activity of essential oils from spice plants: new findings and potential applications. Biomolecules 10(7). https://doi.org/10.3390/biom10070988

Doddawad VG, Jose M, Shivananda S, Shetty A, Shetty A, Ravi MB (2023) Effectiveness of menthol and chlorhexidine mouth rinses on reduction of plaque, gingivitis, and Streptococcus mutans count in Saliva. Int J Pharm Qual Assur 14:105–109. https://doi.org/10.25258/ijpqa.14.1.18

Eccles R (1994) Menthol and related cooling compounds. J Pharm Pharmacol 46(8):618–630. https://doi.org/10.1111/j.2042-7158.1994.tb03871.x

Elsaie L, Mohsen A, Ibrahim I, Mohey-Eddin M, Elsaie M (2016) Effectiveness of topical peppermint oil on symptomatic treatment of chronic pruritus. Clin Cosmet Investig Dermatol 9:333–338. https://doi.org/10.2147/CCID.S116995

Fayed M (2019) Mentha Piperita L. -a promising dental care herb mainly against cariogenic bacteria. Univers J Pharm Res 4:33–38. https://doi.org/10.22270/ujpr.v4i3.271

Fuchs S, Zinn S, Beck T, Mosandl A (1999) Biosynthesis of menthofuran in Mentha x piperita: stereoselective and mechanistic studies. J Agric Food Chem 47(10):4100–4105. https://doi.org/10.1021/jf990439e

Ge J, Liu Z, Zhong Z, Wang L, Zhuo X, Li J, Jiang X, Ye X-Y, Xie T, Bai R (2022) Natural terpenoids with anti-inflammatory activities: potential leads for anti-inflammatory drug discovery. Bioorg Chem 124:105817. https://doi.org/10.1016/j.bioorg.2022.105817

Ghahremani-Nasab M, Del Bakhshayesh AR, Akbari-Gharalari N, Mehdipour A (2023) Biomolecular and cellular effects in skin wound healing: the association between ascorbic acid and hypoxia-induced factor. J Biol Eng 17(1):62. https://doi.org/10.1186/s13036-023-00380-6

Gholamipourfard K, Salehi M, Banchio E (2021) Mentha piperita phytochemicals in agriculture, food industry and medicine: features and applications. S Afr J Bot 141:183–195. https://doi.org/10.1016/j.sajb.2021.05.014

Gill SK, Rossi M, Bajka B, Whelan K (2021) Dietary fibre in gastrointestinal health and disease. Nat Rev Gastroenterol Hepatol 18(2):101–116. https://doi.org/10.1038/s41575-020-00375-4

Golestannejad Z, Gavanji S, Mohammadi E, Motamedi A, Bahrani M, Rezaei F, Larki B, Mojiri A, Bakhtari A (2018) Comparison of antibacterial activity of essential oils of Foeniculum vulgare Mill, Mentha arvensis and Mentha piperita against Streptococcus mutans. Future Nat Prod 4(1):3–13. https://fnp.skums.ac.ir/Article/fnp-67

Gonçalves AC, Portari GV (2021) The B-complex vitamins related to energy metabolism and their role in exercise performance: a narrative review. Sci Sports 36(6):433–440. https://doi.org/10.1016/j.scispo.2020.11.007

Goudarzi MA, Radfar M, Goudarzi Z (2024) Peppermint as a promising treatment agent in inflammatory conditions: a comprehensive systematic review of literature. Phytother Res 38(1):187–195. https://doi.org/10.1002/ptr.8041

Grigoleit H, Grigoleit P (2005) Pharmacology and preclinical pharmacokinetic of peppermint oil. Phytomedicine 12:612–616. https://doi.org/10.1016/j.phymed.2004.10.007

Halmos EP (2017) When the low FODMAP diet does not work. J Gastroenterol Hepatol 32(Suppl 1):69–72. https://doi.org/10.1111/jgh.13701

He Z, Li X, Wang Z, Cao Y, Han S, Li N, Cai J, Cheng S, Liu Q (2023) Protective effects of luteolin against amyloid beta-induced oxidative stress and mitochondrial impairments through peroxisome proliferator-activated receptor γ-dependent mechanism in Alzheimer's disease. Redox Biol 66:102848. https://doi.org/10.1016/j.redox.2023.102848

Hemilä H, Chalker E (2023) Vitamin C reduces the severity of common colds: a meta-analysis. BMC Public Health 23(1):2468. https://doi.org/10.1186/s12889-023-17229-8

Hemilä H, Chalker E (2025) Vitamin C for the common cold and pneumonia. Pol Arch Intern Med 135(1). https://doi.org/10.20452/pamw.16926

Hilfiger L, Triaux Z, Marcic C, Héberlé E, Emhemmed F, Darbon P, Marchioni E, Petitjean H, Charlet A (2021) Anti-Hyperalgesic properties of menthol and Pulegone. Front Pharmacol 12:753873. https://doi.org/10.3389/fphar.2021.753873

Hills JM, Aaronson PI (1991) The mechanism of action of peppermint oil on gastrointestinal smooth muscle: an analysis using patch clamp electrophysiology and isolated tissue pharmacology in rabbit and Guinea pig. Gastroenterology 101(1):55–65. https://doi.org/10.1016/0016-5085(91)90459-X

Hudz N, Kobylinska L, Pokajewicz K, Horčinová Sedláčková V, Fedin R, Voloshyn M, Myskiv I, Brindza J, Wieczorek PP, Lipok J (2023) Mentha piperita: essential oil and extracts, their biological activities, and perspectives on the development of new medicinal and cosmetic products. Molecules 28(21). https://doi.org/10.3390/molecules28217444

Iftinca M, Basso L, Flynn R, Kwok C, Roland C, Hassan A, Defaye M, Ramachandran R, Trang T, Altier C (2020) Chronic morphine regulates TRPM8 channels via MOR-PKCβ signaling. Mol Brain 13(1):61. https://doi.org/10.1186/s13041-020-00599-0

Jarosz M, Taraszewska A (2014) Risk factors for gastroesophageal reflux disease: the role of diet. Prz Gastroenterol 9(5):297–301. https://doi.org/10.5114/pg.2014.46166

Jeyakumar E, Lawrence R, Pal T (2011) Comparative evaluation in the efficacy of peppermint (Mentha piperita) oil with standards antibiotics against selected bacterial pathogens. Asian Pac J Trop Biomed 1(2, Supplement):S253–S257. https://doi.org/10.1016/S2221-1691(11)60165-2

Kashyap N, Kumari A, Raina N, Zakir F, Gupta M (2022) Prospects of essential oil loaded nanosystems for skincare. Phytomed Plus 2(1):100198. https://doi.org/10.1016/j.phyplu.2021.100198

Kazemi A, Iraji A, Esmaealzadeh N, Salehi M, Hashempur MH (2024) Peppermint and menthol: a review on their biochemistry, pharmacological activities, clinical applications, and safety considerations. Crit Rev Food Sci Nutr:1–26. https://doi.org/10.1080/10408398.2023.2296991

Kazemi A, Iraji A, Esmaealzadeh N, Salehi M, Hashempur MH (2025) Peppermint and menthol: a review on their biochemistry, pharmacological activities, clinical applications, and safety

considerations. Crit Rev Food Sci Nutr 65(8):1553–1578. https://doi.org/10.1080/10408398.2023.2296991
Kennedy D, Okello E, Chazot P, Howes M-J, Ohiomokhare S, Jackson P, Haskell-Ramsay C, Khan J, Forster J, Wightman E (2018) Volatile terpenes and brain function: investigation of the cognitive and mood effects of Mentha × Piperita L. essential oil with in vitro properties relevant to central nervous system function. Nutrients 10(8):1029. https://doi.org/10.3390/nu10081029. https://www.mdpi.com/2072-6643/10/8/1029, https://pmc.ncbi.nlm.nih.gov/articles/PMC6116079/pdf/nutrients-10-01029.pdf
Leffingwell J (2009) Cooling ingredients and their mechanism of action. In: Handbook of cosmetic science and technology. Informa Healthcare, pp 661–675. https://doi.org/10.1201/b15273-66
Lemjallad L, Chabir R, Kandri Rodi Y, El Ghadraoui L, Ouazzani Chahdi F, Errachidi F (2019) Improvement of Heliciculture by three medicinal plants belonging to the Lamiaceae Family. Sci World J 2019:2630537. https://doi.org/10.1155/2019/2630537
Li Y, Liu Y, Ma A, Bao Y, Wang M, Sun Z (2017) In vitro antiviral, anti-inflammatory, and antioxidant activities of the ethanol extract of Mentha piperita L. Food Sci Biotechnol 26(6):1675–1683. https://doi.org/10.1007/s10068-017-0217-9
Li XW, Yuan SC, Wang M, Su ZH, Zhao M, Li T, Zhang XM, Xue W, Li WH (2023) Rosmarinic acid ameliorates autoimmune responses through suppression of intracellular nucleic acid-mediated type I interferon expression. Biochem Biophys Res Commun 673:73–80. https://doi.org/10.1016/j.bbrc.2023.05.113
Liu JH, Chen GH, Yeh HZ, Huang CK, Poon SK (1997) Enteric-coated peppermint-oil capsules in the treatment of irritable bowel syndrome: a prospective, randomized trial. J Gastroenterol 32(6):765–768. https://doi.org/10.1007/bf02936952
Lv X, Feng Y, Ma R, Tang Y, Li Y, Cui D, Wu Y (2022) Effects of peppermint essential oil on learning and memory ability in APP/PS1 transgenic mice. Molecules 27(7). https://doi.org/10.3390/molecules27072051
Mahanom H Jr, Azizah A, Dzulkifly M (1999) Effect of different drying methods on concentrations of several phytochemicals in herbal preparation of 8 medicinal plants leaves. Malays J Nutr 5(1):47–54
Mahendran G, Rahman LU (2020) Ethnomedicinal, phytochemical and pharmacological updates on peppermint (Mentha × piperita L.)-A review. Phytother Res 34(9):2088–2139. https://doi.org/10.1002/ptr.6664
Mammari N, Albert Q, Devocelle M, Kenda M, Kočevar Glavač N, Sollner Dolenc M, Mercolini L, Tóth J, Milan N, Czigle S, Varbanov M, Oemonom OBOT (2023) Natural products for the prevention and treatment of common cold and viral respiratory infections. Pharmaceuticals 16(5):662. https://www.mdpi.com/1424-8247/16/5/662, https://pmc.ncbi.nlm.nih.gov/articles/PMC10220542/pdf/pharmaceuticals-16-00662.pdf
Marjani A, Rahmati R, Mansourian AR, Veghary G (2012) Effect of peppermint oil on serum lipid peroxidation and hepatic enzymes after immobility stress in mice. Open Biochem J 6:51–55. https://doi.org/10.2174/1874091x01206010051
Martin SE, Kraft CS, Ziegler TR, Millson EC, Rishishwar L, Martin GS (2023) The role of diet on the gut microbiome, mood and happiness. medRxiv. https://doi.org/10.1101/2023.03.18.23287442
Marwa C, Fikri-Benbrahim K, Ou-Yahia D, Farah A (2017) African peppermint (Mentha piperita) from Morocco: chemical composition and antimicrobial properties of essential oil. J Adv Pharm Technol Res 8(3):86–90. https://doi.org/10.4103/japtr.JAPTR_11_17
McKay DL, Blumberg JB (2006) A review of the bioactivity and potential health benefits of peppermint tea (Mentha piperita L.). Phytother Res 20(8):619–633. https://doi.org/10.1002/ptr.1936
Meamarbashi A (2014) Instant effects of peppermint essential oil on the physiological parameters and exercise performance. Avicenna J Phytomed 4(1):72–78. https://pmc.ncbi.nlm.nih.gov/articles/PMC4103722/pdf/ajp-4-072.pdf
Mesbahzadeh B, Akbari M, Moradikor N, Bayati Zadeh J (2015) The effects of different levels of peppermint alcoholic extract on body-weight gain and blood biochemical parameters of adult male Wistar rats. Electron Physician 7:1376–1380. https://doi.org/10.14661/1376

Moss M, Hewitt S, Moss L, Wesnes K (2008a) Modulation of cognitive performance and mood by aromas of peppermint and ylang-ylang. Int J Neurosci 118:59–77. https://doi.org/10.1080/00207450601042094

Moss M, Steven H, Lucy M, Wesnes K (2008b) Modulation of cognitive performance and mood by aromas of peppermint and Ylang-Ylang. Int J Neurosci 118(1):59–77. https://doi.org/10.1080/00207450601042094

Mucciarelli M, Camusso W, Bertea CM, Maffei M (2001) Effect of (+)-pulegone and other oil components of Mentha x Piperita on cucumber respiration. Phytochemistry 57(1):91–98. https://doi.org/10.1016/s0031-9422(00)00393-9

Nematolahi F, Mohtashamian A, Kaveh G, Sharifi N, Milajerdi A (2024) Effects of Mentha on blood pressure: a GRADE-assessed systematic review and meta-analysis of randomized controlled trials. BMC Complement Med Ther 24(1):406. https://doi.org/10.1186/s12906-024-04701-0

Netzler L, Lovell B (2025) A randomized placebo-controlled clinical trial exploring the short-term cognitive and cerebrovascular effects of consuming peppermint tea: a mediation study. Hum Psychopharmacol Clin Exp 40. https://doi.org/10.1002/hup.70005

Nishi R, Sagiyama K, Fukumoto T, Amitani H, Asakawa A (2025) Heterogeneity in the study of peppermint in aromatherapy on cognition and attention: a critical scoping review. Complement Med Res 32:224–232. https://doi.org/10.1159/000543974. (Heterogenität bei der Untersuchung der Pfefferminz-Aromatherapie für Kognition und Aufmerksamkeit: Ein kritischer Scoping Review)

Oliveira F, Silva E, Matias A, Silva JM, Reis RL, Duarte ARC (2023) Menthol-based deep eutectic systems as antimicrobial and anti-inflammatory agents for wound healing. Eur J Pharm Sci 182:106368. https://doi.org/10.1016/j.ejps.2022.106368

Padalia RC, Verma RS, Chanotiya CS (2011) Variability in volatile Terpenoid compositions of peppermint cultivars and some wild accession from northern India. J Essent Oil Res 23(2):29–33. https://doi.org/10.1080/10412905.2011.9700444

Pelvan E, Karaoğlu Ö, Önder Fırat E, Betül Kalyon K, Ros E, Alasalvar C (2022) Immunomodulatory effects of selected medicinal herbs and their essential oils: a comprehensive review. J Funct Foods 94:105108. https://doi.org/10.1016/j.jff.2022.105108

Pergolizzi JV Jr, Taylor R Jr, LeQuang JA, Raffa RB (2018) The role and mechanism of action of menthol in topical analgesic products. J Clin Pharm Ther 43(3):313–319. https://doi.org/10.1111/jcpt.12679

Piasecki B, Korona-Głowniak I, Kiełtyka-Dadasiewicz A, Ludwiczuk A (2023) Composition and anti-helicobacter pylori properties of essential oils obtained from selected Mentha cultivars. Molecules 28(15). https://doi.org/10.3390/molecules28155690

Quintana Soares Lopes L, Fortes Guerim PH, Maldonado ME, Wagner R, Hadlich Xavier AC, Gutknecht da Silva JL, Bittencourt da Rosa Leal D, de Freitas Daudt N, Christ Vianna Santos R, Kolling Marquezan P (2024) Chemical composition, cytotoxicity, antimicrobial, antibiofilm, and anti-quorum sensing potential of Mentha Piperita essential oil against the oral pathogen Streptococcus mutans. J Toxic Environ Health A 87(20):824–835. https://doi.org/10.1080/15287394.2024.2375731

Raczuk J, Biardzka E, Daruk J (2008) The content of Ca, Mg, Fe and Cu in selected species of herbs and herb infusions. Rocz Panstw Zakl Hig 59(1):33–40. (Zawartość Ca, Mg, Fe i Cu w wybranych gatunkach ziół i ich naparach)

Radivojac A, Bera O, Zeković Z, Teslić N, Mrkonjić Ž, Bursać Kovačević D, Putnik P, Pavlić B (2021) Extraction of peppermint essential oils and lipophilic compounds: assessment of process kinetics and environmental impacts with multiple techniques. Molecules 26(10). https://doi.org/10.3390/molecules26102879

Riachi LG, De Maria CA (2015) Peppermint antioxidants revisited. Food Chem 176:72–81. https://doi.org/10.1016/j.foodchem.2014.12.028

Rohloff J, Dragland S, Mordal R, Iversen TH (2005) Effect of harvest time and drying method on biomass production, essential oil yield, and quality of peppermint (Mentha x piperita L.). J Agric Food Chem 53(10):4143–4148. https://doi.org/10.1021/jf047998s

Rosato A, Carocci A, Catalano A, Clodoveo ML, Franchini C, Corbo F, Carbonara GG, Carrieri A, Fracchiolla G (2018) Elucidation of the synergistic action of Mentha Piperita essential oil with common antimicrobials. PLoS One 13(8):e0200902. https://doi.org/10.1371/journal.pone.0200902

Sajovic J, Meglič A, Glavač D, Markelj Š, Hawlina M, Fakin A (2022) The role of vitamin A in retinal diseases. Int J Mol Sci 23(3). https://doi.org/10.3390/ijms23031014

Saqib S, Ullah F, Naeem M, Younas M, Ayaz A, Ali S, Zaman W (2022) Mentha: nutritional and health attributes to treat various ailments including cardiovascular diseases. Molecules 27(19):6728. https://www.mdpi.com/1420-3049/27/19/6728, https://mdpi-res.com/d_attachment/molecules/molecules-27-06728/article_deploy/molecules-27-06728-v2.pdf?version=1665468033

Sargin SA (2021) Potential anti-influenza effective plants used in Turkish folk medicine: a review. J Ethnopharmacol 265:113319. https://doi.org/10.1016/j.jep.2020.113319

Scavroni J, Boaro CS, Ferreira L (2005) Yield and composition of the essential oil of Mentha piperita L. (Lamiaceae) grown with biosolid. Braz J Plant Physiol 17. https://doi.org/10.1590/S1677-04202005000400002

Schuhmacher A, Reichling J, Schnitzler P (2003) Virucidal effect of peppermint oil on the enveloped viruses herpes simplex virus type 1 and type 2 in vitro. Phytomedicine 10(6–7):504–510. https://doi.org/10.1078/094471103322331467

Seyoum Tola F (2024) The concept of folic acid supplementation and its role in prevention of neural tube defect among pregnant women: PRISMA. Medicine (Baltimore) 103(19):e38154. https://doi.org/10.1097/md.0000000000038154

Shamsi A, Anwar S, Shahbaaz M, Mohammad T, Alajmi MF, Hussain A, Hassan I, Ahmad F, Islam A (2020) Evaluation of binding of Rosmarinic acid with human transferrin and its impact on the protein structure: targeting polyphenolic acid-induced protection of neurodegenerative disorders. Oxid Med Cell Longev 2020:1245875. https://doi.org/10.1155/2020/1245875

Silva WMF, Bona NP, Pedra NS, Cunha KFD, Fiorentini AM, Stefanello FM, Zavareze ER, Dias ARG (2022) Risk assessment of in vitro cytotoxicity, antioxidant and antimicrobial activities of Mentha piperita L. essential oil. J Toxicol Environ Health A 85(6):230–242. https://doi.org/10.1080/15287394.2021.1999875

Singh R, Shushni MAM, Belkheir A (2015) Antibacterial and antioxidant activities of Mentha piperita L. Arab J Chem 8(3):322–328. https://doi.org/10.1016/j.arabjc.2011.01.019

Skalicka-Woźniak K, Walasek M (2014) Preparative separation of menthol and pulegone from peppermint oil (Mentha piperita L.) by high-performance counter-current chromatography. Phytochem Lett 10:xciv–xcviii. https://doi.org/10.1016/j.phytol.2014.06.007

Souza ACR, Vasconcelos AR, Dias DD, Komoni G, Name JJ (2023) The integral role of magnesium in muscle integrity and aging: a comprehensive review. Nutrients 15(24). https://doi.org/10.3390/nu15245127

Spadaccino G, Frabboni L, Petruzzi F, Disciglio G, Mentana A, Nardiello D, Quinto M (2021) Essential oil characterization of Prunus spinosa L., Salvia officinalis L., Eucalyptus globulus L., Melissa officinalis L. and Mentha x piperita L. by a volatolomic approach. J Pharm Biomed Anal 202:114167. https://doi.org/10.1016/j.jpba.2021.114167

Stinson RJ, Morice AH, Sadofsky LR (2023) Modulation of transient receptor potential (TRP) channels by plant derived substances used in over-the-counter cough and cold remedies. Respir Res 24(1):45. https://doi.org/10.1186/s12931-023-02347-z

Swarup V, Ghosh J, Ghosh S, Saxena A, Basu A (2007) Antiviral and anti-inflammatory effects of rosmarinic acid in an experimental murine model of Japanese encephalitis. Antimicrob Agents Chemother 51(9):3367–3370. https://doi.org/10.1128/aac.00041-07

Tayarani-Najaran Z, Talasaz-Firoozi E, Nasiri R, Jalali N, Hassanzadeh M (2013) Antiemetic activity of volatile oil from Mentha spicata and Mentha × piperita in chemotherapy-induced nausea and vomiting. Ecancermedicalscience 7:290. https://doi.org/10.3332/ecancer.2013.290

Terzioglu Bebitoglu B (2020) Frequently used herbal teas during pregnancy-short update. Medeni Med J 35(1):55–61. https://doi.org/10.5222/mmj.2020.69851

Thosar N, Basak S, Bahadure RN, Rajurkar M (2013) Antimicrobial efficacy of five essential oils against oral pathogens: an in vitro study. Eur J Dent 7(Suppl 1):S071–s077. https://doi.org/10.4103/1305-7456.119078

Toumia M, Dhaoui R, Sassi S, Kouraichi C, Bel Haj Ali K, Sekma A, Bakir A, Jaballah R, Yaakoubi H, Youssef R, Zorgati A, Beltaief K, Mezgar Z, Khrouf M, Sghaier A, Jerbi N, Zemni I, Bouida W, Grissa MH et al (2025) Efficacy of a natural herbal topical analgesic versus oral paracetamol in patients with soft tissue injury: a randomized, double-blind, placebo-controlled study. Pain Med. https://doi.org/10.1093/pm/pnaf006

Trif C, Banica AM, Manolache A, Anghel SA, Huţanu DE, Stratulat T, Badea R, Oprita G, Selescu T, Petrescu SM, Sisignano M, Offermanns S, Babes A, Tunaru S (2024) Inhibition of TRPM8 function by prostacyclin receptor agonists requires coupling to Gq/11 proteins. Br J Pharmacol 181(9):1438–1451. https://doi.org/10.1111/bph.16295

Tufarelli V, Ghavami N, Nosrati M, Rasouli B, Kadim IT, Suárez Ramírez L, Gorlov I, Slozhenkina M, Mosolov A, Seidavi A, Ayasan T, Laudadio V (2023) The effects of peppermint (Mentha piperita L.) and chicory (Cichorium intybus L.) in comparison with a prebiotic on productive performance, blood constituents, immunity and intestinal microflora in broiler chickens. Anim Biotechnol 34(7):3046–3052. https://doi.org/10.1080/10495398.2022.2130798

Umezu T, Sano T, Hayashi J (2021) l-Menthol increases extracellular dopamine and c-Fos-like immunoreactivity in the dorsal striatum, and promotes ambulatory activity in mice. PLoS One 16(11):e0260713. https://doi.org/10.1371/journal.pone.0260713

Unalan I, Slavik B, Buettner A, Goldmann WH, Frank G, Boccaccini AR (2019) Physical and antibacterial properties of peppermint essential oil loaded poly (ε-caprolactone) (PCL) electrospun fiber mats for wound healing. Front Bioeng Biotechnol 7:346. https://doi.org/10.3389/fbioe.2019.00346

Uribe E, Marín D, Vega-Gálvez A, Quispe-Fuentes I, Rodríguez A (2016) Assessment of vacuum-dried peppermint (Mentha piperita L.) as a source of natural antioxidants. Food Chem 190:559–565. https://doi.org/10.1016/j.foodchem.2015.05.108

Valdes AM, Walter J, Segal E, Spector TD (2018) Role of the gut microbiota in nutrition and health. BMJ 361:k2179. https://doi.org/10.1136/bmj.k2179

Valussi M, Antonelli M, Donelli D, Firenzuoli F (2021) Appropriate use of essential oils and their components in the management of upper respiratory tract symptoms in patients with COVID-19. J Herb Med 28:100451. https://doi.org/10.1016/j.hermed.2021.100451

Walker JT, Melin JB (1996) Mentha x piperita, Mentha spicata and effects of their essential oils on Meloidogyne in soil. J Nematol 28(4s):629–635. https://pmc.ncbi.nlm.nih.gov/articles/PMC2619730/pdf/629.pdf

Weerts ZZRM, Masclee AAM, Witteman BJM, Clemens CHM, Winkens B, Brouwers JRBJ, Frijlink HW, Muris JWM, De Wit NJ, Essers BAB, Tack J, Snijkers JTW, Bours AMH, de Ruiter-van der Ploeg AS, Jonkers DMAE, Keszthelyi D (2020) Efficacy and safety of peppermint oil in a randomized, double-blind trial of patients with irritable bowel syndrome. Gastroenterology 158(1):123–136. https://doi.org/10.1053/j.gastro.2019.08.026

Xia X, Lin Z, Shao K, Wang X, Xu J, Zhai H, Wang H, Xu W, Zhao Y (2021) Combination of white tea and peppermint demonstrated synergistic antibacterial and anti-inflammatory activities. J Sci Food Agric 101(6):2500–2510. https://doi.org/10.1002/jsfa.10876

Yin Y, Lee SY (2020) Current view of ligand and lipid recognition by the menthol receptor TRPM8. Trends Biochem Sci 45(9):806–819. https://doi.org/10.1016/j.tibs.2020.05.008

Zaia MG, Cagnazzo T, Feitosa KA, Soares EG, Faccioli LH, Allegretti SM, Afonso A, Anibal Fde F (2016) Anti-inflammatory properties of menthol and Menthone in Schistosoma mansoni infection. Front Pharmacol 7:170. https://doi.org/10.3389/fphar.2016.00170

Zhao H, Ren S, Yang H, Tang S, Guo C, Liu M, Tao Q, Ming T, Xu H (2022) Peppermint essential oil: its phytochemistry, biological activity, pharmacological effect and application. Biomed Pharmacother 154:113559. https://doi.org/10.1016/j.biopha.2022.113559

Zhao Q, Zhu L, Wang S, Gao Y, Jin F (2023) Molecular mechanism of the anti-inflammatory effects of plant essential oils: a systematic review. J Ethnopharmacol 301:115829. https://doi.org/10.1016/j.jep.2022.115829

Zheljazkov VD, Cantrell CL, Astatkie T, Hristov A (2010) Yield, content, and composition of peppermint and spearmints as a function of harvesting time and drying. J Agric Food Chem 58(21):11400–11407. https://doi.org/10.1021/jf1022077

Zhou S, Koh H-L, Gao Y, Gong Z-y, Lee EJD (2004) Herbal bioactivation: the good, the bad and the ugly. Life Sci 74(8):935–968. https://doi.org/10.1016/j.lfs.2003.09.035

# Medicinal and Nutritional Importance of *Passiflora incarnata* in Human Health 26

Sanjana Mishra, Siddharth Singh, Kshitij Patel, Prekshi Garg, and Prachi Srivastava

**Abstract**

Neurological disorders are more prevalent nowadays and they pose a major burden on worldwide health. Research studies have revealed that various neurodegenerative, neurodevelopmental and neuropsychiatric disorders are affecting individuals worldwide. This has been a major issue to look up into. There are various drugs prescribed by the physicians to the patients, which show a wide range of side effects. To overcome these side effects of the drug, there is a need to switch for a herbal medicine with least or no side effects. From ancient times, treatments involved *Passiflora incarnata* or Passionflower to treat the neurological disorders like anxiety, sleeplessness, attention-deficit/hyperactivity disorder (ADHD), autism, Parkinson's, Alzheimer's, epilepsy or seizures. Several studies have recently demonstrated its sedative and antianxiety properties. Prolonged stress conditions can lead to mood disorders as well as issues with sleep, and difficulty in focussing. It has also been established that passionflower exhibits beneficial effects on memory functioning and also serves as a neuromodulator, where the GABAergic effects of *Passiflora incarnata* have led to its widespread usage as a sedative and anxiolytic. There were no negative consequences noted, such as memory loss or the collapse of psychometric functions. However, further research is necessary to fully understand the effectiveness and mechanisms of action of *Passiflora* in these contexts. Overall, understanding neurological disorders is crucial for developing effective treatment strategies and improving patient

S. Mishra · S. Singh · K. Patel · P. Srivastava (✉)
Amity Institute of Biotechnology, Amity University Uttar Pradesh, Lucknow Campus, Lucknow, Uttar Pradesh, India
e-mail: psrivastava@amity.edu

P. Garg
BioinfoCore Solutions (OPC) Pvt Ltd, Lucknow, Uttar Pradesh, India

 
M. A. Ansari et al. (eds.), *Medicinal Plants and their Bioactive Compounds in Human Health: Volume 2*, https://doi.org/10.1007/978-981-95-3620-7_26

care. As research advances, new insights may emerge into the causes and potential therapies for these debilitating conditions.

**Keywords**

*Passiflora incarnata* · Neurodevelopmental disorders · Neurodegenerative disorders · Neuropsychiatric disorders · Stress · Anxiety

## 26.1 Introduction

One of the most extensively researched species in the *Passiflora* genus with medicinal qualities is *Passiflora incarnata*. Fruits, flowers and vegetative parts are utilized medicinally (Da Fonseca et al. 2020). It is a great source of quercetin, alkaloids, flavonoids and cyanogenic glycosides. It is also a perennial plant that has an egg-shaped fruit and can reach a height of 10 m. The fruit, which has few calories, is high in calcium, phosphorus, iron and the vitamins A, C, B1 and B2 (Patel 2009). Originating in South America, Australia and South East Asia, the species is being grown for pharmaceutical raw materials. *Passiflora incarnata*, often referred to as maypop, purple passionflower, real passionflower, wild apricot and wild passion vine, is a vigorous perennial vine with climbing or trailing stems. The genus *Passiflora* is the largest in the family Passifloraceae, with approximately 600 species. This genus's species is found in the warm temperate and tropical parts of the New World, but they are far less common in Asia, Australia and tropical Africa (Dhawan et al. 2004).

It is a versatile plant that may be cultivated in a wide range of environments. It can be grown in full sun or partial shade and it is tolerant of a variety of soils. The plant is also relatively drought-tolerant and one of the hardiest species of passionflower. Since ancient times, people have exploited *Passiflora incarnata* to treat a variety of central nervous system (CNS) problems, including anxiety and insomnia (Dhawan et al. 2003). *Passiflora incarnata* has several medicinal uses. The aerial parts of the plant, along with the flowers and fruits, are used medicinally as shown in Fig. 26.1. They are known to have anthelmintic, antispasmodic and anxiolytic properties. Passionflower is also used to treat burns, diarrhoea, painful menstruation, haemorrhoids, neurotic problems, insomnia, morphine dependence and withdrawal symptoms (Janda et al. 2020).

One of the most well-researched species in the *Passiflora* genus with medicinal qualities is *Passiflora incarnata*. The passionflower's main phytochemicals are flavonoids, which include apigenin, luteolin, quercetin and kaempferol, as well as flavonoid glycosides, which include vitexin, isovitexin, orientin and isoorientin (Da Fonseca et al. 2020; Patel 2009). The species has the most isovitexin overall in comparison with other species (Dhawan et al. 2004). The European Medicines Agency officially recognized *Passiflora incarnata* as a medical product, when it released a herbal monograph on the plant's medicinal value (Agency et al. 2014). Clinical studies on the usage of *Passiflora incarnata* revealed no risks to human health (Miyasaka et al. 2007; Movafegh et al. 2008). This plant contains high

**Fig. 26.1** The aerial parts of *Passiflora incarnata* showing therapeutic effects

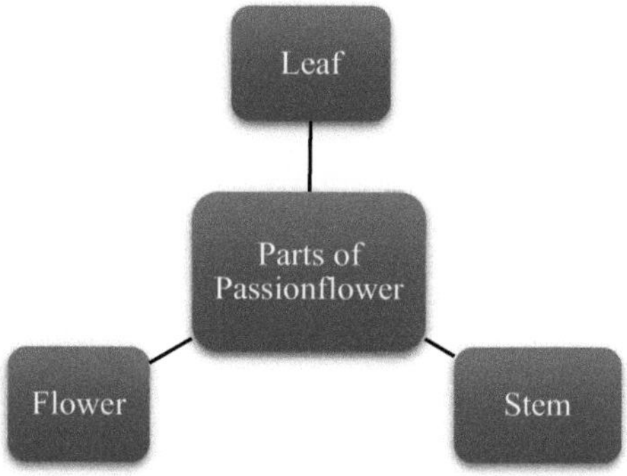


flavonoids, alkaloids, phenolic compounds and cyanogenic glycosides, and due to its action as a monoamine oxidase (MAO) inhibitor, it can have antidepressant properties (Michael et al. 2022). The plant extract has also been shown to improve cognition in Parkinson's disease (PD) and Alzheimer's disease (AD), and there may be a link between these disorders and the extract's antioxidant capabilities (Ingale and Kasture 2017).

The plant species *Passiflora* has shown potential in preliminary studies for managing conditions such as autism spectrum disorder (ASD), attention-deficit/hyperactivity disorder (ADHD), Parkinson's, Alzheimer's and epilepsy. Its potential anxiolytic and sedative properties may also offer a safer alternative to traditional pharmaceutical treatments for neuropsychiatric conditions, including anxiety and insomnia. However, while existing evidence suggests that *Passiflora* could be a valuable tool in managing various neuropsychiatric conditions, neurodegenerative conditions and neurodevelopmental conditions, more extensive studies are needed to confirm these findings and establish it as a reliable treatment option.

## 26.2 Bioactive Compounds as Therapeutics

Neurotoxicity and neural death are caused by cellular and molecular processes that set off the neural malfunction. Neurodegenerative disorders are becoming more prevalent nowadays, and the therapies that are now accessible are aimed at symptom relief (Lima et al. 2018). *Passiflora incarnata* is rich in bioactive compounds as shown in Fig. 26.2. In addition to counteracting the targets that cause neurodegeneration, such as redox imbalance, neuroinflammation, aberrant protein aggregation and a decrease in cognitive and/or motor impairment, vitexin helps to increase neuroprotective factors and pathways. Increased life expectancy and a shifting demography are the primary causes of the rise in neurodegenerative dementias and mobility disorders (Skovronsky et al. 2006). There are presently no medications that can stop the underlying neurodegenerative disorders, although they can alleviate the symptoms of the disease without slowing down or stopping its course. Thus, research into natural compounds and their derivatives as safer and more accessible alternatives

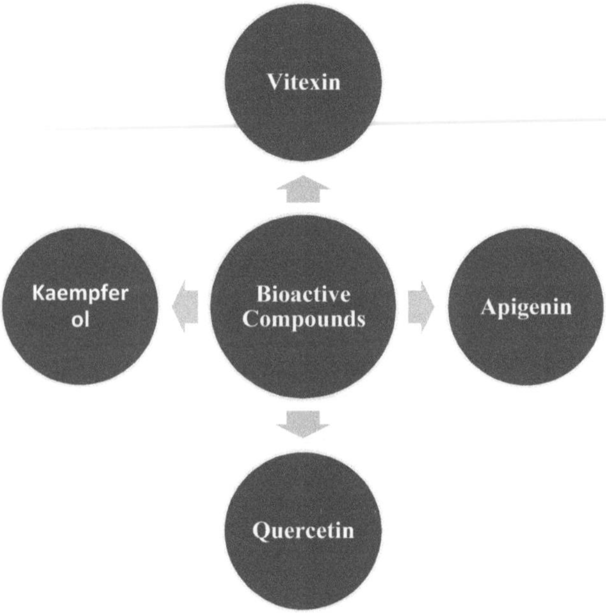


**Fig. 26.2** Illustrating some important bioactive compounds for therapeutics

has been sparked by the hunt for medications that might prevent or postpone neurological loss.

### 26.2.1 Vitexin

The flavonoid known as Vitexin (apigenin-8-C-glycosidic) is present in a number of therapeutic plants, such as *Passiflora incarnata* (Lima et al. 2018). The focus of research on it is because of its pharmacological properties, which include neuroprotective (Borghi et al. 2013; Wang et al. 2015), anti-Alzheimer's (Choi et al. 2014) and antioxidant properties. Additionally, by downregulating apoptotic and proinflammatory signalling pathways and upregulating cell survival pathways, vitexin helps to reduce the amount of brain cell death seen in neurodegenerative disorders and acute fluctuations.

### 26.2.2 Apigenin

Oxidative stress and neuronal or oligodendrocyte cell death are frequently linked to neuroinflammation, a defining feature of many illnesses of the central nervous system. Therefore, pharmacologically targeting neuroinflammation is quite fascinating (Charrière et al. 2024). Nutraceuticals, in particular apigenin, are one treatment alternative. Vegetables, fruits, herbs and some drinks contain apigenin. Apigenin has several biological actions, all of which are equally significant for treating and preventing inflammation. According to several research, apigenin has a variety of pharmacological properties and may be used as a treatment for autoimmune disorders, inflammation, neurological illnesses and even some forms of cancer (Salehi et al. 2019). Research indicates that apigenin provides neuroprotective benefits in animal models by modulating a number of signalling pathways implicated in inflammation, oxidative stress and cell death.

### 26.2.3 Quercetin

Excessive or prolonged neuroinflammation can result in neurotoxicity, which damages the brain and aids in the development and progression of a number of different brain disorders. The potential health advantages of quercetin, such as its anti-inflammatory, antioxidant and anticancer qualities, have been investigated. Additionally, quercetin may improve allergy symptoms and immunological function (Chiang et al. 2023). Furthermore, it has been demonstrated that quercetin possesses neuroprotective and anti-inflammatory qualities. It can also activate (AMP) Adenosine Monophosphate, AMP-activated protein kinase (AMPK), a cellular energy sensor that regulates oxidative stress and inflammation. Quercetin shows promise as a safe and efficient supplementary treatment for neurodegenerative illnesses and other brain disorders (Kuriakose and Xiao 2020) by lowering inflammation and guarding against neuroinflammatory damage.

### 26.2.4 Kaempferol

A flavonoid molecule is present in a variety of fruits, vegetables and medicinal plants. Kaempferol has drawn more attention recently because of its possible neuroprotective benefits in neurological conditions. Various studies show kaempferol's role in neurological conditions, such as stroke, Parkinson's disease, Alzheimer's disease, neuropathic pain and epilepsy (Nezhad Salari et al. 2024). It has been demonstrated that kaempferol alters several signalling pathways, including the (PI3K) Phosphoinositide 3-kinase/Akt, (MAPK) Mitogen-Activated Protein Kinase/(ERK) Extracellular Signal-Regulated Kinase and (NF-κB)Nuclear Factor kappa-light-chain-enhancer of activated B cells pathways (Silva Dos Santos et al. 2021), that are implicated in neurodegeneration and neuroinflammation. Furthermore, by improving neuronal survival, lowering oxidative stress, increasing mitochondrial calcium channel activity, decreasing neuroinflammation, stimulating neurogenesis and boosting cognitive performance, kaempferol shows promise as a treatment.

## 26.3 Neurodevelopmental Disorders

Neurodevelopmental disorders are the group of conditions that affect how your brain functions, which first appear throughout childhood. According to the American Psychiatric Association's *Diagnostic and Statistical Manual of Mental Disorders, Fifth Edition* (DSM-5) published in 2013 (as shown in Fig. 26.3), these problems typically develop in early infancy, before children begin school and can last until adulthood (American Psychiatric Association 2013). The common feature of all of these illnesses is that they have a detrimental impact on a person's functioning in one or more areas of life (personal, social, academic, occupational), depending on the disorder and the deficiencies it has created. They range from moderate limitations that allow people to live relatively normal lives to severe problems that require lifelong care.

### 26.3.1 Autism Spectrum Disorder (ASD)

ASD is defined by persistent deficits in social communication and interaction across multiple contexts, including deficits in social reciprocity, non-verbal communicative behaviours used for social interaction and relationship development, maintenance and comprehension skills. Autism spectrum condition can only be diagnosed

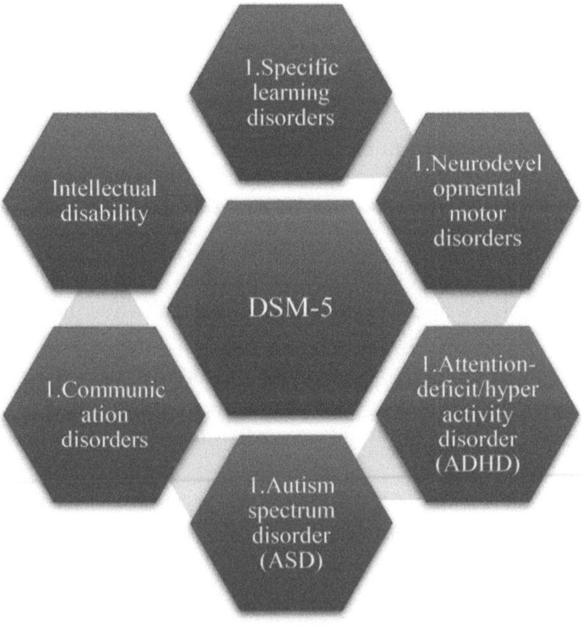


**Fig. 26.3** The DSM-5 (*Diagnostic and Statistical Manual of Mental Disorders*) defines neurodevelopmental disorders into six broad categories

if confined, repetitive patterns of behaviour, interests, or activities exist. It affects over 2% of the pre-school children population and includes an important fraction of the conditions accounting for the heterogeneity of autism. There were 74 research conducted, 26 of which were examined in Asia, 4 in Africa, 25 in Europe, 16 in the United States and three in Australia. Participants in these trials ranged in age from 0 to 27 years. The prevalence of ASD was 0.4% (95% CI: 0.1–1) in Asia, 1% (95% CI: 0.8–1.1) in America, 0.5% (95% CI: 0.2–1) in Europe, 1% (95% CI: 0.3–3.1) in Africa, and 1.7% (95% CI: 0.5–6.1) in Australia (Salari et al. 2022), as shown in Table 26.1.

A study investigated whether *Passiflora* extract could improve autistic-like behaviour in rats exposed to valproic acid during pregnancy. The extract, particularly at 300 mg/kg, significantly improved behavioural problems, and reduced oxidative stress markers and brain cell damage, suggesting its potential as a treatment for autism spectrum disorders (Amini et al. 2023).

### 26.3.2 Attention-Deficit/Hyperactivity Disorder (ADHD)

Neurodevelopmental disorders frequently co-occur with one another, such as individuals with autism spectrum disorder often having intellectual developmental disorder (intellectual disability) and many children with ADHD also having a specific learning disorder. These disorders often co-occur with other mental and behavioural disorders with onset in childhood, such as communication disorders and autism spectrum disorder may be associated with anxiety disorders; ADHD with oppositional defiant disorder; and tics with obsessive-compulsive disorder. Communication disorders include language, speech sound, social (pragmatic) communication disorder and childhood-onset fluency disorders (Tekwu et al. 2013). These disorders have a major influence on an individual's everyday functioning and quality of life. As a result, there is a growing interest in exploring non-pharmaceutical approaches to manage these conditions. One such approach that has gained attention is the use of *Passiflora*, also known as passionflower. *Passiflora* is a plant species that has

**Table 26.1** Analysis of ASD prevalence by continents

| Continent | Study (*n*) | Sample size | Heterogeneity $I^2$% | Begg and Mazumdar test | Prevalence (%) (95% CI) |
|---|---|---|---|---|---|
| Asia | 26 | 7,356,939 | 99.9 | 0.103 | 0.4 (95% CI: 0.1–1) |
| America | 16 | 3,758,240 | 99.6 | 0.137 | 1 (95% CI: 0.8–1.1) |
| Europe | 25 | 17,480,163 | 99.9 | 0.171 | 0.5 (95% CI: 0.2–1) |
| Africa | 4 | 54,326 | 99 | 0.734 | 1 (95% CI: 0.3–3.1) |
| Australia | 3 | 1,563,089 | 99.7 | 0.296 | 1.7 (95% CI: 0.5–6.1) |

*CI* confidence interval

been traditionally used for its potential effects on neurodevelopmental disorders (Tekwu et al. 2013).

ADHD is characterized by impaired levels of inattention, disorganization and/or hyperactivity/impulsivity. Inattention and disorganization are defined as the inability to stay on task, the appearance of not listening and the loss of task-related materials at levels that are inappropriate for age or developmental level. Hyperactivity/impulsivity is characterized by excessive activity, fidgeting, inability to sit still, intruding into other people's activities and unwillingness to wait—symptoms that are out of proportion to age or development level. In Table 26.2, we can see the prevalence of ADHD among US children and adolescents (Xu et al. 2018; Li et al. 2023) as shown below.

A study investigated whether passionflower could be beneficial for children with ADHD. Thirty-four children with ADHD were randomly assigned to take either passionflower or methylphenidate (a common medication for ADHD) for 8 weeks. Both groups showed significant improvement in symptoms based on parent and teacher ratings. While methylphenidate had a higher dropout rate and caused more side effects, passionflower showed similar effectiveness with a better side effect profile. This suggests passionflower could be a potential treatment option for ADHD, but larger studies are needed to confirm these findings (Mohammadi et al. 2005).

Research is ongoing to explore various treatment options for neurodevelopmental disorders, including both pharmaceutical and non-pharmaceutical approaches. *Passiflora*, a plant species traditionally used for its potential effects on these conditions, has shown promising results in preliminary studies for both ASD and ADHD. However, more extensive studies are needed to confirm these findings and establish *Passiflora* as a reliable treatment option. Overall, understanding neurodevelopmental disorders is crucial for providing appropriate support and potential treatment options for individuals affected by these conditions. As research

**Table 26.2** Overall prevalence of ADHD according to age globally

| Year | Overall % | Ages 4–11 | Ages 12–17 |
|---|---|---|---|
| 2021–2022 | 10.5 | 7.5 | 14.6 |
| 2019–2020 | 10.1 | 7.7 | 13.2 |
| 2017–2018 | 10.2 | 7.5 | 13.4 |
| 2015–2016 | 10.2 | 7.7 | 13.5 |
| 2013–2014 | 9.4 | 7.9 | 11.3 |
| 2011–2012 | 9.6 | 7.7 | 12.0 |
| 2009–2010 | 9.0 | 6.9 | 11.9 |
| 2007–2008 | 8.1 | 6.1 | 10.8 |
| 2005–2006 | 7.5 | 6.0 | 9.3 |
| 2003–2004 | 7.3 | 5.8 | 9.3 |
| 2001–2002 | 7.2 | 5.9 | 8.9 |
| 1999–2000 | 6.5 | 5.3 | 8.2 |
| 1997–1998 | 6.1 | 5.3 | 7.2 |

continues to advance, new and improved approaches may emerge to help individuals manage the challenges associated with these disorders and live fulfilling lives.

## 26.4 Neuropsychiatric Disorders

Neuropsychiatric disorders including severe depressive disorders such as insomnia, stress, anxiety and depression are prevalent and, unfortunately, growing concern (Fig. 26.4). Approximately 322 million individuals worldwide suffer from depression, while 264 million people from anxiety (WHO, 2017). Psychotherapy is a potential treatment for mild nervous system dysfunctions; yet, pharmaceutical treatment is necessary for more severe disorders (Cuijpers et al. 2020). Over the duration of last year, the Coronavirus Disease-2019 (COVID-19) epidemic led to subsequent increases in anxiety and depression cases worldwide (Rogers et al. 2020). Although, pharmacotherapy has been effective, there is a chance that it will cause dependency and side effects (Vasileva et al. 2019). Therefore, searching for herbal treatments for neuropsychiatric disorders came into vision (Liu et al. 2015). The perennial plant *Passiflora incarnata* is home to valuable phytochemicals that have been shown to have positive effects on health. Chrysin seems to be the most significant flavonoid of them all because of its neuroprotective properties (Al-Kuraishy et al. 2020).

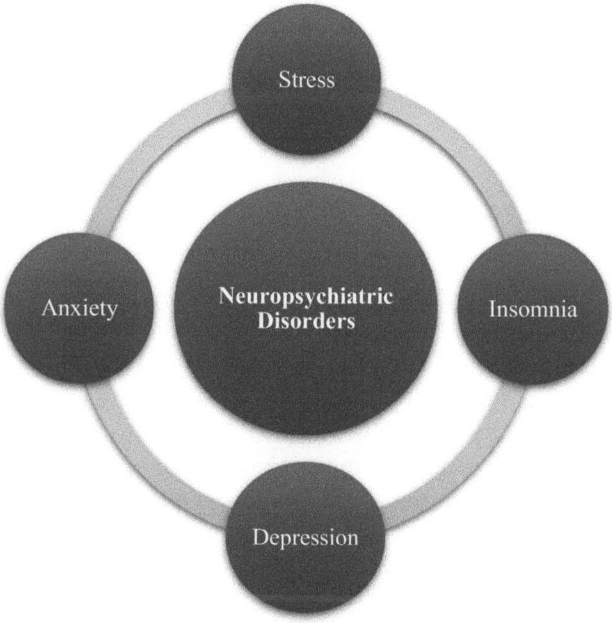


**Fig. 26.4** Various types of neuropsychiatric disorders

### 26.4.1 Stress

Stress is the body's instinctive natural response to both physical and psychological stimuli. Chronic, prolonged stress conditions can lead to mood disorders as well as issues with sleep, and difficulty in focussing or concentration, and memory (Janda et al. 2020). The passionflower is a perennial plant that lasts for longer time duration as compared to plants growing annually. The *Passiflora incarnata* has been determined having major therapeutic effects against several neurological disorders. The passionflower is helpful in the treatment of depression, anxiety and sleeplessness (insomnia) since it relieves a state of tension or stress (Janda et al. 2020). *Passiflora incarnata* has been recognized as a herbal remedy alleviating the adverse effects of stress (Da Fonseca et al. 2020). It has also been established that passionflower exhibits beneficial effects on memory functioning (Kim et al. 2019). *P. incarnata* serves a therapeutic impact in the treatment of degenerative brain disorders, sleeping difficulties and memory loss in those who suffer from chronic insomnia (Miroddi et al. 2013). When it comes to anxiety attacks, restlessness, insomnia and depression, *Passiflora* shows promising results (Kim et al. 2017).

*Passiflora incarnata*'s position as a medicinal product was acknowledged when the European Medicines Agency released a herbal monograph on the plant (European Medicines Agency 2014). There is no health risks associated with using *Passiflora incarnata*, according to clinical investigations (Miyasaka et al. 2007; Movafegh et al. 2008). Some symptoms that are related to neuropsychiatry may be relieved by passionflower. There are no known negative consequences of administering passionflower, such as memory loss or collapse of psychometric functions. *Passiflora incarnata* has an antianxiety effect similar to that of medications like midazolam and oxazepam. As a result, it appears to be a safe and efficient medication to mitigate behaviours associated with depression, anxiety, sleeplessness and stress reactivity (Janda et al. 2020).

### 26.4.2 Insomnia, Anxiety and Depression

Insomnia is becoming more prevalent in the world. For those who suffer from insomnia, it may be quite painful because many of the medications used to treat insomnia may have negative consequences on mental health (Hahn et al. 2014). Insomnia issues are difficult to diagnose because of a variety of social phenomena and lifestyle choices that make it difficult to get a decent night's sleep. Unusual sleep patterns governed by the circadian rhythm can seriously impair a person's quality of life, as well as several bodily metabolic processes (Elbaz et al. 2017) and cognitive functions (Backhaus et al. 2006; Fulda and Schulz 2001; Gorgoni et al. 2013; Joo et al. 2014). The market is always seeing a large influx of prescription medications due to the significance of sleep. Apart from their ability to induce sleep, many of these medicines have adverse effects on the body that include hyperphagia, obesity, diabetes, depression and the inability to recognize aberrant behaviours that might create difficulties (Magee et al. 2010). Traditional herbal therapy, on the other hand, has gained popularity recently as an effective and safer alternative to pharmaceutical treatments for insomnia. It can help lower blood pressure, enhance cognitive function and induce sleep. Passionflower, or *Passiflora incarnata*, has recently drawn attention because of its potentially beneficial and safe sedative, sleep-inducing and antianxiety properties (Aoyagi et al. 1974; Jawna-Zboinska et al. 2016; Kim et al. 2017; Ngan and Conduit 2011; Yurcheshen et al. 2015). *P. incarnata* exhibits a number of CNS-depressing effects, according to reports. Numerous bioactive metabolites (Diniz et al. 2015) of *Passiflora incarnata* have been investigated experimentally and have demonstrated promising safety and effectiveness, along with potential for medicinal use (Ahn et al. 2016).

In North America, passionflower was first used by humans in the late prehistoric era. According to certain examinations, passionflower is just as effective as benzodiazepines while having fewer negative effects (Bokelmann 2022). In Europe, the species *Passiflora incarnata* has long been used in traditional herbal therapy to alleviate anxiety and insomnia. It is used as a sedative tea in North America (Miroddi et al. 2013). *P. incarnata* has a therapeutic impact in the treatment of degenerative brain disorders, difficulty falling asleep and memory loss and also those who suffer from chronic insomnia. Due to its sedative properties, *Passiflora* may help cure insomnia by increasing the likelihood that the individual who is having trouble falling asleep will do so (Miroddi et al. 2013).

### 26.4.3 GABA System Dysfunction

Passionflower and diazepam have been believed to have a similar mode of action since several research indicated that pre-administration of flumazenil, an antagonist of the GABA-A benzodiazepine binding site receptor, attenuates in vivo the anxiolytic effect of *Passiflora incarnata* (Medina et al. 1990). Anxiety and depression disorders are among the numerous neuropsychiatric illnesses linked to Gamma-aminobutyric acid (GABA) system dysfunction. The regulation of the GABA system mediates a number of the pharmacological actions of *Passiflora incarnata*, such as affinity for the GABA-A and GABA-B receptors and effects on GABA absorption. It does not seem feasible that the benzodiazepine site is how *Passiflora incarnata* extract works. Nonetheless, it is conceivable that one of the plant extracts' modes of action is binding to the GABA-site of the GABA-A receptor (Appel et al. 2011).

With growing concerns about the prevalence of conditions like depression, anxiety and insomnia, *Passiflora* offers a potentially safer and more readily available option compared to traditional pharmaceutical treatments. Studies suggest that *Passiflora* possesses several properties beneficial for neuropsychiatric health. Its potential anxiolytic effects show promise in alleviating anxiety symptoms similar to conventional medications but with fewer side effects (Jawna-Zboinska et al. 2016). Additionally, *Passiflora*'s sedative properties may offer relief for individuals struggling with insomnia, promoting better sleep quality without the potential drawbacks associated with sleep medications. As research continues to explore the full potential of *Passiflora incarnata*, it may hold significant promise for improving the lives of individuals struggling with neuropsychiatric disorders, offering a natural and potentially safer approach to managing these conditions.

## 26.5 Neurodegenerative Disorders

Neurodegeneration is the progressive loss of neurons, their structure and/or their functions, which is a major health problem and a key factor in the pathogenesis of various brain illnesses, as neurons are not eternal (Przedborski et al. 2003). Neurons are essential for the effective operation of the human brain, as they serve a crucial function in facilitating communication (van den Heuvel and Sporns 2013). Neurons are distributed throughout the body, not confined to the brain (Pino et al. 2017). In childhood, neural stem cells generate the majority of neurons, but this number significantly diminishes during adulthood (Ganat et al. 2006). Dysfunction of the synapses and neural network and the accumulation of physiologically modified protein variations in the brain are linked to neurodegeneration (Hoover et al. 2010).

Neurodegenerative diseases (NDs) affect millions globally, with age being the biggest risk factor. New research suggests that both genetics and environment can increase the risk of developing these illnesses. Furthermore, even while particular genes (in an individual) responsible for NDs express themselves, their immediate environment has a major impact on the rate and degree of neurodegeneration (Liu et al. 2022). Neurodegenerative diseases, shown in Fig. 26.5, impact various aspects of human functioning, including basic speech, movement, stability, balance, complex tasks like cognitive abilities and bladder and bowel functions, since the brain regulates multiple bodily functions.

### 26.5.1 Parkinson's Disease (PD)

Parkinson's disease is a degenerative neurological condition that causes unsteady walking, tremors, muscle stiffness and issues with balance and coordination (Bartels & Leenders, 2009). PD is brought on by both hereditary and non-genetic factors. Age is thought to be the main PD risk factor (Kouli et al. 2018). Additionally, factors such as high caffeine intake, smoking and exposure to toxins in the environment have been linked to an increased risk of developing Parkinson's disease. However, the exact reasons for this connection are still not fully understood (Priyadarshi et al. 2001). In Parkinson's disease, the loss of cells in the brain's nigrostriatal pathway causes a decrease in dopamine levels in the striatum, leading to the typical motor symptoms. Antioxidant and antiparkinsonian activity of *Passiflora incarnata* leaves delves into the potential antiparkinsonian and antioxidant properties of the butanol extract from the leaves of *P. incarnata*, also known as BEPI (Ingale and Kasture 2017). The findings suggest that BEPI could notably decrease the duration of catalepsy induced by haloperidol and jaw movements induced by tacrine, pointing towards its possible antiparkinsonian benefits. Additionally, BEPI demonstrated a

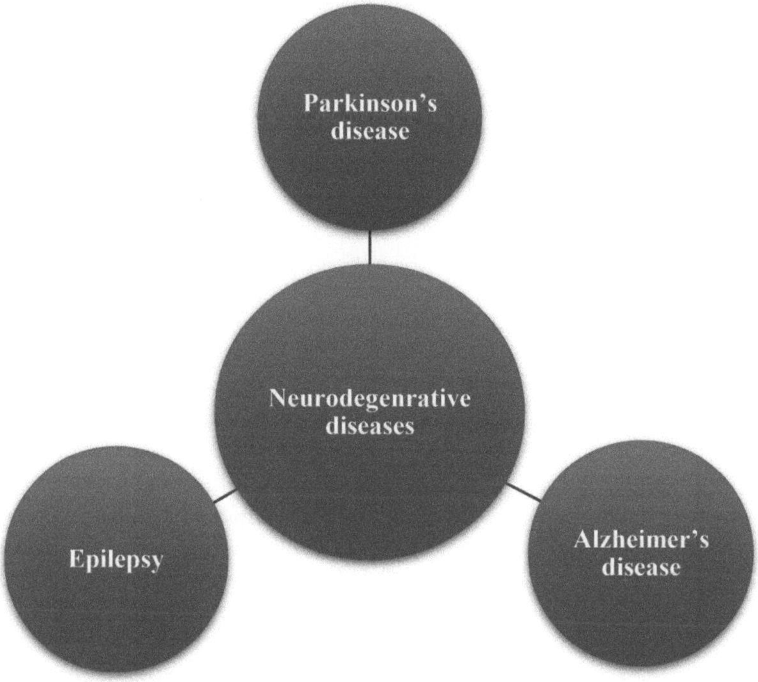


**Fig. 26.5** Some of the most prevalent neurodegenerative disorders that can be cured by *Passiflora incarnata*

significant antioxidant effect (Ingale and Kasture 2014). These findings hint at the possibility that *Passiflora incarnata* might be useful in managing Parkinson's disease, thanks to its neuroprotective and antioxidant attributes.

### 26.5.2 Alzheimer's Disease (AD)

Alzheimer's disease is a progressive neurodegenerative condition for which no cure currently exists. It stands as the leading cause of dementia worldwide. Approximately 40 million individuals are affected by AD, with the majority of cases occurring in people aged 65 years and older. The underlying pathophysiology of this illness is complex and not understood (Querfurth and LaFerla 2010). The main causes of Alzheimer's disease include the hyperphosphorylation of tau proteins, the formation of senile plaques of Aβ, inflammation of brain neurons and increased activity of acetylcholinesterase. These factors lead to a decrease in acetylcholine levels in the brain (Bornstein et al. 2011). Research indicates that the hydroethanolic extract of *Passiflora*, Nitida Kunth, has neuroprotective properties and could provide cognitive benefits in a preclinical model of Alzheimer's-like cognitive impairment caused by streptozotocin (STZ). This extract may improve behavioural assessments and significantly preserve healthy cells in the hippocampal regions CA1 and CA3 (De Almeida et al. 2021). These findings imply that *Passiflora incarnata* and related species might be advantageous in the management of Alzheimer's disease, owing to their neuroprotective and antioxidant characteristics. Nonetheless, additional studies are required to fully comprehend its impact and potential applications in Alzheimer's disease treatment.

### 26.5.3 Epilepsy

Epilepsy, also known as a seizure disorder, is a brain condition causing recurring seizures. Not all seizures are caused by epilepsy, and the term 'epilepsy' simply means the same thing as 'seizure disorders'. It affects around 50 million people of all genders, ethnic backgrounds and ages worldwide and is characterized by recurrent, unprovoked seizures. Symptoms can vary, with some losing awareness, staring blankly, or twitching arms or legs. Epilepsy is diagnosed when at least two unprovoked seizures occur at least 24 h apart. Many traditional medical systems have utilized *Passiflora incarnata* to treat epilepsy (Singh et al. 2012).

Fiest et al. examined 222 population-based studies on the prevalence and incidence of epilepsy worldwide, employing meta-analytic approaches to investigate the factors that contribute to heterogeneity between estimates. Results showed a point prevalence of active epilepsy of 6.38 per 1000 persons, lifetime prevalence of 7.60 per 1000 persons and annual cumulative incidence of 67.77 per 100,000 persons (Fiest et al. 2017).

*Passiflora incarnata* hydroethanolic extract showed significant protection against pentylenetetrazol (PTZ)-induced seizures and post-ictal depression in mice, preserving serotonin and noradrenaline levels (Singh et al. 2012). Unlike diazepam, it improved mood-related outcomes. Pasipay, a similar extract, offered 100% seizure and mortality protection at 0.4 mg/kg. These findings highlight its dual anticonvulsant and antidepressant potential. However, flumazenil and naloxone could suppress Pasipay's anticonvulsant effects. The study suggests that Pasipay could be useful for treatment of absence seizures, possibly due to its effects on GABAergic and opioid systems (Nassiri-Asl et al. 2007). It has been shown that the conventional antiepileptic medication, diazepam, exacerbates the post-ictal depression associated with PTZ-induced seizures. The hydroethanolic extract of *Passiflora incarnata* suppresses these seizures and ameliorates the depression.

While there is currently no cure for most neurodegenerative disorders, ongoing research explores various treatment options to manage symptoms and improve the quality of life for patients. One potential area of interest is the use of *Passiflora*, a plant species with potential neuroprotective and antioxidant properties. Studies suggest that *Passiflora* extracts might be beneficial in managing Parkinson's disease, Alzheimer's disease and epilepsy. However, further research is necessary to fully understand the effectiveness and mechanisms of action of *Passiflora* in these contexts. Overall, understanding neurodegenerative disorders is crucial for developing effective treatment strategies and improving patient care. As research advances, new insights may emerge into the causes and potential therapies for these debilitating conditions.

## 26.6 Conclusion

While research on *Passiflora*'s efficacy is ongoing, existing evidence suggests it could be a valuable tool in managing various neurological diseases. Its safety profile further strengthens its appeal as a potential alternative or complementary therapy. However, it is crucial to consult with a healthcare professional before incorporating *Passiflora* into your routine, especially if using other medications, due to potential interactions. In conclusion, neurodevelopmental and neurodegenerative disorders significantly impact various aspects of life, necessitating ongoing research for effective treatment strategies. However, while existing evidence suggests that *Passiflora* could be a valuable tool in managing various neuropsychiatric conditions, more extensive studies are needed to confirm these findings and establish it as a reliable treatment option. It is also crucial to consult with a healthcare professional before incorporating *Passiflora* into one's routine due to potential interactions with other medications. As such, *Passiflora* represents a promising area of research in the quest for effective, accessible and safer treatments for neurodevelopmental and neurodegenerative disorders.

**Acknowledgements** We extend our heartfelt gratitude to the Amity Institute of Biotechnology, Amity University Uttar Pradesh, Lucknow, India, for providing the necessary infrastructure and resources to facilitate this work. We acknowledge the collaborative and intellectually stimulating environment at Amity University, which played a vital role in the successful completion of this work.

**Conflict of Interest** Authors declare no conflict of interest.

**Consent for Publication** We confirm that all authors have reviewed and approved the manuscript titled 'Medicinal and Nutritional Importance of *Passiflora incarnata* in Human Health' for publication.

We hereby confirm that no permission is required for the figures included in this chapter. The figures are original works created by the authors.

## References

Ahn J, Ahn HS, Cheong JH, Dela Pena I (2016) Natural product-derived treatments for attention-deficit/hyperactivity disorder: safety, efficacy, and therapeutic potential of combination therapy. Neural Plast 2016:1320423

Al-Kuraishy H, Alwindy S, Al-Gareeb A (2020) Beneficial neuro-pharmacological effect of passionflower (Passiflora incarnata L). Online J Neurol Brain Disord 3(5). https://doi.org/10.32474/OJNBD.2020.03.000172

American Psychiatric Association (2013) Diagnostic and statistical manual of mental disorders: DSM-5, 5th edn. American Psychiatric Association, Washington, DC

Amini F, Amini-Khoei H, Haratizadeh S, Setayesh M, Basiri M, Raeiszadeh M, Nozari M (2023) Hydroalcoholic extract of Passiflora incarnata improves the autistic-like behavior and neuronal damage in a valproic acid-induced rat model of autism. J Tradit Complement Med 13(4):315–324. https://doi.org/10.1016/j.jtcme.2023.02.005. PMID: 37396155; PMCID: PMC10310915

Aoyagi N, Kimura R, Murata T (1974) Studies on Passiflora incarnata dry extract. I. Isolation of maltol and pharmacological action of maltol and ethyl maltol. Chem Pharm Bull (Tokyo) 22(5):1008–1013

Appel K, Rose T, Fiebich B, Kammler T, Hoffmann C, Weiss G (2011) Modulation of the γ-aminobutyric acid (GABA) system by Passiflora incarnata L. Phytother Res 25(6):838–843 https://doi.org/10.1002/ptr.3352.Epub 2010 Nov 19. PMID: 21089181

Backhaus J, Junghanns K, Born J, Hohaus K, Faasch F, Hohagen F (2006) Impaired declarative memory consolidation during sleep in patients with primary insomnia: influence of sleep architecture and nocturnal cortisol release. Biol Psychiatry 60(12):1324–1330

Bartels AL, Leenders KL (2009) Parkinson's disease: the syndrome, the pathogenesis and pathophysiology. Cortex 45(8):915–921. https://doi.org/10.1016/j.cortex.2008.11.010. PMID: 19095226

Bokelmann JM (2022) Passionflower. In: Medicinal herbs in primary care. Elsevier, Philadelphia

Borghi SM, Carvalho TT, Staurengo-Ferrari L et al (2013) Vitexin inhibits inflammatory pain in mice by targeting TRPV1, oxidative stress, and cytokines. J Nat Prod 76(6):1141–1149

Bornstein JJ, Eckroat TJ, Houghton JL, Jones CK, Green KD, Garneau-Tsodikova S (2011) Tacrine-mefenamic acid hybrids for inhibition of acetylcholinesterase. MedChemComm 2(5):406–412. https://doi.org/10.1039/c0md00256a

Charrière K, Schneider V, Perrignon-Sommet M, Lizard G, Benani A, Jacquin-Piques A, Vejux A (2024) Exploring the role of apigenin in neuroinflammation: insights and implications. Int J Mol Sci 25(9):5041

Chiang MC, Tsai TY, Wang CJ (2023) The potential benefits of quercetin for brain health: a review of anti-inflammatory and neuroprotective mechanisms. Int J Mol Sci 24(7):6328

Choi JS, Nurul Islam M, Ali MY, Kim EJ, Kim YM, Jung HA (2014) Effects of C-glycosylation on anti-diabetic, anti-Alzheimer's disease and anti-inflammatory potential of apigenin. Food Chem Toxicol 64:27–33

Cuijpers P, Stringaris A, Wolpert M (2020) Treatment outcomes for depression: challenges and opportunities. Lancet Psychiatry 7(11):925–927

Da Fonseca LR, de Rodrigues RA, de Ramos AS, da Cruz JD, Ferreira JLP, de Silva JRA, Amaral ACF (2020) Herbal medicinal products from Passiflora for anxiety: an unexploited potential. ScientificWorldJournal 2020:6598434

De Almeida VL, Silva CG, Campana PRV (2021) Flavonoids of Passiflora: isolation, structure elucidation, and biotechnological application. Stud Nat Prod Chem 71:263–310

Dhawan K, Kumar S, Sharma A (2003) Evaluation of central nervous system effects of Passiflora incarnata in experimental animals. Pharm Biol 41:87

Dhawan K, Dhawan S, Sharma A (2004) Passiflora: a review update. J Ethnopharmacol 94(1):1–23

Diniz TC, Silva JC, Lima-Saraiva SRGD, Ribeiro FPRA, Pacheco AGM, de Freitas RM, Almeida JRGDS (2015) The role of flavonoids on oxidative stress in epilepsy. Oxidative Med Cell Longev 2015:171756

Elbaz I, Zada D, Tovin A, Braun T, Lerer-Goldshtein T, Wang G et al (2017) Sleep-dependent structural synaptic plasticity of inhibitory synapses in the dendrites of hypocretin/orexin neurons. Mol Neurobiol 54(8):6581–6597

European Medicines Agency (2014) Assessment report on Passiflora incarnata L. herba, vol 22. European Medicines Agency, Amsterdam

Fiest KM, Sauro KM, Wiebe S, Patten SB, Kwon CS, Dykeman J et al (2017) Prevalence and incidence of epilepsy: a systematic review and meta-analysis of international studies. Neurology 88(3):296–303

Fulda S, Schulz H (2001) Cognitive dysfunction in sleep disorders. Sleep Med Rev 5(6):423–445

Ganat YM, Silbereis J, Cave C, Ngu H, Anderson GM, Ohkubo Y et al (2006) Early postnatal astroglial cells produce multilineage precursors and neural stem cells in vivo. J Neurosci 26(33):8609–8621

Gorgoni M, D'Atri A, Lauri G, Rossini PM, Ferlazzo F, De Gennaro L (2013) Is sleep essential for neural plasticity in humans, and how does it affect motor and cognitive recovery? Neural Plast 2013:103949

Hahn EA, Wang HX, Andel R, Fratiglioni L (2014) A change in sleep pattern may predict Alzheimer disease. Am J Geriatr Psychiatry 22(11):1262–1271

Hoover BR, Reed MN, Su J, Penrod RD, Kotilinek LA, Grant MK et al (2010) Tau mislocalization to dendritic spines mediates synaptic dysfunction independently of neurodegeneration. Neuron 68(6):1067–1081

Ingale SP, Kasture SB (2014) Antioxidant and antiparkinsonian activity of *Passiflora incarnata* leaves. Orient Pharm Exp Med 14:231–236

Ingale SP, Kasture SB (2017) Protective effect of standardized extract of Passiflora incarnata flower in Parkinson's and Alzheimer's disease. Anc Sci Life 36(4):200–206

Janda K, Wojtkowska K, Jakubczyk K, Antoniewicz J, Skonieczna-Żydecka K (2020) Passiflora incarnata in neuropsychiatric disorders—a systematic review. Nutrients 12(12):3894

Jawna-Zboinska K, Blecharz-Klin K, Joniec-Maciejak I, Wawer A, Pyrzanowska J, Piechal A et al (2016) Passiflora incarnata L. improves spatial memory, reduces stress, and affects neurotransmission in rats. Phytother Res 30(5):781–789

Joo EY, Kim H, Suh S, Hong SB (2014) Hippocampal substructural vulnerability to sleep disturbance and cognitive impairment in patients with chronic primary insomnia: magnetic resonance imaging morphometry. Sleep 37(7):1189–1198

Kim M, Lim HS, Lee HH, Kim TH (2017) Role identification of Passiflora incarnata Linnaeus: a mini review. J Menopausal Med 23(3):156–159

Kim GH, Lim K, Yang HS, Lee JK, Kim Y, Park SK et al (2019) Improvement in neurogenesis and memory function by administration of Passiflora incarnata L. extract applied to sleep disorder in rodent models. J Chem Neuroanat 98:27–40

Kouli A, Torsney KM, Kuan WL (2018) Parkinson's disease: etiology, neuropathology, and pathogenesis. In: Stoker TB, Greenland JC (eds) Parkinson's disease: pathogenesis and clinical aspects [Internet]. Codon Publications, Brisbane

Kuriakose D, Xiao Z (2020) Pathophysiology and treatment of stroke: present status and future perspectives. Int J Mol Sci 21(20):7609

Li Y, Yan X, Li Q, Li Q, Xu G, Lu J et al (2023) Prevalence and trends in diagnosed ADHD among US children and adolescents, 2017–2022. JAMA Netw Open 6(10):e2336872. https://doi.org/10.1001/jamanetworkopen.2023.36872

Lima LKF, Pereira SKS, Junior RDSS, Santos FPDS, Nascimento AS, Feitosa CM et al (2018) A brief review on the neuroprotective mechanisms of vitexin. Biomed Res Int 2018:4785089

Liu L, Liu C, Wang Y, Wang P, Li Y, Li B (2015) Herbal medicine for anxiety, depression and insomnia. Curr Neuropharmacol 13(4):481–493

Liu H, Hu Y, Zhang Y, Zhang H, Gao S, Wang L et al (2022) Mendelian randomization highlights significant difference and genetic heterogeneity in clinically diagnosed Alzheimer's disease GWAS and self-report proxy phenotype GWAX. Alzheimer's Res Ther 14(1):17

Magee CA, Huang XF, Iverson DC, Caputi P (2010) Examining the pathways linking chronic sleep restriction to obesity. J Obes 2010:821710

Medina JH, Paladini AC, Wolfman C, Levi de Stein M, Calvo D, Diaz LE et al (1990) Chrysin (5,7-di-OH-flavone), a naturally-occurring ligand for benzodiazepine receptors, with anticonvulsant properties. Biochem Pharmacol 40(10):2227–2231

Michael HSR, Mohammed NB, Ponnusamy S, Gnanaraj WE (2022) A folk medicine: *Passiflora incarnata* L. phytochemical profile with antioxidant potency. Turk J Pharm Sci 19(3):287–292

Miroddi M, Calapai G, Navarra M, Minciullo PL, Gangemi S (2013) *Passiflora incarnata* L.: ethnopharmacology, clinical application, safety and evaluation of clinical trials. J Ethnopharmacol 150(3):791–804

Miyasaka LS, Atallah AN, Soares BGO (2007) *Passiflora* for anxiety disorder. Cochrane Database Syst Rev (1):CD004518

Mohammadi MR, Akhondzadeh S, Momeni F (2005) *Passiflora incarnata* in the treatment of attention-deficit hyperactivity disorder in children and adolescents. Therapy 2(4):609–614. https://doi.org/10.2217/14750708.2.4.609

Movafegh A, Alizadeh R, Hajimohamadi F, Esfehani F, Nejatfar M (2008) Preoperative oral *Passiflora incarnata* reduces anxiety in ambulatory surgery patients: a double-blind, placebo-controlled study. Anesth Analg 106(6):1728–1732

Nassiri-Asl M, Shariati-Rad S, Zamansoltani F (2007) Anticonvulsant effects of aerial parts of *Passiflora incarnata* extract in mice: involvement of benzodiazepine and opioid receptors. BMC Complement Altern Med 7:26. https://doi.org/10.1186/1472-6882-7-26

Nezhad Salari AM, Rasoulizadeh Z, Shabgah AG, Vakili-Ghartavol R, Sargazi G, Gholizadeh Navashenaq J (2024) Exploring the mechanisms of kaempferol in neuroprotection: implications for neurological disorders. Cell Biochem Funct. https://doi.org/10.1002/cbf.3857

Ngan A, Conduit R (2011) A double-blind, placebo-controlled investigation of the effects of Passiflora incarnata (passionflower) herbal tea on subjective sleep quality. Phytother Res 25(8):1153–1159. https://doi.org/10.1002/ptr.3400

Patel S (2009) *Passiflora incarnata* Linn: a phytopharmacological review. Int J Green Pharm 3(4):277–281. https://doi.org/10.4103/0973-8258.60850

Pino A, Fumagalli G, Bifari F, Decimo I (2017) New neurons in adult brain: distribution, molecular mechanisms and therapies. Biochem Pharmacol 141:4–22. https://doi.org/10.1016/j.bcp.2017.06.052

Priyadarshi A, Khuder SA, Schaub EA, Priyadarshi SS (2001) Environmental risk factors and Parkinson's disease: a meta-analysis. Environ Res 86(2):122–127. https://doi.org/10.1006/enrs.2001.4264

Przedborski S, Vila M, Jackson-Lewis V (2003) Neurodegeneration: what is it and where are we? J Clin Invest 111(1):3–10. https://doi.org/10.1172/JCI17247

Querfurth HW, LaFerla FM (2010) Alzheimer's disease. N Engl J Med 362(4):329–344. https://doi.org/10.1056/NEJMra0909142. Erratum in: N Engl J Med. 2011 Feb 10;364(6):588. PMID: 20107219

Rogers JP, Chesney E, Oliver D, Pollak TA, McGuire P, Fusar-Poli P, Zandi MS, Lewis G, David AS (2020) Psychiatric and neuropsychiatric presentations associated with severe coronavirus infections: a systematic review and meta-analysis with comparison to the COVID-19 pandemic. Lancet Psychiatry 7(7):611–627. https://doi.org/10.1016/S2215-0366(20)30203-0

Salari N, Rasoulpoor S, Rasoulpoor S, Shohaimi S, Jafarpour S, Abdoli N, Khaledi-Paveh B, Mohammadi M (2022) The global prevalence of autism spectrum disorder: a comprehensive systematic review and meta-analysis. Ital J Pediatr 48(1):112. https://doi.org/10.1186/s13052-022-01294-5

Salehi B, Venditti A, Sharifi-Rad M, Kregiel D, Sharifi-Rad J, Durazzo A, Lucarini M, Santini A, Souto EB, Novellino E et al (2019) The therapeutic potential of apigenin. Int J Mol Sci 20(6):1305. https://doi.org/10.3390/ijms20061305

Silva Dos Santos J, Gonçalves Cirino JP, de Oliveira Carvalho P, Ortega MM (2021) The pharmacological action of kaempferol in central nervous system diseases: a review. Front Pharmacol 12:744709. https://doi.org/10.3389/fphar.2021.744709

Singh B, Singh D, Goel RK (2012) Dual protective effect of *Passiflora incarnata* in epilepsy and associated post-ictal depression. J Ethnopharmacol 139(1):273–279. https://doi.org/10.1016/j.jep.2011.11.019

Skovronsky DM, Lee VM-Y, Trojanowski JQ (2006) Neurodegenerative diseases: new concepts of pathogenesis and their therapeutic implications. Annu Rev Pathol 1:151–170. https://doi.org/10.1146/annurev.pathol.1.110304.100113

Tekwu EM, Beng VP, Kuete V (2013) African medicinal plants acting on the reproductive, cardiovascular, and central nervous systems. In: Kuete V (ed) Medicinal plant research in Africa. Elsevier, Amsterdam van den Heuvel MP, Sporns O (2013) Network hubs in the human brain. Trends Cogn Sci 17(12):683–696. https://doi.org/10.1016/j.tics.2013.09.012

Vasileva LV, Ivanovska MV, Murdjeva MA, Saracheva KE, Georgiev MI (2019) Immunoregulatory natural compounds in stress-induced depression: an alternative or an adjunct to conventional antidepressant therapy? Food Chem Toxicol 125:319–332. https://doi.org/10.1016/j.fct.2019.01.005

Wang Y, Zhen Y, Wu X et al (2015) Vitexin protects brain against ischemia/reperfusion injury via modulating mitogen-activated protein kinase and apoptosis signaling in mice. Phytomedicine 22(3):379–384. https://doi.org/10.1016/j.phymed.2014.12.008

WHO Team (Mental Health, Brain Health and Substance Use) (2017) WHO depression and other common mental disorders. Global Health Estimates. WHO/MSD/MER/2017.2https://www.who.int/publications/i/item/depression-global-health-estimates

Xu G, Strathearn L, Liu B, Yang B, Bao W (2018) Twenty-year trends in diagnosed attention-deficit/hyperactivity disorder among US children and adolescents, 1997–2016. JAMA Netw Open 1(4):e181471. https://doi.org/10.1001/jamanetworkopen.2018.1471

Yurcheshen M, Seehuus M, Pigeon W (2015) Updates on nutraceutical sleep therapeutics and investigational research. Evid Based Complement Alternat Med 2015:105256. https://doi.org/10.1155/2015/105256

# Medicinal and Nutritional Importance of *Cinnamomum zeylanicum* in Human Health

27

Akansha Agrwal and Sheetal Mital

**Abstract**

For many ages, people from many civilizations have used cinnamon as a common spice, which is extracted from the internal bark of tropical evergreen trees in the genus "*Cinnamomum*." Apart from its cooking application, "cinnamon" is regarded as a treatment for gynecological, digestive, and pulmonary conditions in traditional Ayurvedic medicine. Almost every element of the cinnamon tree, encompassing the bark, flowers, leaves, fruits, and roots, has a medicinal or culinary use. The volatile oils extracted from the bark, leaves, and root bark exhibit considerable variation in their chemical composition, which implies that their pharmacological property may also vary. The plant's many sections all contain the same range of hydrocarbons in varied amounts, with the main components being camphor (root), eugenol (leaf), and cinnamon (bark). As a result, cinnamon provides a wide variety of oils, each with unique qualities that determine its usefulness to various businesses. For instance, the root, which mostly consists of camphor, is not as valuable commercially as the leaf or bark (Senevirathne et al., J Future Foods 2(2):125–142, 2022). The purpose of this chapter is to present a thorough overview of the scientific literature on the medicinal and nutritional significance of *Cinnamomum zeylanicum* for human health (Błaszczyk et al., Forests 12(5):648, 2021).

**Keywords**

*Cinnamomum zeylanicum* · Antimicrobial · Antiparasitic · Antidiabetic · Antioxidant

A. Agrwal · S. Mital (✉)
Department of Applied Science and Humanities, KIET Group of Institutions, Ghaziabad, Delhi-NCR, Uttar Pradesh, India
e-mail: sheetal.mital@kiet.edu



M. A. Ansari et al. (eds.), *Medicinal Plants and their Bioactive Compounds in Human Health: Volume 2*, https://doi.org/10.1007/978-981-95-3620-7_27

## 27.1 Introduction

Cinnamon is used as a treatment for gynecological, intestinal, and respiratory conditions. Nearly every component from the "cinnamon" tree, such as the roots, leaves, bark, fruits, and flowers, has a therapeutic or culinary purpose. The chemical composition of the volatile oils derived from the bark, leaves, and root barks shows considerable diversity, implying that their pharmacological effects could also vary (Shen et al. 2002). The plant's many sections all contain the same range of hydrocarbons in varied amounts, with the main components being camphor (root), eugenol (leaf), and cinnamon (bark) (Gruenwald et al. 2010; Ranasinghe et al. 2013). Many nations throughout the world classify cinnamon as a medicinal plant and utilize it as a traditional medicine (Gupta 2010).

## 27.2 Antimicrobial Properties

Recent studies indicate that cinnamon has the ability to prevent food spoilage caused by microorganisms (Jessica Elizabeth et al. 2017). Antifungal activities, primarily against *Penicillium*, *Aspergillus*, and *Fusarium*, as well as antibacterial activities, specifically against *Staphylococcus*, *Escherichia*, *Bacillus*, *Salmonella*, and *Pseudomonas*, have been demonstrated (Goñi et al. 2009; Singh et al. 2007). It has been observed that some species of *Klebsiella*, *Clostridium*, *Streptococcus*, *Acinetobacter*, *Enterococcus*, *Proteus*, *Enterobacter*, *Mycobacterium*, *Listeria*, and *Candida* cannot grow when exposed to 500 ppm of "*Cinnamomum zeylanicum*" (CZ) essential oil (EO) (Miyashita and Sadzuka 2013). The lowest bactericidal and inhibitory concentrations of cinnamon essential oils (EOs) were found to be 400–800 mg/L and 200–400 mg/L, respectively, in a study by Becerril et al. (2012). Since they do not trigger the emergence of antibiotic resistance and remain effective for extended periods of time, cinnamon essential oils are the perfect antibacterial material. This result is in line with other studies' findings (Goñi et al. 2009). Strong antibacterial activity of cinnamon has been found in food products. It has been observed that adding 5% cinnamon to pasteurized skim milk exposed to a pulsed electric field reduces *Salmonella typhimurium* cell activity by 52% (Pina-Pérez et al. 2012). Compared to mint, cardamom, and clove essential oils, cinnamon essential oil concentrations of 0.005–0.5% were more effective in preventing the growth of *Streptococcus thermophiles* in yogurt (Bayoumi and Bayoumi 1992). It has been demonstrated that 2 g of cinnamon in white wine inhibits *Helicobacter pylori* more effectively than green tea and raspberry (Lin et al. 2005). When applied to peeled shrimp, agar and alginate bilayer films enhanced with cinnamon essential oil (2 g/100 mL) show antibacterial properties, especially against the pathogenic bacterium *Listeria monocytogenes* (Arancibia et al. 2014). When encapsulated in chitosan, cinnamon essential oil has the potential to greatly reduce the severity and frequency of *Phytophthora drechsleri*-infected cucumber diseases (Mohammadi et al. 2015). Therefore, these findings offer scientific proof that essential oils derived from cinnamon are a potential food preservative source that might stop microbial deterioration and potentially shield against microbially induced foodborne illnesses. This implies that cinnamon may significantly improve the safety of food.

## 27.3 Antiparasitic Effects

The presence of proanthocyanidins (PAC) with a large number of A-type bonds has been reported in cinnamon bark, which makes it rather unique (Fitzpatrick 2013). About 20 essential oils (EOs) using packet tests on unfed *Haemaphysalis longicornis* nymphs were evaluated and it was identified that cinnamon was the most toxic resulting in over 80% mortality. Cinnamon essential oil demonstrated strong acaricidal activity against *Haemaphysalis longicornis*, with the active compound trans-cinnamaldehyde being equally effective across all tick life stages. Additionally, trans-cinnamaldehyde inhibited the reproductive capacity of *H. longicornis*, highlighting its potential as an eco-friendly acaricide (Ruihua et al. 2022). This suggests that cinnamon could be effectively used as a natural acaricide or for controlling ticks and mites, potentially offering an eco-friendly alternative for managing pests in agriculture and food production.

## 27.4 Antioxidant Properties

Antioxidant molecules have a vital role in human health as they protect against disease. Naturally occurring foods contain these substances, and they are also incorporated as additives in the food industry to prevent spoilage. Cinnamon is a source of antioxidant chemicals. This basic flavor has antimicrobial properties. In in vitro experiments, the essential oils extracted from the bark of CZ and eugenol demonstrated extremely potent properties, reducing the production of 3-nitrotyrosine and preventing peroxynitrite-induced lipid peroxidation (Chericoni et al. 2005). Because cinnamon bark infusion contains volatile oil molecules and polyphenols, Ervina et al. demonstrated that it has strong antioxidant properties. Easy extract preparation can be used as a strong antioxidant solution (Ervina et al. 2016; Yashin et al. 2017). At 100 and 200 ppm concentrations, the volatile oils of CZ demonstrated 55.9% and 66.9% antioxidant activity, respectively (Guddadarangavvanahally K. Jayaprakasha et al. 2003). Cinnamon contains natural antioxidants that play a crucial role in the aging process and the development of diseases. Among the various species, *Cinnamomum zeylanicum* and *Cinnamomum cassia* (L.) J. Presl. stand out for their remarkable potential. This type of bark extract contains chemicals that are biologically active, like linalool, trans-cinnamaldehyde, and eugenol. Studies have been conducted to investigate the function of phytonutrients in the stabilization of palm oil. These compounds inhibit the formation of acids, hydrocarbons, ketones, aldehydes, and alcohols. By adding antioxidants, oil oxidation episodes are inhibited, which lowers the amount of hazardous oxidative derivatives and free fatty acids produced (Shahid et al. 2018). In the order of water > acetone > methanol > ethyl acetate > dried fruit extracts of CZ, antioxidant activity was shown (Jayaprakasha et al. 2007). The oxidative process was suppressed by the etheric (0.69 mg), methanolic (0.88 mg), and aqueous (0.44 mg) cinnamon extracts in 68%, 95.5%, and 87.5% of cases, respectively (Mancini-Filho et al. 1998). Cinnamon has ingredients that can be used as health-promoting substances. The antioxidant activity of cinnamon antioxidants is increased when they are encapsulated. Benefits of a diet rich in these chemicals include a reduction in oxidative stress, which is known to induce inflammation (Pisoschi et al. 2021).

## 27.5 Antidiabetic Activities

Diabetes, which is a metabolic illness, leads to a variety of health issues. The most widespread type of diabetes is type 2 diabetes. Synthetic medications are the foundation of traditional medicine; however, because of their bioactive qualities, herbal medications are gaining popularity. The primary tactic is linked to the inhibition of the digestive tract's α-glucosidase and α-amylase enzymes, both are involved in the hydrolysis of biomolecules like carbohydrates. Among the substances detected in the high pressure and decoction method-prepared water extract of *Cinnamomum zeylanicum* were eugenol, (E)-cinnamaldehyde, benzoic acid, trans-cinnamic acid, and *o*-methoxy-cinnamaldehyde. With great promise, these biologically active chemicals isolated using this approach can inhibit α-glucosidase and regulate hyperglycemia (Wariyapperuma et al. 2020). One of the main ingredients in the bark oil that is derived from *Cinnamomum zeylanicum* is cinnamaldehyde. It appears to lower plasma glucose levels more well than metformin, a medication that is frequently prescribed in conventional treatment. Cinnamon oil's bioactive components increase the expression of proteins essential to insulin signaling, glucose transport, and dyslipidemia management. Tissue damage can also result from imbalances in the reaction between oxidative stress and free radicals. Insulin resistance is increased by high levels of oxidative stress. Type 2 diabetes and decreased glucose tolerance are the results of this. This study demonstrated that the bioactive components in cinnamon maintain this equilibrium (Sharma et al. 2020; Tangvarasittichai 2015). According to research examining the effects of different spices, cinnamon's aqueous extract demonstrated an insulin-potentiating effect that surpassed that of the other spices by a factor of 20 (Broadhurst et al. 2000). After separating and analyzing the polyphenol type-A polymers from cinnamon, Anderson et al. discovered that these compounds function as molecules similar to insulin (Anderson et al. 2003). After this characterization, naphthalene methyl ester, a novel chemical derived from derivatives of hydroxycinnamic acid with blood glucose-lowering properties, was discovered (Kim et al. 2006). According to Cao et al. (2007), there was insulin-like activity in the cinnamon aqueous extract that contained polyphenols that had been extracted using "high performance liquid chromatography (HPLC)" (Cao et al. 2007). A significant reduction in the absorption of alanine within the rat gut was observed following the administration of the aqueous extract of cinnamon. Alanine is subsequently converted to pyruvate in the liver, where it plays an important role as a substrate in gluconeogenesis (Kreydiyyeh et al. 2000).

Despite the widespread belief that cinnamon can help with diabetes, a different investigation on postmenopausal diabetic individuals with poor glycemic control found no benefit from further cinnamon supplementation. However, these differences could potentially be attributed to fluctuations in the baseline glucose and lipid levels, as well as the differing dosages of cinnamon used (Vanschoonbeek et al. 2006). According to a recent study, cinnamon with the linalool chemotype at doses of 5, 10, and 20 mg/kg was found to improve insulin secretion, which in turn helped diabetics control their blood sugar levels. It is conceivable that pancreatic β-cells could be protected by the reduction of oxidative stress and the proinflammatory milieu in the pancreas (Lee et al. 2013).

## 27.6 Use of *Cinnamomum* in Diseases

### 27.6.1 Use in Alzheimer's Disease

The deterioration of areas rich in neurons is the root cause of Alzheimer's disease (AD). Acetylcholine and memory performance are connected. Cholinesterase inhibitors are used to treat Alzheimer's disease. They are unable to alter the disease's progress, nevertheless. Other factors can also contribute to AD, such as the existence of senile plaques, which are made of hyperphosphorylated tau protein and amyloid beta peptide fibrils (Huang et al. 2020). By activating enzymes to decrease the generation of amyloid β, oxidative stress can create oligomers Aβ, increasing the likelihood that amyloid plaque will form (Devi et al. 2012). The extract of cinnamon contains polyphenolic compounds that actively prevent oxidative stress enzymes from changing. A reduction in the evaluation of glutathione oxidative stress indicators and malondialdehyde (MDA) levels has demonstrated that *Cinnamomum zeylanicum* offers protective effects against Alzheimer's disease and dementia in rat models of memory impairment induced by scopolamine (Jain et al. 2015). The quantity of cinnamon aldehyde's metabolites determines how active it is. The body has an even distribution of this bioactive substance. A portion of aldehyde is broken down in the liver to produce hippuric acid, and in the plasma to produce methyl cinnamonate and cinnamonyl alcohol. One deciding aspect is how polyphenols interact with the blood barrier (Momtaz et al. 2018).

### 27.6.2 Use in Parkinson's Disease

PD, or Parkinson's disease, is a prevalent neurological condition. Both nonmotor and motor symptoms are present in PD patients. Dopamine levels in the brain decline as a result of neurodegeneration in the black matter, a tissue located in the midbrain, which is linked to Parkinson's disease. One theory about aging and the depletion of vital black blood cells links the body's dopamine levels to a decline. Free radicals and oxidative stress also have an impact. Yet there are numerous indicators of PD. Lewy bodies can be seen inside of the neurons in the pathomorphological picture of Parkinson's disease. They are composed of the protein α-synuclein. The characteristic of Parkinson's disease patients is an aberrant structure of this protein, which results in the secondary build-up of β-amyloid and ultimately in death (Tan et al. 2020). Studies on the therapeutic benefits of cinnamon oil for Parkinson's disease indicate that the extract may influence neurons and slow the advancement of certain neurological conditions. Numerous intracellular pathways and mitochondrial activities can be modulated by cinnamon. Early-onset Parkinson's disease (PD) and partial autosomal recession are caused by mutations in the DJ-1 gene. It is linked to mitochondrial malfunction and oxidative stress (Dolgacheva et al. 2019). Numerous studies have shown that sodium benzoate, an ingredient in cinnamon extract, has neuroprotective properties and can reverse the suppression of the DJ-1 gene in neurons caused by lipopolysaccharide. The components of cinnamon extract control how well apoptosis and mitochondria work. Sodium benzoate-containing cinnamon compounds are promising natural remedies for Parkinson's disease (Angelopoulou et al. 2021).

### 27.6.3 Protecting Against HIV

HIV, Human Immunodeficiency Virus, is caused by a retrovirus that attacks the immune system, specially targeting CD4 cells (T-cells), which are crucial for immune defense. A study was done by Farahpour and Habibi (2012) on mice where *Cinnamomum zeylanicum* (commonly known as Ceylon cinnamon) was found to be beneficial in treating acquired immunodeficiency disorders, specifically HIV/AIDS. This suggests that compounds within cinnamon might have therapeutic effects, potentially supporting immune function or directly inhibiting aspects of the virus. The study involved a broader investigation of 26 different plants used to address HIV/AIDS, indicating an exploration of herbal or natural treatments for immune support or antiviral effects.

Plant-based therapies, including cinnamon, are sometimes researched as complementary treatments to help manage symptoms or support immune health, though they are not replacements for antiretroviral therapy (ART), which is the standard medical treatment for HIV/AIDS. More research would be required to understand the exact mechanisms by which *C. zeylanicum* exerts its effects and to determine its efficacy and safety in a clinical setting.

### 27.6.4 Preventing Multiple Sclerosis (MS)

Multiple sclerosis (MS) is a long-lasting autoimmune condition that influences the central nervous system (CNS), primarily affecting the myelin sheath that protects nerve fibers. In MS, the immune system incorrectly targets the myelin, causing inflammation, the development of scar tissue (sclerosis), and compromised nerve communication. This results in a wide range of symptoms, including muscle weakness, vision problems, and cognitive issues. The condition known as multiple sclerosis (MS) is defined by alternating phases of relapse and remission or a progressive deterioration, depending on the variant of MS. Cinnamon contains beneficial compounds, notably cinnamaldehyde and cinnamic acid, recognized for their anti-inflammatory effects. Such properties may assist in mitigating some of the inflammatory processes linked to MS, which is an autoimmune disease characterized by inflammation in the central nervous system (CNS) (Namjooyan et al. 2014). Research involving animals has indicated that cinnamon may offer neuroprotective benefits by enhancing antioxidant defenses within the central nervous system, which could mitigate oxidative stress—a significant factor in the progression of multiple sclerosis. For example, the compound cinnamaldehyde has demonstrated protective effects on neurons against inflammation in laboratory experiments. Additionally, certain studies propose that cinnamon may exert an immunomodulatory influence, potentially aiding in the regulation of immune responses. However, these findings remain preliminary and have primarily been observed in animal models or cellular studies, rather than in human clinical trials.

### 27.6.5 Reducing the Risk of Cancer

Recent research has shown that cinnamon has antitumor properties, effectively inhibiting the growth of various human cancer cell lines, including those of breast, leukemia, ovarian, and lung cancers. Xiqiang et al. investigated *Cinnamomum cassia* essential oil (CEO)-impregnated chitosan nanoparticles (CS-CEO) for their physicochemical properties, stability, and anticancer effects, conducting evaluations both in vitro and in vivo. The CS-CEO nanoparticles demonstrated enhanced in vitro antitumor activity (52 μg/mL) and, in vivo, significantly inhibited the growth of 4T1 breast cancer cells transplanted into mice. The treatment reduced tumor cell proliferation and induced apoptosis by downregulating the expression of the Ki-67 protein. These findings suggest that chitosan-encapsulated cinnamon oil exhibits superior physical stability and heightened efficacy against the 4T1 breast tumor model, providing a potential reference for incorporating volatile oil components in traditional Chinese medicine applications. Salim et al. (2023) had developed surface-functionalized silver-cinnamon nanoshells (Ag-CNSs) using the pulsed laser ablation in liquid (PLAL) and characterized by X-ray diffractometry (XRD) and energy-dispersive X-ray (EDX) analyses confirming the high crystallinity and elemental composition of the Ag-CNSs. These cinnamon protein-capped Ag nanocrystals are spherical, with an average size of 11.12 ± 1.32 nm, having large surface area, demonstrating enhanced fluorescence and surface plasmon resonance (SPR) absorption. Their anticancer activity and cytotoxicity were tested on liver hepatocellular carcinoma (HepG2) cells and normal (WRL68) cell lines. The nanoparticles showed potent anticancer activity, with a high cytotoxicity level (log $IC_{50} \approx 59.1\%$), suggesting potential for new cancer treatment developments. Additionally, the study revealed that surface functionalization can be used to adjust the anticancer efficacy of Ag-CNSs.

## 27.7 Conclusion and Future Perspective

Cinnamon has historically been appreciated as both a flavorful spice and a natural remedy in various traditional medicinal practices around the world. Existing studies, both in vitro and in vivo, indicate that cinnamon extracts exhibit a variety of beneficial properties, including antioxidant, antimicrobial, antiparasitic, free radical scavenging, and wound-healing effects. These properties may aid in the management and treatment of several medical conditions, such as blood glucose control, serum cholesterol levels, hypertension, hyperlipidemia, gastrointestinal disorders, Alzheimer's disease, and certain cancers. However, since the majority of research has been conducted using animal models, further investigation is required to confirm these effects in human populations, understand their public health significance, and evaluate their safety.

## References

Anderson RA, Broadhurst CL, Polansky MM, Schmidt WF, Khan A, Flanagan VP, Schoene NW, Graves DJ (2003) Isolation and characterization of polyphenol type-A polymers from cinnamon with insulin-like biological activity. J Agric Food Chem 52(1):65–70

Angelopoulou E, Paudel YN, Piperi C, Mishra A (2021) Neuroprotective potential of cinnamon and its metabolites in Parkinson's disease: mechanistic insights, limitations, and novel therapeutic opportunities. J Biochem Mol Toxicol 35(4):e22720

Arancibia M, Giménez B, López-Caballero ME, Gómez-Guillén MC, Montero P (2014) Release of cinnamon essential oil from polysaccharide bilayer films and its use for microbial growth inhibition in chilled shrimps. LWT Food Sci Technol 59(2):989–995

Bayoumi S, Bayoumi S (1992) Bacteriostatic effect of some spices and their utilization in the manufacture of yoghurt. Chem Mikrobiol Technol Lebensm 14(1–2):21–26

Becerril R, Nerín C, Gómez-Lus R (2012) Evaluation of bacterial resistance to essential oils and antibiotics after exposure to oregano and cinnamon essential oils. Foodborne Pathog Dis 9(8):699–705

Broadhurst CL, Polansky MM, Anderson RA (2000) Insulin-like biological activity of culinary and medicinal plant aqueous extracts in vitro. J Agric Food Chem 48(3):849–852

Cao H, Polansky MM, Anderson RA (2007) Cinnamon extract and polyphenols affect the expression of tristetraprolin, insulin receptor, and glucose transporter 4 in mouse 3T3-L1 adipocytes. Arch Biochem Biophys 459(2):214–222

Chericoni S, Prieto JM, Iacopini P, Cioni P, Morelli I (2005) In vitro activity of the essential oil of Cinnamomum zeylanicum and eugenol in peroxynitrite-induced oxidative processes. J Agric Food Chem 53(12):4762–4765

Devi L, Alldred MJ, Ginsberg SD, Ohno M (2012) Mechanisms underlying insulin deficiency-induced acceleration of β-amyloidosis in a mouse model of Alzheimer's disease. PLoS One 7(3):e32792

Dolgacheva LP, Berezhnov AV, Fedotova EI, Zinchenko VP, Abramov AY (2019) Role of DJ-1 in the mechanism of pathogenesis of Parkinson's disease. J Bioenerg Biomembr 51(3):175–188

Ervina M, Nawu YE, Esar SY (2016) Comparison of in vitro antioxidant activity of infusion, extract and fractions of Indonesian Cinnamon (Cinnamomum burmannii) bark. Int Food Res J 23(3):1346–1350

Farahpour MR, Habibi M (2012) Evaluation of the wound healing activity of an ethanolic extract of Ceylon cinnamon in mice. Vet Med 57(1):53–57

Fitzpatrick JL (2013) Global food security: the impact of veterinary parasites and parasitologists. Vet Parasitol 195(3–4):233–248

Goñi P, López P, Sánchez C, Gómez-Lus R, Becerril R, Nerín C (2009) Antimicrobial activity in the vapour phase of a combination of cinnamon and clove essential oils. Food Chem 116(4):982–989

Gruenwald J, Freder J, Armbruester N (2010) Cinnamon and health. Crit Rev Food Sci Nutr 50(9):822–834

Gupta M (2010) Pharmacological properties and traditional therapeutic uses of important Indian spices: a review. Int J Food Prop 13(5):1092–1116

Huang LK, Chao SP, Hu CJ (2020) Clinical trials of new drugs for Alzheimer disease. J Biomed Sci 27(1):1–13

Jain S, Sangma T, Shukla SK, Mediratta PK (2015) Effect of Cinnamomum zeylanicum extract on scopolamine-induced cognitive impairment and oxidative stress in rats. Nutr Neurosci 18(5):210–216

Jayaprakasha GK, Mohan Rao LJ, Sakariah KK (2003) Volatile constituents from Cinnamomum zeylanicum fruit stalks and their antioxidant activities. J Agric Food Chem 51(15):4344–4348

Jayaprakasha GK, Negi PS, Jena BS, Jagan Mohan Rao L (2007) Antioxidant and antimutagenic activities of Cinnamomum zeylanicum fruit extracts. J Food Compos Anal 20(3–4):330–336

Jessica Elizabeth DLT, Gassara F, Kouassi AP, Brar SK, Belkacemi K (2017) Spice use in food: properties and benefits. Crit Rev Food Sci Nutr 57(6):1078–1088

Kim SH, Hyun SH, Choung SY (2006) Anti-diabetic effect of cinnamon extract on blood glucose in db/db mice. J Ethnopharmacol 104(1–2):119–123

Kreydiyyeh SI, Usta J, Copti R (2000) Effect of cinnamon, clove and some of their constituents on the Na+-K+-ATPase activity and alanine absorption in the rat jejunum. Food Chem Toxicol 38(9):755–762

Lee SC, Xu WX, Lin LY, Yang JJ, Liu CT (2013) Chemical composition and hypoglycemic and pancreas-protective effect of leaf essential oil from indigenous cinnamon (Cinnamomum osmophloeum Kanehira). J Agric Food Chem 61(20):4905–4913

Lin YT, Vattem D, Labbe RG, Shetty K (2005) Enhancement of antioxidant activity and inhibition of Helicobacter pylori by phenolic phytochemical-enriched alcoholic beverages. Process Biochem 40(6):2059–2065

Mancini-Filho J, Van-Koiij A, Mancini DAP, Cozzolino FF, Torres RP (1998) Antioxidant activity of cinnamon (Cinnamomum Zeylanicum, Breyne) extracts. Boll Chim Farm 137(11):443–447

Miyashita M, Sadzuka Y (2013) Effect of linalool as a component of Humulus lupulus on doxorubicin-induced antitumor activity. Food Chem Toxicol 53:174–179

Mohammadi A, Hashemi M, Hosseini SM (2015) Chitosan nanoparticles loaded with Cinnamomum zeylanicum essential oil enhance the shelf life of cucumber during cold storage. Postharvest Biol Technol 110:203–213

Momtaz S, Hassani S, Khan F, Ziaee M, Abdollahi M (2018) Cinnamon, a promising prospect towards Alzheimer's disease. Pharmacol Res 130:241–258

Namjooyan F, Ghanavati R, Majdinasab N, Jokari S, Janbozorgi M (2014) Uses of complementary and alternative medicine in multiple sclerosis. J Tradit Complement Med 4(3):145–152

Pina-Pérez MC, Martínez-López A, Rodrigo D (2012) Cinnamon antimicrobial effect against Salmonella typhimurium cells treated by pulsed electric fields (PEF) in pasteurized skim milk beverage. Food Res Int 48(2):777–783

Pisoschi AM, Pop A, Iordache F, Stanca L, Predoi G, Serban AI (2021) Oxidative stress mitigation by antioxidants – an overview on their chemistry and influences on health status. Eur J Med Chem 209:112891

Ranasinghe P, Pigera S, Premakumara GS, Galappaththy P, Constantine GR, Katulanda P (2013) Medicinal properties of 'true' cinnamon (Cinnamomum zeylanicum): a systematic review. BMC Complement Altern Med 13:275

Ruihua Xin, Guibo Wang, Zhengying Qiu, Qian Ma, Saad Ahmad, Feng Yang, Hui Wang, Jiongjie He, Xiaohan Jing, Yan Sun (2022) Screening of essential oils with acaricidal activity against

Haemaphysalis longicornis (Acari:Ixodidae) and analysis of active components. Veterinary Parasitology, Vol. 307–308, 109712. https://doi.org/10.1016/j.vetpar.2022.109712

Salim AA, Bakhtiar H, Dawood DAS, Ghoshal SK (2023) Anticancer and cytotoxicity evaluation of silver-cinnamon nanoshells. Mater Lett 334:133671

Shahid MZ, Saima H, Yasmin A, Nadeem MT, Imran M, Afzaal M (2018) Antioxidant capacity of cinnamon extract for palm oil stability. Lipids Health Dis 17(1):1–8

Sharma S, Mandal A, Kant R, Jachak S, Jagzape M (2020) Is cinnamon efficacious for glycaemic control in type-2 diabetes mellitus? J Pak Med Assoc 70(11):2065–2069

Shen Q, Chen F, Luo J (2002) Comparison studies on chemical constituents of essential oil from Ramulus Cinnamomi and Cortex Cinnamomi by GC-MS. Zhong Yao Cai J Chin Med Mat 25(4):257–258

Singh G, Maurya S, deLampasona MP, Catalan CAN (2007) A comparison of chemical, antioxidant and antimicrobial studies of cinnamon leaf and bark volatile oils, oleoresins and their constituents. Food Chem Toxicol 45(9):1650–1661

Tan EK, Chao YX, West A et al (2020) Parkinson disease and the immune system — associations, mechanisms and therapeutics. Nat Rev Neurol 16(6):303–318

Tangvarasittichai S (2015) Oxidative stress, insulin resistance, dyslipidemia and type 2 diabetes mellitus. World J Diabetes 6(3):456–480

Vanschoonbeek K, Thomassen BJW, Senden JM, Wodzig WKWH, Van Loon LJ (2006) Cinnamon supplementation does not improve glycemic control in postmenopausal type 2 diabetes patients. J Nutr 136(4):977–980

Wariyapperuma NW, Kannangara S, Wijayasinghe YS, Subramanium S, Jayawardena B (2020) In vitro anti-diabetic effects and phytochemical profiling of novel varieties of Cinnamomum zeylanicum (L.) extracts. PeerJ 8:e10070

Yashin A, Yashin Y, Xia X, Nemzer B (2017) Antioxidant activity of spices and their impact on human health: a review. Antioxidants 6(3):70

MIX
Papier aus verantwortungsvollen Quellen
Paper from responsible sources
FSC® C105338
www.fsc.org

If you have any concerns about our products,
you can contact us on
**ProductSafety@springernature.com**

In case Publisher is established outside the EU,
the EU authorized representative is:
**Springer Nature Customer Service Center GmbH**
**Europaplatz 3, 69115 Heidelberg, Germany**

Printed by Libri Plureos GmbH
in Hamburg, Germany